MW00814687

Understanding and Manipulating Excited-State Processes

MOLECULAR AND SUPRAMOLECULAR PHOTOCHEMISTRY

Series Editors

V. RAMAMURTHY

Professor
Department of Chemistry
Tulane University
New Orleans, Louisiana

KIRK S. SCHANZE

Professor
Department of Chemistry
University of Florida
Gainesville, Florida

1. Organic Photochemistry, *edited by V. Ramamurthy and Kirk S. Schanze*
2. Organic and Inorganic Photochemistry, *edited by V. Ramamurthy and Kirk S. Schanze*
3. Organic Molecular Photochemistry, *edited by V. Ramamurthy and Kirk S. Schanze*
4. Multimetallic and Macromolecular Inorganic Photochemistry, *edited by V. Ramamurthy and Kirk S. Schanze*
5. Solid State and Surface Photochemistry, *edited by V. Ramamurthy and Kirk S. Schanze*
6. Organic, Physical, and Materials Photochemistry, *edited by V. Ramamurthy and Kirk S. Schanze*
7. Optical Sensors and Switches, *edited by V. Ramamurthy and Kirk S. Schanze*
8. Understanding and Manipulating Excited-State Processes, *edited by V. Ramamurthy and Kirk S. Schanze*

ADDITIONAL VOLUMES IN PREPARATION

Understanding and Manipulating Excited-State Processes

edited by

V. Ramamurthy
Tulane University
New Orleans, Louisiana

Kirk S. Schanze
University of Florida
Gainesville, Florida

MARCEL DEKKER, INC.   NEW YORK · BASEL

ISBN: 0-8247-0579-3

This book is printed on acid-free paper.

Headquarters
Marcel Dekker, Inc.
270 Madison Avenue, New York, NY 10016
tel: 212-696-9000; fax: 212-685-4540

Eastern Hemisphere Distribution
Marcel Dekker AG
Hutgasse 4, Postfach 812, CH-4001 Basel, Switzerland
tel: 41-61-261-8482; fax: 41-61-261-8896

World Wide Web
http://www.dekker.com

The publisher offers discounts on this book when ordered in bulk quantities. For more information, write to Special Sales/Professional Marketing at the headquarters address
above.

Current printing (last digit):
10 9 8 7 6 5 4 3 2 1

Preface

In Volume 8, we are proud to present 11 chapters by experts who have contributed immensely to photochemistry. This volume covers two areas of photochemistry: (1) understanding excited-state behavior of organic molecules (Chapters 1–3 and 11) and (2) manipulation of the excited-state behavior of molecules via environmental control (Chapters 4–10). Without prior knowledge of the excited-state behavior of a given molecule, one would be at a loss to figure out what to control. At the same time, without proper control of the excited-state behavior of a molecule, the photoreactions are likely to remain complex and less attractive to nonphotochemists. To stress this interdependence, chapters that describe photoreactions of organic molecules as well as strategies adapted to control photoreactions are included in this volume.

Chapters 1 and 2, by Cornelisse and de Haan and Mizuno et al. provide in-depth coverage of photoaddition of excited aromatics to unsaturated molecules. These chapters, along with two others in this series (Fleming and Pincock, Vol. 3, Chap. 6, and Brousmiche et al., Vol. 6, Chap. 1), are exceptional resources in the photochemistry of aromatic molecules.

Chapters 3 and 4, by Orfanopoulos and Lissi et al., emphasize different aspects of the excited-state behavior of $^1O_2$. An in-depth and critical review of the reaction of $^1O_2$ with alkenes provided by Orfanopoulos should be of significant value to newcomers and current practitioners of $^1O_2$ chemistry. Lissi et al. summarize results of the photophysical behavior of $^1O_2$ in various solvents and organized assemblies. The extensive data analysis will be of interest to both photochemists and photobiologists.

In Chapter 5, Tung and his coauthors present a critical review of photoreactions in organized assemblies such as zeolites, nafion membranes, low-density polyethylenes, and vesicles. The chapter provides a flavor of the remarkable control that could be achieved by the use of organized assemblies. Chapters 6 and 7 deal with solid-state (crystal) photochemistry and place the results from the authors' laboratories in the context of existing literature. These reviews and two others in this series (Keating and Garcia-Garibay, Vol. 2, Chap. 5, and Ito, Vol. 3, Chap. 1) are of great value to those interested in manipulating photochemical behavior of molecules in the solid state. These chapters reveal that unprecedented product selectivities could be achieved in the crystalline state. In Chapter 6, Toda and his coauthors who have pioneered the use of host–guest solid complexes to achieve significant chiral induction during photoreactions, elaborate on the methodology with examples. Since the original extensive studies on cinnamic acids by Schmidt's group, coumarins have been investigated at a greater depth by Vishnumurthy's group. These studies have yielded mechanistic details that were unavailable through the studies on cinnamic acids and related systems. In Chapter 7, Vishnumurthy et al. summarize their experimental observations on coumarins and related systems.

Chapter 8, by Ueno and Ikeda, focuses on the photophysics and photochemistry of organic molecules within cyclodextrins. Particular emphasis is placed on the use of modified cyclodextrins as sensors for neutral organic molecules. In this context readers are referred to Volume 7 of the series, which is devoted to sensors and switches.

In Chapter 9, Kumar and Raju provide comprehensive coverage of photophysics of organic molecules incorporated into inorganic layered materials. This chapter illustrates how one could use layered materials such as zirconium phosphates, layered double hydroxides, etc., to control the excited-state behavior of organic, inorganic, and biological molecules. Readers wanting more on this topic, especially clays, are referred to Chapter 2, by Takagi and Shichi, in Volume 5 of this series. The section on chemistry in organized assemblies concludes with Chapter 10, by Mishra. This chapter emphasizes how one can exploit photophysics of organic molecules to probe a medium. The main emphasis has been on probes that undergo medium-dependent excited-state acid–base chemistry. Chapter 10, in addition to two others in this series (Kleinman and Bohne, Vol. 1, Chap. 10, and Bhattacharyya, Vol. 3, Chap. 6), should enable readers to choose appropriate molecules to probe a medium.

Since their discovery photochemists have been attracted to fullerenes and their derivatives. Literature on this topic has exploded during the past decade. In Chapter 11, Mattay and his coauthors summarize the photochemical and photophysical behavior of fullerenes and their derivatives. This chapter, in conjunction with two earlier ones in this series (Maggini and Guldi, Vol. 6, Chap. 4, and Sun, Vol. 1, Chap. 9), should serve readers well for many years to come.

We have particularly enjoyed interacting with the contributors, whose enthusiastic participation enabled the creation of this work. We are indebted to them for distilling more than 2000 references into a comprehensive and critical volume. It is our hope that the chapters in this volume will serve not only as a valuable resource for experts but also as supplementary reading material for graduate students.

V. Ramamurthy
Kirk S. Schanze

Contents

Preface   *iii*
Contributors   *ix*
Contents of Previous Volumes   *xiii*

1. Ortho Photocycloaddition of Alkenes and Alkynes to the Benzene
   Ring   1
   Jan Cornelisse and Rudy de Haan

2. Photocycloaddition and Photoaddition Reactions of Aromatic
   Compounds   127
   Kazuhiko Mizuno, Hajime Maeda, Akira Sugimoto, and
   Kazuhiko Chiyonobu

3. Singlet-Oxygen Ene-Sensitized Photo-Oxygenations:
   Stereochemistry and Mechanisms   243
   Michael Orfanopoulos

4. Singlet-Oxygen Reactions: Solvent and Compartmentalization
   Effects   287
   Eduardo A. Lissi, Else Lemp, and Antonio L. Zanocco

5. Microreactor-Controlled Product Selectivity in Organic
   Photochemical Reactions   317
   Chen-Ho Tung, Kai Song, Li-Zhu Wu, Hong-Ru Li, and Li-Ping Zhang

6. Enantioselective Photoreactions in the Solid State   385
   Fumio Toda, Koichi Tanaka, and Hisakazu Miyamoto

7. Observations on the Photochemical Behavior of Coumarins and
   Related Systems in the Crystalline State   427
   *Kodumuru Vishnumurthy, Tayur N. Guru Row, and
   Kailasam Venkatesan*

8. Supramolecular Photochemistry of Cyclodextrin Materials   461
   Akihiko Ueno and Hiroshi Ikeda

9. Photoactive Layered Materials: Assembly of Ions, Molecules, Metal
   Complexes, and Proteins   505
   Challa V. Kumar and B. Bangar Raju

10. Fluorescence of Excited Singlet-State Acids in Certain Organized
    Media: Applications as Molecular Probes   577
    Ashok Kumar Mishra

11. Photophysics and Photochemistry of Fullerenes and Fullerene
    Derivatives   637
    Jochen Mattay, Lars Ulmer, and Andreas Sotzmann

Index   *753*

Contributors

Kazuhiko Chiyonobu Department of Applied Chemistry, Graduate School of Engineering, Osaka Prefecture University, Sakai, Japan

Jan Cornelisse Department of Chemistry, Leiden University, Leiden, The Netherlands

*Rudy de Haan Department of Chemistry, Leiden University, Leiden, The Netherlands

Hiroshi Ikeda Department of Bioengineering, Graduate School of Biotechnology and Bioscience, Tokyo Institute of Technology, Yokohama, Japan

Challa V. Kumar Department of Chemistry, University of Connecticut, Storrs, Connecticut

Else Lemp Faculty of Chemical and Pharmaceutical Sciences, University of Chile, Santiago, Chile

Hong-Ru Li Technical Institute of Physics and Chemistry, Chinese Academy of Sciences, Beijing, China

---

* Current affiliation: Pink Elephant Netherlands, Zoetermeer, The Netherlands.

Eduardo A. Lissi   Faculty of Chemistry and Biology, University of Santiago, Santiago, Chile

Hajime Maeda   Department of Applied Chemistry, Graduate School of Engineering, Osaka Prefecture University, Sakai, Japan

Jochen Mattay   Department of Chemistry, University of Bielefeld, Bielefeld, Germany

Ashok Kumar Mishra   Department of Chemistry, Indian Institute of Technology Madras, Chennai, India

Hisakazu Miyamoto   Department of Applied Chemistry, Ehime University, Ehime, Japan

Kazuhiko Mizuno   Department of Applied Chemistry, Graduate School of Engineering, Osaka Prefecture University, Sakai, Japan

Michael Orfanopoulos   Department of Chemistry, University of Crete, Crete, Greece

B. Bangar Raju   Department of Chemistry, University of Connecticut, Storrs, Connecticut

Tayur N. Guru Row   Department of Organic Chemistry, Indian Institute of Science, Bangalore, India

Kai Song   Technical Institute of Physics and Chemistry, Chinese Academy of Sciences, Beijing, China

Andreas Sotzmann   Department of Chemistry, University of Bielefeld, Bielefeld, Germany

Akira Sugimoto   Department of Applied Chemistry, Graduate School of Engineering, Osaka Prefecture University, Sakai, Japan

Koichi Tanaka   Department of Applied Chemistry, Ehime University, Ehime, Japan

Fumio Toda   Department of Chemistry, Okayama University of Science, Okayama, Japan

Chen-Ho Tung   Technical Institute of Physics and Chemistry, Chinese Academy of Sciences, Beijing, China

Akihiko Ueno   Department of Bioengineering, Graduate School of Biotechnology and Bioscience, Tokyo Institute of Technology, Yokohama, Japan

Lars Ulmer   Department of Chemistry, University of Bielefeld, Bielefeld, Germany

Kailasam Venkatesan   Department of Organic Chemistry, Indian Institute of Science, Bangalore, India

Kodumuru Vishnumurthy   Department of Organic Chemistry, Indian Institute of Science, Bangalore, India

Li-Zhu Wu   Technical Institute of Physics and Chemistry, Chinese Academy of Sciences, Beijing, China

Antonio L. Zanocco   Faculty of Chemical and Pharmaceutical Science, University of Chile, Santiago, Chile

Li-Ping Zhang   Technical Institute of Physics and Chemistry, Chinese Academy of Sciences, Beijing, China

Contents of Previous Volumes

Volume 1. Organic Photochemistry

1. The Photochemistry of Sulfoxides and Related Compounds
    William S. Jenks, Daniel D. Gregory, Yushen Guo, Woojae Lee, and Troy Tetzlaff

2. The Photochemistry of Pyrazoles and Isothiazoles
    James W. Pavlik

3. Photochemistry of (S-Hetero)cyclic Unsaturated Carbonyl Compounds
    Paul Margaretha

4. Photochemistry of Conjugated Polyalkynes
    Sang Chul Shim

5. Photochemistry and Photophysics of Carbocations
    Mary K. Boyd

6. Regio- and Stereoselective 2,+2 Photocycloadditions
    Cara L. Bradford, Steven A. Fleming, and J. Jerry Gao

7. Photoinduced Redox Reactions in Organic Synthesis
    Ganesh Pandey

8.   Photochemical Reactions on Semiconductor Particles for Organic
      Synthesis
      Yuzhuo Li

9.   Photophysics and Photochemistry of Fullerene Materials
      Ya-Ping sun

10. Use of Photophysical Probes to Study Dynamic Processes in
      Supramolecular Structures
      Mark H. Kleinman and Cornelia Bohne

11. Photophysical and Photochemical Properties of Squaraines in
      Homogeneous and Heterogeneous Media
      Suresh Das, K. George Thomas, and M. V. George

12. Absorption, Fluorescence Emission, and Photophysics of Squaraines
      Kock-Yee Law

Volume 2.   Organic and Inorganic Photochemistry

1.   A Comparison of Experimental and Theoretical Studies of Electron
      Transfer Within DNA Duplexes
      Thomas L. Netzel

2.   Coordination Complexes and Nucleic Acids. Perspectives on Electron
      Transfer, Binding Mode, and Cooperativity
      Eimer Tuite

3.   Photoinduced Electron Transfer in Metal-Organic Dyads
      Kirk S. Schanze and Keith A. Walters

4.   Photochemistry and Photophysics of Liquid Crystalline Polymers
      David Creed

5.   Photochemical Solid-to-Solid Reactions
      Amy E. Keating and Miguel A. Garcia-Garibay

6.   Chemical and Photophysical Processes of Transients Derived from
      Multiphoton Excitation: Upper Excited States and Excited Radicals
      W. Grant McGimpsey

7. Environmental Photochemistry with Semiconductor Nanoparticles
   Prashant V. Kamat and K. Vinodgopal

Volume 3.   Organic Molecular Photochemistry

1. Solid-State Organic Photochemistry of Mixed Molecular Crystals
   Yoshikatsu Ito

2. Asymmetric Photochemical Reactions in Solution
   Simon R. L. Everitt and Yoshihisa Inoue

3. Photochemical cis-trans Isomerization in the Triplet State
   Tatsuo Arai

4. Photochemical cis-trans Isomerization from the Singlet Excited State
   V. Jayathirtha Rao

5. Photochemical Cleavage Reactions of Benzyl-Heteroatom Sigma Bonds
   Steven A. Fleming and James A. Pincock

6. Photophysical Probes for Organized Assemblies
   Kankan Bhattacharyya

**Volume 4.   Multimetallic and Macromolecular Inorganic
               Photochemistry**

1. Metal-Organic Conducting Polymers: Photoactive Switching in Molecular
   Wires
   Wayne E. Jones, Jr., Leoné Hermans, and Biwang Jiang

2. Luminescence Behavior of Polynuclear Metal Complexes of Copper (I),
   Silver (I), and Gold (I)
   Vivian W.-W. Yam and Kenneth K.-W. Lo

3. Electron Transfer Within Synthetic Polypeptides and De Novo Designed
   Proteins
   Michael Y. Ogawa

4. Tridentate Bridging Ligands in the Construction of Stereochemically
   Defined Supramolecular Complexes
   Sumner W. Jones, Micheal R. Jordan, and Karen J. Brewer

5.   Photophysical and Photochemical Properties of Metallo-1,2-enedithiolates
     Robert S. Pilato and Kelly A. Van Houten

6.   Molecular and Supramolecular Photochemistry of Porphyrins and
     Metalloporphyrins
     Shinsuke Takagi and Harou Inoue

Volume 5.   Solid State and Surface Photochemistry

1.   Spectroscopy and Photochemical Transformations of Polycyclic Aromatic
     Hydrocarbons at Silica– and Alumina–Air Interfaces
     Reza Dabestani and Michael E. Sigman

2.   Photophysics and Photochemistry in Clay Minerals
     Katsuhiko Takagi and Tetsuya Shichi

3.   Photochromism of Diarylethenes in Confined Reaction Spaces
     Masahiro Irie

4.   Charge and Electron Transfer Reactions in Zeolites
     Kyung Byung Yoon

5.   Time-Resolved Spectroscopic Studies of Aromatic Species Included in
     Zeolites
     Shuichi Hashimoto

6.   Photo-Oxidation in Zeolites
     Sergey Vasenko and Heinz Frei

Volume 6.   Organic, Physical, and Materials Photochemistry

1.   Photochemistry of Hydroxyaromatic Compounds
     Darryl W. Brousmiche, Alexander G. Briggs, and Peter Wan

2.   Stereoselectivity of Photocycloadditions and Photocyclizations
     Axel G. Griesbeck and Maren Fiege

3.   Photocycloadditions with Captodative Alkenes
     Dietrich Döpp

4. Photo- and Electroactive Fulleropyrrolidines
    Michele Maggini and Dirk M. Guldi

5. Applications of Time-Resolved EPR in Studies of Photochemical
    Reactions
    Hans van Willigen

6. Photochemical Generation and Studies of Nitrenium Ions
    Daniel E. Falvey

7. Photophysical Properties of Inorganic Nanoparticle-DNA Assemblies
    Catherine J. Murphy

8. Luminescence Quenching by Oxygen in Polymer Films
    Xin Lu and Mitchell A. Winnik

Volume 7.   Optical Sensors and Switches

1. Buckets of Light
    Christina M. Rudzinski and Daniel G. Nocera

2. Luminescent PET Signalling Systems
    *A. Prasanna de Silva, David B. Fox, Thomas S. Moody, and
    Sheenagh M. Weir*

3. Sensors Based on Electrogenerated Chemiluminescence
    Anne-Margret Andersson and Russell H. Schmehl

4. From Superquenching to Biodetection: Building Sensors Based on
    Fluorescent Polyelectrolytes
    *David Whitten, Liaohai Chen, Robert Jones, Troy Bergstedt, Peter Heeger,
    and Duncan McBranch*

5. Luminescent Metal Complexes as Spectroscopic Probes of
    Monomer/Polymer Environments
    Alistair J. Lees

6. Photorefractive Effect in Polymeric and molecular Materials
    ALiming Wang, Man-Kit Ng, Qing Wang, and Luping Yu

7. Dynamic Holography in Photorefractive liquid Crystals
    Gary P. Wiederrecht

8. Hologram Switching and Erasing Strategies With Liquid Crystals
   Michael B. Sponsler

9. Novel Molecular Photonics Materials and Devices
   Toshihiko Nagamura

10. Molecular Recognition Events Controllable by Photochemical Triggers or
    Readable by Photochemical Outputs
    Seiji Shinkai and Tony D. James

11. Probing Nanoenvironments Using Functional Chromophores
    Mitsuru Ishikawa and Jing Yong Ye

Understanding and Manipulating Excited-State Processes

1

Ortho Photocycloaddition of Alkenes and Alkynes to the Benzene Ring

Jan Cornelisse and Rudy de Haan
Leiden University, Leiden, The Netherlands

I. INTRODUCTION

Photoexcitation of benzene into its first excited singlet state ($^1B_{2u}$) turns that stable molecule into a very reactive species. The amount of electronic energy in the excited molecule, relative to the ground electronic state, is 112.5 kcal/mol and the six electrons are no longer in pairs filling the three bonding π-orbitals in a closed-shell configuration. The molecule has lost its aromatic stability, is rich in energy, and therefore capable of undergoing chemical reactions not possible in the ground state. For reactions in the ground state of benzene, rather agressive reagents are necessary. Among these reactive species are molecules in their electronically excited states.

Cycloaddition of alkenes to the benzene ring does not occur when both molecules are in their ground electronic states. The reaction can only be brought about by photoexcitation of either of the two addends. Three types of photochemical cycloaddition of alkenes to benzene and its derivatives are presently known. Ortho photocycloaddition, also referred to as 1,2-photocycloaddition or [2 + 2] photocycloaddition, leads to bicyclo[4.2.0]octa-2,4-dienes. Meta photocycloaddition, also referred to as 1,3-photocycloaddition or [2 + 3] photocycloaddition, gives triyclo[3.3.0.0^{2,8}]oct-3-enes, also named 1,2,2a,2b,4a,4b-hexahydrocyclopropa[cd]pentalenes. Para photocycloaddition, also referred to as 1,4-photocycloaddition or [2 + 4] photocycloaddition, results in bicyclo[2.2.2]octa-2,5-

1

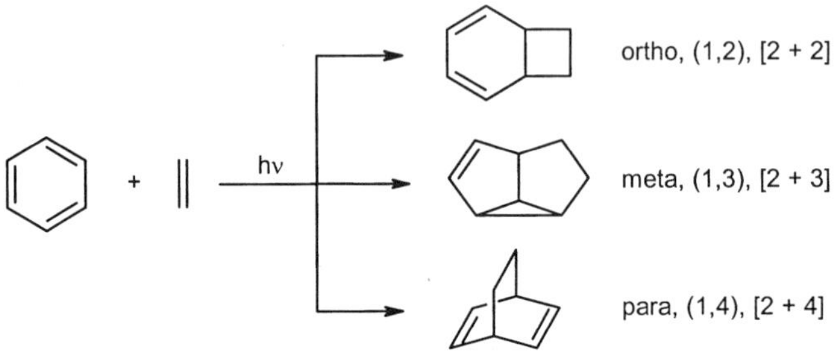

Scheme 1   Ortho, meta, and para photocycloaddition of ethene and benzene.

dienes. The three types are shown in Scheme 1 for the two parent molecules, benzene and ethene.

Ortho photocycloaddition was first reported in a U.S. patent [1] dated September 3, 1957. Irradiation of benzonitrile in the presence of various alkenes resulted in the formation of derivatives of 1-cyanobicyclo[4.2.0]octa-2,4-diene. The first ortho photocycloaddition to benzene was reported in 1959 by Angus and Bryce-Smith [2], who discovered that benzene and maleic anhydride react to form a stable adduct at 60°C under the influence of ultraviolet radiation. This 1:2 adduct was formed from one molecule of benzene and two molecules of maleic anhydride. Two years later, Bryce-Smith and Lodge [3] found that acetylenes could also be photoadded to benzene. The isolated products were cyclooctatetraenes, formed by ring opening of the primarily formed bicyclo[4.2.0]octa-2,4,7-trienes. Since those early years, hundreds of examples of ortho photocycloadditions of alkenes to the benzene ring and many mechanistic investigations have been reported and they will be discussed in this chapter.

Meta photocycloaddition was discovered simultaneously and independently by two groups in 1966. Wilzbach and Kaplan [4] found that the adducts from *cis*-but-2-ene, cyclopentene, and 2,3-dimethylbut-2-ene with benzene are substituted tricyclo[3.3.0.0^{2,8}]oct-3-enes. The adducts were formed by irradiation of solutions (~10%) of the olefins in benzene, at room temperature under nitrogen, with 2537-Å light. Bryce-Smith et al. [5] subjected an equimolar mixture of *cis*-cyclooctene at room temperature or in the solid phase at −60°C to ultraviolet radiation of wavelength 235–285 nm. A mixture of 1:1 adducts was obtained from which the main component (~85%) was readily obtained pure by treatment of the mixture with methanolic mercuric acetate. This 1:1 adduct proved to be a meta photocycloadduct (Scheme 2). The minor nonaromatic adduct (10–15%) could, at that time, not yet be obtained completely free from the meta photocycloadduct; the structure of a rearranged ortho adduct was provisionally assigned to this isomer.

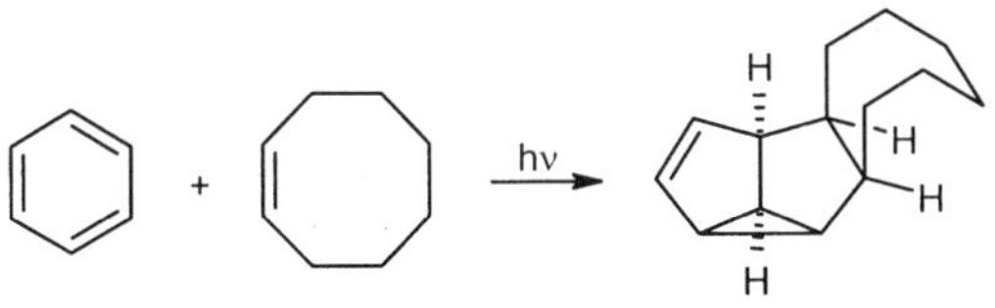

Scheme 2   Meta photocycloaddition of benzene to cyclooctene.

As in the case of the ortho photocycloaddition, the first examples of the meta addition were followed by hundreds of others. A comprehensive review summarizing the state of the art up to 1993 has appeared [6].

Para photocycloaddition of arenes to the benzene ring was first reported in 1971 by Wilzbach and Kaplan [7] as a minor process accompanying ortho and meta photocycloaddition. Since that time, relatively few cases of para photocycloaddition have been described. Para adducts were found as minor products from benzene with cyclobutene [8], *cis*-3,4-dimethylcyclobutene [9], vinylene carbonate [10], 2,3-dihydropyran [11,12], and 1,3-dioxole [13,14] and from α,α,α-trifluorotoluene with vinylene carbonate [15]. Intermolecular para photocycloadducts were major products from the irradiations of benzene and allene [16,17], benzene and cyclonona-1,2-diene [16,17], and from fluorobenzene and cyclopentene [18]. Intramolecular para photocycloadducts were found as major products from the irradiations of phenethyl vinyl ether [19–21] (Scheme 3) and 2,3-dimethyl-6-phenylhex-2-ene [22]. No detailed mechanistic investigations have been published.

In this chapter, we will be concerned with the ortho photocycloaddition of arenes to the benzene ring. This implies that photocycloadditions to larger aromatic systems and to heterocyclic aromatic molecules will not be discussed. The photoadditions of alkynes to benzene and derivatives of benzene, however, are included in this review. The material is organized in sections, according to the pathway that is followed from the ground state of the addends to the ortho photocycloadducts.

There are at least four different routes leading, via photochemical excitation, from ground-state arene and ground-state alkene or alkyne to ortho photocy-

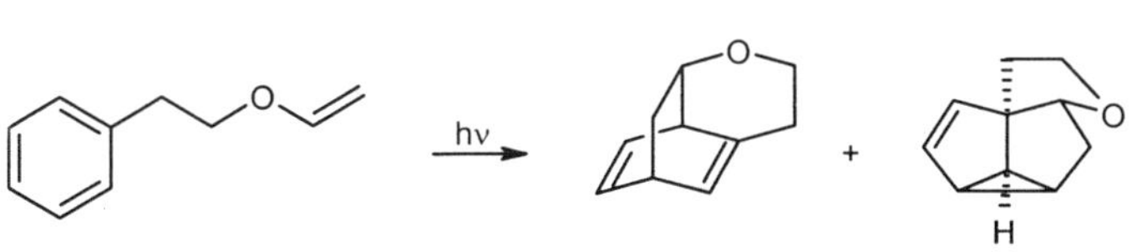

Scheme 3   Para and meta photocycloaddition of phenethyl vinyl ether; $\varphi_{para} = 0.23$, $\varphi_{meta} = 0.006$. (From Ref. 20.)

cloadduct. The pathway that is actually followed depends on the particular combination of arene and alkene:

> Excitation of a ground-state charge-transfer complex between arene and alkene (Sec. II)
>
> Excitation of the alkene or alkyne followed by reaction with ground-state arene (Sec. III)
>
> Excitation of the arene followed by intersystem crossing and reaction with the alkene (Sec. IV)
>
> Excitation of the arene and reaction of the singlet excited arene with the alkene, either directly or via an exciplex (Sec. V)

Most of the ortho photocycloadditions of alkenes and alkynes to the benzene ring that have been reported during the past 40 years have been tabulated in the 4 major sections of this chapter. Many ortho photocycloadducts are unstable under the conditions of irradiation. In some cases, they undergo spontaneous photochemical or thermal rearrangements; in other cases, the investigators add a reagent that turns the unstable ortho adduct into a more stable product that can be isolated and identified. Tables 1–7, for the sake of comparison, list the primary ortho adducts, even if they have never been isolated or detected. In Section VI, the many types of secondary reactions that ortho photocycloadducts can undergo are discussed.

The tables also contain concise information concerning the conditions of irradiation (solvent, concentration, wavelength), product yields and quantum yields, the formation of other products, and types of secondary reactions. Again for the sake of comparison, a few cases have been included in which ortho photocycloaddition was attempted but failed to take place.

Section V describes ortho photocycloadditions of mostly simple alkenes that do not form ground-state complexes with arenes, do not absorb light in the presence of the arenes, and add to the arene in its singlet excited state. Mechanistic investigations, including the search for intermediates (ground-state complexes, excited states, exciplexes, zwitterions), the formulation of empirical rules, and theoretical descriptions of the reaction, have mostly been concerned with this type of ortho addition. Therefore, this section is divided into a number of subsections, each describing a particular aspect of ortho photocycloaddition.

II. ORTHO PHOTOCYCLOADDITION VIA EXCITATION OF A GROUND-STATE CHARGE-TRANSFER COMPLEX BETWEEN ARENE AND ALKENE

Ortho photocycloadditions proceeding via excitation of a ground-state charge-transfer complex have been reported for the combination of benzene and alkylbenzenes with maleic anhydride. The reaction was discovered by Angus and

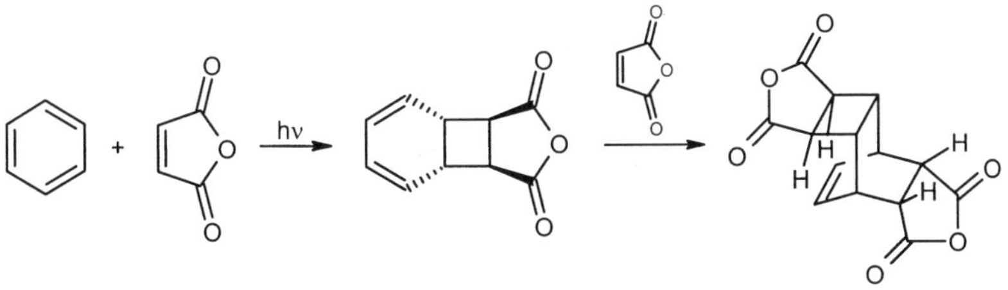

Scheme 4 Photochemical formation of a 1 : 2 adduct from benzene and maleic anhydride.

Bryce-Smith [2] in 1959. They found that benzene and maleic anhydride react to form a stable 1 : 2 adduct under the influence of ultraviolet radiation. The authors proposed that initially an ortho adduct is formed which could normally be expected to revert rapidly to the starting materials, but which would be stabilized by rapid addition of a further molecule of maleic anhydride to the bicyclic diene system. The stereochemistry of the 1:2 adduct was established by Grovenstein et al. [23] (who reported that they had independently discovered the same reaction) and, more firmly, by Bryce-Smith et al. [24] and Pettit et al. [25]. The structure is depicted in Scheme 4.

From the stereochemical structure of the adduct it can be inferred that the initial ortho photocycloaddition occurs with exo stereochemistry, whereas the subsequent Diels–Alder reaction of the ortho adduct with a second molecule of maleic anhydride proceeds with endo stereochemistry.

Schenck and Steinmetz [26] have found that the addition of maleic anhydride to benzene can be photosensitized by benzophenone. The photosensitized reaction also works with toluene, o-xylene, and chlorobenzene. Hardham and Hammond [27,28] have also studied the photosensitized addition and they obtained definitive evidence that this reaction proceeds via the triplet state. They also proposed an ortho adduct as intermediate but were unable to isolate it. Bryce-Smith and Vickery [29] found that phenanthrene reacts with maleic anhydride at the 9,10-positions to give an adduct that does not further react with a molecule of maleic anhydride. This reaction can also be sensitized by benzophenone.

The possible role of a ground-state charge-transfer complex between maleic anhydride and benzene was first mentioned by Angus and Bryce-Smith [30]. It was confirmed later that excitation of a charge-transfer (CT) complex is indeed the primary photochemical step [31]. The ground-state complex formed from maleic anhydride and benzene has absorption maxima around 270 nm [32]. Excitation at this wavelength causes photoaddition to a degree that cannot be explained by excitation of benzene or maleic anhydride alone. The unsensitized addition is only slightly sensitive to the presence of oxygen and the activated intermediate is considered to be of singlet character. The benzophenone-sensitized addition, proceeding via the triplet state of the complex [28], is completely inhibited by oxygen.

Bryce-Smith and Gilbert have shown that toluene, *t*-butylbenzene, chlorobenzene, *o*- and *p*-xylene, and biphenyl all undergo photoaddition to maleic anhydride, yielding 1:2 adducts [33]. All the substituents decrease the rate of formation of adducts. Benzonitrile, nitrobenzene, phenol, methyl benzoate, durene, hexamethylbenzene, naphthalene, and biphenylene fail to undergo the addition; of these, benzonitrile, nitrobenzene, methyl benzoate, and biphenylene do not form charge-transfer complexes. Bradshaw [34] found that various alkylbenzenes give 1:2 adducts with maleic anhydride upon photosensitization with acetophenone.

Rafikov et al. [35] describe a correlation between the electron-donating capacity of substituted benzenes and the efficiency of adduct formation with maleic anhydride. This is only valid if similar compounds are compared. The ionization potentials of benzene and toluene are 9.246 and 8.820 eV, respectively; the yields of adduct formation are 70% and 30%, respectively. In the series of halogenobenzenes, the ionization potentials are as follows: fluorobenzene, 9.195 eV; chlorobenzene, 9.080 eV; bromobenzene, 9.030 eV; the yields of adducts are 7%, 2%, and <1%, respectively. Anisole and diphenyl ether, with ionization potentials of 8.220 and 8.090 eV, respectively, do not give adducts with maleic anhydride. It thus seems that an increase in the electron-donating capacity of the benzene derivative leads to a decrease in the yield of photoadducts.

Irradiation of mixtures of benzene and maleic anhydride in the presence of trifluoroacetic acid has shed more light on the mechanism of the photoaddition [36]. Under these circumstances, formation of the 1:2 adduct is completely suppressed and a new 1:1 adduct, phenylsuccinic anhydride, is obtained. It was concluded from these experiments that excitation of the ground-state complex between benzene and maleic anhydride leads to a zwitterionic intermediate that can be intercepted by protonation (Scheme 5).

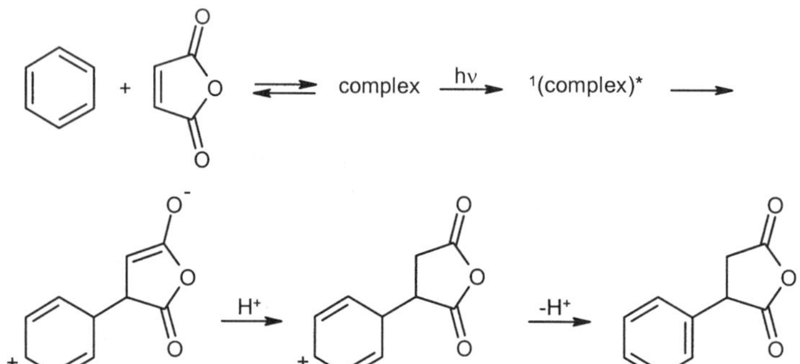

Scheme 5 Formation of phenylsuccinic anhydride by irradiation of benzene and maleic anhydride in the presence of trifluoroacetic acid.

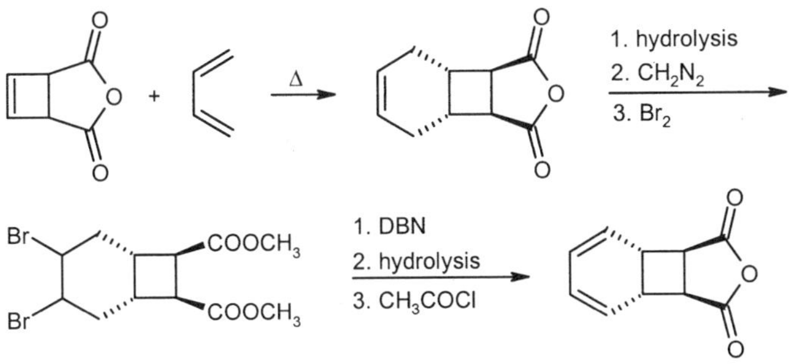

Scheme 6   Preparation of the benzene–maleic anhydride exo–ortho adduct. (From Ref. 39.)

Originally, it was proposed [37,38] that the zwitterionic intermediate would directly undergo the second addition to maleic anhydride (i.e., without first undergoing ring closure to the ortho adduct). This was based on the observation that irradiation of solutions of maleic anhydride together with tetracyanoethylene in benzene, with and without benzophenone as a sensitizer, led to the 1:2 adduct as the only recognizable product: Tetracyanoethylene was recovered almost quantitatively. If the ortho adduct (a diene) would have been formed, it would sup- posedly have undergone a Diels–Alder reaction with the powerful dienophile tetracyanoethylene, competing with maleic anhydride.

Hartman et al. [39,40], however, succeeded in preparing the exo–ortho adduct by Diels–Alder reaction of 1,3-butadiene with cyclobut-3-en-1,2-dicar- boxylic acid anhydride, hydrolysis of the anhydride, esterification, bromination of the double bond, dehydrobromination, ester hydrolysis, and re-formation of the anhydride (Scheme 6).

It was shown that the diene reacted only slowly with tetracyanoethylene, even slower than with maleic anhydride. The reaction of the ortho adduct with maleic anhydride led to the same 1:2 adduct as the photochemical reaction between benzene and maleic anhydride. The previous failure to trap the diene with tetracyanoethylene resulted from an unexpectedly low dienophilic reactivity of this compound toward the ortho adduct in the presence of benzene. Tetracya- noethylene is apparently completely complexed in benzene solution.

Bryce-Smith et al. [41] therefore proposed the mechanism shown in Scheme 7 for the photoaddition of benzene to maleic anhydride.

Both the excited singlet and the triplet state (which can be reached by sen- sitization or by direct irradiation in a heavy-atom solvent) of the complex first form one bond between benzene and maleic anhydride. From the excited singlet state, a zwitterion is formed that can be trapped by protonation. In the absence of protons, it collapses to the ortho adduct. The species formed from the triplet state

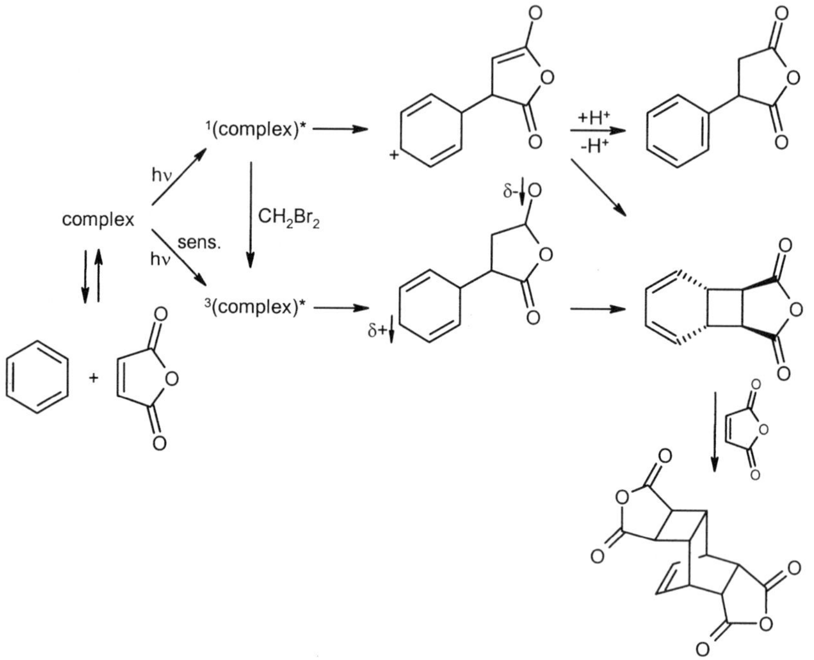

Scheme 7 Mechanism of the direct and the sensitized photoaddition of benzene and maleic anhydride.

has a biradical character; although it may be slightly polarized, it cannot be scavenged by protons. The ortho adduct, once formed, immediately reacts with a second molecule of maleic anhydride.

Mattes and Farid [42] have argued that the formation of phenylsuccinic anhydride need not be regarded as an indication that the ortho cycloadduct is formed via a zwitterionic intermediate. They considered the possibility that charge-transfer excitation of the maleic anhydride–benzene complex leads to the anhydride radical anion and the benzene radical cation. Protonation of the radical anion followed by addition to the benzene radical cation can be considered as an alternative route to phenylsuccinic anhydride.

In the presence of a second dienophile, the ortho adduct can be trapped. Koltzenburg et al. [43] have added duroquinone to a benzene solution of maleic anhydride and obtained a 1:1:1 adduct formed from benzene, maleic anhydride, and duroquinone.

The ground-state complex between benzene and maleic anhydride was found to have the exo configuration. Bryce-Smith and Hems [44] have measured nuclear magnetic resonance chemical shifts of the ethylenic protons of maleic anhydride in carbon tetrachloride and benzene solutions. Large upfield chemical shifts are induced by the benzene solvent. A model in which the hydrogen atoms of the exo-oriented maleic anhydride are located at a mean position 2.55 Å above the plane of the benzene ring provides the best fit with the experimental results.

The benzophenone-sensitized additions of toluene, *o*-xylene, and *p*-xylene each give two thermally stable structural isomers, the ratios of which are dependent on the temperature of the irradiated solution [45]. This behavior may reflect the effect of temperature on equilibria involving the precursor complexes in solution [46] or it may be caused by differing reactivities of the ortho cycloadduct regioisomers with the second molecule of dienophile. At 20°C, the photosensitized addition of maleic anhydride to *p*-xylene gave a 20:1 mixture of A2 and B2 (Scheme 8), whereas at 100°C, this ratio changed to approximately 1:10 [47]. Reduction of the light intensity by 50% caused a change in these ratios. At 20°C, the ratio A2:B2 was 1:10, and at 100°C, this was changed to 3:1. It was concluded that A1 adds maleic anhydride more readily than B1 but is more thermally labile than B1, whereas B1 is photochemically more labile than A1.

Various procedures for large-scale preparation of the 1:2 photoadduct from benzene and maleic anhydride have been published [48–50]. Zhubanov et al. [49] realized a yield of 40% by irradiation of benzene and maleic anhydride in the presence of acetophenone or benzophenone in a quartz reactor and 65% in a Pyrex tube. Increasing the amount of sensitizer above 0.75 mol/mol maleic anhydride did not affect the photoadduct yield. Nurkhodzhaev et al. [50] obtained the highest adduct yield, 76%, in a batch reaction by irradiating the reactants with polychromatic ultraviolet (UV) and visible light at 30°C. With increasing temperature, the reaction rate increased but the adduct yield decreased, due to the occurrence of secondary reactions. An even higher yield, ~91%, was obtained by conducting the reaction under continuous conditions. The optimum reactant ratio was maleic anhydride:acetophenone:benzene = 1.0:75:10.

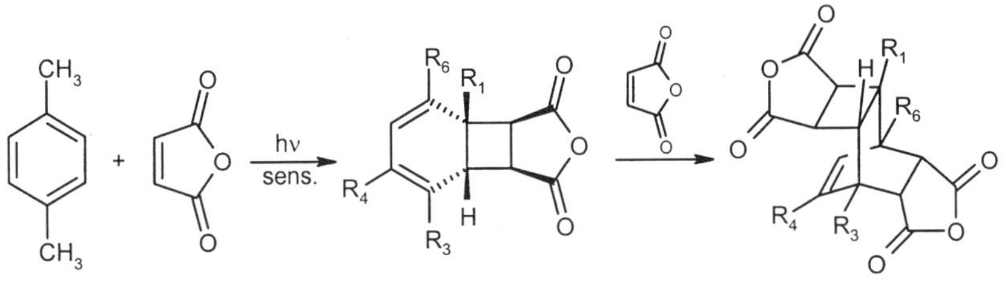

A1: $R_3 = R_6 = CH_3$; $R_1 = R_4 = H$ A2: $R_3 = R_6 = CH_3$; $R_1 = R_4 = H$

B1: $R_1 = R_4 = CH_3$; $R_3 = R_6 = H$ B2: $R_1 = R_4 = CH_3$; $R_3 = R_6 = H$

Scheme 8 Adducts formed by irradiation of *o*-xylene and maleic anhydride (sens. = benzophenone).

The adduct has also been prepared by exposing solutions of maleic anhydride and acetophenone or benzophenone in benzene in tubular glass reactors to solar radiation at various altitudes [51]. The adduct was obtained in high purity.

Ortho photocycloadditions of benzene derivatives to maleic anhydride have been tabulated in Table 1. Only the structures of the primary ortho adducts are given, but these are not the isolated adducts: They always undergo endo [2 + 4] cycloaddition with maleic anhydride, yielding 1:2 adducts. An interesting feature to be seen from Table 1 is that substituents on the benzene (alkyl, phenyl, or halogen) always turn up at the position most remote from the site of addition. In view of the different nature of these substituents, it seems that steric rather than electronic factors are responsible for this regioselectivity.

III. ORTHO PHOTOCYCLOADDITION VIA EXCITATION OF THE ALKENE OR ALKYNE FOLLOWED BY REACTION WITH GROUND-STATE ARENE

Ortho photocycloaddition to benzene of derivatives of acetylene and maleimide proceeds via excitation of the alkyne or the maleimide. A few other alkenes follow the same route to ortho photocycloadducts; among those are dichlorovinylene carbonate and some alkenes in which the double bond is conjugated with a cyano, carbonyl, or phenyl group, which makes it possible to excite them in the presence of the arene.

A. Alkynes

Bryce-Smith and Lodge [3] have discovered that methyl propiolate, dimethyl acetylenedicarboxylate, and phenylacetylene add to benzene upon irradiation at ~50°C. The products are cyclooctatetraenes, formed via initial ortho addition of the acetylene to benzene followed by ring opening (Scheme 9).

The photoreaction of dimethyl acetylenedicarboxylate at room temperature was also reported by Grovenstein and co-workers [57,58]. From acetylene and benzene, only traces of cyclooctatetraene could be detected [59,60].

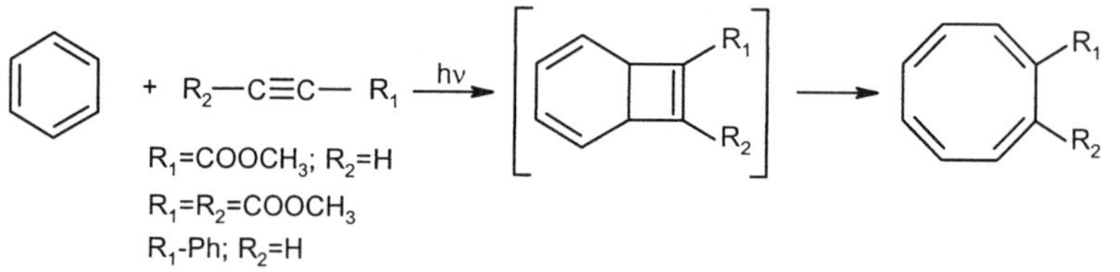

R_1=COOCH$_3$; R_2=H
R_1=R_2=COOCH$_3$
R_1-Ph; R_2=H

Scheme 9   Photocycloaddition of derivatives of acetylene to benzene.

Table 1   Ortho Photocycloaddition of Arenes to Maleic Anhydrides

Arene	Anhydride	Primary adducts	Remarks	Ref.
			solution of maleic anhydride in benzene; primary adduct gives Diels-Alder reaction with MA	2,23,24
			product isolated as Diels-Alder adduct with added MA or tetracyanoethylene	46
			benzophenone-sensitized; product isolated as Diels-Alder adduct with dichloromaleic anhydride	52
			benzophenone-sensitized; products isolated as Diels-Alder adducts with MA; a:b = 1:1 at 20° and 3:2 at 100°	33,34,45
			acetophenone-sensitized; product isolated as Diels-Alder adduct with MA	34
			acetophenone-sensitized; product isolated as Diels-Alder adduct with MA; yield 80%	53
			acetophenone-sensitized; product isolated as Diels-Alder adduct with MA; yield 70%	34,53

Table 1 Continued

Arene	Anhydride	Primary adducts	Remarks	Ref.
n-Bu		n-Bu	acetophenone-sensitized; product isolated as Diels-Alder adduct with MA; yield 78%	53
i-Bu		i-Bu	acetophenone-sensitized; product isolated as Diels-Alder adduct with MA; yield 61%	53
t-Bu		t-Bu	acetophenone-sensitized; product isolated as Diels-Alder adduct with MA; yield 58%	33,34,53
n-Hex		n-Hex	acetophenone-sensitized; product isolated as Diels-Alder adduct with MA; yield 72%	53
Ph		Ph	in cyclohexane at 35°; benzophenone increases rate threefold; product isolated as Diels-Alder adduct with MA	33
F		F	acetophenone or benzophenone-sensitized; yield of Diels-Alder adduct with MA 7–8%	54–56
Cl		Cl	acetophenone or benzophenone-sensitized; yield of Diels-Alder adduct with MA 2–5%	33, 54–56
Br		Br	acetophenone-sensitized; yeild of Diels-Alder adduct with MA 0.66%	55

Table 1 Continued

Arene	Anhydride	Primary adducts	Remarks	Ref.
		a b	benzophenone-sensitized; products isolated as Diels-Alder adducts with MA; a:b = 1:1 at 20° and 2:1 at 100°	33,45
		a b	benzophenone-sensitized; products isolated as Diels-Alder adducts with MA; a:b = 20:1 at 20° and 1:10 at 100°	33,45,47
			product isolated as Diels-Alder adduct with MA	56
			product isolated as Diels-Alder adduct with MA	56
			benzophenone-sensitized; single product isolated as Diels-Alder adduct with MA	45

In the presence of trifluoroacetic acid, irradiation of dimethyl acetylenedi-carboxylate and benzene leads to the formation of dimethyl fumarate and dimethyl maleate in appreciable amounts; in the absence of acid, only traces of these com-pounds were detected [36,61]. Irradiation of dimethyl acetylenedicarboxylate and benzene in the presence of tetracyanoethylene (TCNE) produces a 1:1:1 adduct at the expense of cyclooctatetraene. The photoaddition still proceeds readily behind a filter transmitting at wavelengths >290 nm, where only the acetylene absorbs to a significant degree [60].

It was first proposed [46] that the effect of acid was due to the intermediacy of a zwitterionic species, but later the reaction scheme was revised [61] (Scheme 10).

The principal route of formation of fumarate and maleate is protonation of one of the methoxycarbonyl groups of the bicyclo[4.2.0]oct-2,4,7-triene, followed by proton loss, leading to rearomatization and enolization. The formation of fu-marate and maleate in the absence of acid is ascribed to a photochemical 1,3 H-shift followed by opening of the new cyclobutene ring.

Tinnemans and Neckers [62,63] have published a reinvestigation of the photoaddition of methyl phenylpropiolate to benzene, originally reported by Bryce-Smith et al. [60]. Irradiation with a 450-W medium-pressure mercury lamp through Pyrex produces methyl 5-phenyltetracyclo[$3.3.0.0^{2,4}.0^{3,6}$]oct-7-ene-4-carboxylate (**I**) (65–70%) (Scheme 11).

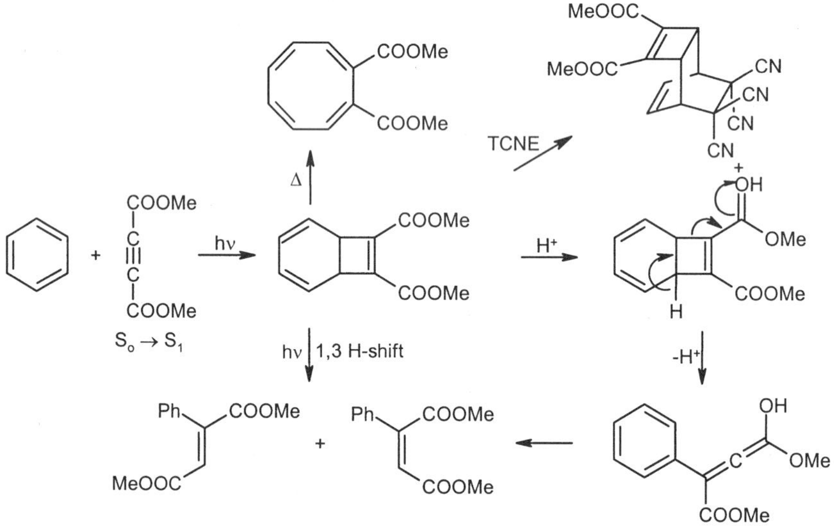

Scheme 10 Mechanism of the photoaddition of dimethyl acetylenedicarboxylate to benzene.

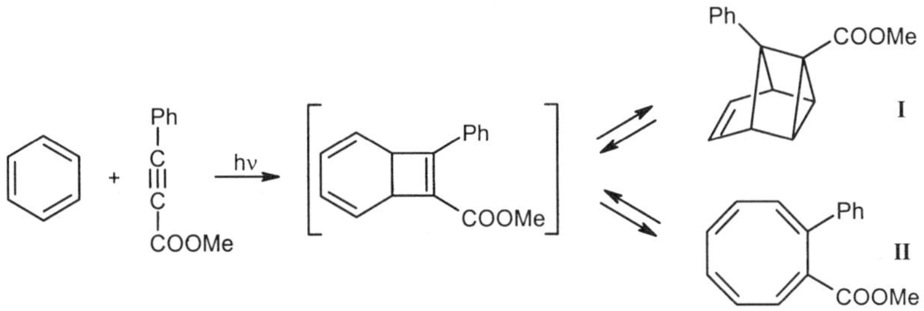

Scheme 11 Photochemical addition of methyl phenylpropiolate to benzene.

However, through quartz at 95% conversion, a mixture of **I** (62%) and 1-carboxymethyl-2-phenylcyclooctatetraene (**II**) (38%) was obtained. Compound **II** could be obtained exclusively (92%) from triplet-sensitized irradiation of **I** with high-energy sensitizers or from the sensitized reaction of methyl phenylpropiolate with benzene. From their experiments with sensitizers, the authors concluded that the primary adduct is formed from triplet alkyne (E_T < 69 kcal/mol^{-1}) and ground-state benzene and that the formation of **I** also proceeds via a triplet state in a two-step radical reaction. Hanzawa and Paquette [64] have also used the photochemical addition of an alkyne to benzene to produce a derivative of tetra-cyclo[3.3.0.0^{2,4}.0^{3,6}]oct-7-ene.

Šket and coworkers [65,66] have irradiated 1-phenyl-2-alkylacetylenes (Ph—C≡C—R; R = H, CH$_3$, n-C$_3$H$_7$, t-C$_4$H$_9$) in the presence of hexafluoroben-zene. They have succeeded in isolating the primary ortho addition products, 1,2,3,4,5,6-hexafluoro-7-phenyl-8-R-bicyclo[4.2.0]oct-2,4,7-trienes in high yield. Upon heating, these compounds were converted into cyclooctatetraenes. The quantum yield of the photoaddition reaction depends on the structure of the acetylene, the concentration of hexafluorobenzene, and the solvent polarity, but not on the concentration of the acetylene. This finding is in agreement with the reaction of excited acetylene with ground-state hexafluorobenzene.

Ortho photocycloaddition of acetylenes to hexafluorobenzene has also been found by Bryce-Smith et al. [67]. Irradiation at 20°C of hexafluorobenzene saturated with propyne gave one major product that could be identified as 1,2,3,4,5,6-hexafluoro-7-methylcyclooctatetraene, apparently formed by ring opening of the primary ortho adduct. When but-2-yne in hexafluorobenzene was irradiated, three 1:1 adducts were formed that could be separated and identified. The ratios were dependent on the time of irradiation and on the temperature in-volved in the workup and analysis procedures. One of the products proved to be the primary ortho adduct, 1,2,3,4,5,6-hexafluoro-7,8 dimethylbicyclo[4.2.0]octa-2,4,7-triene. The second product was the cyclooctatetraene formed by ring open-ing of ortho adduct and the third was a tricyclic compound, formed by 4π ring closure of the butadiene unit in the primary adduct. No evidence for charge-transfer absorption was obtained, but the absorption spectrum of but-2-yne showed the presence of a weak absorption (λ_{max} = 265 nm, ϵ <1) which extended to 310 nm. Indeed, irradiation of mixtures of hexafluorobenzene and but-2-yne at wavelengths >300 nm led to the formation of the ortho adduct. These results are thus consistent with the earlier finding [60] that excitation of the acetylene rather than the arene is involved in these ortho photocycloadditions.

Morrison and co-workers [68,69], however, found that the internal photocycloaddition of 6-phenylhex-2-yne, producing 2-methylbicyclo[6.3.0]undeca-1,3,5,7-tetraene, must proceed via the phenyl excited singlet state. *E*-2-Heptene has a negligible quenching effect on the cycloaddition, yet its isomerization is effectively sensitized by the phenyl triplets of the arylhexyne. The authors state that in their molecule, the site of excitation (the phenyl moiety) is unequivocal [68]. When 6-phenylhex-2-yne is irradiated in hexane saturated with xenon, the yield of addition product is decreased by 53% compared to irradiation in argon-saturated solution [69]. This is due to enhancement of intersystem crossing to the unproductive triplet state. The yield of bicyclo[6.3.0]undecatetraenes is considerably improved when the acetylene bears a trimethylsilyl group [70]. For example, irradiation of 1-trimethylsilyl-3-hydroxy-5-phenylpent-1-yne in hexane gave 2-trimethylsilyl-11-hydroxybicyclo[6.3.0]undeca-1,3,5,7-tetraene in 46% yield (Scheme 12). The reaction could not be sensitized.

Photoaddition of perfluorobut-2-yne to benzene has been accomplished in the vapor phase. The product is 1,2-bis(trifluoromethyl)cycloocta-1,3,5,7-tetraene [71]. In solution, this product undergoes a Diels–Alder reaction with perfluorobut-2-yne.

The photocycloaddition of cyclooctyne to benzene [72], producing bicyclo[6.6.0]tetradeca-1,3,5,7-tetraene can be sensitized (with acetone) and quenched (with piperylene). The unsensitized reaction occurs with very high efficiency (56% yield at 66% conversion). Because transfer of triplet energy from acetone to benzene is improbable, the authors consider the possibility that the acetylene triplet may be the reactive species in the cycloaddition.

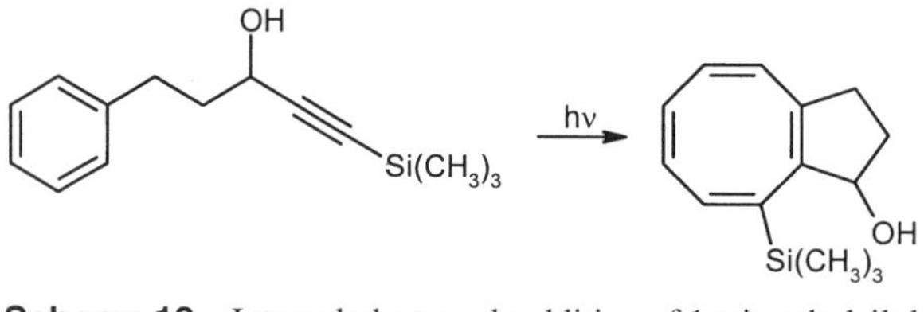

Scheme 12   Internal photocycloaddition of 1-trimethylsilyl-3-hydroxy-5-phenylpent-1-yne.

Atkinson et al. [73] have described the photochemical addition of 3-hexyne and 5-decyne to benzonitrile. The products were 2,3-diethylcyclooctatetraene-1-carbonitrile and 2,3-n-butylcyclooctatetraene-1-carbonitrile, respectively. The ortho addition apparently takes place at positions 1 and 2 of benzonitrile. Subsequent ring opening of the initial bicyclo[4.2.0]octa-2,4,7-trienes leads to the 1,2,3-trisubstituted cyclooctatetraenes.

B. Maleimides

Bradshaw [74] reported in 1966 that maleimide undergoes sensitized photochemical addition to benzene, producing the imide analog of the corresponding maleic anhydride photoadduct. The author observed that the reaction probably proceeds by the addition of electronically excited maleimide to benzene. The photoaddition proved successful also with toluene, t-butylbenzene, and ethylbenzene. Simultaneously, Bryce-Smith and Hems [75] reported that 2:1 photoadducts are formed from maleimide, N-n-butyl-, N-benzyl-, N-o-tolyl-, and N-2,6-xylylmaleimide with benzene. N-Phenyl, N-p-tolyl-, and N-p-methoxyphenylmaleimide did not form photoadducts. Trifluoroacetic acid was found to be virtually without effect on the photoaddition of N-n-butylmaleimide: No phenyl-N-n-butylsuccinimide was detected [36]. It was concluded that a dipolar intermediate is not involved. When N-n-butylmaleimide and benzene were irradiated in the presence of tetracyanoethylene, a 1:1:1 adduct was formed [37].

Although maleimides form ground-state complexes with benzene, such complexes are not responsible for the occurrence of the photoaddition. Their absorption spectra are hidden below the absorption spectrum of benzene and do not extend beyond 270 nm, yet excitation above 280 nm readily leads to photoaddition of maleimide and some of its N-substituted derivatives. The mechanism proposed by Bryce-Smith [38,46] accounts for all of the facts (Scheme 13).

Maleimide also forms 2:1 adducts with acceptor compounds such as benzonitrile, acetophenone, and methyl benzoate [46]. This behavior contrasts with that of maleic anhydride, which neither exhibits charge-transfer absorption with nor photoadds to benzenes bearing strong electron-acceptor substituents. It clearly demonstrates that formation of a ground-state charge-transfer complex is not essential in the formation of 2:1 adducts from maleimide and benzene derivatives.

The efficiency of photoaddition of maleimides to benzene strongly depends on the N-substituent [37,75]. Substituents possessing a π-electron system that can have significant overlap with the nitrogen lone-pair orbital (phenyl, m- and p-tolyl, p-methoxyphenyl) almost completely inhibit the reaction. Maleimides with substituents which either have no π-electron system (H, alkyl) or have a π-electron system which is prevented from significant overlap with the orbital on nitrogen (benzyl, o-tolyl, 2,6-xylyl) show efficient reaction. The lack of reactivity of maleimides with conjugating substituents is considered to result from the

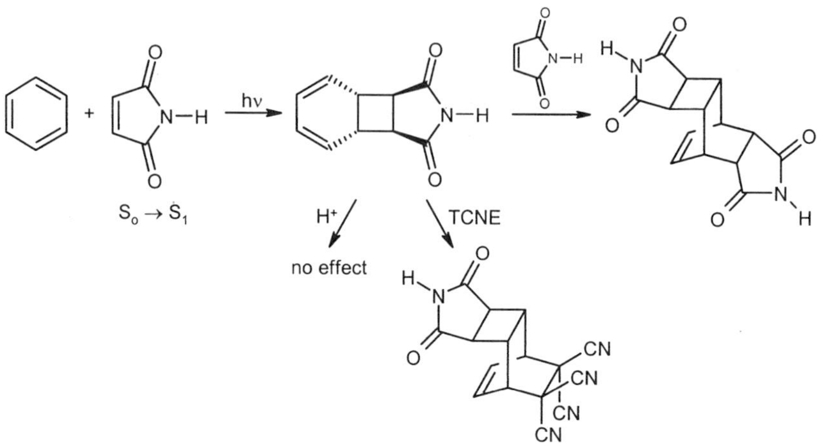

Scheme 13   Mechanism of the photoaddition of maleimide to benzene.

conjugated π-electron system acting as an energy sink [46]. Indeed, the conjugated maleimides have ultraviolet absorption spectra that are shifted considerably to higher wavelengths. This makes it possible to predict whether a particular N-substituted maleimide will be able to form a 2:1 adduct with benzene. Those that are yellow will not react, the pale yellow compounds are of intermediate reactivity, and the colorless ones will react readily.

As stated earlier, charge transfer between maleimide and benzene does not play a significant role in the photoaddition process. In an attempt to increase the importance of charge-transfer excitation, Bryce-Smith et al. [76] have irradiated solutions of maleimide in anisole. The irradiations were performed both in the absence and in the presence of a filter solution excluding radiation of wavelength <327 nm. In both cases, the same mixture of adducts in essentially the same total yield and ratios was obtained. The product mixture consisted of three isomeric 2:1 adducts which result from initial 1,2- 2,3-, and 3,4-addition (ratio 10:10:1) to the aromatic ring followed by Diels–Alder addition of a second molecule of maleimide. The resulting adducts are shown in Scheme 14.

In the presence of 1 M trifluoroacetic acid, formation of the 2:1 adducts was totally suppressed, but no phenylsuccinimides were detected. The authors also studied the effect of the addition of methanol. It was found that upon unfiltered irradiation, methanol greatly promotes formation of the adduct resulting from initial 1,2-addition to anisole. The ratio of 1,2-:2,3-:3,4-addition changed from 10:10:1 in the absence of methanol to 30:10:1 in the presence of 1 M methanol and 60 : 10:1 with 10 M methanol. This selective effect was not observed on irradiation exclusively in the charge-transfer band. The authors conclude that the 1,2-addition favored with anisole arises from the presence of the oxygen atom, with its

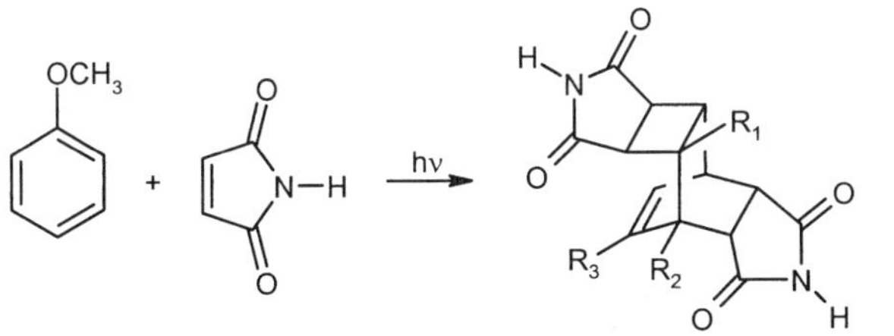

R$_1$=OCH$_3$; R$_2$=R$_3$=H (1,2-addition)

R$_1$=H; R$_2$=OCH$_3$; R$_3$=H (2,3-addition)

R$_1$=R$_2$=H; R$_3$=OCH$_3$ (3,4-addition)

Scheme 14   Photochemical addition of maleimide to anisole.

ability to act as a donor to the maleimide. Such electron donation might involve methanol-promoted complexation which locates the maleimide molecule in proximity to the 1,2 bond, thus favoring addition at this position.

The photoaddition of maleimide to biphenyl has also been reported [77]. The initial addition takes place at positions 3 and 4 of biphenyl and the 2:1 adduct has been obtained in 40% yield.

Wamhoff and Hupe [78] have irradiated 3,4-dibromo-N-methylmaleimide in benzene in the presence and in the absence of a sensitizer (λ < or >313 nm). A mixture of products was formed in which two compounds were detected that had the skeleton of the primary ortho adduct. One of these products had been formed from the primary adduct by the addition of bromine at one of the double bonds of the diene moiety. The second compound had undergone further bromination at the allylic position in the six-membered ring. The photodimer of 3,4-dibromo-N-methylmaleimide, which was also formed, was considered to be the source of the bromine. This is one of the rare instances in which the primary 1:1 adduct of benzene and a maleimide derivative could be isolated, albeit in brominated form.

Pettit et al. [25] have subjected the 2:1 maleimide–benzene adduct to reaction with ammonium hydroxide in water. The solution was kept at room temperature for 14 days and a product was obtained in crystalline form as needles melting above 340°C. This product was identified as a diamide; the imide not connected to the four-membered ring had undergone ammonolysis. An x-ray crystal structure determination of this amide was performed, confirming the stereochemical structure of the maleimide–benzene 2:1 adduct as depicted in Scheme 13. The same authors reported that the 2:1 adduct displayed antitumor activity and was being prepared for eventual clinical trial. The systematic name of the adduct is 3a,3b,4,4a,7a,8,8a,8b-octahydro-4,8-ethenopyrrolo[3′,4′:3,4] cyclobut[1,2f]isoindole-1,3,5,7[2H,6H]-tetrone. Because this is a rather forbidding name for a potential drug, the product was designated mitindomide.

Arkhipova et al. [79] have optimized the synthesis of the adduct. The best yield (65%) was obtained with a benzene solution containing 0.16 M acetophenone and 15 g/L of maleimide and a reaction time of 20 hr at 25°C.

C. Alkenes and Nitriles

As mentioned earlier, maleimide also photoadds to acceptor molecules such as benzonitrile, acetophenone, and methyl benzoate. A similar situation was encountered by Gilbert and Yianni [80], who found that methyl acrylate, methyl methacrylate, acrylonitrile, and methacrylonitrile undergo ortho addition to benzonitrile when irradiated at 254 nm at concentrations where the alkenes absorbed ~60% of the incident radiation. The products are derivatives of 1-cyano- and 3-cyanobicyclo[4.2.0]octa-2,4-dienes, indicating 1,2 and 3,4 attack on the benzonitrile. The location and configuration of the substituents descendant from the alkene (at position 7 or 8) could not be established. No evidence was found for ground-state complexation between the addends, and the formation of excited-state complexes was not considered favorable. The results were rationalized in terms of attack by singlet excited alkene onto ground-state benzonitrile, consistent with predictions [81] from orbital symmetry considerations.

Another reaction in which the excited alkene is the reactive species in ortho photocycloaddition is that of dichlorovinylene carbonate with benzene [82–84] (Scheme 15).

This reaction proceeds only upon sensitization, preferably by acetophenone. The initial reaction is ortho addition of triplet excited dichlorovinylene carbonate to benzene, leading to endo and exo adducts. The exo isomer is quite stable and can be isolated, whereas the endo isomer is very sensitive to further reactions. The authors have considered the possibility of a stepwise reaction, but they prefer a formulation of the reaction mechanism in which an adiabatic reaction leads to triplet excited products. These decay to the ground-state ortho adducts or may

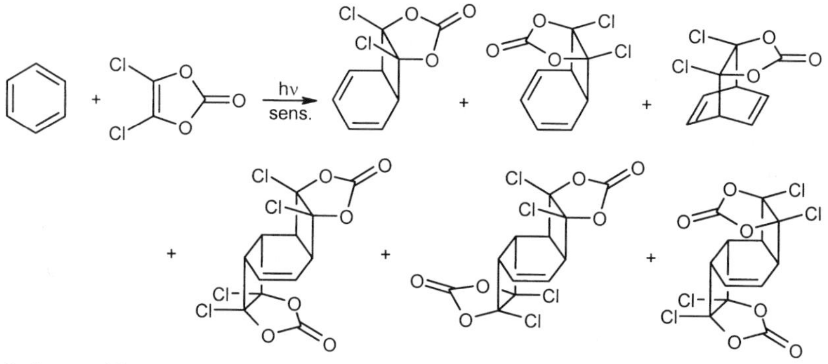

Scheme 15 Ortho photocycloaddition of benzene and dichlorovinylene carbonate.

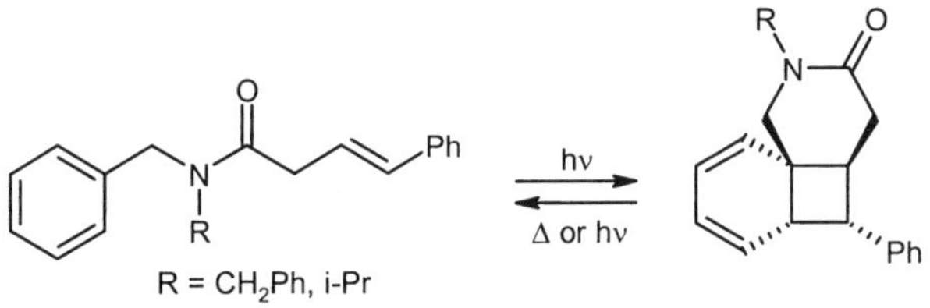

Scheme 16 Ortho photocycloaddition of N-benzylstyrylacetamides. (From Ref. 85.)

undergo a rearrangement to the para addition product. The ortho adducts undergo a second addition of dichlorovinylene carbonate producing the exo–endo, exo–exo, and endo–endo 1:2 adducts.

When N,N-dibenzylstyrylacetamide (Scheme 16, R = CH$_2$Ph) in methanol was irradiated with a low-pressure mercury lamp, the ortho cycloadduct was obtained in 95% yield at 75% conversion [85]. Similarly, irradiation of N-benzyl-N-isopropylstyrylacetamide (Scheme 16, R = iPr) gave the corresponding ortho adduct in 63% yield at 55% conversion. These reactions are presumed to proceed via excitation of the styrene moiety in view of the extinction coefficient at 254 nm ($\epsilon = 2 \times 10^4$ when R = CH$_2$Ph). The absorption and fluorescence spectra of N,N-dibenzylstyrylacetamide are almost the same as those of β-methylstyrene. This indicates the absence of interaction between the styryl group and the phenyl group in both the ground and singlet excited state, although the intervention of nonemissive exciplexes is not excluded.

The carbon–nitrogen triple bond can also undergo ortho photocycloaddition to derivatives of benzene. Al-Jalal et al. [86] found that irradiation of 4-cyanoanisole in acrylonitrile produced three 1 : 1 adducts. Two of these were formed by the addition of the carbon–carbon double bond of acrylonitrile to positions 1,2 and 3,4, respectively, of 4-cyanoanisole. The third product was an aza-cyclooctatetraene, apparently formed by the addition of the carbon–nitrogen triple bond to the arene, followed by ring opening [87].

4-Cyanoanisole adds in the same way to benzonitrile [86]. It could be established that photoexcited 4-cyanoanisole, not photoexcited nitrile, is the species fruitful of reaction [87].

These reactions are mentioned here because of the apparent similarity between the carbon–carbon and the carbon–nitrogen triple bonds, but because the reactive excited state is that of the arene, they do not fall in the category of alkynes and maleimides treated in this section. A somewhat more detailed discussion on photoaddition of nitriles to the benzene ring can be found in Section VI.

Intermolecular and intramolecular ortho photocycloadditions of derivatives of benzene to alkynes are collected in Table 2. Nitriles that give ortho cycloadducts are listed at the end of this table, although mechanistically, as explained, they fall in another category. Table 3 contains the ortho photocycloaddi-

Table 2 Ortho Photocycloaddition of Arenes to Alkynes and Nitriles

Arene	Alkyne	Primary adducts	Remarks	Ref.
	H–C≡C–H		acetylene-saturated benzene under N_2 at 20° (254 nm) $\varphi < 0.001$; product rearranges to cyclooctatetraene	60
			cyclooctyne in benzene (254 nm); yield of ring-opened adduct 56% at 66% conversion	72
	MeOOC–C≡C–H	COOMe	methyl propiolate in benzene under N_2 at 53°; adduct undergoes ring opening	3,59
	t-Bu–C≡C–COOMe	COOMe t-Bu	0.035M alkyne in benzene (254 nm); adduct undergoes intramolecular ring closure	64
	Ph–C≡C–COOMe	COOMe Ph	6.25mM alkyne in benzene; adduct undergoes ring opening as well as ring closure, depending on conditions	60,62,63
	MeOOC–C≡C–COOMe	COOMe COOMe	dimethyl acetylenedicarboxylate in benzene under N_2 at 50–55°; adduct undergoes ring opening; yield 15%	3,57–59
	Ph–C≡C–H	Ph	phenylacetylene in benzene under N_2 at 40°; adduct undergoes ring opening; other products are also formed	3,59
	F_3C–C≡C–CF$_3$	CF$_3$ CF$_3$	benzene and perfluorobutyne in the vapor phase (254 nm); major product is ring-opened adduct	71

Table 2 Continued

Arene	Alkyne	Primary adducts	Remarks	Ref.
CN (benzonitrile)	Et–C≡C–Et	CN, Et, Et (bicyclic adduct)	3-hexyne and benzonitrile in methanol; yield of cyclooctatetraene derivative 8%	73
CN (benzonitrile)	Bu–C≡C–Bu	CN, n-Bu, n-Bu (bicyclic adduct)	5-decyne and benzonitrile in methanol; the cyclooctatetraene derivative is isolated	73
hexafluorobenzene	H–C≡C–CH$_3$	CH$_3$ (fluorinated bicyclic adduct)	hexafluorobenzene saturated with propyne (254 nm); primary adduct undergoes ring opening to cyclooctatetraene	67
hexafluorobenzene	H$_3$C–C≡C–CH$_3$	CH$_3$, CH$_3$ (fluorinated bicyclic adduct)	but-2-yne in hexafluorobenzene (254 nm); primary adduct + cyclooctatetraene + ring-closed product	67
hexafluorobenzene	Ph–C≡C–H	Ph (fluorinated bicyclic adduct)	solution in cyclohexane (254 nm); primary adduct undergoes ring opening to cyclooctatetraene; yield 40%	65,66
hexafluorobenzene	Ph–C≡C–CH$_3$	Ph, CH$_3$ (fluorinated bicyclic adduct)	solution in cyclohexane (254 nm); the primary adduct can be isolated; φ = 0.035 ± 0.005	65,66
hexafluorobenzene	Ph–C≡C–n-Pr	Ph, n-Pr (fluorinated bicyclic adduct)	solution in cyclohexane (254 nm); the primary adduct can be isolated φ = 0.053 ± 0.001	65,66

Table 2 Continued

Arene	Alkyne	Primary adducts	Remarks	Ref.
	Ph—C≡C—t-Bu		solution in cyclohexane (254 nm); the primary adduct can be isolated; $\varphi = 0.133 \pm 0.001$	65,66
	Ph—C≡C—CH₃		solution in cyclohexane (254 nm); the primary adduct undergoes rearrangement; $\varphi = 0.028 \pm 0.002$	66
	Ph—C≡C—t-Bu		solution in cyclohexane (254 nm); the primary adduct can be isolated; $\varphi = 0.028 \pm 0.004$	66
	Ph—C≡C—CH₃		solution in cyclohexane (254 nm); the primary adduct undergoes rearrangement; yield 10.8%	66
	Ph—C≡C—t-Bu		solution in cyclohexane (254 nm); the primary adduct can be isolated; $\varphi = 0.027 \pm 0.002$	66
	Ph—C≡C—CH₃		solution in cyclohexane (254 nm); the primary adduct undergoes rearrangement; yield 19.7%	66
	Ph—C≡C—t-Bu		solution in cyclohexane (254 nm); the primary adduct can be isolated; $\varphi = 0.026 \pm 0.002$	66

Table 2 Continued

Arene	Alkyne	Primary adducts	Remarks	Ref.
			0.03M solution in argon degassed cyclopentane (254 nm); primary adduct opens to cyclooctatetraene; $\varphi = 3.3 \times 10^{-3}$	68
			solution in hexane under N_2 (254 nm); yield of cyclooctatetraene derivative 46%	70
			solution in hexane under N_2 (254 nm); yield to cyclooctatetraene derivative 38%	70
			solution in hexane under N_2 (254 nm); yield of cyclooctatetraene derivative 31%	70
			solution in hexane under N_2 (254 nm); yield to cyclooctatetraene derivative 42%	70
			solution in hexane under N_2 (254 nm); yield of product after ring opening and elimination 55%	70
			solution in hexane under N_2 (254 nm); yield of cyclooctatetraene derivative 50%	70
			solutions in CH_3CN (low pressure Hg lamp); X = H, o-, m-, p-OMe, o-, m-, p-Me, o-, m-, p-Cl; adduct undergoes ring opening	88

Table 2 Continued

Arene	Alkyne	Primary adducts	Remarks	Ref.
			in acetonitrile; yields 15–20% R = H, Me, OMe	89
		a	0.2M arene in acrylonitrile under N_2 (low pressure Hg lamp); (ring-opened) a:b:c = 2:7:1	86,87
		b c		
			ring-opened adduct (azocine) is sole product in ethyl acetate	86,87
			primary adduct undergoes ring opening to azocine derivative	86
		a	in acetonitrile under N_2 (low pressure Hg lamp); ring-opened a:b:c = 1:5:1	90
		b c		

Table 3 Ortho Photocycloaddition of Arenes to Maleimides

Arene	Maleimide	Primary adducts	Remarks	Ref.
(benzene)	N–H maleimide	H H / H H adduct, N–H	maleimide, acetophenone or benzophenone in benzene; primary product forms Diels-Alder adduct with MI; yield 82%	74,75
(benzene)	N–n-Bu maleimide	H H / H H adduct, N–n-Bu	N-n butylmaleimide in benzene; can be sensitized by benzophenone primary product forms Diels-Alder adduct with N-n-BuMI	75
(benzene)	N–CH$_2$Ph maleimide	H H / H H adduct, N–CH$_2$Ph	N-benzylmaleimide in benzene; can be sensitized by benzophenone primary product forms Diels-Alder adduct with N-BnMI	75
(benzene)	N-o-tolyl maleimide	H H / H H adduct	N-o-tolylmaleimide in benzene; can be sensitized by benzophenone primary product forms Diels-Alder adduct with N-TolMI	75
(benzene)	N-xylyl maleimide	H H / H H adduct	N-xylylmaleimide in benzene; can be sensitized by benzophenone primary product forms Diels-Alder adduct with N-XylMI	75
(benzene)	Br, Br dibromomaleimide, N–	H Br / H Br adduct, N–	0.1 M imide in benzene; direct or sensitized irradiation; primary adduct reacts with Br$_2$, but not with the imide	78
(toluene)	N–H maleimide	H H / H H adduct, N–H (a) ; H H / H H adduct, N–H (b)	1% maleimide in toluene (>327 nm); products isolated as Diels-Alder adducts with MI; a:b = 2:1	74,76

Table 3 Continued

Arene	Maleimide	Primary adducts	Remarks	Ref.
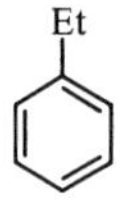 Et	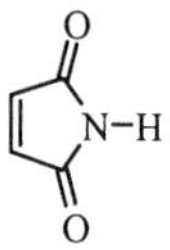	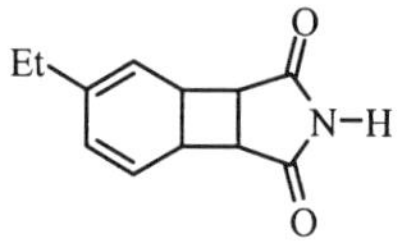 Et (adduct)	maleimide and acetophenone in ethylbenzene; primary product forms Diels-Alder adduct with MI; yield 71%	74
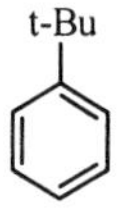 t-Bu	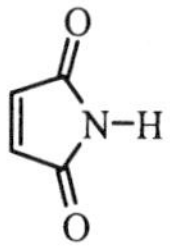	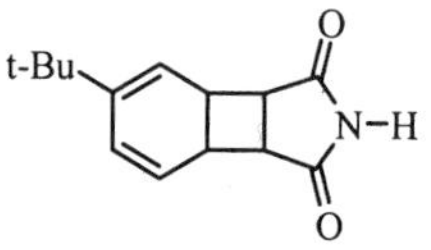 t-Bu (adduct)	maleimide and acetophenone in t-butylbenzene; primary product forms Diels-Alder adduct with MI; yield 35%	74
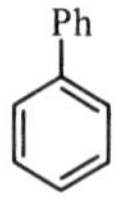 Ph	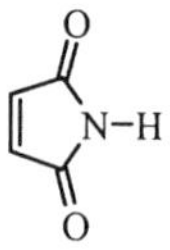	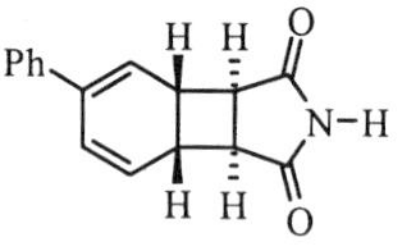 Ph (adduct)	acetophenone-sensitized in acetone/heptane; yield of Diels-Alder adduct with MI ca. 40%	77,91
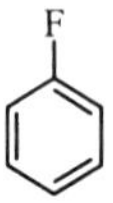 F	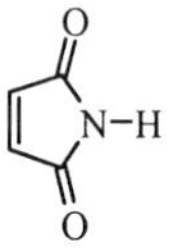	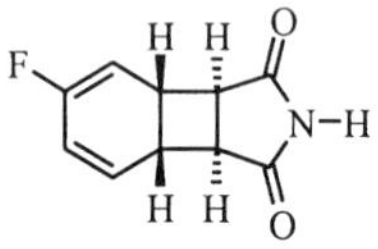 F (adduct)	acetophenone or benzophenone-sensitized; yield of Diels-Alder adduct with MI 80%	55
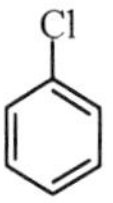 Cl	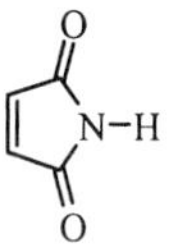	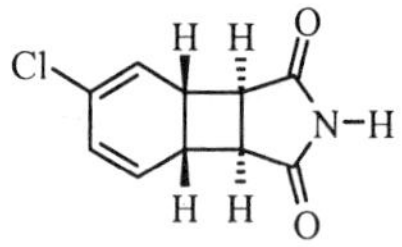 Cl (adduct)	acetophenone or benzophenone-sensitized; yield of Diels-Alder adduct with MI 58%	55
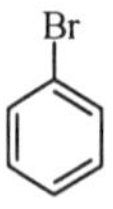 Br	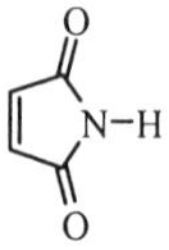	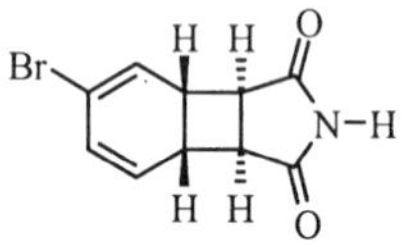 Br (adduct)	acetophenone or benzophenone-sensitized; yield of Diels-Alder adduct with MI 7.6%	55
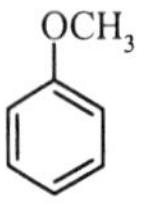 OCH_3	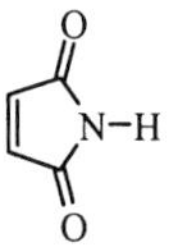	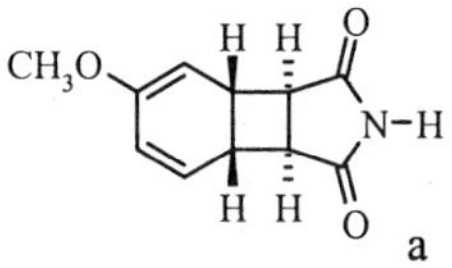 CH_3O (adduct) a	maleimide in anisole; direct irradiation a:b:c = 1:10:10 [76]; sensitized a:b:c = 1:1:1 [92]; sensitized only a (80%) [91]	76,91,92

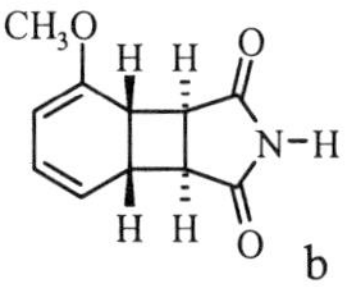

b   c

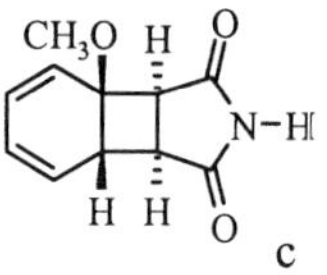

Table 3   Continued

Arene	Maleimide	Primary adducts	Remarks	Ref.
OPh			acetophenone-sensitized; yield of Diels-Alder adduct with MI 68%	91
SPh			acetophenone-sensitized; yield of Diels-alder adduct with MI 68%	91
		a b	acetophenone-sensitized; only traces of b; yield of Diels-Alder adduct with MI 33%	92
OMe OMe			acetophenone-sensitized in acetone/heptane; yield of Diels-Alder adduct with MI 66%	91
OMe OMe			acetophenone-sensitized in acetone/heptane; yield of Diels-Alder adduct with MI 31%	91
			acetophenone-sensitized; yield of Diels-Alder adduct with MI 21%	92

tions of arenes to maleimides. In all cases, as in Table 1, the primary ortho adducts are given although many of them cannot be isolated. The secondary reactions of the primary adducts are treated in Section VI.

IV. ORTHO PHOTOCYCLOADDITION OF TRIPLET-EXCITED AROMATIC KETONES TO ALKENES

Ortho photocycloaddition reactions proceeding via excitation of the arene followed by intersystem crossing to the triplet state and reactions with the alkene have been reported by Wagner [93–103], Gilbert [104,105], Pete [106], and their co-workers.

Most compounds studied have a carbonyl group conjugated with the benzene ring and nearly all reactions are intramolecular cycloadditions. In most cases, the alkenyl moiety is connected to the arene ring via an oxygen atom, ortho or para with respect to the carbonyl group.

In their first report [93], Wagner and Nahm describe the photochemical formation of yellow cyclooctatrienes from *ortho*- and *para*-alkenyloxyacetophenones and *ortho*- and *para*-valerophenones. In another publication [94], they reported further on the process and showed that the final product in most cases is a bicyclo[4.2.0]octa-2,7-diene derivative (Scheme 17).

The initial photochemical reaction is intramolecular ortho cycloaddition leading to bicyclo[4.2.0]octa-2,4-diene derivatives. These normally are not detected because they undergo rapid thermal disrotatory ring opening to all-cis cyclooctatrienes which have large extinction coefficients and absorb a second photon to undergo electrocyclic ring closure to the photostable bicyclo[4.2.0] octa-2,7-dienes.

The kinetics performed on the para-substituted ketones [93] demand two-step, revertible addition of the double bond to the para carbon of the benzene ring. Prompt closure of the 1,4 biradicals yields the ortho cycloadducts.

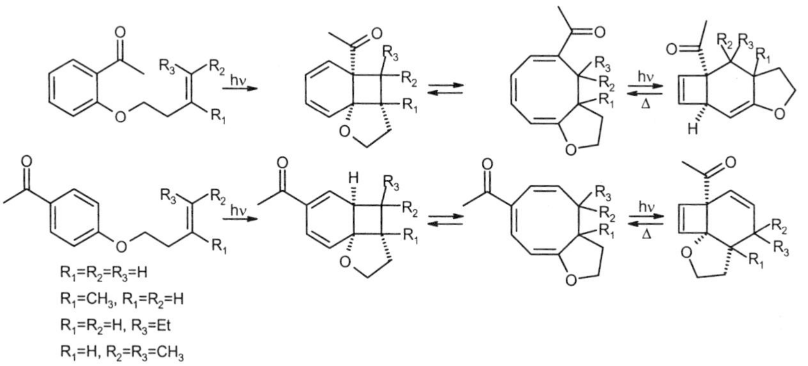

$R_1=R_2=R_3=H$
$R_1=CH_3, R_1=R_2=H$
$R_1=R_2=H, R_3=Et$
$R_1=H, R_2=R_3=CH_3$

Scheme 17   Ortho photocycloaddition of alkenyloxyacetophenones.

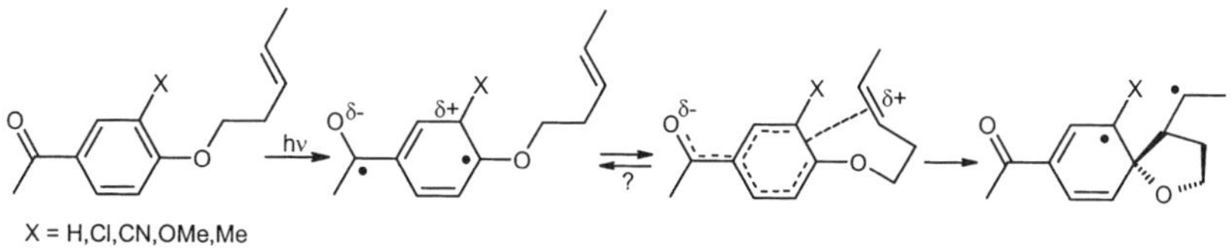

X = H,Cl,CN,OMe,Me

Scheme 18   Mechanism of ortho photocycloaddition of alkenyloxyacetophenones.

Wagner and Sakamoto [95] have studied the triplet decay of *para*-alkeny-loxyacetophenones with a substituent (X = Cl, CN, OMe, Me) at the position ortho to the alkenyloxy side chain. They propose that the cycloaddition reaction involves charge-transfer interaction between donor double-bond and acceptor triplet benzene first, followed by biradical formation (Scheme 18).

Both ortho- and para-substituted ketones react rapidly, whereas their meta isomers give only long-lived triplets that do not add. Alkyl substitution on the tether speeds up the reaction only when it is on the terminus of the double bond. Electron-withdrawing substituents ortho to the unsaturated ether group enhance reactivity, whereas similar electron-donating substituents depress reactivity.

This work has been extended [96] to 1-butenoxy-2-acetonaphthones and 2-butenoxy-1-acetonaphthones. Both types of compounds undergo high-yield internal ortho cycloadditions from their triplet states (Scheme 19).

The effect of substituents at the position ortho to the tether on the regio-chemistry of the addition in *para*-alkenyloxyacetophenones has been examined

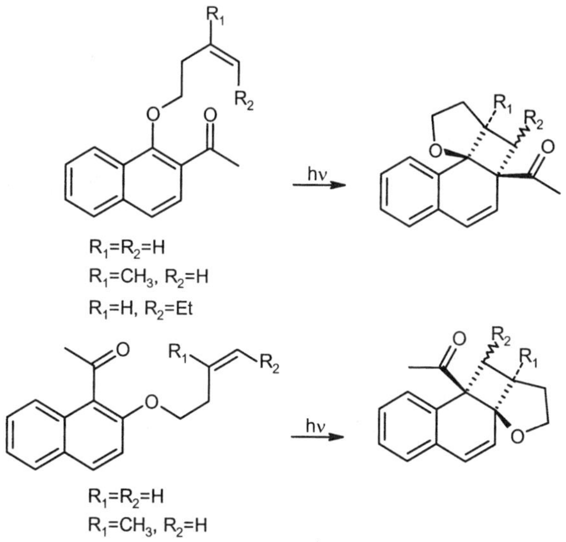

$R_1=R_2=H$
$R_1=CH_3, R_2=H$
$R_1=H, R_2=Et$

$R_1=R_2=H$
$R_1=CH_3, R_2=H$

Scheme 19   Intermolecular photocycloaddition of alkenyloxyacetonaphthones.

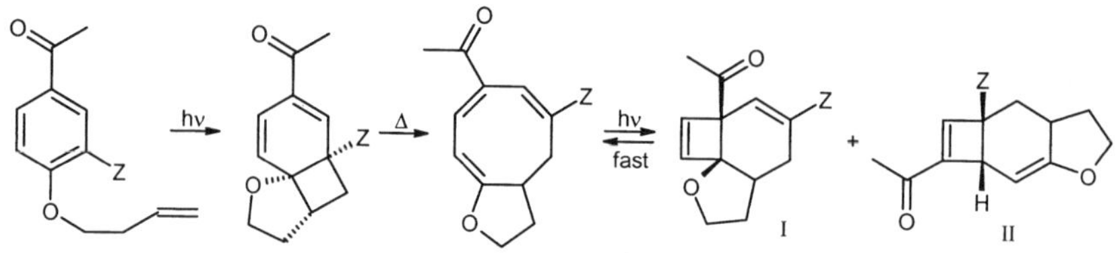

Scheme 20   Ortho addition toward the substituent if Z = CONH$_2$, CN, or CH$_3$.

[97]. When the substituent is CONH$_2$, CN, or CH$_3$, ortho addition occurs toward the substituent (Scheme 20).

From the amide, a >90% yield of a single product **I** (Z = CONH$_2$) is produced. Irradiation of the nitrile produces, in >90% yield, a mixture of **I** (Z = CN) and **II** (Z = CN). Irradiation of the methyl derivative produces mainly I (Z = CH$_3$).

However, if the ortho substituent is OCH$_3$ or SCH$_3$, the regiochemistry of the addition is reversed (Scheme 21). When Z = OCH$_3$, prolonged irradiation leads to a mixture of 80% **III** (Z = OCH$_3$) and 20% **I** (Z = OCH$_3$). From the thiomethoxy derivative only **III** (Z = SCH$_3$) is produced. The explanation offered by the authors is based on the charge-transfer character expected for an exciplex intermediate, so that the double bond shuns electron-rich sites on the benzene ring (such as near methoxy) and is attracted to the more electron-deficient sites near electron-withdrawing groups (Scheme 22).

The diastereoselectivity with which butenoxy acetophenones undergo intramolecular [2 + 2] photocycloadditions has been studied by Wagner and Cheng [98] (Scheme 23).

In every case, R$_1$ and R$_2$ are trans to each other in the major product. R$_2$ is always cis to the cyclobutene ring in the final product. The initial cycloaddition to the benzene ring occurs syn; ring opening proceeds disrotatory and yields a boat-shaped all-cis cyclooctatriene. The photochemical 4π ring closure is also disrotatory; it occurs in such a manner that the five-membered ring is trans to the

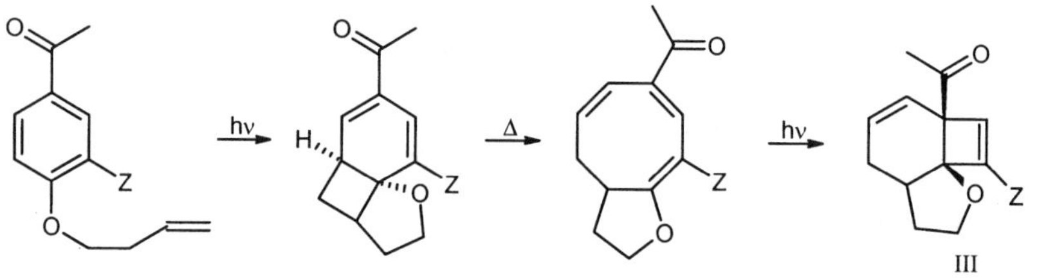

Scheme 21   Ortho addition away from the substituent if Z = OCH$_3$, SCH$_3$.

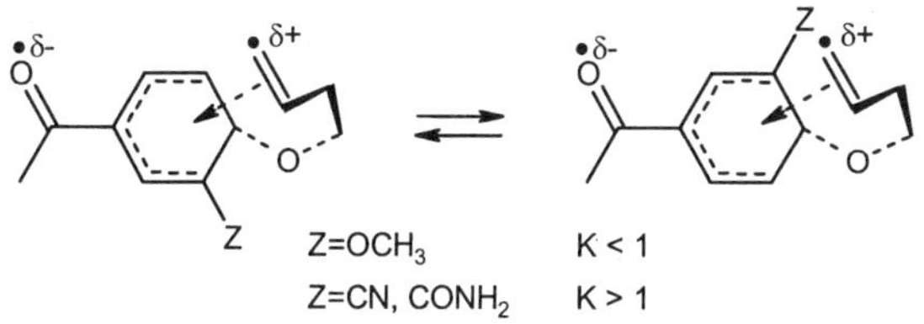
δ-
δ+
O
Z
Z=OCH3
Z=CN, CONH2
K < 1
K > 1

Scheme 22 Substituent-dependent orientation of an exciplex.

cyclobutene ring. The authors presume that the more conjugated diene unit flattens out when excited and the fused five-membered ring then allows the eight-membered ring to pucker in only one direction. When $R_1 = R_2 =$ Me and $R_3 =$ H, the diastereomeric excess of the final tricyclic product shown in Scheme 23 is 80%; with $R_1 =$ iPr, $R_2 =$ Me and $R_3 =$ H; diastereomeric excess is ≥95%.

Wagner and Alehashem [99] have investigated reactions of systems where the alkenyl side chain is anchored to the benzene ring with a methylene group rather than an oxygen atom (Scheme 24).

Changing the anchoring atom from oxygen to carbon does not change the initial photochemistry of the unsaturated ketones. The rates of the subsequent thermal electrocyclic rearrangements are, however, changed substantially. The trifluoroacetyl substituent was chosen to maintain as strong a donor-acceptor interaction in the bicyclooctadienes as possible, as well as to maintain a π,π^* lowest triplet.

Cheng and Wagner [100] have confirmed the intermediacy of a 1,4-biradical by placing a cyclopropyl group onto the vinyl moiety of an alkenyloxyacetophenone (Scheme 25). At 100% conversion, the products **I, II,** and **III** were pro-

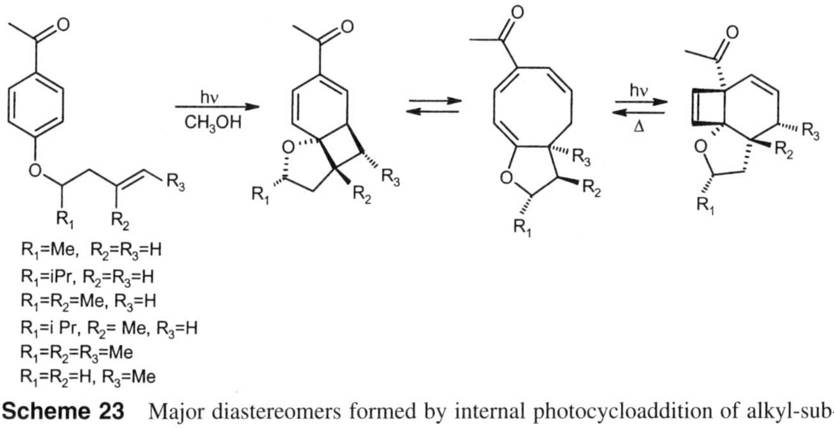
hν
CH3OH
hν
Δ
R1
R2
R3
O
R1=Me, R2=R3=H
R1=iPr, R2=R3=H
R1=R2=Me, R3=H
R1=i Pr, R2= Me, R3=H
R1=R2=R3=Me
R1=R2=H, R3=Me

Scheme 23 Major diastereomers formed by internal photocycloaddition of alkyl-substituted *para*-butenoxyacetophenones.

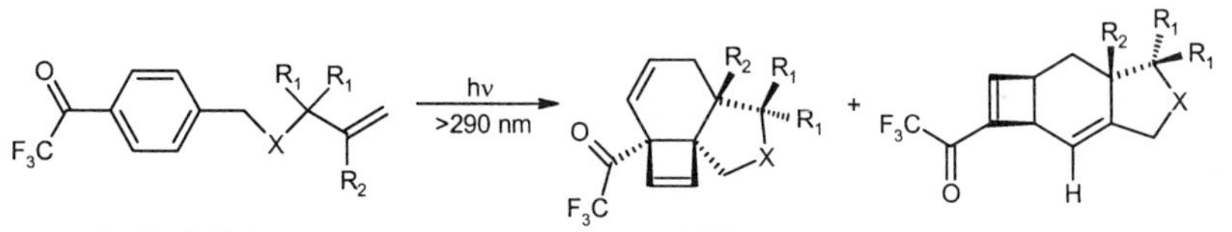

$R_1=R_2=H$; X=O	φ = 0.017	φ = 0.045
$R_1=H$; $R_2=Me$; X=O	φ = 0.021	φ = 0.026
$R_1=Me$; $R_2=H$; X=O	φ = 0.005	φ = 0.009
$R_1=R_2=H$; $X=CH_2$	φ = 0.0034	φ = 0.0064

Scheme 24 Quantum yields of intramolecular photoaddition products from various 5-(4-trifluoroacetylphenyl)-1-pentenes.

duced in the proportion of 5:3:1, respectively, with twice as much of **I** at low conversion. The total quantum yield for formation of compounds **II** and **III** was measured as 0.21. On the basis of product ratios, the quantum yield for formation of **I** drops from 0.50 to 0.26 between 10% and 100% conversion. Assuming 100% efficient formation of the 1,4-biradical from the triplet ketone, the authors conclude that about 70% of the 1,4-biradicals undergo rearrangement and only 30%

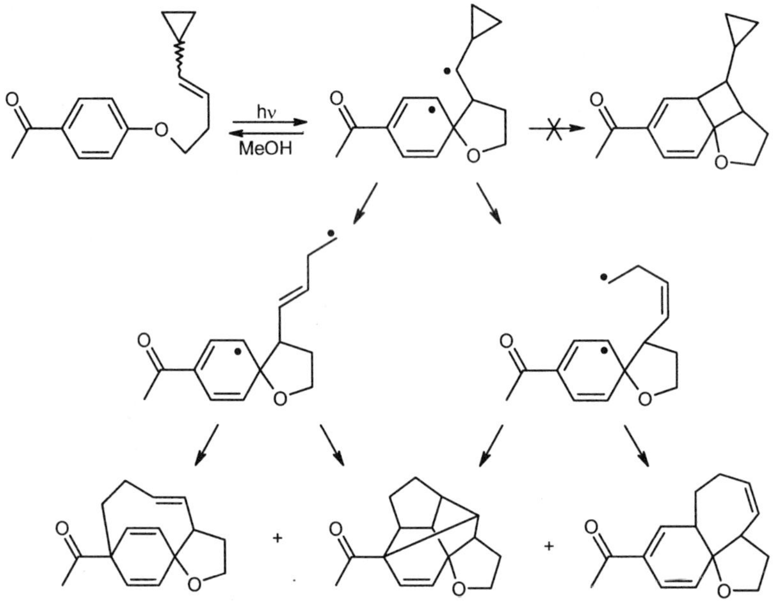

I II III

Scheme 25 Products obtained from irradiation of cyclopropyl-substituted *para*-alkenyloxyacetophenone.

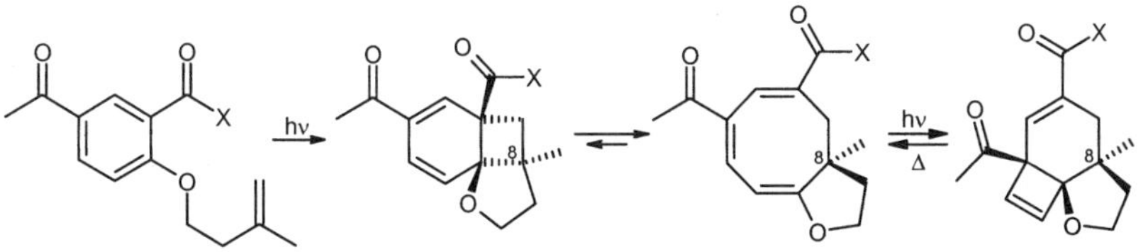

Scheme 26 Diastereoselective intramolecular cycloaddition (X = chiral auxiliary).

revert to the starting ketone. The formation of **II** can be explained by a tandem biradical cyclization process obeying the "rule of five."

Wagner and McMahon [101] have irradiated compounds bearing chiral auxiliaries and observed that high diastereoselectivity can be achieved with photoinduced intramolecular ortho cycloaddition (Scheme 26). Carbon atom 8 is the only permanent new stereocenter created by the ortho photocycloaddition. When X = (2R,5R)-(−)-2,5-dimethylpyrrolidine or (7R)-(+)-camphorsultam, diastereomeric excesses as high as 90% are obtained. The (8S)-cyclooctatriene-sultam derivative could be isolated as a single diastereomer after one recrystallization.

The effect of substituents meta to the tether in *para*-alkenyloxyacetophenones was studied by Smart and Wagner [102] (Scheme 27).

In all cases, 4-acetyl-5-Z-8-methyl-11-oxatricyclo[6.3.0.0^{1,4}]undeca-2,5-dienes (Z = OMe, F, CH$_3$, CF$_3$), resulting from syn addition, were the predominant photoproducts, as determined by direct NMR analysis. Following thermal treatment, the cyclooctatriene and/or cyclohexadiene derivatives were then isolated in 60–70% yield in the ratios shown in Scheme 21. With 2-methyl, 2-methoxy, and 2-trifluoromethyl substitution, the reaction is 100% regioselective syn, addition of the remote double bond occurring only at the 3,4-positions of the benzene ring, not the 4,5. The 2-fluoro derivative alone provides both anti and syn regioisomers in a 15:85 ratio. The 2-cyano derivative did not react. The authors surmised that the

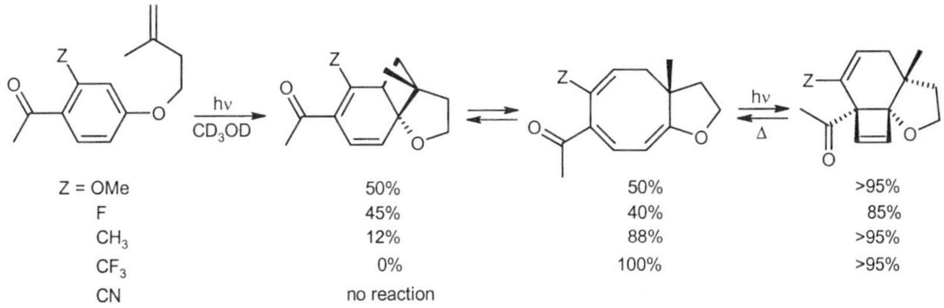

Z = OMe	50%	50%	>95%
F	45%	40%	85%
CH$_3$	12%	88%	>95%
CF$_3$	0%	100%	>95%
CN	no reaction		

Scheme 27 Effect of substituents meta to the tether on intramolecular photocycloaddition of *para*-alkenyloxyacetophenones.

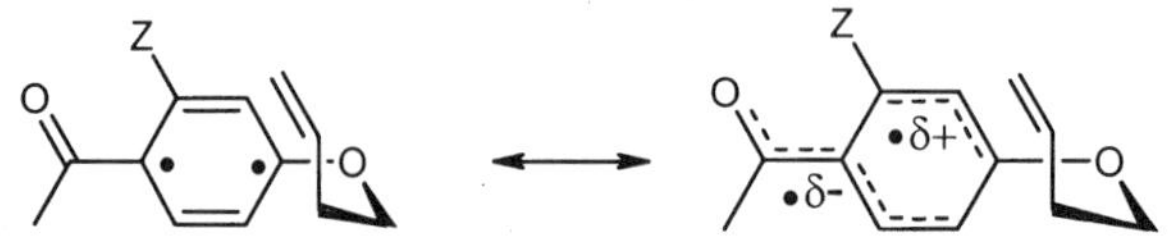

Scheme 28   Polarization of the biradical by electronegative substituents.

electronegative substituents (all but the methyl) increase positive charge on their side of the benzene ring in the triplet, thus attracting the nucleophilic double bond. Semiempirical calculations indeed indicate strong positive charge on the carbon bearing the substituents, in accord with the election distribution depicted in Scheme 28.

Gilbert and co-workers have investigated the intramolecular photochemical reactions of 4-phenoxybut-1-enes bearing an *ortho-*, *para-*, or *meta-*cyano or *ortho-*, *meta-*, or *para-*methoxycarbonyl group [104,105]. The meta–cyano compound upon irradiation produced only trace amounts of an isomer and some polymeric material; 95% of the starting material was recovered. Similarly, the *meta-*methoxycarbonyl derivative proved to be essentially photostable. The *para-*methoxycarbonyl isomer did not undergo any detectable cycloaddition but was rapidly converted into polymeric material. *Ortho–cyano-*, *para–cyano-*, and *ortho-*methoxycarbonyl-4-phenoxy-1-butenes underwent intramolecular ortho photocycloaddition followed by thermal ring opening leading to cyclooctatriene derivatives. These were photochemically converted into photostable 4-oxatricyclo[7.2.0.0^{3,7}]undeca-2,10-dienes (Scheme 29). The primary ortho cycloadducts were not detected.

The formation of the cyclooctatrienes was quenched in the presence of low concentrations of 2,5-dimethylhexa-2,4-diene or 2,3-dimethylbuta-1,3-diene. The photocyclization of the trienes was, however, unaffected by the presence of the dienes. The authors conclude that the primary photoreaction of intramolecular ortho cycloaddition involves the $^3\pi,\pi^*$ state of the arene and the subsequent cyclization of the cycloocta-1,3,5-triene arises from the singlet state.

Wagner and Smart [103] have also demonstrated that not only ketones are capable of undergoing photocycloaddition via the triplet state. Next to ketones,

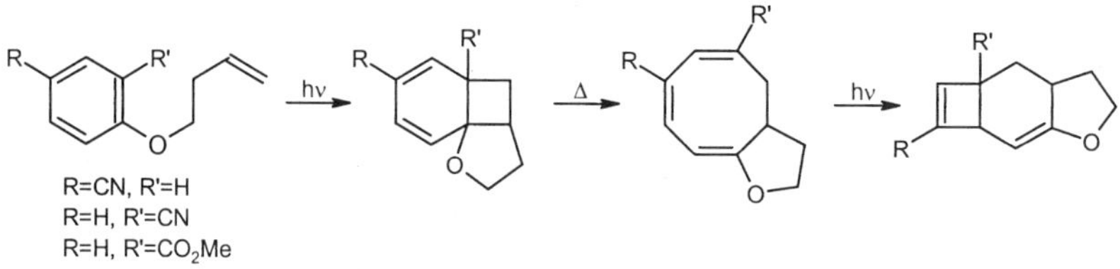

R=CN, R'=H
R=H, R'=CN
R=H, R'=CO$_2$Me

Scheme 29   Intramolecular ortho photocycloaddition of cyano- and methoxycarbonyl-substituted 4-phenoxybut-1-enes.

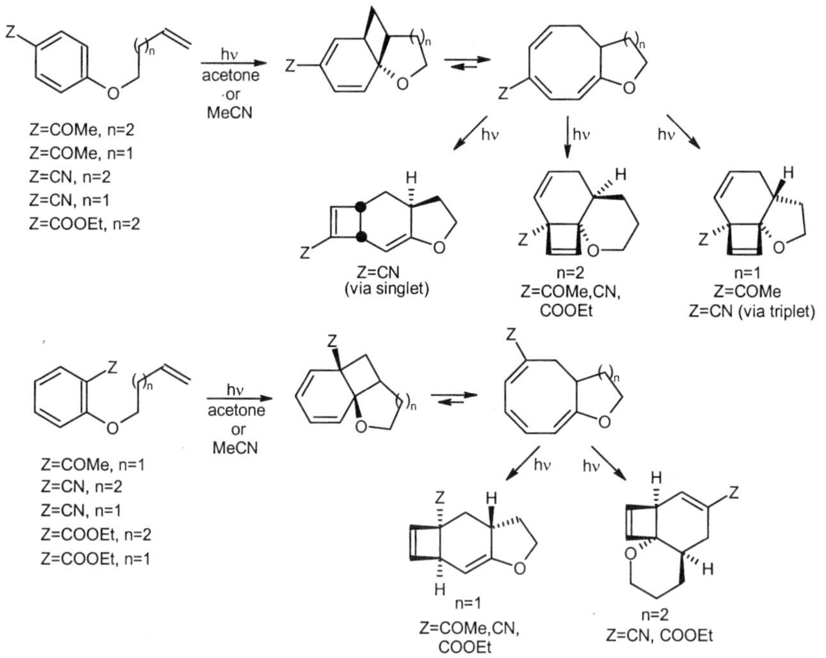

Scheme 30 Intramolecular ortho photocycloaddition of ketones, nitriles, and esters.

they have irradiated benzonitriles and benzoates carrying an alkenyloxy side chain (Scheme 30).

In all cases but one, irradiation at 313 nm of solutions in acetone produced the same products as obtained from irradiation at 254 nm of solutions in acetonitrile. The exception is the para–cyano compound with $n = 1$. Direct as well as sensitized irradiation produces the cyclooctatriene derivative via the ortho adduct. In acetone, however, the cyclooctatriene undergoes sensitized ring closure to give the angular tricyclic product. Direct irradiation gives the linear compound, as had been reported earlier by Gilbert and co-workers [104]. The compound with Z = COOEt and $n = 1$ was also irradiated, but proved to be unreactive, as had also been found by Gilbert et al. [105] for the methyl ester. The compounds with four-atom tethers ($n = 2$) react much slower than those with three-atom tethers, but the initial [2 + 2] cycloaddition nevertheless proceeds efficiently.

Hoffmann and Pete [106] have irradiated O-alk-3-enylsalicylates and obtained products which result from rearrangement reactions of primary ortho adducts (Scheme 31). The authors realized that the linear tricyclic dienes, formed by ring closure of the cyclooctatriene derivatives, are enol ethers which can be converted into acetals by acid-catalyzed addition of an alcohol. This shifts the

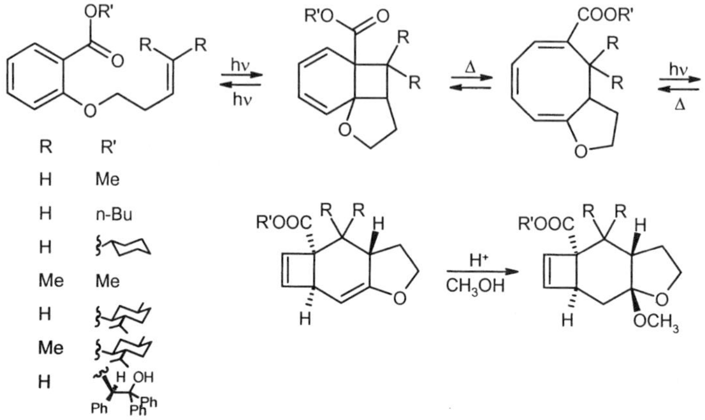

Scheme 31   Ortho photocycloaddition of *O*-alk-3-enylsalicylates.

equilibrium to the product side. Irradiation in cyclohexane, acetonitrile, or methanol in the absence of acid produces complex mixtures, but irradiation in methanol with a catalytic amount of sulfuric acid leads cleanly to the acetals in good yields. The authors do not explicitly mention the intermediacy of a triplet state, but they refer to Wagner's work and state that the ortho adduct is generated from an excited state having a biradical character.

Examples of intermolecular reactions proceeding via triplet excited arenes have also been described. Wagner and Nahm [94] mention in a footnote that irradiation of *p*-methoxyacetophenone in 1-hexene produces a yellow compound with the characteristic NMR spectrum of a cyclooctatriene. Following this lead, Al-Qarawadi et al. [107] have studied intermolecular ortho photocycloaddition reactions of 2- and 4-methoxyacetophenones. These compounds have lowest π,π^* triplet states but with cyclopentene and *cis*-cyclooctene, they gave only minor amounts of complex mixtures. With ethyl vinyl ether and vinyl acetate, however, more successful addition reactions were observed. The results are shown in Scheme 32, in which the primary adducts next to the isolated products are also shown. The authors do not comment on the regiochemical aspects of the reaction—for example the different modes of attack of ethyl vinyl ether and vinyl acetate on *para*-methoxyacetophenone and the different regioselectivities of the reactions of vinyl acetate with the para and the ortho isomer.

The possibility of the intermediacy of the triplet state of benzene itself has been discussed by Atkins et al. [108]. Photoaddition of alkenes to arenes is often accompanied by the formation of dimers of the alkene, a reaction sensitized by triplet benzene. With methyl acrylate and methyl vinyl ketone, however, it was found that the ratio of ortho cycloadducts to alkene dimers increased with the concentration of benzene. Because the yield of T_1 benzene increases with benzene concentration, these results might indicate that ortho photocycloaddition of acry-

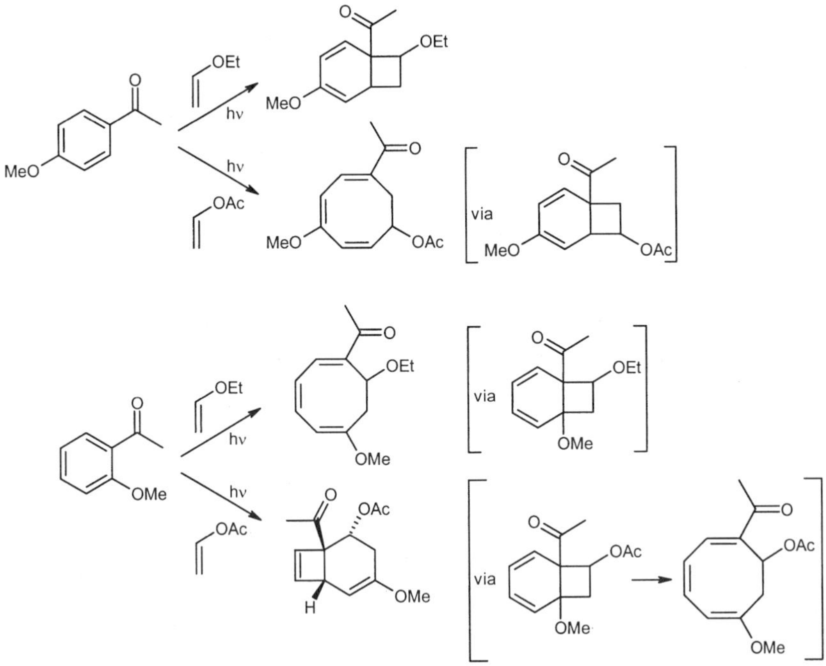

Scheme 32 Ortho photocycloaddition of *p*- and *o*-methoxyacetophenone to ethyl vinyl ether and vinyl acetate.

lates and methyl vinyl ketone is a triplet process and that T_1 benzene in these cases does not sensitize dimer formation.

It is not always easy to ascertain if the addition reaction proceeds via the triplet excited state of the arene. Benzene and its simple derivatives such as anisole and benzonitrile have high triplet energies (benzene, 84 kcal/mol^{-1}; anisole, 81 kcal/mol^{-1}, benzonitrile, 77 kcal/mol^{-1}) which makes sensitization impracticable. Results of quenching experiments are sometimes difficult to interpret, as has become evident from the work of Cantrell. He found [109] that the formation of adducts from benzonitrile and 2,3-dimethylbut-2-ene, vinyl acetate, and 2-methoxypropene in solutions 0.5 *M* in *cis*-1,3-pentadiene occurs at a rate only approximately one-fourth that in the absence of added quencher. Five years later, the author reported [110] that *cis*-1,3-pentadiene itself adds slowly to benzonitrile to give ortho adducts. When a correction was made for the reaction of benzonitrile with the quencher, it became apparent that little, if any, triplet quenching had occurred.

A tabular survey of ortho photocycloadditions proceeding via the triplet excited state of the arene is given in Table 4. As in Tables 1–3, the primary ortho adducts are listed, even if they are not isolated.

Table 4 Ortho Cycloaddition of Triplet-Excited Arenes to Alkenes

Arene	Alkene	Primary adducts	Remarks	Ref.
			dilute solution in CH_3CN (254 nm) or acetone (313 nm); adduct rearranges to NCD*	93,94,103
			degassed solution in CH_3CN (313 nm); adduct rearranges to COT** on GC	93
			dilute solution in CH_3CN (254 nm) or acetone (313 nm); adduct rearranges to NCD	93,94,103
			degassed solution in CH_3CN (313 nm); adduct rearranges to COT on GC	93
			in benzene, CH_3CN or MeOH (313 or 365 nm); GC chemical yield of COT 70-100%	93,94
			degassed solution in CH_3CN (313 nm); adduct rearranges to COT on GC	93
			in benzene, CH_3CN or MeOH (313 or 365 nm); GC chemical yield of COT 70-100%	93,94

Table 4 Continued

Arene	Alkene	Primary adducts	Remarks	Ref.
			degassed solution in CH_3CN (313 nm); adduct rearranges to COT on GC	93
			degassed solution in CH_3CN (313 nm); adduct rearranges to COT on GC	93
			degassed solution in CH_3CN (313 nm); adduct rearranges to COT on GC	93
			degassed solution in CH_3CN (313 nm); adduct rearranges to COT on GC	93
			degassed solution in CH_3CN (313 nm); adduct rearranges to COT on GC	93
			dilute solution in MeOH (>290 nm) diastereomeric excess (primary adduct) ≥ 90%	98
			in benzene, CH_3CN or MeOH (313 or 365 nm); GC chemical yield of COT 70–100%	94

Table 4 Continued

Arene	Alkene	Primary adducts	Remarks	Ref.
			in benzene, CH$_3$CN or MeOH (313 or 365 nm); GC chemical yield of COT 70–100%	94
			in benzene, CH$_3$CN or MeOH (313 or 365 nm); GC chemical yield of COT 70–100%	93,94
			degassed solution in CH$_3$CN (313 nm); adduct rearranges to COT on GC	93
			degassed solution in CH$_3$CN (313 nm); adduct rearranges to COT on GC	93
			degassed solution in CH$_3$CN (313 nm); adduct rearranges to COT on GC	93
			dilute solution in MeOH (>290 nm) diastereomeric excess (COT and NCD) 80%	98
			dilute solution in MeOH (>290 nm) diastereomeric excess COT: 56%, NCD: 41%	98

Table 4 Continued

Arene	Alkene	Primary adducts	Remarks	Ref.
			dilute solution in MeOH (>290 nm) diastereomeric excess COT: 61%, NCD: 67%	98
			dilute solution in MeOH (>290 nm) diastereomeric excess COT: 80%, NCD: 80%	98
			dilute solution in MeOH (>290 nm) diastereomeric excess COT: >95%, NCD: >95%	98
			dilute solution in MeOH (>290 nm) diastereomeric excess primary adduct: 80–95%, NCD: 80–95%	98
			dilute solution in CH_3CN (254 nm) or acetone (313 nm); adduct rearranges to NCD	103–105
			1% w/v in various solvents; <1% conversion after 24h; only trace amounts of isomer	104,105
			dilute solution in CH_3CN 254 nm) or acetone (313 nm); adduct rearranges to NCD	103–105
			0.02M in Ar-flushed benzene (>290 nm); adduct rearranges to NCD	97

Table 4   Continued

Arene	Alkene	Primary adducts	Remarks	Ref.
			0.02M in Ar-flushed benzene (>290 nm); a:b = 80:20	97
			0.02M in Ar-flushed benzene (>290 nm); adduct rearranges to NCD	97
			0.02M in Ar-flushed benzene (>290 nm); adduct rearranges to NCD	97
			0.02M in Ar-flushed benzene (>290 nm); adduct rearranges to two NCDs (>90% yield)	97
			0.01M in MeOH (300 nm)	101
			0.02M in Ar-flushed benzene (>290 nm); adduct rearranges to NCD (>90% yield)	97

For the 0.01M in MeOH (300 nm) entry:

	diastereomeric excess %
X=OH	10
X=(1R,2S,5R)-(−)-menthol	15
X=(S)-(−)-1-phenylethylamine	35
X=(2R,5R)-(−)-2,5-dimethylpyrrolidine	90
X=(7R)-(+)-camphorsultam	90

Table 4 Continued

Arene	Alkene	Primary adducts	Remarks	Ref.
			0.02M in MeOH; adduct rearranges to NCD; regioselectivity >95%	102
			0.02M in MeOH; adduct rearranges to NCD; regioselectivity >95%	102
		a b	0.02M in MeOH; adduct rearranges to NCD; a:b = 44:56	102
			0.02M in MeOH; adduct rearranges to NCD; regioselectivity >95%	102
			0.02M in MeOH; adduct rearranges to NCD; regioselectivity >95%	102
		a b	0.02M in MeOH; adducts rearrange to NCDs; a:b = 85:15	102

Table 4 Continued

Arene	Alkene	Primary adducts	Remarks	Ref.
			in methanol with cat. H_2SO_4 (254 nm); yield of rearranged product 35% at 90% conversion	105,106
			dilute solution in CH_3CN (254 nm) or acetone (313 nm); adduct rearranges to NCD	103
			in methanol with cat. H_2SO_4 (254 nm); yield of rearranged product 77% at 77% conversion	106
			in methanol with cat. H_2SO_4 (254 nm); yield of rearranged product 55% at 81% conversion	106
			in methanol with cat. H_2SO_4 (254 nm); yield of rearranged product 80% at 100% conversion	106
			in methanol with cat. H_2SO_4 (254 nm); yield of rearranged product 80% at 100% conversion	106
			in methanol with cat. H_2SO_4 (254 nm); yield of rearranged product 26% at 80% conversion	106

Table 4 Continued

Arene	Alkene	Primary adducts	Remarks	Ref.
			in methanol with cat. H_2SO_4 (254 nm); yield of rearranged product 60% at 64% conversion	106
			1% w/v in cyclohexane or CH_3CN (254 nm); starting material recovered after 14 days exposure	105
			1% w/v cyclohexane or CH_3CN (254 nm); rapid conversion to polymeric material	104,105
			dilute solution in CH_3CN (254 nm) or acetone (313 nm); photochemically unreactive	103
			adduct rearranges to linear and angular NCDs (φ = 0.017 and 0.045 at 10% conversion) (>290 nm)	99
			adduct rearranges to linear and angular NCDs (φ = 0.021 and 0.026 at 10% conversion) (>290 nm)	99
			adduct rearranges to linear and angular NCDs (φ = 0.005 and 0.009 at 10% conversion) (>290 nm)	99
			primary adduct is major product	99

Table 4 Continued

Arene	Alkene	Primary adducts	Remarks	Ref.
			the primary adduct is the main product (>290 nm)	99
			dilute solution in CH_3CN (254 nm) or acetone (313 nm); adduct rearranges to NCD	94,103
			degassed solution in CH_3CN (313 nm); adduct rearranges to COT on GC	93
			in benzene, CH_3CN or MeOH (313 or 365 nm); GC chemical yield of COT 70–100%	94
			dilute solution in CH_3CN (254 nm) or acetone (313 nm); adduct rearranges to NCD	103
			dilute solution in CH_3CN (254 nm) or acetone (313 nm); adduct rearranges to NCD	103
			dilute solution in CH_3CN (254 nm) or acetone (313 nm); adduct rearranges to NCD	103

Table 4 Continued

Arene	Alkene	Primary adducts	Remarks	Ref.
			dilute solution in CH_3CN (254 nm) or acetone (313 nm); adduct rearranges to NCD	103
	OAc	OAc	254 nm or >290 nm; major adduct is the NCD formed from the primary adduct	107
	OAc	OAc	254 nm or >290 nm; principal 1:1 adduct is the COT formed from the primary adduct	107
	OEt	OEt	254 nm or >290 nm; mixture of three 1:1 adducts; primary adduct could be isolated	107
	OEt	OEt	254 nm or >290 nm; isolated 1:1 adduct is the COT formed from the primary adduct	107

* NCD = A derivative of bicyclo[4.2.0]octa-2,7-diene, a non-conjugated diene.
** COT = A derivative of cycloocta-1,3,5-triene.

V. ORTHO PHOTOCYCLOADDITION OF SINGLET-EXCITED ARENES TO ALKENES

A. Introduction

When there is no ground-state complexation between the addends, when the alkene does not absorb light at longer wavelengths than the arene, and when the arene does not undergo fast intersystem crossing to the triplet state, ortho photocycloaddition usually occurs via interaction between singlet excited arene and ground-state alkene.

The first examples of ortho cycloaddition can be found in a U.S. patent of Ayer and Buchi [1]. Benzonitrile and 2-methylbut-2-ene are reported to yield 7,8,8-trimethylbicyclo[4.2.0]octa-2,4-diene-1-carbonitrile upon irradiation under nitrogen with a mercury resonance arc. Similar reactions, all leading to derivatives of bicyclo[4.2.0]octa-2,4-diene-1-carbonitrile occurred when benzonitrile was irradiated in the presence of 2,4,4-trimethylpent-1-ene, ethyl vinyl ether, vinyl acetate, methyl vinyl ketone, and methyl acrylate. The addend pairs *para*-tolunitrile/oct-1-ene, *ortho*-dicyanobenzene/2-methylbut-2-ene, *para*-dicyanobenzene/but-1-ene, 2,3-dimethylbenzonitrile/propene, and 3,4,5-trimethylbenzonitrile/ethene likewise produced ortho photocycloadducts.

In 1966, two groups simultaneously discovered photochemical addition reactions of simple alkenes to excited benzenes. Wilzbach and Kaplan [4] reported that upon irradiation of 10% solutions of *cis*-but-2-ene, cyclopentene, and 2,3-dimethylbut-2-ene in benzene with light of 254 nm, meta photocycloadducts are formed. Bryce-Smith, et al. [5] irradiated an equimolar mixture of benzene and *cis*-cyclooctene with 254-nm light which yielded a mixture of 1:1 adducts. The main component (~85%) was a 1:1 adduct similar to that found by Wilzbach and Kaplan, a meta photocycloadduct. The minor (10–15%) 1:1 adduct was unstable and underwent thermal isomerization. It was suggested with some reservations that the minor adduct was tetracyclo[6.6.0.0^{2,7}.0^{3,6}]tetradec-4-ene (**I**, Fig. 1). In 1968, this structure was rejected [37], and in 1973, the product was shown to be the ortho adduct, tricyclo[6.6.0.0^{2,7}]tetradeca-3,5-diene [38] (**II**, Fig. 1).

Thus, upon irradiation of benzene in the presence of *cis*-cyclooctene, two different 1:1 addition products are formed. The major product is a meta cycloadduct and the minor one is an ortho cycloadduct.

In Table 5, the ortho adducts are collected, which have been reported to be formed by intermolecular photocycloaddition of benzene and derivatives of benzene to simple alkenes. Ortho photocycloadditions of alkenes to perfluorinated benzenes are separately collected in Table 6. Intramolecular ortho photocycloadditions have been grouped in Table 7. (The many intramolecular ortho photocycloadditions proceeding via the triplet state of the arene moiety are found in Table 4.) All adducts shown in the tables are the primary ortho adducts. Many of these bicyclo[4.2.0]octa-2,4-diene derivatives are, however, unstable and not all of

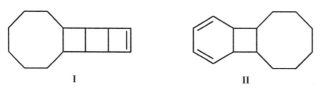

          I   **II**

Figure 1 Originally proposed structure (**I**) and definitive structure (**II**) of the ortho photocycloadduct from benzene and *cis*-cyclooctene.

Table 5 Intermolecular Ortho Photocycloaddition of Arenes to Alkenes

Arene	Alkene	Primary adducts	Remarks	Ref.
(benzene)	(ethene)	(ortho adduct)	50 bar (ca. 12 mol/l) ethene + 1M benzene in CH_2Cl_2 (254 nm); meta adduct is major product; ortho adduct is a major side product	114
(benzene)	(propene)	(ortho adduct)	10% benzene in alkene, $-15°$ (254 nm); $\varphi_{ortho} = 0.06$; $\varphi_{meta} = 0.4$; $\varphi_{para} = 0.01\text{–}0.02$	7,115
(benzene)	(isobutene)	(ortho adduct)	10% benzene in alkene, $-15°$ (254 nm); $\varphi_{ortho} = 0.10$; $\varphi_{meta} = 0.33$; $\varphi_{para} = 0.01\text{–}0.02$	7,115
(benzene)	(2-methyl-2-butene)	(ortho adduct)	1.1M benzene + 3.5M alkene in iso-octane (254 nm); $\varphi_{ortho} = 0.05$; $\varphi_{meta} = 0.18$; $\varphi_{"ene"} = 0.07$	116
(benzene)	(2,3-dimethyl-2-butene)	(ortho adduct)	1.1M benzene + 3.5M alkene in iso-octane (254 nm); $\varphi_{ortho} = 0.25$; $\varphi_{meta} = 0.03$; $\varphi_{"ene"} = 0.03$	7, 115–117
(benzene)	(cyclooctene)	(ortho adduct)	1.1M benzene + 3.5M alkene in iso-octane (254 nm) $\varphi_{ortho} = 0.09$; $\varphi_{meta} = 0.38$	5,37,38, 115,116, 118
(benzene)	(cycloheptene)	(ortho adduct)	1.1M benzene + 3.5M alkene in iso-octane (254 nm) $\varphi_{ortho} = 0.37$; $\varphi_{meta} = 0.28$	115,116
(benzene)	(OEt-vinyl ether)	(ortho adduct, OEt)	1.1M benzene + 3.5M alkene in iso-octane (254 nm); $\varphi_{ortho} = 0.30$; $\varphi_{meta} = 0.16$; φ_{ortho} increases by 20–50% in MeOH and MeCn	12,108, 115,119, 120

Table 5   Continued

Arene	Alkene	Primary adducts	Remarks	Ref.
			ortho:meta cycloaddition rate = 2.7:1	12,108, 120
			similar results to those of n-butyl vinyl ether; structure of ortho adduct not published	12
			equivolume mixture of benzene and alkene; ortho adduct is the only adduct	12
			1.1M benzene +3.5M alkene in iso-octane (254 nm); $\varphi_{ortho} = 0.04$; $\varphi_{meta} = 0.03$	115
			2,3-dihydrofuran in benzene (254 nm); ortho:meta = 7.5:1 two ortho adducts:endo:exo = 5:3	12,120
			1.1M benzene +3.5M alkene (254 nm); in iso-octane $\varphi_{ortho} = 0.70$; in MeCN $\varphi_{ortho} = 0.78$ $\varphi_{meta} + \varphi_{para} < 0.01$	11,12, 108,120, 121
			benzene + dioxole in dioxane (254 nm); $\varphi_{ortho} = 0.21$; $\varphi_{meta} = 0.38$; $\varphi_{para} = 0.04$	12–14, 120,122
			2-Me-1,3-dioxole in benzene (254 nm); two exo-ortho adducts (40%); four meta adducts (40%); para adduct (16%)	12,13, 120

Table 5 Continued

Arene	Alkene	Primary adducts	Remarks	Ref.
			benzene + dioxole in dioxane (254 nm); φ_{ortho} = 0.39; φ_{meta} = 0.12; φ_{para} <0.005	13,14, 122,123
			ratio of ortho:meta adducts >4:1 structures not published	124
			ratio of ortho:meta adducts >9:1 structures not published	124
			ratio of ortho:meta adducts >9:1 structures not published	124
			only ortho adducts and products of ene-reactions structures not published	124
			1,4-dioxene in benzene (254 nm); φ_{ortho} = 0.30	12,13, 108,120, 122
	OAc	OAc	1.1M benzene = 3.5M alkene in iso-octane (254 nm) φ_{ortho} = 0.03; φ_{meta} = 0.22	125,126
	CN	CN CN	1.1M benzene = 3.5M alkene in iso-octane (254 nm); φ_{ortho} = 0.09; endo:exo = 1:5; ZnCl$_2$ and polar solvents promote the reaction	47,61, 108,115, 121, 127–129
	CN	CN a CN b	methacrylonitrile in benzene (254 nm); a:b = 4:5; ZnCl$_2$ and polar solvents promote the reaction	47,128

Table 5 Continued

Arene	Alkene	Primary adducts	Remarks	Ref.
			1.1M benzene = 3.5M alkene in iso-octane (254 nm) φ_{ortho} = 0.02; endo:exo = 1:1	108,115
			methyl acrylate in benzene (1:1) (254 nm); endo:exo = 1:2	108,130
		a b	methyl methacrylate in benzene (1:1) (254 nm); a:b = 1:2	108,130
			methyl (E)-crotonate in benzene (254 nm); rapid E-Z isomerization product mixture may contain all four possible ortho adducts	108,130
			equivolume benzene + alkene (254 nm); two 1:1 adducts, ortho:meta = 1:2.5, = β-Cl-styrene + tetra-Cl-cyclobutanes	131,132
			an aortho adduct of toluene and cis-butene has been isolated and characterized, but the structure has not been published	7

Table 5 Continued

Arene	Alkene	Primary adducts	Remarks	Ref.
	OEt	OEt	ethyl vinyl ether + toluene (1:1) (254 nm); two exo-ortho adducts (ratio 1:1.5) and four meta adducts; ortho:meta = 1:4.5	119,133
			1.1M arene + 3.5M olefin in cyclohexane (254 nm); similar results as with ethyl vinyl ether; no structures published	119
			ratio of ortho:meta adducts 1:4 structures not published	124
			ratio of ortho:meta adducts 1:3 structures not published	124
	OAc	OAc	toluene + vinyl acetate (1:1) (254 nm); mixture of meta adducts +5% of an ortho adduct	47,126
	CN	CN a CN b CN c	toluene + acrylonitrile (1:1) (254 nm); a:b:c = 4:1:2; c is mixture of two regio- or stereoisomers; ZnCl₂ promotes the react ion	47,128

Table 5 Continued

Arene	Alkene	Primary adducts	Remarks	Ref.
			toluene + methacrylonitrile (1:1) (254 nm); several ortho adducts	47
		a b	1M arene + 2M alkene in dioxane (254 nm); ortho:meta = 32:68 at 5% conversion; a:b = 7:1	134
		a b	1M arene + 2M alkene in dioxane (254 nm) ortho:meta = 1:1 at 5% conversion; a:b = 4:1	134
			1M arene + 3.5M alkene in MeOH (254 nm); ortho:meta = 3:7 at 2–5% conversion; structures not published	135
		a b	0.1M arene + 1.0M alkene (254 nm); a:b = 3:2 (cyclohexane); a:b = 20:1 (acetonitrile)	47, 136–138

Table 5 Continued

Arene	Alkene	Primary adducts	Remarks	Ref.
OMe	CN (methacrylonitrile)	a; b; c; d	equimolar arene + alkene (254 nm); a: endo-CN:exo-CN = 3:2 a:b:c:d = 6:3:1:3 (neat) a:b:c:d = 4:2:1:2 (cyclohexane) a:b:c:d = 16:2:1:2 (acetonitrile)	47,137, 139
OMe	CN (crotonitrile)	a	anisole + crotonitrile in MeCN (254 nm); 74% yield of a mixture of stereoisomers	139
OMe	O, OMe (methyl acrylate)	a; b; c	1.1M arene + 3.5 M alkene (254 nm) a:b:c = 55:2.5:12 (cyclohexane) a:b:c = 60:2.5:5 (acetonitrile)	137
OMe	O, OM (methyl methacrylate)	a; b	1.1M arene + 3.5M alkene (254 nm) a:b = 100:33 (cyclohexane) a:b = 200:25 (acetonitrile)	137
OMe	furan	MeO H / H H	20% arene in furan (254 nm); ortho:meta = 1:3	140

Table 5   Continued

Arene	Alkene	Primary adducts	Remarks	Ref.
			benzonitrile + alkene in pentane ($-30°$) (>220 nm); one major product, φ_{ortho} = 0.02, + two minor isomers; yield 40%	109,110
			mixture of benzonitrile and alkene irradiated with Hg resonance arc	1
		a **b**	benzonitrile + alkene in pentane (>220 nm); a major, b minor adduct; φ_{ortho} = 0.18 (254 nm) yield 63%	1,73,109, 110
			benzonitrile + alkene in pentane (>220 nm); 70% ortho, 30% meta adducts; φ_{ortho} = 0.11 (254 nm); yield 27%	109,110, 141
			φ_{ortho}: φ_{meta} = 3:1; structures not published; footnote 21 in ref. 87	87
			benzonitrile + alkene in pentane (>220 nm); yield 36%	110
			benzonitrile + alkene in pentane (>220 nm); φ_{ortho} = 0.16 (254 nm); yield 52%; meta adduct is also formed	109,110

Table 5 Continued

Arene	Alkene	Primary adducts	Remarks	Ref.
	OEt		1.1M arene + 3.5M alkene (254 nm) mixture of stereoisomers	1,73, 119,133
			arene + alkene in cyclohexane (254 nm)	142
	OAc	a 	equimolar arene + alkene (254 nm); a:b + 2:1; plus three minor isomers; φ_{ortho} = 0.15	1,47,109, 110
			mixture of benzonitrile and alkene irradiated with Hg resonance arc	1
	OMe	a b	arene in alkene (with or without cyclohexane and MeCN) (254 nm); alkene absorbs >60% of the light; a (2 isomers) : b (4 isomers) = 2:3	1,80
	OMe	a b	arene in alkene (with or without cyclohexane and MeCN) (254 nm); alkene absorbs >60% of the light; a (3 isomers) : b (3 isomers) = 1:1	80

Table 5 Continued

Arene	Alkene	Primary adducts	Remarks	Ref.
(CN-phenyl)	(CH₂=CH–CN)	(bicyclic adducts, labeled a)	arene in alkene (with or without cyclohexane and MeCN) (254 nm); alkene absorbs >60% of the light; a (3 isomers) : b (2 isomers) = 3:2	80
(CN-phenyl)	(CH₂=C(CH₃)–CN)	(bicyclic adducts, labeled a and b)	arene in alkene (with or without cyclohexane and MeCN) (254 nm); alkene absorbs >60% of the light; a (1 isomer) : b (3 isomers) = 1:3	80
(CN-phenyl)	(cis-dichloroethylene)	(bicyclic adduct)	alkene + arene (>220 nm) yield 35%	109
(CN-phenyl)	(trans-dichloroethylene)	(bicyclic adduct)	equivolume arene + alkene in cyclohexane (254 nm); ortho:meta = 1:5; φ_{ortho} = 0.08	110,132
(CF₃-phenyl)	(cyclopentene)		arene + alkene in cyclohexane (254 nm); ortho:meta = <1:9; structure of ortho adduct not published	141
(CF₃-phenyl)	(1,3-dioxole)	(bicyclic adducts, labeled a and b)	1M arene + 0.5M alkene in dioxane (254 nm); ortho:meta:subst.prod. = 0.8:1.7:0.3; a:b = 1:1	15, 143,144

Table 5 Continued

Arene	Alkene	Primary adducts	Remarks	Ref.
			equivolume arene + alkene in cyclohexane (254 nm); mainly meta addition; minor amount of ortho adduct	132
			20% arene in furan (254 nm); four ortho adducts + one product from [4+4] addition	140
			arene + alkene (254 nm); ortho adduct + 2 minor unidentified 1:1 adducts	145
			arene + alkene (254 nm); ortho adduct comprises ca. 10% of reaction mixture; structure not published	107
			arene + alkene (254 nm); ortho adduct undergoes rearrangement to bicyclo[4.2.0] octa-2,7-diene	107
			arene + alkene (254 nm); ortho adduct undergoes rearrangement to bicyclo[4.2.0] octa-2,7-diene	107
			arene + alkene (solvent not specified) (254 nm); good yields of ortho adduct	107

Table 5 Continued

Arene	Alkene	Primary adducts	Remarks	Ref.
			arene + alkene (solvent not specified) (254 nm); initially 20% ortho, changing to 10%; structures not published; major products are meta adducts	107
			arene + alkene (solvent not specified) (254 nm); only ortho addition	107
			equivolume solution in cyclohexane (254 nm); ortho:meta = 1:14	132
			mixture of benzonitrile and alkene irradiated with Hg resonance arc	1
			arene + alkene (solvent not specified) (254 nm); only ortho addition	107
			0.2M arene in alkene (254 nm); adduct rearranges to a bicyclo [4.2.0]octa-2,7-diene; a minor 1,2 adduct could not be isolated	87,146
		a b	0.2M arene in alkene (254 nm); a:b = 1:1 at low conversion; adducts are photolabile	87,146

Table 5 Continued

Arene	Alkene	Primary adducts	Remarks	Ref.
			0.2M arene in alkene (254 nm); three ortho adducts which could not be obtained pure; meta adduct is major product	87
		a b	0.2M arene in alkene (254 nm); a:b = 1:1.75; in MeCN rate is 4 times as high as in cyclohexane	86,87, 146
		a b	0.2M arene in alkene (254 nm); a:b:azocine = 7:1:2	86,87,90
		a b	0.2M arene + 0.2M alkene in CH_3CN (254 nm); a:b = 4:1	90
			0.2M arene + 0.2M alkene in CH_3CN (254 nm); 65% at <40% arene conversion	90

Table 5 Continued

Arene	Alkene	Primary adducts	Remarks	Ref.
			0.2M arene + 0.2M alkene in CH_3CN (254 nm); 55% at <70% arene conversion	90
			0.2 arene in alkene (254 nm); adduct rearranges to a bicyclo [4.2.0]octa-2,7-diene; also two unknown products (<10%)	87
			22.5 g arene in 50 ml alkene (254 nm); ortho adduct : two meta adducts = 1:6.2	132
			20% arene in furan (254 nm); adduct undergoes HF elimination and rearrangement; a trimer of furan is also formed	140
			20% arene in furan (254 nm); one ortho + two meta adducts; ortho:meta = 1:12	140
			20% arene in furan (254 nm); ortho:meta:subst.prod. = 6:12:7	140
			mixture of benzonitrile and alkene irradiated with Hg resonance arc	1
			mixture of benzonitrile and alkene irradiated with Hg resonance arc	1

Table 5 Continued

Arene	Alkene	Primary adducts	Remarks	Ref.
			arene + alkene in MeCN/MeOH (3:1) (medium pressure Hg arc); ortho adduct (5%) + three meta adducts (together 49%)	147
		a b	dicyanobenzene + alkene + biphenyl in MeCN (medium pressure Hg arc); (rearranged a): b:meta adducts:para adduct = 13:8:44:5	148
			0.2M arene + 0.2 M alkene in cyclohexane (254 nm); adduct rearranges to a cyclooctatriene; 40% at <30% arene conversion	90
			0.2M arene + 0.2 M alkene in cyclohexane (254 nm); 40% at <30% arene conversion	90
			arene + alkene in MeOH or MeCN (254 nm); MeOH: 32% ortho; 49% subst.prod.; MeCN: 63% ortho	139
			arene + alkene in MeOH or MeCN (254 nm); MeOH: only subst.prod. (63%); MeCN: 60% ortho	139

Table 5 Continued

Arene	Alkene	Primary adducts	Remarks	Ref.
			0.1M arene + 1.0M alkene in cyclohexane (254 nm); ortho adduct : alkene dimers = 1.0:0.65	138
			arene dissolved in alkene (254 nm); structures of ortho adducts not published	86
			arene dissolved in alkene (254 nm); structures of ortho adducts not published; azocine is also formed	86
			arene dissolved in alkene (254 nm); structures of ortho adducts not published	86
			arene + alkene (solvent not specified) (254 nm); primary adduct rearranges to a bicyclo[4.2.0]octa-2,7-diene	107
			mixture of benzonitrile and alkene irradiated with Hg resonance arc	1
			mixture of benzonitrile and alkene irradiated with Hg resonance arc	1

Table 5 Continued

Arene	Alkene	Primary adducts	Remarks	Ref.
		a b	0.2M arene + 0.2M alkene in CH$_3$CN (254 nm); (a + cyclooctatriene from a): b : azocine = 5:1:1	90
			arene + alkene in dioxane (254 nm); primary adduct rearranges to a bicyclo[4.2.0]octa-2,7-diene; yield 60%	107
			0.2M arene + 0.2M alkene in CH$_3$CN (254 nm); 70% at <30% arene conversion	90
			arene + alkene (solvent not specified) (254 nm); primary adduct rearranges to a cycloocta-1,3,5-triene	107
			0.2M arene + 0.2M alkene in CH$_3$CN (254 nm); adduct rearranges to a cyclooctatriene; 65% at <50% arene conversion	90
		a b	arene + alkene (solvent not specified) (254 nm); a rearranges to a bicyclo[4.2.0]octa-2,7-diene, b to a cycloocta 1,3,5-triene	107

Table 5 Continued

Arene	Alkene	Primary adducts	Remarks	Ref.
			0.2M arene + 0.2M alkene in CH$_3$CN (254 nm); (a + cyclooctatriene from a): b = 7:1	90
			arene + alkene (solvent not specified) (254 nm); sole ortho adduct, which slowly reverts to starting material	107
			0.2M arene + 0.2M alkene in CH$_3$CN (254 nm); 60% at <25% arene conversion	90
			cyclooctene and trimesic ester in benzene R = Me, Et	149

them have been isolated. In some cases, the irradiations have been performed in the presence of dienophiles such as *N*-phenylmaleimide or tetracyanoethylene, which react with the ortho adducts to form stable Diels–Alder adducts. In the absence of capturing dienophiles, ortho cycloadducts may undergo thermal disrotatory ring opening to cyclooctatriene derivatives. These may, in turn, undergo photochemical 4π disrotatory ring closure to give derivatives of bicyclo[4.2.0]octa-2,7-diene (Scheme 33).

Depending on the reaction conditions and workup procedures, the primary adduct, the triene, the unconjugated diene, or mixtures of these are isolated and identified. Characteristically, after some time of irradiation, a maximum concentration of ortho adduct is reached which decreases upon continued irradiation.

Table 6 Intermolecular Ortho Photocycloaddition of Perfluorinated Arenes to Alkenes

Arene	Alkene	Primary adducts	Remarks	Ref.
			a:b = 85:15 in cyclohexane (254 nm); products undergo ring closure and ring opening $\varphi = 0.080 \pm 0.005$	150,151
			a:b = 82:18 in cyclohexane (254 nm); products undergo ring closure and ring opening	151
			a:b = (46–61):(54–39) dependent on molar ratio in cyclohexane (254 nm); products undergo ring closure and ring opening	152
			a:b = 75:25 in cyclohexane (254 nm); products undergo ring closure and ring opening $\varphi = 0.050 \pm 0.002$	151

Table 6 Continued

Arene	Alkene	Primary adducts	Remarks	Ref.
		a b	a:b = 53:47 in cyclohexane (254 nm); products undergo ring closure and ring opening	67,151, 153
		a b	a:b = 47:53 in cyclohexane (254 nm); products undergo ring closure and ring opening	152
		a b	a:b = 5:95 in cyclohexane, only exo attack to norbornene (254 nm); products undergo ring closure and ring opening $\varphi = 0.054 \pm 0.003$	151,152
		a b	a:b = 47:53 in cyclohexane, only exo attack to norbornadiene (254 nm); products undergo ring closure and ring opening	152,154

Table 6 Continued

Arene	Alkene	Primary adducts	Remarks	Ref.
			exclusive exo-syn attack in cyclohexane (254 nm) products undergo ring closure and ring opening	152
			exclusive syn attack in cyclohexane (254 nm) $\varphi = 0.75 \pm 0.004$	151,155
			exclusive syn attack in cyclohexane (254 nm) $\varphi = 0.097 \pm 0.004$	151,155
	ClFC=CClF		hexa-F-benzene + mixture of 1,2-di-Cl, 1,2-di-F-ethene; adduct not isolated but further transformed into octafluoro cyclooctatetraene	156,157
			in cyclohexane (254 nm); product undergoes further ring closure; yield 65%	158
		a, b, c	a:b:c = 64:16:20 in cyclohexane (254 nm); products undergo further ring closure; yields: a, 37%; (b+c), 26%	158

Table 6 Continued

Arene	Alkene	Primary adducts	Remarks	Ref.
			a:b:c = 34:38:28 in cyclohexane (254 nm); products undergo further ring closure; yields: a, 15%; b, 28%; c, 17%	158
			only one product in cyclohexane (254 nm); product undergoes further ring closure; yield 33%	158
			only one product in cyclohexane (254 nm); product undergoes further ring closure; yield 50%	158
			only one product in cyclohexane (254 nm); product undergoes further ring closure; yield 34%	158
			only one product in cyclohexane (254 nm); product undergoes further ring closure; yield 56%	158
			only one produce in cyclohexane (254 nm); product undergoes further ring closure; yield 48%	158

Table 7 Intramolecular Ortho Photocycloaddition of Arenes to Alkenes

Arene—Alkene	Primary adducts	Remarks	Ref.
		6 mM benzene; $\varphi = 0.023$ (direct irrdn., three other products); $\varphi = 0.26$ (sensitized, one other product)	159
		1% w/v in cyclohexane (254 nm) yield 14% at 15% conversion	112
		5 mM in acetonitrile (medium pressure Hg lamp); rearranged ortho adduct, 58% + one other product, 14% at 85% conversion	160
R=*tert*-butyldimethylsilyl		5 mM in acetonitrile (medium pressure Hg lamp); rearranged ortho adduct, 50–55% + one other product, 17% at 85–90% conversion	160
		0.1% v/v in cyclohexane (254 nm); after 80h, starting material : adduct : unknown isomers = 65:7:9	161
		48 mM in t-butyl methyl ether (high pressure Hg arc); rearranged ortho adduct :meta adducts = ca. 3.5:2.5; total yield 47%	162
		48 mM in t-butyl methyl ether (high pressure Hg arc); rearranged ortho adduct :meta adducts = ca. 1:1	163

Table 7 Continued

Arene—Alkene	Primary adducts	Remarks	Ref.
		0.5% w/v in dioxane (254 nm); ortho:meta = 1:1 at >20% conversion; ortho adduct undergoes rearrangement	164
		5.8 mM in cyclohexane (254 nm); ortho:meta = 1.9:1; ortho adduct undergoes rearrangement	165,166
		30 mM arene + 6 mM H$_2$SO$_4$ in methanol (254 nm); primary adducts undergo acid-catalyzed rearrangement	167,168
		30 mM arene + 6 mM H$_2$SO$_4$ in methanol (254 nm); primary adducts undergo acid-catalyzed rearrangement	167, 168
		30 mM arene + 6 mM H$_2$SO$_4$ in methanol (254 nm); primary adducts undergo acid-catalyzed rearrangement	167, 168

Table 7 Continued

Arene—Alkene	Primary adducts	Remarks	Ref.
		30 mM arene + 6 mM H_2SO_4 in methanol (254 nm); primary adducts undergo acid-catalyzed rearrangement	167, 168
		30 mM arene + 6 mM $H_2 2O_4$ in methanol (254 nm); primary adducts undergo acid-catalyzed rearrangement	167,168
		30 mM arene + 6 mM H_2SO_4 in methanol (254 nm); primary adducts undergo acid-catalyzed rearrangement	167
		0.5% w/v in dioxane (254 nm) rearranged ortho adduct is the sole isolated photoisomer	164
		in tetrahydrofuran (254 nm); only ortho addition; adduct undergoes rearrangement	165,169

Table 7 Continued

Arene—Alkene	Primary adducts	Remarks	Ref.
		30 mM arene + 6 mM H_2SO_4 in methanol (254 nm); primary adducts undergo acid-catalyzed rearrangement	168
		30 mM arene + 6 mM H_2SO4 in methanol (254 nm); primary adducts undergo acid-catalyzed rearrangement	168
		30 mM arene + 6 mM H_2SO4 in methanol (254 nm); primary adducts undergo acid-catalyzed rearrangement	168
		30 mM arene + 6 mM H_2SO_4 in methanol (254 nm); primary adducts undergo acid-catalyzed rearrangement	168

Table 7 Continued

Arene—Alkene	Primary adducts	Remarks	Ref.
		30 mM arene + 6 mM H$_2$SO$_4$ in methanol (254 nm); primary adducts undergo acid-catalyzed rearrangement	168
		30 mM arene + 6 mM H$_2$SO4 in methanol (254 nm); primary adducts undergo acid-catalyzed rearrangement	168
		30 mM arene + 6 mM H$_2$SO$_4$ in methanol (254 nm); primary adducts undergo acid-catalyzed rearrangement	168
		30 mM arene + 6 mM H$_2$SO$_4$ in CH$_3$CN (254 nm); primary adduct undergoes acid-catalyzed rearrangement	170
		30 mM arene + 6 mM H$_2$SO$_4$ in MeCN or MeOH (254 or 300 nm); primary adduct undergoes acid-catalyzed rearrangement	170

Table 7 Continued

Arene—Alkene	Primary adducts	Remarks	Ref.
		30 mM arene + 6 mM H_2SO_4 in methanol (300 nm); primary adduct undergoes acid-catalyzed rearrangement	170
		30 mM arene + 6 mM H_2SO_4 in methanol (300 nm); primary adduct undergoes acid-catalyzed rearrangement	170
		in hexane (254 nm); rearranged ortho adduct, 16% + meta adduct, 38%	171
	R_1 = H or CF_3 R_2 = CF_3 or H	1% w/v in dioxane, cyclohexane or MeCN (254 nm); no meta addition; ortho adduct undergoes rearrangement	172
		55.5 mM in cyclohexane (254 nm); ortho adduct is the only product; yield 65%	111
		6 mM in tetrahydrofuran (254 nm); single photoproduct; yield 23%	169,173

Table 7 Continued

Arene—Alkene	Primary adducts	Remarks	Ref.
		1% w/v in cyclohexane (254 nm) (rearranged ortho):meta = 1:4, changing with time to 1:12; total yield 60%	113
		1% w/v in cyclohexane (254 nm) ortho:meta = 1:2, changing with time to 100% meta; total yield 81%	113
		1% w/v in cyclohexane (254 nm) (rearranged ortho):(unknown prod.) = 2:1, changing with time to 4:1; total yield 17%	113
		1% w/v in cyclohexane (254 nm) (rearranged ortho):meta = 1:2.25, changing with time to 1:2.5; total yield 68%	113
		1% w/v in cyclohexane ortho:meta = 1:2, changing with time to 1:6.25; total yield 95%	113
		in methanol (254 nm); R = CH_2Ph: yield 95% at 75% conversion; R = i-Pr: 63% at 55% conversion	85

Therefore, if ratios are given, these must be considered with great care, especially if irradiation times were long. Ortho adducts formed by intramolecular photocycloaddition are even less stable than those from intermolecular reactions. Only from pentafluorophenyl prop-2-enyl ether [111], 2-methyl-6-(4-fluorophenyl)hex-2-ene [112], and 1-(2-methoxybenzyloxy)-3-methylbut-2-ene [113] have primary ortho adducts reportedly been isolated. The many secondary

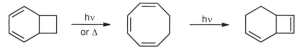

Scheme 33   Thermal ring opening of an ortho adduct to cycloocta-1,3,5-triene and photochemical ring closure of this triene.

reactions that ortho cycloadducts are capable of undergoing will be discussed in Section VI.

The following subsections are devoted to various mechanistic aspects of the ortho photocycloaddition. The possible role of ground-state complexes will be discussed and, subsequently, the intermediate species that are formed or may be formed upon photoexcitation will be treated: the reactive excited state, exciplexes, and zwitterions, biradicals, and ion pairs. Empirical rules, aimed at predicting under what circumstances ortho photocycloaddition (or other modes of addition) may occur, will be discussed next and, finally, the results of theoretical considerations and calculations will be reviewed.

B. Ground-State Complexes

Bryce-Smith et al. [174] have found evidence for stereospecific ground-state interactions between benzene and alkenes using NMR experiments. At that time, the available experimental evidence was that donor alkenes add with endo orientation and acceptor alkenes with exo orientation to the benzene ring [7,24,115,118,129]. NMR spectra (60 MHz) of the alkenes were measured at 35°C using 1.5 mol% alkene solutions in benzene and, for reference, solutions in CCl_4. All donor alkenes (*cis*- and *trans*-but-2-ene, *cis*-di-*t*-butylethylene, cyclopentene, cyclohexene, cycloheptene, *cis*- and *trans*-cyclooctene, norbornene, ethyl vinyl ether, and *cis*- and *trans*-1,2-dimethoxyethylene) showed downfield shifts (~5–8 Hz) of the vinyl protons and comparable upfield shifts of the allylic or alkoxy protons, induced by benzene. The vinyl protons of the acceptor ethylenes (maleic anhydride, methyl vinyl ketone, and acrylonitrile) displayed marked upfield shifts (30–90 Hz), induced by benzene. These findings imply a preferred endo orientation between benzene and the donor alkenes and an even more strongly preferred exo orientation between benzene and the acceptor ethylenes. The authors considered it interesting, and possibly relevant, that these preferred ground-state orientations precisely match the stereospecificities observed in the corresponding ortho photoaddition reactions, although the tendency for ortho photoadditions to occur depends on steric factors and on the magnitude of ionization potential differences between the addends (Ref. 115 and later in this chapter).

The magnitude of the benzene-induced upfield shifts in the acceptor alkenes increases with increasing ionisation potential of the alkene. No such relation was found between the downfield shifts of the donor alkenes and their ionization po-

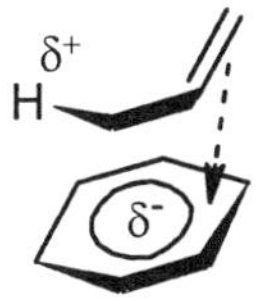

Figure 2   Possible endo ground-state complex between donor alkene and acceptor benzene.

tentials. The authors were surprised that the slight tendency for endo association between benzene and donor ethylenes does not give rise to charge-transfer absorption bands in the ultraviolet. The complex could arise from weak π-bonding involving the allylic or alkoxy hydrogen atoms. These atoms could have a partial positive charge due to charge transfer from the alkene to the aromatic ring (see Fig. 2).

Atkins et al. [130] reported in 1977 that irradiation of mixtures of benzene and methyl acrylate or methyl methacrylate, both acceptors, yields mixtures of endo and exo adducts. A subsequent report from the same groups [120] describes the results of the irradiations of benzene in the presence of ethyl vinyl ether, n-butyl vinyl ether, 2,3-dihydropyran, and 1,4-dioxene. In all these cases, the major products were exo–ortho photocycloadducts. The orientations of these vinyl ethers with respect to benzene, in their loose ground-state associations, were inferred from NMR spectra. For ethyl vinyl ether, n-butyl vinyl ether, 2,3-dihydropyran, and 1,4-dioxene, the vinyl proton resonances were either unaffected by a solvent change from carbon tetrachloride to hexadeuterobenzene or appeared 4–10 Hz downfield, whereas the methyl and/or methylene signals all moved upfield by 10–25 Hz. This implies an endo arrangement of the molecules in the ground state. Thus, the ortho photocycloadducts of vinyl ethers with benzene show exo stereochemistry, even when the ground-state orientation is endo.

Mattay et al. [122] have performed similar experiments with derivatives of 1,3-dioxole as well as with 1,4-dioxene. With benzene, these compounds yield ortho photocycloadducts having the exo configuration. In the NMR spectrum, the olefinic protons of 2,2-dimethyl-1,3-dioxole exhibit an upfield shift by changing the solvent from carbon tetrachloride to benzene; the methyl groups, however, are only slightly shifted. The upfield shifts increase with decreasing temperature. This may be interpreted as a preferred exo orientation between benzene and 2,2-dimethyl-1,3-dioxole in the ground state. With 1,3-dioxole and 2-methyl-1,3-dioxole, no such correspondence between ground-state orientation and photoadduct configuration was found. In these two compounds, the shifts of the olefinic protons are almost equal to those of the other protons and no preferred orientation in the ground state was apparent. 1,4-Dioxene, on the other hand, exhibited a preferred endo orientation in the ground state, but the ortho photocycloadduct has the exo configuration. It was concluded that a "dark complex," formed by ground-state interaction between benzene and alkene, as a direct precursor for the photoadducts was unlikely. A similar conclusion was reached by Gilbert et al. [12], who found that ground-state orientations of the addends as deduced by NMR spectroscopy have little or no influence on the stereochemistry of the ortho photocycloaddition process.

C. Nature of the Excited State

The ortho photocycloadditions of maleic anhydride and its derivatives to benzene and substituted benzenes start with excitation, either by absorption of light or by energy transfer from a triplet sensitizer, of a ground-state complex between the addends (Section II). Ortho photocycloaddition reactions of maleimides and alkynes to benzene and benzene derivatives involve excitation, singlet or triplet, of the alkene or alkyne. Reaction of an excited alkene with ground-state arene has also been reported in a few cases in which the alkene (acrylonitrile, methyl acrylate, and their α-methyl derivatives) absorbs at longer wavelengths than the arene and in which the arene (benzonitrile) is also electron accepting so that ground-state or excited-state charge-transfer complexing is excluded. The photocycloaddition of dichlorovinylene carbonate also proceeds via excitation of the alkene (Section III). The triplet state of the arene is the reactive species in alkenyloxyacetophenones and -butyrophenones, alkenyloxy-substituted benzonitriles and benzoic acid esters, and a few other compounds in which fast intersystem crossing of the singlet excited arene takes place (Section IV). In most other cases, ortho photocycloaddition of benzene and benzene derivatives to simple alkenes is supposed to proceed via the singlet excited state of the arene. Experimental evidence for the involvement of the singlet excited arene in the addition step has been obtained in only a few cases. For most other examples, the singlet mechanism has been adopted simply by analogy.

In one of the earliest reports on ortho photocycloaddition, in which the reaction of benzonitrile with 2-methylbut-2-ene is described, a diradical (triplet) intermediate was proposed [73]. The structure of the product corresponds to the most stable of the four possible diradical intermediates. When benzophenone was added as a sensitizer in an attempt to increase the yield of the photoadduct, only 0.05% of ortho adduct was isolated along with 54% of an oxetane formed by the addition of benzophenone to 2-methylbut-2-ene. In the absence of benzophenone, the ortho adduct was isolated in 63% yield. It is, however, thermally as well as photochemically unstable and reverts to starting materials, supposedly also via a biradical. The authors propose that benzophenone catalyzes bond cleavage of the adduct more efficiently than ortho addition and this would account for the low yield of photoadduct in the presence of benzophenone. From these experiments, no conclusion about the identity of the reactive excited state can be drawn.

Job and Littlehailes [127] have irradiated a 1:1 molar mixture of benzene and acrylonitrile under nitrogen at 0°C and obtained 7-cyanobicyclo[4.2.0] octa-2,4-diene. The reaction did not proceed in the absence of ultraviolet irradiation or in a Pyrex apparatus. The yield of photoadduct is little affected by the presence of air, which, according to the authors, strongly supports the intermediacy of a singlet excited complex. They were, however, unable to detect a UV absorption band of a ground-state complex. However, the yields of adduct were low and the possibility is considered that a low-intensity absorption band is hidden by the benzene spectrum.

According to Wilzbach and Kaplan [7], the retention of olefin configuration in the ortho, meta, and para photocycloadducts of benzene with *cis*- and *trans*-but-2-ene suggests that these reactions are concerted. Because good evidence had already been obtained that meta cycloaddition involves the singlet (B_{2u}) state of benzene [175], the authors consider it likely that this state is also involved in the ortho and para addition, because the relative (initial) yields of the various adducts remain constant over a wide range of concentrations and proportions of the reactants, such that the benzene singlet-triplet ratio would almost certainly vary. Because a concerted ortho photoreaction is not allowed from S_1 benzene plus S_0 alkene [81], an apparently forbidden process occurs with moderate quantum efficiency. The involvement of an excited complex, formed from $^1B_{2u}$ benzene and ground-state olefin, in which mixing of excited states results in relaxation of orbital symmetry restrictions, was proposed as a rationalization.

The rate of ortho photocycloaddition of benzonitrile to 2-methylbut-2-ene, 2-methoxypropene, and vinyl acetate in the presence of *cis*-1,3-pentadiene is only one-fourth that in the absence of quencher [109]. At first, this was taken as an indication that the reaction proceeds via the triplet state. Subsequent studies [110], however, have shown that the quencher reacts with the benzonitrile to give adducts of the bicyclo[4.2.0]octadiene type. When a correction was made for this reaction, it became apparent that little if any triplet quenching was observed. It was, in fact, found that many alkenes, as well as simple dienes, quench benzonitrile fluorescence effectively. Benzonitrile exhibits a fluorescence lifetime in fluid solution at 25°C of 9.6 ns. The quenching rate constants from fluorescence lifetime quenching for 2,3-dimethylbut-2-ene and 2-methylbut-2-ene are 1.06×10^9 L/mol^{-1}/s^{-1} and 0.63×10^9 L/mol^{-1}/s^{-1}, respectively. A quenching rate constant was also obtained from measurements of the dependence of quantum yield of product formation on the concentration of 2,3-dimethylbut-2-ene and found to be 1.3×10^9 L/mol^{-1}/s^{-1}, in reasonable agreement with the value obtained from fluorescence, indicating that the same excited state is involved in both emission and photochemical reaction. The reaction of benzonitrile with 2,3-dimethylbut-2-ene, however, occurs at the cyano group and not at the benzene ring. 2-Methyl-2-butene, on the other hand, adds mainly to the phenyl ring, but for this alkene, no quenching rate constant based on measurements of product quantum yields was reported. From mixtures of both alkenes with benzonitrile, enhanced emission, attributed to exciplexes, was found in the 320–360-nm range, distinct from the benzonitrile fluorescence. The enhancement is moderate in the case of 2-methylbut-2-ene, which adds to the phenyl ring, but enormous in the case of 2,3-dimethylbut-2-ene, which adds to the cyano group. The most likely explanation for this difference in exciplex emission is considered to be a different geometry in the two exciplexes.

Fluorescence quenching experiments have also been performed by Ohashi et al. [139]. The quenching of the fluorescence of *para*-dimethoxybenzene by acrylonitrile as a function of alkene concentration gave a nice linear Stern–Volmer plot with a $k_q\tau$ value of 18.2 M^{-1}, which is taken as an indication that the photochemical reaction proceeds through a singlet excited state of p-dimethoxybenzene. No quenching rate constant from measurements of the quantum yield of product formation as a function of alkene concentration was reported however, so that the possibility that acrylonitrile quenches the emission from the singlet excited state but reacts with the triplet state has not been eliminated. Moreover, *para*-dimethoxybenzene and acrylonitrile do not yield ortho adducts upon irradiation in methanol or acetonitrile [136,138,139]. These experiments, therefore, do not prove the intermediacy of the singlet excited state in the ortho photocycloaddition.

Irradiation of a 1 M solution of 2,2-dimethyl-1,3-dioxole in benzene gives ortho and meta photocycloadducts in the ratio of 2:1 [123]. A dimer of the alkene is also formed in low yield (1–2%). Sensitization experiments using acetophenone ($\lambda > 300$ nm) yield only the dimer. Direct irradiation ($\lambda = 254$ nm) of 2,2-dimethyl-1,3-dioxole in an inert solvent does not lead to formation of the dimer and it is therefore assumed that formation of ortho and meta cycloadducts occurs via the S_1 state of benzene and formation of the dimer of 2,2-dimethyl-1,3-dioxole via sensitization by the T_1 state of benzene. The lowest triplet state of benzene has an energy of 84.3 kcal/mol^{-1}, whereas the energy of the lowest triplet state of acetophenone is 73.7 kcal/mol^{-1}. It is therefore highly unlikely that triplet benzene can be produced by energy transfer from triplet acetophenone and it was therefore not to be expected that ortho photoadducts would be produced in the sensitization experiment. The experiments do, however, show that the dimerization of the alkene is a triplet reaction and it seems logical to assume that if triplet benzene transfers its energy to the alkene, the production of adducts must stem from interaction with the singlet excited state.

The ortho photocycloaddition of 2,3-dihydropyran is a very efficient process [11,12]. The quantum yield, measured with a solution of 1.1 M benzene and 3.5 M 2,3-dihydropyran, is 0.7 in iso-octane and 0.78 in acetonitrile. The higher quantum yield in the more polar solvent supports the proposal that ortho photocycloaddition of alkenes to benzene involves a polar intermediate. It is useful to compare these quantum yields of product formation with the quantum yield of intersystem crossing of benzene. The latter value depends on the concentration of benzene: φ_{isc} = 0.23 at a concentration of 0.11 M in cyclohexane, 0.47 at a concentration of 5.1 M in the same solvent, and 0.56 in neat benzene [176]. The maximum value thus lies below the quantum yield of product formation, implying that the S_1 state of benzene is probably involved in the 1,2 addition. 2,3-Dihydropyran quenches the fluorescence of benzene with k_q values of (3.8–8.99) $\times$ 10^8 L/mol^{-1}/s^{-1} dependent on the dielectric constant of the solvent. No exciplex emission was observed.

The addition of benzene to 2,3-dihydropyran was examined in the presence of cyclopropyl bromide [177] in an attempt to determine the nature of the excited state of benzene responsible for the formation of the ortho adduct [11,12]. The absorption of the bromide at 254 nm was taken into account (90% of the radiation was absorbed by benzene) and a decrease of the rate of formation of 50% in the heavy-atom solvent was observed. Cyclopropyl bromide also quenched the fluorescence of benzene (k_q = 3.82 $\times$ 10^8 L/mol^{-1}/s^{-1}). These data are interpreted as enhanced intersystem crossing of S_1 benzene and the necessary involvement of the S_1 state in formation of the ortho cycloadduct.

Mattay et al. [122] have determined relative quantum yields for the ortho addition of benzene to 1,3-dioxoles and 1,4-dioxene in xenon and in argon atmosphere. The relative quantum yields φ_{rel} = $\varphi^{Xe}/\varphi^{Ar}$ are 0.5 for 1,3-dioxole, 0.3 for 2,2-dimethyl-1,3-dioxole, 0.6 for 2-methyl-1,3-dioxole, and 0.6 for 1,4-dioxene. Because xenon strongly accelerates the singlet–triplet intersystem crossing through the heavy-atom effect, it was concluded that the photoaddition of benzene to the dioxoles and dioxene involves exclusively intermediates of singlet type.

D. Exciplexes

The possibility of an exciplex as an intermediate species in photochemical additions of arenes to alkenes was first put forward by Morrison and Ferree [178] for the intramolecular meta cycloaddition they had discovered. Wilzbach and Kaplan [7], in discussing the intermolecular ortho photocycloaddition, observed that an apparently forbidden process (adduct formation from $^1B_{2u}$ benzene and ground-state olefin) does, in fact, occur with at least moderate quantum efficiency. They proposed that this result can, perhaps, be rationalized by the intervention of an excited complex, formed from $^1B_{2u}$ benzene and ground-state olefin, in which mixing of states results in relaxation of orbital-symmetry restrictions. The possible intermediacy of an exciplex (or even a zwitterion) was also discussed by Bryce-Smith [38]. Exciplex emission from solutions containing an arene and an alkene was first observed by Cantrell [110]. Emission in the 320–360-nm range was observed from mixtures of benzonitrile with 2,3-dimethylbut-2-ene (which undergoes photoaddition to the cyano group) and with 2-methylbut-2-ene (which undergoes ortho photocycloaddition to benzonitrile). A singlet exciplex as inter-mediate in the ortho photocycloaddition was also proposed, but not detected, by Ohashi et al. [139] for the case of anisole and dimethoxybenzenes with cya-noethenes.

In 1977, Scharf and Mattay [123] found that benzene undergoes ortho as well as meta photocycloaddition with 2,2-dimethyl-1,3-dioxole and, subse-quently, Leismann et al. [179,180] reported that they had observed exciplex fluorescence from solutions in acetonitrile of benzene with 2,2-dimethyl-1,3-dioxole, 2-methyl-1,3-dioxole, 1,3-dioxole, 1,4-dioxene, and (Z)-2,2,7,7-tetram-ethyl-3,6-dioxa-2,7-disilaoct-4-ene. The wavelength of maximum emission was around 390 nm. In cyclohexane, no exciplex emission could be detected. No obvious correlation could be found among the ionization potentials of the alkenes, the Stern–Volmer constants of quenching of benzene fluorescence, and the fluorescence emission energies of the exciplexes. Therefore, the observed exciplexes were characterized as "weak exciplexes" with dipole–dipole rather than charge-transfer stabilization. Such exciplexes have been designated as "mixed excimers" by Weller [181].

The alkenes, which give emitting exciplexes with benzene, quench its fluorescence at nearly diffusion-controlled rates. For 1,4-dioxene, 2,2-dimethyl-1,3-dioxole, 2-methyl-1,3-dioxole, and 1,3-dioxole, the quenching rate constants are 35.0×10^8, 81.9×10^8, 82.5×10^8, and 87.3×10^8 L/mol^{-1}/s^{-1}, respectively [122]. For comparison, the quenching rate constants of cyclopentene, cyclohex-ene, and vinylene carbonate were also determined and their values were found to be much smaller: 1.2×10^8, 1.5×10^8, and 3.1×10^8 L/mol^{-1}/$^{-1}$, respectively. The electron-rich alkenes with which exciplex emission was observed give considerable yields of ortho adducts next to meta adducts (1,4-dioxene gives only ortho addition), whereas for the other compounds, no ortho addition has been reported.

A kinetic scheme was proposed [122] with the fluorescent exciplex as precursor of the photoproducts (ortho as well as meta adducts). Quantum yields of adduct formation, exciplex emission, and benzene fluorescence were measured as a function of alkene concentration. The kinetic data fit the proposed reaction scheme. The authors have also attempted to prove the intermediacy of the exciplex in the photoaddition by adding a quencher to the system benzene + 2,2-dimethyl-1,3-dioxole in acetonitrile. It was found that triethylamine quenches the exciplex emission with a Stern–Volmer constant ($K_{SV} = 11.6 \pm 0.4$ L/mol^{-1}) which is nearly identical to that for quenching of the meta photocycloaddition ($K_{SV} = 11.7 \pm 1.5$ L/mol^{-1}). The latter value, however, was not measured in acetonitrile be-cause in that solvent the quantum yield of product formation is rather low; instead, the value measured in dioxane was used, corrected for the difference in viscosity.

Triethylamine not only quenches the fluorescence of the exciplex and the formation of product but it also quenches the fluorescence of benzene. In dioxane, the value of $k_q\tau$ was 86.6 L/mol^{-1} and the equality of the Stern–Volmer constants for quenching of exciplex emission and quenching of photoaddition may not be used as evidence in favor of the intermediacy of an exciplex. If exciplex and photoadduct are formed competitively instead of consecutively, Stern–Volmer constants for quenching of exciplex emission and product formation will also be equal if their common precursor (i.e., S_1 benzene) is quenched.

Leismann et al.[182] have recognized this problem in their publication of 1984, in which they describe a thorough and detailed investigation of the kinetics of formation and deactivation of exciplexes of S_1 benzene or toluene and 1,3-dioxole, 2,2-dimethyl-1,3-dioxole, and 2,2,4-trimethyl-1,3-dioxole. The evolution in time of monomer and exciplex fluorescence after excitation using a nanosecond flash lamp was analyzed, and again it was concluded that the formation of exciplexes is diffusion controlled; their decay proceeds mainly ($\geq 90\%$) via radiationless routes. The polar solvent acetonitrile enhances radiationless deactivation, possibly by promoting radical ion formation. Because decay of benzene fluorescence is essentially monoexponential, dissociation of the exciplex into S_1 benzene and dioxole is negligible.

In experiments with continuous irradiation, Stern–Volmer constants were determined for the quenching of benzene fluorescence by the dioxoles, for the generation of exciplex fluorescence, and for adduct formation. The first two values were found to be identical within experimental error and equal to the values obtained from the flash experiments. The Stern–Volmer constants calculated from adduct formation were, however, much smaller than expected on the basis of the exciplex scheme, and it was concluded that the exciplex hypothesis had to be improved. The authors did this by introducing an additional reaction step in which the exciplex meets a second molecule of dioxole (Scheme 34). Collision leads to quenching of the exciplex with formation of either ground-state reactants or photoadduct. (At this point, no distinction was made between ortho and meta adducts.) The role of the second olefinic partner was presumed to be catalytic. Donation of charge to the cationic site of the exciplex may weaken the solvent stabilization of exciplex and zwitterionic species and thus reduce the solvent-induced quenching of adduct formation. A mechanism involving product formation by direct collapse of the exciplex could, however, not be excluded.

In cyclohexane, no exciplex emission is observed. Leismann et al. [182] have measured the quantum yields of meta and ortho adduct formation from benzene and 2,2-dimethyl-1,3-dioxole in cyclohexane at various concentrations of the

^{1}(Ar.....Ol)* + Ol $\longrightarrow$ adduct + Ol $\qquad$ adduct formation

$\longrightarrow$ 2 Ol + Ar $\qquad$ quenching

Scheme 34 Bimolecular deactivation of the arene olefin. (From Ref. 182.)

alkene in the absence and in the presence of triethylamine (TEA). In plots of $(\varphi_{meta})^{-1}$ versus $[dioxole]^{-1}$, the presence of TEA not only changes the slope but also the intercept. This indicates that TEA quenches a species which is formed from S_1 benzene and dioxole and which is a precursor of the meta adduct. In the plots of $(\varphi_{ortho})^{-1}$ versus $[dioxole]^{-1}$, the slope is different in the presence of TEA but the intercept is the same as in its absence. This might imply that the ortho adduct may be formed directly or via a different short-lived intermediate.

A schematic representation of the possible processes in arene-olefin photocycloaddition according to Leismann et al. [182] is given in Scheme 35. Increasing charge separation in the exciplex, caused by decreasing ionization potentials of the olefins, can lead to zwitterionic intermediates, which are able to form ortho adducts. Increasing solvent polarity can reduce the formation of adducts by propagating the formation of radical ion pairs. The possibility that the ortho adduct may be formed directly or via a different short-lived intermediate was not incorporated in the scheme.

The kinetic results agree with a mechanism involving an exciplex as a reaction intermediate, but it was not explicitly stated that the data are incompatible with a scheme in which exciplex formation and adduct formation are competing instead of consecutive processes.

The 1,3-dioxoles and 1,4-dioxene are the only alkenes with which exciplex formation and decay can be studied in conjunction with photoaddition. The dioxoles give ortho as well as meta photocycloadducts with benzene [13,14,122]. Mattay et al. [15,134,183] have pictured the exciplex as a common precursor to both types. The exciplex is supposed to possess charge-transfer character, from the dioxole to the arene. Formation of the ortho adduct is thought to proceed stepwise as shown in Scheme 36 for anisole and 1,3-dioxole

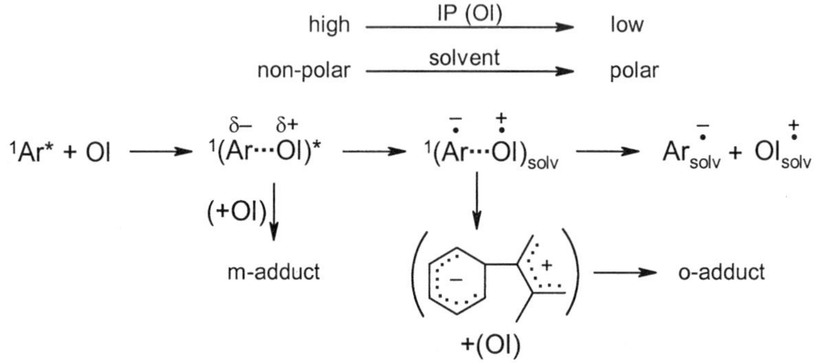

Scheme 35 Formation of the arene-olefin (Ar-Ol) exciplex and its deactivation by charge-transfer effects induced by changes of olefin ionization potential (IP) and solvent polarity according to Leismann et al. (From Ref. 182.)

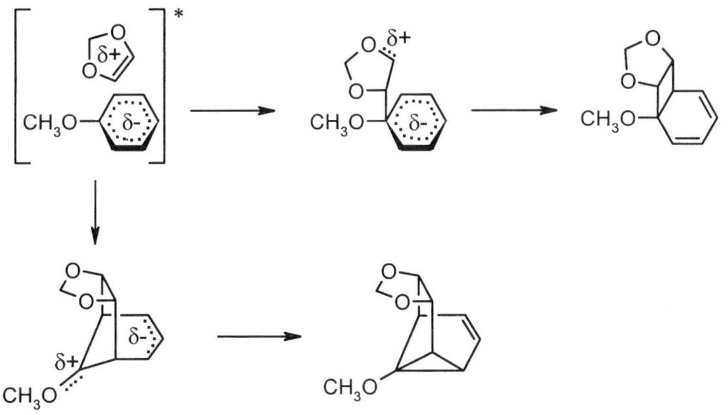

Scheme 36   Formation of ortho and meta adducts via a common exciplex. (From Ref. 183.)

[183]. Another route leads to the meta adduct via a zwitterionic structure (or a polarized biradical).

Although the relative orientations of the two addends must be different for ortho and meta addition, it is conceivable that both processes should proceed via the same exciplex. One may speculate that the exciplex does not have one favorite rigid geometry, but that it is in a double-minimum energy well on the excited-state potential-energy surface, with the minima separated by a small barrier.

In the case of anisole and 1,3-dioxole, two ortho adducts are formed [134], exo and endo, in a ratio of 7:1. Because the ortho adduct is proposed to arise via stepwise addition, formation of both stereoisomers can be accounted for if one assumes that rotation around the first-formed bond is possible. The predominance of exo–ortho over endo–ortho is ascribed to repulsion between the lone pairs on the oxygen atoms and the negative charge residing in the six-membered ring [183]. 2-Methyl-1,3-dioxole gives two exo–ortho adducts with benzene, next to four meta adducts [12,13,120]. One ortho adduct is formed via the approach with the methyl group pointing toward the benzene ring and the other with the methyl group away from benzene. The involvement of at least two different exciplexes seems inevitable if exciplexes are intermediates. Exciplex emission has been observed with this system [122,179,180], but there is no evidence concerning dual exciplexes.

From the evidence thus far available, it cannot be concluded that exciplexes are intermediate species in the ortho photocycloaddition of arenes to alkenes. In those cases in which exciplex emission has been observed, there exists no definitive proof that the exciplex is able to undergo bond formation leading to the adduct. For the numerous cases in which no exciplexes have been detected, of course no statements about their intermediacy can be made.

E. Zwitterions, Biradicals, and Ion Pairs

Irradiation (254 nm) of 2,3-dimethylbut-2-ene in benzene (10% v/v) yields three adducts, which were identified as an ortho cycloadduct, a meta cycloadduct, and the product of an "ene" reaction [7,115–117] (Scheme 37).

The quantum yields [115,116] (1.1 M alkene + 3.5 M benzene in isooctane) of the products are as follows: ortho, 0.25; meta, 0.03; "ene," 0.03. In preparative irradiations, the relative proportions were found to be dependent on both reactant concentration and irradiation time. The use of a 10% solution of the alkene in benzene led to a photostationary ratio of ortho : meta : "ene" = 1:4:8. The ortho adduct is photolabile and was identified via its Diels–Alder adduct with maleic anhydride [116,117]. It could be shown that the "ene" product was not formed in a concerted reaction. Irradiation of the alkene in hexadeuterobenzene gave two products, one with hydrogen at C-4 cis and one with hydrogen trans with respect to the substituent. This proves at the same time that hydrogen-atom or proton transfer can take place within an intermediate species, diradical, exciplex or zwitterion. The identity of the intermediate was investigated by performing a study of solvent effects in cyclohexane, ether, methanol, and acetonitrile. The yield of ortho adduct was ~70–80% greater in both methanol and acetonitrile than in cyclohexane or ether. The yield of the "ene" product was increased by ~35% in methanol, but the yield in acetonitrile was similar to that in the nonpolar solvents. Irradiation in the presence of deuteromethanol (MeOD) led to 20% deuterium incorporation in the 4-position of the cyclohexadiene moiety. Proton transfer is clearly involved in the formation of the "ene" product, and the proton-transfer step evidently follows the initial addition of the alkene to the benzene ring.

The experimental data are consistent [116] with the intermediate depicted in Scheme 38 or at least with a corresponding polarized exciplex capable of acting as a proton acceptor. For a singlet species, as in the present case, little or no diradical character is likely. The delocalization shown in Scheme 38 involves an interaction between centers formally separated by an sp^3 hybridized carbon atom. The tendency for a hypothetical diradical to polarize in a preferred sense is also shown in the scheme. The stability of such a singlet intermediate species of the major left-hand canonical type (Scheme 38) would be favored by dimethylation at the positive pole, and the apparent failure of propene and the but-2-enes to un-

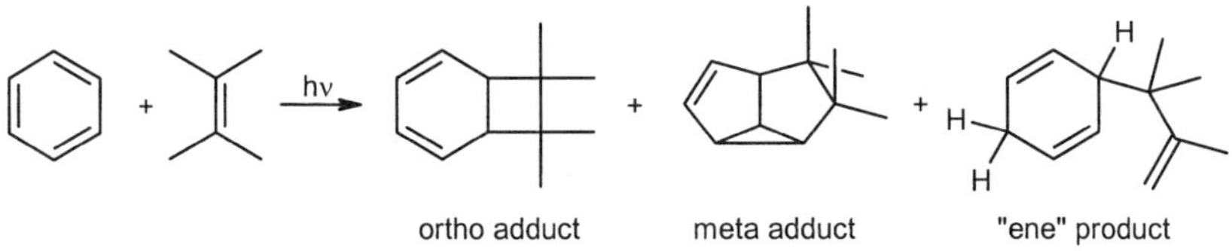

           ortho adduct   meta adduct   "ene" product

Scheme 37   Photoreaction of benzene and 2,3-dimethylbut-2-ene.

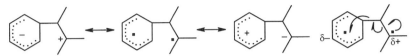

Scheme 38   Polarized biradical (or zwitterion) as intermediate in the formation of the "ene" product (and the ortho adduct) from 2,3-dimethylbut-2-ene and benzene.

dergo "ene" addition may in part be attributed to the absence of this structural feature.

Collapse of the polarized diradical/zwitterion would lead to the ortho cycloadduct. Stereospecificity in ortho addition by this route would then be governed not by concertedness in the usual sense, but rather by coulombic interaction between the charged centers [38].

Bryce-Smith et al. [61] have examined the effect of acid and solvent polarity on the ortho cycloaddition of acrylonitrile to benzene. Irradiation (254 nm) of the two compounds in cyclohexane or of solutions of acrylonitrile in benzene yields a mixture of ortho adducts in which the endo:exo ratio is ~5 (cf. Ref. 47; endo:exo = 1:5). Irradiation in methanol or acetonitrile strongly promotes the formation of these adducts but the endo:exo ratio is unchanged, and no other 1:1 adducts could be detected. Exactly the same results were obtained upon irradiation of an equimolar mixture of benzene and acrylonitrile in the presence and absence of 10% trifluoroacetic acid, indicating that the promoting effect of methanol is more a result of increased dielectric constant than of acidity. The effects of solvent polarity indicate the involvement of an exciplex or formally bonded intermediate having markedly polar character, whereas the insensitivity to proton donors implies that this intermediate is insufficiently dipolar in character and/or of too short a lifetime relative to ring closure for it to be intercepted by protonation. The authors excluded a polar intermediate formed by full charge transfer from S_1 benzene to S_0 acrylonitrile followed by bond formation (a zwitterion) because they found it hard to believe that a fully developed carbanionic center (on the CH_2 group of the acrylonitrile moiety) could escape protonation in the presence of trifluoroacetic acid. It was therefore considered likely that the degree of dipolar character which develops in the course of the ortho cycloaddition is sufficient to escape the restrictions imposed by orbital-symmetry considerations on a fully concerted homopolar ortho cycloaddition of S_1 benzene to an S_0 alkene, but insufficient to permit interception by a proton donor.

Ohashi et al. [136,139] have proposed a scheme describing the formation of ortho adducts and substitution products from anisole and the three dimethoxybenzenes with acrylonitrile, methacrylonitrile, and crotonitrile (Scheme 39). Here, the ortho cycloadduct is supposed to be formed directly from an encounter complex or exciplex, whereas the substitution product arises via formation of an ion pair from the complex, followed by protonation of the radical anion and radical

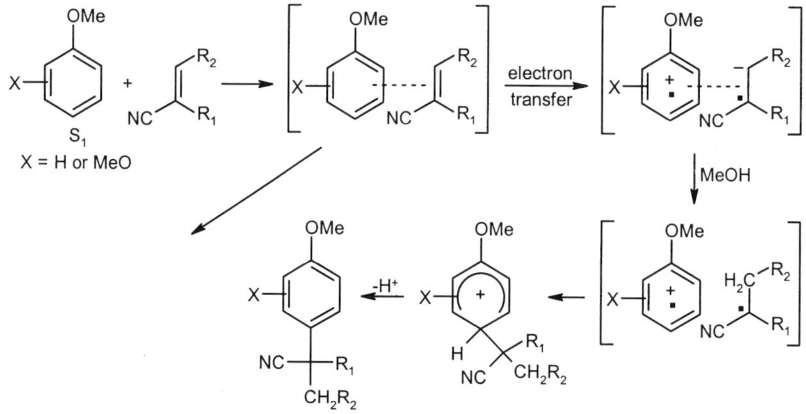

Scheme 39 Formation of ortho photocycloadducts and substitution products from anisole and dimethoxybenzenes with acrylonitrile ($R_1 = R_2 = H$), methacrylonitrile ($R_1 = Me$; $R_2 = H$), and crotonitrile ($R_1 = H$; $R_2 = Me$).

combination. This mechanism explains the observed shift from mainly addition product to mainly substitution product in changing the solvent from acetonitrile to methanol. The scheme also rationalizes the experimental finding that a deuterium atom becomes incorporated in the CH_2R_2 group of the substitution product when the reaction is carried out in MeOD [136].

In this respect, it is of interest that the intramolecular ortho photocycloaddition of (E)-6-(2-methoxyphenyl)-5,5-dimethyl-2-hexenenitrile proceeds with retention of (E)-olefin geometry in acetonitrile from the first excited singlet state of the anisole chromophore [160] (Scheme 40). The ortho adduct undergoes further

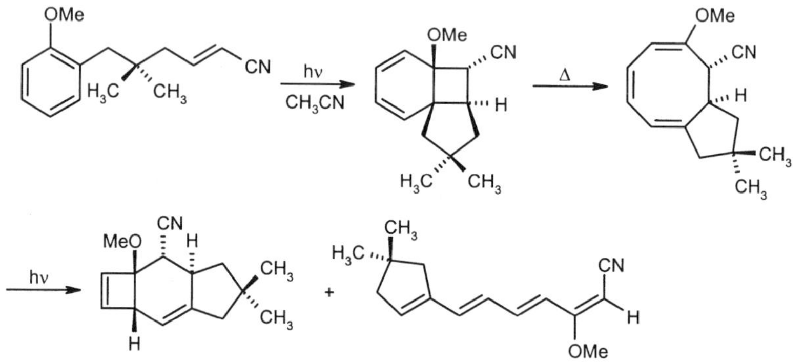

Scheme 40 Intramolecular photocycloaddition of (E)-6-(2-methoxyphenyl)-5,5-dimethyl-2-hexenenitrile.

thermal and photochemical rearrangements. At 85% conversion, the yield of the tricyclic diene is 58% and that of the tetraene is 14%.

Upon irradiation of α,α,α-trifluorotoluene and 1,3-dioxole in dioxane, two ortho cycloadducts and two meta cycloadducts are formed, next to small amounts of substitution products and some unidentified products. Mattay et al. [15] have proposed a mechanism in which an exciplex may be the common precursor of both types of adducts (Scheme 41). Formation of one bond within the exciplex gives rise to zwitterions in which the negative charge is optimally stabilized by the CF_3 group. The formation of the second C—C bond then leads to the 1-substituted endo–and exo–ortho adducts. The meta adducts, which both have the exo configuration, may be formed either by simultaneous formation of two bonds in the exciplex or by an alternative bond formation in the zwitterion that also leads to the endo–ortho adduct.

Similar mechanisms involving exciplexes and zwitterions were proposed for the formation of ortho adducts from anisole and 1,3-dioxoles [134,183] and from anisole with acrylonitrile [183]. For the latter reaction, Mattay has proposed a zwitterion in which the terminal carbon atom of the double bond of acrylonitrile

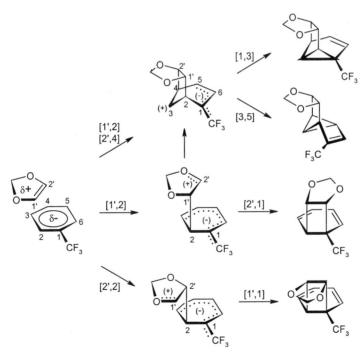

Scheme 41   Formation of ortho and meta cycloadducts from α,α,α-trifluorotoluene and 1,3-dioxole via exciplex and zwitterions.

is bonded to an ortho position of anisole. This structure seems unlikely because it does not correspond to the substitution product that is formed in methanol (cf. Scheme 39) and to the result of the reaction in MeOD.

F. Empirical Rules: Ortho Versus Meta Photocycloaddition

In discussing the first examples of ortho photocycloaddition of simple mono-olefins to benzene [7,117], Bryce-Smith [38] addressed the question how stereospecific 1,2 (and 1,4) photochemical cycloadditions of mono-olefins to benzene can occur under conditions which appear to involve S_1 (benzene) ($^1B_{2u}$) and S_0 (olefin) when concerted 1,2 (and 1,4) additions of S_0 olefin to S_1 benzene are forbidden on orbital-symmetry grounds. At that time, evidence had already been obtained that polar intermediates are important. In the photoadditions of cis- and trans-cyclooctenes to benzenes, the quantum yield for ortho addition is much higher for the trans olefin ($\varphi = 0.39$ at 3.8 M olefin + 1.1 M benzene) than for the cis isomer ($\varphi = 0.10$). The ratio $\varphi_{ortho}/\varphi_{meta}$ is 1.2 for trans-cyclooctene and 0.26 for cis-cyclooctene. Of these two olefins, trans-cyclooctene is the stronger electron donor (the first ionization potentials are 8.56 and 8.81 eV for the trans and cis isomers, respectively) and has the greater tendency to undergo ortho cycloaddition to benzene. Toluene, being a weaker acceptor [ionization potential (IP) = 8.82 eV], gives only meta cycloaddition with cis- or trans-cyclooctene, whereas the stronger acceptor benzonitrile (IP = 9.71 eV) undergoes ortho addition with simple olefins.

On the basis of these observations, Bryce-Smith et al. [115] introduced a rule stating that for addition to benzene, $\varphi_{ortho} > \varphi_{meta}$ when 9.6 eV < IP (alkene) <8.65 eV. They concluded that if this rule is correct, ortho addition of ethylenes to S_1 benzene necessarily involves an element of charge transfer to or from the ethylene. Indeed, a marked effect of polar solvents (methanol or acetonitrile) in promoting the ortho addition of benzene to ethyl vinyl ether and tetramethylethene was observed: φ_{ortho} increased by 20–50%, whereas φ_{meta} was unaffected. One exception to this rule was found by Heine and Hartmann [10], who discovered that vinylene carbonate (IP = 10.08 eV) undergoes mainly meta photocycloaddition to benzene, accompanied by some para addition. Bryce-Smith and Gilbert [46] commented that their rule referred to quantum yields and not chemical yields, whereas no quantum yields were given for the vinylene carbonate additions. Moreover, quantum yield measurements should be made at low conversions because most ortho cycloadducts are photolabile.

The most simple combination of arene and alkene, benzene + ethene, was studied by Mirbach et al. [114]. Their experiments were carried out in a UV autoclave fitted with a low-pressure mercury lamp under elevated olefin pressure in various solvents. The major product was the meta photocycloadduct, tricyclo[3.3.0.0^{2,8}]oct-3-ene. After short irradiation times, the major side products bicyclo[4.2.0]oct-2,4-diene and cycloocta-1,3,5-triene, formed by ortho cycloaddition and subsequent ring opening were identified. At higher conversions, these products were present in only small amounts, but now their photoproducts could be identified. In dichloromethane at 27 bar olefin pressure, the quantum yield for formation of meta adduct, φ_{meta}, is 0.08 and the quantum yield of formation of other photoproducts, φ_{others}, 0.07. At a pressure of 50 bar, the values are $\varphi_{meta} = 0.11$ and $\varphi_{others} = 0.08$. Thus, although a substantial amount of ortho adduct is found, the preferred mode of addition of benzene and ethene is meta, in spite of the high ionization potential of ethene (10.5 eV). The photoreaction of benzene with propene, studied by the same authors, was rather clean. Except for a small amount of polymer, no side products were detectable. The ionization potential of propene is 9.73 eV, and although the preference for meta addition of ethene appears to form an exception to the ionization potential rule, the relative behavior of ethene and propene is in accord with this rule.

Cis-cyclooctene undergoes both ortho and meta addition to hexafluorobenzene [67], as it does to benzene. There is a significant difference in ionization potential between the aromatic compound (IP = 10.4 eV) and the olefin (8.75 eV) and ortho addition was to be expected. The ratio $\varphi_{ortho}/\varphi_{meta}$ was 0.60 (cf. 0.24 for the corresponding addition to benzene). However, the authors had expected a somewhat higher value and were concerned about the failure of cyclopentene (IP = 9.02 eV) to undergo meta photocycloaddition to hexafluorobenzene [150]. They therefore proposed that electron affinities might provide a better correlation with the photochemical behavior in these systems than ionization potentials, a suggestion that had also been made by Mirbach et al. [114].

Mattay et al. [13,14,122] have also observed deviations from the ionization potential rule. They have studied the photocycloadditions of benzene to 1,3-dioxole (Scheme 42) and some of its derivatives.

The ionization potential of 1,3-dioxole is 8.56 eV and thus outside the region where meta addition is expected to dominate over ortho addition. Yet, the quantum yield of meta addition (exo + endo) is 0.38 and that of ortho addition 0.21. With 2-methyl-1,3-dioxole and benzene, 40% meta adducts, 40% ortho adducts, and 16% para adduct are found [13].

In spite of these deviations, the ionization potential rule still has considerable predictive value, although it may be a little too restricted to cover all the addends [116,184]. A slight modification of the rule was introduced by Gilbert et al. [12] who proposed that it might be more meaningful to relate the relative efficiencies of the meta and ortho photocycloaddition with the difference in ionization potential only within a series of structurally very similar alkenes. It was also recognized that ionization potentials relate to properties of the ground state of the reactants rather than of the excited state [185]. Nevertheless, the IP rule retains its predictive value in a series in which various arenes are irradiated in the presence of the same alkene [133]. Ethyl vinyl ether and benzonitrile (IP = 10.02 eV) yield

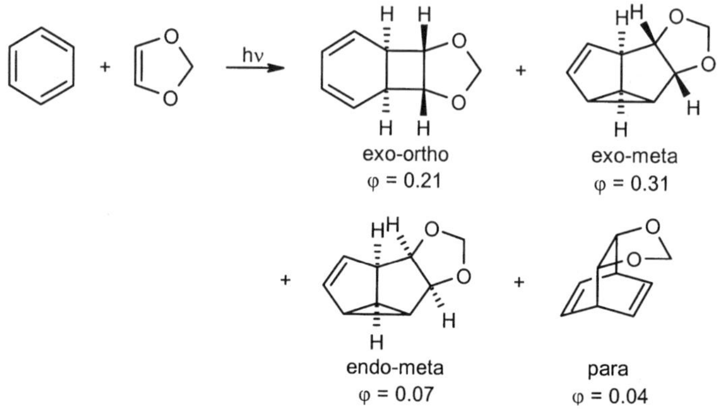

Scheme 42 Photocycloaddition of benzene to 1,3-dioxole. The quantum yields [13,122] were measured at 254 nm in dioxane at 20°C with 1.1 M benzene and a dioxole concentration at which the quantum yield reached the maximum value.

regioisomers and stereoisomers of ortho cycloadducts; with ethyl vinyl ether and benzene (9.24 eV), the ortho:meta ratio is 2:1, with toluene (8.82 eV), it is 1:4.5, and with anisole (8.54 eV), ethyl vinyl ether gives only regioisomers and stereoisomers of a meta cycloadduct. In this series, in which the mode of addition shifts from exclusively ortho to exclusively meta, the difference in ionization potential between the two addends changes from 1.24 to 0.06 eV.

Ohashi et al. [128] found that the yields of ortho photoaddition of acrylonitrile and methacrylonitrile to benzene and that of acrylonitrile to toluene are considerable increased when zinc(II) chloride is present in the solution. This was ascribed to increased electron affinity of (meth)acrylonitrile by complex formation with $ZnCl_2$ and it confirmed the occurrence of charge transfer during ortho photocycloaddition. This was further explored by investigating solvent effects on ortho additions of acceptor olefins and donor arenes [136,139]. Irradiation of anisole and acrylonitrile in acetonitrile at 254 nm yielded a mixture of stereoisomers of 1-methoxy-8-cyanobicyclo[4.2.0]octa-2,4-diene as a major product. A similar reaction occurred in ethyl acetate. However, irradiation of a mixture of anisole and acrylonitrile in methanol under similar conditions gave the substitution products 4-methoxy-α-methylbenzeneacetonitrile (49%) and 2-methoxy-α-methylbenzeneacetonitrile (10%) solely (Scheme 43).

When the irradiation was performed in MeOD, incorporation of one deuterium atom into the α-methyl groups could be demonstrated. The reaction proceeds via photoinduced electron transfer from the excited anisole to acrylonitrile, possible in an exciplex [139]. The radical anion of acrylonitrile abstracts a proton from the solvent and the resulting radical combines with the radical cation of anisole forming a σ-complex that finally loses a proton.

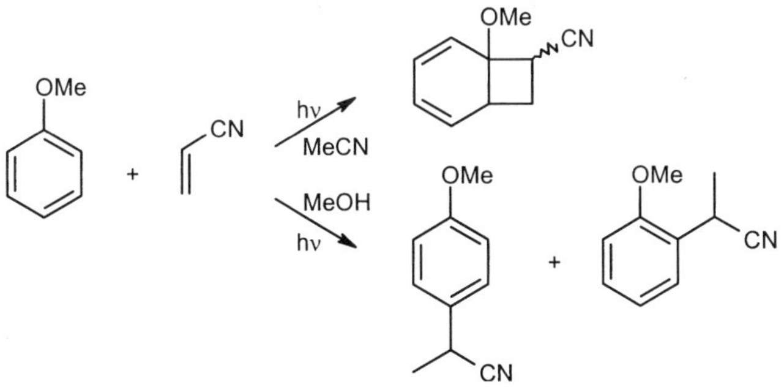

Scheme 43   Photoreaction of anisole with acrylonitrile in acetonitrile and in methanol.

Similar results were obtained [139] with the three dimethoxybenzenes and acrylonitrile, methacrylonitrile, and crotonitrile. The amounts of substitution products decrease in the order acrylonitrile (49%) > methacrylonitrile (45%) > crotonitrile (6%), which agrees with the electron affinities of these compounds. Simultaneously, the amount of addition product increases: acrylonitrile, 0%; methacrylonitrile, 38%; crotonitrile, 67%. In the series of anisole and the dimethoxybenzenes with crotonitrile, the amount of substitution products decrease in the order *ortho*- and *para*-dimethoxybenzene > *meta*-dimethoxybenzene > anisole, which is just the reverse of the order of their oxidation potentials. Ohashi et al. [139] have attempted to relate the photochemical behavior of these systems to the free enthalpy of electron transfer in the excited state as calculated with the Rehm–Weller equation, $\Delta G = E(D/D^{+}) - E(A^{-}/A) - e_0^2/\varepsilon R - \Delta E_{00}$.

Al-Jalal and Gilbert [138] have also examined the systems anisole/acrylonitrile and *para*-dimethoxybenzene/acrylonitrile. In marked contrast to the results reported by Ohashi et al. [136,139], it was found that from anisole and acrylonitrile in methanol, the major product is still the ortho cycloadduct; the relative quantum yields of ortho adduct, para-substitution product, and ortho-substitution product being 22:6:3. With *para*-dimethoxybenzene and acrylonitrile in both methanol and acetonitrile, large quantities of insoluble polymer were found, but no substitution products, again in contrast to the earlier reports. The photoreactions of this system were investigated in a range of solvents and it was found that in cyclohexane, iso-octane, ethyl acetate, dioxane, and *tert*-butyl alcohol, the ortho adduct was formed, next to alkene dimers. The hydrocarbon solvents gave the highest yields of ortho cycloadducts. Irradiation in isopropyl alcohol, ethanol, or methanol yielded only polymer and in acetonitrile alkene dimers were formed next to polymers. The authors have no logical explanation for the marked differences between their results and those described by Ohashi et al. [136,139]. There seems to be agreement, however, on the general mechanism. Ortho addition occurs directly from the excited-state complex, whereas in more polar solvents, the substitution products arise via the sequence electron transfer within the complex, proton transfer from the solvent, and radical combination.

Mattay et al., having discovered exciplex emission from solutions of benzene and 1,3-dioxole [122], continued their investigations with a study on selectivity and charge transfer in photoreactions of α,α,α-trifluorotoluene with 1,3-dioxole and some of its derivatives, and with vinylene carbonate and dimethylvinylene carbonate [15,143,144]. α,α,α-Trifluorotoluene and 1,3-dioxole upon irradiation yield three types of products: ortho cycloadducts, meta cycloadducts, and so-called substitution products (Scheme 44). The products are formed in the ratio ortho adducts:meta adducts:substitution products = 0.8:1.7:0.3. The substitution reaction (which is really an addition of a C—F bond to the double bond of 1,3-dioxole, but named substitution in order to distinguish it from the ortho addition [186] is supposed to start with electron transfer from 1,3-dioxole to excited α,α,α-trifluorotoluene. The radical anion then releases a fluoride ion, which adds to the 1,3-dioxole radical cation. Radical combination then leads to the product.

In view of the occurrence of electron transfer and also in view of the observations made by Bryce-Smith and Gilbert (on which the ionization potential rule is based), Mattay et al. [15,143,144] have proposed a relationship between the mode of reaction (ortho addition, meta addition, substitution) and the free enthalpy of electron transfer between the reaction partners. The free enthalpy was calculated using the Rehm–Weller equation

$$\Delta G = F[E_{1/2}^{ox}(D) - E_{1/2}^{red}(A)] - \Delta E_{0-0} + \Delta E_{coul}$$

in which F is the Faraday constant, 9.648×10^4 C/mol^{-1}; $E_{1/2}^{ox}(D)$ is the oxidation potential of the donor (in V, measured in acetonitrile), $E_{1/2}^{red}(A)$ is the reduction potential of the acceptor (in V, measured in acetonitrile), ΔE_{0-0} is the excitation en-

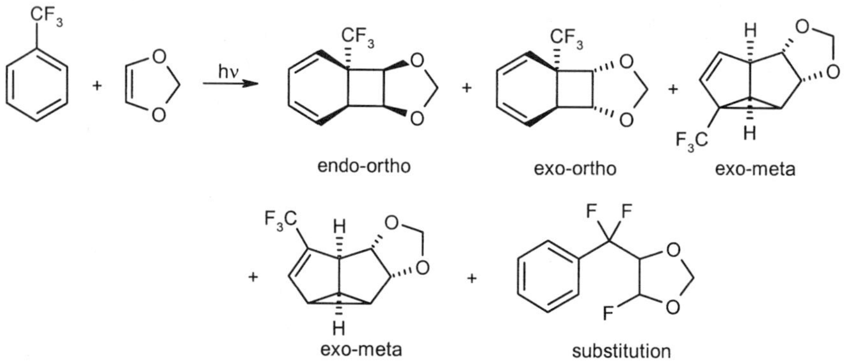

Scheme 44   Photoreaction of α,α,α-trifluorotoluene with 1,3-dioxole.

ergy (in J/mol^{-1}), and ΔE_{coul} is the Coulomb interaction energy (in J/mol^{-1}) of D^+ and A^-, in a given solvent where

$$\Delta E_{coul} = \frac{e^2 N}{4\pi\varepsilon_0 a}\left(\frac{1}{\varepsilon} - \frac{2}{37.5}\right)$$

with $e = 1.602 \times 10^{-19}$ C, $N = 6.023 \times 10^{23}$ mol^{-1}, $\varepsilon_0 = 8.854 \times 10^{-12}$ F/m^{-1}, ε is the dielectric constant of the solvent, and a is the encounter distance (7 Å). The Coulomb interaction energy pertains to a D^+/A^- ion pair at encounter distance a.

For the series of 1,3-dioxoles and vinylene carbonates with α,α,α-trifluorotoluene, the results shown in Table 8 were obtained.

Use of the Rehm–Weller equation made it possible to predict the direction of charge transfer in a given system and to establish empirical correlations between the ΔG value for electron transfer and the mode of the photoreaction (meta addition, ortho addition, and/or substitution).

For photocycloaddition, to benzene the following conclusions were drawn from this empirical correlation [124]. Olefins with poor electron-donor or poor electron-acceptor abilities yield mainly meta adducts with benzene (i.e., if $\Delta G >$ 1.4–1.6 eV, all other olefins yield mainly ortho adducts). Even ethene, which had seemed to behave exceptionally, fits into this correlation provided that it acts as the acceptor. The transition area from ortho to meta cycloaddition (i.e., the ΔG region where ortho:meta = 1:1) is relatively large (~0.2 eV). This is considered not to be surprising because the ΔG correlation is based on many different types of olefins. When only ΔG values for derivatives of 1,3-dioxole and for 1,4-dioxene were used, the transition area was narrowed to 0.03 eV. Not only ethene but also vinylene carbonate now fit into the correlation. According to the ionization potential rule, this compound should give only ortho photocycloaddition with benzene. Mattay's empirical rule predicts mainly meta addition, which is indeed found experimentally.

Table 8 Free Enthalpies of Electron Transfer and Mode of Reaction of α,α,α-Trifluorotoluene With Various Olefins

Olefin	$E_{1/2}$ (V)	ΔG (eV)	Substitution	Cycloaddition
2,2,4,5-Tetramethyl-1,3-dioxole	+0.42	−0.23	+	—
4,5-Dimethyl-1,3-dioxole	+0.48	−0.17	+	—
2,2-Dimethyl-1,3-dioxole	+0.69	+0.04	+	—
1,3-Dioxole	+0.76	+0.11	+	Ortho, meta
Dimethylvinylene carbonate	+1.47	+0.82	−	Meta
Vinylene carbonate	−2.64	+1.72	−	Meta

Source: Ref. 144.

A correlation between free enthalpy of electron transfer and mode of the photoreaction was also constructed for addition of alkenes to benzonitrile. Four areas could be differentiated: Full electron transfer, leading to substitution, is only observed if $\Delta G < 0$ eV; cycloaddition to the cyano group occurs if $0 < \Delta G < \sim 0.4$ eV. All olefins for which $\Delta G > \sim 0.4$ eV preferentially undergo cycloaddition to the aromatic ring, ortho cycloaddition if $\Delta G < \sim 1.7$ eV and meta cycloaddition if $\Delta G > \sim 1.7$ eV.

The correlations between free enthalpies of electron transfer and the mode of the reactions are empirical and therefore may be restricted to one series of molecules owing to similar structural features of intermediates (i.e., absolute values of ΔG may change in some series) [183].

G. Theory

Orbital-symmetry relationships for thermal and photochemical concerted cycloadditions to the benzene ring were published in 1969 by Bryce-Smith [81]. From correlations between low-lying excited states of starting materials and products, it was concluded that photochemical ortho addition is allowed:

- From S_1 or T_1 alkene and ground-state benzene
- From S_2 or T_1 benzene and ground-state alkene
- Following charge-transfer excitation with either the alkene or benzene as donor

The ortho cycloaddition is thermally forbidden in a suprafacial–suprafacial manner and the photochemical reaction is forbidden with S_1 benzene and ground-state alkene. On the basis of these considerations, it could be understood that the ortho addition had only been observed with systems where the alkene is the lowest excited singlet species (as with maleimides [37,74,75] or where either the alkene or the arene has marked acceptor properties (the only examples known at that time were benzene–acrylonitrile [127] and benzonitrile + a mono-olefin [1,73]. Benzene–acrylonitrile and benzonitrile–olefin systems do not display charge-transfer absorption, but charge transfer could well follow excitation. Bryce-Smith further stated that irradiation of benzene in the presence of simple mono-olefins normally provides B_{2u} (S_1) benzene as the lowest excited singlet species, which leads to meta rather than ortho addition, but the latter process might, in principle, be able to occur under conditions where a B_{1u} (S_2) state of benzene is populated.

The molecular orbitals of benzene are schematically represented in Fig. 3. The first excited state of benzene cannot be described by one electron configuration, due to the degeneracy of the highest occupied molecular orbitals (HOMOs) and the lowest unoccupied molecular orbitals (LUMOs). The S_1 state of benzene (B_{2u}) can be represented as $\phi_2\phi_4 - \phi_3\phi_5$ and the S_2 or T_1 state (B_{1u}) as $\phi_2\phi_5 - \phi_3\phi_4$.

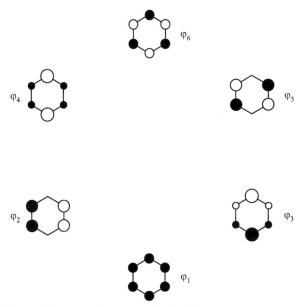

Figure 3   The molecular π orbitals of benzene.

The correlation diagram [38,46,187] for ortho addition of benzene to ethene is given in Fig. 4. Orbital symmetries are with respect to a plane of symmetry (σ), which is retained during the transition from benzene plus ethene to bicyclo[4.2.0]octa-2,4-diene.

From the molecular orbital (MO) correlation diagram, it follows that the configuration $\phi_2\phi_5$ of benzene correlates with the first excited state of the product. This configuration corresponds to an S_2 or T_1 state of benzene. When the lowest singlet transition of benzene is higher in energy than that of the alkene, the correlation diagram changes in such a way that the π_1 level is above the HOMOs (ϕ_2 and ϕ_3) and π_2 below the LUMOs (ϕ_4 and ϕ_5) of benzene. In Fig. 4, π_2 could then correlate with π'_2, ϕ_5 with π'_4, π_1 with π'_3 and ϕ_2 with π'_1. A concerted ortho addition is then seen to be allowed with ground-state benzene plus lowest excited singlet alkene. This situation is reached when the ethylene bears substituents which lower its S_1 state below that of S_1 benzene, as is the case with maleimide.

If the energy levels are such that $\pi_1 \rightarrow (\phi_4,\phi_5)$ is the lowest energy transition, the alkene acts as donor. If $(\phi_2,\phi_3) \rightarrow \pi_2$ is the lowest energy transition, benzene will be the donor. In these cases, photoexcitation can, in principle, proceed by charge transfer. Bryce-Smith and Gilbert [187] have remarked that for charge-transfer absorption to occur to a significant extent, it is probably necessary that benzene and the ethene form a suitably oriented "dark" molecular complex in

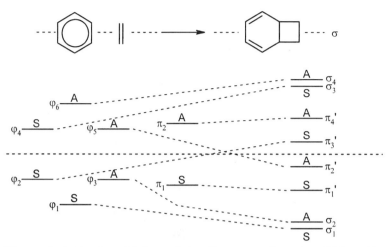

Figure 4 Molecular orbital correlation diagram for the ortho addition of benzene to ethene. A plane of symmetry (σ) is retained during the transition from benzene plus ethene to bicyclo[4.2.0]octa-2,4-diene.

solution so that some degree of interpenetration of orbitals of the donor and acceptor molecules is realized prior to photoexcitation (as in the case of maleic anhydride [46]). If such complexation is not possible, there remains the possibility of "contact" charge transfer. However, charge transfer can also occur after photoexcitation, either to or from the photoexcited molecule, which produces, according to the authors [187], an exciplex intermediate. As mentioned earlier, benzene + acrylonitrile and benzonitrile + mono-olefin show no direct charge-transfer absorption. From later studies, it has become clear that in most cases, no "dark" or "contact" charge-transfer complexes exist prior to excitation [12,122].

Houk [188] has given a qualitative explanation of the selectivities in ortho, meta, and para cycloadditions of singlet excited substituted benzenes to alkenes using frontier molecular orbitals. In Fig. 5, the frontier molecular orbitals of benzene together with the π HOMO and π^* LUMO orbitals of ethene are represented. The orbitals are labeled according to their symmetries with respect to a plane perpendicular to the benzene ring and passing through the position of attachment of a substituent (the bottom carbon).

The lowest excited singlet state of benzene is the negative combination of transitions involving orbitals of opposite symmetry (A $\rightarrow$ S* and S $\rightarrow$ A*). In the ortho approach of ethylene to benzene, stabilization may be achieved through the interaction of the benzene A orbital with the ethene π HOMO or of the benzene A* orbital with the ethene π^* LUMO. Of less significance are the interactions between the benzene S* orbital with the ethene π HOMO and between the benzene S orbital with the ethene π^* LUMO.

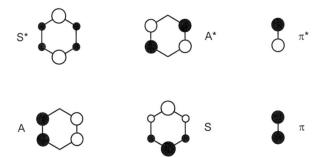

Figure 5   The frontier molecular orbitals of benzene and ethene.

For the meta approach in the sense shown in Fig. 6, the most important interactions are those between the benzene S orbital and the ethene π HOMO and between the benzene A* orbital and the ethene π* LUMO. Interactions of less significance are between the benzene A orbital and the ethene π* LUMO and between the benzene S* orbital and the ethene π HOMO.

In the para approach, there are only weak interactions, such as S with π* and S* with π. This weak excited-state interaction implies that the para addition is photochemically forbidden.

In the ortho and meta additions, there are stabilizing frontier orbital interactions in the excited state. The interaction of the SA* configuration stabilizes the meta complex (through S–π and A*–π*) more than the ortho (only A*–π*) and the AS* configuration stabilizes only the ortho complex (through A–π). According to the correlation diagram given earlier, the ortho addition is forbidden from S_1 benzene with S_0 ethene, but Houk only considers the interactions at the beginning of the reaction path and the correlation of the benzene ϕ_2 orbital with the π'_3 orbital of the butadiene system is not taken into account.

Upon comparing the distances (Fig. 6) between the benzene carbon atoms which are involved in the addition with the bond length of ethene (1.34 Å), it can be clearly seen that overlap is most favored in the case of ortho approach and this will contribute to the likelihood of occurrence of this mode of addition.

Ab initio calculations by Houk [188] using the STO-3G basis set with ground-state ethene and ground-state benzene in parallel planes at 2.5 Å clearly re-

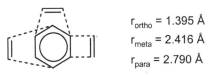

r_{ortho} = 1.395 Å

r_{meta} = 2.416 Å

r_{para} = 2.790 Å

Figure 6   Distances between the benzene carbon atoms involved in ortho, meta, or para-cycloaddition.

veal the greater orbital overlap in the ortho than in the meta approach by the greater changes in the frontier orbitals upon the ortho approach. These calculations also show the strong mixing of the benzene A orbital with the ethene π LUMO in the ortho approach, stabilizing the AS* configuration. For the meta approach, the SA* configuration is highly stabilized by the mixing of the benzene S orbital with the ethene π HOMO and the A* of benzene with the π* LUMO of ethene.

The frontier molecular orbitals of an electron-deficient alkene are stabilized with respect to those of ethene. In this situation the interaction between benzene S or A orbitals with the alkene π HOMO will diminish, but because the alkene π* LUMO is now appreciably below benzene S* and A*, stabilization of a complex will occur due to charge transfer. An electron from benzene A* will then be in the lower-energy orbital, which is the bonding combination of A* and π* for the ortho as well as for the meta complex. Ortho addition will be favored over meta addition due to better orbital overlap. Additional stabilization of the ortho complex can occur by interaction of S with π*. With an electron-rich alkene, there will be charge transfer from the alkene to the benzene. The π HOMO of the alkene is now higher in energy than the S and A orbitals of benzene. The interaction of the A orbital with the π HOMO will result in stabilization of an ortho complex due to electron transfer from the π orbital to a lower-energy orbital, which is a bonding combination of A and π. An ortho complex is again more stabilized by orbital overlap of A with π than a meta complex, which has less orbital overlap of S with π. S*–π interaction provides additional stabilization for the ortho approach.

Houk concluded on the basis of this frontier orbital theory that the ortho addition is favored when the alkene is either a better donor or a better acceptor than benzene. The author qualifies this conclusion as simply the Bryce-Smith generalization, derived in a slightly different way [188].

The 1,2 and 3,4 regioselectivity of the ortho addition of an alkene to an arene substituted with a strong donor substituent is, according to Houk, dominated by interaction of the benzene S orbital with the alkene π HOMO, because the A*–π* interaction would result in a 2,3 ortho adduct. For an arene substituted with a strong donor, the donor substituent, rather than the approaching alkene, will determine the symmetry of the orbitals [188].

With regard to stereoselectivity, Houk observes that the endo–ortho approach is destabilized by secondary antibonding orbital interactions between the alkene substituents and the secondary positions of the benzene.

From the orbital correlation diagram derived by Bryce-Smith [38], it was deduced that the ortho cycloaddition is forbidden from the lowest excited singlet state of benzene and the ground state of ethene. Van der Hart et al. [189] have constructed molecular orbital and state correlation diagrams for the ortho photocycloaddition of benzene to ethene. The molecular orbital correlation diagram differs from that given by Bryce-Smith, because natural correlations have been used. From a topological point of view, it seems less desirable to correlate the π

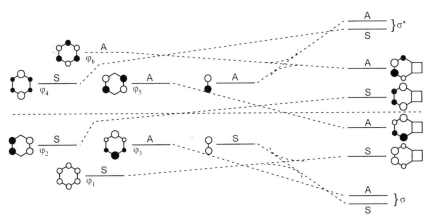

Figure 7 Natural molecular orbital correlation diagram for the ortho addition of benzene to ethene.

and π^* orbitals of ethene with the orbitals of the butadiene moiety of the adduct. In the correlation diagram of Van der Hart et al., the ethene orbitals are correlated with the new σ and σ^* orbitals. The correlations of φ_1 and φ_6 of benzene have been changed accordingly. This correlation diagram is shown in Fig. 7. The electron configurations $\varphi_2\varphi_4$ and $\varphi_3\varphi_5$, both involved in the first excited state of benzene, combined with the ground-state electronic configuration of ethene, correlate with highly excited electronic configurations of the product. The first excited state of the reactants is therefore correlated with a high-lying excited state of the adduct. On the basis of this correlation, the ortho photocycloaddition of benzene to ethene has to be considered forbidden, as observed earlier by Bryce-Smith and Gilbert [38,46,187].

The configuration $\varphi_2\varphi_5$ of benzene, combined with the ground-state electronic configuration of ethene, correlates with the electronic configuration of the first excited state of the product. Thus, the second excited state of the reactants correlates with the first excited state of the product. The ground state of the reactants correlates with a doubly excited state of the product and the ground state of the product correlates with a doubly excited state of benzene plus ground-state ethene. On the basis of these correlations, Van der Hart et al. [189] assume a barrier in the potential energy surface of the ground state and a reactive minimum in the potential energy surface of the first excited state.

Two possible state correlation diagrams [189] for the ortho addition of benzene to ethene are shown in Fig. 8.

The occurrence of a reactive minimum in the potential energy surface of the first excited state makes the ortho photocycloaddition an allowed reaction if that minimum is accessible.

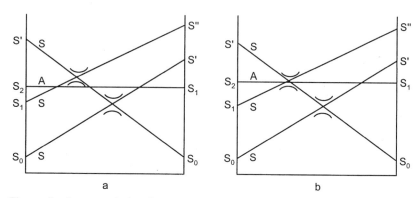

Figure 8   State correlation diagrams for the ortho cycloaddition of ethylene to benzene.

Because S_1 of the reactants correlates with a high-lying excited state (S'') of the reactants, a barrier may be expected before the funnel is reached (Fig. 8a). The second excited state of the reactants (S_2), however, correlates with S_1 of the products. The resulting level crossing will decrease the barrier. It is also possible (Fig. 8b) that the double excited electronic configurations involved in the avoided crossing leading to the funnel mix with the electronic configurations of the first excited state of benzene at an early stage of the reaction, because they have the same symmetry.

If charge transfer dominates the interaction, the contribution of the electronic configuration of the charge-transfer complex to the electronic configurations of the first excited state considerably lowers the potential energy. This will affect the potential energy surface during the addition process from start to funnel. Assumed barriers become lower and pathways to funnels which were inaccessible without charge transfer become passable.

Coulombic forces will determine the regioselectivity of the ortho addition [189]. In the charge-transfer complexes of monosubstituted benzenes with alkenes, the charge (positive or negative) on the arene is largely located at the carbon atoms ipso and (to a lesser extent) para to the substituent. The carbon atoms of the alkene double bond will preferentially be located in the neighborhood of either the ipso carbon or (to a lesser extent) the para carbon atom of the monosubstituted benzene. This would explain the 1,2 and 3,4 selectivity in the ortho photocycloaddition.

Clifford and co-workers [190] have performed complete active space self-consistent field (CASSCF) *ab initio* calculations on the photocycloaddition reactions of benzene and ethene. An eight-electron, eight-orbital active space involving the π-orbitals of the benzene and ethene moieties was used. The geometries were optimized using the 4-31G basis set, and the energies were recomputed at the 6-31G* level.

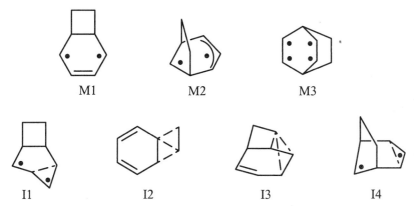

Figure 9 Minima (M1–M3) on the S_1 surface and conical intersections S_1/S_0 (I1–I4) for benzene + ethene.

From the data obtained in the computations, it became clear that structures M1–M3 (Fig. 9) are minima on the excited state surface and that I1–I4 are conical intersections. The conical intersection I3 is the only one with lower energy than S_1 of benzene + ethene at the S_0 geometry. No transition state could be found along a reaction path leading to this structure. The formation of a bond between benzene and ethene thus occurs without a barrier to yield a low-energy conical intersection. Decay from this point may form either the ortho, meta, or para products. The molecular mechanics with valence bond (MMVB) method was employed to study the dynamics of decay from the conical intersection. Ground-state steepest descent lines (SDL) were followed from 12 equally spaced geometries around a "circle" (of radius 0.02 Å) centered on the conical intersection I3. It was found that there exists a channel in one direction leading to the meta addition product. SDL from the intersection in all other directions terminate at the ortho adduct. No SDL leading to dissociation and to the para addition product were seen at all. This implies that the meta product formation channel is narrower and less deep than the ortho. This is considered to be simply a reflection of the greater stability of the S_0 ortho minimum. The major role of substituent effects may be to move the conical intersection I3 closer to the ortho or meta side by stabilizing the incipient biradical structures such as I1 or I4.

VI. SECONDARY REACTIONS OF ORTHO PHOTOCYCLOADDUCTS

Many ortho photocycloadducts are thermally and photochemically labile compounds. When a mixture of isomeric adducts is formed from a particular combination of arene and alkene, its composition is often time dependent. Ratios of regioisomeric or stereoisomeric ortho adducts and ratios of ortho versus meta adducts therefore often vary with the time of irradiation and are sometimes difficult to interpret. Conclusions about factors that influence regioselectivity and stereoselectivity and the mode of addition ideally have to be based on quantum yields rather than product yields. Moreover, these quantum yields should be measured at very low degrees of conversion. Analytical methods for determining product ratios and techniques for separation and purification (in particular, gas chromatography) should be used with great care in view of the thermal lability of the ortho adducts.

If the alkene, from which the ortho photocycloadduct is derived, is a powerful dienophile, a thermal Diels–Alder reaction between adduct and alkene may take place during the irradiation so that a 2:1 alkene:arene adduct is formed. Such adducts are quite stable and can easily be purified. Their structure contains all the regiochemical and stereochemical information of the primary 1:1 ortho adduct. This Diels–Alder reaction occurs spontaneously with the alkenes maleic anhydride, maleimide, and their derivatives, as discussed in the sections of this chapter that deal with photoreactions of ground-state charge-transfer complexes (maleic anhydrides) and reactions of excited alkenes with ground-state arenes (maleimides). Ortho adducts formed from other alkenes have often been identified via their reaction with a good dienophile.

A complex reaction mixture (twelve 1:1 adduct isomers) was formed upon irradiation of ethyl vinyl ether and 4-cyanoanisole [87]. Two major adducts, together comprising approximately 70%, were formed at <50% conversion. These 1:1 adducts reacted readily with N-phenylmaleimide (in refluxing toluene) and were isolated and separated as their Diels–Alder adducts. The structures of these 1:1:1 adducts could be assigned on the basis of their spectroscopic data, and from these structures, the primary ortho adducts could be identified as 1-cyano-4-methoxy-8-ethoxy- and 1-methoxy-4-cyano-7-ethoxybicyclo[4.2.0]octa-2.4-diene (Scheme 45).

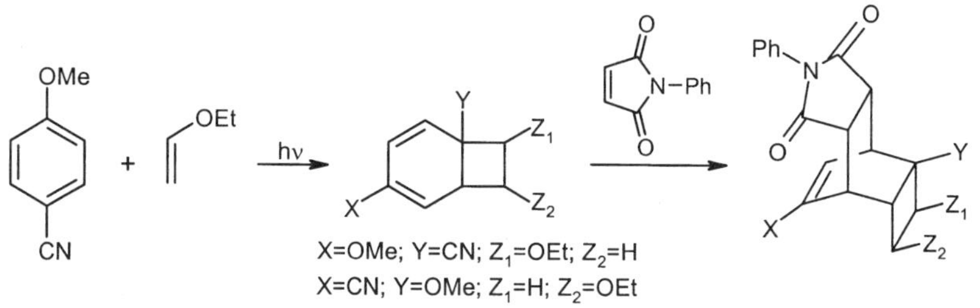

X=OMe; Y=CN; Z_1=OEt; Z_2=H
X=CN; Y=OMe; Z_1=H; Z_2=OEt

Scheme 45 Formation of ortho photocycloadducts from 4-cyanoanisole and ethyl vinyl ether and reaction of the adducts with N-phenylmaleimide.

The Diels–Alder reaction with *N*-phenylmaleimide has frequently been used for the separation, purification, and structure determination of ortho photocycloadducts [12,47,86,90,108,116,126,132,133,138]. Other dienophiles that have been successfully employed in Diels–Alder reactions with ortho adducts are *N*-(para-bromophenyl)maleimide [116,120], maleimide [116,118,127], maleic anhydride [127,191], tetracyanoethylene [11], and dimethyl acetylenedicarboxylate [73,127]. The Diels–Alder product of *N*-(*para*-bromophenyl)maleimide with the exo–ortho adduct formed from 1,4-dioxene and benzene [120] and the Diels–Alder product of maleimide with the endo–ortho adduct from *cis*-cyclooctene and benzene [118] were obtained in crystalline form and their structures could be determined by means of x-ray diffraction.

The primary ortho adducts formed from benzene derivatives and acetylenes are derivatives of bicyclo[4.2.0]octa-2,4,7-triene. These products usually are not isolated but isomerize during the irradiation to cyclooctatetraenes [58,59,68,72]. From combinations of hexafluorobenzene and pentafluoroalkoxybenzenes with various disubstituted acetylenes, however, the isolation of relative stable primary ortho adducts has been reported [65–67]. In Scheme 46, the products of the photochemical reaction of hexafluorobenzene with but-2-yne are shown [67].

In this particular case, the primary adduct not only undergoes ring opening but also ring closure to a tricyclic compound, a reaction that is more often seen in perfluorobenzene derivatives than in benzene derivatives. The authors propose that the ring opening of the ortho adduct can be activated by light as well as by heat. Because the ring junction in the ortho adduct must be cis, the ring opening reaction, if concerted, must proceed in a disrotatory fashion. This could formally be viewed as a disrotatory four-electron photochemical ring opening of a cyclobutene or as a disrotatory six-electron thermal ring opening of a cyclohexadiene. For the reverse reaction, however, only a photochemical path is indicated. In

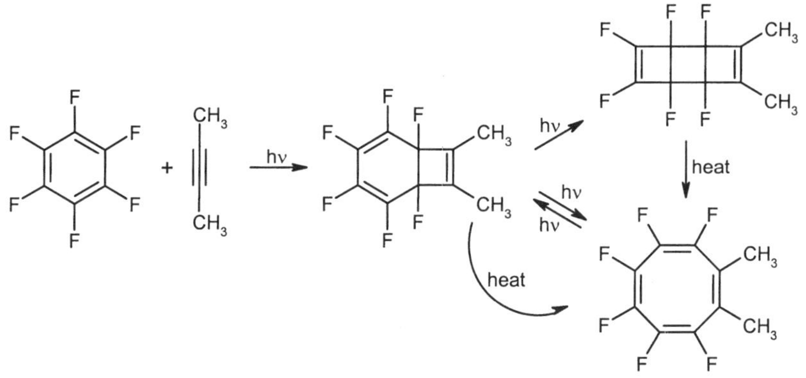

Scheme 46   Ortho photocycloaddition of hexafluorobenzene and but-2-yne.

principle, five different hexafluorodimethylbicyclo[4.2.0]octa-2,4,7-trienes can be formed by ring closure of the cyclooctatetraene derivative, but apparently only one, the original adduct, is found.

Upon irradiation of dimethyl acetylenedicarboxylate in benzene, however, a small amount of dimethyl phthalate (and presumably also acetylene) is produced, next to dimethyl cyclooctatetraene-1,2-dicarboxylate [36,61]. Upon irradiation of dimethyl cyclooctatetraene-1,2-dicarboxylate in benzene, dimethyl phthalate was the only product formed. This reaction was markedly catalyzed by trifluoroacetic acid. The reactions shown in Scheme 47 rationalize this interesting discovery.

In this case, only a thermal pathway for the formation of the cyclooctatetraene is proposed, in agreement with many other publications (for references, see Sec. II and Table 3). Tinnemans and Neckers [62], however, describe the ring opening of the ortho adduct from methyl phenylpropiolate and benzene as well as the reverse reaction as photochemical processes. The formation of this ortho adduct can also be accomplished in a xanthone-sensitized photoreaction [63], and in that case, the authors consider the ring opening as a thermal or a triplet-sensitized reaction and the reverse reaction as one proceeding via the singlet.

Not only carbon–carbon triple bonds but also carbon–nitrogen triple bonds can undergo ortho photocycloaddition to derivatives of benzene. The reaction was discovered by Al-Jalal et al. [86] who found that irradiation (254 nm) of 4-cyanoanisole in acrylonitrile produced three 1:1 adducts in the ratio of 7:1:2. Two of these (ratio 7:1) were formed by the addition of the carbon–carbon double bond of acrylonitrile to positions 1,2 and 3,4, respectively, of 4-cyanoanisole. The third product was an azocine derivative, apparently formed by the addition of the carbon–nitrogen triple bond to positions 2,1 of the arene, followed by ring opening to an azacyclooctatetraene [87] (Scheme 48).

4-Cyanoanisole adds in the same fashion to benzonitrile, methyl 4-methoxybenzoate to acrylonitrile [86], and 2-methyl-4-cyanoanisole to acryloni-

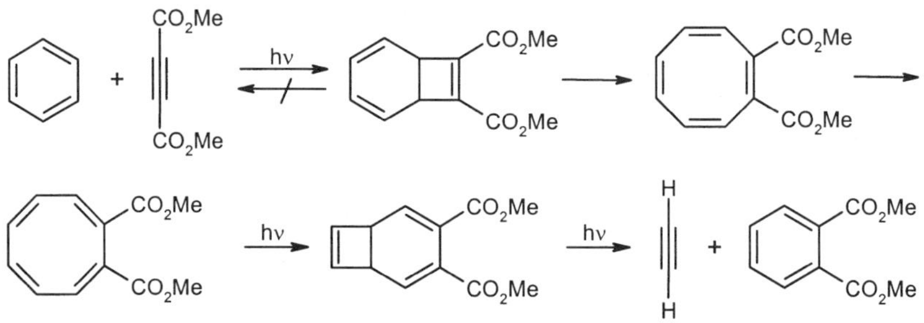

Scheme 47 Formation of dimethyl phthalate from the ortho photocycloadduct of benzene and dimethyl acetylenedicarboxylate. (From Ref. 61.)

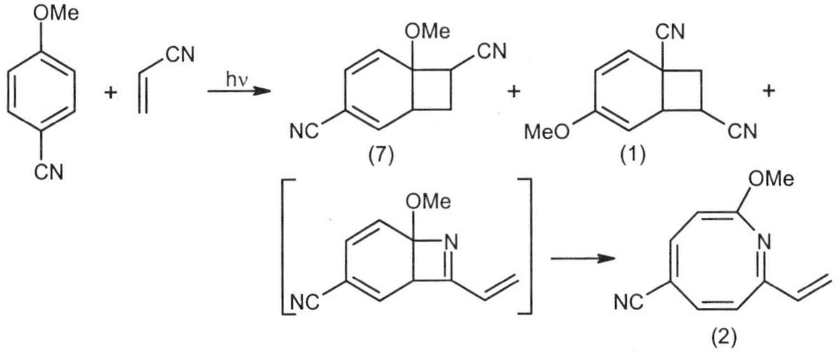

Scheme 48 Ortho photocycloaddition of 4-cyanoanisole and acrylonitrile.

trile [90]. In all cases, the nitrogen atom of the cyano group adds to the arene carbon atom bearing the methoxy group. The primary ortho adducts could not be isolated. From a study of the effect of addend concentration on the formation of the azocine, it was deduced that photoexcited 4-cyanoanisole, not photoexcited nitrile, is the species fruitful of reaction [87]. It was first assumed that nitrile addition to the benzene ring only occurs for arenes have electron-donor and electron-acceptor groups in a para relationship, but Al-Jalal [88,89] has found that the reaction also takes place when phenol, 4-methyl-, or 4-methoxyphenol are irradiated in the presence of benzonitrile and also between phenol and *ortho-*, *meta-*, or *para*-methoxy-, methyl- or chloro-substituted benzonitriles. In these cases, hydroxyazocines are formed which tautomerize to cyclic amides.

One of the secondary reactions that ortho adducts from alkenes and benzene or benzene derivatives may undergo is formation of a tetraene, by complete ring opening. The process was first described by Atkinson et al. [73] in their report on the ortho photocycloaddition of benzonitrile to 2-methylbut-2-ene (Scheme 49). They did not detect the tetraenes among the irradiation products, but found it when they pyrolyzed or photolyzed the ortho adduct. Pyrolysis at 128°C gave 53% alkene, 6% benzonitrile, and 41% tetraene; photolysis caused almost complete reconversion (90%) to arene and alkene, whereas formation of tetraene occurred to the extent of 8–10%. Most likely a mixture of cis and trans isomers of 2-methyl-

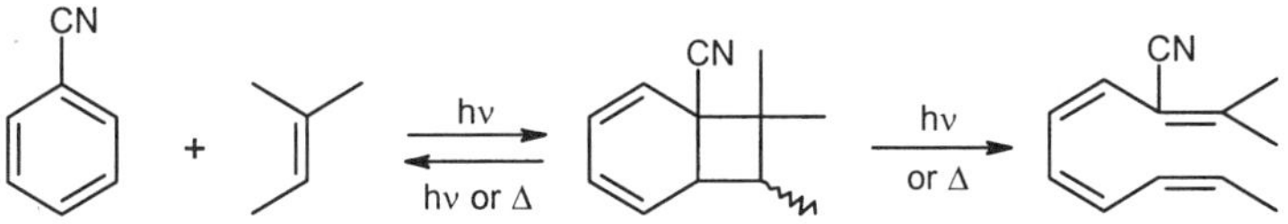

Scheme 49 Ortho photocycloaddition of benzonitrile and 2-methylbut-2-ene and secondary reactions of the adduct.

3-cyano-2,4,6,8-decatetraene is formed. The ring opening reactions were considered to be radical reactions, but no explanation for the contrast between photolysis and pyrolysis could be offered.

Ring opening of ortho adducts may also have occurred in the experiments reported by Perrins and Simons [192], who irradiated benzene, toluene, and anisole in the presence of dichloroethenes (all isomers), trichloroethene, and tetrachloroethene. The products were chlorine-substituted linear tetraenes, but no ortho adducts were found. An analogous reaction occurred with benzonitrile and 2-methylbut-2-ene, but, in contrast to the results reported by Atkinson et al. [73], the ortho adduct was not detected.

The formation of tetraenes from ortho adducts was also reported by Kaupp and Jostkleigrewe [193], who have indicated that it may proceed thermally as well as photochemically. Bryce-Smith et al. [116], however, only mention the photochemical formation of a tetraene from the ortho photocycloadduct of benzene and 2,3-dimethylbut-2-ene. A 1-ethoxy-2-cyanoocta-1,3,5,7-tetraene was observed by Gilbert et al. [133] who considered it to be the thermal product of the ring opening of the ortho adduct from ethyl vinyl ether and benzonitrile. The proportion of tetraenes slightly increased with the age of the sample, as was evident from the appearance of absorptions in the 290–330-nm region of the ultraviolet absorption spectrum. Ring opening leading to tetraenes has also been found to occur with ortho adducts from methacrylonitrile and benzene [47], and in this case, they were found upon prolonged irradiation of the mixture of addends.

In the formation of tetraenes from bicyclo[4.2.0]octa-2,4-dienes, two bonds are broken. This may occur in one concerted reaction which can be regarded as a retro [2 + 2] cycloaddition. It is also possible that the central bond, being part of a cyclohexadiene system, is the first one to break in a thermal, concerted disrotatory process that leads to a 1,3,5-cyclooctatriene derivative. Ring opening of the cyclooctatriene then might take place photochemically, again disrotatory, to produce a tetraene. This two-step sequence was first observed by Mirbach et al. [114] in their study of the photocycloaddition of the two parent molecules benzene and ethene. The same explanation for the formation of a tetraene was given by Nuss et al. [160] in their report on the intramolecular ortho photocycloaddition of (E)-6-(2-methoxyphenyl)-5,5-dimethyl-2-hexenenitrile (see Scheme 40).

The formation of a derivative of cyclooctatriene had earlier been observed by Bryce-Smith et al. [153] during their investigation of the photoaddition of cis-cyclooctene to hexafluorobenzene. One of the products of this reaction is 2,3,4,5,6,7-hexafluorobicyclo[6.6.0]tetradeca-2,4,6-triene, formed by ring opening of the primary ortho adduct. The ortho adducts derived from hexafluorobenzene and its derivatives seem especially prone to undergo this isomerization (see below), but for derivatives of benzene itself, the reaction was not often reported prior to 1987. The study by Mirbach et al. [114] has been referred to earlier and Atkins et al. [120] mention in a note that endo–ortho adducts might be converted into presumably more stable exo–ortho adducts via ring opening to a cycloocta-triene followed by ring closure with the other stereochemistry. Such a rearrange-ment was not actually observed in this study, but Mattay et al. [134] have found that their endo– and exo–ortho adducts from anisole and 2,2-dimethyl-1,3-dioxole are indeed thermally interconvertible. During the photochemical addition, the endo isomer is formed only in small amounts as evidenced by high-perfor-mance liquid chromatography analysis of the crude reaction mixture. After workup via chromatography or distillation, the exo:endo ratio is 1.2:1. The rearrangement was found to be totally reversible up to 80°C. Two possible mech-anisms for the interconversion were considered. One of the newly formed bonds might break in a heterolytic fashion whereupon the resulting zwitterion might undergo rotation around the remaining single bond followed by ring closure yielding the stereoisomer. The alternative mechanism is disrotatory ring opening of the ortho adduct to a cyclooctatriene derivative followed by disrotatory closure of the hexatriene system in the other direction.

A cyclooctatriene was formed upon pyrolysis at high temperature (up to 350°C) of the exo–ortho adduct from benzene and 1,4-dioxene [12]; the position of the double bonds, however, was not as expected for a ring-opened ortho adduct. The ortho adduct from benzene and 1,1-dimethoxyethene [12] is an acetal and with a trace amount of acid in methanol solution it is converted into cycloocta-2,4,6-trien-1-one (Scheme 50).

From 1987 onward, many examples of rearrangements of ortho photocy-cloadducts to derivatives of cyclooctatriene and subsequent reactions of these compounds have been reported. The same year marks the start of a series of in-vestigations on intramolecular ortho photocycloadditions. This may not be a co-incidence, because ortho adducts formed by intramolecular photoaddition seem especially prone to undergo rearrangement, although it has also frequently been observed in adducts arising from intermolecular addition.

Almost all triplet-state photocycloadditions studied by Wagner's group [93–103] yield ortho adducts that undergo thermal ring opening to a cycloocta-triene followed by photochemical 4π ring closure. Many examples of these reactions are shown in the section devoted to additions via the triplet state. Scheme 51 schematically illustrates the basic reactions for intermolecular as well as intramolecular cycloadducts. A substituted benzene has been chosen for the intermolecular reaction in order to demonstrate the two modes of ring closure.

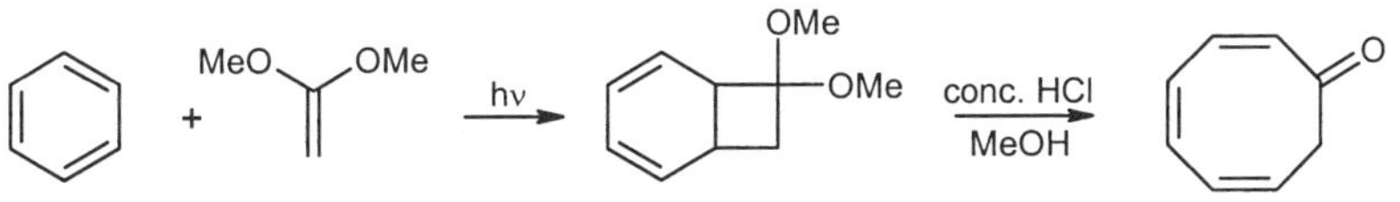

Scheme 50 Photocycloaddition of benzene to 1,1-dimethoxyethene and acid hydroly-sis of the product.

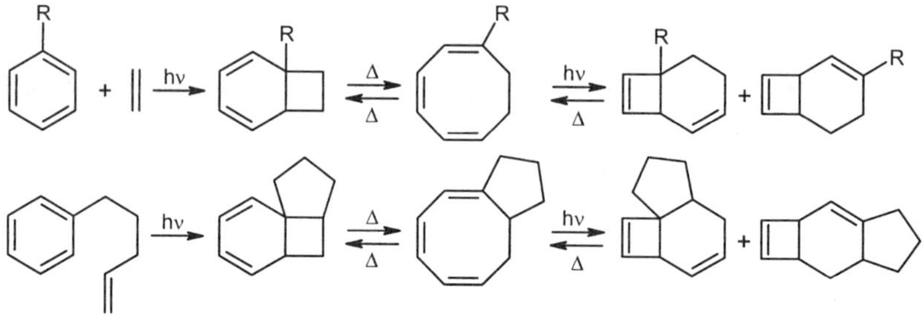

Scheme 51   Intermolecular and intramolecular ortho photocycloaddition followed by thermal ring opening and photochemical ring closure.

The formation of the cyclooctatriene derivative is a thermal disrotatory six-electron ring opening of a cyclohexadiene. The reaction is reversible and the position of the equilibrium depends on the position and the nature of substituents. In many of the cases studied by Wagner's group, the equilibrium lies strongly to the right and the primary ortho adduct is seldom isolated. This is certainly the case when the atom connecting the phenyl ring with the side chain is oxygen, but when it is carbon, the cyclohexadiene structure may be favored over its cyclooctatriene isomer [99]. The cyclooctatrienes undergo a photochemical disrotatory ring closure of a four π-electron unit, and because there are two such units, two products, derivatives of bicyclo[4.2.0]octa-2,7-diene, are possible. In the intramolecular case, this leads to a linear and an angular tricyclic molecule. In many instances, one of these two is formed preferentially, depending on the substituents. The nonconjugated dienes are often quite stable, but they may revert to the cyclooctatrienes in a thermal reaction. This is probably not a concerted ring opening and whether or not it occurs is again substituent-dependent. Especially substituents forming a donor-acceptor pair on the bridgehead carbon atoms promote this ring opening. Interestingly, the isomeric nonconjugated dienes are also interconvertible, via a thermal Cope rearrangement. This turns one isomer into the other (in the intramolecular case, the linear adduct into the angular isomer) or, if a single diastereomer of the ortho adduct is formed, one isomer into the epimer of the other [99].

Several articles have been published on the intramolecular ortho photocycloaddition of substituted 4-phenoxybut-1-enes [104,105,164–166]. In most cases studied, ortho adducts are converted into derivatives of bicyclo[4.2.0]octa-2,7-diene. This was also found [113] for ortho adducts derived from *ortho*-methyl- and *ortho*-methoxy-3-benzyloxyprop-1-enes. In these cases, the angular products are formed exclusively, which is attributed to steric properties of the cyclooctatriene.

Keese and co-workers [162,163] have constructed molecules with the intention of producing [5.5.5.5]fenestranes via intramolecular meta photocy-

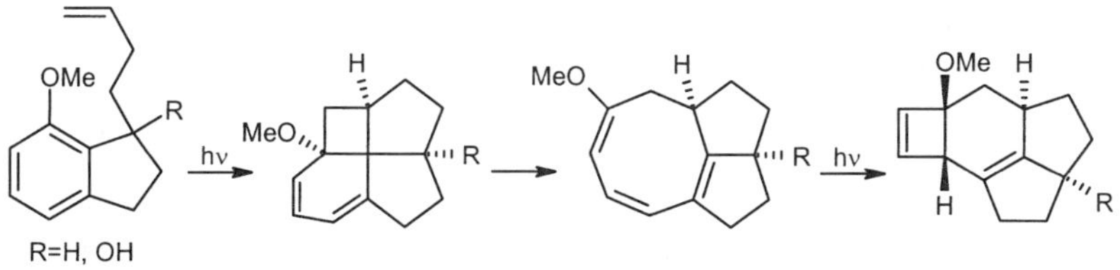

Scheme 52   Intramolecular ortho photocycloaddition of 1-(but-3-enyl)-7-methoxyindane and 1-(but-3-enyl)-7-methoxyindan-1-ol.

cloaddition. Irradiation did indeed produce meta photocycloadducts, but these were accompanied by rearranged ortho adducts (Scheme 52).

Avent et al. [171] have utilized the intramolecular meta photocycloaddition reaction of 3-(phenoxydimethylsilyl)prop-1-ene to prepare a molecule with the gelsemine skeleton. In addition to meta cycloaddition, the starting material undergoes intramolecular ortho photocycloaddition. The ortho adduct undergoes ring opening to a cyclooctatriene which is subsequently photochemically converted into a mixture of the linear and the angular nonconjugated dienes. Irradiation of 3-[3-(trifluoromethyl)phenoxydimethylsilyl]prop-1-ene does not yield the expected meta cycloadducts [172]; instead, ortho addition occurs and the products are probably the linear bicyclo[4.2.0]octa-2,7-diene derivatives.

As mentioned earlier, the rearrangement of ortho photocycloadducts into cyclooctatrienes and subsequently into bicyclo[4.2.0]octa-2,7-dienes is not limited to intramolecular ortho additions. The ortho adduct from 2-cyanoanisole and ethyl vinyl ether [146] could not be detected but underwent facile ring opening to 1-methoxy-6-cyano-7-ethoxycycloocta-1,3,5-triene. The lability of the ortho adduct is probably due to the presence of a methoxy and a cyano group at the bridgehead carbon atoms. The cyclooctatriene was photochemically converted into only one of two possible isomers, 3-methoxy-5-ethoxy-6-cyanobicyclo[4.2.0]octa-2,7-diene. This process can be reverted thermally. Ortho photocycloadducts derived from 4-cyanoanisole and acrylonitrile or allyl cyanide undergo the same type of rearrangement [87]. Thermal ring opening to cyclooctatrienes, but not the subsequent photochemical ring closure, has also been observed with ortho adducts from 4-cyano-2-methylanisole, 4-cyano-2-methoxyanisole, 4-cyano-3-methoxyanisole and 1,2-dimethoxybenzene with acrylonitrile [90]. From a study of the photoadditions of the methylbenzonitriles, 2,3-, 2,6-, 2,4-, 3,4-, and 3,5-dimethoxybenzonitriles and 2- and 4-methoxyacetophenones with ethyl vinyl ether and vinyl acetate [107], it has become especially clear that the thermal ring opening of the primary ortho photoadduct and the photochemical reactivity of the cyclooctatriene derivative depend markedly on the nature and the position of the substituents. From some combinations of arene and alkene, the primary ortho adducts were isolated; some gave principally cyclooctatrienes and, in other cases, the bicyclo[4.2.0]octa-2,7-diene derivatives were the stable products.

In the formation of cyclooctatrienes from ortho adducts, the bond common to the cyclohexadiene and cyclobutane units is broken. A completely different way of breaking the cyclobutane ring was discovered by Garcia et al. [140]. Irradiation of 2-fluorobenzonitrile and furan gave a single product, of which the molecular mass was equal to the sum of the masses of the addends minus that of HF. The product was shown to be 2-(2-furyl)benzonitrile, which was considered to originate from antiperiplanar elimination of HF from the exo–ortho cycloadduct and ring opening of the resulting cyclobutene. However, no evidence for either intermediate was obtained (Scheme 53).

The first investigations by Bryce-Smith et al. [46,67,153] on ortho photocycloaddition of an alkene to hexafluorobenzene have revealed yet another secondary reaction of ortho photocycloadducts. Irradiation of a solution of hexafluorobenzene in *cis*-cyclooctene leads to the rapid formation of seven adducts of which six were identified: (**i**) the exo–meta adduct, (**ii**) a product that can be formed from the meta adduct by a thermal 1,5 H-shift but which apparently is also a primary product, (**iii**) an ortho adduct of which the configuration could not be established, (**iv**) a cyclooctatriene derivative formed by thermal ring opening of the ortho adduct, and (**v**) and (**vi**) two stereoisomers of 2,3,4,5,6,7-hexafluorotetracyclo[6.6.0.0^{2,7}.0^{3,6}]tetradec-4-ene. The experiment was repeated 9 years later by Šket et al. [151] with the important difference that cyclohexane was used as a diluent. The meta adduct (**i**) and its formal rearrangement product (**ii**) were not found. One ortho adduct (**iii**), the cyclooctatriene (**iv**), and the two tetracyclic products (**v**) and (**vi**) could be identified and their stereochemistry determined. From their results, the authors concluded that a second ortho adduct with the alternative stereochemistry must also have been formed. They also performed experiments in which the influence of the solvent on the course of the reaction was studied and found that the difference between their results and those of Bryce-

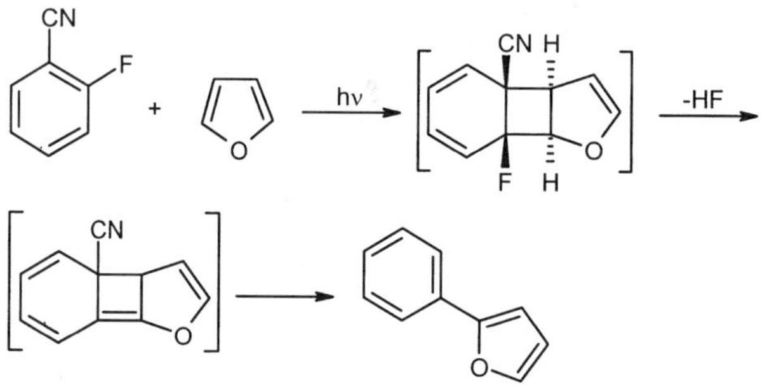

Scheme 53 Ortho photocycloaddition of 2-fluorobenzonitrile with furan and further reactions of the adduct.

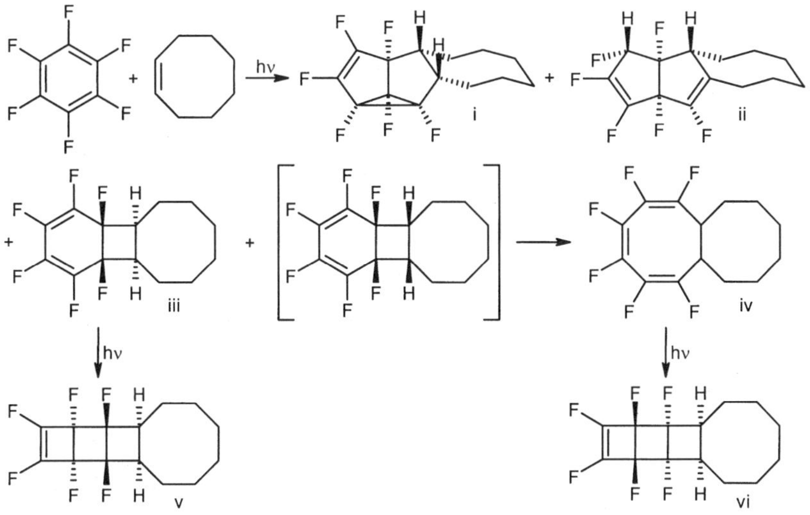

Scheme 54   Photocycloaddition of hexafluorobenzene and *cis*-cyclooctene. (From Refs. 67,151.)

Smith et al. are likely to be caused by solvent effects. In Scheme 54, the results of both investigations are combined.

The exo–ortho adduct (**iii**) is stable enough to be isolated [151]. Upon irradiation, it is stereospecifically converted into the tetracyclic product (**v**). The endo–ortho adduct, however, is unstable at room temperature. It undergoes ring opening, thus forming the cyclooctatriene derivative (**iv**). The latter is stereospecifically converted into the tetracyclic product (**vi**) upon irradiation. Similar results were obtained with cyclopentene [150,151], cyclohexene [151], and cycloheptene [151]. The product distribution depends on the ring size of the alkene: The exo:endo ratio changes from 85:15 for cyclopentene to 53:47 for cyclooctene. Ortho photocycloaddition of hexafluorobenzene to norbornene [151,152] and norbornadiene [152,154] yields the same types of adducts and their rearrangement products. In both cases, the attack on the bicyclic alkene is exclusively exo. From benzonorbornadiene and hexafluorobenzene only the endo–ortho adduct is formed (with exo stereochemistry at the norbornene skeleton), and like the other endo adducts, this undergoes further photochemical ring closure [152]. From indene and 1,2-dihydronaphthalene with hexafluorobenzene, only endo–ortho photocycloadducts are formed [151,155] and these do not undergo further reactions under the conditions used. The photoadditions of cyclohexa-1,4-diene and cycloocta-1,5-diene to hexafluorobenzene yield the same types of adducts and products of rearrangement as the corresponding cycloalkenes [152]. Ortho photocycloaddition of cyclopentene to 2,3,4,5,6-pentafluoroanisole,

-ethoxybenzene, -isopropoxybenzene, and -*tert*-butoxybenzene, of cycloheptene and cyclooctene to 2,3,4,5,6-pentafluoroanisole and of norbornene to 2,3,4,5,6-pentafluoroanisole and -ethoxybenzene yield only tetracyclic (pentacyclic in the case of norbornene) products, formed by secondary reactions of the ortho photoadducts [158]. Only in the case of cyclopentene with 2,3,4,5,6-hexafluoroanisole could the unstable primary ortho adduct be detected in the irradiation mixture on the basis of its NMR spectrum.

3-(Pentafluorophenoxy)prop-1-ene upon irradiation in cyclohexane remarkably yields an intramolecular ortho adduct by the addition of the double bond to the ipso,ortho positions of the benzene ring [111]. This reaction is exceptional, because the alkene and the arene are connected by a two-atom bridge.

Photocycloaddition of 1,2-dichloro-1,2-difluoroethene to hexafluorobenzene produces an unstable ortho adduct that during the irradiation undergoes further ring closure, yielding 3,4-dichloro-1,2,3,4,5,6,7,8-octafluorotricyclo[4.2.0.0^{2,5}]oct-7-ene [156,157]. This was further elaborated into perfluorocyclooctatetraene [156] and into its valence isomer, perfluorobicyclo[4.2.0] octa-2,4,7-triene [157].

Examples of nonfluorinated ortho photocycloadducts undergoing ring closure of the 1,3-diene unit are few and far between. The photoadducts formed in benzene solution from trimesic acid triester and cyclooctene have been reported to undergo this reaction [149] (Scheme 55).

Hoffmann and Pete [167,168] have discovered intramolecular ortho photocycloaddition of 3-alkenyloxyphenols. The efficiency of the reaction increases if it is carried out in the presence of added acid and the primary adducts undergo acid-catalyzed rearrangements. An example is given in Scheme 56.

The ratio of benzocyclobutenes formed from A and B is 1:9 (yield 50% at 83% conversion). The reaction proceeds similarly with 3-((3-methylbut-3-enyl)oxy)phenol (ratio 1:5) and with 3-(pent-4-enyloxy)phenol (ratio 1:2), but it fails with 3-(hex-5-enyloxy)phenol. Understandably, when there is a 2-methyl group on the arene, the product from A is not formed.

When the terminal carbon atom of the double bond bears two methyl groups, different results are obtained [167,168] (Scheme 57). Ortho adduct P follows the same course as adduct A in Scheme 56; the ensuing benzocy-

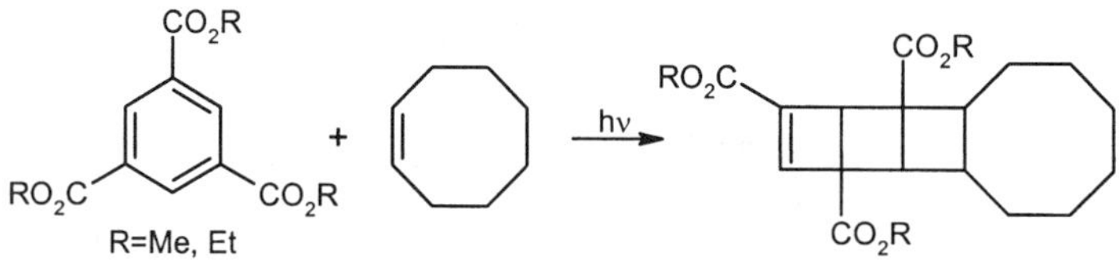

Scheme 55 Photocycloaddition of trimesic acid triesters to cyclooctene.

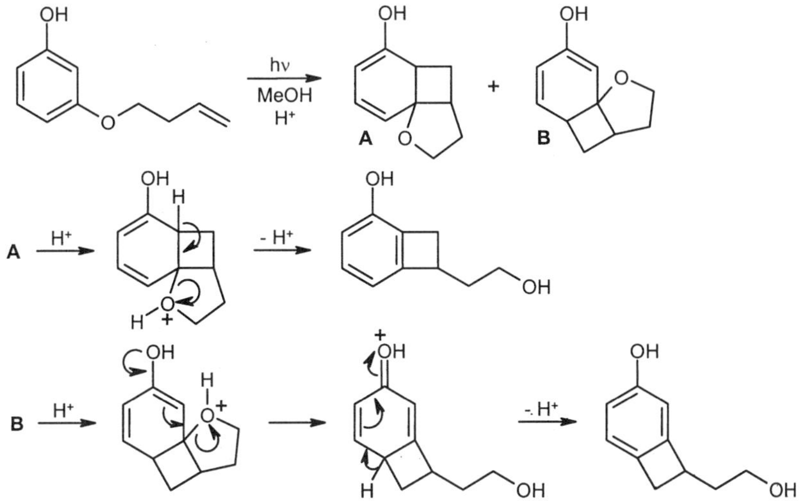

Scheme 56 Acid-catalyzed rearrangement of ortho photocycloadducts from 3-(but-3-enyl-oxy)phenol.

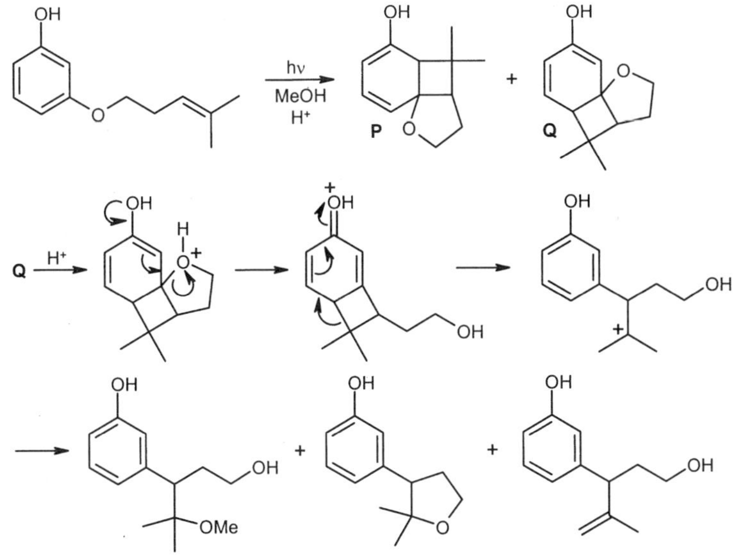

Scheme 57 Acid-catalyzed rearrangement of ortho photocycloadducts from 3-((4-methyl-pent-3-enyl)oxy)phenol.

clobutene is obtained in 13% yield at 73% conversion. Ortho adduct Q, how-ever, follows a path completely different from that of B. A tertiary carbocation is created by breaking of one of the cyclobutane bonds. This cation is captured by methanol (37%) and undergoes intramolecular ring closure with the nucle-ophilic hydroxyl group (10%), or loses a proton (10%). Similar results were ob-tained with terminal ethyl groups on the alkene, but with isopropyl groups, the reaction fails.

Compounds in which the hydroxyl group at the meta position is replaced by an alkoxy group also undergo photochemical ortho addition and acid-catalyzed rearrangement of the adducts [168].

Hoffmann and Pete [170] have also investigated the photochemical behavior of derivatives of methyl benzoate bearing a hydroxy or alkoxy sub-stituent at one meta position and an alkenyloxy side chain at the other meta position. In contrast to most other intramolecular ortho photocycloadditions, the addition in these molecules takes place at the positions ortho and meta with respect to the side chain, instead of ipso and ortho (Scheme 58).

When X=R=H, irradiation in acetonitrile (254 nm) in the presence of small quantities of sulfuric acid yields only **I** (yield 47% at 50% conversion). Likewise, when X=Me and R=H, only compound **I** is obtained (57% at 254 nm and 65% at 300 nm). Replacement of acetonitrile by methanol as solvent affords completely different results: compound **II** and **III** are formed in yields of 64% and 16% at 82% conversion. With X=Me and R=Me at 300 nm in MeOH, yields of **II** and **III** are somewhat lower, 38% and 10%, respectively, at 100% conversion (a complex mixture of by-products is also formed). Similar results are obtained with X=H and R=Me.

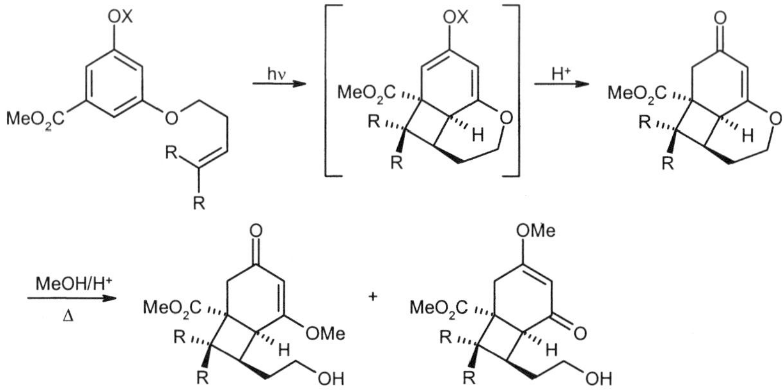

Scheme 58 Intramolecular ortho photocycloaddition of 3-alkenyloxy-5-hydroxy- and 3-alkenyloxy-5-alkoxybenzoic acid methyl esters.

REFERENCES

1. Ayer, D. E.; Buchi, G. H. U.S. patent 2,805,242 (1957); *Chem. Abstr.* **1958,** *52,* 2904.
2. Angus, H. J. F.; Bryce-Smith, D. *Proc. Chem. Soc.* **1959,** 326–327.
3. Bryce-Smith, D.; Lodge, J. E. *Proc. Chem. Soc.* **1961,** 333–334.
4. Wilzbach, K. E.; Kaplan, L. *J. Am. Chem. Soc.* **1966,** *88,* 2066–2067.
5. Bryce-Smith, D.; Gilbert, A.; Orger, B. H. *J. Chem. Soc., Chem. Commun.* **1966,** 512–514.
6. Cornelisse, J. *Chem. Rev.* **1993,** *93,* 615–669.
7. Wilzbach, K. E.; Kaplan, L. *J. Am. Chem. Soc.* **1971,** *93,* 2073–2074.
8. Srinivasan, R. *IBM J. Res. Dev.* **1971,** *15,* 34–40.
9. Srinivasan, R. *J. Am. Chem. Soc.* **1972,** *94,* 8117–8124.
10. Heine, H.-G.; Hartmann, W. *Angew. Chem.* **1975,** *87,* 708–708.
11. Gilbert, A.; Taylor, G. *Tetrahedron Lett.* **1977,** 469–472.
12. Gilbert, A.; Taylor, G. N.; bin Samsudin, M. W. *J. Chem. Soc., Perkin Trans. 1* **1980,** 869–876.
13. Mattay, J.; Leismann, H.; Scharf, H.-D. *Chem. Ber.* **1979,** *112,* 577–599.
14. Mattay, J.; Runsink, J.; Leismann, H.; Scharf, H.-D. *Tetrahedron Lett.* **1982,** *23,* 4919–4922.
15. Mattay, J.; Runsink, J.; Gersdorf, J.; Rumbach, T.; Ly, C. *Helv. Chim. Acta* **1986,** *69,* 442–455.
16. Bryce-Smith, D.; Foulger, B. E.; Gilbert, A. *J. Chem. Soc., Chem. Commun.* **1972,** 664–665.
17. Berridge, J. C.; Forrester, J.; Foulger, B. E.; Gilbert, A. *J. Chem. Soc., Perkin Trans. 1* **1980,** 2425–2434.
18. Bryce-Smith, D.; Dadson, W. M.; Gilbert, A. *J. Chem. Soc., Chem. Commun.* **1980,** 112–113.
19. Gilbert, A.; Taylor, G. *J. Chem. Soc., Chem. Commun.* **1978,** 129–130.
20. Gilbert, A.; Taylor, G. N. *J. Chem. Soc., Chem. Commun.* **1979,** 229–230.
21. Gilbert, A.; Taylor, G. N. *J. Chem. Soc., Perkin Trans. 1* **1980,** 1761–1768.
22. Ellis-Davies, G. C. R.; Gilbert, A.; Heath, P.; Lane, J. C.; Warrington, J. V.; Westover, D. L. *J. Chem. Soc., Perkin Trans. 2* **1984,** 1833–1841.
23. Grovenstein, E., Jr.; Rao, D. V.; Taylor, J. W. *J. Am. Chem. Soc.* **1961,** *83,* 1705–1711.
24. Bryce-Smith, D.; Vickery, B.; Fray, G. I. *J. Chem. Soc. (C)* **1967,** 390–394.
25. Pettit, G. R.; Paull, K. D.; Herald, C. L.; Herald, D. L.; Riden, J. R. *Can. J. Chem.* **1983,** *61,* 2291–2294.
26. Schenck, G. O.; Steinmetz, R. *Tetrahedron Lett.* **1960,** *21,* 1–8.
27. Hammond, G. S.; Hardham, W. M. *Proc. Chem. Soc.* **1963,** 63–64.
28. Hardham, W. M.; Hammond, G. S. *J. Am. Chem. Soc.* **1967,** *89,* 3200–3205.
29. Bryce-Smith, D.; Vickery, B. *Chem. Ind.* **1961,** 429.
30. Angus, H. J. F.; Bryce-Smith, D. *J. Chem. Soc.* **1960,** 4791–4795.
31. Bryce-Smith, D.; Lodge, J. E. *J. Chem. Soc.* **1962,** 2675–2680.
32. Bryce-Smith, D.; Connett, B. E.; Gilbert, A. *J. Chem. Soc. (B)* **1968,** 816–822.
33. Bryce-Smith, D.; Gilbert, A. *J. Chem. Soc.* **1965,** 918–924.

34. Bradshaw, J. S. *J. Org. Chem.* **1966**, *31*, 3974–3976.
35. Rafikov, S. R.; Tolstikov, G. A.; Naletova, G. P.; Shaikhrazieva, V. Sh.; Vshivtseva, N. S.; Tal'vinskii, A. V. *Zhur. Org. Khim.* **1979**, *15*, 2073–2075.
36. Bryce-Smith, D.; Deshpande, R.; Gilbert, A.; Grzonka, J. *J. Chem. Soc., Chem. Commun.* **1970**, 561–562.
37. Bryce-Smith, D. *Pure Appl. Chem.* **1968**, *16*, 47–63.
38. Bryce-Smith, D. *Pure Appl. Chem.* **1973**, *34*, 193–212.
39. Hartmann, W.; Heine, H.-G.; Schrader, L. *Tetrahedron Lett.* **1974**, 883–886.
40. Hartmann, W.; Heine, H.-G.; Schrader, L. *Tetrahedron Lett.* **1974**, 3101–3104.
41. Bryce-Smith, D.; Deshpande, R. R.; Gilbert, A. *Tetrahedron Lett.* **1975**, 1627–1630.
42. Mattes, S. L.; Farid, S. *Acc. Chem. Res.* **1982**, *15*, 80–86.
43. Koltzenburg, G.; Fuss, P. G.; Mannsfield, S.-P.; Schenck, G. O. *Tetrahedron Lett.* **1966**, 1861–1866.
44. Bryce-Smith, D.; Hems, M. A. *J. Chem. Soc. (B)* **1968**, 812–815.
45. Bryce-Smith, D.; Gilbert, A. *J. Chem. Soc., Chem. Commun.* **1968**, 19–21.
46. Bryce-Smith, D.; Gilbert, A. *Tetrahedron* **1977**, *33*, 2459–2489.
47. Gilbert, A.; Yianni, P. *Tetrahedron* **1981**, *37*, 3275–3283.
48. Grovenstein, Jr., E.; Rao D. V.; Taylor, J. W. in *Organic Photochemical Syntheses,* Srinivasan, R.; Roberts, T. D.; Cornelisse, J., eds., Wiley: New York; **1976**, Vol. 2, 97–98.
49. Zhubanov, B. A.; Berzhanova, S. K.; Kutzhanov, R. T.; Belekhov, S. A. *Izv. Nats. Akad. Nauk Resp. Kaz., Ser. Khim.* **1993**, *6*, 69–72; *Chem. Abstr.* **1995**, *123*, 227,751.
50. Nurkhodzhaev, Z. A.; Kutzhanov, R. T.; Samazhanova, K. B.; Zhubanov, B. A. *Izv. Akad. Nauk Kaz. SSR, Ser. Khim.* **1990**, *1*, 51–54; *Chem. Abstr.* **1990**, *113*, 7100.
51. Berzhanova, S. K.; Belekhov, S. A.; Nazarbekova, M. T.; Kutzhanov, R. T.; Zhubanov, B. A.; Almabekov, O. A. *Izv. Akad. Nauk Resp. Kaz., Ser. Khim.* **1992**, 39–42; *Chem. Abstr.* **1992**, *117*, 91,314.
52. Vermont, G. B.; Riccobono, P. X.; Blake, J. *J. Am. Chem. Soc.* **1965**, *87*, 4024–4026.
53. Zhubanov, B. A.; Almabekov, O. A.; Ismailova, Zh. M. *Zhur. Org. Khim.* **1981**, *17*, 996–999.
54. Shaikhrazieva, V. Sh.; Tal'vinskii, E. V.; Tolstikov, G. A. *Zhur. Org. Khim.* **1971**, *7*, 2225.
55. Shaikhrazieva, V. Sh.; Tal'vinskii, E. V.; Tolstikov, G. A.; Shakirova, A. M. *Zhur. Org. Khim.* **1973**, *9*, 1452–1458.
56. Kravtsova, V. D.; Almabekov, O. A.; Zhubanov, B. A. *Izv. Akad. Nauk Kaz. SSR, Ser. Khim.* **1980**, *3*, 62–67; *Chem. Abstr.* **1981**, *94*, 3781.
57. Grovenstein, E., Jr.; Rao, D. V. *Tetrahedron Lett.* **1961**, 148–150.
58. Grovenstein, Jr., E.; Campbell, T. C.; Shibata, T. *J. Org. Chem.* **1969**, *34*, 2418–2428.
59. Bryce-Smith, D.; Lodge, J. E. *J. Chem. Soc.* **1963**, 695–701.
60. Bryce-Smith, D.; Gilbert, A.; Grzonka, J. *J. Chem. Soc., Chem. Commun.* **1970**, 498–499.
61. Bryce-Smith, D.; Gilbert, A.; Al-Jalal, N.; Deshpande, R. R.; Grzonka, J.; Hems, M. A.; Yianni, P. *Z. Naturforsch. B* **1983**, *38*, 1101–1112.

62. Tinnemans, A. H. A.; Neckers, D. C. *J. Am. Chem. Soc.* **1977,** *99,* 6459–6460.
63. Tinnemans, A. H. A.; Neckers, D. C. *Tetrahedron Lett.* **1978,** 1713–1716.
64. Hanzawa, Y.; Paquette, L. *Synthesis* **1982,** 661–662.
65. Šket, B.; Zupan, M. *J. Am. Chem. Soc.* **1977,** *99,* 3504–3505.
66. Šket, B.; Zupančič, N.; Zupan, M. *Tetrahedron* **1984,** *40,* 3795–3804.
67. Bryce-Smith, D.; Gilbert, A.; Orger, B. H.; Twitchett, P. J. *J. Chem. Soc., Perkin Trans. 1* **1978,** 232–243.
68. Lippke, W.; Ferree, W. Jr.; Morrison, H. *J. Am. Chem. Soc.* **1974,** *96,* 2134–2137.
69. Morrison, H.; Nylund, T.; Palensky, F. *J. Chem. Soc., Chem. Commun.* **1976,** 4–5.
70. Pirrung, M. C. *J. Org. Chem.* **1987,** *52,* 1635–1637.
71. Liu, R. S. H.; Krespan, C. G. *J. Org. Chem.* **1969,** *34,* 1271–1278.
72. Miller, R. D.; Abraitys, V. Y. *Tetrahedron Lett.* **1971,** 891–894.
73. Atkinson, J. G.; Ayer, D. E.; Büchi, G.; Robb, E. W. *J. Am. Chem. Soc.* **1963,** *85,* 2257–2263.
74. Bradshaw, J. S. *Tetrahedron Lett.* **1966,** 2039–2042.
75. Bryce-Smith, D.; Hems, M. A. *Tetrahedron Lett.* **1966,** 1895–1899.
76. Bryce-Smith, D.; Gilbert, A.; Halton, B. *J. Chem. Soc., Perkin Trans. 1* **1978,** 1172–1175.
77. Shaikhrazieva, V. Sh.; Tal'vinskii, E. V.; Tolstikov, G. A. *Zhur. Org. Khim.* **1974,** *10,* 665.
78. Wamhoff, H.; Hupe, H.-J. *Chem. Ber.* **1978,** *111,* 2677–2688.
79. Arkhipova, I. A.; Zhubanov, B. A.; Saidenova, S. B. *Izv. Akad. Nauk Kaz. SSR, Ser. Khim.* **1972,** *22,* 44–48; *Chem. Abstr.* **1972,** *77,* 100,908.
80. Gilbert, A.; Yianni, P. *Tetrahedron Lett.* **1982,** *23,* 4611–4614.
81. Bryce-Smith, D. *J. Chem. Soc., Chem. Commun.* **1969,** 806–808.
82. Scharf, H.-D.; Klar, R. *Tetrahedron Lett.* **1971,** 517–520.
83. Scharf, H.-D.; Klar, R. *Chem. Ber.* **1972,** *105,* 575–587.
84. Scharf, H.-D.; Leismann, H.; Erb, W.; Gaidetzka, H. W.; Aretz, J. *Pure Appl. Chem.* **1975,** *41,* 581–600.
85. Aoyama, H.; Arata, Y.; Omote, Y. *J. Chem. Soc., Chem. Commun.* **1990,** 736–737.
86. Al-Jalal, N.; Drew, M. G. B.; Gilbert, A. *J. Chem. Soc., Chem. Commun.* **1985,** 85–86.
87. Al-Jalal, N.; Gilbert, A.; Heath, P. *Tetrahedron* **1988,** *44,* 1449–1459.
88. Al-Jalal, N. A. *J. Heterocyclic Chem.* **1990,** *27,* 1323–1327.
89. Al-Jalal, N. A. *J. Chem. Res., Synop.* **1989,** 110–111.
90. Al-Jalal, N.; Gilbert, A. *Recl. Trav. Chim. Pays-Bas* **1990,** *109,* 21–25.
91. Shaikhrazieva, V. Sh.; Tal'vinskii, E. V.; Tolstikov, G. A. *Zhur. Org. Khim.* **1978,** *14,* 1522–1529.
92. Deutsch, H. M.; Gelbaum, L. T.; McLaughlin, M.; Fleischmann, T. J.; Earnhart, L. L.; Haugwitz, R. D.; Zalkow, L. H. *J. Med. Chem.* **1986,** *29,* 2164–2170.
93. Wagner, P. J.; Nahm, K. *J. Am. Chem. Soc.* **1987,** *109,* 4404–4405.
94. Wagner, P. J.; Nahm, K. *J. Am. Chem. Soc.* **1987,** *109,* 6528–6530.
95. Wagner, P. J.; Sakamoto, M. *J. Am. Chem. Soc.* **1989,** *111,* 8723–8725.
96. Wagner, P. J.; Sakamoto, M. *J. Am. Chem. Soc.* **1989,** *111,* 9254–9256.
97. Wagner, P. J.; Sakamoto, M.; Madkour, A. E. *J. Am. Chem. Soc.* **1992,** *114,* 7298–7299.

98. Wagner, P. J.; Cheng, K.-L. *Tetrahedron Lett.* **1993,** *34,* 907–910.
99. Wagner, P. J.; Alehashem, H. *Tetrahedron Lett.* **1993,** *34,* 911–914.
100. Cheng, K.-L.; Wagner, P. J. *J. Am. Chem. Soc.* **1994,** *116,* 7945–7946.
101. Wagner, P. J.; McMahon, K. *J. Am. Chem. Soc.* **1994,** *116,* 10,827–10,828.
102. Smart, R. P.; Wagner, P. J. *Tetrahedron Lett.* **1995,** *36,* 5131–5134.
103. Wagner, P. J.; Smart, R. P. *Tetrahedron Lett.* **1995,** *36,* 5135–5138.
104. Cosstick, K. B.; Drew, M. G. B.; Gilbert, A. *J. Chem. Soc., Chem. Commun.* **1987,** 1867–1868.
105. Al-Qaradawi, S. Y.; Cosstick, K. B.; Gilbert, A. *J. Chem. Soc., Perkin Trans. 1* **1992,** 1145–1148.
106. Hoffmann, N.; Pete, J.-P. *Tetrahedron Lett.* **1995,** *36,* 2623–2626.
107. Al-Qaradawi, S.; Gilbert, A.; Jones, D. T. *Recl. Trav. Chim. Pays-Bas* **1995,** *114,* 485–491.
108. Atkins, R. J.; Fray, G. I.; Gilbert, A.; bin Samsudin, M. W.; Steward, A. J. K.; Taylor, G. N. *J. Chem. Soc., Perkin Trans. 1* **1979,** 3196–3202.
109. Cantrell, T. S. *J. Am. Chem. Soc.* **1972,** *94,* 5929–5931.
110. Cantrell, T. S. *J. Org. Chem.* **1977,** *42,* 4238–4245.
111. Šket, B.; Zupančič, N.; Zupan, M. *Tetrahedron* **1986,** *42,* 753–754.
112. Neijenesch, H. A.; De Ruiter, R. J. P. J.; Ridderikhoff, E. J.; Van den Ende, J. O.; Laarhoven, L. J.; Van Putten, L. J. W.; Cornelisse, J. *J. Photochem. Photobiol. A.: Chem.* **1991,** *60,* 325–343.
113. Blakemore, D. C.; Gilbert, A. *J. Chem. Soc., Perkin Trans. 1* **1992,** 2265–2270.
114. Mirbach, M. F.; Mirbach, M. J.; Saus, A. *Tetrahedron Lett.* **1977,** 959–962.
115. Bryce-Smith, D.; Gilbert, A.; Orger, B.; Tyrrell, H. *J. Chem. Soc., Chem. Commun.* **1974,** 334–336.
116. Bryce-Smith, D.; Foulger, B.; Forrester, J.; Gilbert, A.; Orger, B. H.; Tyrell, H. M. *J. Chem. Soc., Perkin Trans. 1* **1980,** 55–71.
117. Bryce-Smith, D.; Foulger, B. E.; Gilbert, A.; Twitchett, P. J. *J. Chem. Soc., Chem. Commun.* **1971,** 794–795.
118. Tyrrell, H. M.; Wolters, A. P. *Tetrahedron Lett.* **1974,** 4193–4196.
119. Gilbert, A.; Taylor, G. *J. Chem. Soc., Chem. Commun.* **1977,** 242–243.
120. Atkins, R. J.; Fray, G. I.; Drew, M. G. B.; Gilbert, A.; Taylor, G. N. *Tetrahedron Lett.* **1978,** 2945–2948.
121. Gilbert, A.; Heritage, T. W.; Isaacs, N. S. *J. Chem. Soc., Perkin Trans. 2* **1992,** 1141–1144.
122. Mattay, J.; Leismann, H.; Scharf, H.-D. *Mol. Photochem.* **1979,** *9,* 119–156.
123. Scharf, H.-D.; Mattay, J. *Tetrahedron Lett.* **1977,** 401–404.
124. Mattay, J. *Tetrahedron* **1985,** *41,* 2405–2417.
125. Gilbert, A.; bin Samsudin, M. W. *Angew. Chem.* **1975,** *87,* 540–541.
126. Gilbert, A.; bin Samsudin, M. W. *J. Chem. Soc., Perkin Trans. 1* **1980,** 1118–1123.
127. Job, B. E.; Littlehailes, J. D. *J. Chem. Soc. (C)* **1968,** 886–889.
128. Ohashi, M.; Yoshino, A.; Yamazaki, K.; Yonezawa, T. *Tetrahedron Lett.* **1973,** 3395–3396.
129. Atkins, R. J.; Fray, G. I.; Gilbert, A. *Tetrahedron Lett.* **1975,** 3087–3088.
130. Atkins, R. J.; Fray, G. I.; Gilbert, A.; bin Samsudin, M. W. *Tetrahedron Lett.* **1977,** 3597–3600.

131. Cornelisse, J.; Gilbert, A.; Rodwell, P. W. *Tetrahedron Lett.* **1986,** *27,* 5003–5006.
132. Gilbert, A.; Heath, P.; Rodwell, P. W. *J. Chem. Soc., Perkin Trans. 1* **1989,** 1867–1873.
133. Gilbert, A.; Taylor, G. N.; Collins, A. *J. Chem. Soc., Perkin Trans. 1* **1980,** 1218–1224.
134. Mattay, J.; Runsink, J.; Piccirilli, J. A.; Jans, A. W. H.; Cornelisse, J. *J. Chem. Soc., Perkin Trans. 1* **1987,** 15–20.
135. Osselton, E. M.; Eyken, C. P.; Jans, A. W. H.; Cornelisse, J. *Tetrahedron Lett.* **1985,** *26,* 1577–1580.
136. Ohashi, M.; Tanaka, Y.; Yamada, S. *J. Chem. Soc., Chem. Commun.* **1976,** 800.
137. Gilbert, A.; Yianni, P. *Tetrahedron Lett.* **1982,** *23,* 255–256.
138. Al-Jalal, N.; Gilbert, A. *J. Chem. Research (S)* **1983,** 266–267.
139. Ohashi, M.; Tanaka, Y.; Yamada, S. *Tetrahedron Lett.* **1977,** 3629–3632.
140. Garcia, H.; Gilbert, A.; Griffiths, O. *J. Chem. Soc., Perkin Trans. 2* **1994,** 247–252.
141. Osselton, E. M.; Cornelisse, J. *Tetrahedron Lett.* **1985,** *26,* 527–530.
142. Mattay, J.; Runsink, J.; Heckendorn, R.; Winkler, T. *Tetrahedron* **1987,** *43,* 5781–5789.
143. Mattay, J.; Runsink, J.; Rumbach, T.; Ly, C.; Gersdorf, J. *Proceedings Xth IUPAC Symposium on Photochemistry, Interlaken, Switzerland,* 1984, 257–258.
144. Mattay, J.; Runsink, J.; Rumbach, T.; Ly, C.; Gersdorf, J. *J. Am. Chem. Soc.* **1985,** *107,* 2557–2558.
145. Bryce-Smith, D.; Dadson, W. M.; Gilbert, A.; Orger, B. H.; Tyrrell, H. M. *Tetrahedron Lett.* **1978,** 1093–1096.
146. Gilbert, A.; Heath, P. *Tetrahedron Lett.* **1987,** *28,* 5909–5912.
147. Arnold, D. R.; McManus, K. A.; Du, X. *Can. J. Chem.* **1994,** *72,* 415–429.
148. De Lijser, H. J. P.; Cameron, T. S.; Arnold, D. R. *Can. J. Chem.* **1997,** *75,* 1795–1809.
149. Katsuhara, Y.; Nakamura, T.; Shimizu, A.; Shigemitsu, Y.; Odaira, Y. *Chem. Lett.* **1972,** 1215–1218; *Chem. Abstr.* **1973,** *78,* 83,885.
150. Šket, B.; Zupan, M. *J. Chem. Soc., Chem. Commun.* **1977,** 365–366.
151. Šket, B.; Zupančič, N.; Zupan, M. *J. Chem. Soc., Perkin Trans. 1* **1987,** 981–985.
152. Šket, B.; Zupančič, N.; Zupan, M. *Bull. Chem. Soc. Jpn.* **1989,** *62,* 1287–1291.
153. Bryce-Smith, D.; Gilbert, A.; Orger, B. H. *J. Chem. Soc., Chem. Commun.* **1969,** 800–802.
154. Šket, B.; Zupan, M. *Tetrahedron Lett.* **1977,** 2811–2814.
155. Šket, B.; Zupan, M. *J. Chem. Soc., Chem. Commun.* **1976,** 1053.
156. Lemal, D. M.; Buzby, J. M.; Barefoot, A. C., III; Grayston, M. W.; Laganis, E. D. *J. Org. Chem.* **1980,** *45,* 3118–3120.
157. Waldron, R. F.; Barefoot III, A. C.; Lemal, D. M. *J. Am. Chem. Soc.* **1984,** *106,* 8301–8302.
158. Šket, B.; Zupan, M. *Tetrahedron* **1989,** *45,* 6741–6748.
159. Zimmerman, H. E.; Bunce, R. A. *J. Org. Chem.* **1982,** *47,* 3377–3396.
160. Nuss, J. M.; Chinn, J. P.; Murphy, M. M. *J. Am. Chem. Soc.* **1995,** *117,* 6801–6802.
161. De Haan, R.; De Zwart, E. W.; Cornelisse, J. *J. Photochem. Photobiol. A: Chem.* **1997,** *102,* 179–188.
162. Zhang, C.; Bourgin, D.; Keese, R. *Tetrahedron* **1991,** *47,* 3059–3074.

163. Mani, J.; Schüttel, S.; Zhang, C.; Bigler, P.; Müller, C.; Keese, R. *Helv. Chim. Acta* **1989,** *72,* 487–495.
164. De Keukeleire, D.; He, S.-L.; Blakemore, D.; Gilbert, A. *J. Photochem. Photobiol. A: Chem.* **1994,** *80,* 233–240.
165. Busson, R.; Schraml, J.; Saeyens, W.; Van der Eycken, E.; Herdewijn, P.; De Keukeleire, D. *Bull. Soc. Chim. Belg.* **1997,** *106,* 671–676.
166. Van der Eycken, E.; De Keukeleire, D.; De Bruyn, A. *Tetrahedron Lett.* **1995,** *36,* 3573–3576.
167. Hoffmann, N.; Pete, J.-P. *Tetrahedron Lett.* **1996,** *37,* 2027–2030.
168. Hoffmann, N.; Pete, J.-P. *J. Org. Chem.* **1997,** *62,* 6952–6960.
169. Saeyens, W.; De Keukeleire, D.; Herdewijn, P.; De Bruyn, A. *Biomed. Chromatogr.* **1997,** *11,* 79–80.
170. Hoffmann, N.; Pete, J.-P. *Tetrahedron Lett.* **1998,** 5027–5030.
171. Avent, A. G.; Byrne, P. W.; Penkett, C. S. *Org. Lett.* **1999,** *1,* 2073–2075.
172. Amey, D. M.; Gilbert, A.; Jones, D. T. *J. Chem. Soc., Perkin Trans. 2* **1998,** 213–218.
173. Saeyens, W.; Busson, R.; Van der Eycken, J.; Herdewijn, P.; De Keukeleire, D. *J. Chem. Soc., Chem. Commun.* **1997,** 817–818.
174. Bryce-Smith, D.; Gilbert, A.; Tyrrell, H. M. *J. Chem. Soc., Chem. Commun.* **1974,** 699–700.
175. Morikawa, A.; Brownstein, S.; Cvetanovič, R. J. *J. Am. Chem. Soc.* **1970,** *92,* 1471–1476.
176. Cundall, R. B.; Pereira, L. C.; Robinson, D. A. *Chem. Phys. Lett.* **1972,** *13,* 253–256.
177. Fleming, R. H.; Quina, F. H.; Hammond, G. S. *J. Am. Chem. Soc.* **1974,** *96,* 7738–7741.
178. Morrison, H.; Ferree, W. I., Jr.; *J. Chem. Soc., Chem. Commun.* **1969,** 268–269.
179. Leismann, H.; Mattay, J.; Scharf, H.-D. *J. Photochem.* **1978,** *9,* 338.
180. Leismann, H.; Mattay, J. *Tetrahedron Lett.* **1978,** 4265–4268.
181. Weller, A. *The Exciplex,* Gordon, M.; Ware, W. R., eds.; Academic Press: London, 1975, 23–36.
182. Leismann, H.; Mattay, J.; Scharf, H.-D. *J. Am. Chem. Soc.* **1984,** *106,* 3985–3991.
183. Mattay, J. *J. Photochem.* **1987,** *37,* 167–183.
184. Gilbert, A. *Pure Appl. Chem.* **1980,** *52,* 2669–2682.
185. Bryce-Smith, D.; Gilbert, A.; Mattay, J. *Tetrahedron* **1986,** *42,* 6011–6014.
186. Mattay, J. *Tetrahedron* **1985,** *41,* 2393–2404.
187. Bryce-Smith, D.; Gilbert, A. *Tetrahedron* **1976,** *32,* 1309–1326.
188. Houk, K. N. *Pure Appl. Chem.* **1982,** *54,* 1633–1650.
189. Van der Hart, J. A.; Mulder, J. J. C.; Cornelisse, J. *J. Photochem. Photobiol. A: Chem.* **1995,** *86,* 141–148.
190. Clifford, S.; Bearpark, M. J.; Bernardi, F.; Olivucci, M.; Robb, M. A.; Smith, B. R. *J. Am. Chem. Soc.* **1996,** *118,* 7353–7360.
191. Berridge, J.; Bryce-Smith, D.; Gilbert, A. *J. Chem. Soc., Chem. Commun.* **1974,** 964–966.
192. Perrins, N. C.; Simons, J. P. *J. Chem. Soc., Chem. Commun.* **1967,** 999–1000.
193. Kaupp, G.; Jostkleigrewe, E. *Angew. Chem.* **1976,** *88,* 812–818.

2

Photocycloaddition and Photoaddition Reactions of Aromatic Compounds

Kazuhiko Mizuno, Hajime Maeda,
Akira Sugimoto, and
Kazuhiko Chiyonobu
Osaka Prefecture University, Sakai, Japan

I. INTRODUCTION

Photocycloaddition and photoaddition can be utilized for new carbon–carbon and carbon–heteroatom bond formation under mild conditions from synthetic viewpoints. In last three decades, a large number of these photoreactions between electron-donating and electron-accepting molecules have been appeared and discussed in the literature, reviews, and books [1–10]. In these photoreactions, a variety of reactive intermediates such as excimers, exciplexes, triplexes, radical ion pairs, and free-radical ions have been postulated and some of them have been detected as transient species to understand the reaction mechanism. Most of reactive species in solution have been already characterized by laser flash photolysis techniques, but still the prediction for the photochemical process is hard to visualize. In preparative organic photochemistry, the dilemma that the transient species including emission are hardly observed in the reaction system giving high chemical yields remains in most cases [11,12].

   This chapter deals primarily with the intermolecular and intramolecular photocycloaddition of unsaturated compounds to aromatic rings and the intermolecular and intramolecular photoaddition of nucleophiles, electrophiles, and radical species to the activated aromatic ring. The photochemistry of carbonyl compounds including α,β-unsaturated ketones are also interesting and important, but they are not treated here because they are discussed in other reviews [13–17]. Previously, McCullough has reviewed the photocycloaddition of aromatic compounds [18]. We have also reviewed the photoaddition and photocycloaddition reactions via photoinduced electron transfer [19]. Moreover, many reviews about the photocycloaddition and photoaddition to aromatic rings have been already appeared in recent years [11,12,20–37]. Therefore, we will mainly focus on the recent development of the photocycloaddition to aromatic rings via exciplexes and the photoaddition to aromatic rings via radical ions.

II. GENERATION OF REACTIVE SPECIES IN THE ELECTRON-DONOR–ACCEPTOR SYSTEM

The reactive species generated by the photoexcitation of organic molecules in the electron-donor–acceptor systems are well established in last three decades as shown in Scheme 1. The reactivity of an exciplex and radical ion species is discussed in the following sections. The structure–reactivity relationship for the exciplexes, which possess infinite lifetimes and often emit their own fluorescence, has been shown in some selected regioselective and stereoselective photocycloadditions. However, the exciplex emission is often absent or too weak to be identified although the exciplexes are postulated in many photocycloadditions [11,12]. The different reactivities among the contact radical ion pairs (polar exciplexes), solvent-separated radical ion pairs, and free-radical ions as ionic species

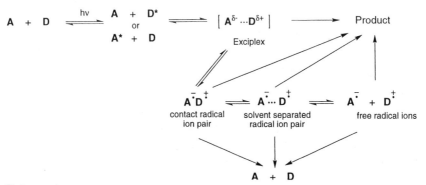

Scheme 1

have been also discussed from the synthetic and mechanistic viewpoints in the photoinduced electron-transfer reactions [19].

We will discuss briefly the reactive species such as an exciplex and radical ion species generated by the excitation of organic molecules in the electron-donor (D)–acceptor (A) system. An exciplex is produced usually in nonpolar solvents by an interaction of an electronically excited molecule D* (or A*) with a ground-state molecule A (or D). It is often postulated as an important intermediate in the photocycloaddition between D and A. In the case of D = A, an excimer is formed as an excited reactive species to cause photodimerization. In some cases, a termolecular interaction of an exciplex with another D or A generates a triplex, which is also a reactive intermediate for photocycloaddition. The evidence for the formation of excimers, exciplexes, and triplexes are shown in the fluorescence quenching. Excimer and exciplex emission is, in some cases, observed and an emission of triplex rarely appears.

Radical ion species ($D^{+\cdot}$, $A^{-\cdot}$) are usually produced by one-electron transfer from D* to A or D to A* in polar solvents. The calculated values for the free-energy change using the Rehm–Weller equation (Eq. 1) predict a photoinduced electron-transfer process between D and A [38]

$$\Delta G = E^{ox}(D) - E^{red}(A) - E_{0\text{-}0} - \frac{e^2}{\varepsilon r} \tag{1}$$

where $E^{ox}(D)$, $E^{red}(A)$, and $E_{0\text{-}0}$ are the oxidation potential of D, the reduction potential, and the 0–0 excitation energy of the molecule to be excited, respectively. The term $e^2/\varepsilon r$ expresses the coulombic interaction energy between $D^{+\cdot}$ and $A^{-\cdot}$ in a solvent of dielectric constant ε of distance r apart. When ΔG is negative, the formation of radical ion pair takes place as an exothermic process. These radical ion species, which have an ionic nature, can be utilized as reactive intermediates in the photoinduced electron-transfer reactions. The unavoidable problem in this system is a rapid back-electron transfer from $A^{-\cdot}$ to $D^{+\cdot}$. This problem is solved in some cases by using additives such as aromatic hydrocarbons (ArH) and metal salts [19]. The addition of salts into the D–A system suppresses a back-electron transfer and the photoinduced electron-transfer reaction is accelerated [39]. The ArH-sensitized photoreaction, named redox photosensitization or cosensitization, also affect the photoinduced electron-transfer reactions (Scheme 2). In the primary process of the photoinduced electron-transfer reaction, ArH becomes the radical cation ($ArH^{+\cdot}$) via one-electron transfer from ArH* to A (or from ArH to A*). This radical cation ($ArH^{+\cdot}$) interacts with D to generate $D^{+\cdot}$, which is a real reactive species. In most cases, ArH is finally recovered. The role of these aromatic hydrocarbons is discussed as redox photosensitization and cosensitization [19,40–42].

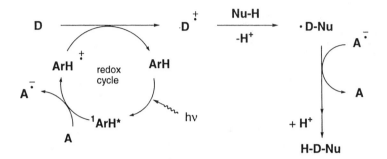

D : donor   **A** : acceptor   **ArH** : aromatic hydrocarbon

Nu-H : nucleophile
Scheme 2

III. PHOTOCYCLOADDITION TO AROMATIC RINGS

A. Classification

The photocycloaddition of alkenes to benzene rings can be classified into three types: $(2 + 2)$ (*ortho*), $(3 + 2)$ (*meta*), and $(4 + 2)$ (*para*) photocycloaddition, depending on the substituents, the reaction media, and the additives, as shown in Scheme 3. In general, *para* addition is very limited, and *ortho* and *para* adducts are more labile thermally and photochemically than the *meta* adducts. The $(2 + 2)$ (*ortho*) photocycloadducts, bicyclo[4.2.0]octa-2,4-dienes **1,** give thermally the ring-opened 1,3,5-cyclooctatrienes **4** and their following photochemical ring clo-

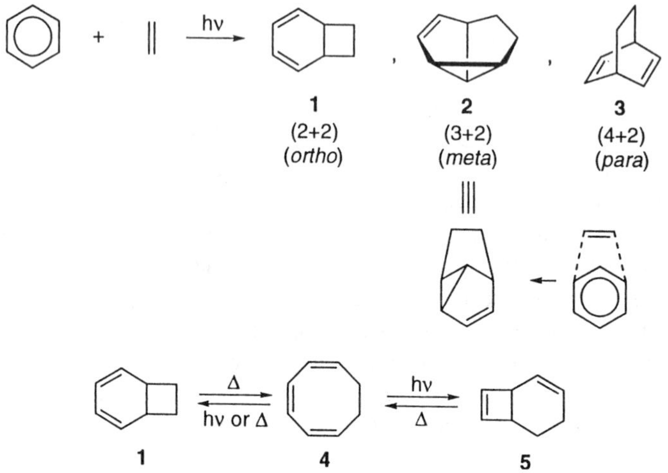

	1	**2**	**3**
	(2+2)	(3+2)	(4+2)
	(*ortho*)	(*meta*)	(*para*)

Scheme 3

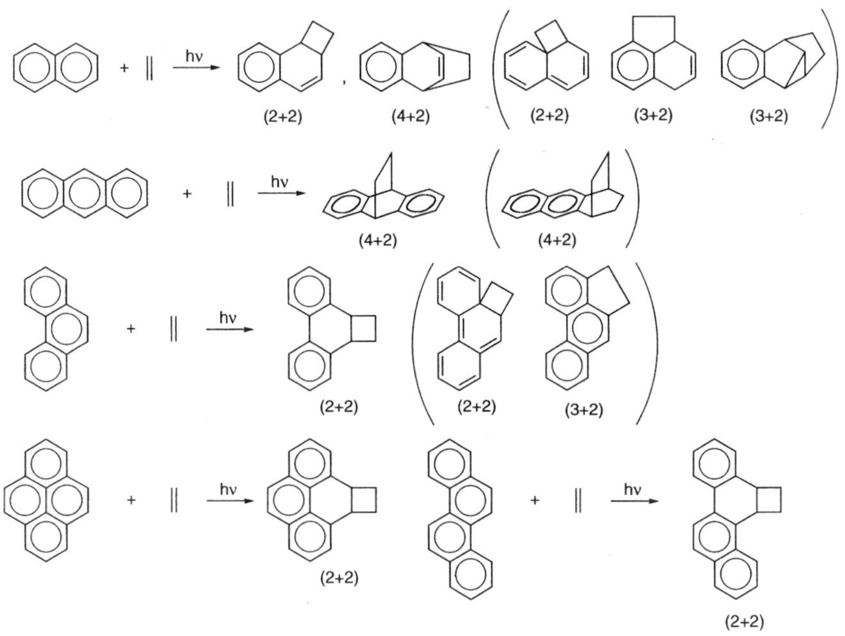

Scheme 4

sure often produces the bicyclo[4.2.0]octa-2,7-dienes **5**. The patterns of the photocycloaddition of alkenes to naphthalene rings are shown in Scheme 4; most of them are (2 + 2) cycloaddition at the 1,2-position, and some (4 + 2) addition at the 1,4-position, (3 + 2) addition at unusual 1,3- and 1,8-positions, and unusual (2 + 2) addition at the 1,8a-position have been shown in the literature. The photocycloaddition of polycyclic aromatic compounds such as anthracenes, phenanthrenes, pyrenes, chrysenes, and so on occurs usually at the expected reactive positions such as 9,10-positions of anthracenes and phenanthrenes, the 4,5-position of pyrenes, and 5,6-position of chrysenes. However, in some cases, the photoreactions take place at unprecedented positions such as 1,4-positions of anthracenes and 10,10a- and 1,10-positions of phenanthrenes. Specific examples of these photoreactions will be mentioned in the following subsections.

The photocycloaddition of cyclic and acyclic 1,3-alkadienes or furans to aromatic rings has been shown in some cases. Not only symmetrically allowed (4 + 4) photocycloaddition but also (4 + 2) photocycloaddition to benzene, naphthalene, and anthracene rings have been reported. These results are also shown in the following subsections.

Caldwell proposed a predictor of reactivity in allowed photodimerization and photocycloaddition by using a simple equation [43]. A paradigm for the prediction of reactivity in the allowed (2 + 2) and (4 + 4) photocyclodimerization and photocycloaddition via excited singlet states is developed. The quantitative algorithm

$$\gamma(r_c) = \frac{E_T^A + E_T^B - E_S^A}{c^2} \tag{2}$$

is derived, where $\gamma(r_c)$ is the resonance integral for end-on interaction of the carbon $2p$ orbital at distance r_c, E_T^A and E_T^B are triplet energies of reactants A and B, E_S^A is the singlet energy of the excited reactant A, and c^2 is the sum of HOMO and LUMO orbital coefficient products over reacting positions:

$$c^2 = 2({}^A C_1^{HOMO\ B} C_1^{HOMO} + {}^A C_2^{HOMO\ B} C_2^{HOMO})$$
$$+ 2({}^A C_1^{LUMO\ B} C_1^{LUMO} + {}^A C_2^{LUMO\ B} C_2^{LUMO}) \tag{3}$$

the algorithm correlates a substantial body of known photoreactivities with considerable success. Favorable features for high reactivity are high singlet energy, low triplet energy, and high frontier orbital density at reacting positions. The reactivity is simply classified by the calculated values of $\gamma(r_c)$: For $\gamma(r_c) < 20$ kcal/mol, (2 + 2) and (4 + 4) photodimerization and photocycloaddition proceed smoothly. In the range of $\gamma(r_c)$ of ~20–25 kcal/mol, reactivity appears to be moderate. In the photocyclodimerization and photocycloaddition, the substrates with $\gamma(r_c) > 25$ kcal/mol have not been reported. This equation is convenient for simple prediction for reactivity. However, as Caldwell pointed out, it is necessary to consider modification to incorporate charge-transfer effects and excited complex formation such as excimer and exciplex.

B. Intermolecular Photocycloaddition to Benzene Rings

Several reviews have been appeared about the synthetic utility and mechanistic details of photocycloadditions to aromatic rings [18,20–23]. In this subsection, we will deal with mainly the examples of last two decades. Photocycloaddition of alkenes to benzene rings proceeds at *ortho, meta,* and *para* positions as described earlier. It has been suggested that alkenes and arenes having either electron-releasing or electron-accepting substituents favor the *ortho* process, whereas relatively simple alkenes and arenes undergo *meta* addition [24,44].

Excellent regioselectivity and stereoselectivity has been achieved in each photocycloaddition mode [45–48]. Regiochemistry and stereochemistry in the *meta* process is decided by the orientation of the addends in the exciplex and by stabilization of biradical intermediates having a change transfer (CT) character (**6**) by the substituents on the arene. Intermolecular *meta* cycloaddition of arenes with cycloalkenes proceeds with *endo* selectivity (**7**) (Scheme 5). In the *ortho*-process, selectivities can be controlled mainly by the substituents on the reactants.

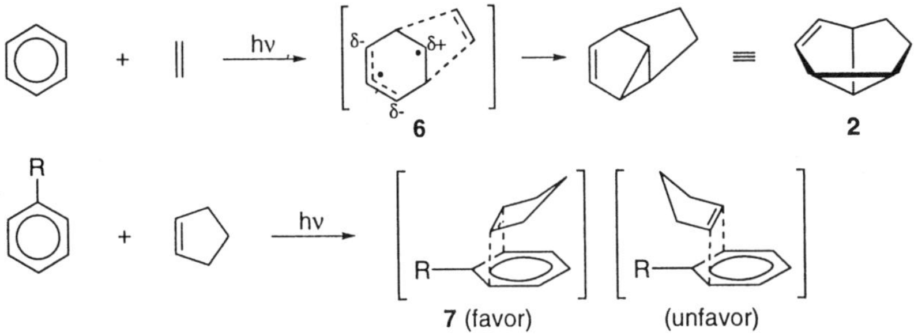

Scheme 5

In recent years, these donor–acceptor correlations to the reaction mode [49], substituent effects to stabilize *meta* intermediates [50–52], *exo/endo* selectivity [53–55], kinetics by means of fluorescence study [56] and deuterium isotope effect [57,58], formation and deactivation studies of exciplex [59,60], influence of pressure [61], and theoretical calculations [62–64] have been extensively studied.

Photocycloaddition of 1,3-dienes to benzene ring forms a mixture of *meta* (3 + 2) and *para*-(4 + 4) adducts accompanied by 2 : 1 and 2 : 2 adducts, and the yields are recognized to be low in general [65–68]. Recently, Gilbert reinvestigated the photoreaction of arenes with 1,3-dienes [69]. Various mono-substituted benzenes react with 2,3-dimethyl-1,3-butadiene to give a mixture of (2 + 2), (3 + 2), and (4 + 4) adducts; however, the (4 + 4) photocycloadduct **8** is essentially the sole product from benzonitrile and the diene, which was investigated in Takamuku et al.'s earlier study [70].

Gilbert also reported that the photocycloaddition of furan to benzonitrile gave only the 2,5-2′,6′ (*meta*) adduct (**9**) as a sole product, although the donor–acceptor interaction between addends is relatively high [71,72] (Scheme 6). Unsubstituted benzene reacts with furan to give a mixture of (4 + 4), (4 + 3), (3 + 2), and (2 + 2) adducts by 254-nm irradiation [73,74].

Al-Jalal and co-workers showed that nitrile groups undergo photocycloaddition to benzene rings to give azocine derivatives (**10**) [75–77] in a manner analogous to that of acetylene [78–84] (Scheme 7).

Photoreaction of hexafluorobenzene (**11**) with cycloalkenes proceeds at only the *ortho* mode with *exo* selectivity (**12** > **13**) [85] (Scheme 8). The *exo* selectivity decreased with increasing the ring size of cycloalkenes. However, the photoreaction of **11** with indene or dihydronaphthalene predominates the *endo* adduct due to the π-stacking. The ether **14** reacts regioselectivity and stereoselectively to produce **15–17** [86].

Ultraviolet (UV) irradiation of uracil derivative (**18**) in *p*- and *m*-xylene (**19**, **20**) in the presence of trifluoroacetic acid gave novel cycloaddition products (**20**,

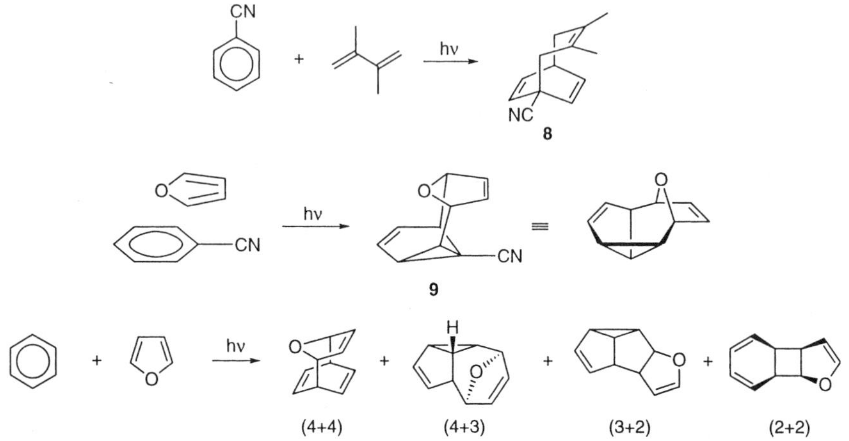

Scheme 6

21) [87] (Scheme 9). On the contrary, a similar irradiation in frozen benzene at −15 to −20°C affords different types of photocycloadduct **21–23** [88].

Pentafluoropyridine is known to undergo photocycloaddition with ethylene [89,90] and cycloalkenes [91] at the 3,4-position of the pyridine ring. Recently, Sakamoto et al. reported that metacrylonitrile adds to the 2,3-position of 2-alkoxy-3-cyano-4,6-dimethylpyridines (**24a,b**) to give 7-azabicyclo[4.2.0]octa-2,7-di-enes (**25a,b**) [92] (Scheme 10). This work was developed to the photocycloaddition of furan to pyridine ring as a first example (Scheme 10) [93].

C. Intermolecular Photocycloaddition to Polycyclic Aromatic Rings

Photocycloaddition of a variety of unsaturated compounds to unsubstituted naphthalene ring has been well investigated in 1960s and 1970s: the (2 + 2) cycload-

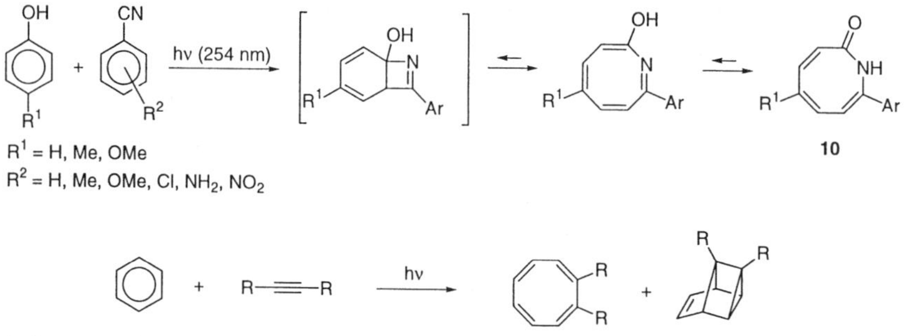

R^1 = H, Me, OMe
R^2 = H, Me, OMe, Cl, NH$_2$, NO$_2$

Scheme 7

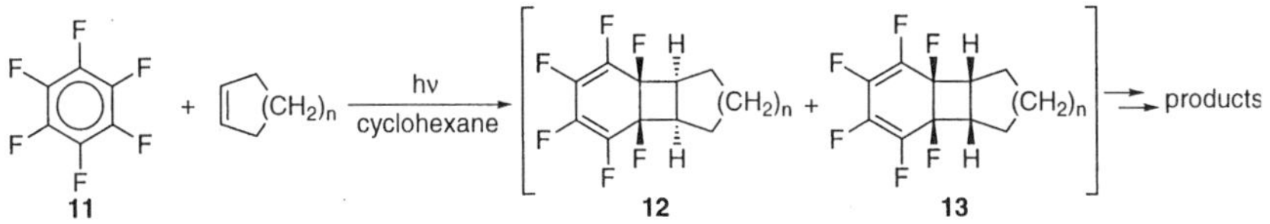

	12		13
n = 1	85	:	15
n = 2	82	:	18
n = 3	75	:	25
n = 4	53	:	47

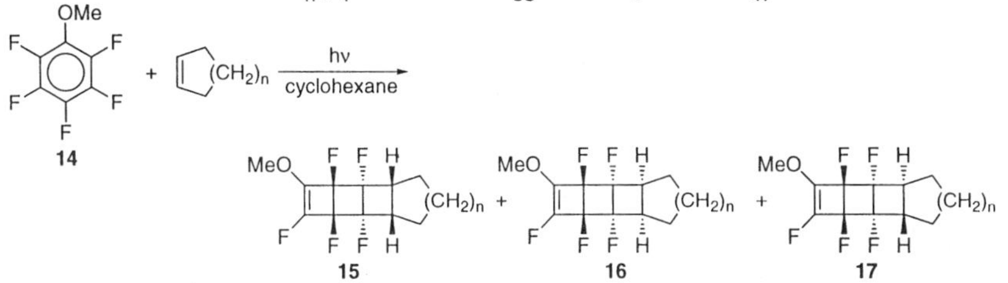

	15		16		17
n = 1	100	:	0	:	0
n = 3	64	:	16	:	20
n = 4	34	:	38	:	28

Scheme 8

dition of acrylonitrile [94,95], the (2 + 2 + 2) cycloaddition of diarylacetylenes [96] and dimethyl acetylenedicarboxylate [97], and the (4 + 4) addition of acyclic dienes [67] and cyclic dienes [98,99]. More chemoselective and efficient reactions have been developed by using 1- and 2-cyanonaphthalenes. Pac first reported that 1-cyano- and 2-cyano-naphthalenes [100–105] react with furan to give a (4 + 4)

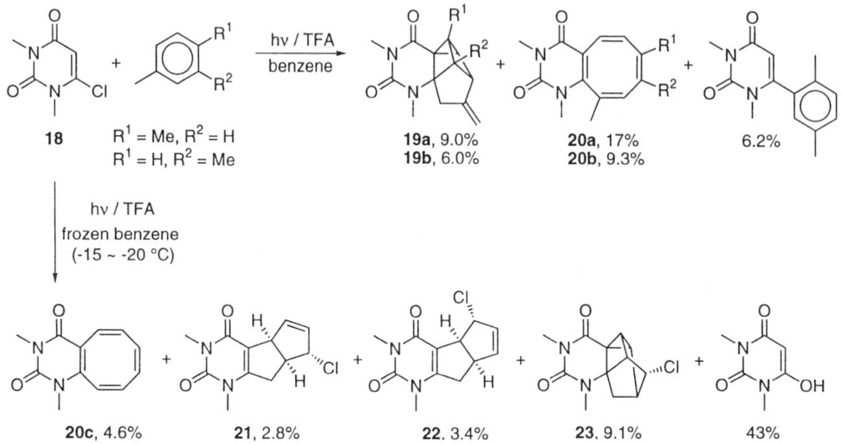

Scheme 9

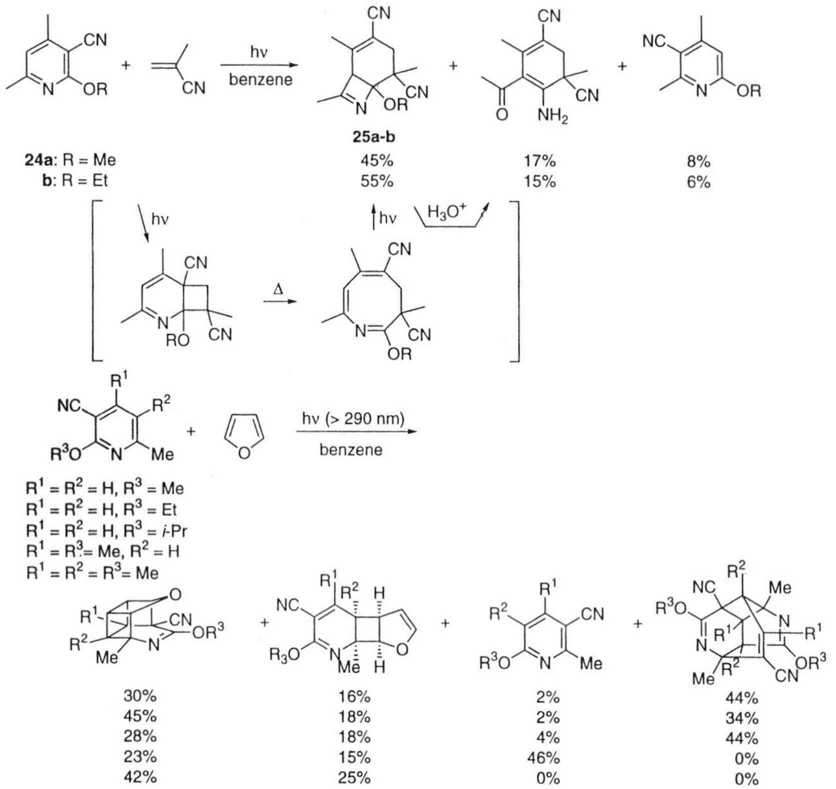

Scheme 10

cycloadduct **26** and a caged compound **27,** respectively (Scheme 11). The different regioselectivity is elucidated by the difference of the dipole moments in the excited states.

In the photoreaction of 1-cyanonaphthalene with phenyl vinyl ethers, the cycloadducts having *endo*-phenoxy group are much more favorably produced than *exo* adducts [106–108]. The photocycloaddition of *cis-* and *trans*-phenoxy-1-propenes to 1-cyanonaphthalene, which was reported by Mizuno and Pac, proceeded in a stereospecific manner. The *endo* selectivity is not observed in the photocycloaddition of alkyl propenyl ethers, although the stereospecific photocycloaddition proceeds in a highly regioselective manner [109,110]. Regioselective photocycloaddition of trimethylsilyl enol ethers can lead to the Michael-type alkylation on the naphthalene ring [111].

Irradiation of 2-cyanonaphthalene and alkyl vinyl ethers at 313 nm exclusively gives an *endo* (2 + 2) cycloadduct, whereas the irradiation >280 nm for a

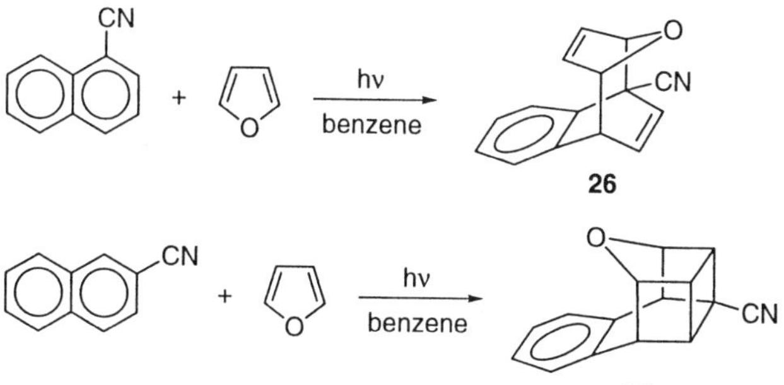

Scheme 11

longer time results in formation of a cyclobutene (**28**) as a main product [112–114] (Scheme 12).

Photoreaction of cyanonaphthalenes with tetramethylethylene leads to cyclobutane formation in nonpolar solvents via exciplex intermediates, whereas in polar media electron transfer occurs [115–117]. 1-Azetine is formed in the photoreaction of highly electron-donating alkene with 1-cyanonaphthalene [118].

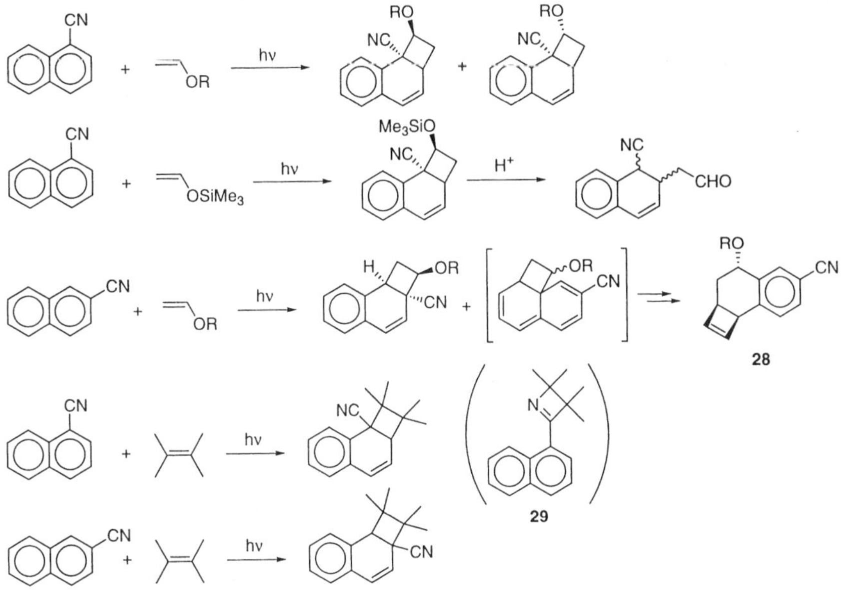

Scheme 12

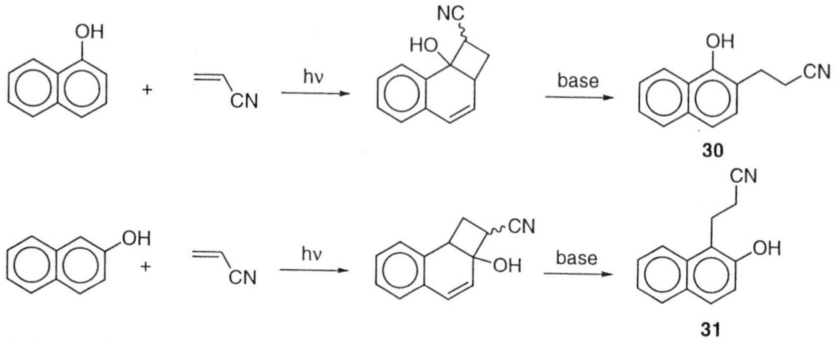

Scheme 13

Irradiation of naphthols or their trimethylsilyl ethers with acrylonitrile gives the photocycloadduct, which is converted to cyanoethyl-substituted naphthols (**30,31**) by treatment with a base [111,119] (Scheme 13).

Irradiation of naphthalene and *trans*-cyclooctene yielded the *cis* isomer and three 1 : 1 adducts, whereas *cis*-cyclooctene did not give any product under the same reaction conditions [120] (Scheme 14).

Mizuno reported that the photoreaction of 1-naphthylmethyltrimethylsilane (**32a**) with acrylonitrile (**33a**) gave *endo* (**34a**) and *exo* (**35a**) photocycloadducts in a 15 : 1 ratio [121] (Scheme 15). However, employing **33b** and **33c** in the reaction with **32a**, *endo* selectivity decreased. The photoreaction of germyl derivative (**32b**) gave similar results, but the photoreaction of 1-naphthylmethyldisilane (**32c**) afforded **34c** in a good yield with high regioselectivities and *endo* selectivities. Irradiation of stannyl derivative (**32d**) in the presence of **33a** gave 1,2-bis(1-naphthylmethyl)ethane as a major product, via the photocleavage of the C-Sn bond. *Endo* selectivity decreased in the photoreaction of 2-naphthylmethyltrimethylsilane with **33a** [121] (Table 1).

Suginome developed a useful ring expansion reaction through (2 + 2) photocycloaddition–β-scission sequences [122,123]. Photolysis of **37,** which is derived in three steps from the regioselective photocycloadducts **36,** gives benzocyclooctenone derivatives (**38**). The treatment of benzocyclooctenones with base affords benzohomotropones (**39**) (Scheme 16).

Photocycloadditions of alkenes to 1- and 2-naphthols are useful for the synthesis of benzobicyclo[4.2.0]octandiene derivatives [119,122,124–126]. Odaira

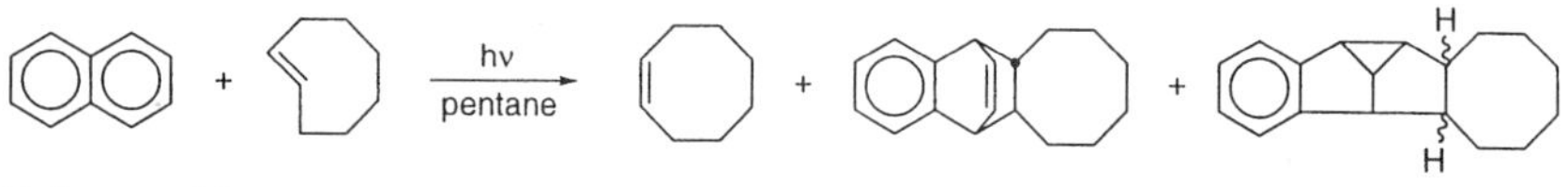

Scheme 14

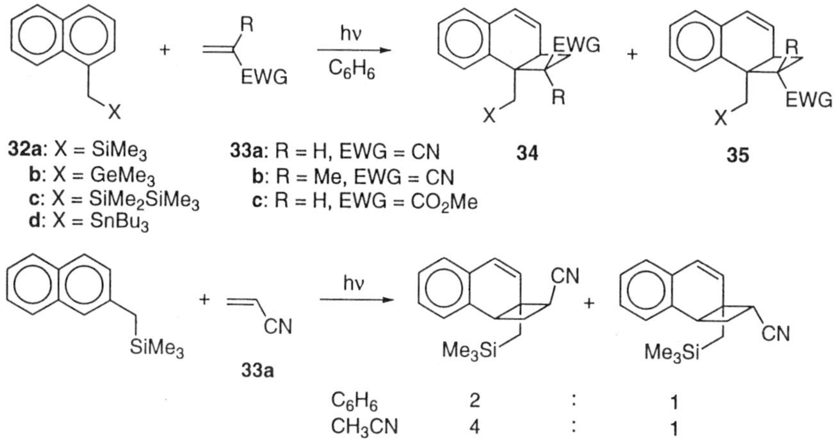

32a: X = SiMe$_3$
  b: X = GeMe$_3$
  c: X = SiMe$_2$SiMe$_3$
  d: X = SnBu$_3$

33a: R = H, EWG = CN   **34**
  b: R = Me, EWG = CN
  c: R = H, EWG = CO$_2$Me

           35

C$_6$H$_6$	2 :	1
CH$_3$CN	4 :	1

Scheme 15

and his co-workers reported (2 + 2) photocycloaddition of 2-naphthol derivatives (**40**) to ethylene in the presence of aluminum trichloride to give cycloadducts **42** by means of the photoexcitation of the complexes **41** [127] (Scheme 17).

    McCullough and co-workers first reported the photocycloaddition of acrylonitrile (**33a**) to naphthalene in 1970 [94,95]. They systematically investigated the photocycloaddition of **33a** to mono- and di-methylnaphthalenes; however, the yields and selectivity were not high and usually a complex mixture was produced [128] (Scheme 18).

    1,4-Adduct from naphthalene was also obtained by the reaction of octafluoronaphthalene (**43**) with indene [129] (Scheme 19). Photoirradiation of cyclo-

Table 1   Photocycloaddition of **33a–c** to Naphthalene Derivates **32a–d**

							Ratio of products		
32	X	**33**	R	EWG	Time/h	Yield/%	**34**	:	**35**
a	SiMe$_3$	a	H	CN	4	50	15	:	1
		b	Me	CN	32	20	2	:	1
		c	H	CO$_2$Me	12	10	1	:	1
b	GeMe$_3$	a	H	CN	5	40	15	:	1
		b	Me	CN	35	39	3	:	1
		c	H	CO$_2$Me	13	0	—		
c	SiMe$_2$SiMe$_3$	a	H	CN	3	27	(Only **34**)		
d	SnMe$_3$	a	H	CN	5	0	—		

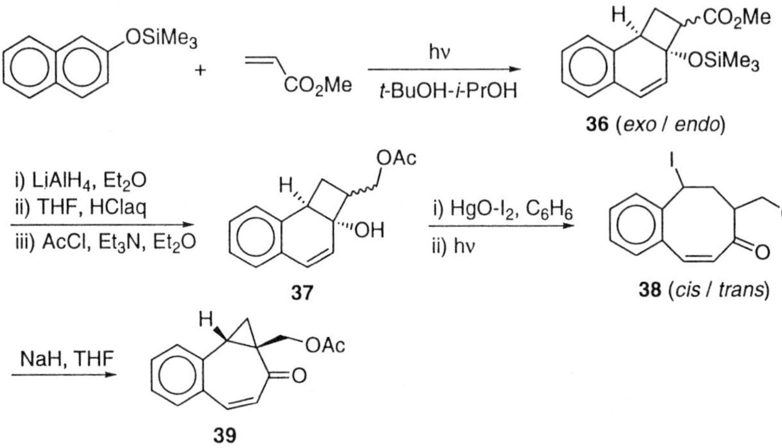

Scheme 16

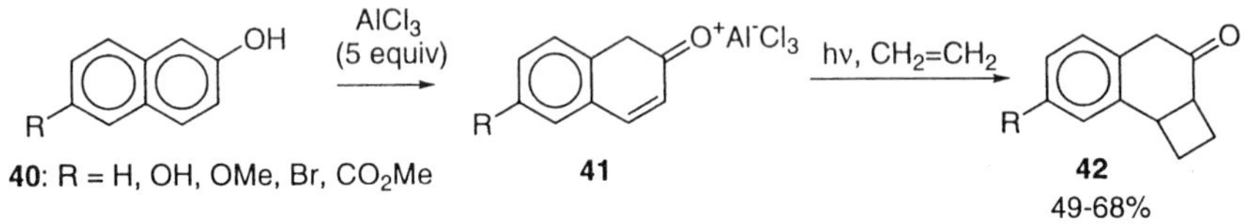

Scheme 17

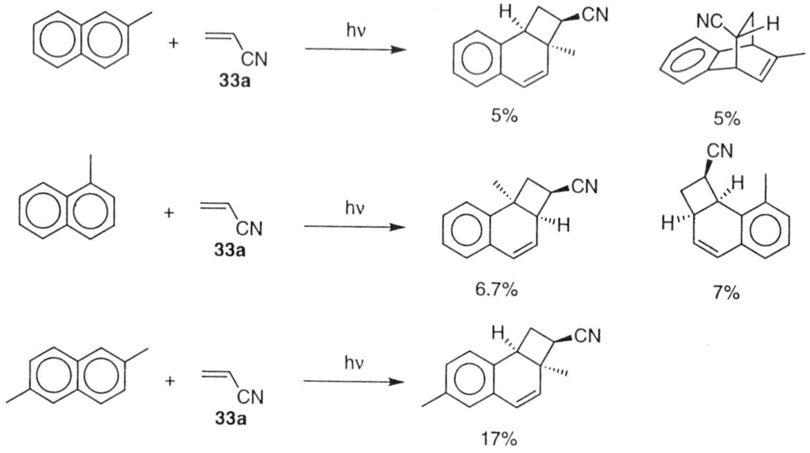

Scheme 18

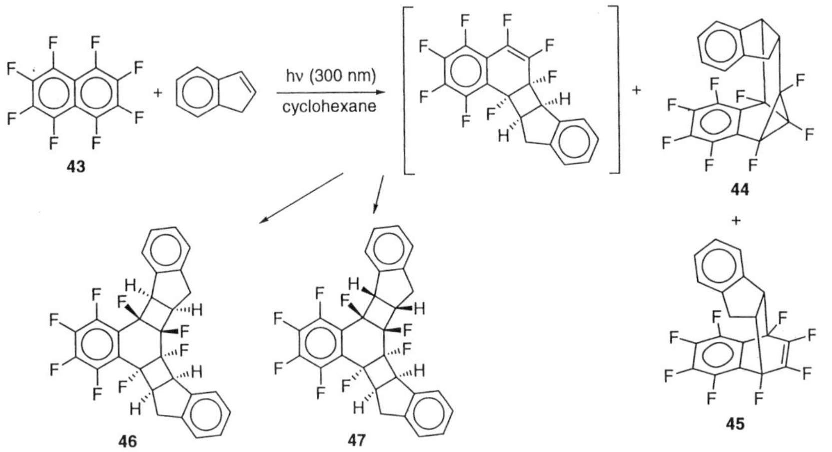

Scheme 19

hexane solution containing **43** and indene with a 300-nm light gave 1 : 2 (2 + 2) adducts (**46** and **47**), a (3 + 2) adduct (**44**), and a (4 + 2) adduct (**45**) in a 39 : 16 : 35 : 10 ratio. It is concluded that all 1,2*endo*, 1,3*endo*, and 1,4*endo* adducts are the primary products.

Photochemical synthesis of a 1,4-difluorobenzene-naphthalene biplanemer (**50**) and its photochemical and physical properties were investigated by Kimura et al. [130] (Scheme 20). The compound **50** was prepared through the electrolytic oxidative decarboxylation of the (4 + 4) photocycloadduct **49** between naphthalene and 3,6-difluoro-1,2-dihydrophthalic anhydride (**48**). The direct irradiation of **50** gave 1,4-difluorobenzene and the excited singlet and triplet states of naph-

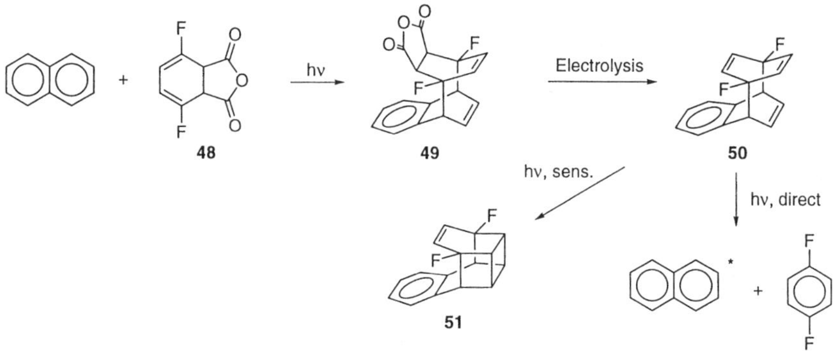

Scheme 20

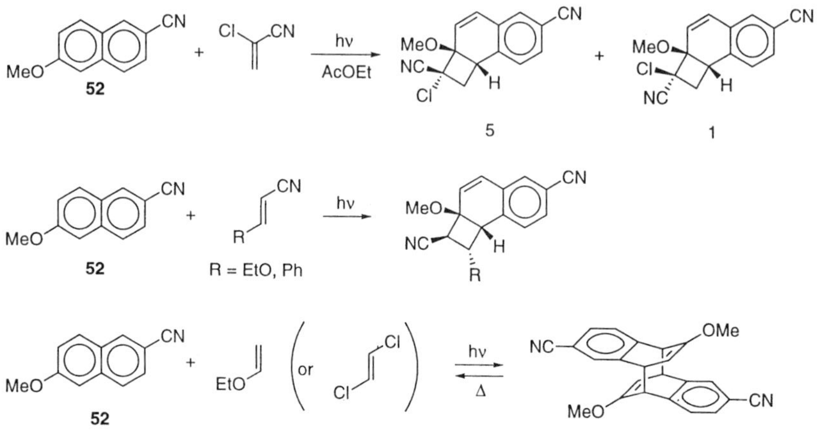

Scheme 21

thalene at 77 K in EPA (ether:pentane:alcohol -5:5 = 2). The triplet-sensitized photoreaction affords a caged compound (**51**) via the biplanemer **50**.

Al-Jalal reported that the photocycloaddition of 1-cyano-4-methoxynaph-thalene with alkenes commonly occurs at the 1,2 bonds of the naphthalene ring [126,131]. However, when 2-cyano-6-methoxynaphthalene (**52**) was treated with 2-chloroacrylonitrile, (*E*)-ethoxyacrylonitrile, or (*E*)-cinnamonitrile, the (2 + 2) cycloadducts showed a different regiochemistry [132,133] (Scheme 21). Irradiation of **52** in the presence of ethyl vinyl ether or (*E*)-dichloroethene did not give a cross-adduct, but gave a photodimer of **52**.

The photoreaction of 1-cyanonaphthalene with norbornadiene in benzene yields four kinds of (2 + 2) cycloadducts [134]. The addition occurs exclusively on the *exo* face of norbornadiene. In more polar solvents, the reaction becomes more selective, favoring the *exo* face to the naphthalene ring at the 1,2-position.

Photoexcited 1-cyanonaphthalene adds to 2-morphorinoacrylonitrile in (2 + 2) mode both at 1,2- and 7,8-positions. Ketone **55** is produced by hydrolysis of the initial cycloadduct **54** [135] (Scheme 22).

The photoreactions of mono- and di-cyanonaphthalene derivatives with al-lyltrimethylsilane have been reported by Mizuno and Albini, independently [136,428,433,436]. The product ratios of photocycloadducts via exciplexes and the allylated products via electron transfer were dependent on the solvent polarity and additives. The intramolecular photocycloaddition of the reductive allylated products efficiently gives tricyclic compounds in high yields [428,433,436] (Scheme 23).

Albini et al. investigated the photocycloadditions of 1- and 2-cyanonaph-thalenes with 1,3-dienes [137]. Irradiation of 1-cyanonaphthalene in the presence of 2,3-dimethyl-1,3-butadiene in cyclohexane gave a complex mixture containing

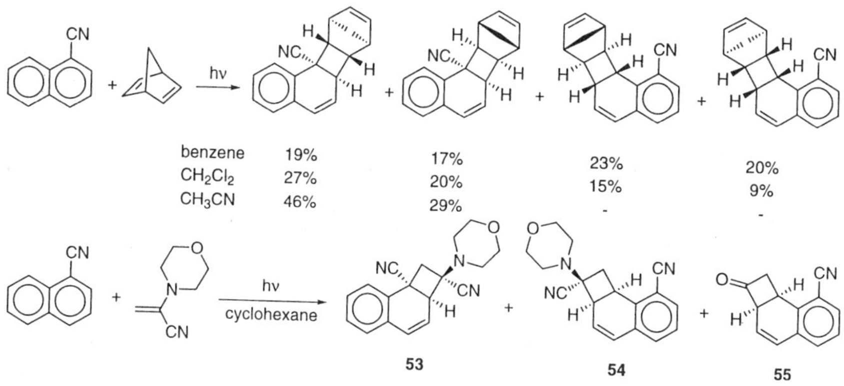

Scheme 22

$(2 + 2)$ adducts (**56, 57**), $(4 + 4)$ adducts (**58, 59**), and an eight-membered product (**61**) and its dimers (**62, 63**) (Scheme 24). The product **61** was formed by ring expansion of the cycloadduct **60** at the 1,8a-position of naphthalene. Irradiation of 1-cyanonaphthalene with cyclohexa-1,3-diene leads to 1 : 1 adducts (**64, 65**) with excellent *endo* selectivity. Another product obtained was a $(4 + 4)$ cycloadduct (**66**). From the reaction with 2,5-dimethylhexa-2,4-diene, a $(2 + 2)$ cycloadduct

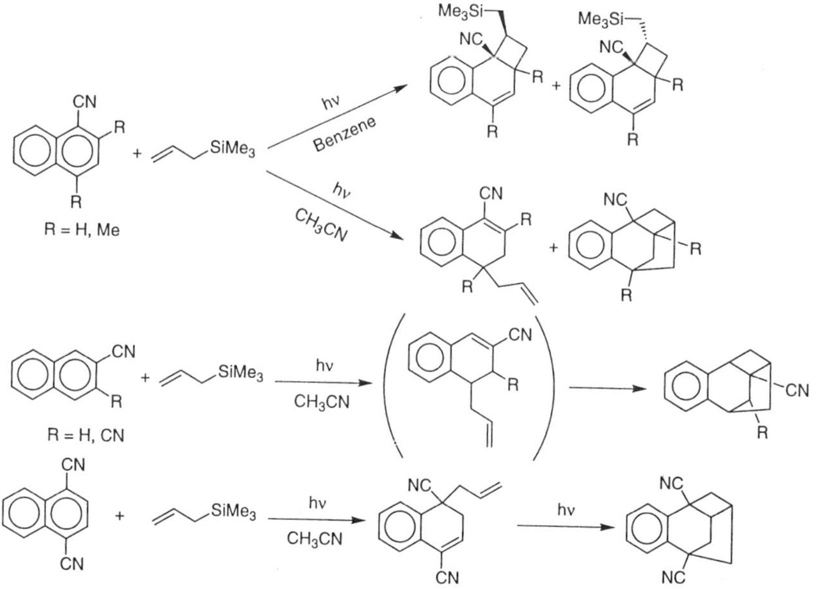

Scheme 23

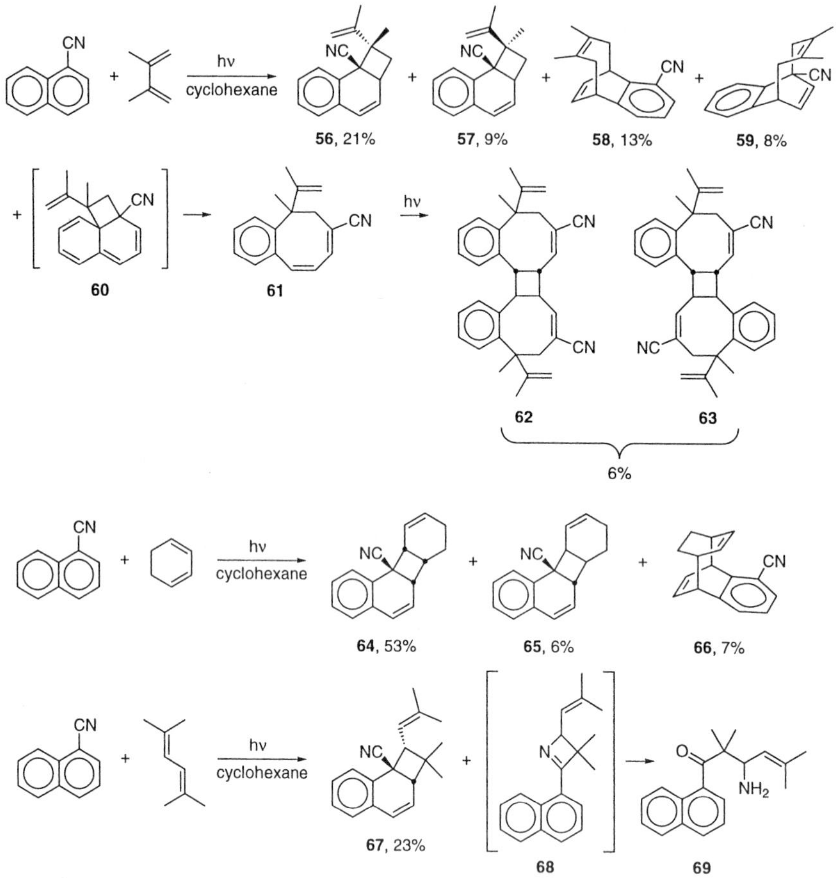

Scheme 24

(**67**) and an aminoketone (**69**) were obtained. Product **69** was obtained from the hydrolysis of the azetine (**68**).

Irradiation of 2-cyanonaphthalene in the presence of 2,3-dimethyl-1,3-buta-diene yielded (4 + 4) cycloadducts (**70, 71**) onto the naphthalene ring (Scheme 25). The main product from the reaction of 2-cyanonaphthalene with cyclohexa-1,3-diene is the (2 + 2) adduct (**72**), whereas a minor product is the (4 + 4) adduct (**73**). The reaction with 2,5-dimethyl-2,4-hexadiene leads again to a (2 + 2) cycloadduct (**74**) and aminoketone (**75**).

To prevent secondary processes due to facile thermal reaction, low-temper-ature irradiations of cyanonaphthalenes with cyclohexa-1,3-diene through a Pyrex filter were conducted by Noh et al. [138]. A dichloromethane solution of 1-

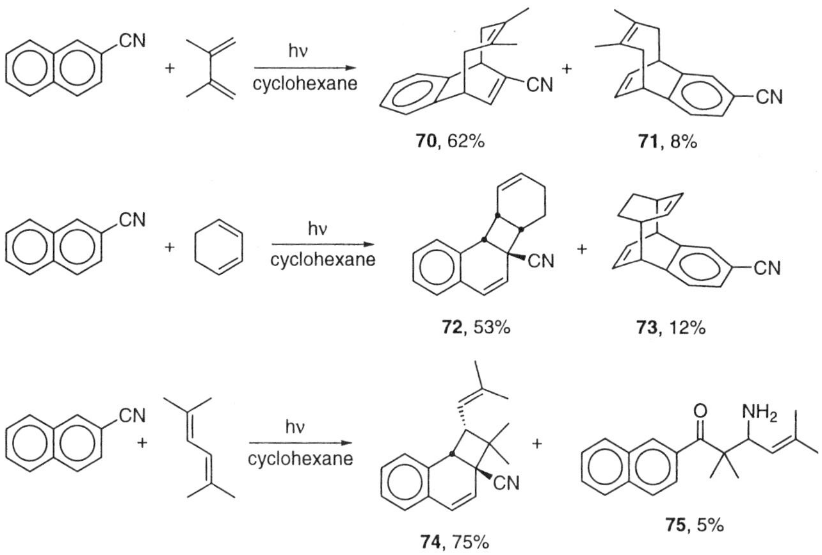

Scheme 25

cyanonaphthalene and cyclohexa-1,3-diene at $-78°C$ was irradiated for 4 hr, and two (2 + 2) adducts (**76, 77**) were isolated in 13% and 58% yields, respectively (Scheme 26). Similar irradiation of 2-cyanonaphthalene and cyclohexa-1,3-diene for 5 hr gave a complex mixture, containing (2 + 2) and (4 + 4) 1 : 1 adducts (**78, 79**) and a 1 : 2 adduct (**80**) in 16%, 52%, and 8% yields, respectively. The low-temperature ¹H-NMR (nuclear magnetic resonance) study and xanthone-sensitized photoreaction suggested that the primary major products are the *exo*-(4 + 4) adducts and *syn*-(2 + 2) adducts.

By the low-temperature irradiation experiments, Noh and Kim reinvestigated the photocycloaddition of furan to 1-cyanonaphthalene [139], which was first reported by Pac et al. [100,101]. Irradiation of a mixture through Pyrex filter at $-78°C$ yielded exclusively *endo*-(4 + 4) cycloadduct (*endo*-**26**) with a small amount of *syn*-(2 + 2) cycloadduct. Formation of **81** was proposed by Cope rearrangement of *exo*-(4 + 4) cycloadduct (*exo*-**26**)

Mizuno investigated the diastereoselectivity (*de*) in the reaction of (*l*)-menthyl 2-naphthoate with furan [140]. The *de* of the caged products **83** was highest in the reaction in pentane solution at ambient temperature (Scheme 27). The *de* was decreased with increasing polarity of solvents or by use of benzene or toluene. The *de* increased with decreasing the concentration of furan, or lowering the reaction temperature. Furthermore, the photoreaction of 1-phenylethyl ester (**84**) with furan gave up to 30% *de*.

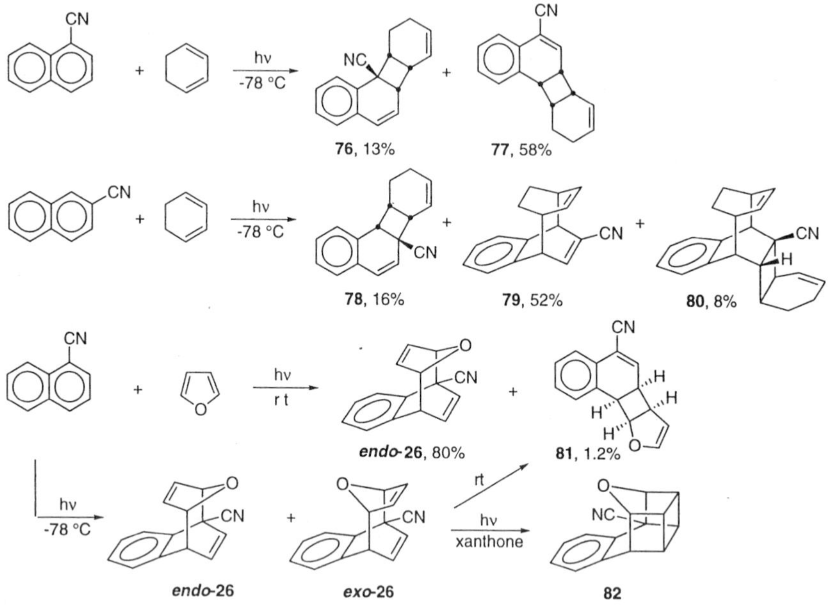

Scheme 26

Chow et al. developed the photocycloaddition of acetylacetone (**87**) with naphthols (**86a, 88a**) and their methyl ethers (**86b, 88b**) [141] (Scheme 28). The cyclization occurs regiospecifically to give diketones with low quantum efficiency. Whereas the excitation of the naphthol component does not take place during photoreaction, the excitation of the acetylacetone component leads to the formation of photocycloadducts.

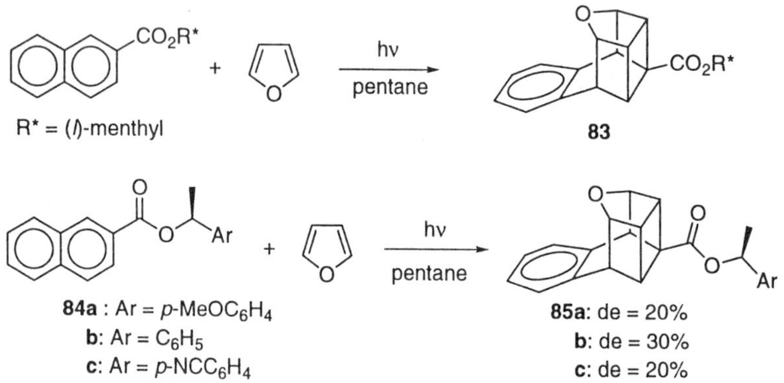

Scheme 27

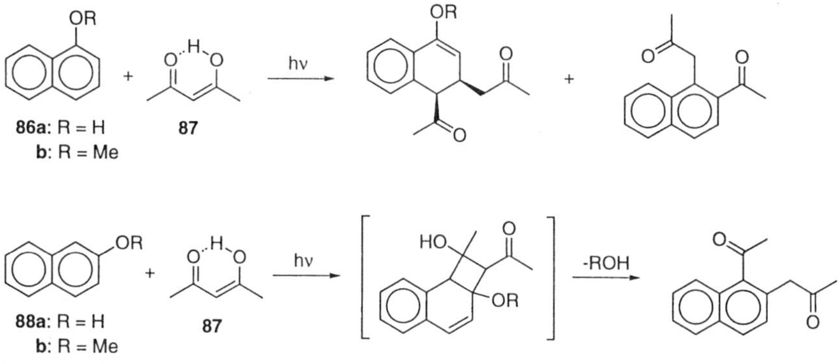

Scheme 28

In the case of methyl 1-naphthoate, the addition of **87** occurred at the 7,8-position of naphthoate to give **89** via **88,** which undergoes intramolecular cycloaddition involving three π-bonds (two C=O groups and one C=C group) to give **90** [142,143] (Scheme 29). The final rearrangement is probably initiated from the lowest triplet state of styrene moiety in **89** by a radical mechanism. Irradiation of methyl 2-naphthoate in the presence of **87** arose during photocycloaddition to the 1,2-position of the naphthyl ring to give **91** and **92** in ~1 : 1 ratio. The excited state of **92** should be able to undergo an oxa-di-π-methane rearrangement or a similar pathway to form a strained oxetane intermediate **93** and a subsequent ring opening to give **94.** Conversion of a nonemissive exciplex to **91** and **92** is enhanced by the addition of sulfuric acid [144].

In the photoreaction of cyanonaphthalenes, the regiochemistry was the same with the methyl naphthoate, but the final products were different [145].

Although photodimerization of 2-alkoxynaphthalenes [146–151] or 2-cyanonaphthalenes [152–154] was well described, unprecedented photodimerization of 1-substituted naphthalenes was recently reported by Noh et al. [155]. Upon irradiation of 1-cyano- or 1-methoxy-naphthalene at −78°C, *syn*-(2 + 2) cyclodimer (**102**) was isolated; however, no cyclodimer was found in the irradiation of 1-methyl-, 1-methoxy-, or unsubstituted naphthalene (Scheme 30). By the xanthone-sensitized photoreaction and low-temperature NMR study, **103** was found to be produced through Cope rearrangement of *endo*-(4 + 4) cyclodimer (**101**) in the cases of 1-cyano- and 1-methoxy-naphthalene.

Albini et al. reported that the cross-dimerization of 9-cyanoanthracene with substituted naphthalenes and anthracenes takes place through a significantly polar intermediates and is affected by solvents and substituents [156]. The combination of 9-cyanoanthracene and unsubstituted anthracene in CH_3CN gave the best result (Scheme 31). Photoirradiation of 9-cyanoanthracene with 1-methylnaphthalene,

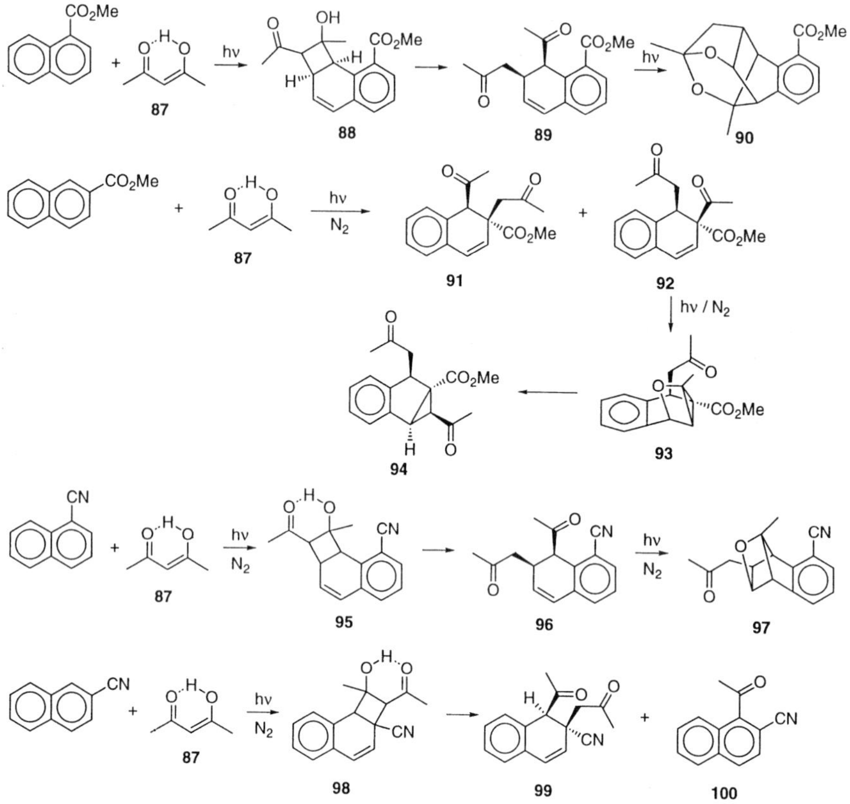

Scheme 29

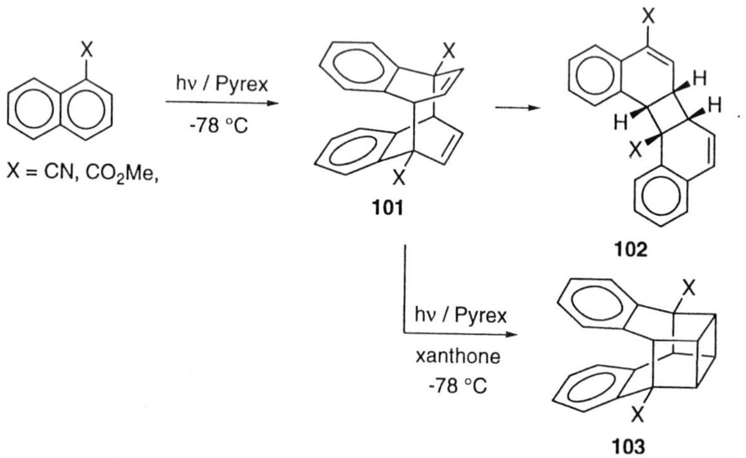

X = CN, CO₂Me,

Scheme 30

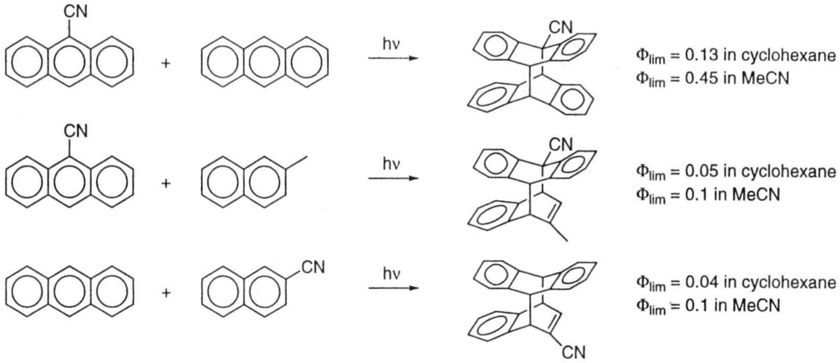

$\Phi_{lim} = 0.13$ in cyclohexane
$\Phi_{lim} = 0.45$ in MeCN

$\Phi_{lim} = 0.05$ in cyclohexane
$\Phi_{lim} = 0.1$ in MeCN

$\Phi_{lim} = 0.04$ in cyclohexane
$\Phi_{lim} = 0.1$ in MeCN

Scheme 31

2-methoxynaphthalene, and 2-cyanonaphthalene gave only the homodimer of 9-cyanoanthracene. The cycloreversion process from the homodimer and heterodimer of anthracenes and naphthalenes is essentially adiabatic [157].

Throughout the irradiation of **104** with >440-nm light in benzene, (4 + 2), (4 + 4), and (6 + 6) photodimerized products **105–108** are formed in a constant ratio of about 20 : 2 : 2 : 1 [158] (Scheme 32). The (4 + 2) cycloadduct **110** was obtained by the photochemical Diels–Alder-type dimerization of 9-(phenylethynyl)anthracene **109** [159]. The photochemical (4 + 4) dimerization of **111** does not proceed, but three types of dimerized product do, whereas the photoreaction of 9-vinylanthracene gave the (4 + 4) cycloadduct as the major product [160].

Photodimerizations of 9-substituted anthracenes in solution are efficient, usually giving rise to head-to-tail dimers rather than head-to-head dimers. In an attempt to conduct head-to-head photodimerization by using noncovalent bonding, diamines are used as a linker molecule to connect two acid molecules in solid-state photoreaction [161,162]. Double salt **112b** gave a small amount of the head-to-tail dimer **113,** but no head-to-head dimer **114** (Scheme 33). In contrast, the photoreaction of double salt **112a** proceeded efficiently, producing the head-to-head dimer **114** as a sole product. Double salt **112c** afforded a similar result, but the yield for **114** was considerably low and a small amount of **113** was detected as a by-product.

The photocycloaddition of 9-substituted anthracenes incorporated within Nafion membranes was investigated by Tung and Guan [163]. Whereas irradiation of **115a–e** in homogeneous solutions mainly gave rise to their head-to-tail photocyclodimers, photoirradiation of **115a–e** incorporated within Nafion membranes almost exclusively resulted in the head-to-head photocyclodimers (Scheme 34). Furthermore, the head-to-head photocyclodimers are significantly more stable in Nafion membranes than in the homogeneous solutions. These observations

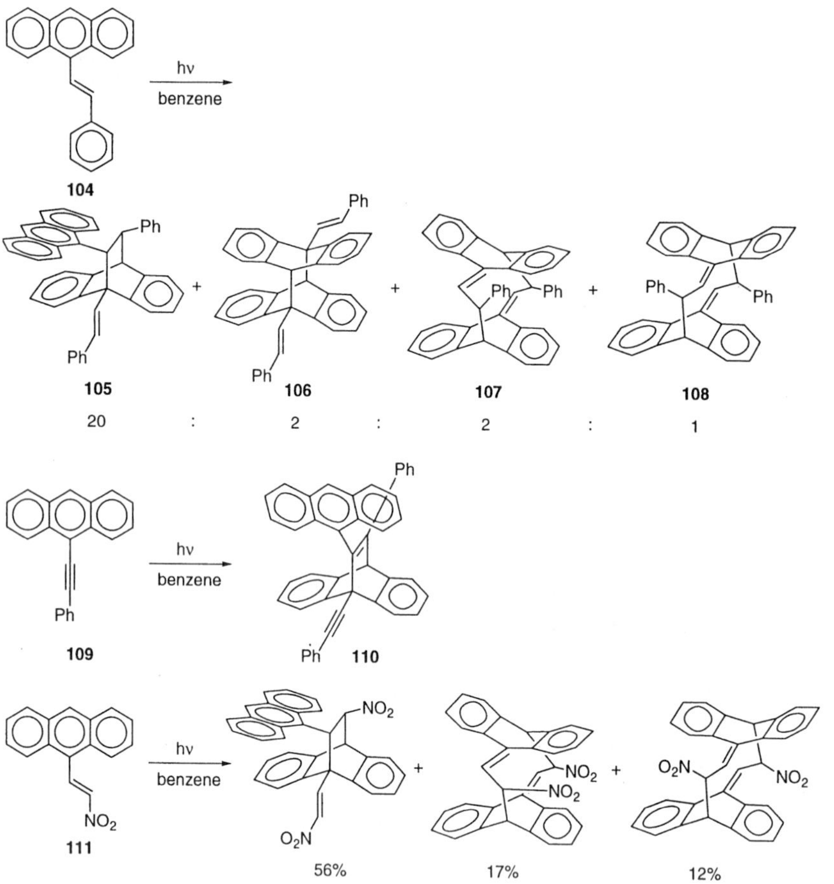

Scheme 32

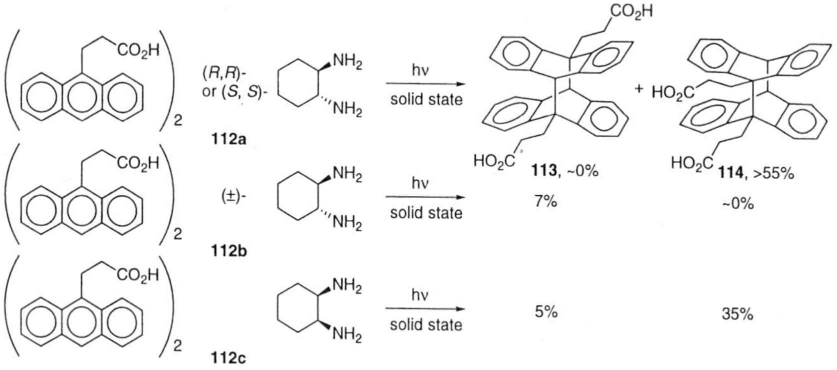

Scheme 33

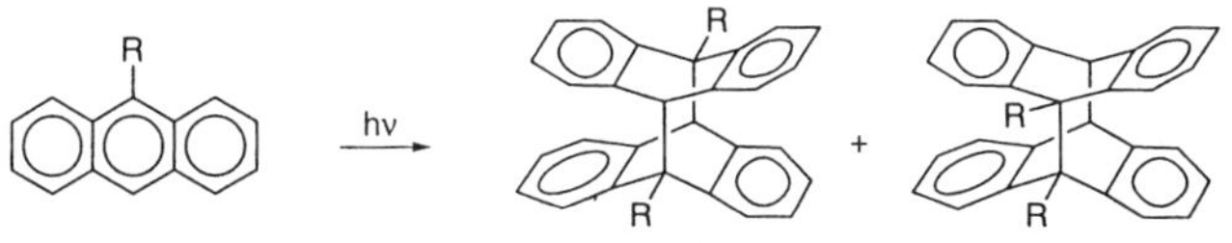

	(h-t)		(h-h)	
115a: R = CH$_2$N$^+$(CH$_3$)$_3$Br$^-$	90	:	10	in Et$_2$O
	0	:	100	in Nafion membrane
b: R = CH$_2$COO$^-$Na$^+$	83	:	17	in Et$_2$O
	7	:	93	in Nafion membrane
c: R = CH$_2$OH	84	:	16	in Et$_2$O
	6	:	94	in Nafion membrane
d: R = COCH$_3$	100	:	0	in Et$_2$O
	6	:	94	in Nafion membrane
e: R = Me	93	:	7	in Et$_2$O
	92	:	8	in Nafion membrane

Scheme 34

can be explained in terms of the preorientation of the substrate molecules in the inverse micellelike clusters of Nafion.

Noh and Lim reported asymmetrical photodimerization of methyl anthroate [164]. Irradiation of a diethyl ether solution of methyl anthroate through a uranium glass filter ($\lambda > 330$ nm) gave the 1,4-10',9' and 1,4-9',10' cyclodimers **117** and **118** as well as the normal 9,10-10',9' cyclodimer **116** (Scheme 35). This is the first example of intermolecular photodimerization involving the 1,4-9',10' positions of *meso*-substituted anthracenes.

Anthracene–benzene biplanemers (**119**) were prepared through the electrolytic oxidative decarboxylation of (4 | 4) photocycloadducts between substituted anthracenes and 1,2-dihydrophthalic anhydrides [165,166] (Scheme 36). Chemiluminescence was observed for **119** only in the solid state at > 120°C, but not in a liquid phase, whereas **119** (X = F or Cl, Y = Z = H) were not chemiluminescent [167]. Efficient chemiluminescence was observed in the photochlorereversion of all biplanemers tested in both phases.

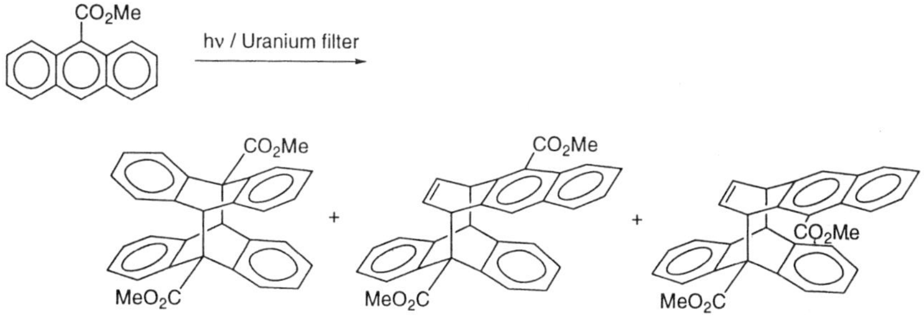

116, 83.6%   **117**, 8.2%   **118**, 2.1%

Scheme 35

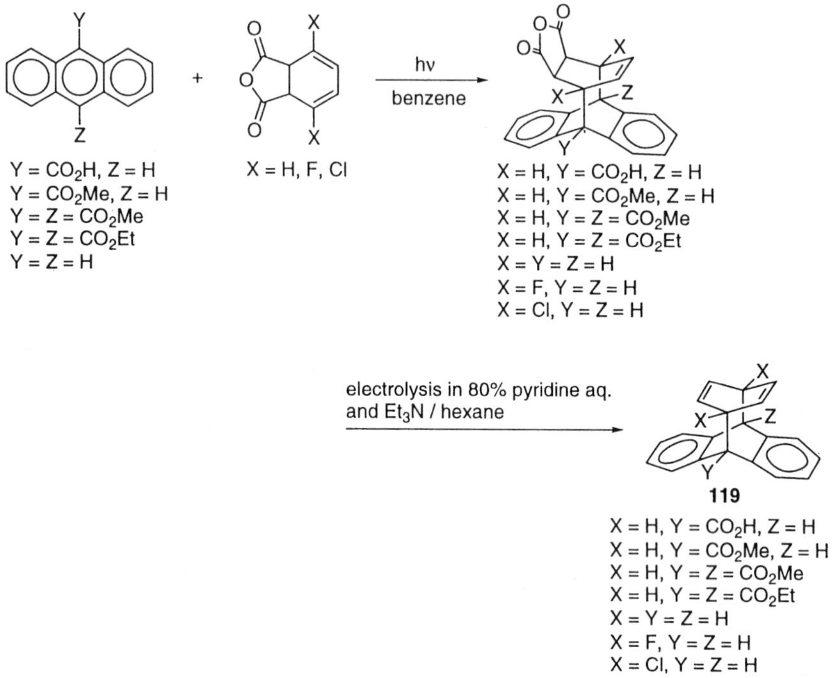

Y = CO$_2$H, Z = H
Y = CO$_2$Me, Z = H
Y = Z = CO$_2$Me
Y = Z = CO$_2$Et
Y = Z = H

X = H, F, Cl

X = H, Y = CO$_2$H, Z = H
X = H, Y = CO$_2$Me, Z = H
X = H, Y = Z = CO$_2$Me
X = H, Y = Z = CO$_2$Et
X = Y = Z = H
X = F, Y = Z = H
X = Cl, Y = Z = H

electrolysis in 80% pyridine aq.
and Et$_3$N / hexane

119

X = H, Y = CO$_2$H, Z = H
X = H, Y = CO$_2$Me, Z = H
X = H, Y = Z = CO$_2$Me
X = H, Y = Z = CO$_2$Et
X = Y = Z = H
X = F, Y = Z = H
X = Cl, Y = Z = H

Scheme 36

Photocycloaddition of anthracenes to 1,3-dienes gives (4 + 4), (4 + 2), and (2 + 2) photocycloadducts, extensively studied by Yang and colleagues [168–179], Kaupp and colleagues [180–187], and others [188,189] mostly in the 1970s. Irradiation of 9,10-dichloroanthracene in the presence of 2,5-dimethyl-2,4-hexadiene in benzene at 25°C gives a single (4 + 2) adduct (**123**) [168,172] (Scheme 37). The singlet and triplet biradical intermediates **121** and **122** from the singlet exciplex **120** are involved for the formation of **123** from the fluorescence quenching experiments, the transient absorption measurements, and some kinetic treatments [190,191].

Farid et al. reported the formation of two types of (4 + 2) photocycloadduct of 9,10-dicyanoanthracene (DCA) with 3-carbomethoxy-1,2-diphenylpropene [192,193]. The product ratio depends on the solvent polarity. In benzene, *exo*-**125** is selectively obtained via exciplex (Scheme 38). In acetonitrile, *endo*-**125** is obtained as a sole product via the radical ion pair. Photochemical reactions of DCA with 1,2-diarylcyclopropanes gave (4 + 3) cycloadducts [194,195]. In degassed acetonitrile solution, (4 + 3) photocycloaddition occurred to give *cis* and *trans* cycloadducts in a 3 : 1 ratio in good chemical yields, although the quantum yields

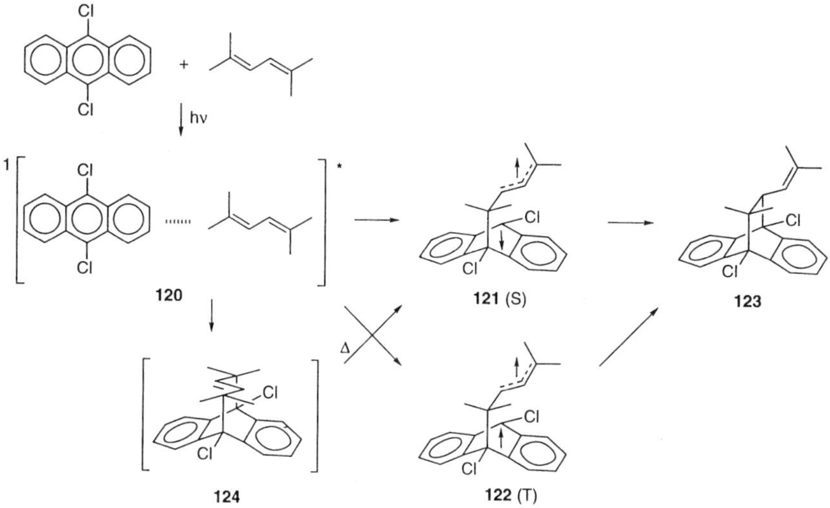

Scheme 37

were not high (Φ = 0.002–0.01). This photocycloaddition did not occur in benzene.

Intermolecular ($2\pi + 2\pi$) photocycloaddition to the 9,10-position of the phenanthrene ring has been extensively studied in the last three decades. Unsubstituted phenanthrene reacts with electron-poor olefins such as dichloroethene,

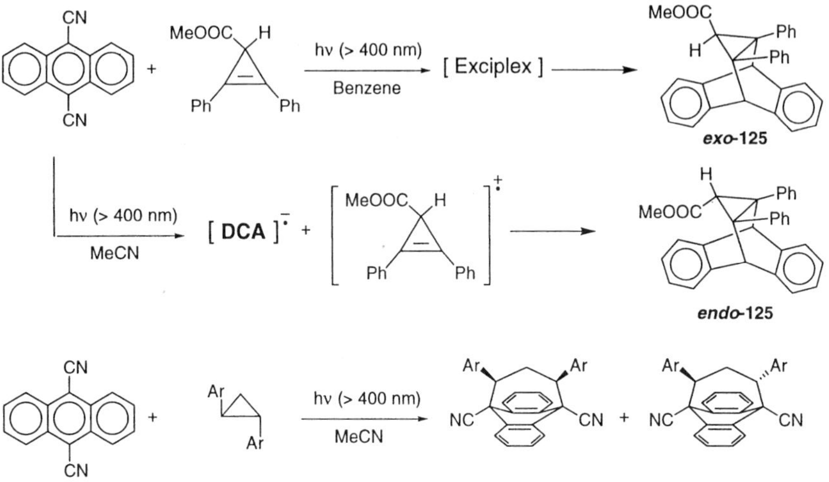

Scheme 38

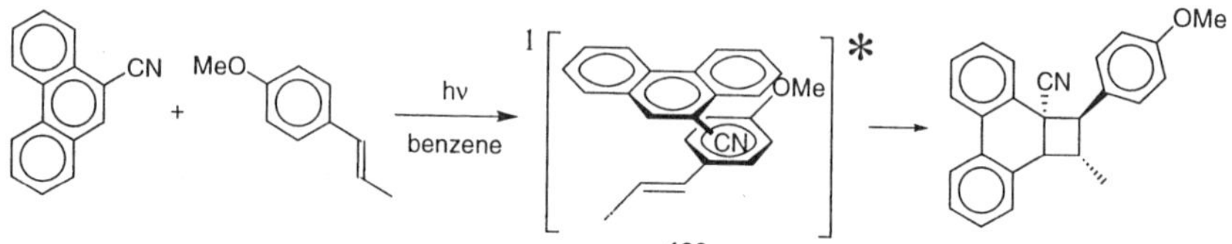

Scheme 39

dimethyl maleate, and fumarate at the 9,10-position of phenanthrene [196–200]. The reaction of 9-cyanophenanthrene with vinyl ethers proceeds in a regioselective and stereoselective manner [201,202] Caldwell showed that in the 9-cyanophenanthrene–anethole system, the emitting singlet exciplex 126 is the precursor of cycloadducts 127 [203–205] (Scheme 39). Caldwell's and Ford's groups investigated in detail the intermediacy and behavior of singlet and triplet exciplexes in photoreactions of phenanthrene with various alkenes [11,206–211].

Kubo et al. reported that the irradiation (>400 nm) of N-methyl-9,10-phenanthrenedicarboximide and allyltrimethylsilane in benzene gave two isomeric cyclobutanes 128 and 129, which were derived from the triplet state of the phenanthrenedicarboximide [212] (Scheme 40). On the contrary, photoreaction in acetonitrile afforded allylated products.

Lewis et al. examined the photocycloaddition of 9-methoxycarbonylphenanthrene with 2,3-dimethyl-2-butene in the absence or presence of a Lewis acid [213]. The conformations of the free and complexed molecules in the ground states have been investigated by means of NMR and Gaussian 88 calculations. In the absence of a Lewis acid, a second unidentified adduct and the photodimer are observed as well as the cycloadduct at the 9,10-position (Scheme 41). In the presence of a Lewis acid, neither the second adduct nor the photodimer is detected, and the yield of the photocycloadduct is inferior to that obtained in the absence of Lewis acid. From the fluorescence quenching experiments, it seems likely that the photocycloaddition occurs via a triplet mechanism, whereas the photocycloaddition of their Lewis acid complexes occurs via a singlet mechanism (Table 2).

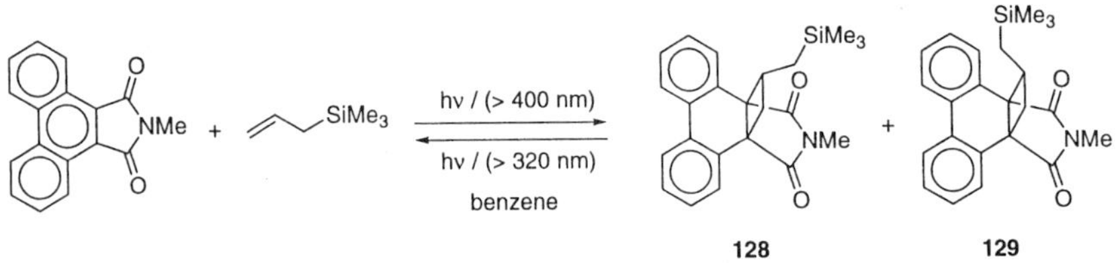

Scheme 40

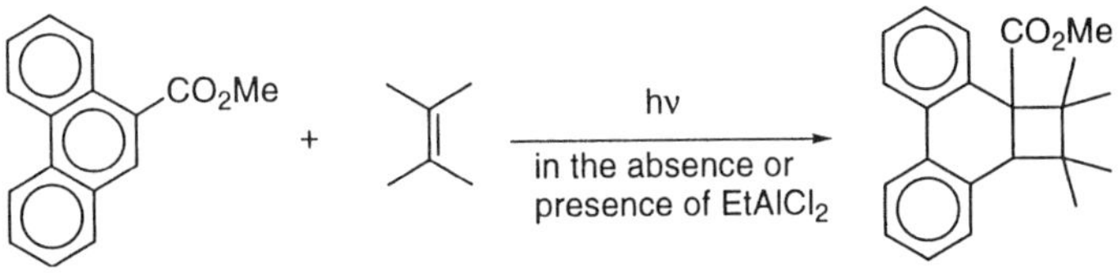

Scheme 41

Mizuno et al. reported stereospecific photocycloaddition of electron-defi-cient arylalkenes to 9-cyanophenanthrene via an exciplex [214]. Irradiation of a benzene solution containing 9-cyanophenanthrene and methyl *trans*-cinnamate gave exclusively the $(2\pi + 2\pi)$ photocycloadduct **E-130a** (Scheme 42). Pro-longed irradiation gave a mixture of **E-130a** along with a small amount of the dimer of 9-cyanophenanthrene. Similar irradiation of methyl *cis*-cinnamate af-forded stereospecifically **Z-130a** in an initial stage of the reaction. The photocy-cloaddition of *E*- and *Z*-cinnamonitriles with 9-cyanophenanthrene gave **E-130b** and **Z-130b** in a stereospecific manner. In all the adducts obtained, the phenyl group lies always at the *endo* position. The photochemical and thermal cyclore-version of **E-130a–b** and **Z-130a–b** predominantly afforded *trans*- and *cis*-alkenes, respectively, with the formation of 9-cyanophenanthrene. Moreover, the product ratio of **E-130a** to **Z-130a** was enhanced by the addition of Lewis acids such as $BF_3 \cdot Et_2O$ and $AlCl_3$.

Irradiation of lactone **131** with (*E*)-anethole in benzene gave cyclobutane **132** together with *p*-methoxybenzaldehyde [215] (Scheme 43). Similar irradiation of lactone **133** gave both cyclobutane **134** and an olefin **136** derived from an oxe-tane precursor **135**.

Photoirradiation of phenanthrene with acetylacetonatoboron difluoride **137** in ether or dioxane irreversibly forms a fluorescent exciplex which is the precur-sor to the product *cis*-**138** [216] (Scheme 44). Trace amount of *trans*-**138** was also obtained. In contrast, the excitation of **137** in the presence of phenanthrene gives neither exciplex emission nor any products.

Table 2 Fluorescence Quenching 9-Substituted Phenanthrenes by 2,3-Dimethyl-2-Butene

9-Substituent	$k_q\tau$ (M^{-1})	$10^{-8}k_q$ (M^{-1}/sec)
CO_2Me	0.1	0.1
$CO_2Me \cdot EtAlCl_2$	39	57.0
$CONH_2$	0.7	0.6
$CONH_2 \cdot BF_3$	70	74.0

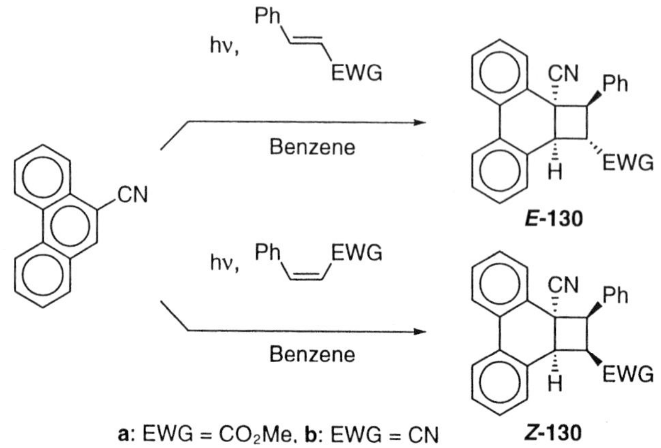

a: EWG = CO$_2$Me, **b**: EWG = CN

Scheme 42

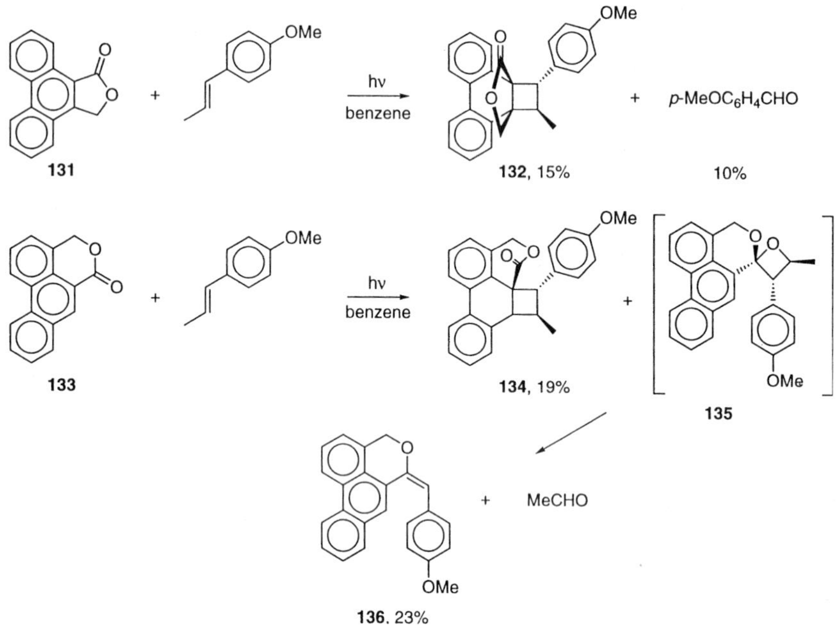

131

132, 15%

10%

p-MeOC$_6$H$_4$CHO

133

134, 19%

135

136. 23%

+ MeCHO

Scheme 43

137

cis-138

trans-138

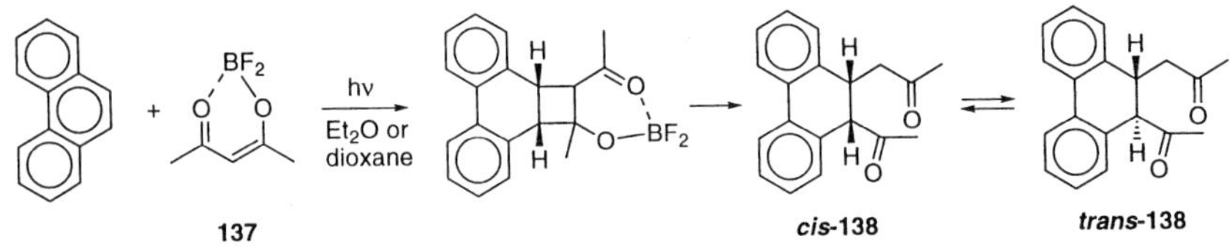

Scheme 44

156

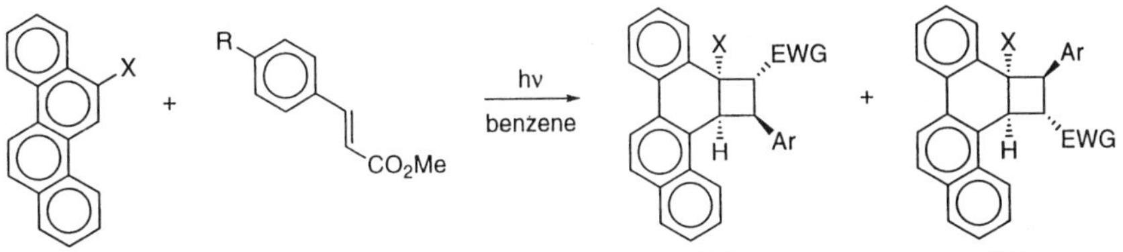

Scheme 45

X = H	R = H	139	140
	R = H	46%	10%
	R = MeO	60%	13%
	R = CN	<1%	<1%
	R = Ph	24%	10%
X = CN	R = MeO	53%	5%

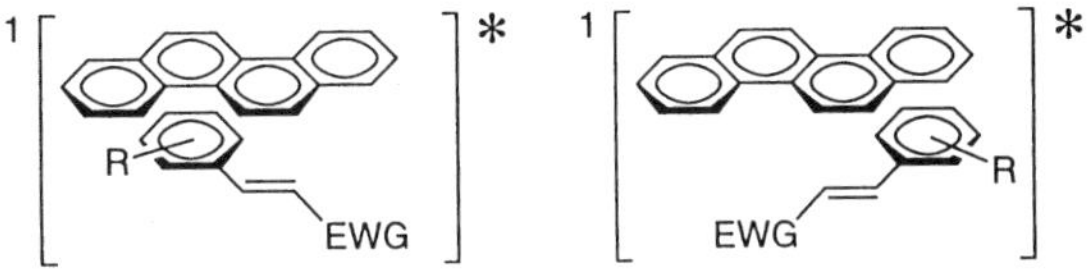

Photocycloaddition of electron-deficient arylalkenes to chrysene stereose-lectively occurred at the 5,6-position of the chrysene ring to give two ($2\pi + 2\pi$) photocycloadducts [217]. Irradiation of a benzene solution containing chrysene and methyl *trans*-cinnamate gave two regioisomeric ($2\pi + 2\pi$) photocy-cloadducts (**139** and **140**) in a stereoselective manner. A methoxy group on *para* position of methyl *trans*-cinnamate accelerated the reaction; however, introduc-tion of *p*-cyano group caused the almost recovery of chrysene. Reaction with methyl *p*-phenylcinnamate gave poor selectivity of **139/140**. The photoreaction of 6-cyanochrysene with methyl *p*-methoxycinnamate proceeded in >10 : 1 selec-tivity. Regioselectivity and stereoselectivity were explained by two possible struc-tures of sandwich-type singlet exciplexes **141** and **142** (Scheme 45). Aromatic groups of the electron-deficient arylalkenes might favor overlapping on the naph-thalene ring than on the benzene ring. The highly regioselective photocycloaddi-tion of 6-cyanochrysene to methyl *p*-methoxycinnamate have been achieved by the combination of π–π overlapping and favorable dipole moment in the exciplex state.

($2\pi + 2\pi$) Photocycloaddition of cyclohexa-1,3-diene to pyrene proceeds via triplet pyrene to give cycloadducts [218]. Photoreaction of pyrene with elec-tron-deficient arylalkenes afforded 1 : 1 and 1 : 2 adducts in high yields in a stere-ospecific and *endo* selective manner [219,220]. For prolonged irradiation, a 1 : 2 cycloadduct **147** was precipitated. A phenyl group lies at the *endo* position in the

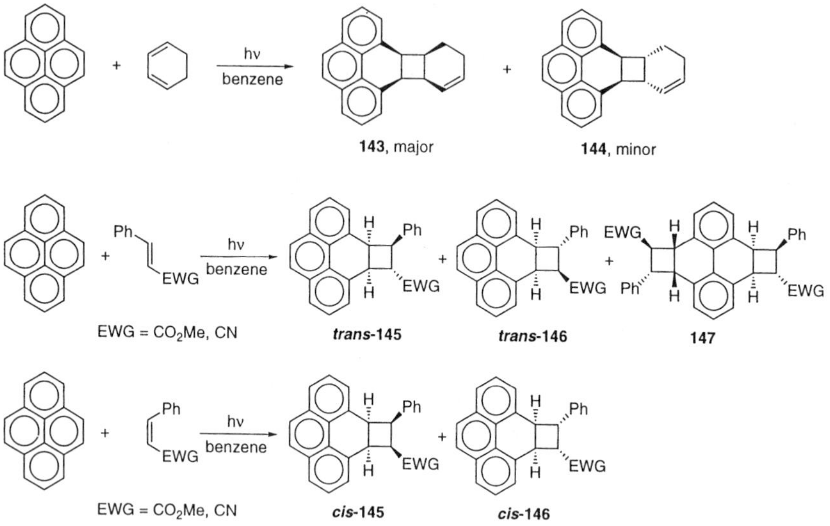

143, major   **144**, minor

EWG = CO₂Me, CN   *trans*-145   *trans*-146   147

EWG = CO₂Me, CN   *cis*-145   *cis*-146

Scheme 46

major 1 : 1 adducts (**trans-145** and **cis-145**) and 1 : 2 adduct (**147**) (Scheme 46). Electron-deficient alkenes having no aryl substituent such as acrylonitrile and methyl acrylate and electron-donating alkenes such as 2,3-dimethyl-2-butene and ethyl vinyl ether did not add to pyrene under the same reaction conditions. Sandwich-type singlet exciplexes between pyrene and arylalkenes are proposed as reactive intermediates.

Photocycloadditions of more condensed aromatic hydrocarbons such as benz[*a*]anthracene, naphthacene, dibenz[*a,h*]anthracene, dibenz[*a,j*]anthracene, dibenz[*a,c*]anthracene, with 1,3-cyclohexadiene, gave (4 + 4), (4 + 2), and (2 + 2) cycloadducts [176,177,221,222] (Scheme 47).

D. Intermolecular Photocycloaddition Assisted by Hydrogen-Bonding

Photocycloaddition reactions have been used as key transformations in many organic syntheses to construct organic compounds having unique structures that are hardly accessible by other methods. However, their stereoselectivity is not necessarily high, and many efforts have been done to attain the highly regioselective and stereoselective photocycloadditions. They are discussed in terms of the electronic nature of substrates, the steric repulsions between substituents, and the conformational restrictions of intramolecular reactions.

Recently, the photocycloaddition assisted by hydrogen-bonding substituents such as hydroxyl and amino groups frequently interact with other functional groups or heteroatoms [223–225]. The hydrogen-bond-assisted photocy-

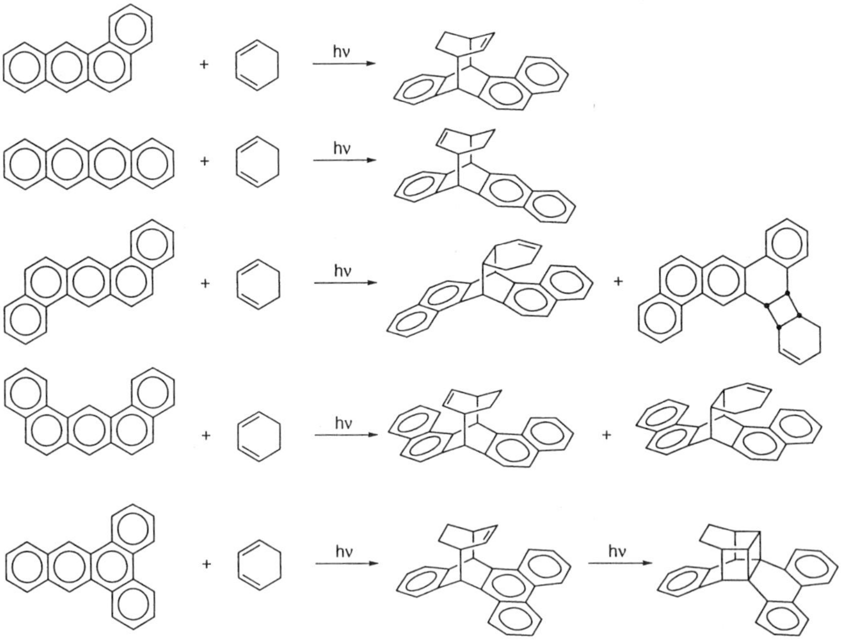

Scheme 47

cloaddition has been classified into two types: One is the hydrogen-bonding in the ground states followed by their excitation. The other one is no interaction in the ground states, but the hydrogen-bonding appears in the excited states. Each interaction should cause the highly stereoselective photocycloadditions.

Previously, a variety of (2 + 2) photocycloaddition of alkenes to naphthalene rings have been reported as discussed in the previous subsections. Very recently, Yokoyama and Mizuno have reported the highly regioselective and stereoselective (2 + 2) photocycloaddition between 4-substituted 1-cyanonaphathalenes (**148**) and 1-substituted 3-methyl-2-butenes (**149**) assisted by hydrogen-bonding in the ground states as shown in Scheme 48 [226].

Table 3 summarizes the substituent effect of the photocycloaddition of **148** to **149** in benzene solution. When X and Y can form a hydrogen bond, the *endo* isomers (**150**) were preferentially formed over the *exo* isomers (**151**) (entries 3–5). However, if the combination of X and Y was unable to form a hydrogen bond, no selectivity was observed (entries 1, 2, and 11). Solvent and temperature effects support the hydrogen-bonding in the photocycloaddition of **148c** to **149a**. At room temperature, photocycloaddition in benzene (entry 3) and in CH_2Cl_2 (entry 4) gave **150ca** preferentially over **151ca** in the ratio of 3 : 1. Use of more polar solvents such as CH_3CN and MeOH decreased the selectivity.

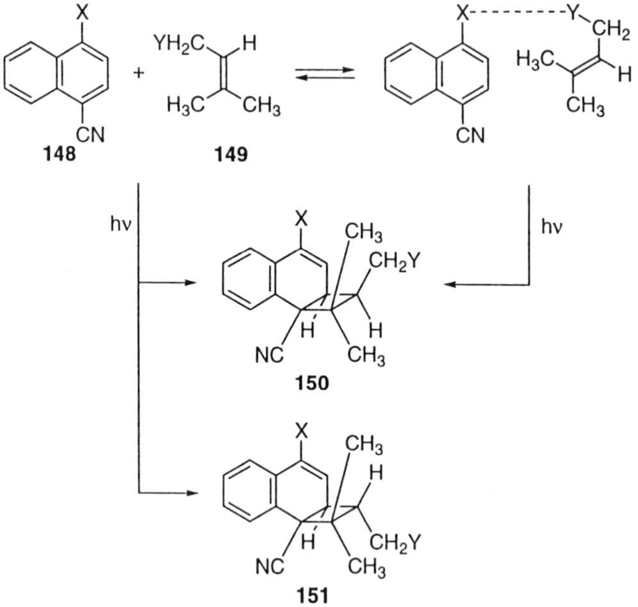

Scheme 48

Table 3 Regioselectivities on the Photocycloaddition of **148** and **149**

	Reactants				Cycloadducts			
Entry	–X	–Y	Solvent	Temp.	150 (*endo*)	151 (*exo*)	Yield	150/151
1	–H (**148a**)	–OH (**149a**)	Benzene	RT[a]	**150aa**	**151aa**	22	0.72
2	–CH$_3$ (**148b**)	149a	Benzene	RT	**150ba**	**151ba**	22	0.84
3	CH$_2$OH (**148c**)	149a	Benzene	RT	**150ca**	**151ca**	42	3.3
4	148c	149a	CH$_2$Cl$_2$	RT	**150ca**	**151ca**	33	3.4
5	148c	149a	CH$_2$Cl$_2$	–20°C	**150ca**	**151ca**	28	6.6
6	148c	149a	CH$_2$Cl$_2$	–60°C	**150ca**	**151ca**	73	13
7	148c	149a	CH$_3$CN	RT	**150ca**	**151ca**	39	1.6
8	148c	149a	CH$_3$OH	RT	**150ca**	**151ca**	28	1.0
9	–CH$_2$OCH$_3$ (**148d**)	149a	Benzene	RT	**150da**	**151da**	47	2.4
10	148c	–OCH$_3$ (**149b**)	Benzene	RT	**150cb**	**151cb**	22	1.7
11	148d	149b	Benzene	RT	**150db**	**151db**	18	0.95

[a] RT = room temperature.

The effect of temperature was examined in CH_2Cl_2. Lowering the reaction temperature to $-20°C$ increased the ratio of **150ca : 151ca** (entry 5), and irradiation at $-60°C$ gave a highly selective formation of *endo* isomer (**150ca : 151ca** = 13 : 1) in a high yield (73%). The contribution of the hydrogen bond to the *endo* selectivity is also confirmed by low-temperature ^{1}H-NMR (Fig. 1). At 20°C, mix-

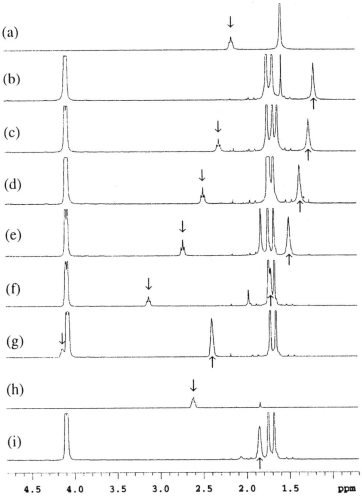

Figure 1 Variable-temperature ^{1}H-NMR spectra of **148c** and **149a** (300 MHz, CD_2Cl_2). ↓ and ↑ indicate the resonances of hydroxylic protons of **148c** and **149a,** respectively. (a) **148c** (27 mM) at $+20°C$; (b) **149a** (60 mM) at $+20°C$; (c–g) a mixture of **148c** (27 mM) and **149a** (48 mM) at $+20°C$ (c), 0°C (d), $-20°C$ (e), $-40°C$ (f), and $-60°C$ (g); (h) **148c** (27 mM) at $-60°C$; (i) **149a** (60 mM) at $-60°C$.

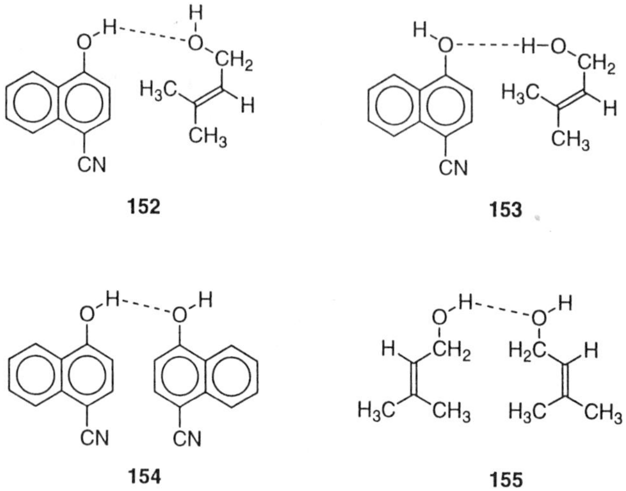

Scheme 49

ing of **148c** and **149a** led to downfield shifts of their hydroxylic proton resonances, indicating that the type of hydrogen bonds in **152** and **153** were favored over those in **154** and **155** (Scheme 49).

When the mixture is cooled, the hydroxylic proton resonances shift to a lower field, and the extent of their shifts at $-60°C$ is larger than those of individual solutions. This result suggests that lowering the temperature increased the dominance of **152** and **153** in the ground state and that photocycloadditions via such a hydrogen-bonded complex produce the *endo* adducts selectively.

Chiyonobu and Mizuno also found a specific example in the photocycloaddition to aromatic rings assisted by hydrogen-bonding in the excited states [227]. Irradiation of 1-cyanonaphthalene (1-CN) with furfuryl methyl ether in benzene gives two isomeric (4 + 4) cycloadducts in a 4 : 1 ratio [100,101]. However, similar photoreaction of 1-CN with furfuryl alcohol affords only one isomer exclusively. The low-temperature ^{1}H-NMR experiments in the latter case does not show any hydrogen-bonding interaction in the ground state compared to that in the 1-cyano-4-hydroxymethylnaphthalene with prenol system. On the other hand, the fluorescence quenching rate of 1-CN by furfuryl alcohol in cyclohexane is five times faster than that by furfuryl methyl ether. In addition, the exciplex emission between 1-CN and furfuryl alcohol is observed to be much stronger than that between 1-CN and furfuryl methyl ether (Scheme 50).

The photocycloaddition of furan derivatives to 2-cyanonaphthalene (2-CN) affords caged products, which was first reported by Pac et al. [100–105]. The product ratio in the photoreaction of 2-CN with 2-methylfuran is almost 1 : 1.

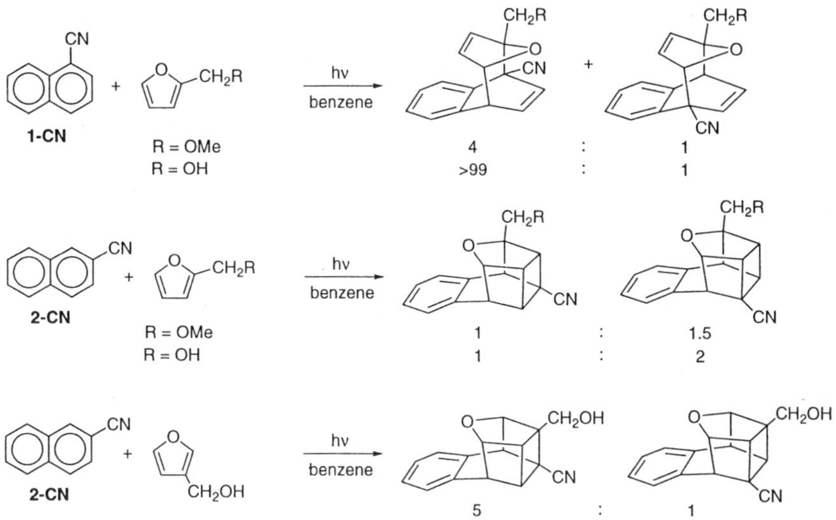

Scheme 50

Similar irradiation of 2-CN with furfuryl alcohol and furfuryl methyl ether gives the corresponding caged products in almost the same ratios (2 : 1). However, the photoreaction of 3-furylmethanol to 2-CN gives two isomeric cage products in a 5 : 1 ratio.

These results clearly show that the stereoselectivity in the photocycloaddition of 1-CN and 2-CN with 2- and/or 3-furylmethanol is reasonably explained by the hydrogen-bonding in the excited states. Probably, a partial electron transfer from furylmethanols (F-OH) to the excited singlet states of cyanonaphthalenes (ArCN) contributes the formation of polar exciplexes [ArCN$^{\delta-}$F–OH$^{\delta+}$], where ArCN$^{\delta-}$ interacts with more acidic OH group of F–OH$^{\delta-}$ to produce hydrogen bonds in the excited states.

E. Intramolecular Photocycloaddition to Benzene Rings

Intramolecular photocycloaddition of alkenes to benzene rings is a synthetically useful method for the construction of polycyclic compounds including natural products [23,25]. In 1969, Morrison and Ferree reported the first example of intramolecular 1,3-photocycloaddition of *cis*-6-phenylhex-2-ene (**156**) [228,229] (Scheme 51).

In general, intramolecular photocycloaddition between simple benzene and simple alkenes, in which any electron-releasing or electron-withdrawing groups are absent, proceeds at 1,3 (*meta*) positions [23]. The products obtained by the *meta* addition is most commonly divided into three types: 1,3-addition-2,4-clo-

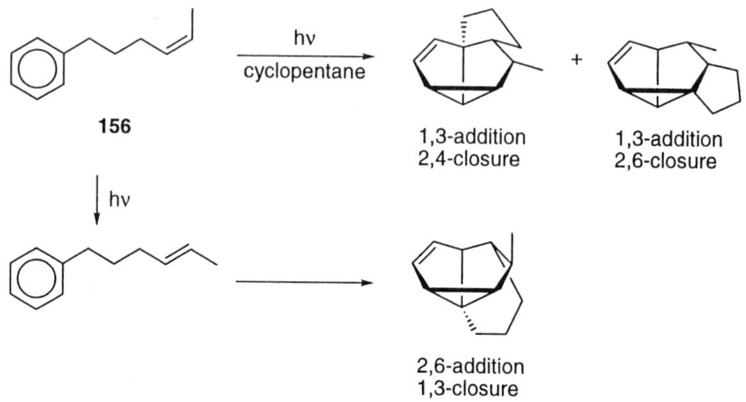

Scheme 51

sure, 1,3-addition-2,6-closure, and 2,6-addition-1,3-closure. Along with the substitution pattern, the number of isomers increases.

Gilbert et al. investigated intramolecular photocycloaddition of more than 20 kinds of 5-arylpent-1-ene substituted on the phenyl ring and/or chain. They concluded that *ortho*-, *meta*-, and *para*-methyl groups and *ortho*-methoxy group promote both efficiency and selectivity of *meta* cycloaddition, and *para*-methoxy, *para*-cyano, and *para*-acetyl groups inhibit the intramolecular cyclization itself [230,231]. As for the substitution of methylene chain, 2,3,5-trimethyl and 1,1,2-trimethyl substitution depress both efficiency and selectivity. These results are reasonably explained by the preferred conformations of the bichromophores. Cornelisse et al. reported that the efficiency of the intramolecular photocycloaddition depends on the substituted position of the methoxy group, where **157** and **158** give intramolecular 2,6 and 1,3 adducts, but **159** and **160** do not give any cycloadducts [232] (Scheme 52). They explained these results by the repulsive interaction between the *endo* oxygen atom and the negative charge which develops in the excited phenyl ring upon approach of the internal alkenes. In addition, the *para*-flu-

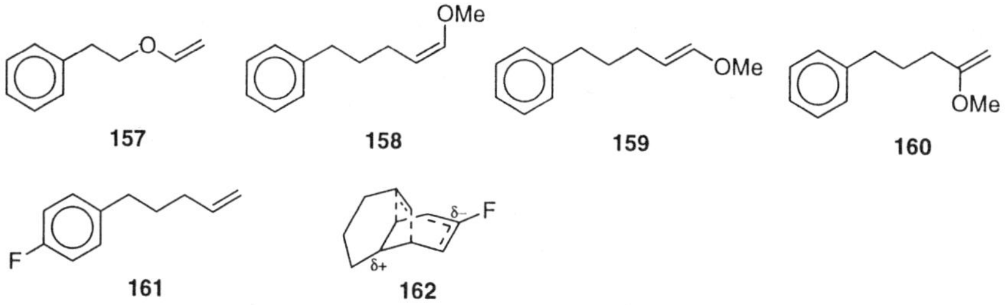

Scheme 52

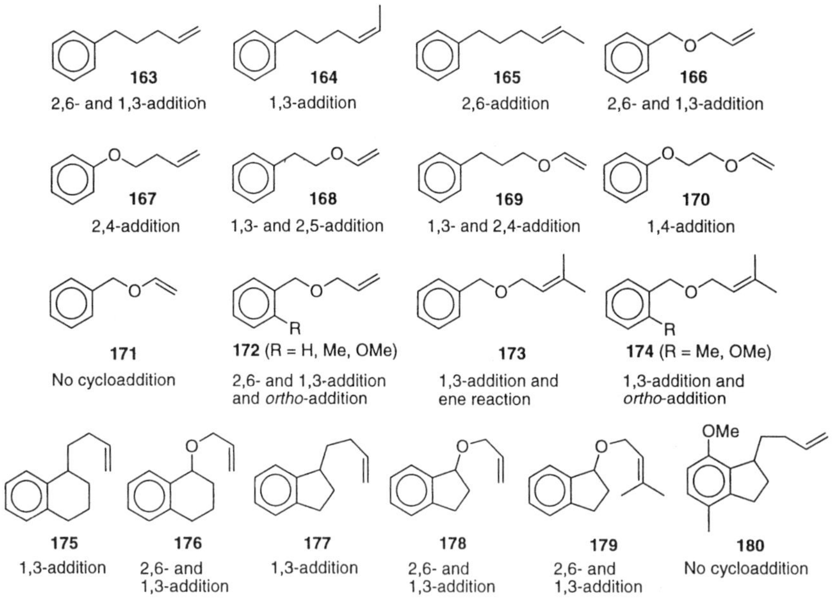

163
2,6- and 1,3-addition

164
1,3-addition

165
2,6-addition

166
2,6- and 1,3-addition

167
2,4-addition

168
1,3- and 2,5-addition

169
1,3- and 2,4-addition

170
1,4-addition

171
No cycloaddition

172 (R = H, Me, OMe)
2,6- and 1,3-addition
and *ortho*-addition

173
1,3-addition and
ene reaction

174 (R = Me, OMe)
1,3-addition and
ortho-addition

175
1,3-addition

176
2,6- and
1,3-addition

177
1,3-addition

178
2,6- and
1,3-addition

179
2,6- and
1,3-addition

180
No cycloaddition

Scheme 53

oro derivative (**16!**) gives selectively to the 2,6 adduct, due the stabilization of the zwitterionic intermediate (**162**) [233].

To clarify the effects of linking chains, Gilbert et al. investigated the photocycloaddition of **163–180**, which produce various types of adducts [231,234–237] (Scheme 53). In some cases, not only the *meta* addition but also *ortho* and *para* addition and ene reaction took place.* The observed regiochemistry was rationalized in terms of the stabilization of zwitterionic intermediates and the steric constraints.

Morrison reported that *trans*-decalin (**181**) is photochemically inert, but the *cis* isomer (**182**) undergoes a singlet-derived, intramolecular *meta* cycloaddition to give **183** (Scheme 54). Compound **183** efficiently photocycloeliminates to give a carbene-derived product **184** [238].

Intermolecular *meta* photocycloaddition of cycloalkenes to benzene rings occurs with marked *endo* stereoselectivity, although the *exo* process would be favored on purely steric grounds. Because of intramolecular anomalous stereoselectivity, the *exo* process is often feasible [55] (Scheme 55).

---

* In some cases, the formation of *para* adduct is doubtful. For example, Gilbert reported the formation of *para* adduct from **168** [234–236]; however, supplementary experiments done by Cornelisse concluded the observed product by Gilbert had been *meta* adduct [232].

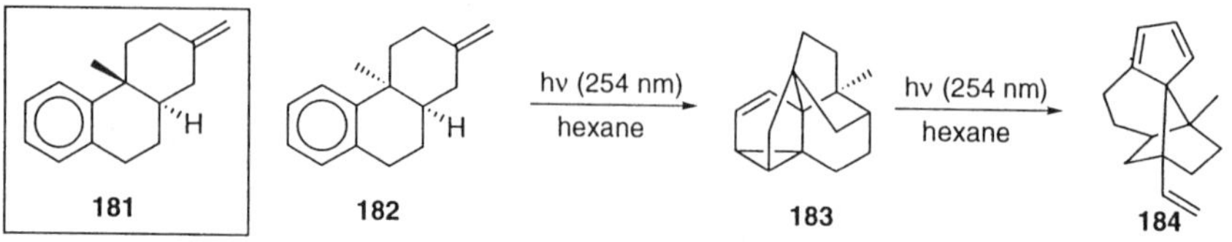

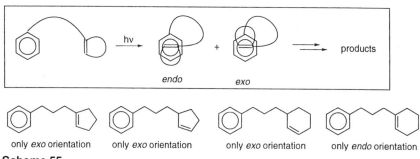

Scheme 54

Sugimura et al. investigated diastereoselectivity in the photocycloaddition of arene and alkene connecting by optically active 2,4-pentanediol [239]. High diastereofacial differentiation was observed at the addition step and sufficient regiocontrol of the subsequent ring-closure step.

In recent investigations about the intramolecular *meta*-addition processes, the effects of heteroatoms in the linking chains are obvious. Blakemore and Gilbert reported the photoreaction of **185** linked by the amide group [240]. Photoinduced intramolecular electron transfer can be inhibited by *N*-acetyl or *N*-carbomethoxy groups. Fluorescence is then observed from the arene chromophore and intramolecular *meta* photocycloaddition occurs through high selectivity.

Blakemore and Gilbert found that the SiMe$_2$ unit induces high selectivity in the *meta* photocycloaddition [241]. Irradiation (254 nm) of the silicon-tethered bichromophore **186** gives only one photoisomer **187** in 60% yield (Scheme 56). The ability of silicon to stabilize a β-positive center acts to stabilize the developing positive charge in the ipso carbon; hence, the photocycloaddition directs to the 2,6-position of the benzene ring rather than the 1,3-addition.

Amey et al. proposed the Si–F interaction in the photocycloaddition of silicon-tethered bichromophore **188** by the fact that **189** was the sole product and **190** was not obtained [242,243] (Scheme 57). On the contrary, similar irradiation of siloxy derivative (**191**) gave only an *ortho* adduct (**192**). Moreover, removal of the trifluoromethyl substituent gave rise to the formation of the *meta* cycloadduct (**193**) as the major product [244].

Scheme 55

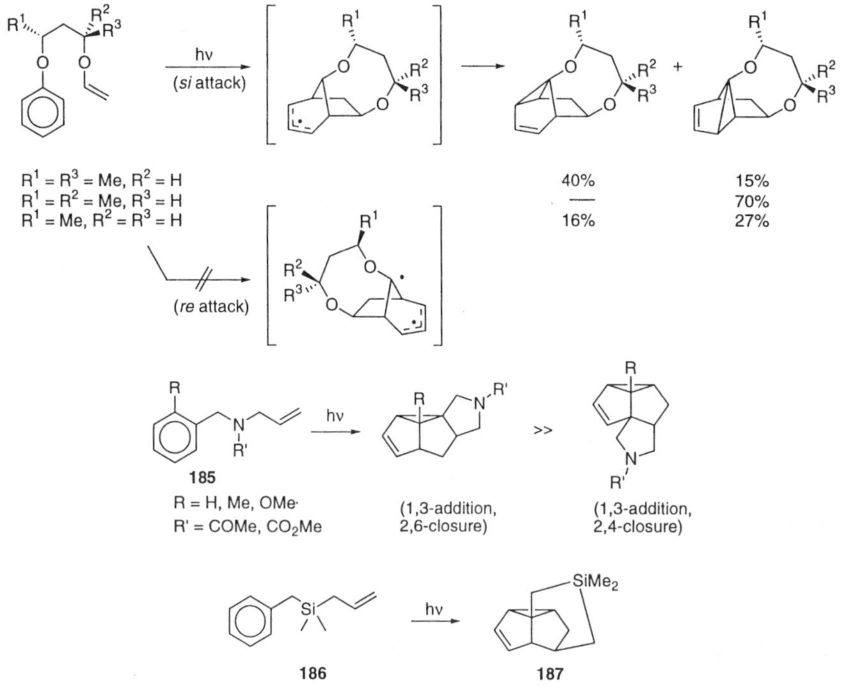
hν
(si attack)
(re attack)
$R^1 = R^3 = Me, R^2 = H$
$R^1 = R^2 = Me, R^3 = H$
$R^1 = Me, R^2 = R^3 = H$
40% 15%
— 70%
16% 27%
hν
185
R = H, Me, OMe
R' = COMe, CO_2Me
(1,3-addition, 2,6-closure)
>>
(1,3-addition, 2,4-closure)
hν
$SiMe_2$
186
187

Scheme 56

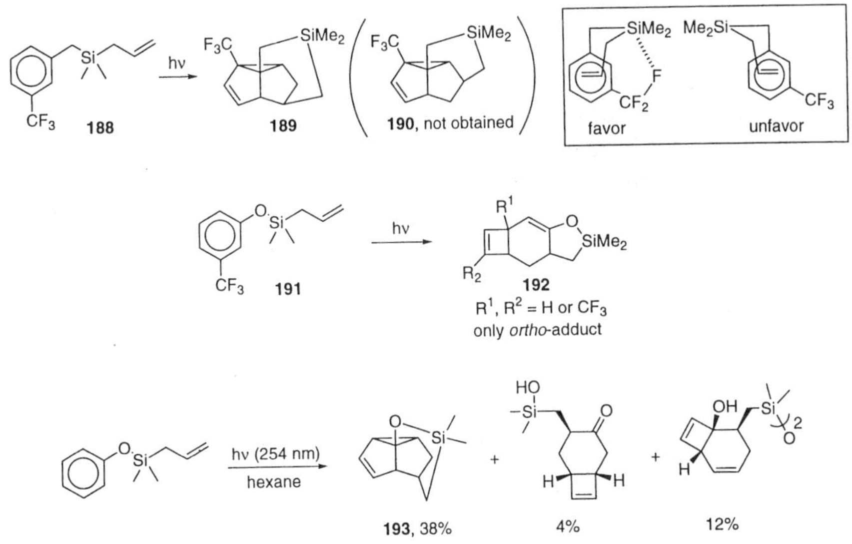
hν
F_3C
$SiMe_2$
188
189
190, not obtained
favor
unfavor
191
192
R^1, R^2 = H or CF_3
only ortho-adduct
hν (254 nm)
hexane
193, 38%
4%
12%

Scheme 57

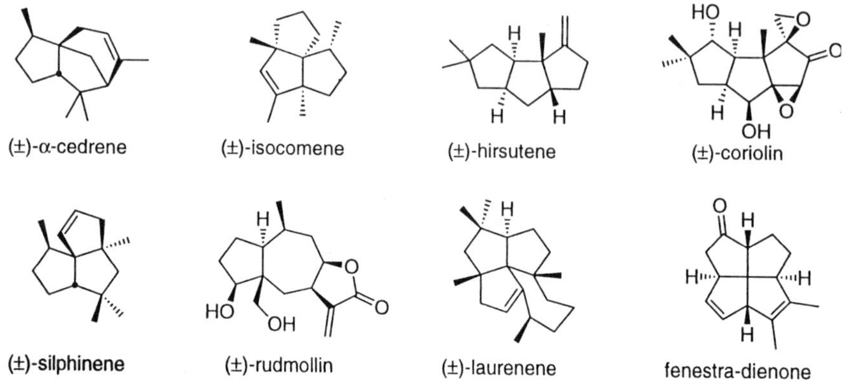

(±)-α-cedrene   (±)-isocomene   (±)-hirsutene   (±)-coriolin

(±)-silphinene   (±)-rudmollin   (±)-laurenene   fenestra-dienone

Scheme 58

These intramolecular meta-addition processes were utilized in the key steps for the total synthesis of α-cedrene [245], isocomene [246], hirsutene [247], coriolin [248], silphinene [249], rudmollin [250], laurenene [251], and fenestranes [252–254], which were synthesized by Wender's and Keese's groups (Scheme 58).

Although relatively simple alkenes add to benzene rings intramolecularly as shown earlier, donor–acceptor pair of addends tends to cycloadd at the *ortho* position. Sket et al. [255] and Gilbert et al. [256] clarified that the pentafluorophenyl ring and the *p*-cyanophenyl ring suffer intramolecular *ortho* addition by the internal alkenyl groups in their earlier works. Intramolecular *ortho* photocycloaddition of **194** proceeded efficiently to give **195** as a sole product, but the irradiation of the sulfur analog **196** led to C—S bond cleavage (Scheme 59). In the case of **197**, the primary photocycloadduct (**198**) is thermally labile and yields the cycloocta-triene derivative **199**, which undergoes electrocyclic reaction to give **200**.

Gilbert's and other groups investigated the effects of substituents and linking chains in the intramolecular *ortho* addition on 254-nm irradiation in aliphatic hydrocarbon solvents. The effect of substituents were investigated in detail by the photoirradiation of substituted 4-phenoxy-1-enes (**201**) (Scheme 60). A cyano substituent in the *ortho* or *para* position of **201** promotes the efficiency of *ortho* cycloaddition to give 4-oxatricyclo[7.2.0.0^{3,7}]undeca-2,10-dienes (**202**) via the photolabile intramolecular *ortho* cycloadducts and 11-oxabicyclo[6.3.0]undeca-1,3,5-trienes [257]. The *meta* isomer of **201** reacts inefficiently, and only the *ortho* isomer in the methoxycarbonyl series undergoes the photocycloaddition. Methyl substitution on the arene units of **203** does not affect the regiochemistry, but the presence of the *ortho*-methoxy group inhibits the *meta* process and only the *ortho* addition occurs [258]. Alkyl groups attached to the allyl carbon of **204** does not influence significantly the modes of selectivity [259]. Moreover, the amide chain of **205** can also inhibit the *meta* process [260].

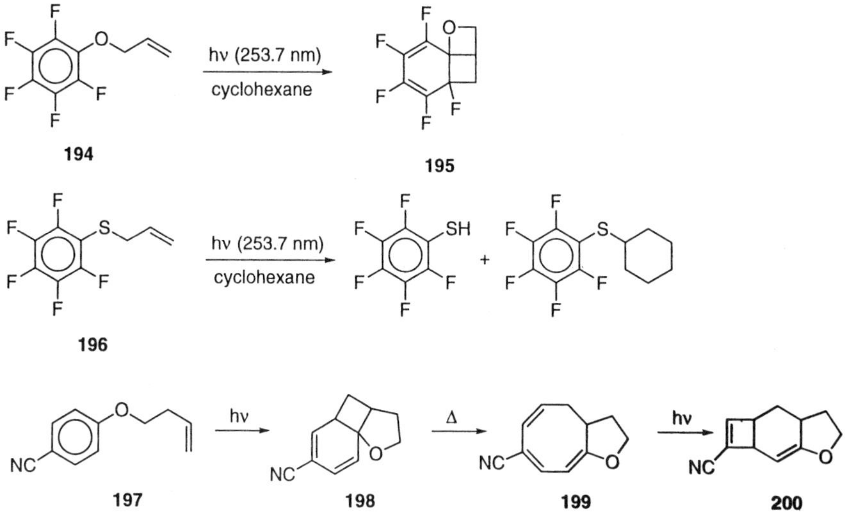

194 → **195**

hv (253.7 nm), cyclohexane

196 → hv (253.7 nm), cyclohexane

Scheme 59

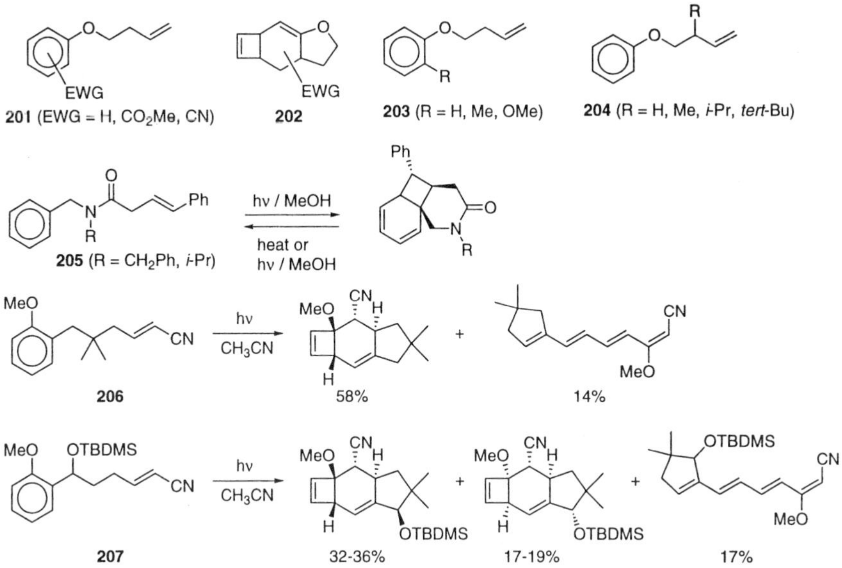

197 → hv → **198** → Δ → **199** → hv → **200**

201 (EWG = H, CO$_2$Me, CN)

202

203 (R = H, Me, OMe)

204 (R = H, Me, i-Pr, tert-Bu)

205 (R = CH$_2$Ph, i-Pr)   hv / MeOH ⇌ heat or hv / MeOH

206   hv / CH$_3$CN   58%   14%

207   hv / CH$_3$CN   32–36%   17–19%   17%

Scheme 60

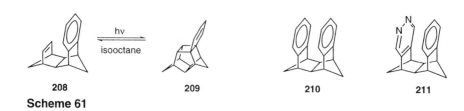

208   209   210   211

Scheme 61

As a result, cycloaddition between excited aryl groups and simple alkyl-substituted olefins generally leads to the formation of *meta* cycloadducts, whereas *ortho* cycloaddition predominates from the singlet state when excited charge transfer is energetically favorable. In fact, direct irradiation of donor–acceptor bichromophores **206** and **207** affords no *meta* cycloadduct [261].

To prevent the *meta* photocycloaddition, the use of rigid "face-to-face" skeletons is also an effective strategy. Acetone-sensitized excitation of **208** gave only **209** reversibly [262–264]; reactions of **210** and **211** gave similar results [265,266] (Scheme 61). X-ray analysis of **210** showed the distance of the aromatic carbons which react is 3.04 Å [263].

Hoffman and Pete suggested the presence of acid can increase both chemoselectivity and efficiency, and the role of acids was investigated in detail. For example, irradiation of *O*-alk-3-enylsalicylates (**212**) at 254 nm yields tricyclic compounds (**213**) through **214–216** [267] (Scheme 62). The efficiency of the intramolecular photocycloaddition increased and the reaction mixture was greatly simplified when the reaction was carried out in methanol in the presence of a catalytic amount of acid. Irradiation of 3-alkenyloxyphenols (**217**) in a methanolic solution yields benzocyclobutenes (**218** and **219**) [268–270]. The efficiency increases if the reaction was carried out in the presence of H_2SO_4.

In the last 15 years, Wagner and his co-workers have developed the intramolecular *ortho* photocycloaddition of alkenes to the benzene ring of the acetophenone chromophore. In these photoreactions, the ring opening products of *ortho* photocycloadducts, 1,3,5-cyclooctatriene derivatives (**220**), and their secondary photocycloadducts, tricyclic compounds (**221**), were often produced [271,272] (Scheme 63). The triplet exciplexes generated from π,π^*-triplet states of acetophenone derivatives were postulated as the reactive intermediates [273].

By incorporating a cyclopropylcarbinyl radical clock, Cheng and Wagner have confirmed the intermediacy of a 1,4-biradical **222,** which cyclized very slowly [274] (Scheme 64).

The effects of 2- or 3-substituents of the acetophenone chromophore toward regioselectivity were examined by use of **223a–e** [275] and **227a–d** [276] (Scheme 65). As for the 2-substitution, the electron-releasing groups facilitate the reaction to give only 3,4-adducts (**224–226**), whereas the electron-withdrawing

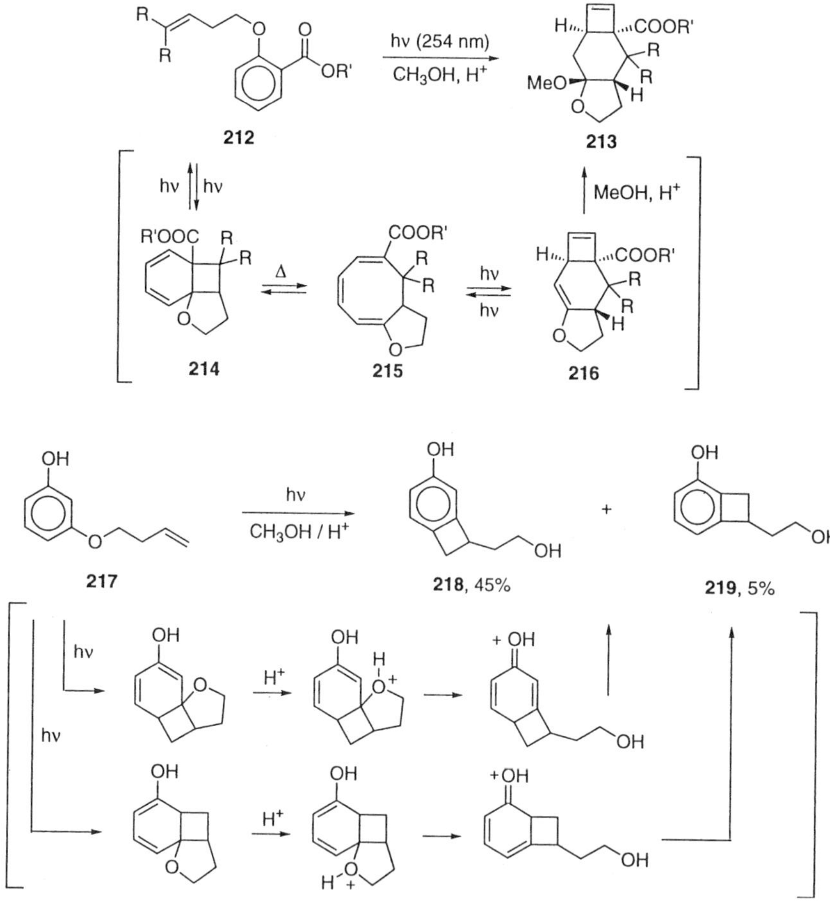

Scheme 62

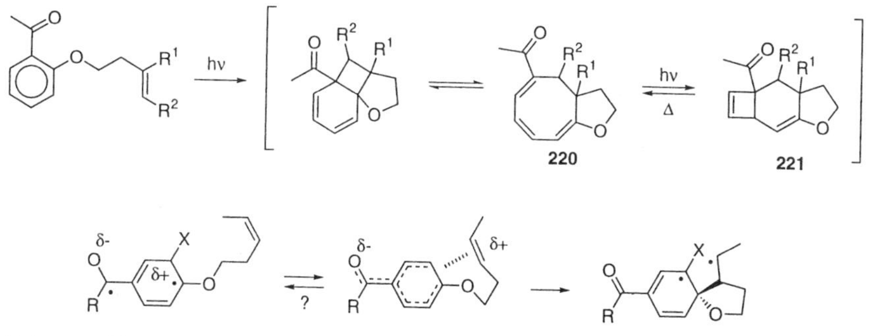

Scheme 63

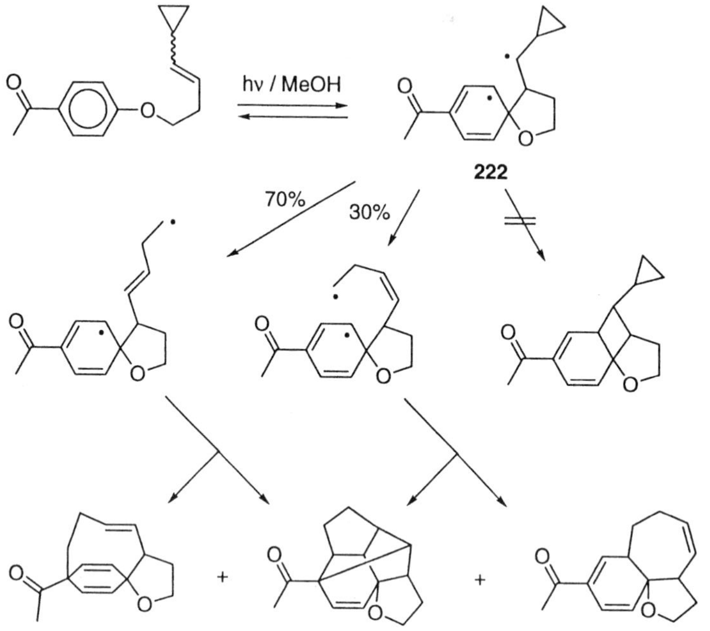

Scheme 64

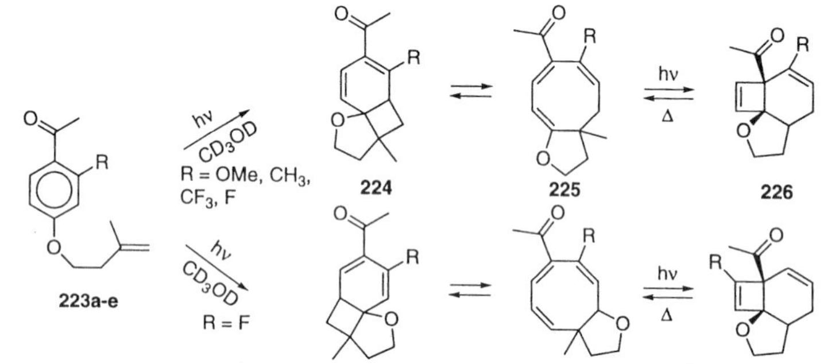

a: R = OMe   b: R = F   c: R = CH₃   d: R = CF₃   e: R = CN

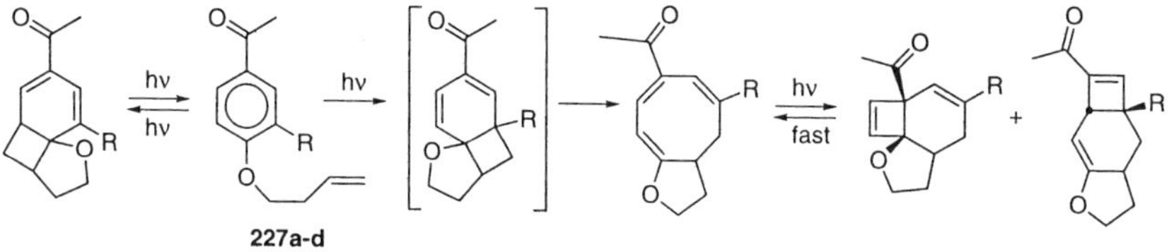

a: R = OMe   b: R = CONH₂ c: R = CN   d: R = CH₃   e: R = SCH₃

Scheme 65

ones give poor yields of products: 2-Trifluoromethyl derivative reacts very sluggishly and the 2-cyano derivative does not react. The 2-fluoro derivative alone provides both 3,4 and 4,5 adducts in a 85 : 15 ratio. A similar tendency is observed in the reaction of 3-substituted derivatives **227a–d.** 3-Methoxy group promotes the 4,5-addition while amide, cyano, methyl, thiomethyl groups favored 3,4-addition. These selectivities toward 4,5- and 3,4-addition reflect inductive effects both on the initial triplet state cycloaddition and on the competing thermal and photochemical reactions.

Stereoselectivity of some methyl substitution on the alkenyl group was examined by the series **228** [277] (Scheme 66). Irradiation of methanol solutions of **228** at >290 nm resulted in the clean conversion of each reactant into a mixture of two diastereomers of **231.** The **231** are then converted thermally (25–100°C) to equilibrium mixtures with **229** and **230.** High diastereomeric excesses (R^2 is *trans* to R^1 or R^3) in **229–231** are always observed. This fact indicates that the photoinduced electrocyclization to a cyclobutene **231** of one diene unit of the bicyclo[6.3.0]undecatriene intermediate **230** undergoes disrotatory photoclosure in only one of two possible directions.

For the investigation of the chain length, photoreactions of **232** were examined under both direct (254 nm in CH_3CN) and triplet-sensitized irradiation (313 nm in acetone) [278]. In **233b,d** (n = 1), the 6/5 ring junction is *cis;* in contrast, *trans* 6/6 fusions are observed in the structures of **234a,c,e.** Nitrile **232d** also forms a linear tricycloundecadiene (**235d**), but **232f** is unreactive.

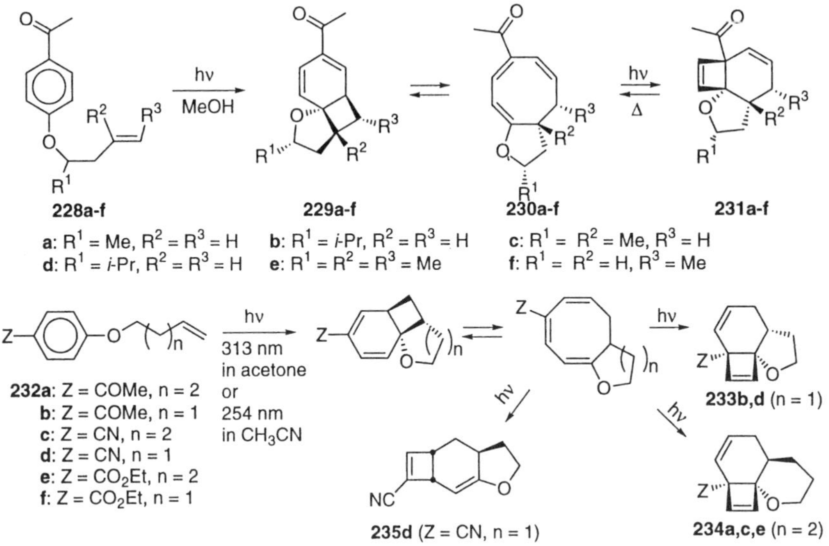
hν
MeOH
hν
Δ
228a-f
229a-f
230a-f
231a-f
a: R¹ = Me, R² = R³ = H
d: R¹ = i-Pr, R² = R³ = H
b: R¹ = i-Pr, R² = R³ = H
e: R¹ = R² = R³ = Me
c: R¹ = R² = Me, R³ = H
f: R¹ = R² = H, R³ = Me
hν
313 nm
in acetone
or
254 nm
in CH₃CN
232a: Z = COMe, n = 2
b: Z = COMe, n = 1
c: Z = CN, n = 2
d: Z = CN, n = 1
e: Z = CO₂Et, n = 2
f: Z = CO₂Et, n = 1
233b,d (n = 1)
235d (Z = CN, n = 1)
234a,c,e (n = 2)

Scheme 66

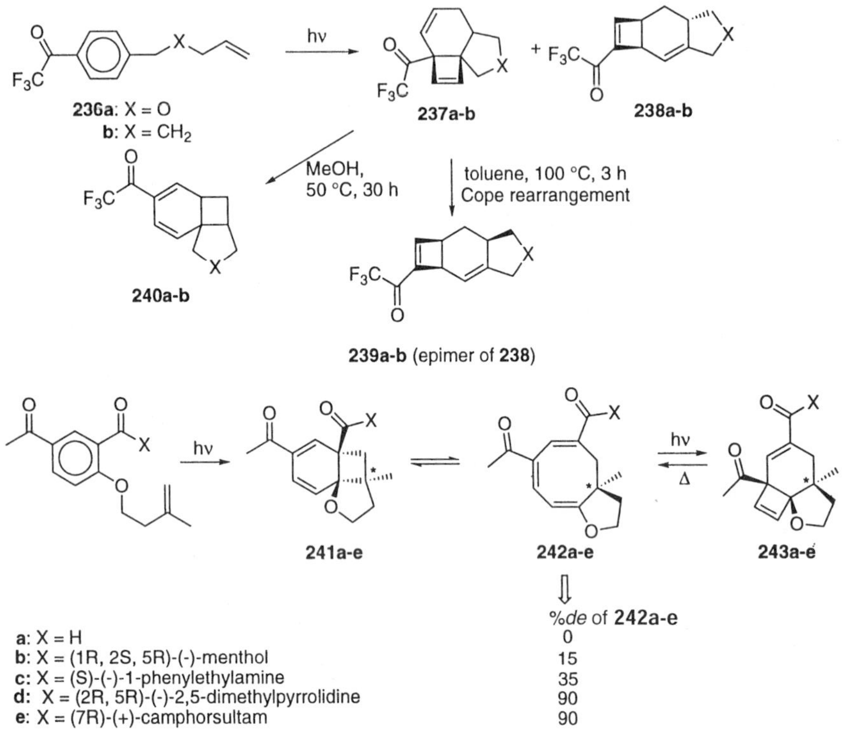

a: X = H
b: X = (1R, 2S, 5R)-(-)-menthol
c: X = (S)-(-)-1-phenylethylamine
d: X = (2R, 5R)-(-)-2,5-dimethylpyrrolidine
e: X = (7R)-(+)-camphorsultam

Scheme 67

Ortho cycloaddition of trifluoroacetyl derivatives (**236a–b**) proceeds in a similar way to obtain **237a–b** and **238a–b** [279] (Scheme 67). Compounds **237a–b** undergo low-temperature Cope rearrangement to generate epimers of **238a–b** (**239a–b**) and also open to equilibrium mixtures of bicycloundecatrienes and tricycloundecatrienes. These results come from strong donor–acceptor inter-action due to the trifluoroacetyl group.

The diastereodifferentiation of **241–243** has been examined [280]. The orig-inal cycloadduct **241** converts thermally to **242**, which then photocyclizes to **243**. Cyclobutenes **243** undergo thermal reversion to **242**. In methanol, (8S)-**242e** is formed in only 30% *de* at 20% conversion, but the *de* increases to 90% as the re-action completed. The reason for this enhancement of *de* with conversion was ex-plained by kinetic resolution, whereby photoepimerization of (8R)-**242e** occurs to give (8S)-**242e**, which slowly cyclizes to **243e**.

This intramolecular *ortho* photocycloaddition is used in the cyclization of the chromanone derivative (**244**) to give oxapentacyclotetradecanedione (**246**) via sec-ondary photorearrangement of the photocycloadduct (**245**) [281,282] (Scheme 68).

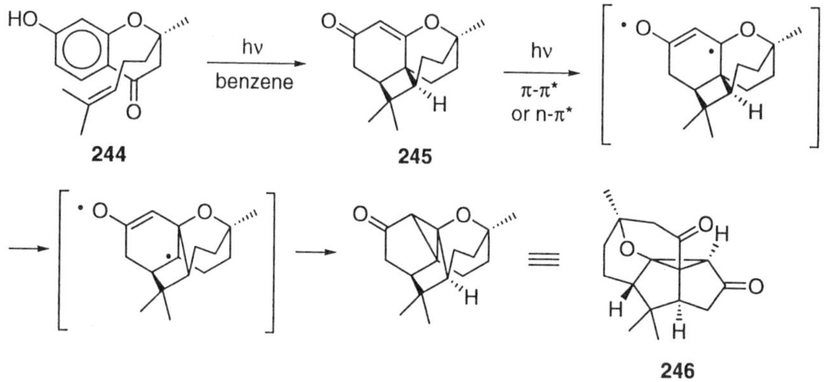

Scheme 68

The isomeric cage products **248–250** are formed from **247** via intramolecular benzene–alkene photocycloadditions between chloranil and 1,1-diarylethenes, induced by the triplet excited enedione moiety [283] (Scheme 69).

The intermolecular *ortho* photocycloaddition of acetylenes to benzene has provided routes to cyclooctatetraenes [78–84]. Intramolecular photocycloaddition of alkynes to aromatic rings is also investigated. Morrison et al. reported that photolysis of 6-phenyl-2-hexyne in hexane solution using 254-nm light leads to the formation of bicyclo[6.3.0]undecatetraene (**251** and/or **252**) [284] (Scheme 70). Pirrung prepared various bicyclo[6.3.0]undecatetraenes (**254a–e**) by the in-

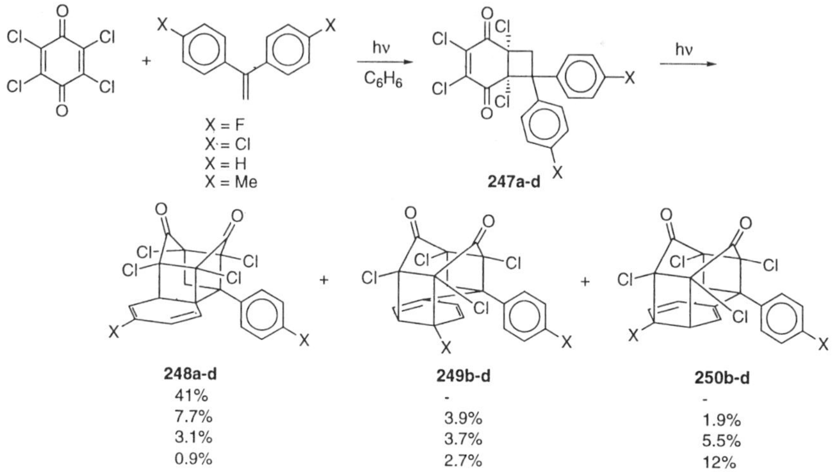

248a-d	249b-d	250b-d
41%	-	-
7.7%	3.9%	1.9%
3.1%	3.7%	5.5%
0.9%	2.7%	12%

Scheme 69

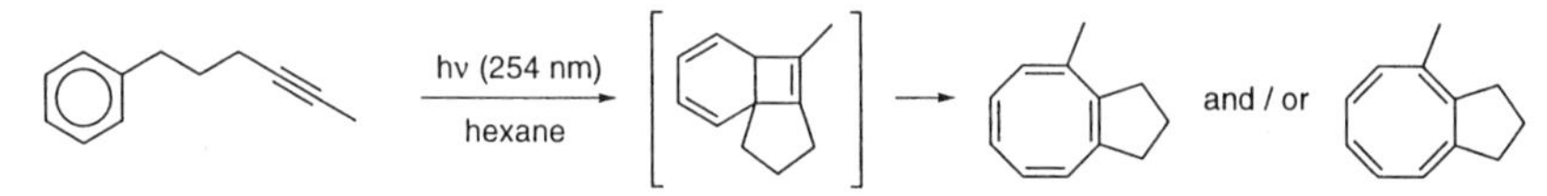

251   252

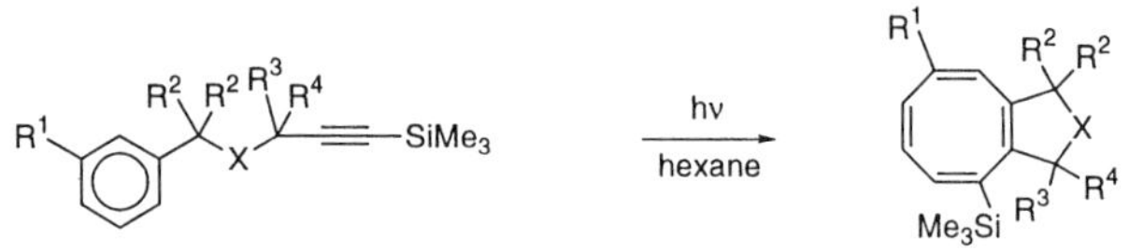

254a–e

253a: $R^1 = R^2 = R^4 = H$, $R^3 = OH$, $X = CH_2$ 46%
   b: $R^1 = Me$, $R^2 = R^4 = H$, $R^3 = OH$, $X = CMe_2$ 38%
   c: $R^1 = MeO$, $R^2 = R^4 = H$, $R^3 = OH$, $X = CMe_2$ 31%
   d: $R^1 = R^2 = R^3 = R^4 = H$, $X = O$ 42%
   e: $R^1 = H$, $R^2 = R^4 = H$, $R^3 = Me$, $X = CH_2$ 50%

Scheme 70

tramolecular photocycloaddition of silyl-substituted alkynes (**253a–e**) in good yields [285].

F. Intramolecular Photocycloaddition to Polycyclic Aromatic Rings

Typical intramolecular (2 + 2) photocycloaddition of electron-rich alkenes to the 1-cyanonaphthalene ring via an exciplex has been reported by McCullough and Gilbert, independently. McCullough reported intramolecular photocycloaddition of bichromophoric molecules **255, 258,** and **259,** in which the chromophores are linked by a three-atom ether chain [286,287] (Scheme 71). The fluorescence of the 1-cyanonaphthalene chromophore of **255, 258,** and **259** is efficiently quenched by alkenes and the typical intramolecular exciplex emissions are observed. Photocycloaddition of these 1-cyanonaphthalenes in benzene occurs at 1,2-, 3,4-, or 5,6-positions of the naphthalene ring, respectively.

Gilbert and his co-workers reported that the photocycloaddition of 1-cyanonaphthalenes (**260** and **262**) gave **261** and **263,** independently [288] (Scheme 72). In addition, they prepared and photoirradiated a variety of compounds that connected naphthalene and alkenes by methylene, ether, and ester tethers at various lengths [289]. Among them, the reaction of **264** gave the best result, whereas the 2-substituted derivative (**265**) did not give any product.

Wagner and Sakamoto reported the first example of the intramolecular (2 + 2) photocycloaddition of simple alkenes to the excited triplet state of naphthalene [290]. Cyclizations of **266** and **267** are entirely triplet-state reactions: The formation of cycloadducts is sensitized by Michler's ketone and is quenched by conju-

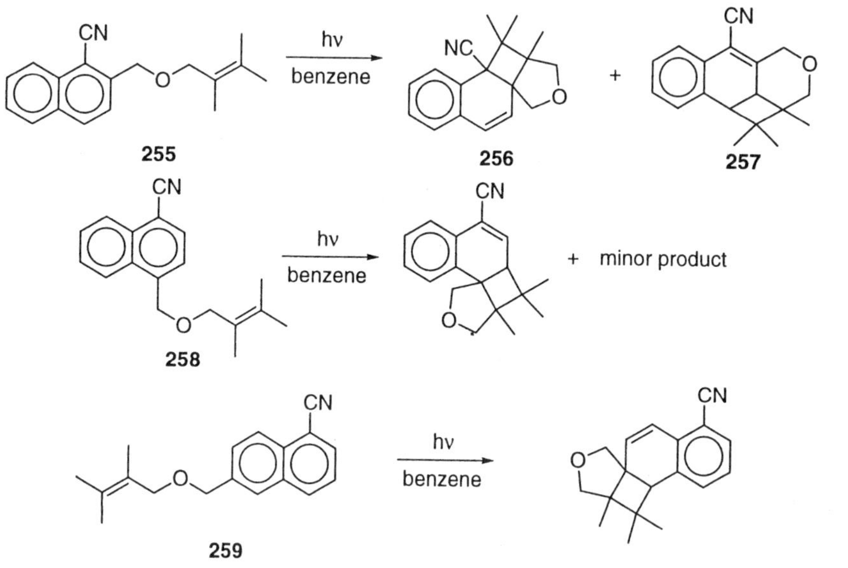

Scheme 71

gated dienes or *trans*-stilbene (Scheme 73). The rate differences between **266a–c** and **267a–c** suggest that biradical closure from **267** is less efficient than that from **266** in the rate-determining step for cyclization.

Kubo et al. reported the photoreaction of imide (**268**) with 2,3-dimethylbutadiene to give naphthazepinediones (**269,270**), which gave rise to the in-

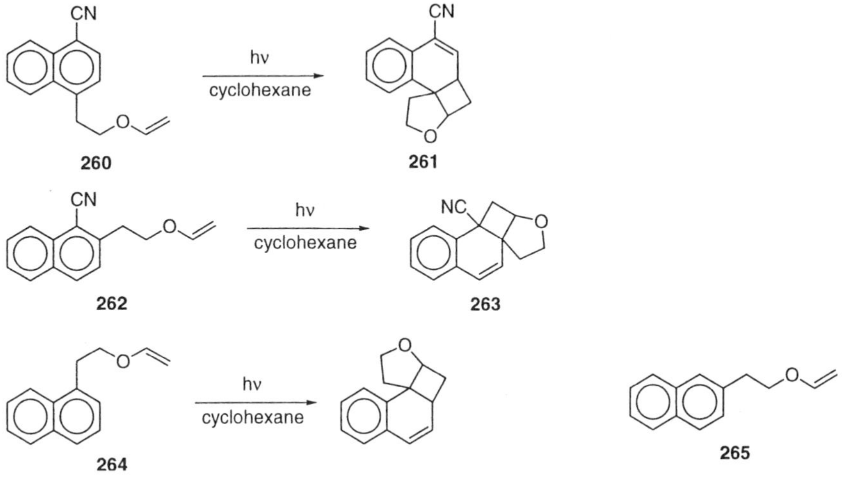

Scheme 72

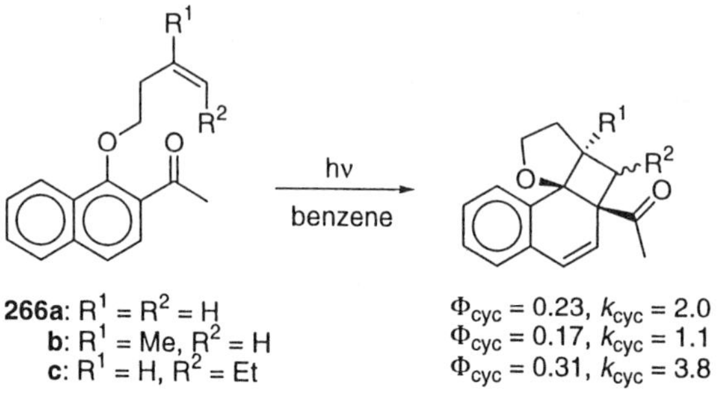

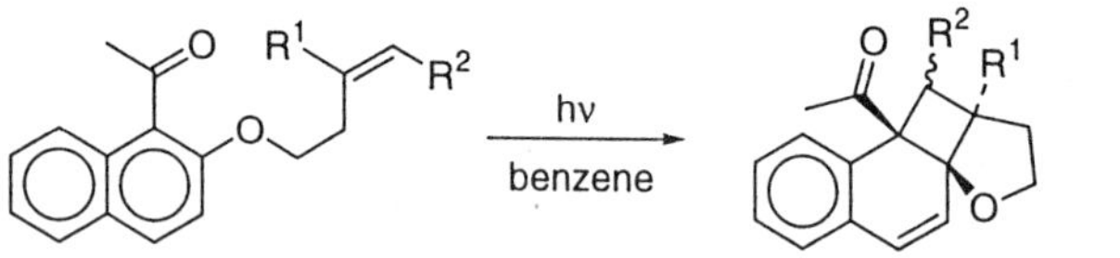

266a: $R^1 = R^2 = H$
  b: $R^1 = Me, R^2 = H$
  c: $R^1 = H, R^2 = Et$

$\Phi_{cyc} = 0.23, k_{cyc} = 2.0$
$\Phi_{cyc} = 0.17, k_{cyc} = 1.1$
$\Phi_{cyc} = 0.31, k_{cyc} = 3.8$

267a: $R^1 = R^2 = H$
  b: $R^1 = Me, R^2 = H$
  c: $R^1 = H, R^2 = Et$

$\Phi_{cyc} = 0.01, k_{cyc} = 0.03$
$\Phi_{cyc} = 0.04, k_{cyc} = 0.07$
(no cycloadduct)

Scheme 73

tramolecular photocycloaddition of alkenes to the naphthalene ring [291] (Scheme 74).

Dittami et al. reported that the alkene side chain in **271** cycloadded to internal naphthalene ring as secondary photoprocess during the course of photocyclization of aryl vinyl sulfide (**271**) [292,293] (Scheme 75).

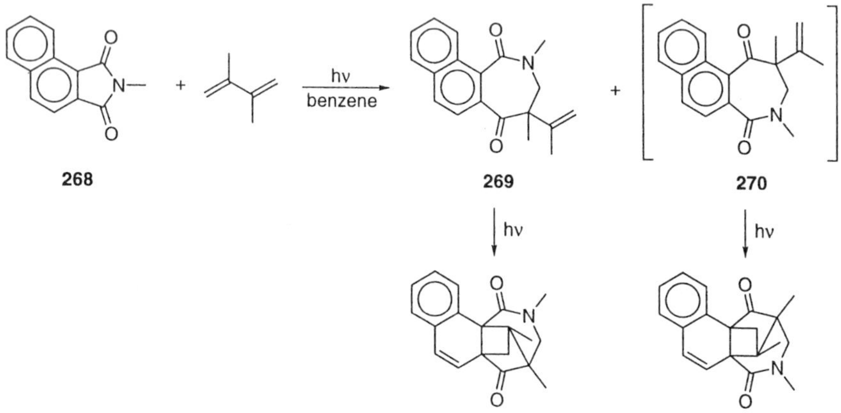

268   **269**   **270**

Scheme 74

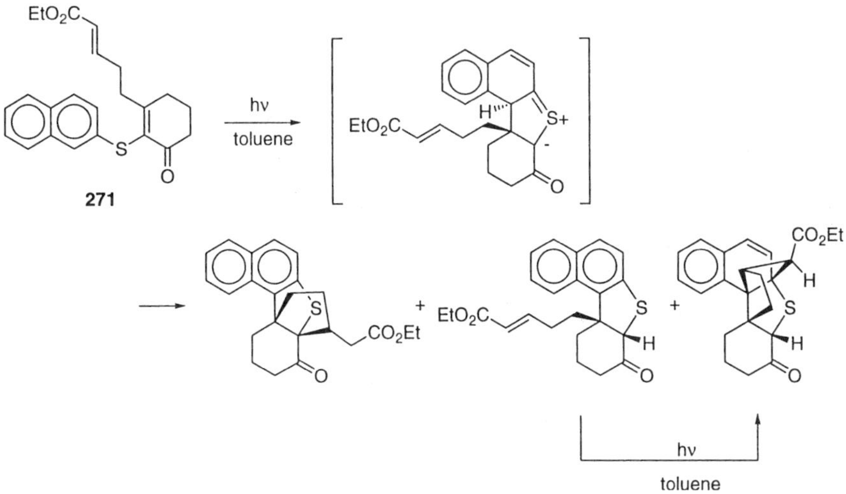

Scheme 75

Prinzbach and co-workers studied the photocycloaddition of a rigid compound in which naphthalene and alkene were placed [294]. Distances between naphthyl β-carbons and olefinic carbons of **272** are 2.911 and 2.944 Å. Direct or acetone-sensitized photoexcitation induced an efficient *meta* addition to the naphthalene ring to give **273,** the *ortho* addition did not occur, although it is not clear whether the *ortho* process occurred (Scheme 76).

Kohmoto et al. examined the photocyclization of *N*-cinnamoyl-1-naphthamides (**274a–f**) in benzene solution and in the solid state [295] (Scheme 77). Intramolecular $(2\pi + 2\pi)$ and $(2\pi + 4\pi)$ photocycloaddition of styryl group to the naphthalene ring was observed in benzene solution, whereas **275a–c** were also obtained in the solid state.

Mizuno et al. reported the site-selective photocycloaddition of **255** in the presence of Eu(III) salt [296] (Scheme 78). In the photoreaction of **255a** in benzene, the cycloadduct **256a** at the 1,2-position on the naphthalene ring predomi-

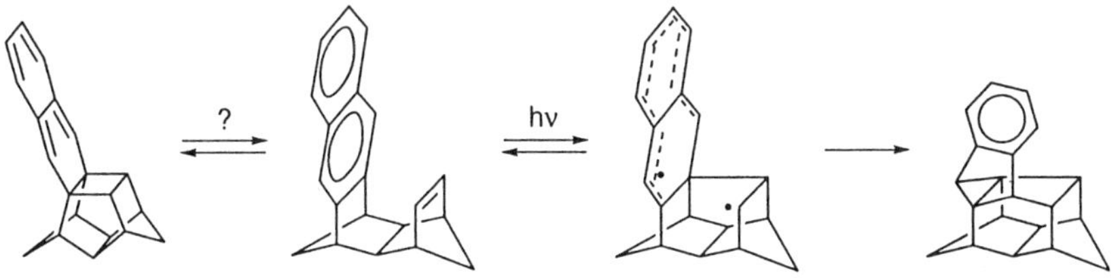

Scheme 76

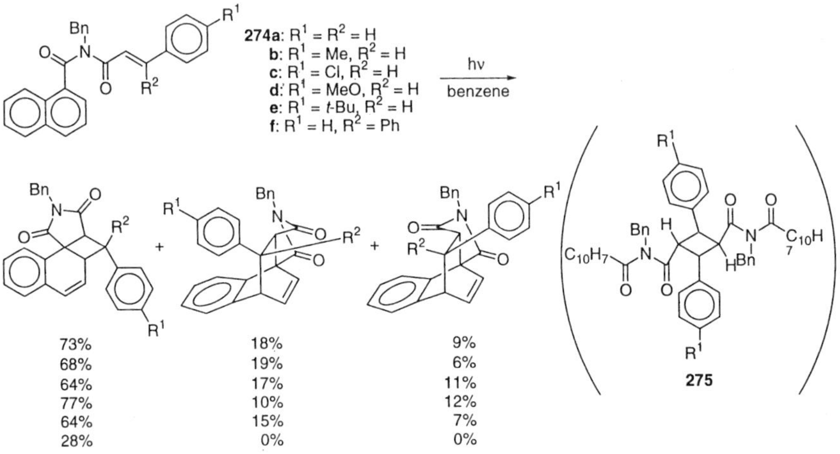

Scheme 77

nantly obtained in the initial stage. However, prolonged irradiation afforded the cycloadduct **257a** at the 3,4-position as a major product [296,297]. Under the same irradiation conditions, the isolated **256a** efficiently cycloreversed to **255a,** but **257a** did not. The addition of Eu(hfc)$_3$ (**279,** hfc = tris[3-(heptafluoropropyl-hydroxymethylene)-d-camphorato]) into the reaction system suppressed not only the formation of **257a** but also the cycloreversion of **256a** [296]. In the reactions of **255b–c, 256b–c** were regioselectively obtained. In the case of **255d–e,** photo-cycloaddition did not occur in benzene, but in acetonitrile, **257d–e** were slowly produced, and the formation of **257d–e** was accelerated by the addition of Mg(ClO$_4$)$_2$ (Table 4).

The solvent effect in the photoreaction of **255c** clearly showed that the formation of **257c** strongly correlates with the polarity of solvent, but the formation of **256c** did not [297]. From these facts, the formation of 3,4-adducts **257** is explained by the charge-transfer contribution of the singlet exciplex, whereas the 1,2-adducts **256** are formed via normal exciplex with less charge-transfer na-

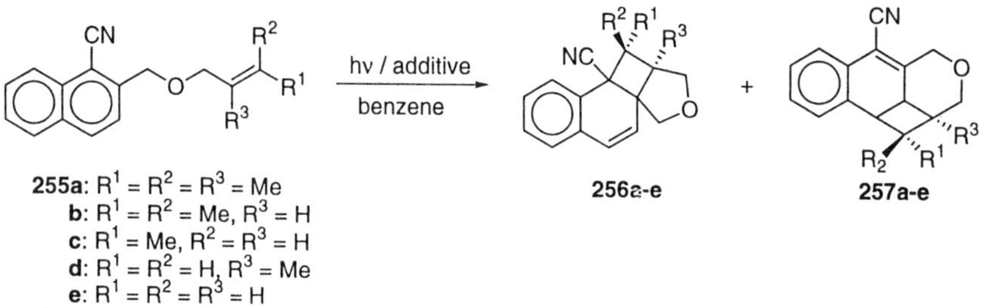

255a: R^1 = R^2 = R^3 = Me
   b: R^1 = R^2 = Me, R^3 = H
   c: R^1 = Me, R^2 = R^3 = H
   d: R^1 = R^2 = H, R^3 = Me
   e: R^1 = R^2 = R^3 = H

256a–e   **257a–e**

Scheme 78

Table 4 Salt and Solvent Effects on the Photocycloaddition of **255a** and **255e**

255	Solvent[a]	Irradiation time (hr)	Additive ($\times 10^{-3} M$)	Yields[b]/(%) 256	257
255a	Benzene	3	None	48	16
255a	Benzene	20	None	41	44
255a	Benzene	3	Eu(hfc)$_3$ (**261**) (5.0)	64	3
255e	Benzene	5	None	0	0
255e	Acetonitrile	5	None	0	45
255e	Acetonitrile	5	Mg(ClO$_4$)$_2$ (5.0)	0	75

Note: **255** = $3.0 \times 10^{-2} M$.
[a] 5 mL.
[b] Gas–liquid chromatography yields based on **255** used.

ture (Scheme 79). Molecular orbital (MO) calculation of 1-cyanonaphthalene and 1-cyano-2-methylnaphthalene supports the fact where anionic charges spread to the 3,4-position and the spin density is located at the 4-position (Table 5).

Nishiyama and Mizuno investigated the enantioselectivity by the addition of chiral sources **276–280** into the photoreaction of **255a**; however, enantiomer excess (*ee*) was very low [297] (Scheme 80, Table 6). The diastereodifferentiation of compound **281** that connects 1-cyanonaphthalene and an alkene by di-(*l*)-menthyl malonate gave **282** in 14% *de;* moreover, the addition of Lewis acid at low temperature increased the *de* to 17% (Table 7).

In the case of 1-methoxycarbonylnaphthalene-bearing alkenyl group (**283**), medium-ring-size lactone (**284**) was obtained by unusual 1,9-hydrogen abstrac-

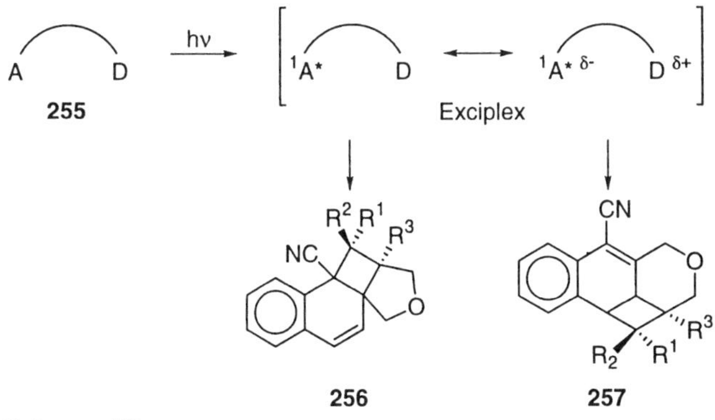

Scheme 79

Table 5 UHF-PM3 Molecular Orbital Calculations on
the Radical Anions of Cyanonaphthalenes

Cyanonaphthalene	Position	Charge	Spin density
	1	-0.2365	0.2708
	2	-0.0895	0.2410
	3	-0.1576	-0.2238
	4	-0.2307	0.5822
	1	-0.2305	0.2427
	2	-0.0782	0.2793
	3	-0.1364	-0.2588
	4	-0.2306	0.5848

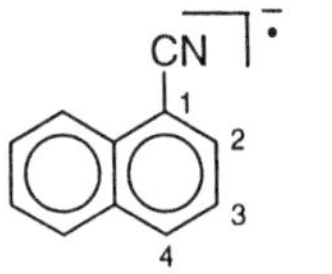

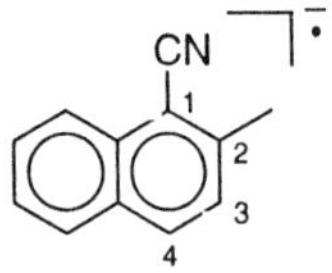

tion of methoxycarbonyl group accompanied by the formation of normal (2π + 2π) photocycloadducts (**285,286**) [298] (Scheme 81).

Mizuno and colleagues examined the intramolecular photocycloaddition of naphthylmethyl furfuryl ether derivatives [299]. Irradiation of a benzene solution of cyano- and methoxycarbonyl derivatives (**287a–b**) gave intramolecular (4π + 4π) photocycloadducts (**288a–b**) in high yields. In the case of unsubstituted **287c,** two kinds of (4π + 4π) cycloadduct (**288c, 289c**) and a caged compound (**290c**) were obtained (Scheme 82). It is noteworthy that the intermolecular photocycloaddition of furan to 1-methoxycarbonyl- and unsubstituted naphthalene did not proceed at all.

An intriguing intramolecular (3 + 2) photocycloaddition of alkenyl methyl 1,4-naphthalenedicarboxylates developed by Kubo [300] will be discussed later.

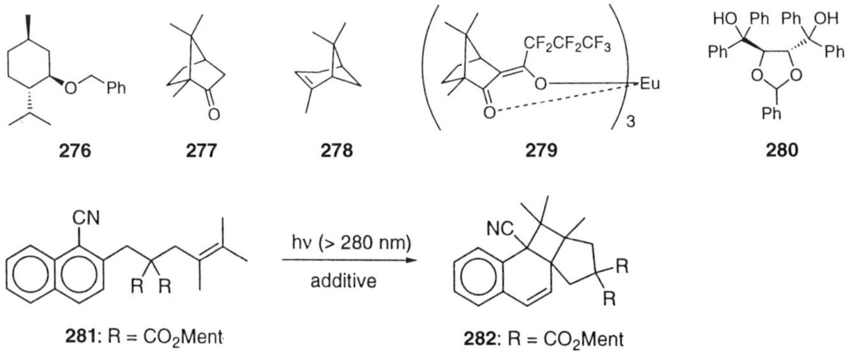

| 276 | 277 | 278 | 279 | 280 |

281: R = CO$_2$Ment   **282**: R = CO$_2$Ment

Scheme 80

Table 6 Additive and Solvent Effects on Photocyclization of **255a**

Entry	Solvent	Additive	Irradiation Time (hr)	Temp. (°C)	Yield (%)[a] 256a	Yield (%)[a] 257a	ee of 256a[b] (%)
1	Cyclohexane	**276** $(3.0 \times 10^{-1} M)$	1.0	RT[c]	55	8	0.8
2	Cyclohexane	**277** $(3.0 \times 10^{-1} M)$	1.3	RT	23	2	0.4
3	Cyclohexane	**278** $(3.0 \times 10^{-1} M)$	1.3	RT	28	2	0.1
4	Pentane	**276** $(3.0 \times 10^{-1} M)$	4.0	−30	19	1	0.5
5	Benzene	**279** $(5.0 \times 10^{-3} M)$	3.0	RT	42	1	0.8
6	Benzene	**280** $(3.0 \times 10^{-1} M)$	1.5	RT	40	7	1.1
7	Toluene	**280** $(3.0 \times 10^{-1} M)$	4.0	−30	24	2	0.2
8	**276**	None	1.5	RT	19	2	0.4
9	**278**	None	1.0	RT	62	7	1.2

Note: **255a** $= 3.0 \times 10^{-2} M$.
[a] Determined by gas–liquid chromatography.
[b] Determined by high-performance liquid chromatography.
[c] RT = room temperature.

Table 7 Additive and Temperature Effects on the Photocyclization of **281** in $CDCl_3$

Entry	Additive	Temp. (°C)	Irradiation Time (hr)	Yield[a] (%)	de[a] (%)
1	None	RT[b]	1.0	57	< 5
2	Al(OPr-i)$_3$ $(9.0 \times 10^{-2} M)$	RT	1.0	59	< 5
3	Ti(OPr-i)$_4$ $(9.0 \times 10^{-2} M)$	RT	1.0	48	< 5
4	None	−20	3.0	36	14
5	Al(OPr-i)$_3$ $(9.0 \times 10^{-2} M)$	−20	3.0	28	13
6	Ti(OPr-i)$_4$ $(9.0 \times 10^{-2} M)$	−20	3.0	15	17

Note: **281** $= 3.0 \times 10^{-2} M$.
[a] Determined by ^{1}H-NMR (270-MHz) spectrum.
[b] RT = room temperature.

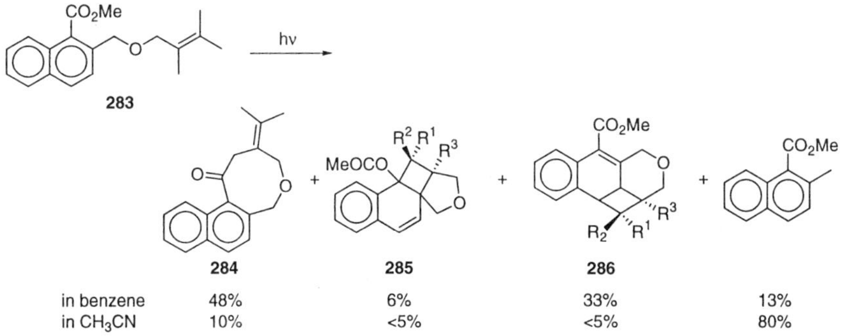

	284	**285**	**286**	
in benzene	48%	6%	33%	13%
in CH$_3$CN	10%	<5%	<5%	80%

Scheme 81

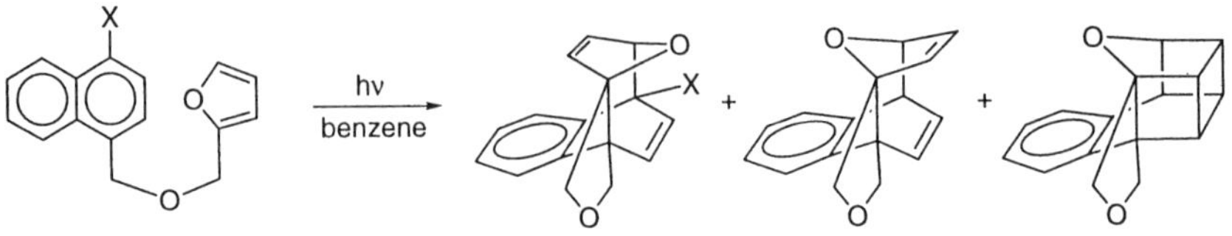

287a: X = CN
  b: X = CO₂Me
  c: X = H

	288a-c	289c	290c
	quant.	—	—
	70%	—	—
	40%	25%	7%

Scheme 82

Photochemical (4 + 2) cycloaddition of anthracene and dimethyl fumarate or maleate have been reported by Kaupp [301]. The reaction requires a high concentration of dimethyl fumarate or maleate owing to the short lifetime of the excited singlet state of anthracene and its facile dimerization. However, irradiation of a benzene solution of (9-anthryl)methyl methyl fumarate or maleate resulted in the intramolecular (4 + 2) photocycloaddition efficiently [302]. Asymmetric induction in this intramolecular photocycloaddition of **291b–c** was also investigated (Scheme 83).

The intramolecular photocycloaddition of the styryl group to the phenanthrene ring linked by an ester has been reported by Sakuragi et al. [303–306]. Although the precedented cyclobutanes (**293b–d**) are obtained in some cases, the oxetanes formed from the reaction between the ester carbonyl group and the styryl group are primarily obtained (Scheme 84). The oxetanes **294b–d** are slowly decomposed to **295b–d** and acetaldehyde. The products dihydropyranol (**298b**) and tetrahydroxepinol (**298c**) can be rationalized by the initial formation of acetals, followed by polar rearrangements and dehydration.

Mizuno et al. investigated an unusual intramolecular photocycloaddition of 9-cyano-10-phenanthrylmethyl cinnamyl ether **299a** giving two kinds of cycloadduct, **300a** and **301a** [307]. Product **300a** was the initial product and the prolonged irradiation afforded **301a** (Scheme 85). The isolated **300a** was smoothly cy-

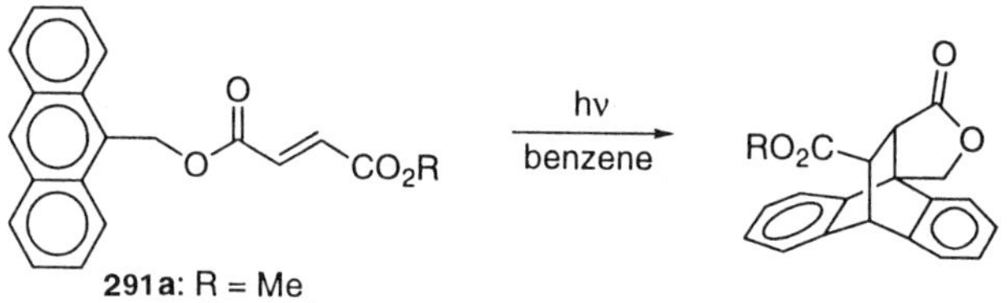

291a: R = Me
  b: R = (-)-bornyl
  c: R = (-)-menthyl
  c: R = (-)-menthyl

	rt	9% de
	rt	35% de
	-20 °C	56% de

Scheme 83

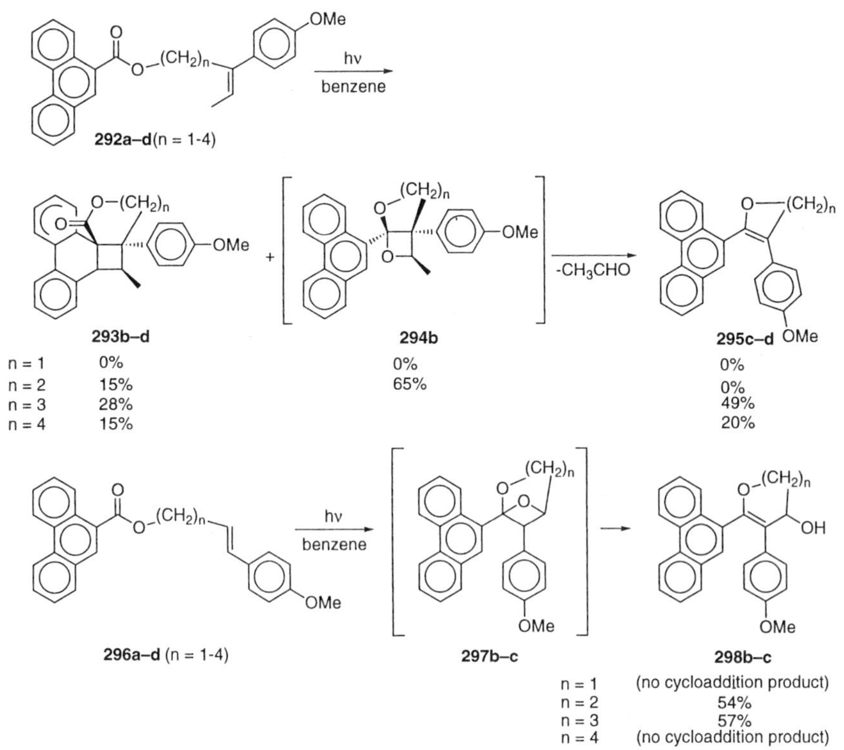

292a–d(n = 1-4)

293b–d

	293b–d	**294b**	**295c–d** OMe
n = 1	0%	0%	0%
n = 2	15%	65%	0%
n = 3	28%		49%
n = 4	15%		20%

296a–d (n = 1-4)   **297b–c**   **298b–c**

n = 1	(no cycloaddition product)
n = 2	54%
n = 3	57%
n = 4	(no cycloaddition product)

Scheme 84

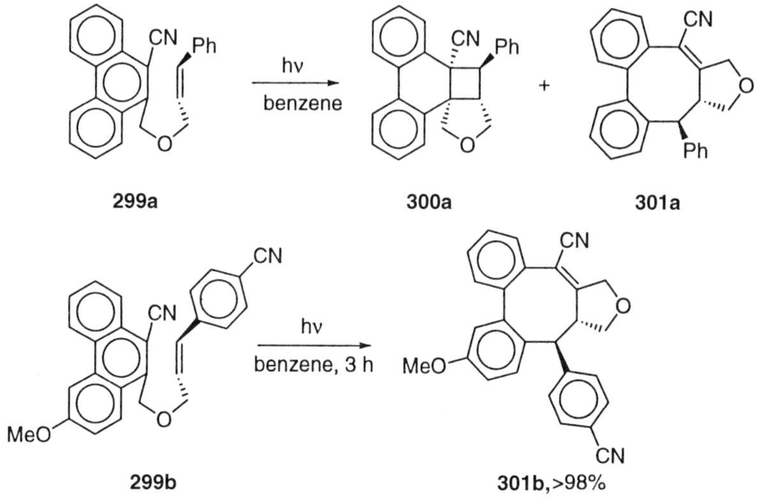

299a   **300a**   **301a**

299b   **301b,>98%**

Scheme 85

cloreversed to **299a** in these reaction conditions. The formation of **300a** and **301a** was not sensitized by the addition of Michler's ketone and was not quenched by 2-methyl-1,3-butadiene. The relative intensity of the fluorescence of phenanthrene chromophore was quenched intramolecularly by the internal cinnamyl group.

From these results, two sandwich-type exciplexes are postulated as reactive intermediates via the excited singlet state of phenanthrenes. The normal (2π + 2π) photocycloadduct **300a** is produced from the more favorable type of exciplex (**302a**) (Scheme 86). Another type of exciplex (**303a**) affords strained ($2\pi + 2\pi$) photocycloadduct **304a** at the 10,10a-position of phenanthrene. Thermal ring expansion of **304a** affords the eight-membered compound **301a.** This is the first example of the photocycloaddition at the 10,10a-position of phenanthrene.

The electron-releasing group on the 3-position of the phenanthrene ring and electron-withdrawing group on the *para* position of the phenyl ring of **299** accelerated the formation of the eight-membered compound **301**. Indeed, the reaction of the 3-methoxy-*p*-cyano derivative **299b** gave **301b** almost quantitatively.

Regioselective and site-selective intramolecular photocycloaddition of the cinnamyl group to the pyrene ring has been recently reported by Mizuno and his co-workers [308]. The photoirradiation of a benzene solution containing **305a–c** gave intramolecular ($2\pi + 2\pi$) photocycloadducts **306a–c** (Scheme 87). On the other hand, irradiation of *ortho*-substituted derivative **307** afforded **308**. The regioselective photocycloaddition is reasonably explained by the sandwich-type exciplexes. Photoreaction of the *meta* derivative (*trans*-**309**) resulted in the brief photoisomerization to the *cis* isomer (*cis*-**309**) and did not give any cycloadduct. Although the cycloreversion from **306** to **305** is included in the former reaction, **308** was not converted to **307** at all.

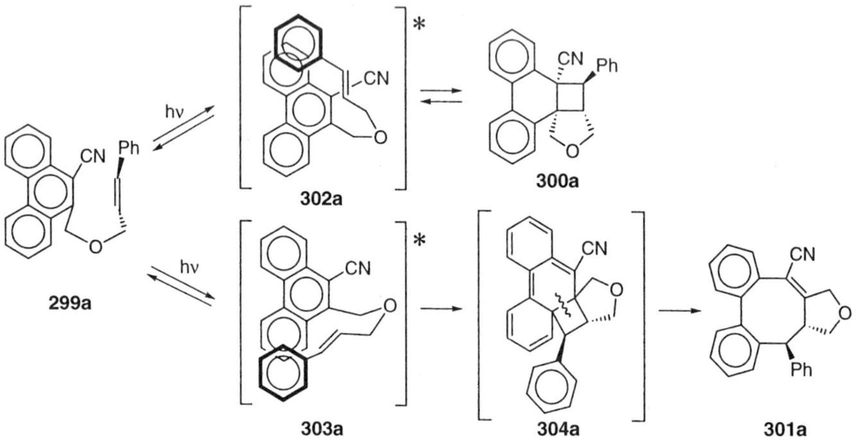

Scheme 86

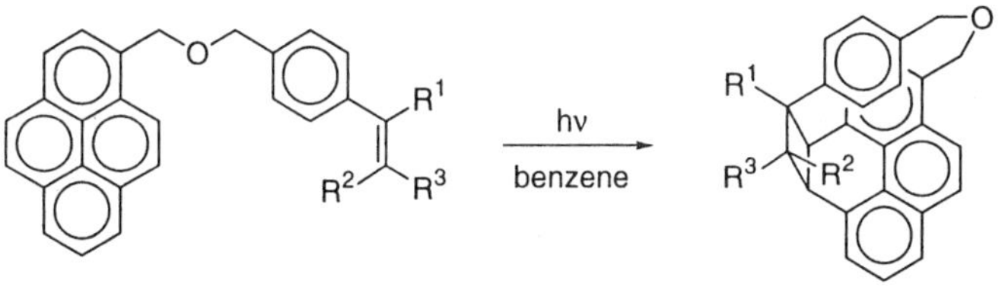

305 a: R^1 = R^2 = H, R^3 = CO$_2$Me
b: R^1 = R^2 = H, R^3 = CN
c: R^1 = CN, R^2 = H, R^3 = Ph

306 a, 83%
b, 81%
c, 22%

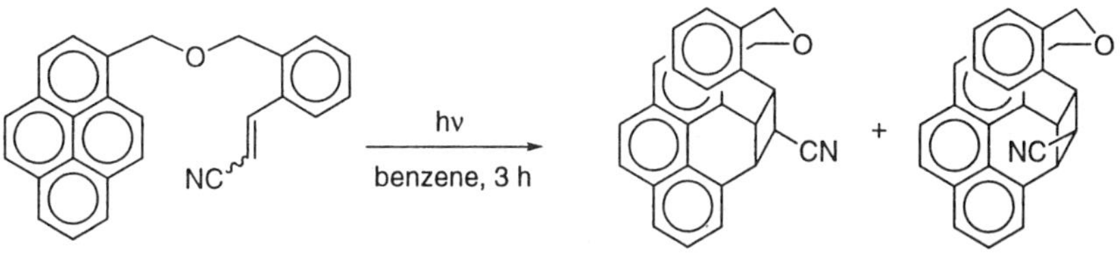

307 (trans / cis = 78 / 22), 0.03 M

trans-308, 75%   **cis-308**, 11%

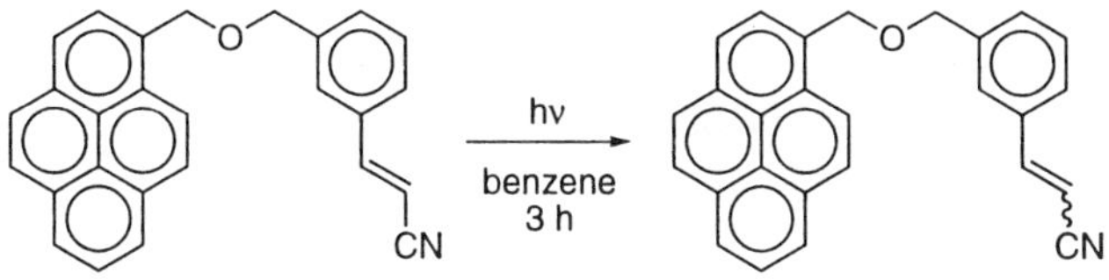

trans- 309, 0.03 M

92% (trans/cis = 93 / 7)

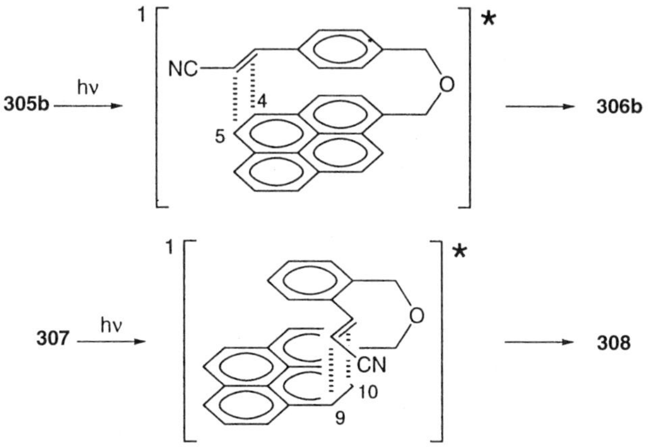

Scheme 87

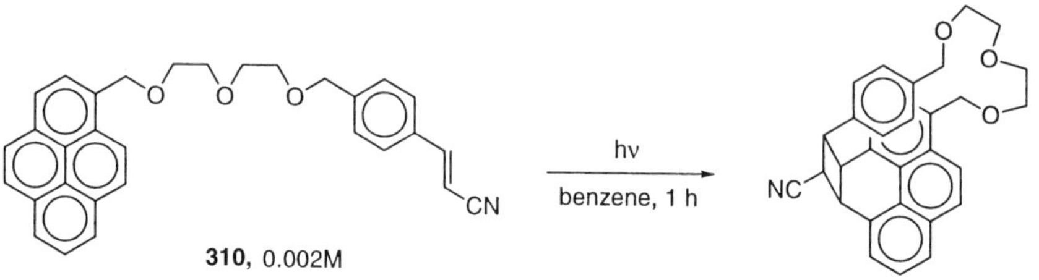

310, 0.002M

311, 67%

Scheme 88

The photoreaction of **310** that connects pyrene and cinnamonitrile by diethylene glycol linkage gave only the photocycloadduct **311** at the 9,10-position of pyrene [309] (Scheme 88).

G. Intramolecular Photocycloaddition and Photodimerization of Two Aromatic Rings

Compared with the typical photodimerization of condensed aromatics such as anthracenes, benzene is inert to photodimerization even if two benzene rings are connected intramolecularly.* Photoirradiation of higher strained [2.2]paracyclophane (**312**) with a low-pressure mercury lamp at $-196°C$ gave rise to photocleavage to give p-quinodimethane (**313**) without any photodimerized product [311] (Scheme 89).

Misumi and co-workers reported the first example of photodimerization of benzene rings in their photoinduced desulfurization study from **314b** [312,313]. Upon irradiation, three quadruple-layered cyclophanes **314a–c** undergo photodimerization of two inner benzenes to give good yields of cage compounds **315a–c**, which reverse to **314a–c**. Prinzbach reported the two benzene rings of **210** suffer face-to-face stacking to be enabled for photodimerization of benzene rings [272–276].

Recently, Shinmyozu et al. reported that the photolysis of [3₄](1,2,3,4)cyclophane (**317**) and [3₄](1,2,3,5)cyclophane (**320**) in H_2O-saturated CH_2Cl_2 afforded polycyclic caged diols **319** and **322** via hexaprismane derivatives **318** and **321** [314,315] (Scheme 90).

[2.2](1,4)-naphthalenophane **323,** whose *anti* form is favored more than the *syn* form [316], undergoes the intramolecular photoisomerization to give dibenzoequinene (**324**) [317] (Scheme 91). This (4 + 2) photoreactions giving **324** would be geometrically favored where intramolecular (4 + 4) photodimerization

---

* Intramolecular excimer formation of diphenyl and triphenyl alkanes was investigated in detail. For example, see Ref. 310.

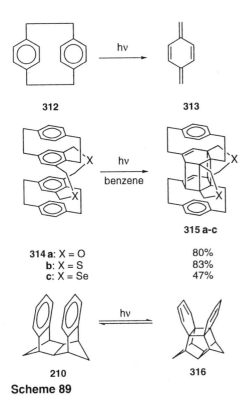

312   313

315 a-c

314 a: X = O   80%
    b: X = S   83%
    c: X = Se   47%

210   316

Scheme 89

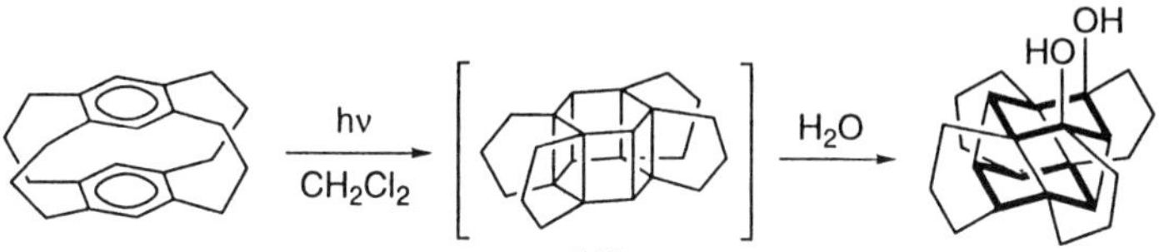

317   318   319

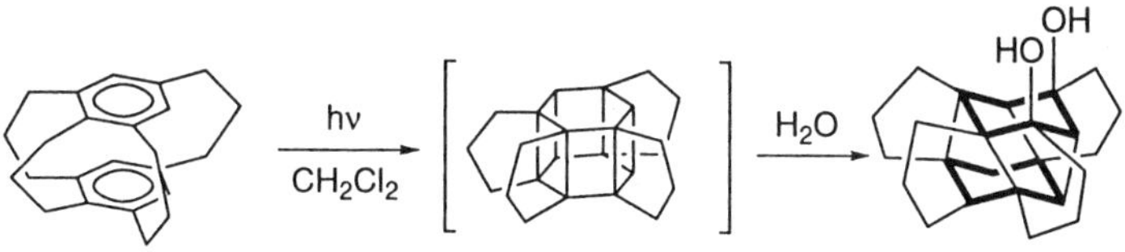

320   321   322

Scheme 90

from the *anti* form is kinetically preferred. In fact, irradiation of **323** at low temperature followed by ozonization resulted in the formation of **326,** but prolonged irradiation without ozonolysis gave a thermally more stable escape product **324** [318]. Kaupp and Zimmermann clarifed that **325** is thermally labile. At room temperature, **325** undergoes extremely rapid rearomatization to regenerate **323** ($t_{1/2} =$ 76 sec at 20°C) [318]. Kaupp also reported that the lower-temperature irradiation at 254 nm in ether/ethanol produces **327,** which is detectable by UV absorption [311].

Intramolecular excimer formation of 1,*n*-dinaphthylalkanes has been investigated in detail [319,320]. Chandross and Dempster reported the first example of intramolecular photodimerization of 1,3-di(1-naphthyl)propane (**328**), which was discovered during their intramolecular excimer study [321,322]. Upon irradiation of **328** in methylcyclohexane, **328** formed **329,** which is thermally unstable to give **330** (Scheme 92). Both **329** and **330** are converted substantially to **328** upon irradiation with 254-nm light. Similar irradiation of 1,3-di(2-naphthyl)propane did not give any photocycloadduct. The authors assumed that the intramolecular cyclomerization occurred via an intramolecular excimer. Davidson and Whelan reported further evidence for this assumption [320].

On the other hand, Put and DeSchryver reported that the reaction of di-(1-naphthylmethyl) ether leads to both the *endo* and *exo*-cyclomers **332** and **334** [323]. The *endo* cyclomer **332** smoothly rearranges to **333,** which is photolyzed to the starting **331.**

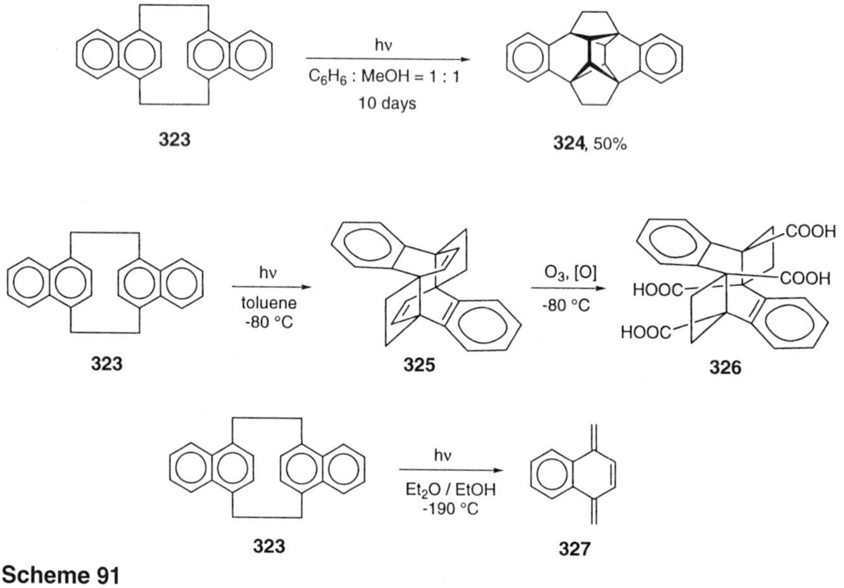

323            $\xrightarrow[\substack{C_6H_6 : MeOH = 1 : 1 \\ 10\ days}]{h\nu}$ **324,** 50%

323     $\xrightarrow[\substack{toluene \\ -80\ °C}]{h\nu}$ 325 $\xrightarrow[\substack{O_3,\ [O] \\ -80\ °C}]{}$ 326

323     $\xrightarrow[\substack{Et_2O\ /\ EtOH \\ -190\ °C}]{h\nu}$ 327

Scheme 91

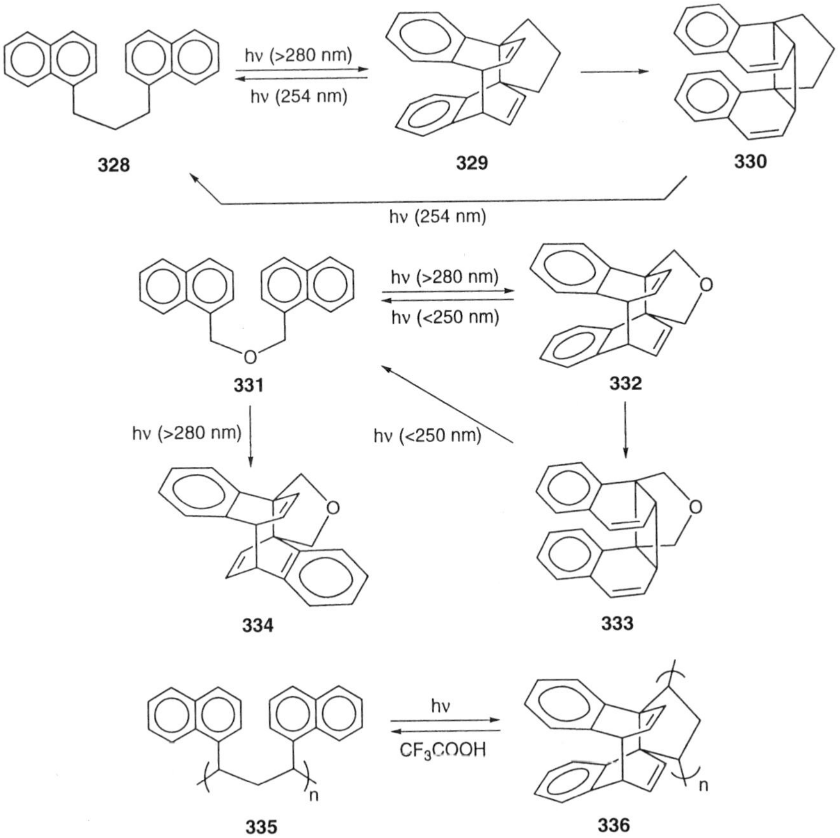

Scheme 92

Irie and co-workers studied the intramolecular photocycloaddition of naphthyl groups in poly(1-vinylnaphthalene) (**335**) in cyclohexane, benzene, and dichloromethane [338]. In cyclohexane, the reaction proceeds via first-order kinetics to a high conversion of 70%, whereas in dichloromethane, the reaction levels off at a very low conversion of 20%. Quenching and sensitizing experiments proved that the triplet mechanism is predominant in cyclohexane, whereas in dichloromethane, a singlet mechanism is predominant. The addition of trifluoroacetic acid to the illuminated sample restores the initial absorbance.

Irradiation of poly(ethylene glycol) labeled at the chain terminal with two 2-naphthyl groups **337** in NaY zeolite leads to formation of intramolecular photocyclomers **338** to the exclusion of intermolecular products [325] (Scheme 93). These results are explained in terms of the compartmentalization of the guest molecules in the supercages of NaY zeolite. Thus, this work demonstrates the util-

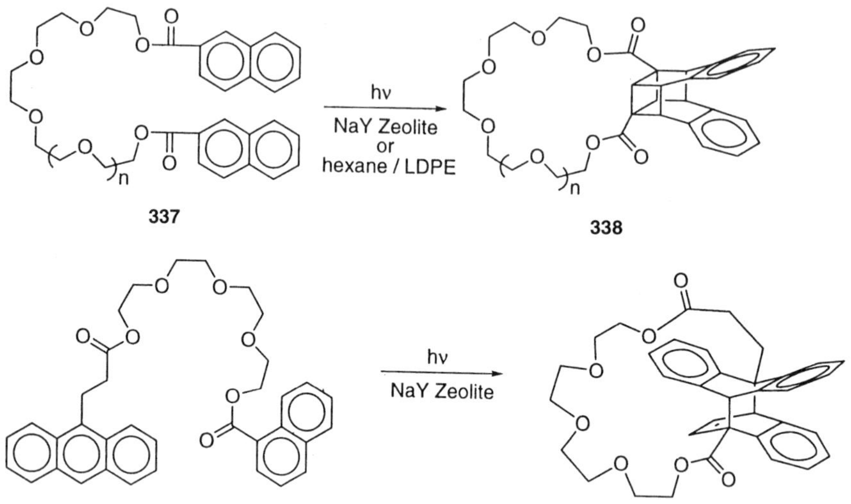

Scheme 93

ity of the micropores of zeolites to promote the formation of large-ring compounds in synthetically useful quantities. Of particular interest is the enhanced formation of the cross-photocyclomers **340** of anthracene and naphthalene derivatives in high yields. This process is not observed in homogeneous solutions.

The photocycloaddition of **337** has been investigated in hexane solutions and in low-density polyethylene (LDPE) films [326]. At loading levels less than 1×10^{-2} mol/f-film, irradiation of **337** in LDPE leads to the formation of **338** to the exclusion of intramolecular products again. On the other hand, irradiations in hexane containing 1×10^{-3} M of **337** produce large amounts of oligomer materials. The difference between the results in solution and the film is explained in terms of the compartmentalization of the guest molecules in the LDPE amorphous regions.

Intramolecular photocycloaddition of naphthalene and anthracene has been studied by Chandross and Schiebel [327]. At concentrations above 10^{-3} M, bimolecular photodimerization of the anthracene occurs in deaerated methylcyclohexane solution. In contrast, irradiation of much more dilute (~ 2×10^{5} M) solutions resulted in the formation of intramolecular adduct **342** (Scheme 94). Bouas-Laurent et al. showed that the CH$_2$—O—CH$_2$ link is more efficient than the (CH$_2$)$_3$ chain in bridging the two chromophores [328]. Irradiation of diethyl ether or methylcyclohexane solution of **343** (5×10^{-5} M) with a high-pressure mercury lamp and liquid filter ($\lambda > 335$ nm) gave a single photoadduct **344**, which was isolated quantitatively. The quantum yield of **344** is 10 times higher than that of **342**.

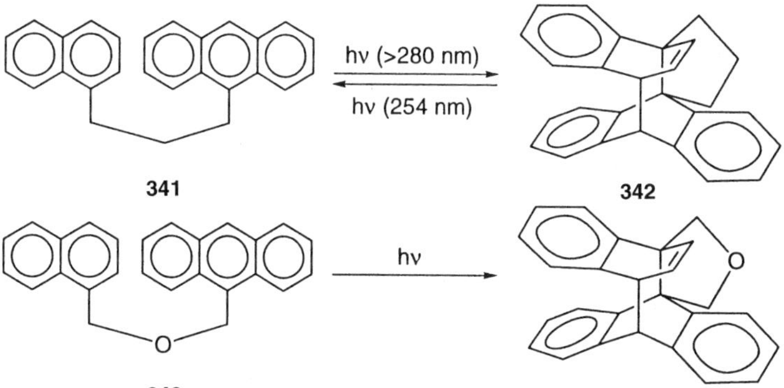

Scheme 94

Intramolecular (4 + 4) photodimerization of two anthryl groups has been summarized in excellent reviews written by Bouas-Laurent et al. and Becker [329,330]. Until 1980, intramolecular photocyclomerization of various bianthryl compounds in which two 9-anthryl groups are linked by ethylene, methylene, hydroxymethyl, carbonate, amide, and azo linkages **345a–j** and [2.2]anthracenophanes **347a–b** had been investigated [301,331–346] (Scheme 95). Photoreaction of all these compounds gave $(4\pi + 4\pi)$ photocycloadducts at the 9,10-position of the anthracene rings, with concomitant photocycloreversion at shorter-wavelength irradiation, or the thermal reversion. The so-called "$n = 3$ rule," which represents suitable excimer conformation and photodimerization, is not valid in the case of dianthrylalkanes, whose photodimerizations are possible even in linking by C1 chains (**345d–f**) [335,339]. Photochemical cyclomerization of **345c** ($n = 3$) is less efficient than that of **345a** ($n = 2$) [330,346].

Becker et al. prepared a series of 10-substituted and 10,10′-disubstituted 1,2-di(9-anthryl)ethanes **349** and their photochemical properties have been studied by determining the quantum yields for intramolecular $(4\pi + 4\pi)$ photocyclomerization and the quantum yields of fluorescence in both cyclohexane solutions [347] (Scheme 96). The quantum yields for the intramolecular cyclization of monosubstituted 1,2-dianthrylethanes are higher than those for disubstituted analogs (see Table 8). Bulky substituents increase the fluorescence efficiencies and decrease the quantum yields for cycloaddition. The rate of the cycloreversion is enhanced by the addition of trifluoroacetic acid.

Bouas-Laurent and colleagues studied the photocyclomerization of bis-9-anthrylmethyl ether **351a** and its derivatives **351b–f** [348] (Scheme 97) and concluded that the three-membered ether linkage enable more relevant conformations for efficient photocyclomerization than the trimethylene chain (see Table 9).

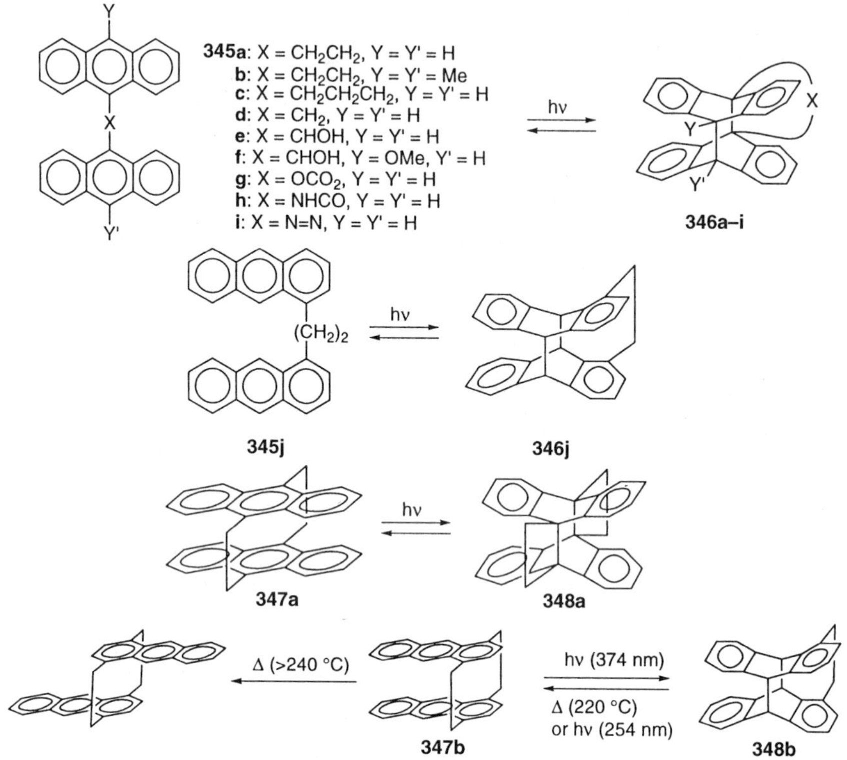
345a: X = CH2CH2, Y = Y' = H
b: X = CH2CH2, Y = Y' = Me
c: X = CH2CH2CH2, Y = Y' = H
d: X = CH2, Y = Y' = H
e: X = CHOH, Y = Y' = H
f: X = CHOH, Y = OMe, Y' = H
g: X = OCO2, Y = Y' = H
h: X = NHCO, Y = Y' = H
i: X = N=N, Y = Y' = H
hν
346a–i
(CH2)2
hν
345j
346j
hν
347a
348a
Δ (>240 °C)
hν (374 nm)
Δ (220 °C)
or hν (254 nm)
347b
348b

Scheme 95

Table 8 Quantum Yields for Intramolecular Dimerization and Fluorescence Quantum Yields of **350**

R	R'	Φ_{dimer}	Φ_f	R	R'	Φ_{dimer}	Φ_f
H	H	0.26	0.20	Me	Me	0.14	0.16
H	Me	0.26	0.15	MeO	MeO	0.11	0.30
H	MeO	0.24	0.14	MeO	AcO	0.068	0.45
H	AcO	0.14	0.40	AcO	AcO	0.015	0.74
H	Ph	0.034	0.61	n-Bu	n-Bu	$<<0.015$	0.70
				Ph	Ph	$<<0.015$	0.79

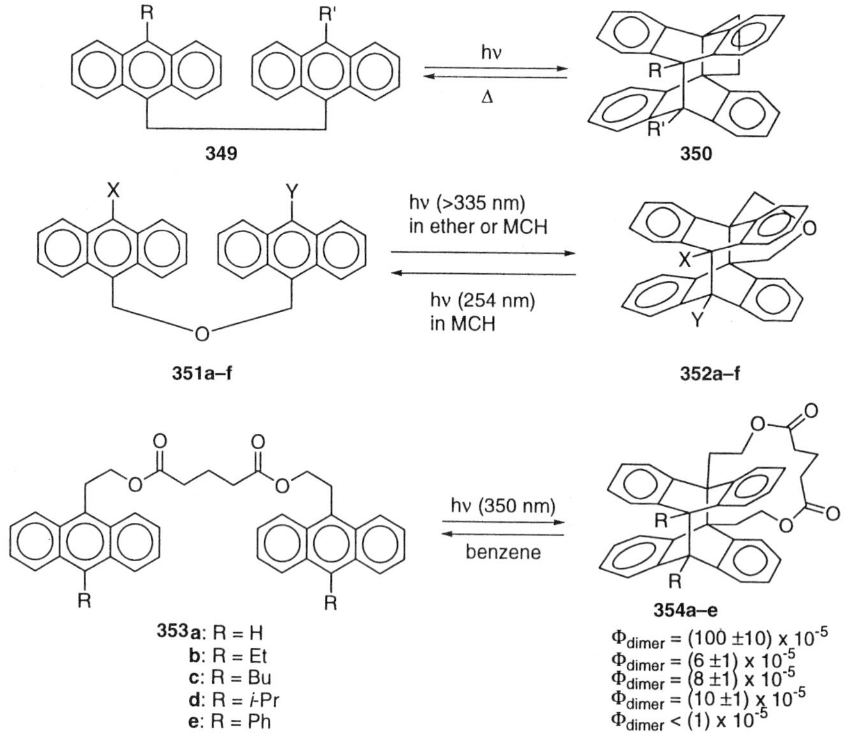

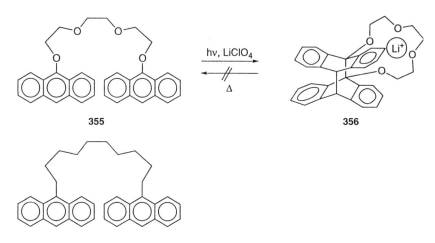

349 **350**

hν

Δ

351a–f **352a–f**

hν (>335 nm)
in ether or MCH

hν (254 nm)
in MCH

hν (350 nm)

benzene

353a: R = H
b: R = Et
c: R = Bu
d: R = *i*-Pr
e: R = Ph

354a–e

$\Phi_{dimer} = (100 \pm 10) \times 10^{-5}$
$\Phi_{dimer} = (6 \pm 1) \times 10^{-5}$
$\Phi_{dimer} = (8 \pm 1) \times 10^{-5}$
$\Phi_{dimer} = (10 \pm 1) \times 10^{-5}$
$\Phi_{dimer} < (1) \times 10^{-5}$

Scheme 96

hν, LiClO₄

Δ

355 **356**

357

Scheme 97

Table 9 Intramolecular Photodimerization and Photocycloreversion Quantum Yields for **351a–f** and **352a–f** in Methylcyclohexane at 20°C

Compounds	X	Y	$\Phi_{351\to352}$ ($\lambda = 366$ nm)	$\Phi_{352\to351}$ ($\lambda = 254$ nm)
351a–352a	H	H	0.32	0.64
351b–352b	Me	H	0.32	0.38
351c–352c	Ph	H	0.19	0.27
351d–352d	Br	H	0.18	0.28
351e–352e	Me	CN	0.25	a
351f–352f	Ph	CN	0.16	a

[a] Thermal instability of **352e–f** prevents the determination of $\Phi_{352\to351}$.

The substituent effect of 10,10′-positions of bis[2-(9-anthryl)ethyl] glutarates **353a–e,** which have more flexible chains, was investigated by irradiation with 350-nm light in dilute (10^{-4} *M*) deoxyganated benzene [349]. The quantum yields for the dimerization (Φ_{dimer}) falls with the steric bulk of the substituents.

Desvergne and Bouas-Laurent showed that the conformational properties of a polyoxyethylene chain linking two anthracene molecules provide an easy way to generate complexing agents by intramolecular photoaddition [350]. The quantum yields of photocyclomerization of **355** and **357** are determined in degassed benzene (Table 10). Product **355** photocyclomerizes with a high quantum yield in benzene or ether, but the photocycloisomer is not thermally stable and rapidly reverts to **355** ($t_{1/2} \approx 3$ min). In contrast, irradiation of **357** in ether or benzene only leads to recovered starting material. Irradiation of **355** in the presence of 2–5 equivalents of LiClO$_4$ gives **356** with a similar quantum yield, which is called the "cation-locked" cycloadduct. Because it is thermally stable up to melting point (206–210°C), it is poorly soluble.

The quantum yields in the photocyclomerization of **358–361** in methylcyclohexane at 20°C in the presence of LiClO$_4$ showed that the efficiency of the cyclomerization was dependent on the length of the chain and the presence of an

Table 10 Quantum Yields for Intramolecular Dimerization of **355** and **357**

	Φ_{dimer}		Φ_{dimer} (LiClO$_4$)	
	Benzene	Et$_2$O	Benzene	Et$_2$O
355	0.32	0.27	0.29	0.29
357	$<10^{-4}$	—	—	—

Table 11 Quantum Yields for Intramolecular Photodimerization

Compound[a]		Φ_{dimer}
An–O–O–O–An	358	0.045
An–O–O–O–O–An	359	0.17
An–O–O–O–O–An	360	0.19
An–O–O–O–O–An	355	0.26
An⌇An	361	0.0006
An⌇An	362	0.08
An–O⌇O–An	363	0.22
An–O⌇An	364	0.20

[a] An = 9-anthryl.

oxygen atom in the linking chains [351] (see Table 11). The photocyclomerization of **362–364** in methylcyclohexane is accelerated by the introduction of an oxygen atom into the 4-methylene chain [352].

The presence of the $SiMe_2$ group strongly hindered the formation of photocyclomers **366** and **368** where the bonds were formed between 9,10- and 1′,4′-positions [352,353]. Irradiation of di(9-anthryl)dimethylsilane **369** gave a single photocyclomer **370** by intramolecular (4 + 4) photodimerization without the formation of silacyclopropane **371** [354] (Scheme 98).

Triplet-state reactivity of di(9-anthryl)ethane **345a** was examined by the irradiation in the presence of biacetyl [355]. Photoexcitation of biacetyl in a benzene solution containing **345a** filtered by potassium chromate aqueous solution (λ > 420 nm) gave unprecedented Diels–Alder type adduct **372** (Scheme 99). Because the phosphorescence of biacetyl is quenched by **345a** in a nearly diffusion-controlled rate, the sensitized photochemical isomerization is explicable in terms of triplet energy transfer. Isomerization of **373** sensitized by biacetyl gave (4 + 2) adducts **374** and **375** in a similar manner. The ground-state geometry of **377** is characterized by partially overlapping aromatic moieties, as is evident from its broadened UV-absorption spectrum. Upon photoexcitation by biacetyl sensitization, **377** undergoes (4 + 4) cycloaddition with high quantum yield to give **378** and **379.** In contrast, the molecular geometry of **380** is characterized by *anti* conformation. Biacetyl-sensitized photoreaction of **380** leads to the exclusive formation of **381,** whereas the direct irradiation of **380** gives **382.**

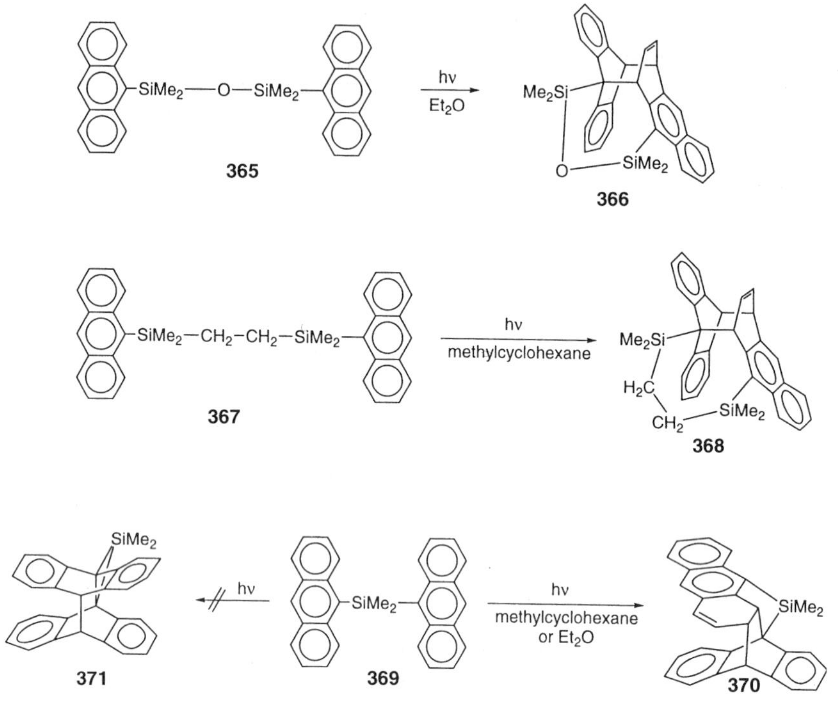

Scheme 98

Quantum yields of various bianthryl compounds **345c, 383a–l** bearing a ketone carbonyl or hydroxy group under direct irradiation in cyclohexane or methylcyclohexane are shown in Table 12 [356]. All these compounds gave intramolecular $(4\pi + 4\pi)$ photocycloadducts at their 9,10–9′,10′-positions. Chemical yields are between 80% and 90%. The observed differences in cyclization efficiency are attributable to the differences in ground-state conformations. Only in the case of di-9-anthryl ketone **384,** direct irradiation gave $(4 + 2)$ adduct **385** in low quantum yield [357] (Scheme 100). The course of photoisomerization of **373** under direct irradiation is dependent on the substrate concentration, affording mainly $(4 + 2)$ cyclomers **374** and **375** in dilute soution, but $(4 + 4)$ cyclomer **376** in concentrated solution.

Photoirradiation of a tetrahydrofuran (THF) solution of calix[4]arene **386** having two anthryl groups produces **387** (Scheme 101). Encapsulation of alkali metal ions into the ionophoric cavities of **386–387** was investigated by Shinkai and co-workers [358]. Compound **386** itself showed the poor ion affinity and the poor ion selectivity, whereas **387** showed the much improved ion affinity and the sharp Na^+ selectivity.

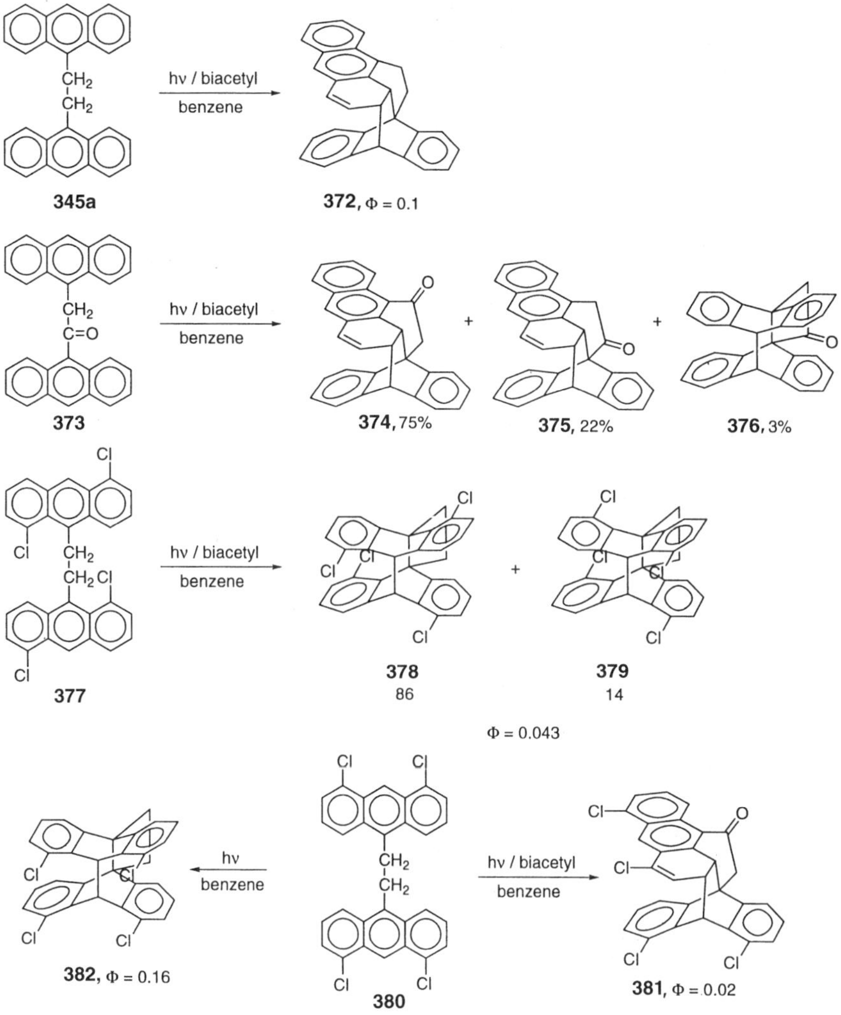

Scheme 99

Red and purple photochromic molecules **388a–389a** and **388b–389b** undergo intramolecular photocycloaddition on irradiation with white light to form pale yellow and colorless adducts **390a–b,** respectively, which undergo the reverse reactions thermally or on exposure to UV light [359] (Scheme 102).

One molecule of **391** displays a positive cooperative effect in binding two sodium cations when **391** is treated with sodium perchlorate, whereas **391** generates a 1 : 1 complex with K$^+$ [360]. The photocyclomerization was demonstrated

Table 12 Fluorescence Quantum Yield and
Quantum Yield for Intramolecular Dimerization

Compound[a]		Φ_f	Φ_{dimer}
An⌒⌒An	**345c**	0.47	0.14
An(OH)⌒An	**383a**	0.27	0.14
An⌒(OH)⌒An	**383b**	0.48	0.046
An⌒(OH)⌒An	**383c**	0.24	0.021
An⌒(OH)⌒An	**383d**	0.15	0.23
An⌒(OH)⌒An	**383e**	0.07	0.23
An⌒(O)⌒An	**383f**	0.009	0.25
An⌒(O)⌒An	**383g**	0.007	0.22
An⌒(O)⌒An	**383h**	<0.001	0.40
An⌒(O)⌒An	**383i**	<0.001	0.19
An⌒(O)⌒An	**383j**	<0.001	0.17
An⌒(O)⌒An	**383k**	<0.001	0.65
An⌒(O)⌒An	**383l**	<0.001	0.72

[a] An = 9-anthryl.

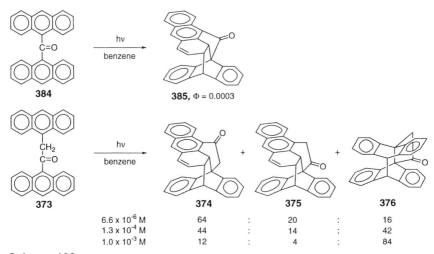

384 → **385**, Φ = 0.0003

373 →

	374		**375**		**376**
6.6 x 10⁻⁶ M	64	:	20	:	16
1.3 x 10⁻⁴ M	44	:	14	:	42
1.0 x 10⁻³ M	12	:	4	:	84

| | 6.6×10^{-6} M | 64 | : | 20 | : | 16 |

Scheme 100

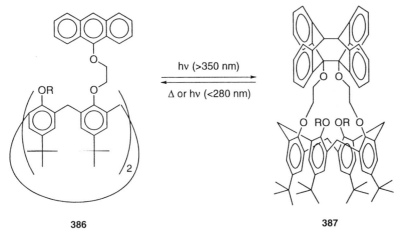

hv (>350 nm)

Δ or hv (<280 nm)

386   **387**

Scheme 101

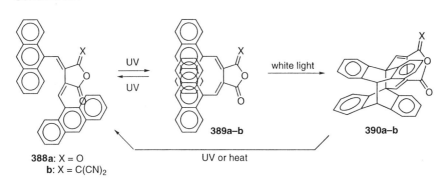

UV / UV

white light

UV or heat

388a: X = O
b: X = C(CN)₂

389a–b   **390a–b**

Scheme 102

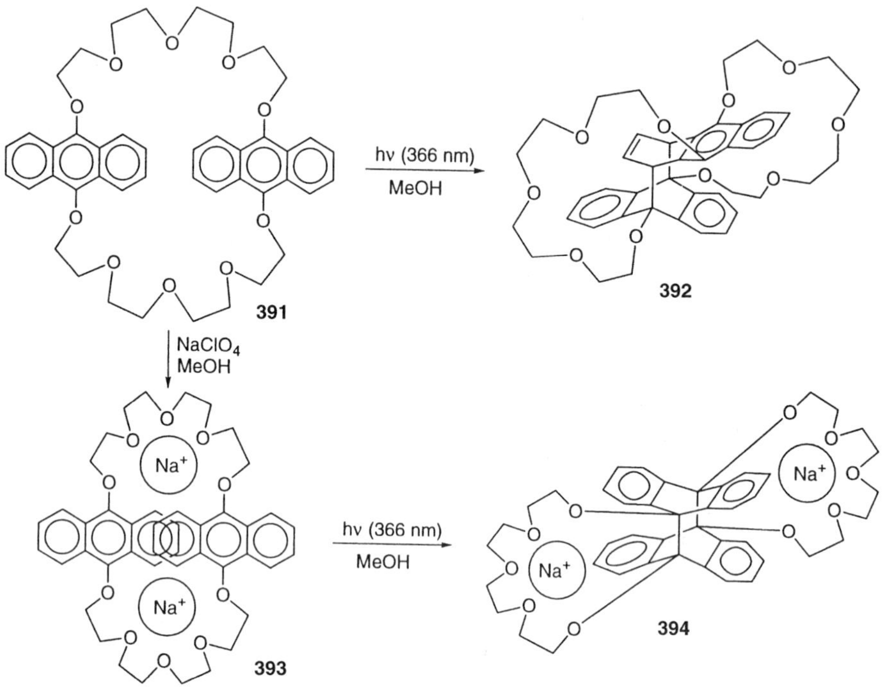

Scheme 103

to be regiospecifically directed by Na$^+$ (Scheme 103). In methanol, the free receptor **391** exclusively generates the 9,10–1′,4′-photoadduct **392,** whereas the sodium biscoronate leads to the 9,10–9′,10′ photoadduct **394.**

   Intramolecular photocyclomerization of **395a–b** was investigated in NaY zeolite [325]. Photoirradiation of **395a** included in NaY zeolite only gave h–h photocyclomer **396a,** but the irradiation of **395b** yields both h–h and h–t photocyclomers **396b, 397b** with the latter as the major product (Scheme 104). In these conditions, the two terminal aryl groups of one molecule of **395a–b** are included within one supercage of the zeolite, and these photocyclomers are too large to escape from the zeolite supercage through a 7.4-Å window.

H.   Unusual Photocycloaddition to Aromatic Rings

Photochemical (2 + 2) and (4 + 4) photocycloaddition is well investigated and recent advances in this field easily predict the patterning of the photoreactivity. The intermolecular and intramolecular photocycloaddition of alkenes to benzene ring also has a citizenship in this field. On the other hand, most people believe the photocycloaddition of unsaturated compounds to naphthalene ring takes place at

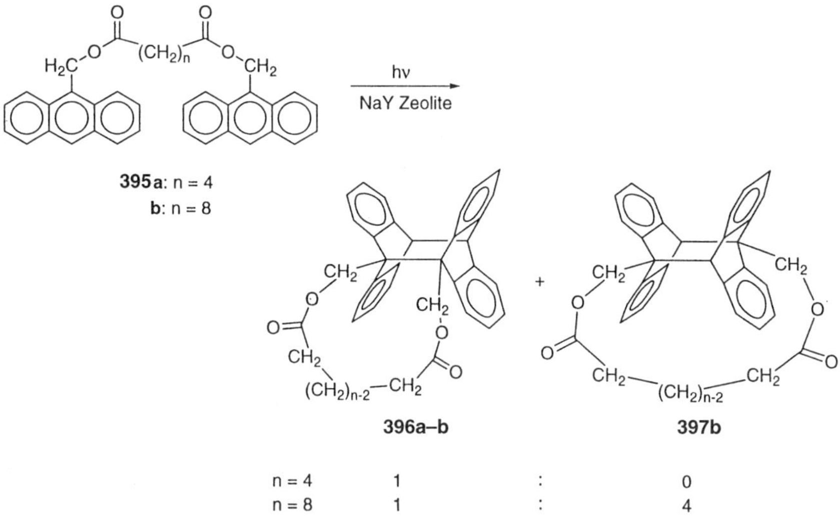

Scheme 104

1,2- and/or 1,4-positions. The most active sites of phenanthrene and anthracene derivatives are predicted at 9,10-positions of both compounds.

However, Kubo et al. found the unusual (3 + 2) intermolecular photocycloaddition of alkenes to the naphthalene ring at the 1,8-position. The same group also reported the (3 + 2) photocycloaddition of alkenes to phenanthrene ring at the 8,9-position.

Irradiation of benzene, ether, acetonitrile solutions containing dimethyl 1,4-naphthalenedicarboxylate **398a** and alkenes, such as isobutene, 1-hexene, cyclopentene, and styrene, leads (3 + 2) photocycloaddition to give 1,8-adducts efficiently [361]. This unusual photocycloaddition proceeds stereospecifically with retention of the stereochemistry of the alkenes, via the formation of an exciplex **401** and synchronous two-bond formation to give a zwitterionic intermediate **402** followed by proton transfer (Scheme 105). Stereospecificity and fluorescence quenching studies support the singlet mechanism. A ground-state interaction between dimethyl 1,4-naphthalenedicarboxylate **398a** and alkenes is absent by UV spectra. Application of the Rehm–Weller equation suggests that an electron transfer between these substrates would be endothermic, hence unlikely. In the exciplex, the alkene moiety may have some radical cationic nature and the diester moiety may have some radical anionic character. The reaction of methyl 4-cyano-1-naphthalenecarboxylate and 1,4-dicyanonaphthalene also proceeds in a similar manner [362].

(3 + 2) Photocycloadditions of 1,2-, 1,3- and 2,3-dicyanonaphthalenes to alkenes proceed to the 1,8-, 4,5-, and 1,8-positions of the naphthalene ring of di-

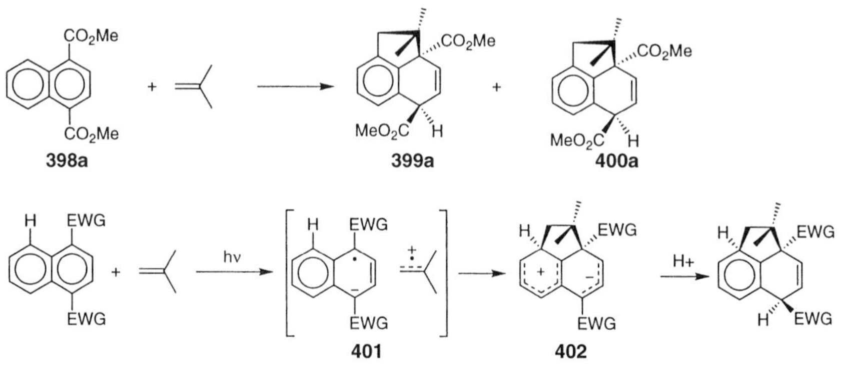

Scheme 105

cyanonaphthalenes, respectively [363,364] (Scheme 106). Observed addition sites of the alkenes is rationalized on the basis of the spin and charge densities of the radical anions of dicyanonaphthalenes.

Diastereoselectivity in the $(3 + 2)$ photocycloaddition of di-$(-)$-menthyl, di-$(-)$-8-phenylmenthyl, and di-$(-)$-bornyl derivatives (**398b–d**) is largely dependent on the concavity of the auxiliary, steric bulk of the substituents of the alkenes, and reaction temperature [365] (Scheme 107). The highest *de* value was obtained when 8-phenylmenthyl ester **398c** was employed, although the yield of the adducts was relatively low. The bornyl ester **398d** showed poor diastereoselectivity. The effect of the chiral auxiliary seems to be reasonably explained in terms of the concavity of the auxiliary. When α-methylstyrene was used as the alkene, two stereoisomers of $(3 + 2)$ adducts were obtained in lower yields than

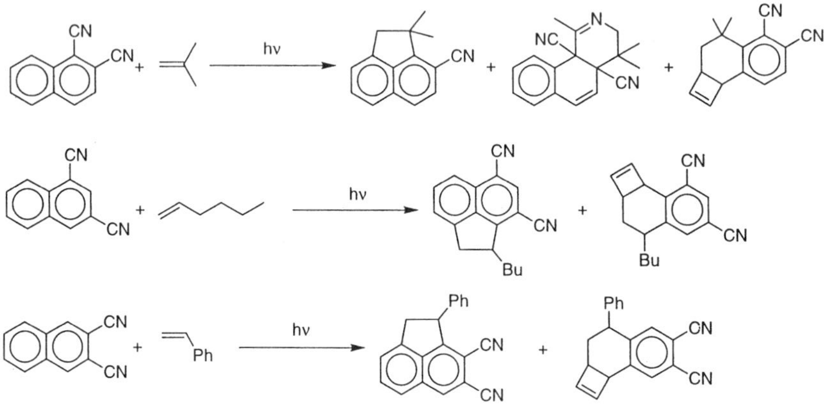

Scheme 106

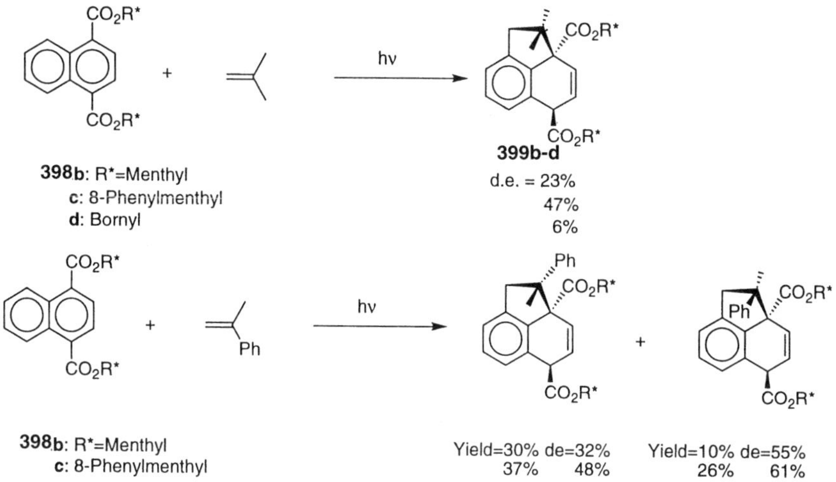

Scheme 107

398b: R*=Menthyl
  c: 8-Phenylmenthyl
  d: Bornyl

399b-d
d.e. = 23%
47%
6%

398b: R*=Menthyl
  c: 8-Phenylmenthyl

Yield=30% de=32%
37% 48%

Yield=10% de=55%
26% 61%

that of isobutene, but with higher *de* values. The *de* increases considerably up to 62% with the decrease of the reaction temperature.

Photoreactions of 9,10-dicyanophenanthrene with 1,3-butadiene gave a novel (3 + 2) photocycloadduct at the 1,10-position of the phenanthrene ring (**403**) to one double bond of 1,3-butadiene and a (2 + 2) cycloadducts at the 9,10-position (**404, 405**) [366] (Scheme 108). Compound **403** was formed from the singlet excited state of 9,10-dicyanophenanthrene, and **404–405** were from triplet state of 9,10-dicyanophenanthrene. This is the first example of the (3 + 2) photocycloaddition to the phenanthrene ring.

An intriguing intramolecular (3 + 2) photocycloaddition of alkenyl methyl 1,4-naphthalenedicarboxylates has been developed by Kubo et al. [300]. Intramolecular (3 + 2) photocycloaddition of alkenyl methyl 1,4-naphthalene dicarboxylates, which contain remote alkene moieties proceeded largely dependent on the chain lengths. On irradiation of benzene or acetonitrile solutions of **406a–c,** no reaction was observed despite its chain length. On the other hand, in the photoreactions of **407a–e,** intramolecular (3 + 2) photocycloadditions proceeded

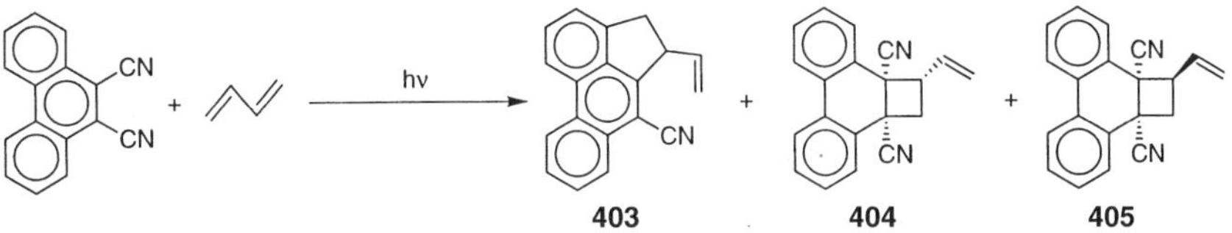

403 **404** **405**

Scheme 108

only in the case of $n = 3$–5 to give adducts **408b–d** in moderate yields (Scheme 109). Reactions of styrene derivatives **409a–d** are completely controlled by the chain lengths. Thus, in the case of $n = 2$, an exclusive intramolecular oxetane formation was observed to give **410a** in almost quantitative yield. On the contrary, in the cases of $n = 3$–5, only **411b–d** were obtained.

Unprecedented intramolecular photocycloaddition of styrene chromophore to the 10,10a-position of phenanthrene was recently reported by Mizuno et al. as discussed in previous section (Schemes 85 and 86) [307].

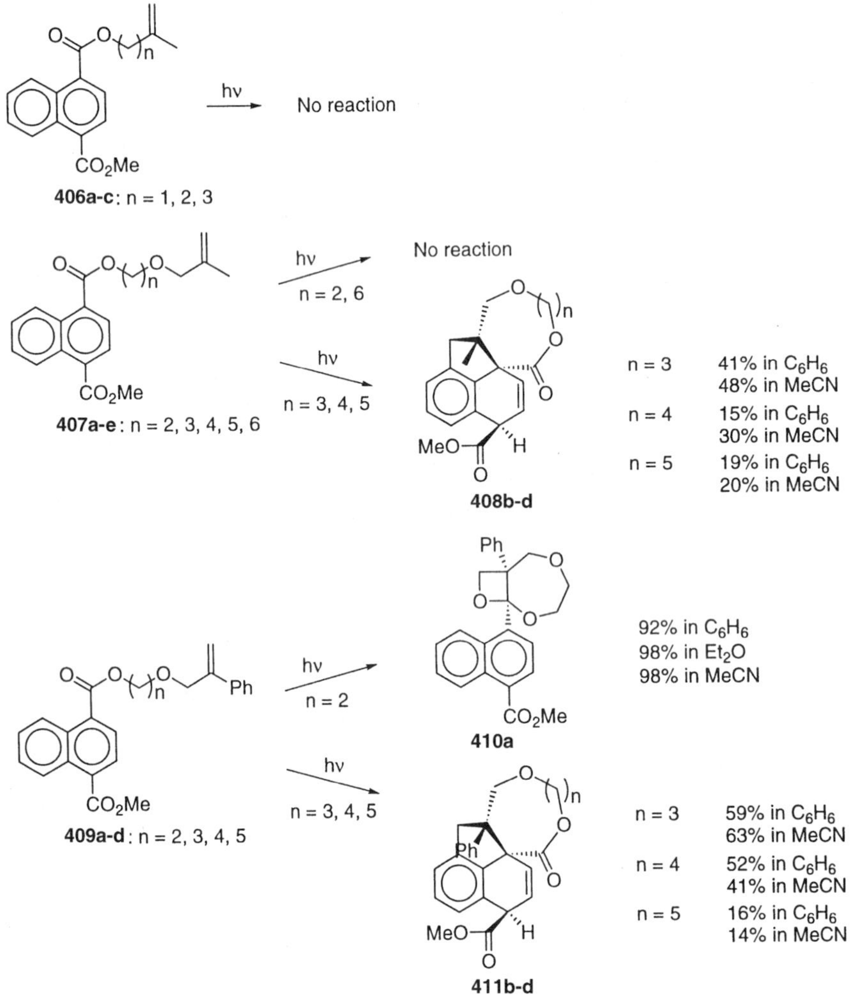

Scheme 109

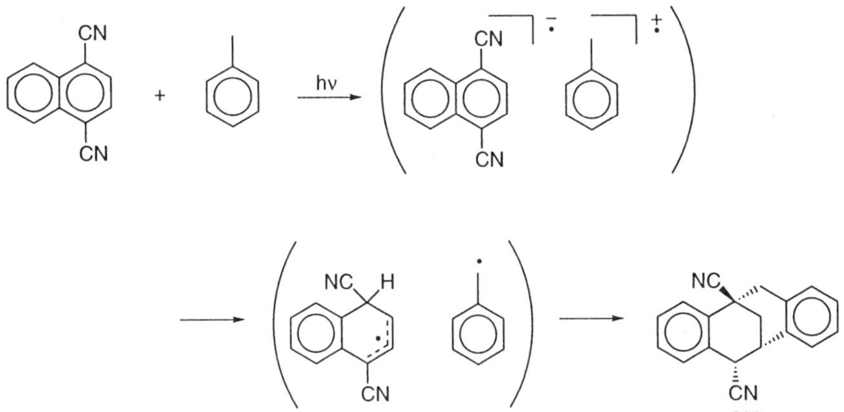

412

Scheme 110

Synthesis of tetracyclic compounds by the photoreaction of cyanoarenes with benzylic donors via photoinduced electron transfer is one of the unusual (3 + 3) photocycloaddition reactions (Scheme 110). It is not treated here and we concede to Albini et al.'s excellent review [36].

IV. PHOTOADDITION TO AROMATIC RINGS

A. Classification

The photoaddition to aromatic ring is simply classified in Table 13. Direct attack of nucleophiles such as cyanide anion and carbanion to aromatic rings in the excited species has been recognized by $S_{RN}2$ reactions [Eq. (4)], which were discussed in the past three decades [27,28]. We will discuss briefly some examples in this section, including intramolecular photocyclization such as photo-Smiles rearrangement.

Photoinduced electron-transfer reactions generate the radical ion species from the electron-donating molecule to the electron-accepting molecules. The radical cations of aromatic compounds are favorably attacked by nucleophiles [Eq. (5)]. On the contrary, the radical anions of aromatic compounds react with electrophiles [Eq. (6)] or carbon radical species generated from the radical cations [Eq. (7)]. In some cases, the coupling reactions between the radical cations and the radical anions directly take place [Eq. (8)] or the proton transfer from the radical cation to the radical anion followed by the radical coupling occurs as a major pathway. In this section, we will mainly deal with the intermolecular and intramolecular photoaddition to the aromatic rings via photoinduced electron transfer.

Table 13 Classification of Photoaddition to Aromatic Rings

Reactive species	Addend	Reaction type
$M \xrightarrow{h\nu} M^*$		
M^*	N_u^-	(4)
$D + A \xrightarrow{h\nu} D^{\bullet+} + A^{\bullet-}$		
$D^{\bullet+}$	N_u^-	(5)
$A^{\bullet-}$	E^+	(6)
$A^{\bullet-}$	$R^{\bullet}$	(7)
$A^{\bullet-}$	$D^{\bullet+}$	(8)

Note: M = molecule; N_u^- = nucleophile; D = electron donor; A = electron acceptor; E^+ = electrophile; R = radical; L = substituent; EDG = electron-donating group; EWG = electron-withdrawing group.

B. Intermolecular Photoaddition of Amines to Aromatic Rings

Photosubstitution of aromatic compounds by nucleophiles usually proceeds via the addition–elimination mechanism. Photochemical direct substitution of aromatic compounds by ammonia and primary and secondary amines were reviewed by Cornelisse and Havinga in 1975 [27]. The photosubstitution of 1-methoxy-4-nitronaphthalene with amines has been investigated by Bunce et al., as shown in Scheme 111 [367]. In this photoreaction, the primary amines are replaced by a nitro group via an S_N2Ar^* process. On the other hand, the secondary amines are substituted with a methoxy group via a photoinduced electron transfer from amines to the naphthalene derivatives. Cantos et al. have reported that the photosubstitution of 4-nitroanisole with n-hexylamine gives a methoxy-substituted product via an electron-transfer process, whereas the reaction with ethyl glycinate proceeds via an S_N2Ar^* process to give a nitro-group-substituted product (Scheme 111) [368]. The photo-Smiles rearrangement as an intramolecular process, which was first reported by Matsui et al., proceeds via a similar pathway (Scheme 112) [369]. The photorearrangements of 3-nitro- and 4-nitro-phenyl 2-aminoethyl ether and 4-nitrophenyl 2-anilinoethyl ether have been investigated by Wubbels and Mutai, in-

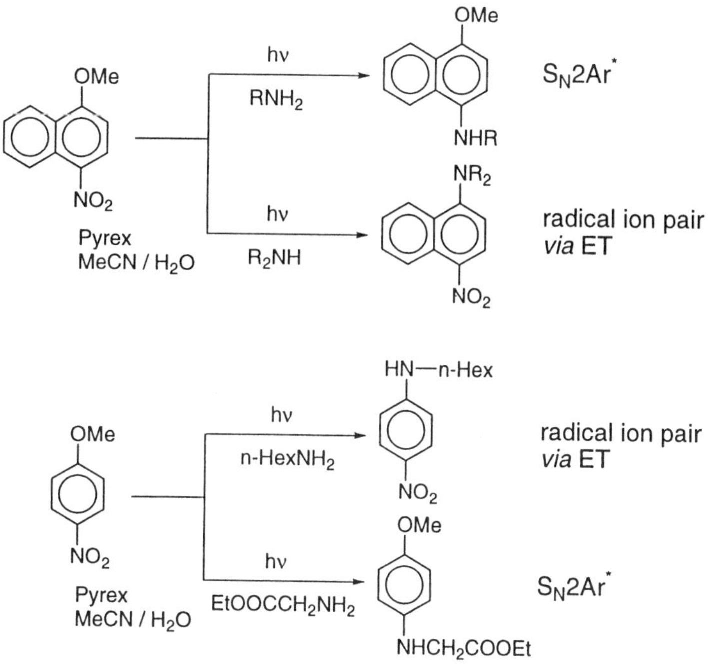

Scheme 111

dependently [370–372]. The reactivity depends on the position of substituents. The latter photoreaction proceeds via a photoinduced electron-transfer mechanism. Recently, the nucleophilic aromatic substitution has been applied to the photochemical synthesis of aniline and aromatic amides via the regioselective amination of *m*-dinitrobenzene assisted by the fluoride anion (Scheme 112) [373,374].

Photoinduced electron-transfer reaction of aromatic compounds with amines is one of the most fundamental reactions in the electron-donor–acceptor systems, which was recently reviewed by Lewis [35]. Because of the low oxidation potentials of the amines, the photoinduced one-electron transfer from the amines to the excited singlet states of aromatic hydrocarbons (^{1}ArH*) readily occurs to give the radical cations of amines and the radical anions of aromatic compounds even in the less polar solvents.

The fluorescence quenching experiments of aromatic hydrocarbons by tertiary amines, including *N,N*-dialkylanilines, in less polar solvents show the typical exciplex emissions [382–384], but products are not obtained or inefficiently produced. On the other hand, in polar solvents such as acetonitrile or methanol, the photoinduced electron transfer from the amines to ^{1}ArH* efficiently occurs to give the addition products. Interestingly, some primary and secondary aliphatic and aromatic amines caused the photoinduced electron transfer even in nonpolar solvents.

The photoamination by use of ammonia and some primary amines such as nucleophiles to the radical cations has been extensively developed by Yasuda and Shima [30,32]. The key point in this photoreaction is that the oxidation potentials

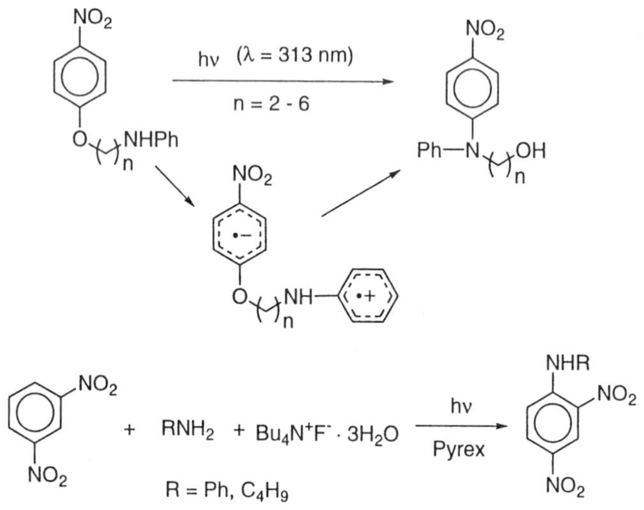

Scheme 112

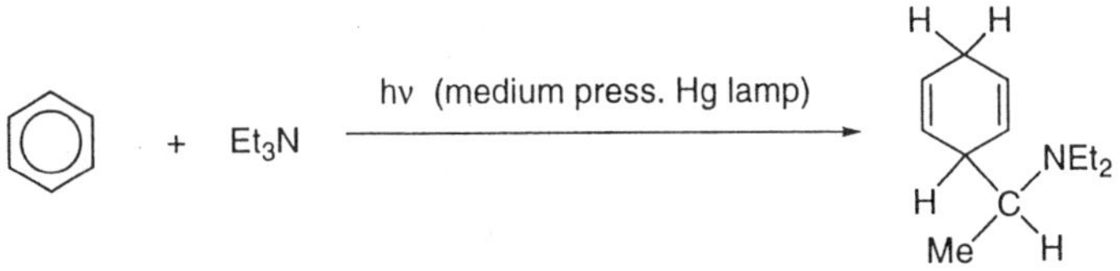

Scheme 113

of ammonia and primary amines are not so low. Therefore, the radical cations of electron-donating molecules generated from the photoinduced electron transfer are not reduced by these amines. Although there have been several reviews of the photochemistry of aromatic compounds with amines via electron transfer [30,32,35], we will focus on the photoaddition of amines to aromatic rings in this section.

The 1,4-photoaddition of aliphatic amines with benzene via photoinduced electron transfer was first reported by Bryce-Smith more than 30 years ago [375–378]. In the photoreaction of triethylamine with benzene, the proton transfer from the radical cation of triethylamine to the radical anion of benzene is proposed as a probable pathway (Scheme 113). In the case of tertiary amines, the photoaddition is accelerated by the addition of methanol or acetic acid as a proton source. Similar photoaddition of diethyl ether to benzene takes place assisted by trifluoroacetic acid, where methanol is not affective [379]. In these photoreactions, α-hydrogen next to the heteroatom moves to the radical anion of benzene as a proton, followed by radical ccoupling to give 1,4-addition products. Similar photoaddition of amines to the benzene ring has been reported by Ohashi et al. [380,381].

The fluorescence of anthracene in benzene is efficiently quenched by N,N-dimethylaniline and a strong exciplex emission appears in a longer wavelength than the emission of anthracene [382–384]. However, the addition product was not obtained at all, except the (4 + 4) anthracene dimer (Scheme 114). In contrast, the addition product and reductive dimerization product of dimethylaniline to the anthracene ring are produced via photoinduced electron transfer, which was first reported by Pac and Davidson [385–387]. In the case of N-methylaniline, some addition products are obtained both in nonpolar and polar solvents [386–389].

Photoaddition of N-methylaniline to phenantherene ring via C—O cleavage also occurs to give N-, o-, and p-(phenanthrylmethyl)-substituted N-methylanilines, as shown in Scheme 115 [390]. It is explained in terms of one-electron transfer from N-methylaniline to the phenanthrene derivatives, followed by proton transfer and the radical cross-coupling.

Nucleophilic attack of amines to the radical cations of aromatic compounds [Eq. (6)] is much more favorable than the direct attack to the aromatic rings bearing electron-donating substituents or unsubstituted aromatic hydrocarbons [Eq.

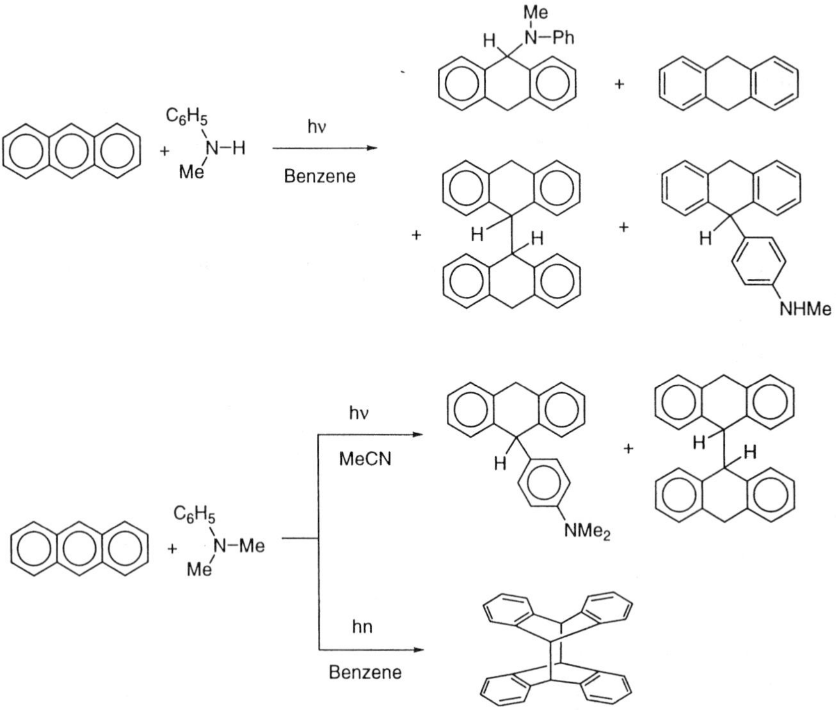

Scheme 114

(4)]. Yasuda, Pac, and Shima reported the photoamination of aromatic compounds by use of ammonia and primary amines as nucleophiles in the presence of *m*-dicyanobenzene as an electron acceptor [30,32]. Irradiation of an acetonitrile–water (9 : 1) solution containing phenanthrene and *m*-dicyanobenzene in the presence of ammonia or primary amines gives 9-amino-9,10-dihydrophenanthrene deriva-

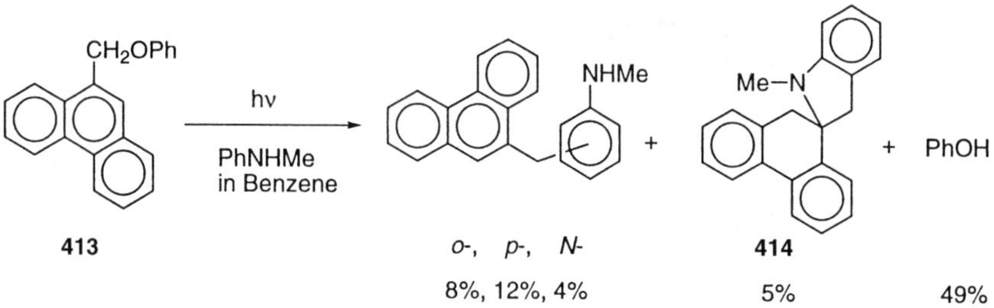

413

o-, p-, N-

8%, 12%, 4%

414

5%

+ PhOH

49%

Scheme 115

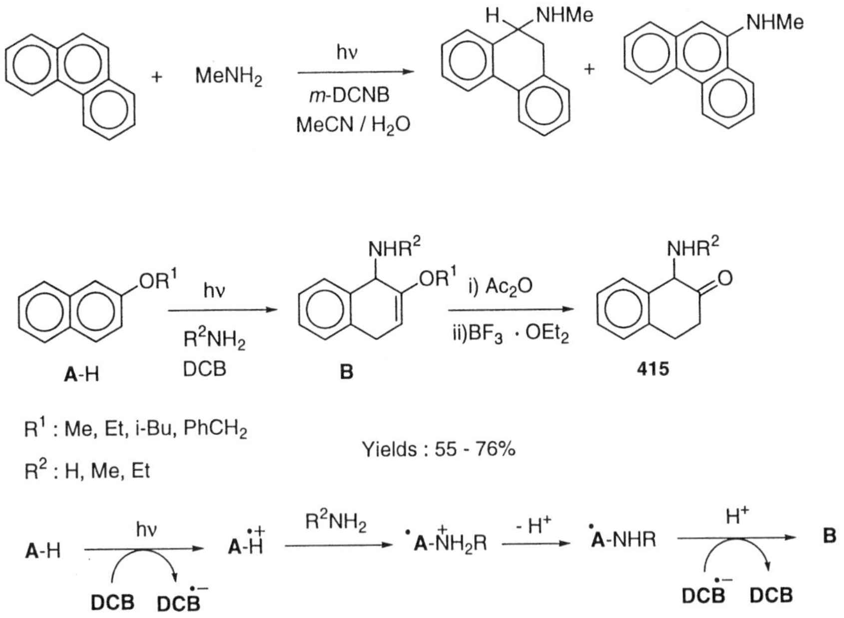

R^1 : Me, Et, i-Bu, PhCH$_2$

R^2 : H, Me, Et

Yields : 55 - 76%

Scheme 116

tives in high yields [391,392] (Scheme 116). This is a general method for the amination of electron-donating substrates, including arylalkenes and arylcyclopropanes [393–395]. For example, the photoamination of 2-naphthol derivatives by this method is applied to the synthesis of 1-amino-2-tetralone derivatives [396].

C. Intermolecular Photoaddition of Other Nucleophiles

Nucleophilic addition to the radical cations of aromatic hydrocarbons in the presence of p-dicyanobenzene generated through a photoinduced electron transfer has been developed to the photo-Birch reduction using sodium borohydride, photocyanation, and photophosphorylation [Eq. (6)]. Irradiation of an aqueous acetonitrile solution containing phenanthrene and sodium borohydride in the presence of p-dicyanobenzene as an electron acceptor afforded 9,10-dihydrophenanthrene in a high yield [397,398] (Scheme 117). Similar photoreduction of aromatic hydrocarbons such as naphthalene, anthracene, and their derivatives occurs smoothly to give the corresponding dihydroaromatic compounds in high yields. The radical cations of aromatic hydrocarbons were also attacked by cyanide anion and trialkyl phosphite to produce the reductive cyanation and phosphonation products [399–401].

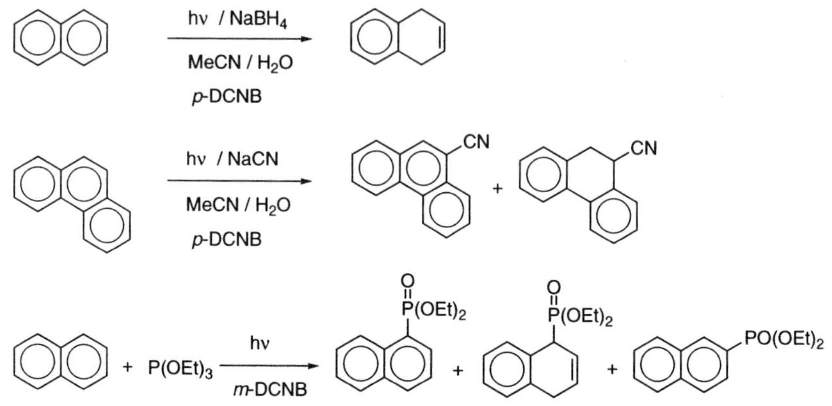

Scheme 117

D. Intermolecular Photoaddition of Electrophiles

The radical anions of electron-deficient aromatic compounds and aromatic hydrocarbons, which are generated by photoinduced electron transfer, can be protonated by protic solvents or by the radical cations of amines to produce their neutral radicals [Eq. (7)]. Dispropotionation of the radicals yields the reduction products. The radical anions of 1,1-diphenylethene in the presence of 1,4-dimethoxynaphthalene as an electron donor is also protonated by protic solvents to give Markownikoff adducts [402,403] (Scheme 118).

Photoinduced carboxylation of aromatic hydrocarbons via their radical anions in the presence of tertiary amines has been reported by Tazuke and Ozawa [404]. Similar photofixation of carbon dioxide on styrene has been reported Toki et al. [405]. In these photoreactions, the anionic part of the radical anions are trapped by electrophiles such as proton and carbon dioxide (Scheme 119).

Mizuno and his co-workers reported the photooxygenation of 9,10-dicyanoanthracene (DCA) in the presence of biphenyl to give anthraquinone and

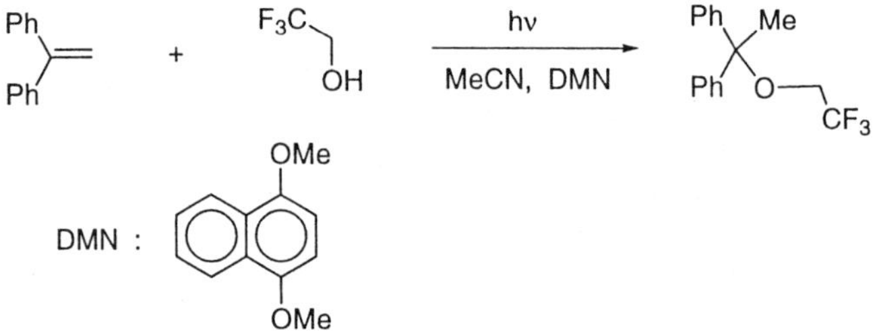

Scheme 118

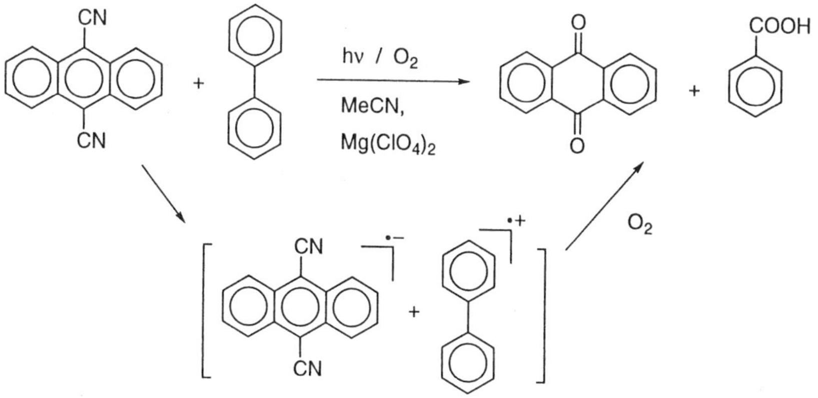

Scheme 119

benzoic acid, where both of the radical anion of DCA and the radical cation of biphenyl are attacked by molecular dioxygen [406–408]. Although the detailed mechanism is not clarified yet, probably in the initial stage both of the anion and radical sites of DCA$^{\cdot-}$ are oxygenated to give anthraquinone.

E. Intermolecular Photoaddition of Radical Species

Previously, Ohashi and his co-workers reported the photosubstitution of 1,2,4,5-tetracyanobenzene (TCNB) with toluene via the excitation of the charge-transfer complex between TCNB and toluene [409]. The formation of substitution product is explained by the proton transfer from the radical cation of toluene to the radical anion of TCNB followed by the radical coupling and the dehydrocyanation. This type of photosubstitution has been well investigated and a variety of examples are reported. Arnold reported the photoreaction of *p*-dicyanobenzene (*p*-DCB) with 2,3-dimethyl-2-butene in the presence of phenanthrene in acetonitrile to give 1-(4-cyanophenyl)-2,3-dimethyl-2-butene and 3-(4-cyanophenyl)-2,3-dimethyl-1-butene [410,411]. The addition of methanol into this reaction system affords a methanol-incorporated product. This photoreaction was named the photo-NOCAS reaction (photochemical nucleophile–olefin combination, aromatic substitution) by Arnold. However, a large number of nucleophile-incorporated photoreactions have been reported as three-component addition reactions via photoinduced electron transfer [19,40,113,114,201,410–425]. Some examples are shown in Scheme 120.

Photoaddition and substitution of electron-deficient aromatic compounds such as *o*-dicyanobenzene (*o*-DCNB), *p*-DCNB, and TCNB by use of group 14 organometallic compounds are classified to the reaction of the radical anions of electron-deficient aromatic compounds with carbon radical species generated

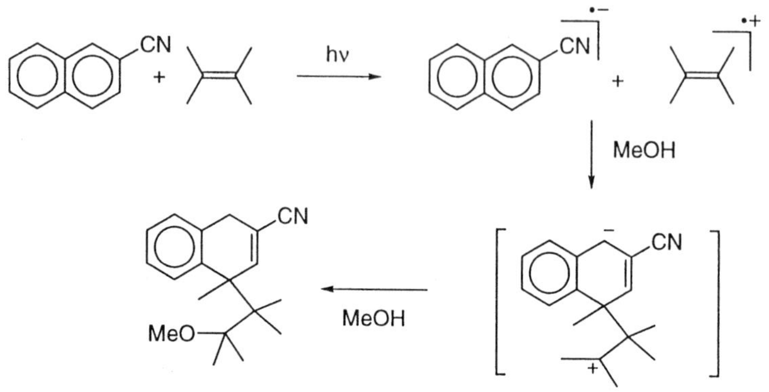

Scheme 120

from the carbon–metal bond cleavage of the radical cations of group 14 organometallic compounds [Eq. (7)] [426–459]. Mizuno et al. first reported the photosubstitution of o-DCNB and p-DCNB using allyltrimethylsilane, giving 4-allylbenzonitrile exclusively [426–434]. They applied a variety of photoaddition and photosubstitution of electron-deficient aromatic compounds by the use of group 14 organometallic compounds. Irradiation of mono- and di-cyanonaphthalene derivatives with allyltrimethylsilanes afforded the tricyclic compounds in a regioselective and stereoselective manner, which are produced by the intramolecular (2 + 2) photocycloaddition of the reductive allylation products (Scheme 121). The nucleophile-assisted photocleavage of C—Si and Si—Si bonds of the radical cations of organosilicon compounds has been suggested by several groups. Schuster reported the photosubstitution of dicyanoaromatic compounds by use of tetraalkylborate via photoinduced electron transfer [460,461].

Nakadaira et al. have reported the photoalkylation and photosilylation of polycyanobenzenes by the use of disilanes [438–440]. They also reported that photosubstitution of TCNB by use of $(n\text{-Bu})_3\text{SnSiMe}_3$ (Scheme 122). In this photoreaction, the Sn—C bond of the radical cation is cleaved in preference to the Sn—Si and Si—C bonds to generate an n-butyl radical. Similar photosubstitution of 2,3-dicyanopyradines has been reported by Mizuno and his co-workers [437].

Albini and his co-workers have reported a new type of photoreaction of 1,4-dicyanonaphthalene with toluene and its related compounds to give the substitution products, the reductive arylmethylation products, and 1,3-photoaddition products [135,456–465] (Scheme 123). They reviewed a series of photoreactions and discussed a mechanism in detail [36]. They also reported the photosubstitution of TCNB by use of 2,2-dimethyl-1,3-dioxolane and 1-cyanoadamantane.

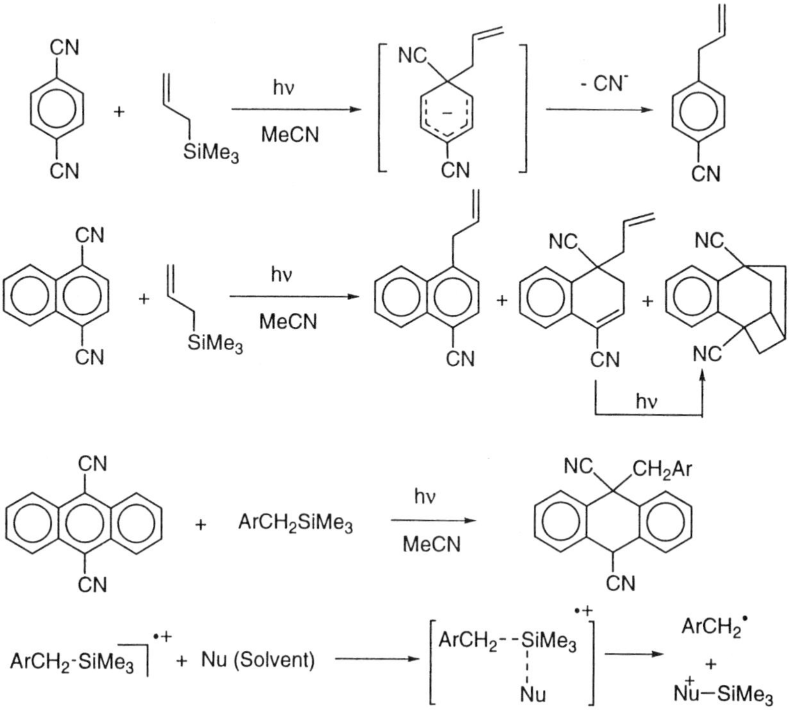

Scheme 121

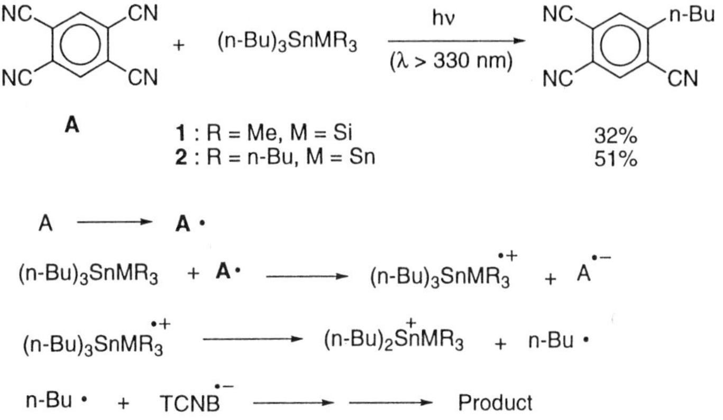

| | 1 : R = Me, M = Si | 32% |
| | 2 : R = n-Bu, M = Sn | 51% |

$$A \longrightarrow A \cdot$$

$$(n\text{-Bu})_3SnMR_3 + A\cdot \longrightarrow (n\text{-Bu})_3SnMR_3^{\bullet+} + A^{\bullet-}$$

$$(n\text{-Bu})_3SnMR_3^{\bullet+} \longrightarrow (n\text{-Bu})_2\overset{+}{Sn}MR_3 + n\text{-Bu}\cdot$$

$$n\text{-Bu}\cdot + TCNB^{\bullet-} \longrightarrow \longrightarrow Product$$

Scheme 122

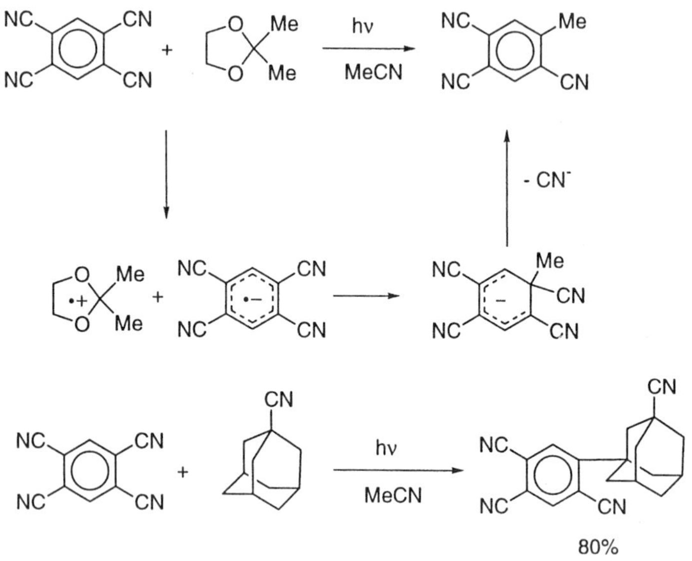

Scheme 123

F. Intramolecular Photocyclization Giving Azaheterocyclic Compounds

Intramolecular photocyclization is a useful method for the preparation of a variety of polycyclic compounds in a few steps. The photocyclization is classified into two types: One type is that intramolecular photocyclization bearing both electron-donating and electron-accepting chromophores has been initiated by the formation of intramolecular exciplex followed by electron transfer. When an intramolecular photoinduced electron transfer occurs from an amino group as an electron donor to an excited aromatic compound as an electron acceptor, a radical cation of the amino group and a radical anion of the aromatic compound are produced. The proton transfer from the radical cation of amino group to the radical anion of aromatic compound generates intramolecular 1,n-biradicals followed by the radical coupling, giving azaheterocyclic compounds. The other one is the intramolecular nucleophilic attack by amino group to the radical cation of aromatic ring in the presence of an electron acceptor, as discussed in Section IV.E.

The intramolecular photocyclization of 1-dimethylamino-3-phenylpropane to give a *meta* cycloadduct via photoinduced electron transfer was first reorted by Bryce-Smith et al. in 1973 [377]. Yoon and Mariano reported the intramolecular photocyclization of **416** in the presence of DCA to give **417** and **418** via the radical cation of amido moiety followed by desilylation (Scheme 124).

Sugimoto and his co-workers have systematically reported the intramolecular photocyclization of aromatic hydrocarbons such as naphthalene, phenan-

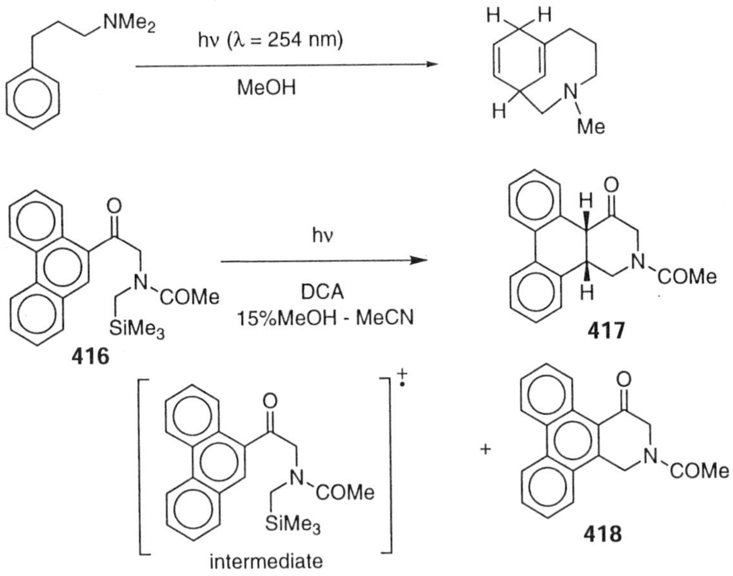

Scheme 124

threne, and anthracene derivatives having anilinoalkyl groups tethered by methylenes, ethers, and esters [390,466–476]. Irradiation of **419** in benzene efficiently affords the spiro compounds **420** in good yields, although the exciplex emission was not observed in these secondary amine derivatives [466]. The thermal or acid-catalyzed cycloreversion reaction occurs to produce the starting compounds **419** in quantitative yields (Scheme 125).

The reactivities of 9-(N-phenylaminoalkyl)phenanthrenes **421** and **424** depend on the length of methylene chain. In the case of $n = 1$, photo-Claisen-type rearranged products **422** are mainly produced [467] (Scheme 126). 9-Methylene-9,10-dihydrophenanthrene **423** is exclusively obtained in a 60% isolated yield

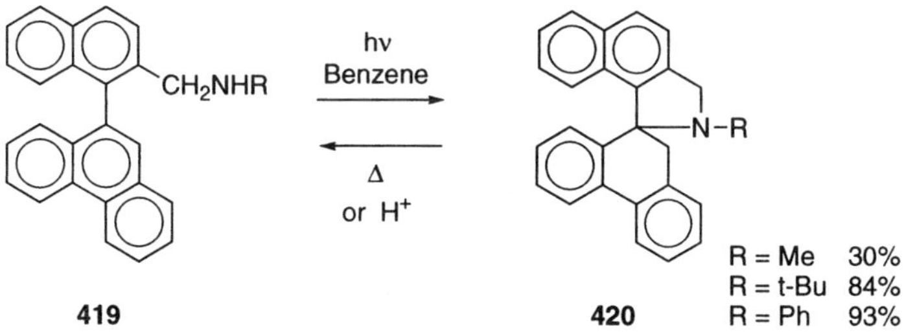

	R = Me	30%
	R = t-Bu	84%
	R = Ph	93%

419   **420**

Scheme 125

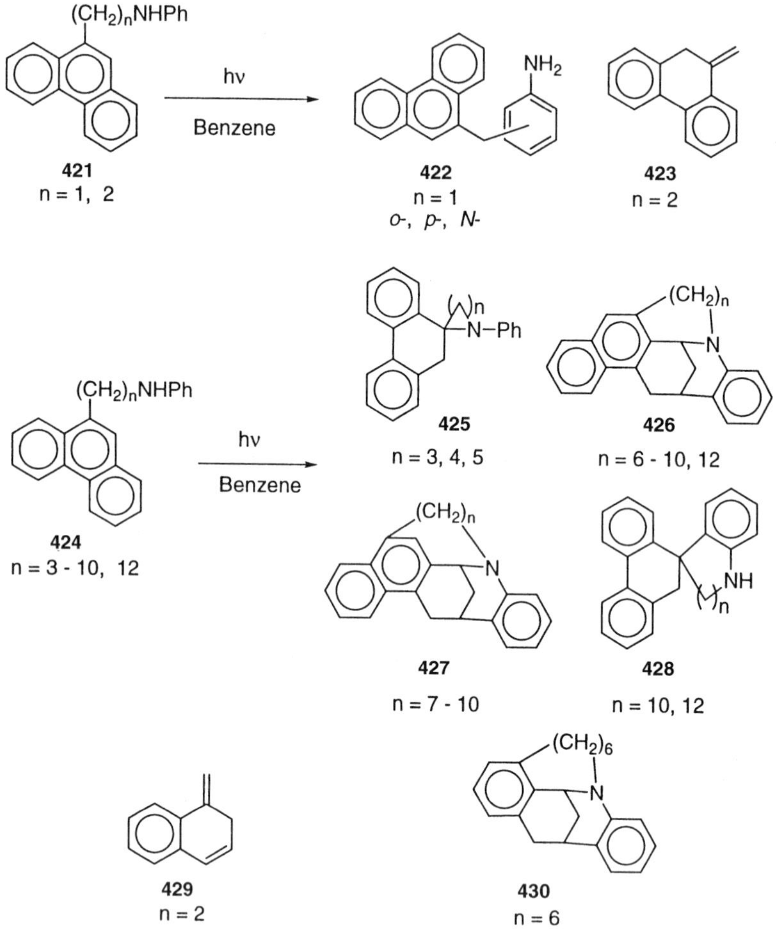

Scheme 126

from **421** ($n = 2$) [468,469]. Similarly, 1-methylene-1,2-dihydronaphthalene **429** is also obtained from 1-(N-phenylethyl)naphthalene [469]. Spiro compounds **425** and **428** are obtained in the photoreactions of **424** ($n = 3$–5, 10, 12). However, **424** ($n = 6$–10, 12) and 1-(N-phenylhexyl)naphthalene give tricyclic azaheterocyclic compounds **426** ($n = 6$–10, 12), **427** ($n = 7$–10), and **430**, respectively [470–472].

Sugimoto also reported the intramolecular photocyclization of **431–434**, giving a variety of tricyclic azaheterocyclic compounds and their related compounds (Scheme 127). The compounds **435** and **436** tethered by ether caused the C—O bond cleavage as a major pathway accompanying some cyclized products,

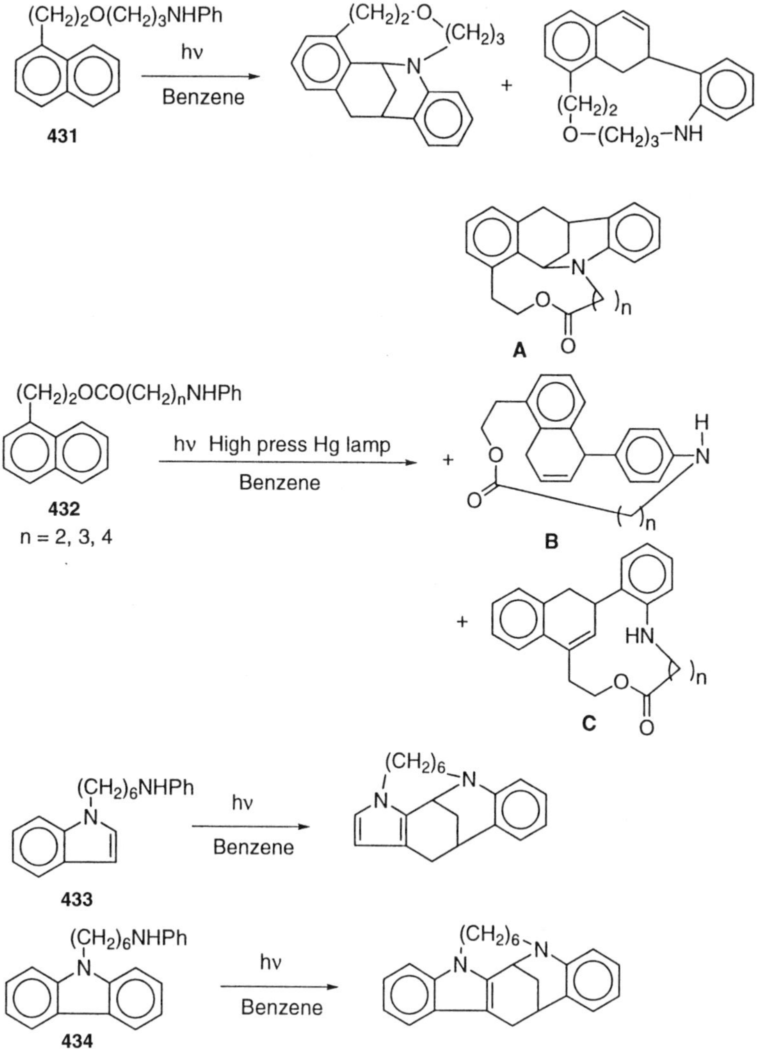

Scheme 127

as shown in Scheme 128. The structures of most of these photoproducts are determined by X-ray crystallography.

Lewis, Pandey, and Yasuda and co-workers have independently reported the intramolecular photocyclization of **437–439** via nucleophilic attack of amino groups to the aromatic ring in the presence of an electron acceptor [477–480]. The

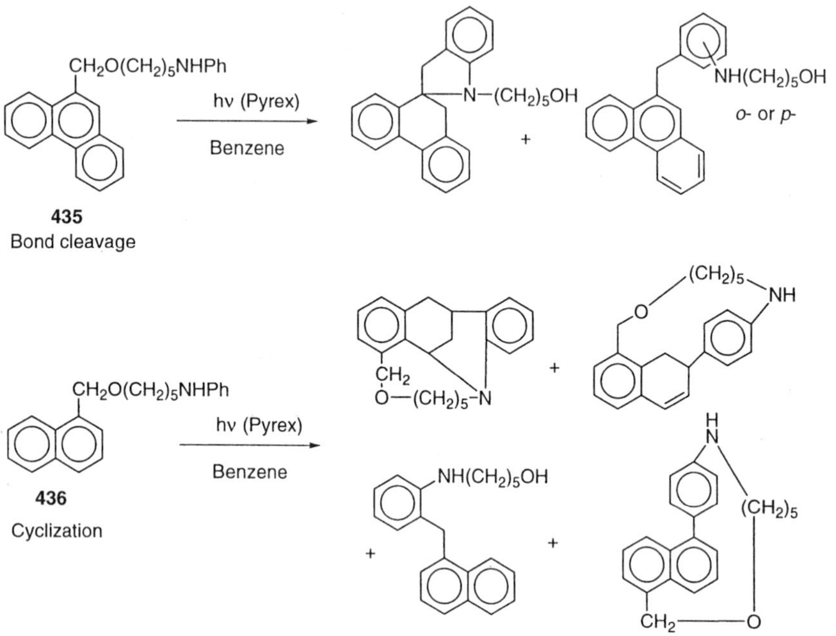

435
Bond cleavage

436
Cyclization

Scheme 128

photocyclization did not proceed in the absence of an electron acceptor (Scheme 129).

The intramolecular photoinduced electron transfer reaction of N-(o-chlorobenzyl)aniline **440** in the presence of sodium hydroxide in aqueous acetonitrile afforded, 9,10-dihydrophenanthridine and its dimer, which is reasonably explained by dechlorination from the radical anion of chlorobenzene chromophore followed by the cyclization (Scheme 130) [481]. Similar photocyclization 9-(ω-anilinoalkyl)-10-bromophenantherens **441** takes place to give spiro compounds, cyclized products, and reduction products dependent on the methylene chain length. The efficient intramolecular photocyclization occurs when the methylene tether is $n = 3$ [476] (Scheme 131).

G. Some Miscellaneous Intramolecular Photocyclization

Photo-NOCAS reactions of p-dicyanobenzene with 2-methylpropene in acetonitrile afforded novel 3,4-dihydroisoquinoline derivatives, as shown in Scheme 132 [482]. This photoreaction is initiated by a single electron transfer from olefin to p-dicyanobenzene. Acetonitrile as a nucleophile combined with the alkene radical cation and the resulting radical cation adds to the radical anion of 1,4-dicyanobenzene. Cyclization to the *ortho* position of phenyl group followed by loss

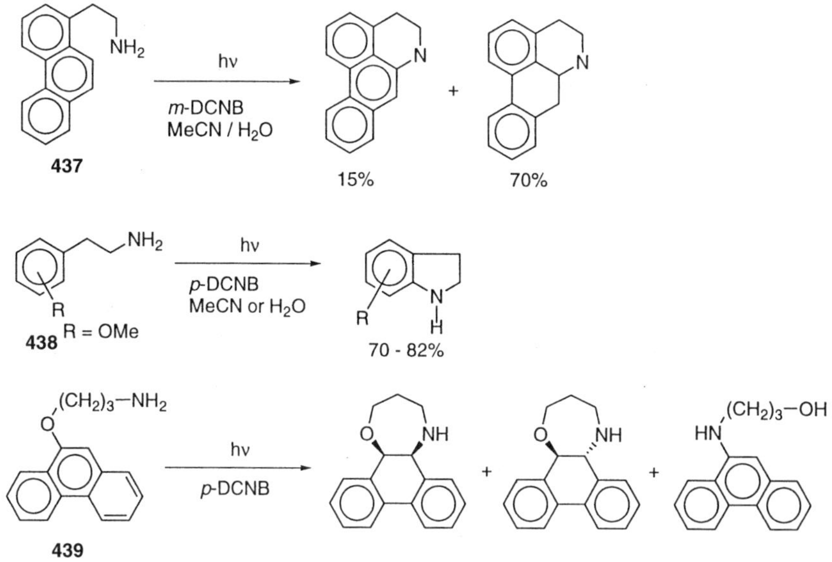

437

15%   70%

438 R = OMe

70 - 82%

439

Scheme 129

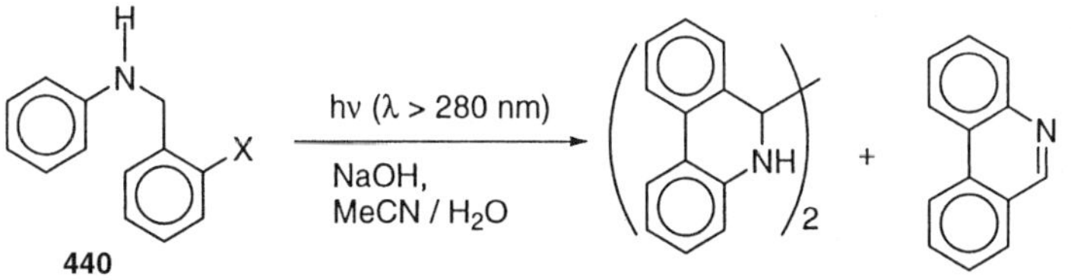

440

Scheme 130

441
n = 3 - 6, 10

n = 3, 4, 5   n = 3, 4

n = 5, 6, 10

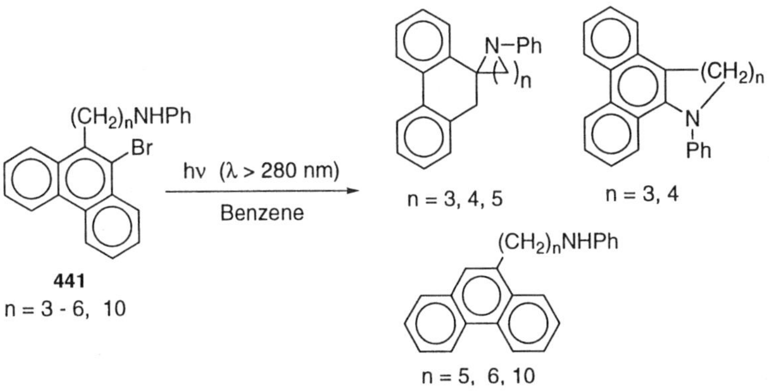

Scheme 131

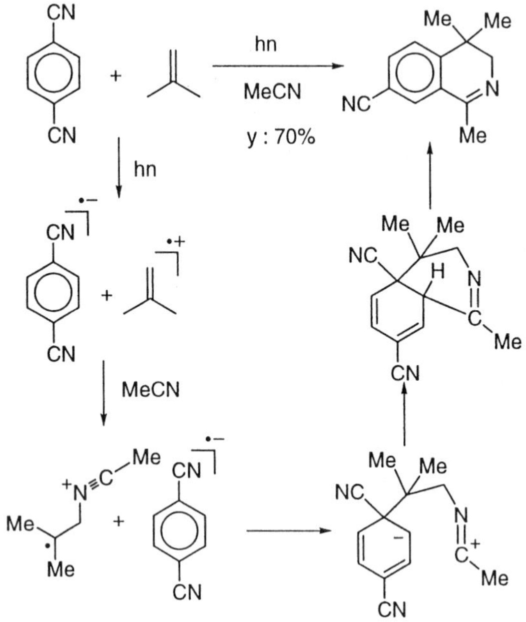

Scheme 132

of HCN gives the dihydroisoquinoline derivatives. The Ritter-type photoaddition of cyano group to cationic center generated by photoinduced electron transfer followed by cyclization has recently been reported as several examples in the 1990s [483–489].

Hirano et al. reported on the stereoselective cyclization to give tetralin derivatives using the phenanthrene-*p*-dicyanobenzene sensitizer system. Pandey independently reported the intramolecular photocyclization of methoxybenzene derivatives bearing silyl enol ether chromophore via their heterodimer radical cations in the presence of 1,4-dicyanonaphthalene gave benzo-annulated cyclic ketones in 70% yields [490] (Scheme 133).

V. CONCLUSION

Intermolecular and intramolecular photocycloaddition and photoaddition to aromatic rings in the electron-donor and electron-acceptor systems were discussed in this chapter. The highly stereoselective and regioselective photocycloaddition is a synthetically useful method for the construction of polycyclic carbon-skeleton compounds, including natural products. New aspects for the stereoselective intermolecular and intramolecular photocycloaddition reactions via exciplexes in less

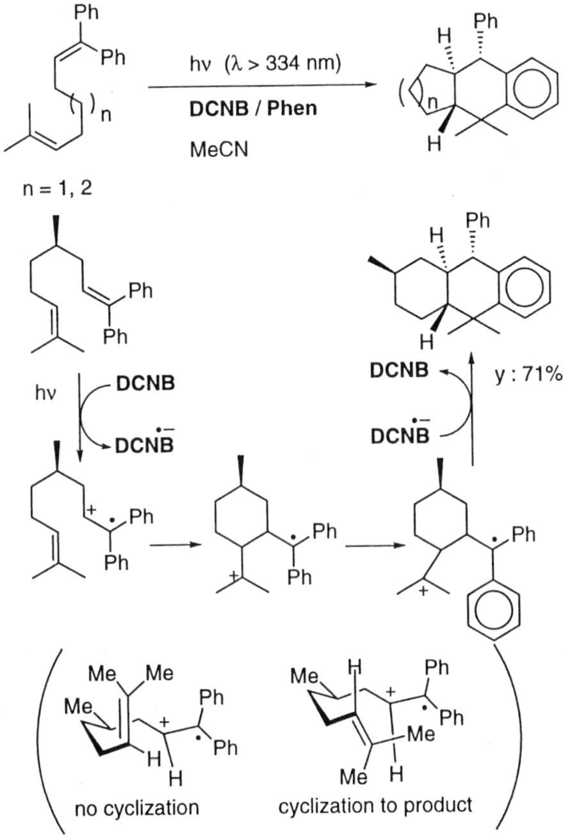

Scheme 133

polar solvents have been described depending on the effects of solvents, substituents, additives, and reaction temperature. The introduction of the functional group to the aromatic rings via photoinduced electron transfer giving new carbon–carbon and carbon–heteroatom bonds under mild neutral conditions based on the reactivities of the radical ion species is also discussed.

It will be interesting to follow the developments on the highly efficient and more stereoselective photocycloaddition and photoaddition to aromatic rings from the viewpoints for the synthesis of more complex compounds, including natural products. In addition, the chiral induction in the excited states should be more attractive projects in the near future. Although some excellent reviews about the asymmetric photochemical reactions have been reported in recent years [490–492], the highly enantioselective or diastereoselective photocycloaddition and photoaddition have been reported in only limited cases.

ACKNOWLEDGMENTS

K. M. would like to thank Dr. Chyongjin Pac (Kawamura Institute of Technology), Emeritus Professor Yoshio Otsuji (Osaka Prefecture University), and Professor Richard A. Caldwell (University of Texas at Dallas) for their stimulating discussions. We are also grateful to our co-workers mentioned in the literature. Financial support was partially provided by a Special Coordination Funds for Promoting Science and Technology (Leading Research Utilizing Potential of Regional Science and Technology) of the Ministry of Education, Culture, Sports, Science, and Technology of the Japanese Government.

REFERENCES

1. Turro, N. J. *Modern Molecular Photochemistry,* The Benjamin/Cummings Publishing Company: Menlo Park, CA, 1978.
2. Kavarnos, G. J. *Fundamentals of Photoinduced Electron Transfer,* Wiley-VCH: New York, 1993.
3. Gordon, M.; Ware, W. R., eds. *The Exciplex,* Academic Press: New York, 1975.
4. Horspool, W. M. ed. *Synthetic Organic Photochemistry,* Plenum Press: New York, 1984.
5. Horspool, W. M.; Song, P.-S., eds. *CRC Handbook of Organic Photochemistry and Photobiology,* CRC Press: New York, 1995.
6. Fox, M. A.; Chanon, M., eds. *Photoinduced Electron Transfer Part A–D,* Elsevier: Amsterdam, 1988.
7. Mattay, J., ed. *Photoinduced Electron Transfer I–V,* Topics in Current Chemistry Vols. 156, 158, 159, and 163, Springer-Verlag: Berlin, 1990–1992.
8. Mattay, J., ed. *Electron Transfer I–II,* Topics in Current Chemistry, Vols. 169, and 177, Springer-Verlag: Berlin, 1994, 1996.
9. Mariano, P. S., ed. *Advances in Electron Transfer Chemistry* London, 1991–1999, Vols. 1–6.
10. Ramamurthy, V.; Schanze, K. S., eds. *Molecular and Supramolecular Photochemistry,* Marcel Dekker: New York, 1997–1999, Vols. 1–4.
11. Caldwell, R. A.; Creed, D. *Acc. Chem. Res.* **1980,** *13,* 45–50.
12. Sakurai, H.; Pac, C. *Mem. Inst. Sci. Ind. Res., Osaka Univ.* **1980,** *37,* 59–78.
13. Schuster, D. I.; Lem. G.; Kaprinidis, N. A. *Chem. Rev.* **1993,** *93,* 3–22.
14. Müller, F.; Mattay, J. *Chem. Rev.* **1993,** *93,* 99–117.
15. Oppolzer, W. *Acc. Chem. Res.* **1982,** *15,* 135–141.
16. Bach, T. *Synthesis* **1998,** 683–703.
17. Dell, C. P. *J. Chem. Soc., Perkin Trans. 1* **1998,** 3873–3905.
18. McCullough, J. J. *Chem. Rev.* **1987,** *87,* 811–860.
19. Mizuno, K.; Otsuji, Y. *Topics Curr. Chem.* **1994,** *169,* 301–346.
20. Gilbert, A. in *Synthetic Organic Photochemistry,* Horspool, W. M., ed.; Plenum Press: New York, 1984, 1–60.
21. Wender, P. A.; Dore, T. M. in *CRC Handbook of Organic Photochemistry and Photobiology,* Horspool, W. M., Song, P.-S., eds., CRC Press: New York, 1995, 280–290.

22. Fleming, A.; Bradford, C. L.; Gao, J. J. in *Molecular and Supramolecular Photochemistry,* Ramamurthy, V., Schanze, K. S., eds., Marcel Dekker: New York, 1997, Vol. 1 (Organic Photochemistry), 187–243.
23. Cornelisse, J. *Chem. Rev.* **1993,** *93,* 615–669.
24. Mattay, J. *Tetrahedron* **1985,** *41,* 2405–2417.
25. De Keukeleire, D.; He, S. -L. *Chem. Rev.* **1993,** *93,* 359–380.
26. Cohen, S. G.; Parola, A.; Parsons, G. H., Jr. *Chem. Rev.* **1973,** *73,* 141–161.
27. Cornelisse, J.; Havinga, E. *Chem. Rev.* **1975,** *75,* 353–388.
28. Cornelisse, J. in *CRC Handbook of Organic Photochemistry and Photobiology,* Horspool, W. M.; Song, P-S., eds., CRC Press: New York, 1995, 250–265.
29. Mattay, J. *Synthesis,* **1989,** 233–252.
30. Yasuda, M.; Shima, K. *Rev. Heteroatom Chem.* **1991,** *4,* 27–47.
31. Roth, H. D. *Topics Curr. Chem.* **1992,** *163,* 131–245.
32. Yasuda, M.; Shima, K. *Trends Org. Chem.* **1993,** *4,* 291–302.
33. Yoon, U. C.; Mariano, P. S.; Givens, R. S.; Atwater III, B. W. *Advances in Electron Transfer Chemistry,* **1994,** *4,* 117–205.
34. Hintz, S.; Heidbreder, A.; Mattay, J. *Topics Curr. Chem.* **1996,** *177,* 77–124.
35. Lewis, F. D. *Adv. Electron Transfer Chem.* **1996,** *5,* 1–39.
36. Albini, A.; Fasani, E.; Freccero, M. *Adv. Electron Transfer Chem.* **1996,** *5,* 103–140.
37. Schmittel, M.; Burghart, A. *Angew. Chem. Int. Ed. Engl.,* **1997,** *36,* 2550–2589.
38. Rehm, d.; Weller, A. *Isr. J. Chem.* **1970,** *8,* 259–271.
39. Santamaria, J. in *Photoinduced Electron Transfer, Part B,* Fox, M. A.; Chanon, M., eds., Elsevier: Amsterdam, 1988, 483–540.
40. Majima, T.; Pac, C.; Nakasone, A.; Sakurai, H. *J. Am. Chem. Soc.* **1981,** *103,* 4499–4508.
41. Pac, C. *Pure Appl. Chem.* **1986,** *58,* 1249.
42. Mizuno, K.; Tamai, T.; Sugimoto, A.; Maeda, H. in *Advances in Electron Transfer Chemistry,* Mariano, P. S., ed., JAI Press: London, 1999, Vol. 6, 131–165.
43. Caldwell, R. A. *J. Am. Chem. Soc.* **1980,** *102,* 4004–4007.
44. Mattay, J. *Tetrahedron* **1985,** *41,* 2393–2404.
45. Gilbert, A.; Taylor, G. *Tetrahedron Lett.* **1977,** 469–472.
46. Atkins, R. J.; Fray, G. I.; Drew, M. G. B.; Gilbert, A.; Taylor, G. N. *Tetrahedron Lett.* **1978,** 2945–2948.
47. Atkins, R. J.; Fray, G. I.; Gilbert, A.; bin Samsudin, M. W.; Steward, A. J. K.; Taylor, G. N. *J. Chem. Soc., Perkin Trans. 1* **1979,** 3196–3202.
48. Gilbert, A.; Yianni, P. *Tetrahedron* **1981,** *37,* 3275–3283.
49. Al-Qaradawi, S.; Gilbert, A.; Jones, D. T. *Recl. Trav. Chim. Pays-Bas* **1995,** *114,* 485–491.
50. Drew, M. G. B.; Gilbert, A.; Heath, P.; Mitchell, A. J.; Rodwell, P. W. *J. Chem. Soc., Chem. Commun.* **1983,** 750–751.
51. Cornelisse, J.; Gilbert, A.; Rodwell, P. W. *Tetrahedron Lett.* **1986,** *27,* 5003–5006.
52. Gilbert, A.; Heath, P.; Rodwell, P. W. *J. Chem. Soc., Perkin Trans. 1* **1989,** 1867–1873.
53. Weber, G.; Runsink, J.; Mattay, J. *J. Chem. Soc., Perkin Trans. 1* **1987,** 2333–2338.
54. Mattay, J.; Rumbach, T.; Runsink, J. *J. Org. Chem.* **1990,** *55,* 5691–5696.

55. Bryce-Smith, D.; Drew, M. G. B.; Fenton, G. A.; Gilbert, A.; Proctor, A. D. *J. Chem. Soc., Perkin Trans.* 2 **1991,** 779–784.
56. Osselton, E. M.; van Dijk-Knepper, J. J.; Cornelisse, J. *J. Chem. Soc., Perkin Trans.* 2 **1988,** 1021–1025.
57. de Vaal, P.; Lodder, G.; Cornelisse, J. *Tetrahedron Lett.* **1985,** *26,* 4395–4398.
58. de Vaal, P.; Lodder, G.; Cornelisse, J. *Tetrahedron* **1986,** *42,* 4585–4590.
59. Mattay, J.; Runsink, J.; Leismann, H.; Scharf, H.-D. *Tetrahedron Lett.* **1982,** *23,* 4919–4922.
60. Leismann, H.; Mattay, J.; Scharf, H.-D. *J. Am. Chem. Soc.* **1984,** *106,* 3985–3991.
61. Gilbert, A.; Heritage, T. W.; Isaacs, N. S. *J. Chem. Soc., Perkin Trans.* 2 **1992,** 1141–1144.
62. van der Hart, J. A.; Mulder, J. J. C.; Cornelisse, J. *1. Photochem. Phtobiol., A: Chem.* **1991,** *61,* 3–13.
63. Clifford, S.; Bearpark, M. J.; Bernardi, F.; Olivucci, M.; Robb, M. A.; Smith, B. R. *J. Am. Chem. Soc.* **1996,** *118,* 7353–7360.
64. Bearpark, M. J.; Deumal, M.; Robb, M. A.; Vreven, T.; Yamamoto, N.; Olivucci, M.; Bernardi, F. *J. Am. Chem. Soc.* **1997,** *119,* 709–718.
65. Koltzenburg, G. Kraft, K.; *Tetrahedron Lett.* **1966,** 389–395.
66. Kraft, K.; Koltzenburg, G. *Tetrahedron Lett.* **1967,** 4357–4362.
67. Kraft, K.; Koltzenburg, G. *Tetrahedron Lett.* **1967,** 4723–4728.
68. Berridge, J. C.; Forrester, J.; Foulger, B. E.; Gilbert, A. *J. Chem. Soc., Perkin Trans. 1* **1980,** 2425–2434.
69. Gilbert, A.; Griffiths, O. *J. Chem. Soc., Perkin Trans. 1* **1993,** 1379–1384.
70. Okumura, K.; Takamuku, S.; Sakurai, H. *J. Chem. Soc. Jpn., Ind. Chem. Sec.* **1969,** *72,* 200–203.
71. Gilbert, A.; Rodwell, P. *J. Chem. Soc., Chem. Commun.* **1985,** 1057–1058.
72. Gilbert, A.; Rodwell, P. W. *J. Chem. Soc., Perkin Trans. 1* **1990,** 931–935.
73. Berridge, J. C.; Gilbert, A.; Taylor, G. N. *J. Chem. Soc., Perkin Trans. 1* **1980,** 2174–2178.
74. Garcia, H.; Gilbert, A.; Griffiths, O. *J. Chem. Soc., Perkin Trans.* 2 **1994,** 247–252.
75. Al-Jalal, N.; Drew, M. G. B.; Gilbert, A. *J. Chem. Soc., Chem. Commun.* **1985,** 85–86.
76. Al-Jalal, N. *J. Chem. Res. (S)* **1989,** 110–111.
77. Al-Jalal, N. A. *J. Heterocycl. Chem.* **1990,** *27,* 1323–1327.
78. Bryce-Smith, D.; Lodge, J. E. *J. Chem. Soc.* **1963,** 695–701.
79. Grovenstein, E., Jr.; Campbell, T. C.; Shibata, T. *J. Org. Chem.* **1969,** *34,* 2418–2428.
80. Bryce-Smith, D.; Gilbert, A.; Grzonka, J. *J. Chem. Soc., Chem. Commun.* **1970,** 498–499.
81. Šket, B.; Zupan, M. *J. Am. Chem. Soc.* **1977,** *99,* 3504–3505.
82. Tinnemans, A. H. A.; Neckers, D. C. *J. Am. Chem. Soc.* **1977,** *99,* 6459–6460.
83. Tinnemans, A. H. A.; Neckers, D. C. *Tetrahedron Lett.* **1978,** 1713–1716.
84. Hanzawa, Y.; Paquette, L. *Synthesis* **1982,** 661–662.
85. Šket, B.; Zupančič, N.; Zupan, M. *J. Chem. Soc., Perkin Trans. 1* **1987,** 981–985.
86. Šket, B.; Zupan, M. *Tetrahedron* **1989,** *45,* 6741–6748.
87. Ohkura, K.; Kanazashi, N.; Okamura, K.; Date, T.; Seki, K. *Chem. Lett.* **1993,** 667–670.

88. Ohkura, K.; Noguchi, Y.; Seki, K. *Chem. Lett.* **1997,** 99–100.

89. Barlow, M. G.; Brown, D. E.; Haszeldine, R. N. *J. Chem. Soc., Chem. Commun.* **1977,** 669–670.

90. Barlow, M. G.; Brown, D. E.; Haszeldine, R. N. *J. Chem. Soc., Perkin Trans. 1* **1978,** 363–365.

91. Šket, B.; Zupančič, N.; Zupan, M. *J. Org. Chem.* **1982,** *47,* 4462–4464.

92. Sakamoto, M.; Sano, T.; Takahashi, M.; Yamaguchi, K.; Fujita, T.; Watanabe, S. *Chem. Commun.* **1996,** 1349–1350.

93. Sakamoto, M.; Kinbara, A.; Yagi, T.; Takahashi, M.; Yamaguchi, K.; Mino, T.; Watanabe, S.; Fujita, T. *J. Chem. Soc., Perkin Trans. 1* **1999,** 171–177.

94. Bowman, R. M.; McCullough, J. J. *Chem. Commun.* **1970,** 948–949.

95. Bowman, R. M.; Chamberlain, T. R.; Huang, C. W.; McCullough, J. J. *J. Am. Chem. Soc.* **1970,** *92,* 4106–4108.

96. Teitei, T.; Collin, P. J.; Sasse, W. H. F. *Aust. J. Chem.* **1972,** *25,* 171–182.

97. Grovenstein, E. Jr.; Campbell, T. C.; Shibata, T. *J. Org. Chem.* **1969,** *34,* 2418–2428.

98. Yang, N. C.; Libman, J. *J. Am. Chem. Soc.* **1972,** *94,* 9228–9229.

99. Kimura, M.; Sagara, S.; Morosawa, S. *J. Org. Chem.* **1982,** *47,* 4344–4347.

100. Pac, C.; Sugioka, T.; Sakurai, H. *Chem. Lett.* **1972,** 39–42.

101. Sugioka, T.; Pac, C.; Sakurai, H. *Chem. Lett.* **1972,** 667–668.

102. Kan, K.; Kai, Y.; Yasuoka, N.; Kasai, N. *Bull. Chem. Soc. Jpn.* **1979,** *52,* 1634–1636.

103. Sugioka, T.; Pac, C.; Sakurai, H. *Chem. Lett.* **1972,** 791–792.

104. Pac, C.; Sakurai, H. *Chem. Lett.* **1976,** 1067–1072.

105. Majima, T.; Pac, C.; Sakurai, H. *Bull. Chem. Soc. Jpn.* **1978,** *51,* 1811–1817.

106. Pac, C.; Mizuno, K.; Sugioka, T.; Sakurai, H. *Chem. Lett.* **1973,** 187–188.

107. Pac, C.; Sugioka, T.; Mizuno, K.; Sakurai, H. *Bull. Chem. Soc. Jpn.* **1973,** *46,* 238–243.

108. Pac, C.; Mizuno, K.; Sakurai, H. *Nihon Kagaku Kaishi* **1984,** 110–118.

109. Mizuno, K.; Pac, C.; Sakurai, H. *J. Chem. Soc., Chem. Commun.* **1974,** 648–649.

110. Yasuda, M.; Pac, C.; Sakurai, H. *Bull. Chem. Soc. Jpn.* **1980,** *53,* 502–507.

111. Pac, C.; Mizuno, K.; Okamoto, H.; Sakurai, H. *Synthesis* **1978,** 589–590.

112. Mizuno, K.; Pac, C.; Sakurai, H. *J. Chem. Soc., Chem. Commun.* **1973,** 219–220.

113. Mizuno, K.; Pac, C.; Sakurai, H. *J. Chem. Soc., Perkin Trans. 1* **1975,** 2221–2227.

114. Mizuno, K.; Pac, C.; Sakurai, H. *J. Org. Chem.* **1977,** *42,* 3313–3315.

115. Cantrell, T. S. *J. Am. Chem. Soc.* **1972,** *94,* 5929–5931.

116. McCullough, J. J.: Miller, R. C.; Fung, D.; Wu, W. -S. *J. Am. Chem. Soc.* **1975,** *97,* 5942–5943.

117. McCullough, J. J.; Miller, R. C.; Wu, W. -S. *Can. J. Chem.* **1977,** *55,* 2909–2915.

118. Yang, N. C.; Kim, B.; Chiang, W.; Hamada, T. *J. Chem. Soc., Chem. Commun.* **1976,** 729–730.

119. Akhtar, I. A.; McCullough, J. J. *J. Org. Chem.* **1981,** *46,* 1447–1450.

120. Inoue, Y.; Nishida, K.; Ishibe, K.; Hakushi, T.; Turro, N. J. *Chem. Lett.* **1982,** 471–474.

121. Mizuno, K.; Nakanishi, K.; Yasueda, M.; Miyata, H.; Otsuji, Y. *Chem. Lett.* **1991,** 2001–2004.

122. Suginome, H.; Liu, C. F.; Tokuda, M.; Furusaki, A. *J. Chem. Soc., Perkin Trans. 1* **1985,** 327–329.
123. Suginome, H.; Itoh, M.; Kobayashi, K. *J. Chem. Soc., Perkin Trans. 1* **1988,** 491–496.
124. Chamberlain, T. R.; McCullough, J. J. *Can. J. Chem.* **1973,** *51,* 2578–2589.
125. Suginome, H.; Liu, C. F.; Tokuda, M. *J. Chem. Soc., Chem. Commun.* **1984,** 334–335.
126. Al-Jalal, N. A. *J. Photochem. Photobiol. A* **1990,** *51,* 429–436.
127. Ue, M.; Kinugawa, M.; Kakiuchi, K.; Tobe, Y.; Odaira, Y. *Tetrahedron Lett.* **1989,** *30,* 6193–6194.
128. McCullough, J. J.; McMurry, T. B.; Work, D. N. *J. Chem. Soc., Perkin Trans. 1* **1991,** 461–464.
129. Zupancic, N.; Sket, B. *J. Chem. Soc., Perkin Trans. 1* **1992,** 179–180.
130. Okamoto, H.; Kimura, M.; Satake, K.; Morosawa, S. *Bull. Chem. Soc. Jpn.* **1993,** *66,* 2436–2439.
131. Al-Jalal, N. A. *Gazz. Chim. Ital.* **1994,** *124,* 205–207.
132. Al-Jalal, N. A.; Pritchard, R. G.; McAuliffe, C. A. *J. Chem. Res. (S)* **1994,** 452.
133. Al-Jalal, N. A. *J. Chem. Res. (S)* **1995,** 44.
134. Weng, H.; Roth, H. D. *Tetrahedron Lett.* **1996,** *37,* 4895–4898.
135. Döpp, D.; Erian, A. W.; Henkel, G. *Chem. Ber.* **1993,** *126,* 239–242.
136. Mella, M.; Fasani, E.; Albini, A. *J. Org. Chem.* **1992,** *57,* 6210–6216.
137. Albini, A.; Fasani, E.; Giavarini, F. *J. Org. Chem.* **1988,** *53,* 5601–5607.
138. Noh, T.; Kim, D.; Kim, Y. -J. *J. Org. Chem.* **1998,** *63,* 1212–1216.
139. Noh, T.; Kim, D. *Tetrahedron Lett.* **1996,** *37,* 9329–9332.
140. Mizuno, K.; Konishi, G.; Chiyonobu, K. *unpublished result.*
141. Chow, Y. L.; Buono-Core, G. E.; Liu, X. -Y.; Itoh, K.; Qian, P. *J. Chem. Soc., Chem. Commun.* **1987,** 913–915.
142. Chow, Y. L.; Liu, X. -Y. Hu, S. *J. Chem. Soc., Chem. Commun.* **1988,** 1047–1048.
143. Chow, Y. L.; Liu, X. -Y. *Can. J. Chem.* **1991,** *69,* 1261–1272.
144. Chow, Y. L.; Johansson, C. I. *Can. J. Chem.* **1994,** *72,* 2011–2020.
145. Chow, Y. L.; Buono-Core, G. E.; Zhang, Y. -H.; Liu, X. -Y. *J. Chem. Soc., Perkin Trans. 2* **1991,** 2041–2045.
146. Bradshaw, J. S.; Hammond, G. S. *J. Am. Chem. Soc.* **1963,** *85,* 3953–3955.
147. Selinger, B. K.; Sterns, M. *J. Chem. Soc., Chem. Commun.* **1969,** 978–979.
148. Bradshaw, J. S.; Nielsen, N. B.; Rees, D. P. *J. Org. Chem.* **1968,** *33,* 259–261.
149. Collin, P. J.; Sugowdz, G.; Teitei, T.; Wells, D.; Sasse, W. H. F. *Aust. J. Chem.* **1974,** *27,* 227–230.
150. Teitei, T.; Wells, D.; Sasse, W. H. F. *Tetrahedron Lett.* **1974,** 367–370.
151. Teitei, T.; Wells, D.; Spurling, T. H.; Sasse, W. H. F. *Aust. J. Chem.* **1978,** *31,* 85–96.
152. Mattingly, T. W.; Lancaster, J. E.; Zweig, A. *J. Chem. Soc., Chem. Commun.* **1971,** 595–596.
153. Albini, A.; Giannantonio, L. *J. Org. Chem.* **1984,** *49,* 3862–3863.
154. Albini, A.; Fasani, E.; Gamba, A. *J. Photochem. Photobiol., A: Chem.* **1988,** *41,* 215–225.
155. Noh, T.; Jeong, Y.; Kim, D. *J. Chem. Soc., Perkin Trans. 1* **1998,** 2501–2504.
156. Albini, A.; Fasani, E.; Faiardi, D. *J. Org. Chem.* **1987,** *52,* 155–157.

157. Albini, A.; Fasani, E. *J. Am. Chem. Soc.* **1988,** *110,* 7760–7763.
158. Becker, H. -D.; Andersson, K.; Sandros, K. *J. Org. Chem.* **1985,** *50,* 3913–3916.
159. Becker, H. -D.; Andersson, K. *J. Photochem.* **1984,** *26,* 75–77.
160. Becker, H. -D.; Sörensen, H.; Sandros, K. *J. Org. Chem.* **1986,** *51,* 3223–3226.
161. Ito, Y.; Fujita, H. *J. Org. Chem.* **1996,** *61,* 5677–5680.
162. Ito, Y.; Olovsson, G. *J. Chem. Soc., Perkin Trans. 1* **1997,** 127–133.
163. Tung, C. -H.; Guan, J.-Q. *J. Org. Chem.* **1998,** *63,* 5857–5862.
164. Noh, T.; Lim, H. *Chem. Lett.* **1997,** 495–496.
165. Kimura, M.; Okamoto, H.; Kura, H.; Okazaki, A.; Nagayasu, E.; Satake, K.; Morosawa, S.; Fukuzawa, M.; Abdol-Halim, M.; Cowan, D. O. *J. Org. Chem.* **1988,** *53,* 3908–3911.
166. Kimura, M.; Kura, H.; Nukada, K.; Okamoto, H.; Satake, K.; Morosawa, S. *J. Chem. Soc., Perkin Trans. 1* **1988,** 3307–3310.
167. Kimura, M.; Okamoto, H.; Kashino, S. *Bull. Chem. Soc. Jpn.* **1994,** *67,* 2203–2212.
168. Yang, N. C.; Libman, J. *J. Am. Chem. Soc.* **1972,** *94,* 1405–1406.
169. Yang, N. C.; Libman, J.; Barrett, L., Jr.; Hui, M. H.; Loeschen, R. L. *J. Am. Chem. Soc.* **1972,** *94,* 1406–1408.
170. Yang, N. C. V.; Neywick, C. V.; Srinivasachar, K. *Tetrahedron Lett.* **1975,** 4313–4316.
171. Yang, N. C.; Shold, D. M.; McVey, J. K. *J. Am. Chem. Soc.* **1975,** *97,* 5004.
172. Yang, N. C.; Srinivasachar, K.; Kim, B.; Libman, J. *J. Am. Chem. Soc.* **1975,** *97,* 5006–5008.
173. Yang, N. C.; Srinivasachar, K. *J. Chem. Soc., Chem. Commun.* **1976,** 48–49.
174. Yang, N. C.; Yates, R. L.; Masnovi, J.; Shold, D. M.; Chiang, W. *Pure Appl. Chem.* **1979,** *51,* 173–180.
175. Yang, N. C.; Shold, D. M. *J. Chem. Soc., Chem. Commun.* **1978,** 978.
176. Yang, N. C.; Masnovi, J.; Chiang, W. *J. Am. Chem. Soc.* **1979,** *101,* 6465–6466.
177. Yang, N. C.; Shou, H.; Wang, T.; Masnovi, J. *J. Am. Chem. Soc.* **1980,** *102,* 6652–6654.
178. Yang, N. C.; Chen, M.-J.; Chen, P.; Mak, K. T. *J. Am. Chem. Soc.* **1982,** *104,* 853–855.
179. Yang, N. C.; Chen, M.-J.; Chen, P. *J. Am. Chem. Soc.* **1984,** *106,* 7310–7315.
180. Kaupp, G. *Angew. Chem., Int. Ed. Engl.* **1972,** *11,* 718–719.
181. Kaupp, G.; Dyllick-Brenzinger, R.; Zimmerman, I. *Angew. Chem., Int. Ed. Engl.* **1975,** *14,* 491–492.
182. Kaupp, G. *Liebigs Ann. Chem.* **1977,** 254–275.
183. Kaupp, G.; Grüter, H.-W. *Angew. Chem., Int. Ed. Engl.* **1979,** *18,* 881–882.
184. Kaupp, G.; Grüter, H.-W. *Chem. Ber.* **1980,** *113,* 1458–1471.
185. Kaupp, G.; Schmitt, D. *Chem. Ber.* **1981,** *114,* 1567–1571.
186. Kaupp, G.; Grüter, H. W.; Teufel, E. *Chem. Ber.* **1983,** *116,* 630–644.
187. Kaupp, G.; Teufel, E. *Chem. Ber.* **1980,** *113,* 3669–3674.
188. Mizuno, K.; Pac, C.; Sakurai, H. *J. Chem. Soc., Perkin Trans. 1* **1974,** 2360–2364.
189. Fukazawa, Y.; Fujihara, T.; Usui, S.; Shiobara, Y.; Kodama, M. *Tetrahedron Lett.* **1986,** *37,* 5621–5624.
190. Smothers, W. K.; Meyer, M. C.; Saltiel, J. *J. Am. Chem. Soc.* **1983,** *105,* 545–555.
191. Saltiel, J.; Dabestani, R.; Sears, D. F., Jr.; McGowan, W. M.; Hilinsky, E. F. *J. Am. Chem. Soc.* **1995,** *117,* 9129–9138.

192. Farid, S.; Brown, K. A. *J. Chem. Soc., Chem. Commun.* **1976,** 564–565.
193. Brown-Wensley, K. A.; Mattes, S. L.; Farid, S. *J. Am. Chem. Soc.* **1978,** *100,* 4162–4172.
194. Ichinose, N.; Mizuno, K.; Hiromoto, Z.; Otsuji, Y. *Tetrahedron Lett.* **1986,** *27,* 5619–5620.
195. Mizuno, K.; Ichinose, N.; Otsuji, Y. *J. Org. Chem.* **1992,** *57,* 1855–1860.
196. Miyamoto, T.; Mori, T.; Odaira, Y. *Chem. Commun.* **1970,** 1598–1599.
197. Kaupp, G. *Angew. Chem., Int. Ed. Engl.* **1973,** *12,* 765–767.
198. Caldwell, R. A. *J. Am. Chem. Soc.* **1973,** *95,* 1690–1692.
199. Ried, W.; Schinzel, H.; Schmidt, A. H. *Chem. Ber.* **1980,** *113,* 255–260.
200. Hacker, N. P.; McOmie, J. F. W.; Meunier-Piret, J.; Van Meerssche, M. *J. Chem. Soc., Perkin Trans. 1* **1982,** 19–23.
201. Mizuno, K.; Pac, C.; Sakurai, H. *Chem. Lett.* **1973,** 309–310.
202. Mizuno, K.; Pac, C.; Sakurai, H. *J. Am. Chem. Soc.* **1974,** *96,* 2993–2994.
203. Caldwell, R. A.; Smith, L. *J. Am. Chem. Soc.* **1974,** *96,* 2994–2996.
204. Caldwell, R. A.; Ghali, N. I.; Chien, C. -K.; DeMarco, D.; Smith, L. *J. Am. Chem. Soc.* **1978,** *100,* 2857–2863.
205. Caldwell, R. A.; Mizuno, K.; Hansen, P. E.; Vo, L. P.; Frentrup, M.; Ho, C. D. *J. Am. Chem. Soc.* **1981,** *103,* 7263–7269.
206. Creed, D.; Caldwell, R. A. *J. Am. Chem. Soc.* **1974,** *96,* 7369–7371.
207. Farid, S.; Hartman, S. E.; Doty, J. C.; Williams, J. L. R. *J. Am. Chem. Soc.* **1975,** *97,* 3697–3702.
208. Caldwell, R. A.; Creed, D. *J. Am. Chem. Soc.* **1977,** *99,* 8360–8362.
209. Caldwell, R. A.; Creed, D. *J. Phys. Chem.* **1978,** *82,* 2644–2652.
210. Creed, D.; Caldwell, R. A.; Ulrich, M. M. *J. Am. Chem. Soc.* **1978,** *100,* 5831–5841.
211. Caldwell, R. A.; Creed, D.; DeMarco, D. C.; Melton, L. A.; Ohta, H.; Wine, P. H. *J. Am. Chem. Soc.* **1980,** *102,* 2369–2377.
212. Kubo, Y.; Imaoka, T.; Shiragami, T.; Araki, T. *Chem. Lett.* **1986,** 1749–1752.
213. Lewis, F. D.; Barancyk, S. V.; Burch, E. L. *J. Am. Chem. Soc.* **1992,** *114,* 3866–3870.
214. Mizuno, K.; Caldwell, R. A.; Tachibana, A.; Otsuji, Y. *Tetrahedron Lett.* **1992,** *33,* 5779–5782.
215. Itoh, H.; Maruyama, S.; Fujii, Y.; Senda, Y.; Sakuragi, H.; Tokumaru, K. *Bull. Chem. Soc. Jpn.* **1993,** *66,* 287–293.
216. Chow, Y. L.; Wu, S. P.; Ouyang, X. *J. Org. Chem.* **1994,** *59,* 421–428.
217. Maeda, H.; Waseda, S.; Mizuno, K. *Chem. Lett.* **2000,** 1238–1239.
218. Kimura, M.; Nukada, K.; Satake, K.; Morosawa, S. *J. Chem. Soc., Perkin Trans. 1* **1984,** 1431–1433.
219. Williams, J. L. R.; Farid, S. Y.; Doty, J. C.; Daly, R. C.; Specht, D. P.; Searle, R.; Borden, D. G.; Chang, H. J.; Martic, P. A. *Pure Appl. Chem.* **1977,** *49,* 523–538.
220. Mizuno, K.; Maeda, H.; Inoue, Y.; Sugimoto, A.; Vo, L. P.; Caldwell, R. A. *Tetrahedron Lett.* **2000,** *41,* 4913–4916.
221. Yang, N. C.; Masnovi, J.; Chiang, W. -L.; Wang, T.; Shou, H.; Yang, D. -D. H. *Tetrahedron* **1981,** *37,* 3285–3300.
222. Yang, N. C.; Gan, H.; Kim, S. S.; Masnovi, J. M.; Rafalko, P. W.; Ezell, E. F.; Lenz, G. R. *Tetrahedron Lett.* **1990,** *31,* 3825–3828.

223. Sydnes, L. K.; Hansen, K. I.; Oldroyd, D. L.; Weedon, A. C.; Jørgensen, E. *Acta Chem. Scand.* **1993,** *47,* 916–924.
224. Bach, T.; Bergmann, H.; Harms, K. *J. Am. Chem. Soc.* **1999,** *121,* 10,650–10,651.
225. Adam, W.; Peter, K.; Peter, E. M.; Stegmann, V. R. *J. Am. Chem. Soc.* **2000,** *122,* 2958–2959.
226. Yokoyama, A.; Mizuno, K. *Org. Lett.* **2000,** *2,* 3457–3459.
227. Chiyonobu, K.; Mizuno, K., *unpublished data.*
228. Morrison, H.; Ferree, W. I., Jr. *Chem. Commun.* **1969,** 268–269.
229. Ferree, W., Jr.; Grutzner, J. B.; Morrison, H. *J. Am. Chem. Soc.* **1971,** *93,* 5502–5512.
230. Ellis-Davies, G. C. R.; Gilbert, A.; Heath, P.; Lane, J. C.; Warrington, J. V.; Westover, D. L. *J. Chem. Soc., Perkin Trans. 2* **1984,** 1833–1841.
231. Blakemore, D. C.; Gilbert, A. *J. Chem. Soc., Perkin Trans. 1* **1992,** 2265–2270.
232. De Zwart, E. W.; De Haan, R.; Cornelisse, J. *J. Photochem. Photobiol. A: Chem.* **1994,** *77,* 161–168.
233. Ellis-Davies, G. C. R.; Cornelisse, J. *Tetrahedron Lett.* **1985,** *26,* 1893–1896.
234. Gilbert, A.; Taylor, G. *J. Chem. Soc., Chem. Commun.* **1978,** 129–130.
235. Gilbert, A.; Taylor, G. N. *J. Chem. Soc., Chem. Commun.* **1979,** 229–230.
236. Gilbert, A.; Taylor, G. N. *J. Chem. Soc., Perkin Trans. 1* **1980,** 1761–1768.
237. Ellis-Davies, G. C. R.; Cornelisse, J.; Gilbert, A. *Tetrahedron Lett.* **1988,** *29,* 4319–4322.
238. Pallmer, M.; Morrison, H. *J. Org. Chem.* **1980,** *45,* 798–803.
239. Sugimura, T.; Nishiyama, N.; Tai, A.; Hakushi, T. *Tetrahedron: Asymmetry* **1994,** *5,* 1163–1166.
240. Blakemore, D. C.; Gilbert, A. *Tetrahedron Lett.* **1994,** *35,* 5267–5270.
241. Blakemore, D. C.; Gilbert, A. *Tetrahedron Lett.* **1995,** *36,* 2307–2310.
242. Amey, D. M.; Blakemore, D. C.; Drew M. G. B.; Gilbert, A.; Heath, P. *J. Photochem. Photobiol. A: Chem.* **1997,** *102,* 173–178.
243. Amey, D. M.; Gilbert, A.; Jones, D. T. *J. Chem. Soc., Perkin Trans. 2* **1998,** 213–218.
244. Avent, A. G.; Byrne, P. W.; Penkett, C. S. *Org. Lett.* **1999,** *1,* 2073–2075.
245. Wender, P. A.; Howbert, J. J. *J. Am. Chem. Soc.* **1981,** *103,* 688–690.
246. Wender, P. A.; Dreyer, G. B. *Tetrahedron* **1981,** *37,* 4445–4450.
247. Wender, P. A.; Howbert, J. J. *Tetrahedron Lett.* **1982,** *23,* 3983–3986.
248. Wender, P. A.; Howbert, J. J. *Tetrahedron Lett.* **1983,** *24,* 5325–5328.
249. Wender, P. A.; Ternansky, R. J. *Tetrahedron Lett.* **1985,** *26,* 2625–2628.
250. Wender, P. A.; Fisher, K. *Tetrahedron Lett.* **1986,** *27,* 1857–1860.
251. Wender, P. A.; von Geldern, T. W.; Levine, B. H. *J. Am. Chem. Soc.* **1988,** *110,* 4858–4860.
252. Mani, J.; Keese, R. *Tetrahedron* **1985,** *41,* 5697–5701.
253. Mani, J.; Schüttel, S.; Zhang, C.; Bigler, P.; Müller, C.; Keese, R. *Helv. Chim. Acta* **1989,** *72,* 487–495.
254. Zhang, C.; Bourgin, D.; Keese, R. *Tetrahedron* **1991,** *47,* 3059–3074.
255. Šket, B.; Zupančič, N.; Zupan, M. *Tetrahedron* **1986,** *42,* 753–754.
256. Cosstick, K. B.; Drew, M. G. B.; Gilbert, A. *J. Chem. Soc., Chem. Commun.* **1987,** 1867–1868.

257. Al-Qaradawi, S. Y.; Cosstick, K. B.; Gilbert, A. *J. Chem. Soc., Perkin Trans 1* **1992,** 1145–1148.
258. De Keukeleire, D.; He, S.-L.; Blakemore, D.; Gilbert, A. *J. Photochem. Photobiol. A: Chem.* **1994,** *80,* 233–240.
259. Van der Eycken, E.; De Keukeleire, D.; De Bruyn, A. *Tetrahedron Lett.* **1995,** *36,* 3573–3576.
260. Aoyama, H.; Arata, Y.; Omote, Y. *J. Chem. Soc., Chem. Commun.* **1990,** 736–737.
261. Nuss, J. M.; Chinn, J. P.; Murphy, M. M. *J. Am. Chem. Soc.* **1995,** *117,* 6801–6802.
262. Fessner, W.-D.; Prinzbach, H.; Rihs, G. *Tetrahedron Lett.* **1983,** *24,* 5857–5860.
263. Prinzbach, H.; Sedelmeier, G.; Krüger, C.; Goddard, R.; Martin, H.-D.; Gleiter, R. *Angew. Chem., Int. Ed. Engl.* **1978,** *17,* 271–272.
264. Fischer, G.; Beckmann, E.; Prinzbach, H.; Rihs, G.; Wirz, J. *Tetrahedron Lett.* **1986,** *27,* 1273–1276.
265. Sedelmeimer, G.; Fessner, W.-D.; Grund, C.; Spurr, P. R.; Fritz, H.; Prinzbach, H. *Tetrahedron Lett.* **1986,** *27,* 1277–1280.
266. Mathew, T.; Keller, M.; Hunkler, D.; Prinzbach, H. *Tetrahedron Lett.* **1996,** *37,* 4491–4494.
267. Hoffmann, N.; Pete, J.-P. *Tetrahedron Lett.* **1995,** *36,* 2623–2626.
268. Hoffmann, N.; Pete, J.-P. *Tetrahedron Lett.* **1996,** *37,* 2027–2030.
269. Hoffmann, N.; Pete, J.-P. *J. Org. Chem.* **1997,** *62,* 6952–6960.
270. Hoffmann, N.; Pete, J.-P. *Tetrahedron Lett.* **1998,** *39,* 5027–5030.
271. Wagner, P. J.; Nahm, K. *J. Am. Chem. Soc.* **1987,** *109,* 4404–4405.
272. Wagner, P. J.; Nahm, K. *J. Am. Chem. Soc.* **1987,** *109,* 6528–6530.
273. Wagner, P. J.; Sakamoto, M. *J. Am. Chem. Soc.* **1989,** *111,* 8723–8725.
274. Cheng, K.-L.; Wagner, P. J. *J. Am. Chem. Soc.* **1994,** *116,* 7945–7946.
275. Smart, R. P.; Wagner, P. J. *Tetrahedron Lett.* **1995,** *36,* 5131–5134.
276. Wagner, P. J.; Sakamoto, M.; Madkour, A. E. *J. Am. Chem. Soc.* **1992,** *114,* 7298–7299.
277. Wagner, P. J.; Cheng, K.-L. *Tetrahedron Lett.* **1993,** *34,* 907–910.
278. Wagner, P. J.; Smart, R. P. *Tetrahedron Lett.* **1995,** *36,* 5135–5138.
279. Wagner, P. J.; Alehashem, H. *Tetrahedron Lett.* **1993,** *34,* 911–914.
280. Wagner, P. J.; McMahon, K. *J. Am. Chem. Soc.* **1994,** *116,* 10,827–10,828.
281. Kalena, G. P.; Pradhan, P.; Banerji, A. *Tetrahedron Lett.* **1992,** *33,* 7775–7778.
282. Kalena, G. P.; Pradhan, P. P.; Swaranlatha, Y.; Singh, T. P.; Banerji, A. *Tetrahedron Lett.* **1997,** *38,* 5551–5554.
283. Xue, J.; Xu, J.-W.; Yang, L.; Xu, J.-H. *J. Org. Chem.* **2000,** *65,* 30–40.
284. Lippke, W.; Ferree, W. Jr.; Morrison, H. *J. Am. Chem. Soc.* **1974,** *96,* 2134–2137.
285. Pirrung, M. C. *J. Org. Chem.* **1987,** *52,* 1635–1637.
286. McCullough, J. J.; MacInnis, W. K.; Lock, C. J. L.; Faggiani, R. *J. Am. Chem. Soc.* **1980,** *102,* 7780–7782.
287. McCullough, J. J.; MacInnis, W. K.; Lock, C. J. L.; Faggiani, R. *J. Am. Chem. Soc.* **1982,** *104,* 4644–4658.
288. Gilbert, A.; Heath, P.; Kashoulis-Koupparis, A.; Ellis-Davies, G. C. R.; Firth, S. M. *J. Chem. Soc., Perkin Trans. 1* **1988,** 31–36.
289. Kashoulis, A.; Gilbert, A.; Ellis-Davis, G. *Tetrahedron Lett.* **1984,** *25,* 2905–2908.
290. Wagner, P. J.; Sakamoto, M. *J. Am. Chem. Soc.* **1989,** *111,* 9254–9256.

291. Kubo, Y.; Toda, R.; Araki, T. *Bull. Chem. Soc. Jpn.* **1987,** *60,* 429–431.

292. Dittami, J. P.; Nie, X.-Y.; Buntel, C. J.; Rigatti, S. *Tetrahedron Lett.* **1990,** *31,* 3821–3824.

293. Dittami, J. P.; Nie, X. Y.; Nie, H.; Ramanathan, H.; Buntel, C.; Rigatti, S.; Bordner, J.; Decosta, D. L.; Williard, P. *J. Org. Chem.* **1992,** *57,* 1151–1158.

294. Thiergardt, R.; Keller, M.; Wollenweber, M.; Prinzbach, H. *Tetrahedron Lett.* **1993,** *34,* 3397–3400.

295. Kohmoto, S.; Kobayashi, T.; Nishio, T.; Iida, I.; Kishikawa, K.; Yamamoto, M.; Yamada, K. *J. Chem. Soc., Perkin Trans. 1* **1996,** 529–535.

296. Mizuno, K.; Konishi, S.; Takata, T.; Inoue, H. *Tetrahedron Lett.* **1996,** *37,* 7775–7778.

297. Yoshimi, Y.; Konishi, S.; Maeda, H.; Mizuno, K., submitted for publication.

298. Mizuno, K.; Konishi, S.; Yoshimi, Y.; Sugimoto, A. *Chem. Commun.* **1998,** 1659–1660.

299. Chiyonobu, K.; Konishi, G.; Inoue, Y.; Mizuno, K., *submitted for publication.*

300. Kubo, Y.; Adachi, T.; Miyahara, N.; Nakajima, S.; Inamura, I. *Tetrahedron Lett.* **1998,** *39,* 9477–9480.

301. Kaupp, G. *Angew. Chem., Int. Ed. Engl.* **1972,** *11,* 313–314.

302. Okada, K.; Samizo, F.; Oda, M. *Tetrahedron Lett.* **1987,** *28,* 3819–3822.

303. Sakuragi, H.; Tokumaru, K.; Itoh, H.; Terakawa, K.; Kikuchi, K.; Caldwell, R. A.; Hsu, C.-C. *J. Am. Chem. Soc.* **1982,** *104,* 6796–6797.

304. Sakuragi, H.; Tokumaru, K.; Itoh, H.; Terakawa, K.; Kikuchi, K.; Caldwell, R. A.; Hsu, C.-C. *Bull. Chem. Soc. Jpn.* **1990,** *63,* 1049–1057.

305. Sakuragi, H.; Tokumaru, K.; Itoh, H.; Terakawa, K.; Kikuchi, K.; Caldwell, R. A.; Hsu, C.-C. *Bull. Chem. Soc. Jpn.* **1990,** *63,* 1058–1061.

306. Tanaka, N.; Yamazaki, H.; Sakuragi, H.; Tokumaru, K. *Bull. Chem. Soc. Jpn.* **1994,** *67,* 1434–1440.

307. Mizuno, K.; Nakashima, R.; Maeda, H.; Sugimoto, A., *unpublished data.*

308. Maeda, H.; Sugimoto, A.; Mizuno, K. *Org. Lett.* **2000,** *2,* 3305–3308.

309. Maeda, H.; Mizuno, K., *unpublished data.*

310. Hirayama, F. *J. Chem. Phys.* **1965,** *42,* 3163–3171.

311. Kaupp, G. *Angew. Chem., Int. Ed. Engl.* **1976,** *15,* 442–443.

312. Higuchi, H.; Takatsu, K.; Otsubo, T.; Sakata, Y.; Misumi, S. *Tetrahedron Lett.* **1982,** *23,* 671–672.

313. Higuchi, H.; Kobayashi, E.; Sakata, Y.; Misumi, S. *Tetrahedron* **1986,** *42,* 1731–1739.

314. Sentou, W.; Satou, T.; Yasutake, M.; Lim, C.; Shinmyozu, T. *Eur. J. Org. Chem.* **1999,** 1223–1231.

315. Lim, C.; Yasutake, M.; Shinmyozu, T. *Tetrahedron Lett.* **1999,** *40,* 6781–6784.

316. Wasserman, H. H.; Keehn, P. M. *J. Am. Chem. Soc.* **1969,** *91,* 2374–2375.

317. Wasserman, H. H.; Keehn, P. M. *J. Am. Chem. Soc.* **1967,** *89,* 2770–2772.

318. Kaupp, G.; Zimmermann, I. *Angew. Chem., Int. Ed. Engl.* **1976,** *15,* 441–442.

319. Chandross, E. A.; Dempster, C. J. *J. Am. Chem. Soc.* **1970,** *92,* 3586–3593.

320. Davidson, R. S.; Whelan, T. D. *J. Chem. Soc., Chem. Commun.* **1977,** 361–362.

321. Chandross, E. A.; Dempster, C. J. *J. Am. Chem. Soc.* **1970,** *92,* 703–704.

322. Chandross, E. A.; Dempster, C. J. *J. Am. Chem. Soc.* **1970,** *92,* 704–706.

323. Todesco, R.; Gelan, J.; Martens, H.; Put, J.; Boens, N.; De Schryver, F. C. *Tetrahedron Lett.* **1978,** 2815–2818.

324. Kamijo, T.; Irie, M.; Hayashi, K. *Bull. Chem. Soc. Jpn.* **1978,** *51,* 3286–3289.

325. Tung, C.-H.; Wu, L.-Z.; Yuan, Z.-Y.; Su, N. *J. Am. Chem. Soc.* **1998,** *120,* 11,594–11,602.

326. Tung, C.-H.; Yuan, Z.-Y.; Wu, L.-Z.; Weiss, R. G. *J. Org. Chem.* **1999,** *64,* 5156–5161.

327. Chandross, E. A.; Schiebel, A. H. *J. Am. Chem. Soc.* **1973,** *95,* 611–612.

328. Desvergne, J.-P.; Bitit, N.; Castellan, A.; Bouas-Laurent, H. *J. Chem. Soc., Perkin Trans. 2* **1983,** 109–114.

329. Bouas-Laurent, H.; Castellan, A.; Desvergne, J.-P. *Pure Appl. Chem.* **1980,** *52,* 2633–2648.

330. Becker, H.-D. *Pure Appl. Chem.* **1982,** *54,* 1589–1604.

331. Golden, J. H. *J. Chem. Soc.* **1961,** 3741–3748.

332. Koutecký, J.; Paldus, J. *Tetrahedron* **1963,** *19*(Suppl. 2), 201–221.

333. Livingston, R.; Wei, K. S. *J. Am. Chem. Soc.* **1967,** *89,* 3098–3100.

334. Applequist, D. E.; Lintner, M. A.; Searle, R. *J. Org. Chem.* **1968,** *33,* 254–259.

335. Weinshenker, N. M.; Greene, F. D. *J. Am. Chem. Soc.* **1968,** *90,* 506.

336. Toyoda, T.; Otsubo, I.; Otsubo, T.; Sakata, Y.; Misumi, S. *Tetrahedron Lett.* **1972,** 1731–1734; see also Toyoda, T.; Kasai, N.; Misumi, S. *Bull. Chem. Soc. Jpn.* **1985,** *58,* 2348–2356.

337. Kaupp, G. *Liebigs Ann. Chem.* **1973,** 844–878.

338. Cristol, S. J.; Perry, J. S., Jr. *Tetrahedron Lett.* **1974,** 1921–1923.

339. Applequist, D. E.; Swart, D. J. *J. Org. Chem.* **1975,** *40,* 1800–1804.

340. Hayashi, T.; Mataga, N.; Sakata, Y.; Misumi, S.; Morita, M.; Tanaka, J. *J. Am. Chem. Soc.* **1976,** *98,* 5910–5913.

341. Ferguson, J.; Morita, M.; Puza, M. *Chem. Phys. Lett.* **1976,** *42,* 288–292.

342. Hayashi, T.; Suzuki, T.; Mataga, N.; Sakata, Y.; Misumi, S. *Chem. Phys. Lett.* **1976,** *38,* 599–601.

343. Hayashi, T.; Suzuki, T.; Mataga, N.; Sakata, Y.; Misumi, S. *J. Phys. Chem.* **1977,** *81,* 420–423.

344. Bergmark, W. R.; Jones, G., II; Reinhardt, T. E.; Halpern, A. M. *J. Am. Chem. Soc.* **1978,** *100,* 6665–6673.

345. Anderson, B. F.; Ferguson, J.; Morita, M.; Robertson, G. B. *J. Am. Chem. Soc.* **1979,** *101,* 1832–1840.

346. Castellan, A.; Desvergne, J.-P.; Bouas-Laurent, H. *Chem. Phys. Lett.* **1980,** *76,* 390–397.

347. Becker, H.-D.; Elebring, T.; Sandros, K. *J. Org. Chem.* **1982,** *47,* 1064–1068.

348. Castellan, A.; Lacoste, J.-M.; Bouas-Laurent, H. *J. Chem. Soc., Perkin Trans. 2* **1979,** 411–419.

349. Fox, M. A.; Britt, P. F. *J. Phys. Chem.* **1990,** *94,* 6351–6360.

350. Desvergne, J.-P.; Bouas-Laurent, H. *J. Chem. Soc., Chem. Commun.* **1978,** 403–404.

351. Desvergne, J.-P.; Bitit, N.; Bouas-Laurent, H. *J. Chem. Res. (S)* **1984,** 214–215.

352. Desvergne, J.-P.; Bitit, N.; Castellan, A.; Webb, M.; Bouas-Laurent, H. *J. Chem. Soc., Perkin Trans. 2* **1988,** 1885–1894.

353. Felix, G.; Lapouyade, R.; Bouas-Laurent, H.; Clin, B. *Tetrahedron Lett.* **1976,** 2277–2278.
354. Daney, M.; Vanucci, C.; Desvergne, J.-P.; Castellan, A.; Bouas-Laurent, H. *Tetrahedron Lett.* **1985,** *26,* 1505–1508.
355. Becker, H.-D.; Andersson, K. *Tetrahedron Lett.* **1985,** *26,* 6129–6132.
356. Becker, H.-D.; Amin, K. A. *J. Org. Chem.* **1989,** *54,* 3182–3188.
357. Becker, H.-D.; Hansen, L.; Andersson, K. *J. Org. Chem.* **1986,** *51,* 2956–2961.
358. Deng, G.; Sakaki, T.; Nakashima, K.; Shinkai, S. *Chem. Lett.* **1992,** 1287–1290.
359. Heller, H. G.; Ottaway, M. J. *J. Chem. Soc., Chem. Commun.* **1995,** 479–480.
360. Marquis, D.; Desvergne, J.-P.; Bouas-Laurent, H. *J. Org. Chem.* **1995,** *60,* 7984–7996.
361. Kubo, Y.; Inoue, T.; Sakai, H. *J. Am. Chem. Soc.* **1992,** *114,* 7660–7663.
362. Kubo, Y.; Noguchi, T.; Inoue, T. *Chem. Lett.* **1992,** 2027–2030.
363. Kubo, Y.; Yoshioka, M.; Kiuchi, K.; Nakajima, S.; Inamura, I. *Tetrahedron Lett.* **1999,** *40,* 527–530.
364. Kubo, Y.; Kiuchi, K.; Inamura, I. *Bull. Chem. Soc. Jpn.* **1999,** *72,* 1101–1108.
365. Kubo, Y.; Yoshioka, M.; Nakajima, S.; Inamura, I. *Tetrahedron Lett.* **1999,** *40,* 2335–2338.
366. Kubo, Y.; Kusumoto, K.; Nakajima, S.; Inamura, I. *Chem. Lett.* **1999,** 113–114.
367. Bunce, N. J.; Cater, S. R.; Scaiano, J. C.; Johnston, L. J. *J. Org. Chem.* **1987,** *52,* 4214–4223.
368. Cantos, A.; Marquet, J.; Moreno-Mañas, M.; González-Lafont, A.; Lluch, J. M.; Bertrán, J. *J. Org. Chem.* **1990,** *55,* 3303–3310.
369. Matsui, K.; Maeno, N.; Suzuki, S. *Tetrahedron Lett.* **1970,** 1467–1469.
370. Wubbels, G. G.; Winitz, S.; Whitaker, C. *J. Org. Chem.* **1990,** *55,* 631–636.
371. Mutai, K.; Kajii, Y.; Nakagaki, R.; Obi, K. *Tetrahedron Lett.* **1996,** *37,* 505–508.
372. Nakagaki, R.; Mutai, K. *Bull. Chem. Soc. Jpn.* **1996,** *69,* 261–274.
373. Huertas, I.; Gallardo, I.; Marquet, J. *Tetrahedron Lett.* **2000,** *41,* 279–281.
374. Cervera, M.; Marquet, J. *Tetrahedron Lett.* **1996,** *37,* 7591–7594.
375. Bellas, M.; Bryce-Smith, D.; Gilbert, A. *Chem. Commun.* **1967,** 263–264, 862–863.
376. Bryce-Smith, D.; Clarke, M. T.; Gilbert, A.; Klunklin, G.; Mannig, C. *Chem. Commun.* **1971,** 916–918.
377. Bryce-Smith, D.; Gilbert, A.; Klunklin, G. *J. Chem. Soc., Chem. Commun.* **1973,** 330–331.
378. Bellas, M.; Bryce-Smith, D.; Clarke, M. T.; Gilbert, A.; Klunkin, G.; Krestonosich, S.; Manning, C.; Wilson, S. *J. Chem. Soc. Perkin 1,* **1977,** 2571–2580.
379. Bryce-Smith, D.; Cox, G. B. *Chem. Commun.* **1971,** 915–916.
380. Ohashi, M.; Tsujimoto, K.; Furukawa, Y. *Chem. Lett.* **1977,** 543–546.
381. Ohashi, M.; Miyake, K.; Tsujimoto, K. *Bull. Chem. Soc. Jpn.* **1980,** *53,* 1683–1688.
382. Birks, J. B. *Photophysics of Aromatic Molecules,* Wiley-InteScience: New York, 1970.
383. Stevens, B. *Adv. Photochem.* **1971,** *8,* 161–226.
384. Weller, A. *Pure Appl. Chem.* **1968,** *16,* 115–123.
385. Pac, C.; Sakurai, H. *Tetrahedron Lett.* **1969,** 3829–3832.
386. Yasuda, M.; Pac, C.; Sakurai, H. *Bull. Chem. Soc. Jpn.* **1981,** *54,* 2352–2355.
387. Davidson, R. S. *Chem. Commun.* **1969,** 1450–1451.

388. Yang, N. C.; Libman, J. *J. Am. Chem. Soc.* **1973,** *95,* 5783–5784.

389. Yang, N. C.; Shold, D. M.; Kim, B. *J. Am. Chem. Soc.* **1976,** *98,* 6587–6596.

390. Sugimoto, A.; Kimoto, S.; Adachi, T.; Inoue, H. *J Chem. Soc. Perkin Trans. 1* **1995,** 1459–1466.

391. Yasuda, M.; Yamashita, T.; Matsumoto, T.; Shima, K.; Pac, C. *J. Org. Chem.* **1985,** *50,* 3667–3669.

392. Yasuda, M.; Yamashita, T.; Shima, K. *J. Org. Chem.* **1987,** *52,* 753–759.

393. Yasuda, M.; Matsuzaki, Y.; Shima, K.; Pac, C. *J. Chem. Soc., Perkin Trans. 2,* **1988,** 745–751.

394. Kojima, R.; Shiragami, T.; Shima, K.; Yasuda, M.; Majima, T. *Chem. Lett.* **1997,** 1241–1242.

395. Yasuda, M.; Kojima, R.; Ohira, R.; Shiragami, T.; Shima, K. *Bull. Chem. Soc. Jpn.* **1998,** *71,* 1655–1660.

396. Yamashita, T.; Tanabe, K.; Yamano, K.; Yasuda, M.; Shima, K. *Bull. Chem. Soc. Jpn.* **1994,** *67,* 246–250.

397. Mizuno, K.; Okamoto, H.; Pac, C.; Sakurai, H. *J. Chem. Soc., Chem. Commun.* **1975,** 839–840.

398. Yasuda, M.; Pac, C.; Sakurai, H. *J. Org. Chem.* **1981,** *46,* 788–792.

399. Mizuno, K.; Pac, C.; Sakurai, H. *J. C. S. Chem. Commun.* **1975,** 553.

400. Yasuda, M.; Pac, C.; Sakurai, H. *J. Chem. Soc. Perkin Trans. 1,* **1981,** 746–750.

401. Yasuda, M.; Yamashita, T.; Shima, K. *Bull. Chem. Soc. Jpn.* **1990,** *63,* 938–940.

402. Arnold, D. R.; Maroulis, A. J. *J. Am. Chem. Soc.* **1977,** *99,* 7355–7356.

403. Maroulis, A. J.; Arnold, D. R. *Synthesis,* **1979,** 819–820.

404. Tazuke, S.; Ozawa, H. *J. Chem. Soc., Chem. Commun.* **1975,** 273–274.

405. Toki, S.; Hida, S.; Takamuku, S.; Sakurai, H. *Nippon Kagaku Kaishi,* **1984,** 152–157.

406. Mizuno, K.; Ichinose, N.; Tamai, T.; Otsuji, Y. *Tetrahedron Lett.* **1985,** *26,* 5823–5826.

407. Mizuno, K.; Tamai, T.; Nakanishi, I.; Ichinose, N.; Otsuji, Y. *Chem. Lett.* **1988,** 2065–2068.

408. Tamai, T.; Mizuno, K.; Hashida, I.; Otsuji, Y. *Photochem. Photobiol.* **1991,** *54,* 23–29.

409. Yoshino, A.; Ohashi, M.; Yonezawa, T. *Chem. Commun.* **1971,** 97.

410. Borg, R. M.; Arnold, D. R.; Cameron, T. S. *Can. J. Chem.* **1984,** *62,* 1785–1802.

411. Arnold, D. R.; Snow, M. S. *Can. J. Chem.* **1988,** *66,* 3012–3026.

412. Pac, C.; Nakasone, A.; Sakurai, H. *J. Am. Chem. Soc.* **1977,** *99,* 5806.

413. Rao, V. R.; Hixson, S. S. *J. Am. Chem. Soc.* **1979,** *101,* 6458–6459.

414. Mizuno, K.; Ogawa, J.; Otsuji, Y. *Chem. Lett.* **1981,** 741–744.

415. Mazzocchi, P. H.; Somich, C.; Edwards, M.; Morgan, T.; Ammon, H. L. *J. Am. Chem. Soc.* **1986,** *108,* 6828–6829.

416. McCullough, J. J.; Wu, W. S. *J. Chem. Soc., Chem. Commun.* **1972,** 1136–1137.

417. Lewis, F. D.; DeVoe, R. J.; MacBlane, D. B. *J. Org. Chem.* **1982,** *47,* 1392–1397.

418. Lewis, F. D.; DeVoe, R. J. *Tetrahedron,* **1982,** *38,* 1069–1077.

419. Arnold, D. R.; Du, X. *J. Am. Chem. Soc.* **1989,** *111,* 7666–7667.

420. Kubo, Y.; Suto, M.; Araki, T.; Mazzocchi, P. H.; Klingler, L.; Shook, D.; Somich, C. *J. Org. Chem.* **1986,** *51,* 4404–4411.

421. Maruyama, K.; Kubo, Y. *J. Org. Chem.* **1985**, *50*, 1426–1435.
422. Maruyama, K.; Kubo, Y.; Machida, M.; Oda, K.; Kanaoka, Y.; Fukuyama, K. *J. Org. Chem.* **1978**, *43*, 2303–2304.
423. Mariano, P. S.; Stavinoha, J. L.; Pepe, G.; Meyer, E. F. *J. Am. Chem. Soc.* **1978**, *100*, 7114–7116.
424. Mazzocchi, P. H.; Minamikawa, S.; Wilson, P. *J. Org. Chem.* **1985**, *50*, 2681–2684.
425. Yamada, S.; Kimura, Y.; Ohashi, M. *J. Chem. Soc., Chem. Commun.* **1977**, 667–668.
426. Mizuno, K.; Ikeda, M.; Otsuji, Y. *Tetrahedron Lett.,* **1985**, *26*, 461–464.
427. Nakanishi, K.; Mizuno, K.; Otsuji, Y. *Bull. Chem. Soc. Jpn.* **1993**, *66*, 2371–2379.
428. Mizuno, K.; Terasaka, K.; Ikeda, M.; Otsuji, Y. *Tetrahedron Lett.,* **1985**, *26*, 5819–5822.
429. Mizuno, K.; Terasaka, K.; Yasueda, M.; Otsuji, Y. *Chem. Lett.* **1988**, 145–148.
430. Mizuno, K.; Nakanishi, K.; Otsuji, Y. *Chem. Lett.* **1988**, 1833–1836.
431. Mizuno, K.; Kobata, T.; Maeda, R.; Otsuji, Y. *Chem. Lett.* **1990**, 1821–1824.
432. Tamai, T.; Mizuno, K.; Hashida, I.; Otsuji, Y. *Chem. Lett.* **1992**, 781–784.
433. Mizuno, K.; Nishiyama, T.; Terasaka, K.; Yasuda, M.; Shima, K.; Otsuji, Y. *Tetrahedron,* **1992**, *48*, 9673–9686.
434. Mizuno, K.; Nakanishi, K.; Kobata, T.; Sawada, Y.; Otsuji, Y. *Chem. Lett.* **1993**, 1349–1352.
435. Tamai, T.; Mizuno, K.; Hashida, I.; Otsuji, Y. *Bull. Chem. Soc. Jpn.* **1993**, *66*, 3747–3754.
436. Nishiyama, T.; Mizuno, K.; Otsuji, Y.; Inoue, H. *Tetrahedron,* **1995**, *51*, 6695–6706.
437. Mizuno, K.; Konishi, G.; Nishiyama, T.; Inoue, H. *Chem. Lett.* **1995**, 1077–1078.
438. Kyushin, S.; Matsuda, Y.; Matsushita, K.; Nakadaira, Y.; Ohashi, M. *Tetrahedron Lett.* **1990**, *31*, 6395–6398.
439. Kyushin, S.; Nakadaira, Y.; Ohashi, M. *Chem. Lett.* **1990**, 2191–2194.
440. Kyushin, S.; Ehara, Y.; Nakadaira, Y.; Ohashi, M. *J. Chem. Soc., Chem. Commun.* **1989**, 279–280.
441. Kubo, Y.; Todani, T.; Inoue, T.; Ando, H.; Fujiwara, T. *Bull. Chem. Soc. Jpn.* **1993**, *66*, 541–549.
442. Mizuno, K.; Nishiyama, T.; Takahashi, N.; Inoue, H. *Tetrahedron Lett.* **1996**, *37*, 2975–2978.
443. Sulpizio, A.; Albini, A.; d'Alessandro, N.; Fasani, E.; Pietra, S. *J. Am. Chem. Soc.* **1989**, *111*, 5773–5777.
444. d'Alessandro, N.; Fasani, E.; Mella, M.; Albini, A. *J. Chem. Soc. Perkin Trans. 2,* **1991**, 1977–1980.
445. d'Alessandro, N.; Albini, A.; Mariano, P. S. *J. Org. Chem.* **1993**, *58*, 937–942.
446. Fasani, E.; d'Alessandro, N.; Albini, A.; Mariano, P. S. *J. Org. Chem.* **1994**, *59*, 829–835.
447. Dinnocenzo, J. P.; Farid, S.; Goodman, J. L.; Gould, I. R.; Todd, W. P. Mattes, S. L. *J. Am. Chem. Soc.* **1989**, *111*, 8973–8975.
448. Todd, W. P.; Dinnocenzo, J. P.; Farid, S.; Goodman, J. L.; Gould, I. R. *Tetrahedron Lett.* **1993**, *34*, 2863–2866.

449. Lin, X.; Kavash, R. W.; Mariano, P. S. *J. Org. Chem.* **1996,** *61,* 7335–7347.
450. Khim, S.-K.; Cederstrom, E.; Ferri, D. C.; Mariano, P. S. *Tetrahedron,* **1996,** *52,* 3195–3222.
451. Hasegawa, E.; Brumfield, M. A.; Mariano, P. S. Yoon, U.-C. *J. Org. Chem.* **1988,** *53,* 5435–5442.
452. Yoon, U. C.; Mariano, P. S. *Acc. Chem. Res.* **1992,** *25,* 233–240.
453. Kim, J.-M.; Hoegy, S. E.; Mariano, P. S. *J. Am. Chem. Soc.* **1995,** *117,* 100–105.
454. Yoon, U. C.; Kim, D. U.; Lee, C. W.; Choi, Y. S.; Lee, Y.-J.; Ammon, H. L.; Mariano, P. S. *J. Am. Chem. Soc.* **1995,** *117,* 2698–2710.
455. Kako, M.; Morita, T.; Torihara, T.; Nakadaira, Y. *J. Chem. Soc., Chem. Commun.* **1993,** 678–680.
456. Mella, M. Fasani, E.; Albini, A. *J. Photochem. Photobiol. A* **1992,** *65,* 383–389.
457. Fukuzumi, S.; Kuroda, S.; Tanaka, T. *J. Chem. Soc., Chem. Commun.* **1986,** 1553–1554.
458. Fukuzumi, S.; Kitano, T.; Mochida, K. *Chem. Lett.* **1989,** 2177–2180.
459. Lan, J. Y.; Schuster, G. B. *J. Am. Chem. Soc.* **1985,** *107,* 6710–6711.
460. Lan, J. Y.; Schuster, G. B. *Tetrahedron Lett.* **1986,** *27,* 4261–4264.
461. Albini, A.; Fasani, E.; Sulpizio, A. *J. Am. Chem. Soc.* **1984,** *106,* 3562–3566.
462. Albini, A.; Fasani, E. *Tetrahedron,* **1982,** *38,* 1027–1034.
463. Mella, M.; Fasani, E.; Albini, A. *J. Org. Chem.* **1992,** *57,* 3051–3057.
464. Mella, M.; Freccero, M.; Soldi, T.; Fasani, E.; Albini, A. *J. Org. Chem.* **1996,** *61,* 1413–1422.
465. Mella, M.; Freccero, M.; Albini, A. *J. Chem. Soc., Chem. Commun.* **1995,** 41–42.
466. Sugimoto, A.; Sumida, R.; Tamai, N.; Inoue, H.; Otsuji, Y. *Bull. Chem. Soc. Jpn.* **1981,** *54,* 3500–3504.
467. Sugimoto, A.; Kimoto, S.; Inoue, H. *J. Chem. Res. (S)* **1996,** 506–507.
468. Sugimoto, A.; Yoneda, S. *J. Chem. Soc., Chem. Commun.* **1982,** 376–377.
469. Sugimoto, A.; Yamano, J.; Yasueda, M.; Yoneda, S. *J. Chem. Soc. Perkin Trans. 1* **1988,** 2579–2584.
470. Sugimoto, A.; Sumi, K.; Urakawa, K.; Ikemura, M.; Sakamoto, S.; Yoneda, S.; Otsuji, Y. *Bull. Chem. Soc. Jpn,* **1983,** *56,* 3118–3122.
471. Sugimoto, A.; Fukada, N.; Adachi, T.; Inoue, H. *J. Chem. Soc. Perkin Trans. 1* **1995,** 1597–1602.
472. Sugimoto, A.; Fukada, N.; Adachi, T.; Inoue, H. *J. Chem. Res. (S)* **1996,** 252–253.
473. Sugimoto, A.; Yamano, J.; Suyama, K.; Yoneda, S. *J. Chem. Soc. Perkin Trans. 1* **1989,** 483–487.
474. Sugimoto, A.; Hayashi, C.; Omoto, Y.; Mizuno, K. *Tetrahedron Lett.* **1997,** *38,* 3239–3242; *unpublished data.*
475. Sugimoto, A.; Maruyama, H.; Mizuno, K., *unpublished data.*
476. Sugimoto, A.; Hiraoka, R.; Fukada, N.; Kosaka, H.; Inoue, H. *J. Chem. Soc. Perkin Trans. 1* **1992,** 2871–2875.
477. Lewis, F. D.; Reddy, G. D.; Cohen, B. E. *Tetrahedron Lett.* **1994,** *35,* 535–538.
478. Lewis, F. D.; Reddy, G. D. *Tetrahedron Lett.* **1990,** *31,* 5293–5296.
479. Pandey, G.; Sridhar, M.; Bhalerao, U. T. *Tetrahedron Lett.* **1990,** *31,* 5373–5376.
480. Yasuda, M.; Shiomori, K.; Hamasuna, S.; Shima, K.; Yamashita, T. *J. Chem. Soc. Perkin Trans. 2* **1992,** 305–310.

481. Mizuno, K.; Pac, C.; Sakurai, H. *Bull. Chem. Soc. Jpn.,* **1973,** *46,* 3316–3317.

482. de Lijser, H. J. P.; Arnold, D. R. *J. Org. Chem.* **1997,** *62,* 8432–8438 and references cited therein.

483. Evans, T. R.; Wake, R. W. Jaenicke, O. in *The Exciplex,* Gordon, M.; Ware W. R., eds., Academic Press: New York, **1975,** 345–358.

484. Arnold, D. R.; Du, X. *Can. J. Chem.* **1994,** *72,* 403–414.

485. Zona, T. A.; Goodman, J. L. *J. Am. Chem. Soc.* **1993,** *115,* 4925–4926.

486. Weng, H.; Sheik, Q.; Roth, H. D. *J. Am. Chem. Soc.* **1995,** *117,* 10,655–10,661; Herbertz, T.; Roth, H. D. *J. Am. Chem. Soc.* **1996,** *118,* 10,954–10,962.

487. de Lijser, H. J. P.; Arnold, D. R. *J. Chem. Soc. Perkin Trans. 2* **1997,** 1369–1380.

488. Mizuno, K.; Nire, K.; Sugita, H.; Otsuji, Y. *Tetrahedron Lett.* **1993,** *34,* 6563–6566.

489. Ishii, H.; Yamaoka, R.; Imai, Y.; Hirano, T.; Maki, S.; Niwa, H.; Hashizume, D.; Iwasaki, F.; Ohashi, M. *Tetrahedron Lett.* **1998,** *39,* 9501–9504.

490. Everitt, S. R. L.; Inoue, Y. V. in *Molecular and Supramolecular Photochemistry* Ramamurthy, K. S. Schanze, eds., Marcel Dekker: New York, **1999,** 71–130.

491. Inoue. Y. *Chem. Rev.* **1992,** *92,* 741–770.

492. Rau, H. *Chem. Rev.* **1983,** *83,* 535–547.

3

Singlet-Oxygen Ene-Sensitized Photo-Oxygenations: Stereochemistry and Mechanisms

Michael Orfanopoulos

University of Crete, Crete, Greece

I. GENERAL ASPECTS

Since its discovery by Schele and Priestly more than two centuries ago, molecular oxygen has moved up and down on the scale of research interest [1]. Until that time, the theory of phlogistron (with negative weight) was developed and used to explain the oxidation of metals and the burning of organic compounds. The fall of this theory was followed by a period of exciting discovery of the role of oxygen in life processes. In the period between 1928 and 1935, Mulliken [2] interpreted the paramagnetic nature of molecular oxygen as the result of two outer electrons with parallel spins. Childe and Mecke [3] spectroscopically identified the $^1\Sigma_g^+$ higher-energy electronic state, and Herzberg [4] discovered another singlet excited state $^1\Delta_g$. These two excited states ($^1\Sigma_g^+$, $^1\Delta_g$) of the molecular oxygen had an excess energy from the ground state equal to 37.5 and 22.5 kcal/mol, respectively.

In 1939, Kautsky [5,6] reported experimental evidence for a "metastable reactive oxygen molecule" and the intermediate in dye-sensitized photo-oxygenation reactions. Kautsky's results were extensive and his conclusions basically correct based on what is known today. However, his proposal was rejected by many reputable scientists (e.g., see Ref. 6) and it was never accepted during his lifetime.

Although singlet oxygen was discovered 70 years ago, until the early 1960s it was considered to be a molecular species of passing astrophysical interest and rather limited as a research subject. It was the independent work of Foote and Wexler [7] and Corey and Taylor [8] that brought singlet oxygen back into the mainstream of chemical research. They opened up a gigantic new field of organic reaction chemistry. This field of singlet molecular oxygen chemistry has been developed rapidly since then. Today, the physics, chemistry, biology, and medicine of singlet oxygen are actively explored areas of research interest.

The chemical and photochemical generation of singlet oxygen, its ready detection, and its unusual but stereocontrolled chemical reactivity have made this research subject remarkably attractive [9–13]. Most of photosensitized oxygenations are now well established to invovle the $^1\Delta_g$ excited singlet state of molecular oxygen. It has a sufficiently long solution lifetime (milliseconds to microseconds) to be responsible for chemical reactions.

The most convenient way of generating singlet oxygen is the dye photosensitization, as shown in Eqs. (1)–(3).

$$\text{Sens} \xrightarrow{\text{visible light}} \text{Sens}^* \tag{1}$$

$$\text{Sens}^* + {}^3O_2 \longrightarrow \text{Sens} + {}^1O_2 \tag{2}$$

$${}^1O_2 + \text{substrate} \longrightarrow \text{oxygenated products} \tag{3}$$

These photooxygenations represent one of the most important hydrocarbon functionalization reactions available to the synthetic organic chemist.

Singlet-oxygen studies have being reported to such diverse areas as chemiluminescence [14], photocarcinogenity [15], ozonolysis [16], photodynamic action [17], peroxide decomposition [7], photosynthesis [14], air pollution [18], metallocatalyzed oxygenation reactions [19,20], synthetic applications [21], and polymer degradation [10].

The reaction of singlet oxygen with carbon–carbon double bonds can be classified into three categories:

1. The [4 + 2] cycloaddition to conjugated dienes or anthracenes to yield endoperoxides

$$\tag{4}$$

2. The reaction with enol ethers, enamines, and electron-rich alkenes, without allylic hydrogens of proper orientation, to yield 1,2-dioxetanes

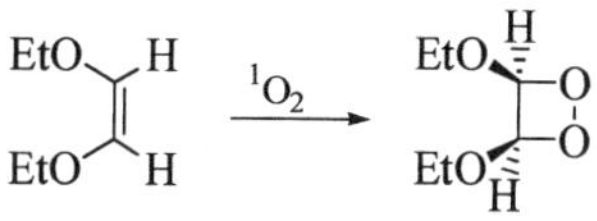

$$\text{(5)}$$

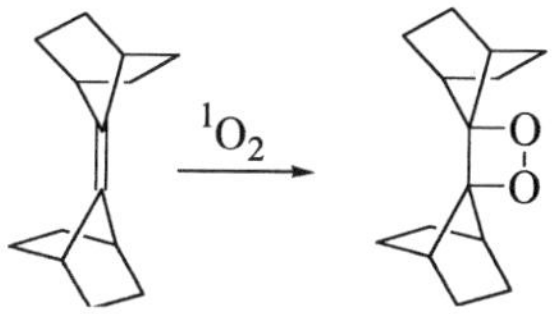

$$\text{(6)}$$

3. The ene or "Shenck reaction" with alkenes to form allylic hydroperoxides

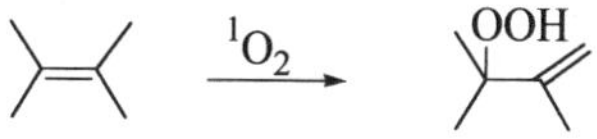

$$\text{(7)}$$

The purpose of this chapter is to summarize the recent stereochemical and mechanistic features of the ene photo-oxygenation of alkenes and related substrates by singlet oxygen.

II. THE ENE (SCHENCK) REACTION

A. Potential Mechanism

The ene reaction of singlet oxygen with alkenes was originally discovered [22,23] by Schenck in 1943. It is of synthetic [21,24], biochemical [25–27], and environmental significance [25,28] (Scheme 1). Over the years, the mechanism of this re-

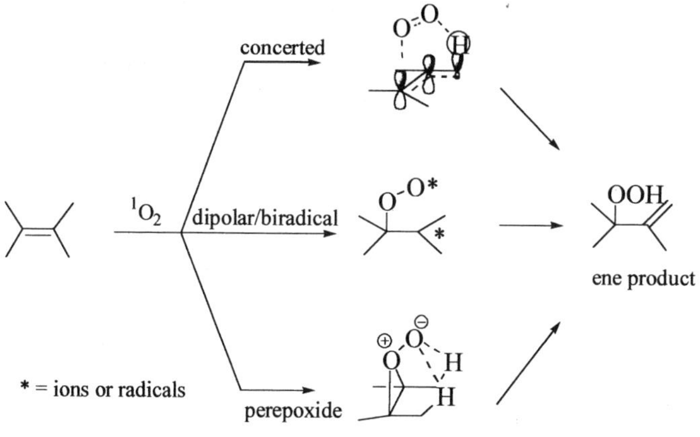

Scheme 1 Proposed mechanisms.

action was the subject of controversy, the main question being whether the reaction is concerted or involves intermediates.

A concerted mechanism in which the characteristic bond shifts take place through the cyclic transition state has been favored for some time [29,30]. The lack of substituent and solvent effects on product distribution [31] and the lack of direct evidence of any intermediates were given as support for the concerted ene mechanism.

Zwitterions [32] and biradicals [33] have been also proposed as intermediates. These mechanistic possibilities were insufficient for rationalizing the lack of correlation between the reaction rates and solvent polarity [14,31]. An early theoretical study by Goddard [33] favored a biradical intermediate. However, isotope effect studies [34], the lack of Markovnikov-type directing effects [14,31], and the fact that radical scavengers have no effect on the reaction eliminated a biradical mechanism for the singlet-oxygen-alkene ene reaction.

Today, the perepoxide intermediate proposed earlier by Sharp [35] seems to be the most popular mechanism that found support from results derived from kinetic isotope effects, stereochemical studies, trapping experiments [36–38], and theoretical calculations [39–43].

Dioxetanes have been also proposed [44] as intermediates. However, the pathway through dioxetanes has been found to give only carbonyl cleavage products rather than rearrange to allylic hydroperoxides [45].

The perepoxide intermediate was established by using Stephenson's tetramethylethylene-d_6 isotope effect test [46,47]. For example, substantial intramolecular isotope effects were found in reactions of $^1O_2$ with cis-related methyl and deuteriomethyl groups in tetramethylethylenes and cis-2-butene, but only a small or negligible isotope effect was observed with trans groups (Scheme 2).

Based on these results, a concerted pathway where all methyl groups should be equally competitive was excluded and a stepwise mechanism with rate-determining formation of the perepoxide-like intermediate was established.

Some researchers prefer an intermediate exciplex instead of the perepoxide [48–50]. Because the geometry of the intermediate is well defined from the isotope effects and the stereochemical studies, if an intermediate exciplex is formed, its geometrical features must resemble those of a perepoxide.

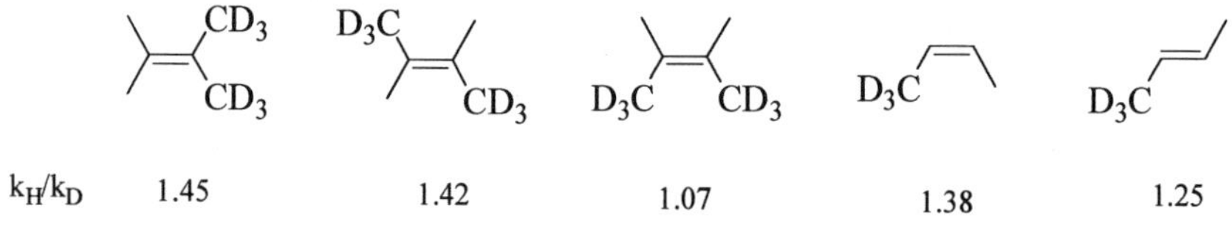

| k_H/k_D | 1.45 | 1.42 | 1.07 | 1.38 | 1.25 |

Scheme 2 Intramolecular isotope effects of $^1O_2$ with tetramethylethylenes and 2-butenes.

   Although no direct experimental evidence was reported for a perepoxide intermediate, episulfoxide [51], an analogous compound of perepoxide, has been prepared and rearranged at 25°C to the allylic sulfenic acid:

$$
\underset{-30\,^\circ C}{\xrightarrow{[O]}} \quad \underset{25\,^\circ C}{\xrightarrow{}} \quad \xrightarrow{\text{SOH}} \quad \xrightarrow{(PhO)_3P} \quad \text{SH} \qquad (8)
$$

   This could be the sulfur counterpart of a singlet-oxygen ene reaction.

B. Stereochemistry: A Suprafacial Ene Process

The first detailed stereochemical investigation of an ene reaction was reported in 1980 [52]. In this study the correlation between the stereochemistry of C—O bond-making to C—H(D) bond-breaking in the ene reaction was investigated. This intriguing experiment was performed by combining stereoisotopic studies in the photo-oxygenation reaction of the optically and isomerically pure olefins **1** and **2**:

$$
\begin{array}{cc}
\text{H}\text{—Ph} & \text{H}\text{—Ph} \\
\text{D} & \text{D} \\
\textbf{1} & \textbf{2}
\end{array} \qquad (9)
$$

   Before proceeding to a discussion of relatively complex stereochemical results, it is considered helpful at this point to illustrate the analytical technique with the perprotio olefin **3**:

$$
\underset{\textbf{3}}{\xrightarrow[\text{H}]{\text{—Ph}}} \quad \xrightarrow[2)\,[\text{H}]]{1)\,{}^1O_2} \quad \underset{R\text{-}\textbf{3a}}{\overset{\text{HO}\quad \text{H}_2}{\underset{\text{H}_1}{\text{—Ph}}}} \quad + \quad \underset{S\text{-}\textbf{3b}}{\overset{\text{HO}\quad \text{H}_2}{\underset{\text{H}_1}{\text{—Ph}}}} \qquad (10)
$$

Rose Bengal-sensitized oxidation of **3** yields two products **R-3a** and **S-3b**. Hydrogens H_1 and H_2 of the *R*- and *S*-enantiomers resonate in two distinguishable doublets in the vinyl region of the ^{1}H-NMR (nuclear magnetic resonance) spectrum. Enantiomers *R* and *S* are indistinguishable in an achiral environment. However, upon the addition of the chiral shift reagent Eu(tfc)$_3$, each doublet becomes a set of two doublets of equal intensity corresponding to the *R*- and *S*-enantiomers. Therefore, the enantiomeric ratio (*R/S*) of the products can be determined by examination of either the methyl or the olefinic diastereotopic protons. With this stereochemical understanding, the stereoisotopic analysis of the photo-oxygenation of optically active olefins **1** and **2** was performed.

Optically active olefin (Z)-(S)-$(-)$-1-deuterio-3,4-dimethyl-1-phenyl-2-pentene, **1** (Scheme 3), was photo-oxidized in acetone at $-10°C$. The tertiary allylic hydroperoxide was obtained in high yield and only in the trans configuration.

The approach of oxygen from the upper face of the olefin **1** would form the R tertiary hydroperoxide, and necessitates D abstraction in order to produce trans olefin. The approach from the lower face would form the S hydroperoxide with H abstraction (Scheme 3). These possibilities are designated R_H and S_D. Proton integration in the vinyl region of the ^{1}H-NMR spectrum of the product revealed no isotopic discrimination (i.e., $D/H = k_H/k_D = 1.01$). The stereoisomeric–isotopic relationship R_H and S_D can be obtained by NMR examination of the olefinic protons in the presence of a chiral shift reagent. It can be shown that for the R-enantiomer, only H appears at the vinyl position C_1, whereas the S-enantiomer has only D remaining (see Scheme 3). The amounts of R_D and S_H recorded correlate accurately with the known optical and geometric purity of the starting material. At the same time, the ratio of the diastereomeric methyl signals in the presence of $Eu(tfc)_3$ represents the enantiomeric ratio at the C_3 of the alcohol. This ratio was found to be $S/R = 1.0$. The observed stereoisomeric–isotopic relationship indicates that although the reaction is highly stereospecific, it does not show any isotopic discrimination ($k_H/k_D = 1.0$). For the latter reason, a concerted pathway for this reaction is eliminated.

These results were completely consistent with a rate-determining attack on the double bond to give an intermediate in which the stereochemistry about the olefinic bond is retained. Such a pathway is shown in Scheme 4. In such an hypothesis, an equivalent attack on the upper and the lower olefin faces would be expected, leading to equal yields of the two intermediates. Subsequent decomposition of these perepoxide intermediates would give the observed stereoiso-

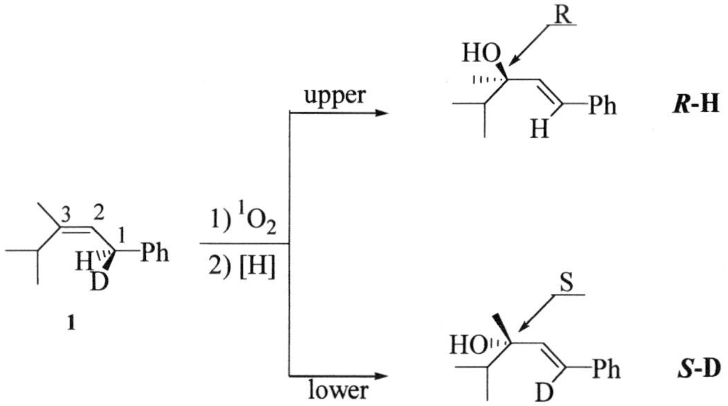

Scheme 3 Stereochemistry of the photo-oxygenation of **1**.

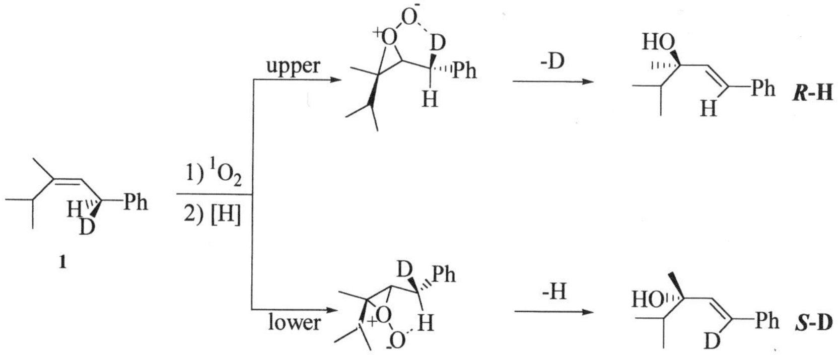

Scheme 4 Proposed mechanism of the sensitized photo-oxygenation of alkene **1.**

meric–isotopic relationship and no deuterium isotope effect would be observed because the rate-determining step would no longer involve abstraction of H or D.

In Scheme 4, the perepoxide intermediate is used for illustrative purposes. An exciplex between the olefin and oxygen with similar stereochemical requirements with that of a perepoxide might also accommodate these results.

Photo-oxygenation of the optically active olefin **2** [Eq. (11)], the (E)-isomer of **1,** gave very interesting results:

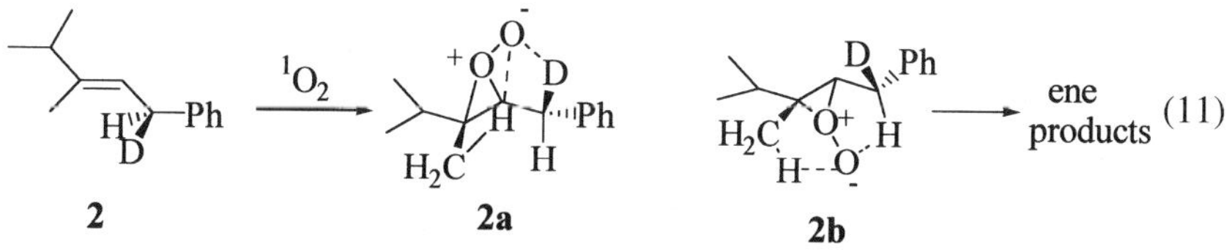

This olefin is the key substrate, which tests all the known stereoisotopic observations made thus far [52]. Because the methyl group is placed on the reactive disubstituted side of the olefin, a cis competition (Scheme 2) with the chiral, monodeuterated center would be expected.

As indicated in Eq. (11), the **2a** intermediate has the choice of abstracting D from the chiral center or H from the reactive methyl group leading to an isotope effect of 1.4 (cis competition, Scheme 2). On the other hand, intermediate **2b** can react with either the H of the chiral center or the H of the methyl group (noncompeting isotopes). The experimental isotopic analysis of this system showed $k_H/k_D \approx 1.2$, which is exactly the predicted average of 1.4 and 1.0. The enantiomeric analysis gave an R/S value of 1.25. Again, the enantiomeric ratio correlates with the isotopic ratio.

In conclusion, the ene reaction of singlet oxygen with simple alkenes is a highly stereospecific suprafacial process. It proceeds via an irreversible formation

of perepoxide or an exciplex intermediate with similar stereochemical requirements.

C. Site Selectivity With Trisubstituted Alkenes

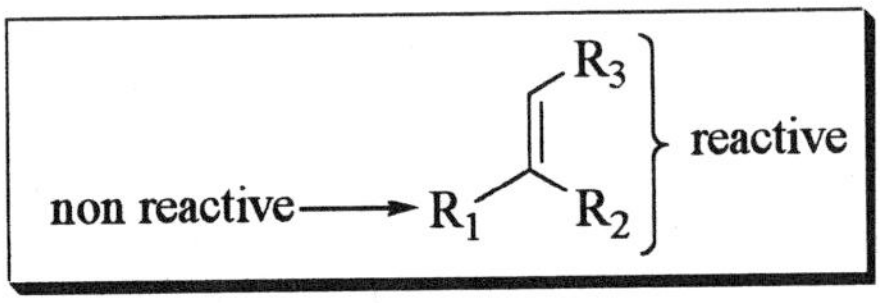

The regioselectivity of singlet-oxygen–olefin reactions went unrecognized for
more than 20 years of mechanistic study. It was generally recognized that methyl
and methylene hydrogens are reactive and that isopropyl C—H and certain con-
formationally inaccessible hydrogen atoms are not.

Photo-oxidation of trimethylethylene was frequently used as evidence to
support or contradict various mechanistic possibilities for the ene reaction. For example, the equal amounts of photo-oxidized products (olefin **4**) led to the conclusion that the ene reaction proceeds without any regioselectivity (Scheme 5).

The lack of Markovnikov directing effects in this system has been interpreted as evidence against a biradical or ionic intermediate. Because product **4a**
results from H abstraction from either methyl group (a) or (b) of **4** [Eq. (12)], the
relative reactivity of these groups was not known.

The stereospecific deuterium labeling and subsequent photo-oxidation of the
olefin **5** and similar olefins **6** and **7** revealed the hidden regioselectivity of the
singlet-oxygen–ene reaction.

A strong preference of H abstraction from the more substituted side of the
double bond was found [53,54], as is illustrated in Scheme 6. Reviewed in light of
this principle, earlier product distributions obtained in the photo-oxidation of a

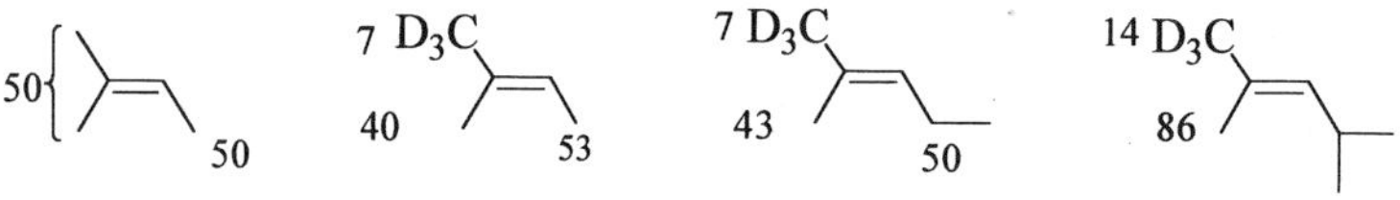

Scheme 5   Site selectivity of the photo-oxygenation of deuterated olefins (cis effect).

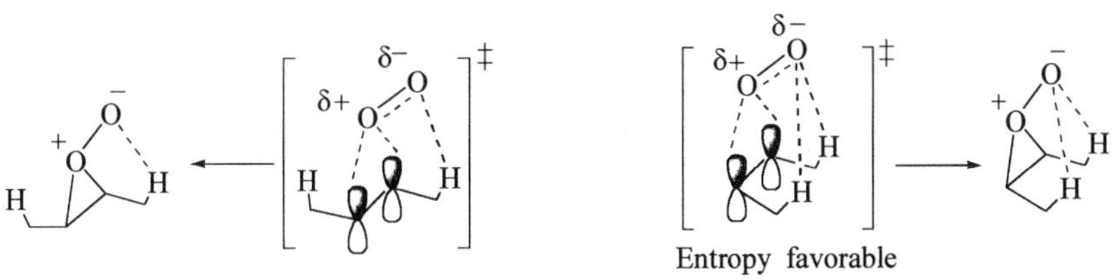

Scheme 6 Transition states of *cis*- and *trans*-butenes with $^1O_2$.

wide variety of olefins fall into a predictive pattern, the only notable exception being 1-methyl-cyclohexene. Selective examples are shown in Table 1 and demonstrate clearly that the overall reactivity of a given C—H bond is greater on the more crowded side of the trisubstituted olefin (cis effect). The interaction between the incoming oxygen and allylic hydrogens of the more crowded side of the olefin highly stabilizes the transition state of a perepoxide or an exciplex formation.

Schuster and co-workers [55,56] demonstrated that trans olefins show distinctly lower normalized (negative) entropies of activation $\Delta S\ddagger$ for the ene reaction than cis and suggested that a reversible exciplex can be formed, followed by an allylic hydrogen–oxygen interaction in the rate-determining step. In the reaction of *cis*-butene, the activation entropy $\Delta S\ddagger$ is less negative by 10 e.u. than that of the *trans*-butene, whereas the activation enthalpies are very similar. The considerable difference in the activation entropies was attributed to the fact that

Table 1 Site Selectivity of the Photo-Oxygenation of Trisubstituted Alkenes (cis Effect)[a]

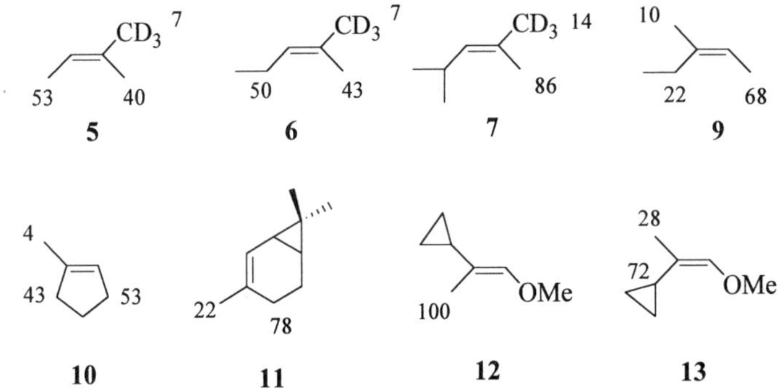

[a]Numbers indicate percentage of double bond formation in the ene adducts.

transition states in the case of cis olefins require more organization because of the simultaneous interaction of the incoming oxygen with two allylic hydrogens, as shown in Scheme 6.

Gorman et al. found that the reaction of singlet oxygen with a variety of substrates have negative activation enthalpies and concluded that a reversible formed exciplex is the intermediate [49,57]. They pointed out that none of the experiments clearly distinguish between formation of a perepoxide or a geometrically similar exciplex.

A simple way to rationalize the site selectivity is the formation of a perepoxide between olefin and singlet oxygen as the rate-determining step of this reaction. The formation of this intermediate is irreversible at least within the time scale of the ene reaction. The low activation energy (~10 kcal/mol) provides support for the assumption of irreversibility. This complex is formulated in such a way that O_a is over the monosubstituted side and O_b over the disubstituted side of the olefin (Scheme 7).

A crisscross flexibility of $^1O_2$ in such a complex was reported previously by Frimer and Bartlett [58,59]. Carbon–oxygen bond formation and C—H bond-breaking would then follow such an intermediate. Because the formation of the C—O bond requires closer approach to the olefin than the C—H bond-breaking, a transition state favoring C–O_a formation (less hindered O) over C—O_b (more hindered O) and subsequent C—H bond-breaking would be sufficient to rationalize the regioselectivity of the singlet-oxygen–trisubstituted olefin reaction. Stephenson suggested [60] that an interaction between the lowest unoccupied molecular orbital (LUMO) of the oxygen and the highest occupied molecular orbital (HOMO) of the olefin stabilizes the transition state of perepoxide formation. This is particularly true for the case of *cis*-2-butene, where the bonded orbitals of the allylic hydrogen contribute to the HOMO of the olefin, thus generating a system similar to the Ψ_3 state of butadiene. The interaction of this system with the π^* LUMO of the oxygen is favorable. For the *trans* isomer, the contribution of the bonded orbitals of the allylic C—H bonds to the HOMO of the olefin is approximately half, thus making the interaction with the oxygen LUMO less favorable. The future and possible properties of this complex will be discussed in light of more experimental results in the following sections.

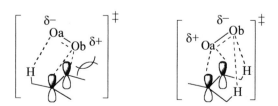

Scheme 7   Transition states of $^1O_2$–alkene ene reaction.

The remarkable site selectivity was rationalized by Houk and co-workers [61] in terms of rotational-barrier differences within the alkyl groups of the double bond. For example, STO-G6 semiempirical calculations, showed that the cis methyl groups in 2-methyl-2-butene have a lower (approximately 1 kcal/mol) rotational barrier than the trans methyl group. Therefore, the cis allylic hydrogens adopt a perpendicular conformation to the double-bond plane easier than the trans allylic hydrogens. According to this postulate, the lower the calculated rotational barrier, the higher the reactivity of the alkyl group. This mechanism was challenged by more recent results indicating that rotational barriers do not dictate the regioselectivity [62] of this reaction. This postulate will be discussed in the light of additional experimental results in Section III.B.

D. With Allylic Alcohols and Amines

The high diastereoselectivity observed in the photo-oxygenation of certain chiral allylic alcohols [63,64] and amines [65,66] and derivatives was rationalized in terms of hydroxyl or amino group coordination with the incoming oxygen. The synergy of oxygen–hydroxyl coordination as well as the 1,3 allylic strain exhibits, in nonpolar solvents, high selectivity for threo allylic hydroperoxides. The *syn/anti* selectivity of the photo-oxygenation of deuterium labeled alcohols **14-E** and **14-Z** was controlled by this effect.

The usual site selectivity was observed in a nonpolar solvent, whereas in polar solvents, such as methanol, where a hydrogen-bonding with the hydroxyl is efficient, the regioselectivity reverses (Table 2) [67,68]. Photo-oxygenation of

Table 2 *Syn/anti* Selectivity in the Photo-Oxygenation of Allylic Alcohols **14-E** and **14-Z**

Solvent	Syn/anti selectivity	Substrate
CCl$_4$	75/25	**14-E**
benzene	73/27	**14-E**
CH$_3$CN	41/59	**14-E**
MeOH	33/67	**14-E**
CCl$_4$	72/28	**14-Z**
CH$_3$CN	40/60	**14-Z**

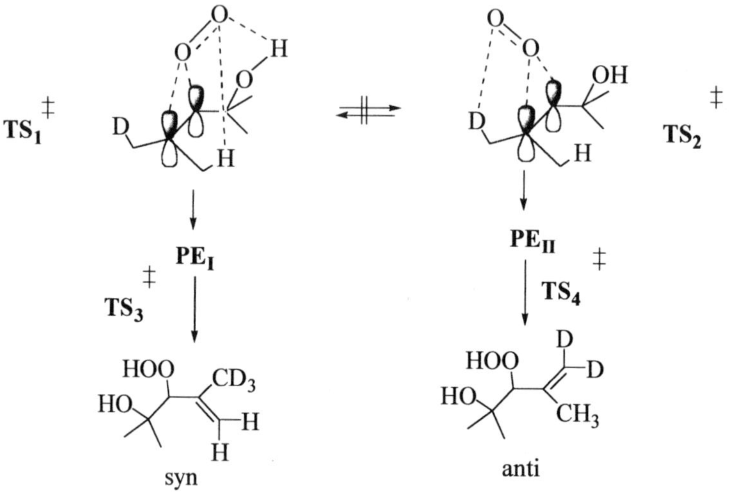

Scheme 8 Mechanism of the photo-oxygenation of allylic alcohol **14-E**.

14-E and **14-Z** in CCl_4 and CH_3CN showed similar *syn* selectivity. This result indicates that the *syn/anti* product selectivity is independent on the specific labeling of the methyl groups. Therefore, the perepoxide formation occurs in the rate-determining step. The mechanistic rational of the observed regioselectivity derived from examination of the possible reaction transition states is shown in Scheme 8. In TS_1, leading to the major *syn* product, the hydrogen-bonding interactions between the hydroxyl functionality and the incoming oxygen—in nonpolar solvents—are expected to be stronger than those in TS_2, where this interaction is absent. However, in polar solvents, the hydroxyl functionality of the allyl alcohol interacts with the solvent; thus, the ability to coordinate with the incoming oxygen is considerably reduced. In this case, the efficiency of hydroxyl–oxygen coordination depends on solvent polarity. For example, in methanol, where the hydroxyl group is fully coordinated by the solvent, the selectivity reverses, favoring the *anti* product by 67% (*syn/anti*= 33 : 67). In this case, the selectivity is controlled exclusively by steric factors.

E. *Anti* "Cis Effect" Selectivity

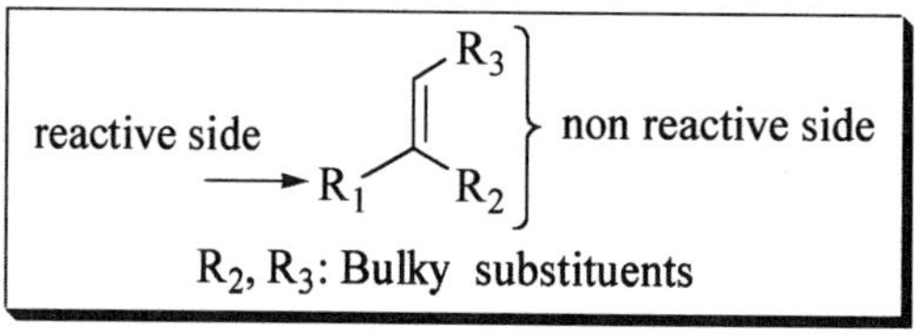

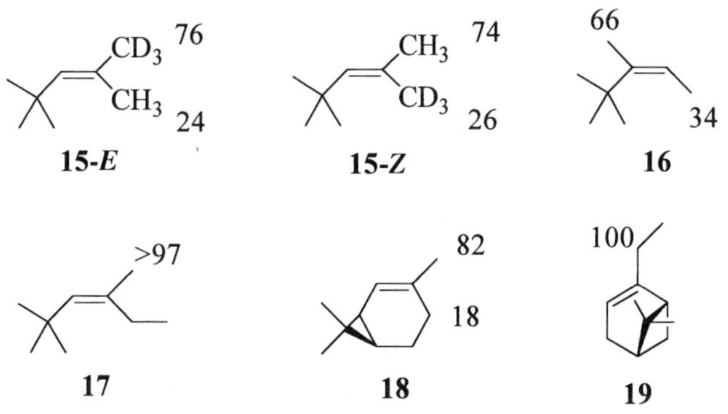

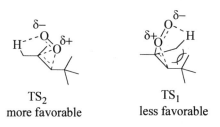

Scheme 9

In trisubstituted alkenes with the cis alkyl groups highly crowded, the site selectivity (Sec. II.C) does not apply. In those cases, usually only one allylic hydrogen is available in the more crowded side, not sufficient to stabilize the proper transition state that could produce *syn* products. Instead, the major product is now produced by abstraction of the allylic hydrogen in the less substituted side of the double-bond "*anti* selectivity" [69]. These results are summarized in Scheme 9. Photo-oxygenation of trisubstituted acyclic **15–17** [69] and cyclic **18** [70] and **19** [9] alkenes illustrates impressively a strong preference for hydrogen abstraction on the less substituted side of the double bond.

Examination of the possible transition states leading to the major (*anti*) and minor (*syn*) product provides reasonable mechanistic rational into the *anti* selectivity. In TS$_1$, leading to the minor (*syn*) product, the nonbonded interactions involving the large *tert*-butyl group and the incoming oxygen are expected to be stronger than those in TS$_2$, where this steric interaction is absent (see Scheme 10).

The most impressive *anti* selectivity is demonstrated in olefin **17**, where only the *trans* methyl hydrogens react. This can be attributed to the fact that in TS$_1$, apart from the 1,3 nonbonded interactions of the *tert*-butyl group with the

δ−
H---O $\delta+$ O

$\delta+$ O O δ− H

TS$_2$
more favorable

TS$_1$
less favorable

Scheme 10   Transition states in the ene reaction of $^1O_2$ with olefin **15**.

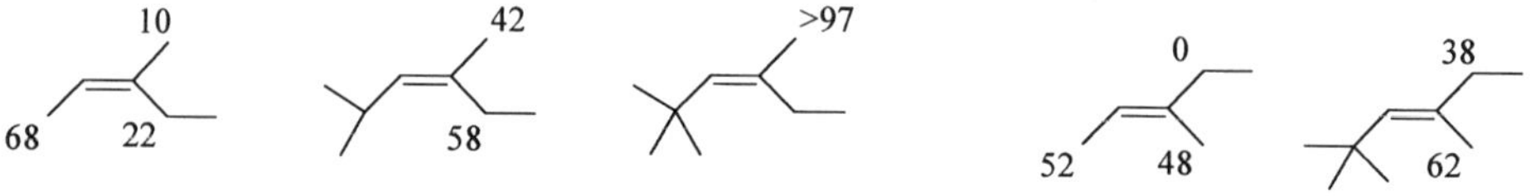

Scheme 11

oxygen, the cis configuration of the newly forming double bond places the methyl groups *cis* to each other.

The *anti* selectivity increases as the disubstituted side of the double bond becomes more crowded (Scheme 11). This is illustrated with the trisubstituted alkenes **20, 21,** and **17** [69]. Alkene **20** shows the normal "cis effect" selectivity where only the 10% of the anti ene adduct is formed. However, as the size of the cis alkyl substituent increases from methyl in **20** to isopropyl in **21** and *tert*-butyl in **17**, the *anti* selectivity increases from 10% to 42% to >97%, respectively. The same trend is also noted in substrate **23**. A substantial deviation from "cis effect" selectivity is observed by replacing one methyl group in **22** to a *tert*-butyl group in **23**. The totally unreactive methylene hydrogens in **22** (cis effect) become reactive in **23**, producing the exo ene adduct in 38% yield.

In conclusion, the *anti* selectivity for hydrogen abstraction of the ene reaction of trisubstituted olefins is related (1) to the degree of crowdedness of the more substituted side of the olefin, (2) to the nonbonded interactions in the new double-bond formation, and (3) to the lack of interaction of oxygen with two allylic hydrogens.

F. *Syn* Selectivity in Phenyl-Substituted Alkenes

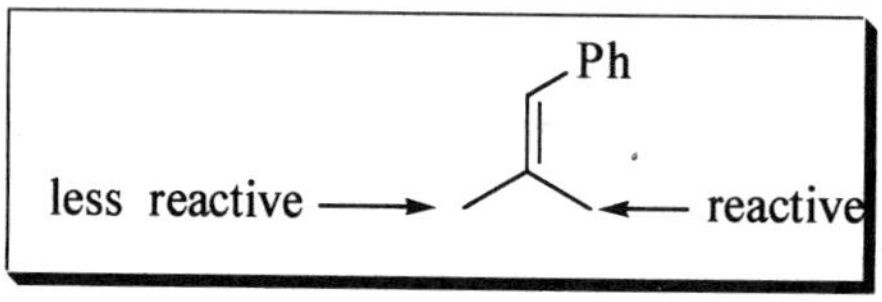

The phenyl ring of styrene substrates directs a *syn* selectivity in their ene reactions with singlet oxygen [71]. This effect was demonstrated by the photo-oxygenation of β,β-dimethyl styrene. This substrate, a part of the ene product, produces the 1,2-dioxetane and two diastereomeric diendoperoxides [72,73], as shown in Scheme 12. The stereospecific labeling of the *anti* methyl by deuterium in compound **24** to produce substrate **25** was required in order to study the *syn/anti* stereoselectivity of the ene products.

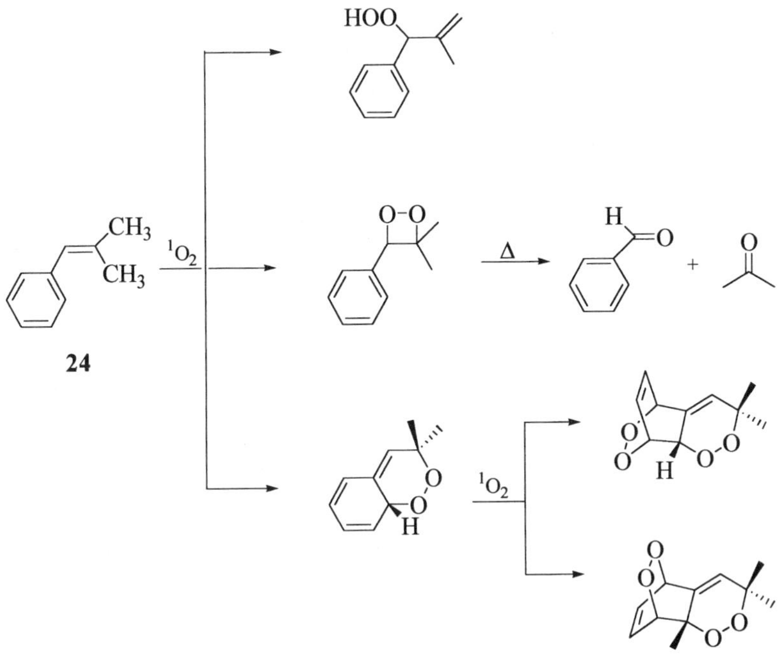

Scheme 12   Photo-oxygenation products of β,β-dimethylstyrene.

The ene products in different solvents showed a preference for hydrogen abstraction from the methyl *syn* to the phenyl group. The magnitude of this selectivity depends on solvent polarity. On increasing the solvent polarity, a substantial increase in the amount of *syn* product occurs (Table 3).

The intermolecular isotope effect of **24** with its deuterated analog **26** in chloroform was negligible ($k_H/k_D = 1.00 \pm 0.02$) [71]. As in other trisubstituted alkenes, this result was interpreted in terms of irreversible formation of a perepoxide intermediate. (See Scheme 13.)

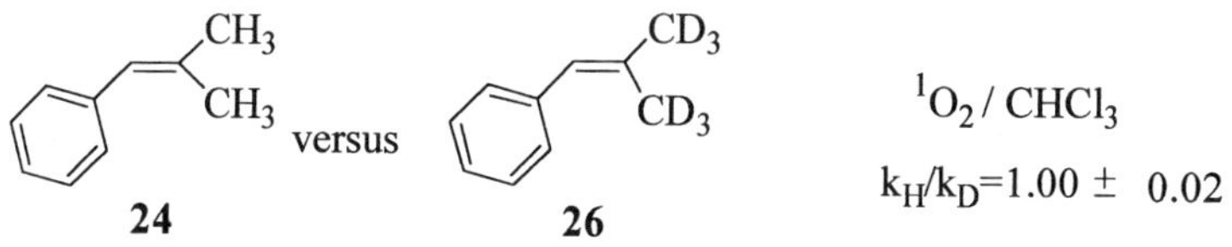

Scheme 13   Intermolecular isotope effects in the photo-oxygenation of alkenes **24** versus **26.**

Table 3 *Syn/anti* Stereoselectivity in the Ene Reaction of
$^1O_2$ With β,β-Dimethylstyrene

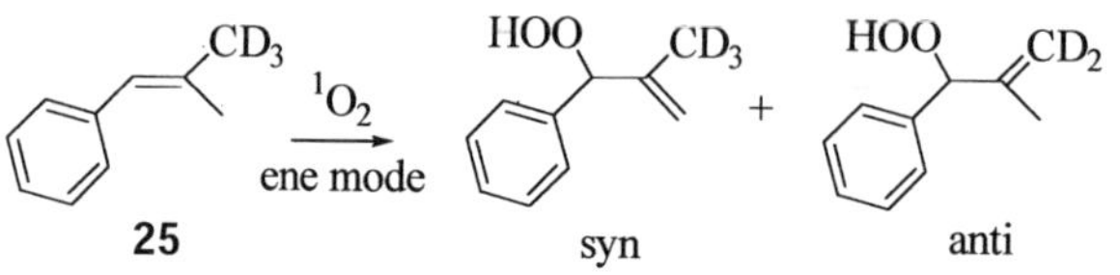

Solvent	Syn/anti selectivity[a]
CCl$_4$	56/44
CH$_3$CN	71/29
MeOH	82/18

[a] syn = H- abstraction, anti = D- abstraction.

A mechanistic possibility that accounts for the observed *syn* selectivity is shown in Scheme 14. In TS$_1$, the incoming oxygen is oriented toward the more substituted side of the double bond; there is only one allylic hydrogen interaction with singlet oxygen and TS$_1$ leads to the major ene product. In TS$_2$, leading to the minor *anti* product, singlet oxygen again interacts with only one allylic hydrogen. Therefore, the extra stabilization of TS$_1$ (major product) versus TS$_2$ (minor

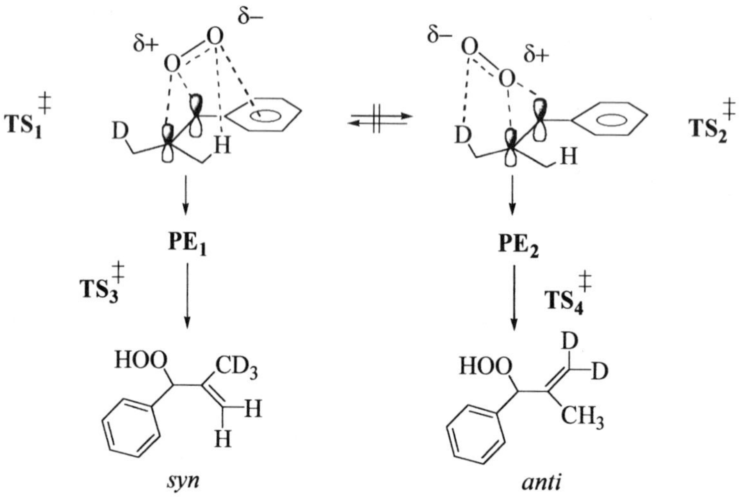

Scheme 14   Mechanism of photo-oxygenation of β,β-dimethylstyrene.

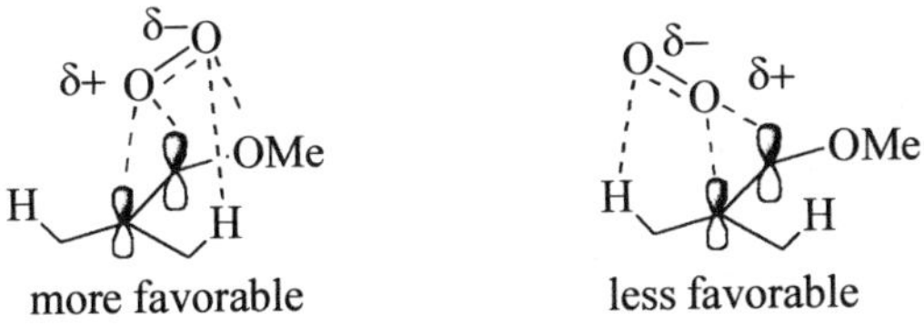

Scheme 15

product) must arise from "positive" interactions of singlet oxygen with the phenyl ring. In TS_1, the electron-donating ability of the phenyl ring stabilizes the partial positive charge which is developing on the benzylic carbon of the double bond. Simultaneously, the negative charge which is developing into the oxygen during the formation of the perepoxide, is stabilized by the partially positive phenyl group. Thus, the overall effect stabilizes the *syn* transition state TS_1 better than the anti TS_2, where this effect is absent.

The "cis effect" observed also in trisubstituted enol ethers [74,75] is well rationalized by a similar mechanism (Scheme 15). In this case, the electron-donating group that stabilizes the partially charged double-bond carbon of the perepoxide in the *syn* transition state is the alkoxy moiety. Therefore, the favorable interactions of the singlet oxygen with the phenyl or alkoxy substituents direct the orientation of the perepoxide intermediate.

III.   REGIOSELECTIVITY

A.   With Cis and Trans Alkenes

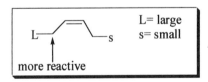

The reaction of singlet oxygen with cis and trans alkenes shows an unexpected regioselectivity [76] for hydrogen abstraction on the large alkyl group of the double bond. This remarkable and synthetically useful selectivity applies to both cis and trans nonsymmetrically substituted alkenes. The regioselective photo-oxygenation reactions of some cis nonsymmetrical alkenes and their regio-limitations are shown in Table 4.

For example, when L (larger group) is isopropyl or *tert*-butyl or phenyl, and s (smaller group) is hydrogen, compounds **27, 29** and **33,** respectively, the preferential abstraction of allylic hydrogen adjacent to the larger group is greater than 70%. In this case, the methyl hydrogens are statistically six to seven times

Table 4 Regioselectivity in the Photo-Oxygenation of Cis Alkenes[a]

L⌒═⌒s →($^1O_2$)→ L⌒═⌒(OOH)⌒s (Major product) + HOO⌒═⌒s (Minor product)

L = large substituent
s = small substituent

Major product

Minor product

Substrate	Large side %	Small side %
27	70	30
28	65	35
29	73	27
30	75	25
31	70	30
32	56	44 (Ph)
33 (Ph)	70	30
34 (Ph)	49	51
35 (Ph, Ph)	75	25
36 (Ph, Ph, Ph)	95	5
37 (Ph)	11	89

[a]Numbers indicate percentage of double bond formation in the ene adducts.

more reactive than the s-methylene hydrogens. In substrate **36**, where L = triphenyl group, the regioselectivity is nearly exclusive (95%) on the L-substituted side. As the size of the s group becomes larger, the regioselectivity toward the L substituent decreases. This is demonstrated with substrates **32** and **34**, where the preferential hydrogen abstraction is only slightly different on the two sides of the double bond. For example, when L and s are phenyl and isopropyl, respectively (substrate **34**), competition for the two allylic sides leads to nearly equal hydrogen abstraction from the two methylene sides. This result indicates further that nonbonding interactions play a more important role than conjugation with the π system of the phenyl ring in the transition state of this reaction.

The regioselectivity was rationalized [76] mainly on steric interactions in the possible transition states of the photooxygenation reaction (Scheme 16). In the transition state TS_2 leading to the major product, the nonbonding interactions involving the large group are smaller than those of the transition state TS_1 leading to the minor product. Because TS_2 is expected to have lower energy than TS_1, the C—O bond next to the larger substituent should be weakened more than the other C—O bond to the smaller group.

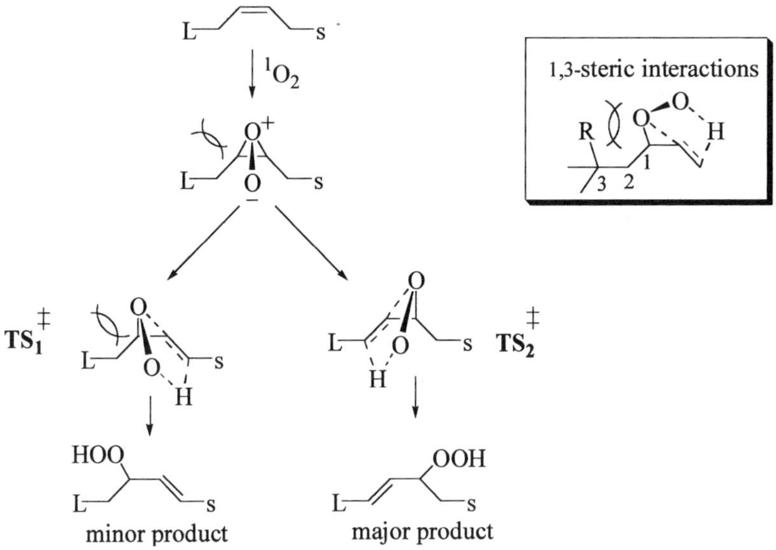

Scheme 16 Regioselectivity of the photo-oxygenation reaction of cis-alkenes.

Trans alkenes show similar regioselectivity on their photo-oxygenation reactions. However, their reactivity toward singlet oxygen is much less than that of the corresponding cis alkenes [61].

B. Effect of a Bulky Allylic Substituent

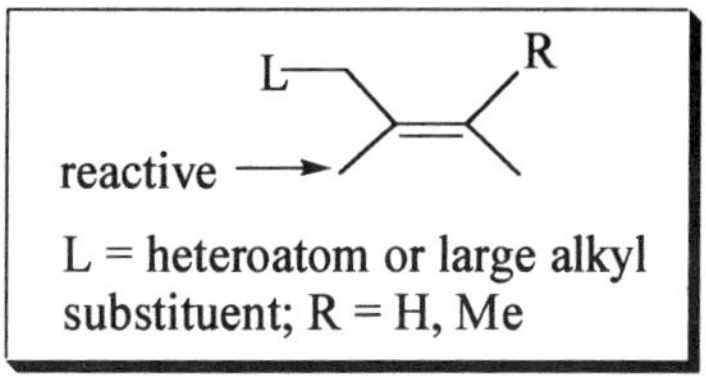

The reaction of singlet oxygen with alkenes shows general preference for hydrogen abstraction from the group that is geminal to the larger substituent of the double bond. For example substrates produced by replacement of an allylic hydrogen in tetramethylethylene with a series of functional groups (sulfides, sulfoxides, sulfones, cyano, halides, etc.) undergo selective photo-oxygenations with a surprising geminal selectivity with respect to the allylic functionality [77]. Some examples are summarized in Table 5.

Table 5   Geminal Selectivity in the Photo-Oxygenation of Alkenes Bearing an Allylic Functionality[a]

[a]Numbers indicate percentage of double bond formation in the ene adducts.

The various explanations that were provided to rationalize the observed regioselectivity involved (a) electronic repulsions between the lone pairs of the heteroatoms and the negatively charged oxygen of the perepoxide, (b) anchimeric assistance by the allylic substituent leading to regioselective opening of the possible perepoxide intermediate [Eq. (13)],

(13)

and (c) rotational barrier differences within the alkyl groups of the double bond.

The geminal selectivity holds even in the reactions of singlet oxygen with simple alkyl-substituted alkenes [78,79]. Photo-oxygenation of a series of alkyl- and phenyl-substituted olefins shows a strong preference for hydrogen abstraction on the methyl group that is geminal to the larger alkyl or aryl substituent of the alkene. These results are summarized in Table 6.

Disubstituted alkene **42** impressively illustrates this point. Similarly, trisubstituted and tetrasubstituted alkenes with the bulky alkyl substituent in the allylic position show preferential hydrogen abstraction from the geminal methyl group. Alkenes **43–46** give, again, as the major product, the ene adduct with the double bond on the methyl that is geminal to the larger alkyl group. This selectivity is demonstrated again in symmetrical olefin **47,** where only the methyl hydrogens react. The same trend was also noted in cyclic alkenes **48–50** [80].

Examination of the possible transition states leading to the major and the minor product in the reaction of $^1O_2$ with alkene **46** (Scheme 17) provides a new insight into the geminal selectivity. In transition state TS$_3$, leading to the minor or absent product, the nonbonding interactions involving the large *t*-butyl group and

Table 6 Geminal Selectivity on the Photo-Oxygenation of Alkenes Bearing a Large Alkyl Substituent at the Allylic Position[a]

L = large alkyl substituent   Major product   Minor product
R = H, Me

[a]Numbers indicate percentage of double bond formation in the ene adducts.

the methyl group, which are placed in a cis configuration, are expected to be stronger than those in transition states TS_1, TS_2, and TS_4, where this steric interaction is absent. Because of 1,3 nonbonded interactions, transition state TS_1 leading to the major product is expected to have lower energy than TS_2 and TS_4.

An alternative explanation of the geminal selectivity was based on a previous computational model proposed by Houk et al. [61]. According to this model, the lower the calculated rotational barrier, the higher the reactivity of the alkyl group.

For example, molecular mechanics calculations (MM2) showed that the methyl group geminal to the neopentyl group in 2,3,5,5-tetramethyl-2-hexene (**51**, Scheme 18) has the lowest rotational barrier and is the most reactive [79]. Furthermore, the ethyl group in 2,3-dimethyl-2-pentene (**52**) has a much higher rotational barrier (5.76 kcal/mol) than the methyl groups and is totally inactive to $^1O_2$. Similar trends hold with 2-methyl-2-butene (**4**). However, the barrier to rotation does not always predict the regioselectivity of the ene reaction of $^1O_2$ with alkenes. As it was shown later [62], it is the nonbonded interactions in the isomeric

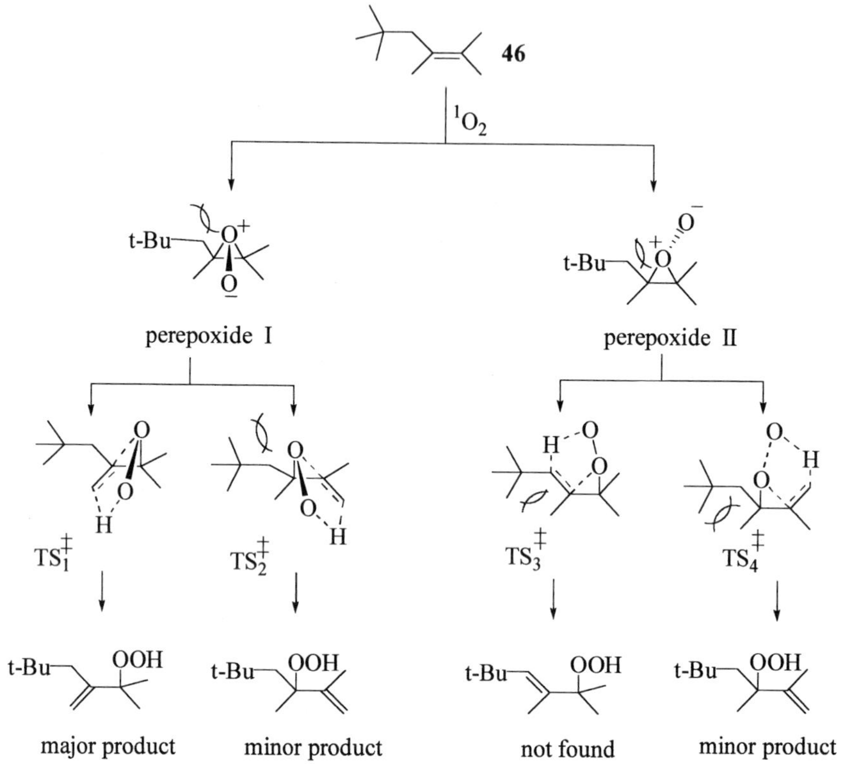

Scheme 17 Mechanism of the geminal selectivity on the photo-oxygenation of alkenes bearing a large alkyl substituent in the allylic position.

transition states that control product formation, and the barriers to rotation are rather irrelevant.

The calculated rotational barrier values for the allylic methyls, with the HF-STO-3G method [62], in a series of trisubstituted alkenes and the experimentally observed ene regioselectivity of a series of selective substrates are shown in Table 7.

For alkenes **15-E** and **15-Z**, the *syn* methyl groups have lower rotational barriers than the corresponding *anti* ones by 0.5 kcal/mol. This is in the opposite

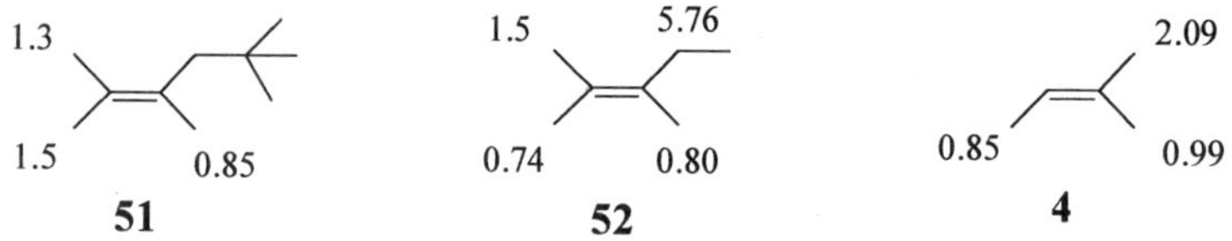

Scheme 18 Calculated rotational barriers (in kcal/mol) by MM2.

Table 7 Relative Yields of the Ene Products and Rotational Barriers of the Methyl Groups

	a	b		a	b
15-E (structure with CD$_3$)	76	1.63	**16** (structure)	66	2.91
	24	1.11		34	0.91
15-Z (structure with CD$_3$)	74	1.63	**53** (Ph structure)	64	1.45
	26	1.11		36	1.22
7 (structure with CD$_3$)	14	1.64			
	86	0.40			

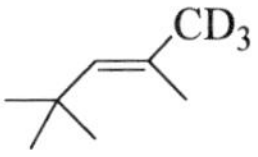

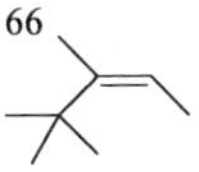

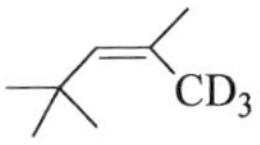

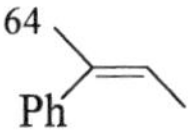

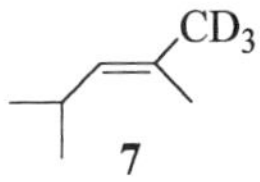

[a] Percentage of double bond formation in the ene adducts.
[b] Barriers to rotation (kcal/mol).

direction to the proposed theoretical model. However, for alkene **7**, there is a correlation between rotational barriers and ene reactivity. Alkenes **16** and **53** also demonstrate impressively that there is no correlation between ene reactivity and rotational barriers. Again, in these examples, the more reactive trans methyl group is the one which shows the higher rotational barrier in contrast to the predictions of the proposed theoretical model. These results indicate that there is not always a correlation between the reactivity of the methyl groups and their rotational barriers. It was also emphasized [62] that according to the Curtin–Hammett principle, the rotational barriers are irrelevant to the product distribution because their values are too small (0.5–2.0 kcal/mol) compared to the activation energies of the reactions (6–13 kcal/mol).

C. Geminal Selectivity With Respect to a Bulky Vinyl Substituent

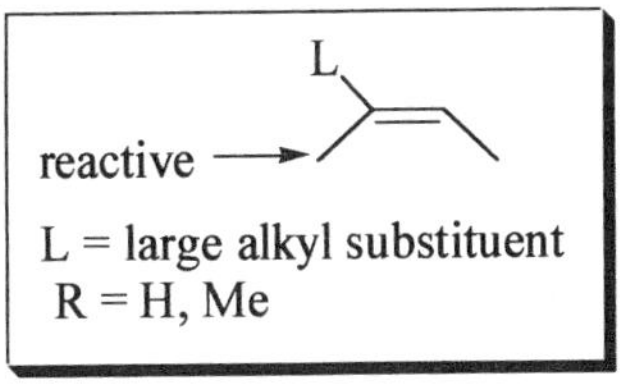

reactive → (structure)
L = large alkyl substituent
R = H, Me

Geminal selectivity is also observed in cases where the bulky substituent is in the vinylic position [66,78,80]. The examples are shown in Table 8. The presence of a para-phenyl substituent does not alter the geminal selectivity. This is demon-

Table 8 Geminal Selectivity With Respect to a Bulky Vinyl Substituent

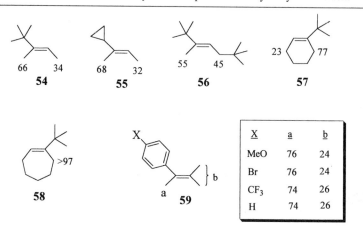

X	a	b
MeO	76	24
Br	76	24
CF$_3$	74	26
H	74	26

strated with substrate **59,** where the selectivity is insensitive to substitution on the phenyl ring. This result indicates that nonbonding interactions play a more important role than electronic effects of the para-substituted phenyl ring in determining the stability of the transition state of the product-determining step of the reaction [66,76]. For these styrene-type substrates, although the active stereosize of the phenyl group is smaller than the *t*-butyl group in **54,** the geminal selectivity is higher. This may be attributed to the fact that the forming double bond is in conjugation with the phenyl group.

For cycloalkenes, the variation in the percentage of geminal regioselectivity is probably due to the combination of both effects: (a) the different conformational arrangements of the allylic hydrogens in the ring systems as proposed earlier [81] and (b) the 1,3 nonbonded interactions between the large groups and the oxygen.

On the basis of similar arguments previously proposed, transition state TS$_1$ is expected to have lower energy than TS$_2$ and thus account for the geminal selectivity of alkenes with the large L group in the vinyl position (see Scheme 19).

D. Allylic Silanes, Stannanes, and Germanes

Me$_3$Si—/═\ ⟵ reactive

In the reaction of $^1O_2$ with allylic silanes [82–84], a different regioselectivity pattern was observed when compared to their carbon analogs (Sec. III.A). In this case

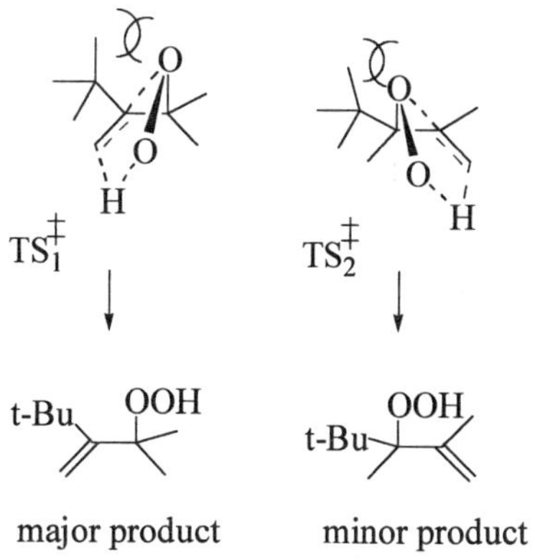

TS$_1^{\ddagger}$ TS$_2^{\ddagger}$

t-Bu   OOH   OOH
                  t-Bu

major product   minor product

Scheme 19

hydrogen abstraction occurs at the side of the double bond that is opposite to the silicon functionality. Also, in addition to trans, significant amounts of ene products are formed with a cis configuration. It is interesting to compare the regioselectivity in the photooxygenation of alkyl substituted alkenes with allylic silanes. The C—Si bond length is 1.89 Å, almost 25% longer than that of the C—C bond. Although the Me$_3$Si group is bulkier than the t-butyl, the 1,3 nonbonded interactions must be far less important, because the methyl groups lie away from the oxygen of the perepoxide. For example, replacement of the t-butyl for trimethyl silyl group in substrates **63** and **31** to produce **60** and **61** (Table 9) leads to very different regioselectivity, as well as stereoselectivity (mixtures of E and Z allylic hydroperoxides). The few studies of photo-oxygenation of some cyclic allylic germanes [85] indicate a similar regioselectivity trend to the silicon analogs, but the directing effect of germanium is less pronounced.

Table 9   Regioselectivities in the Photo-Oxygenation of Allylic Silanes, Alkenes, and Germanes

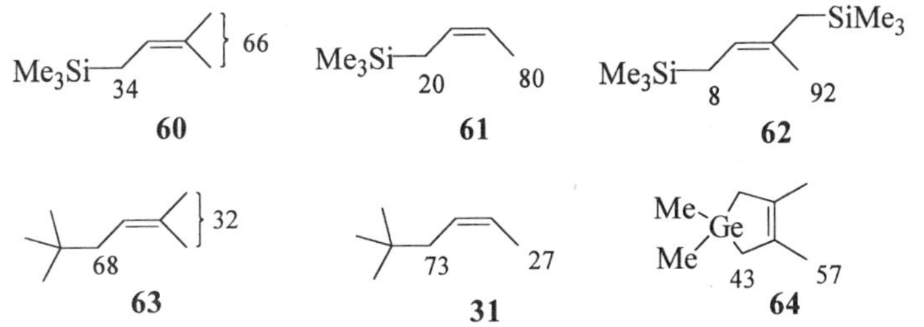

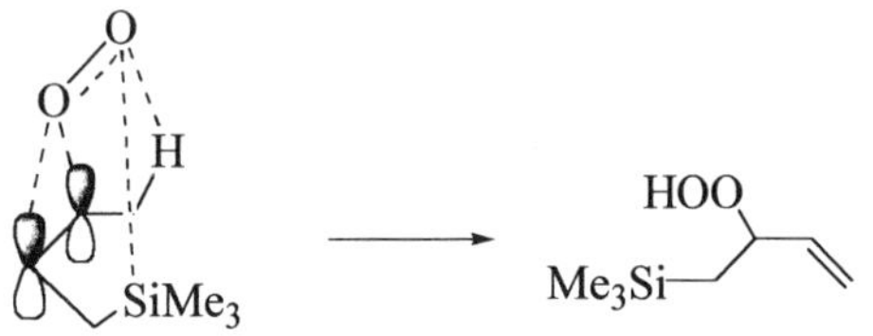

Scheme 20

To explain the regioselectivity trend in the photo-oxygenation of allylic silanes, it was proposed that an interaction between the negatively charged oxygen of the perepoxide and the silicon atom controls the abstraction of allylic hydrogen atoms (Scheme 20). The formation of cis ene adducts [86] was attributed either to the intermediacy of a zwitterionic intermediate or to a nonconcerted pathway. A perepoxide intermediate was considered unlikely.

Apart from the ene products, photo-oxygenation of allylic stannanes produces stannyldioxolanes [87–89]. Dioxolanes probably derive from ring opening of the intermediate perepoxide with 1,2-migration of the stannyl group to form a zwitterion:

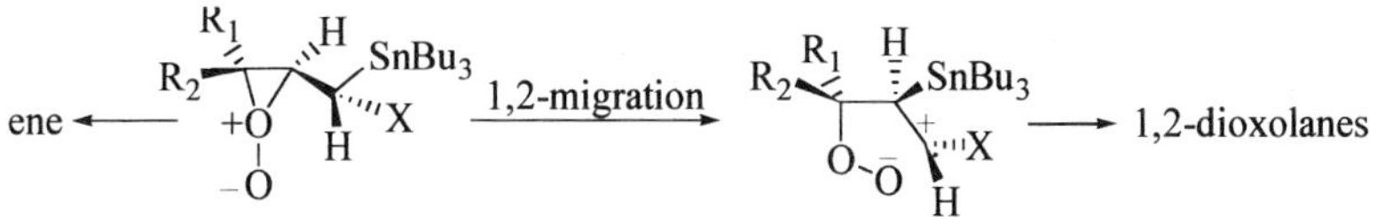

The ene regioselectivity is surprising, with the trialkyltin group directing allylic hydrogen abstraction next to it, in contrast to the regioselectivity observed in allylic silanes [90], as seen in substrates **65** and **66** in Scheme 21. Although the tertiary allylic hydrogen seems to have the proper geometry for ab-

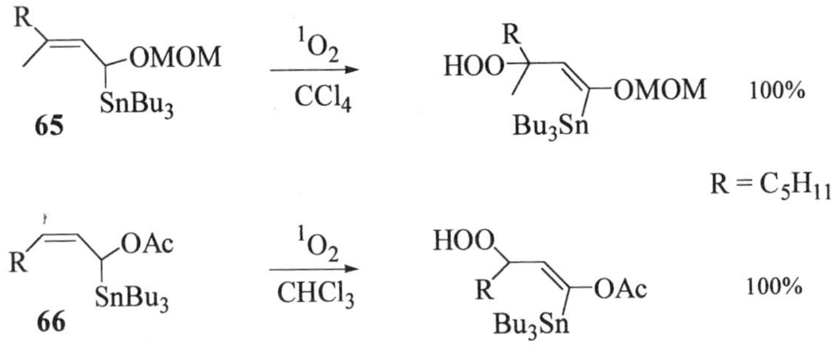

Scheme 21 Regioselectivity in the photo-oxygenation of alkenes bearing a trialkyltin substituent in allylic position.

straction in the perepoxide intermediate, additional factors may control this un-expected selectivity.

E. Induced Geminal Regioselectivity from Electron-Withdrawing Group at the α-Position

X = CHO, COOH, COMe, SOR, C=NtBu, CN

For alkenes bearing an electron-withdrawing group at the α-position, such as aldehyde [91], carboxylic acid [92], ester [93], ketone [94,95], imine [96,97], sulfoxide [98], and cyano [99], a high degree of geminal selectivity has been demonstrated. The results are summarized in Table 10.

Table 10 Germinal Selectivity in the Photo-Oxygenation of Functionalized Alkenes[a]

major product

[a]Numbers indicate percentage of double bond formation in the ene adducts.

Many mechanisms had been proposed in the past to rationalize this selectivity (trioxanes, perepoxide, exciplex, dipolar, or biradical intermediates); however, it is now generally accepted [100] that the mechanism proceeds through an intermediate exciplex which has the structural requirements of a perepoxide. This assumption is supported (a) by the lack of stereoselectivity in the reactions with chiral oxazolines [101] and tiglic acid esters [102], (b) by the comparison of the diastereoselectivity of dialkyl-substituted acrylic esters [103] with structurally similar nonfunctionalized alkenes, (c) by intermolecular isotope effects [104] in the photo-oxygenation of methyl tiglate, and (d) by solvent effects on regioselectivity [105].

It was proposed that in the hydrogen-abstraction step, the perepoxide opens preferentially at the C—O bond next to the unsaturated moiety. Due to the forthcoming conjugation in the adduct, this transition state is favorable, as exemplified in the photo-oxygenation of methyl tiglate in Scheme 22.

The 1,3 nonbonded interactions which control the regioselectivity in the photooxygenation of non-functionalized alkenes (see Secs. III.A and III.B), do not contribute significantly to the geminal selectivity of alkenes bearing an electron withdrawing group at the α-position. As seen in substrate **75** (Table 10), no reactivity of the methylene hydrogens next to the *t*-butyl group was found. This result indicates that the non-bonded interactions between the *t*-butyl group and the oxygen atom are less important than the forthcoming double bond-carbonyl conjugation in the ene adduct.

F. The Effect of β-Electron-Withdrawing and β-Hydroxyl Groups on Regioselectivity

The presence of an electron-withdrawing substituent at the β-position with respect to the double bond causes various trends in the ene regioselectivity [106] (Table

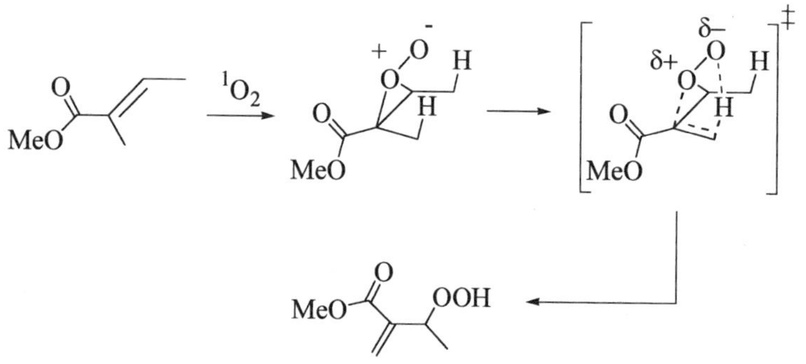

Scheme 22

Table 11 Regioselectivity in the Photo-Oxygenation of Functionalized Alkenes at the β-Position[a]

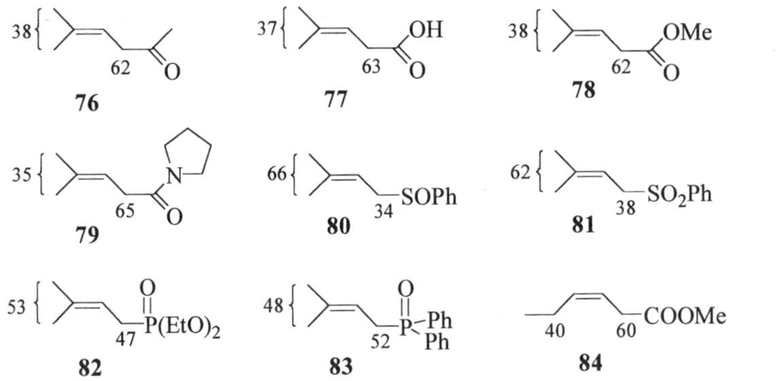

[a]Numbers indicate percentage of double bond formation in the ene adducts.
[b]For previous reports on the regioselectivity of **76** see Refs. 107,108.
[c]For substrate **80** see Ref. 109.

11). For example, in the carbonyl derivatives (**76–79**), the regioselectivity is invariable and the methylene hydrogen atoms are approximately five times as reactive as those of the methyl groups. The forthcoming conjugation of the double bond with the functionality shows moderate regioselectivity compared to the corresponding α,β-isomers (Table 10).

In substrates **80–83**, where the substituents are sulfoxide, sulfonyl, phosphate, or phosphine oxide, the reactivity next to the allylic substituent decreases significantly, although these substituents are quite bulky. This behavior was attributed to the electronic repulsions between the highly polarized oxygen atoms of the S—O and P—O bonds and the negative oxygen atom of the perepoxide, which direct the intermediate to abstract a hydrogen from the methyl group (Scheme 23). The electronic repulsions probably do not force singlet oxygen to form the inter-

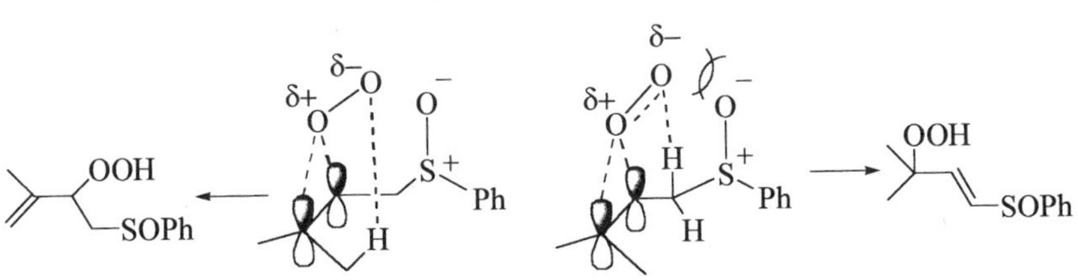

Scheme 23

mediate perepoxide on the less substituted side of the olefin, because the cis disubstituted β,γ-unsaturated ester (**84**) exhibits very similar reactivity to the trisubstituted substrate **78**. Analogous unfavorable repulsions [110–112] have been proposed to control the diastereoselectivity in the photo-oxygenation of chiral acylated allylic amines, sulfoxides, carbonyl, and carboxylic acid derivatives. The electronic control in the stereoselectivity of the electrophilic addition to alkenes is well documented [113–115].

The hydroxyl group at the allylic position has a significant effect on the *syn/anti* methyl stereoselectivity [67,68] and the diastereoselectivity [63,64] of the photo-oxygenation ene reaction (see Sec. II.B). To assess the effect of the hydroxyl at the more remote homoallylic position, the reaction of $^1O_2$ with the geminal dimethyl trisubstituted homoallylic alcohols (**85, 86, 89**) and the cis disubstituted **90** was examined in nonpolar solvents [116]. The regioselectivity trend was compared with that of the structurally similar trisubstituted alkenes (**87, 88, 91**) [105]. The results are summarized in Table 12.

For alkenes **87, 88,** and **91,** the regiochemistry is solely dependent on the steric hindrance of the allylic substituent. It is obvious that the regioselectivity trend for the homoallylic alcohols **85, 86,** and **89** is different from that of compounds **87, 88,** and **91,** respectively, although the hydroxyl group exerts approximately the same steric hindrance as a methyl group.

The increased percentage of hydrogen abstraction from the geminal methyl groups can be attributed either to the fact that hydroxyl coordination to the negatively charged oxygen of the perepoxide directs hydrogen abstraction from the methyl group (seven-membered ring transition state) or more likely due to electronic repulsions between a hydroxyl group and an oxygen atom in the perepoxide intermediate which occurs in a six-membered ring transition state,

Table 12   Regioselectivity in the Photo-Oxygenation of Alkenes and Homoallylic Alcohols[a]

[a]Numbers indicate percentage of double bond formation in the ene adducts.

analogous to that shown in Scheme 23. This assumption is further supported by the regioselectivity trend of substrate **90.** In this case, although steric hindrance is approximately the same on both sides of the double bond [76], an ene product ratio of 60/40 was found.

G. Twisted 1,3-Dienes: Allylic Versus Allene Hydroperoxide Formation

Selectivity was observed in the photo-oxygenation of significantly twisted 1,3-dienes which cannot adopt a reactive [4 + 2] conformation. In these substrates, a vinylic hydrogen atom is properly aligned in a perpendicular conformation to the olefinic plane and competes with the allylic hydrogens of the methyl group for ene product formation. Abstraction of the vinylic hydrogen leads to the formation of allene hydroperoxides. Some *trans*-β-ionone derivatives have been reported [117–120] to afford small amounts of allene hydroperoxides by vinylic hydrogen abstraction. Recently, Katsumura and co-workers reported [121–123] that several twisted 1,3-dienes afford allene hydroperoxides in surprising high yields (Scheme 24).

This selectivity was attributed to the perpendicular geometry of the vinylic hydrogen atom to the olefinic plane. In such a conformation, the vinylic hydrogen is "activated," considering the large σ*–π interactions between the vinyl

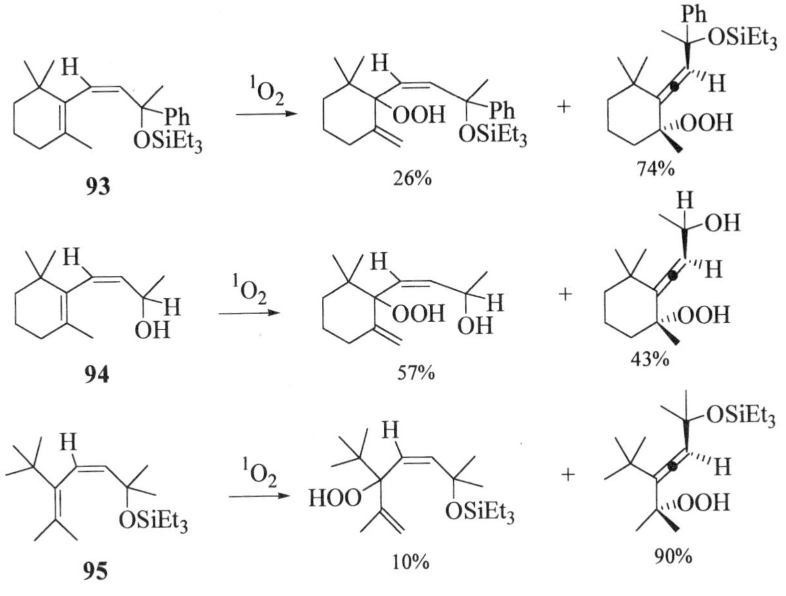

Scheme 24   Regioselectivity in the photo-oxygenation of twisted 1,3-dienes.

π (C-C) bond- σ* (C-H) bond interaction

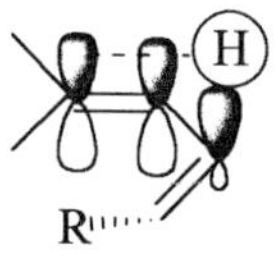

Scheme 25

C—H bond and the reacting C=C double bond (see Scheme 25). For example, in substrates **93** and **95** the major ene adducts (74% and 90%, respectively) are formed by vinylic hydrogen atom abstraction, whereas the minor ene adducts (26% and 10%, respectively) are produced by allylic hydrogen abstraction. The percentage of allene hydroperoxide is lower (43%) in the photo-oxidation of **94.** In this substrate, the steric hindrance at the stereogenic center is lower than in **93** or **95,** and the diene is therefore less twisted and the σ*–π interaction less profound, or by analogy, the vinylic hydrogen atom is less perpendicular to the double bond.

H. Regioselective Self-Sensitized Oxygenation of Fullerene Adducts

It was shown recently that fullerene C_{60} is a potent photosensitizer [124–128]. Its triplet excited state is produced with quantum yield close to unity, and by energy transfer to molecular oxygen, it produces large quantities of $^1O_2$. In the case that a fullerene C_{60} adduct bears an oxidizable group, in the presence of oxygen and light, self-sensitized oxidation has been reported [129–131]. Only a few reports so far in the literature deal with the ene regioselectivity of self-sensitized oxygenation of fullerene adducts. For example, the photo-oxygenation of the adduct **96** produced by the [2 + 2] photochemical cycloaddition of 2,5-dimethyl-2,4-hexadiene to C_{60} [132] affords a moderate threo/erythro diastereoselectivity of 55/45. This result is in contrast with that of chiral nonfunctionalized olefins, which afford higher diastereomeric ratios [133]. In the case of self-sensitized ene photo-oxygenation of fullerene derivatives **97** and **98,** abstraction of the hydrogen adjacent to the fullerene substituent is predominant [134]. This type of selectivity reminiscent the regioselectivity for hydrogen abstraction on the large alkyl group of the double bond of cis or trans alkenes, discussed in Section III.A. It is interesting to note that the preferential hydrogen abstraction from the fullerene side of the double bond (substrates **97** and **98**) is even higher than that of substrates **29, 30** (Table 4), and **87** (Table 11), where L (large group) is *t*-butyl. Apart from the large group nonbonded interactions, some other electronic effects may also contribute to the observed regioselectivity. The second case deals with the ene photoxygenation of several adducts prepared by the thermal [4 + 2] cycloaddition of conjugated dienes to C_{60} (Table 13) [135].

Table 13 Regioselectivity in the Self-Sensitized Oxygenation of Fullerene Adducts[a]

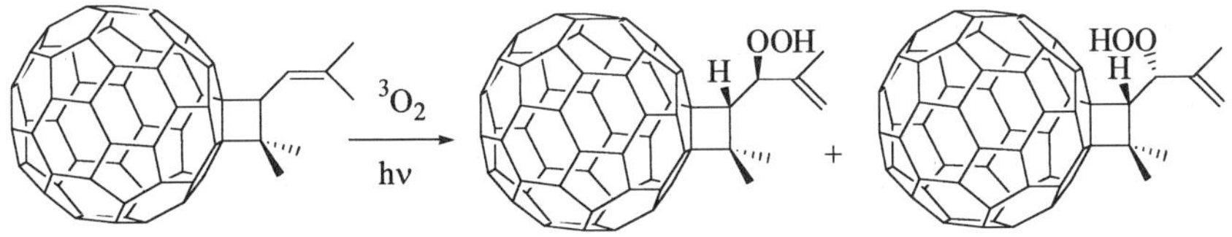

96   Diastereoselectivity 55/45

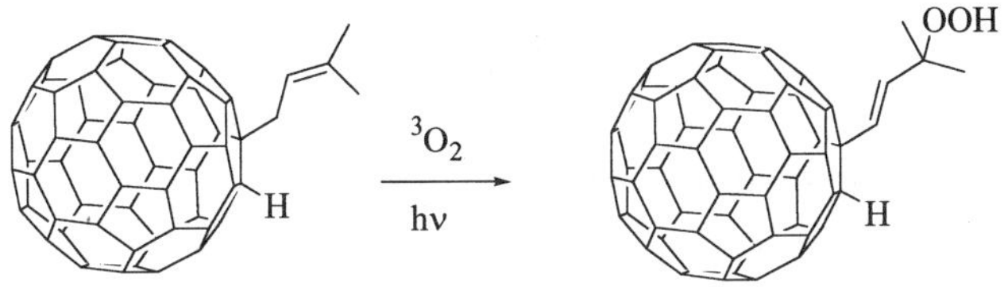

97   major product

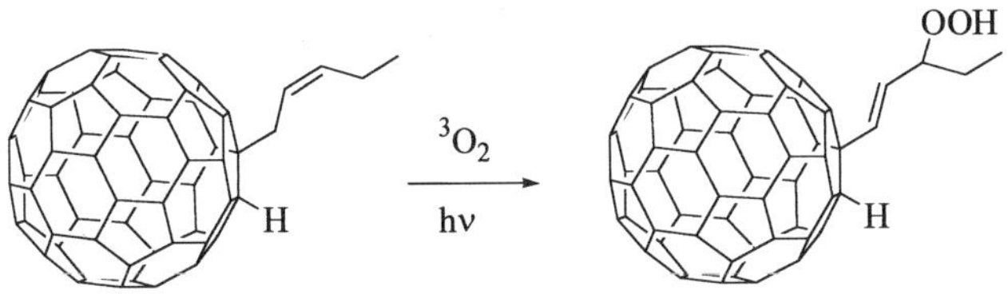

98   major product

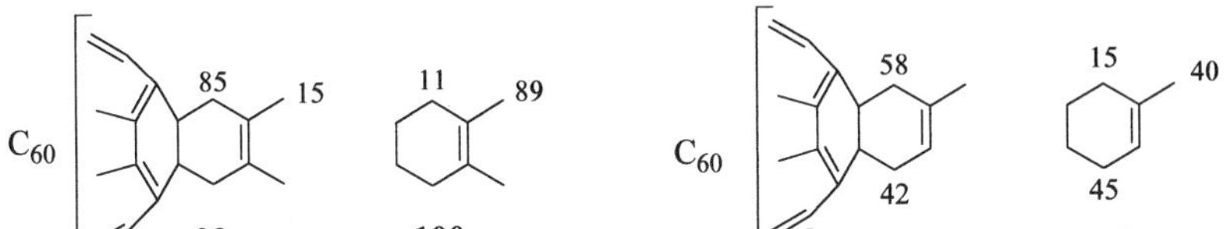

[a]Numbers indicate percentage of double bond formation in the ene adducts.

For substrates **99** and **101,** the methylene hydrogens next to fullerene are more reactive than the methyls (compare substrates **99** to **100,** and **101** to **102**). The strong preference for the formation of the perepoxide *syn* to the fullerene, which leads to the major ene products, was attributed to the ideal interaction of singlet oxygen with the methylene allylic hydrogen atoms during the formation of the *syn* perepoxide. The methylene hydrogens adopt the preferable perpendicular conformation to the olefinic double bond [81]. However, additional factors such as favorable electrostatic or electronic interactions may also contribute to the observed regioselectivity.

IV. REGIOSELECTIVITY WITHIN ZEOLITES

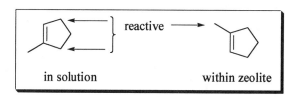

The photo-oxygenation reactions of alkenes on dye-supported zeolites have recently received considerable interest [136]. When both sensitizer and alkene were incorporated within the zeolite cavities, a remarkable ene regioselectivity was observed. Recently, Ramamurthy has shown an interesting variation of regioselectivity upon irradiation of several cyclic and acyclic alkenes absorbed in thionin-supported Y-zeolites compared to that in solution [137–139]. For example, the photo-oxygenation of 1-methylcycloalkenes **5** in zeolite Na-Y gave predominantly the allylic hydroperoxide derived by hydrogen abstraction from the methyl group. However, in solution, preferential hydrogen abstraction from the most substituted side of the double bond (cis effect) gave the opposite selectivity. These results are summarized in Table 14.

In the reaction with geminal dimethyl trisubstituted alkene **104** (Scheme 27), the methylene hydrogen atoms showed little or negligible reactivity depending on the cation of the zeolite [140]. These results demonstrate clearly that regioselectivity within zeolites is different to that found in solution.

Three mechanistic possibilities were proposed (Scheme 26) in order to rationalize the remarkable zeolite ene regioselectivity. In model 1, stereochemical nonbonded interactions within the cavity of zeolite place the larger alkyl chain of the alkene in such a conformation that methylene hydrogens are geometrically unaccessible and therefore unreactive. In model 2, polarization of the alkene double bond by zeolite cationic interactions causes electrophilic oxygen attack to the disubstituted carbon. Thus, hydrogen abstraction occurs at the geminal methyls, leading to secondary hydroperoxide. Finally, in model 3, Clennan, based on regioselectivities of tetrasubstituted alkenes [141,142], and Stratakis, based on experimental and computational work [143], independently proposed that stereoelectronic effects among the zeolite cations, the double bond, and the particular structure of alkenes affect the stability of the transition states, leading to the most stable perepoxide.

Table 14 Regioselectivity of Ene Photo-Oxygenations in Solution and Within Zeolite Na-Y

	major	minor
In solution	major	minor
within zeolite Na-Y	traces	>97 %

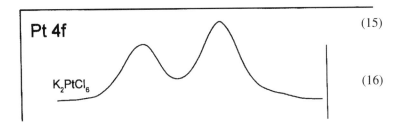

In solution	50-80%	50-20%
within zeolite Na-Y	0-40%	60-100%

More specifically, it was found that with geminal dimethyl-trisubstituted alkenes, the methyl group in the less substituted side of the double bond became significantly more reactive within zeolite compared to the solution. This result is illustrated in Scheme 27.

The asymmetric photooxygenation of alkenes in chiral-supported zeolites was also reported [144,145]:

$$
\text{Pt 4f} \qquad\qquad (15)
$$

$$
\text{K}_2\text{PtCl}_6 \qquad\qquad (16)
$$

Upon irradiation of alkene **103** in a dye-exchanged zeolite Na-Y in the presence of a chiral media [(+)-ephedrin] and molecular oxygen, a small enantioselectivity of 15% enantiomeric excess (*ee*) was found. This first experiment of an asymmetric ene photo-oxygenation reaction within the zeolite cavity is interesting, promising, and merits further investigation.

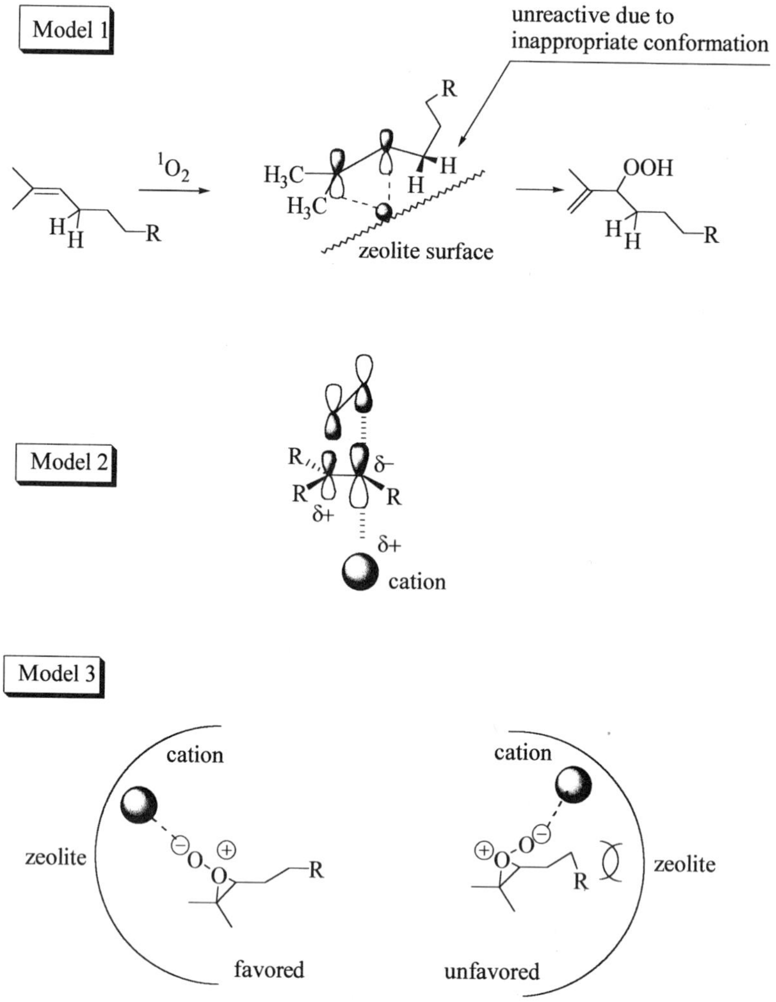

Scheme 26 Proposed mechanisms of the alkene photo-oxygenations in zeolites.

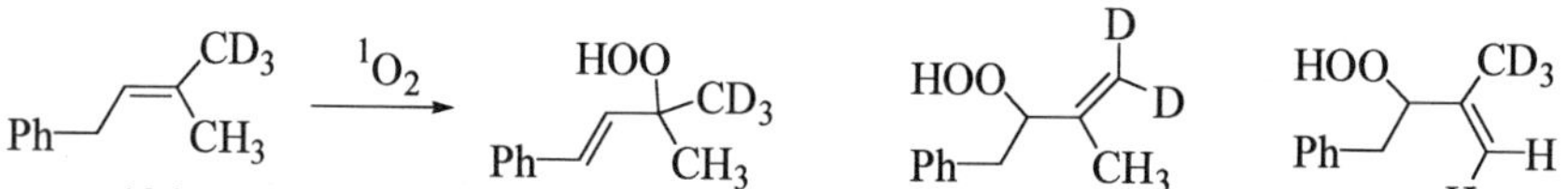

in solution (CCl$_4$)	60%	4%	36%
within zeolite (hexane)	<1%	62%	38%

Scheme 27 Photo-oxygenation of **104** in solution and within zeolite.

V. SOLVENT EFFECTS IN THE ENE REGIOSELECTIVITY

Earlier studies have shown that the rate of the $^1O_2$ ene reaction with alkenes showed negligible dependence on solvent polarity [146,147]. A small variation in the distribution of the ene products of this reaction by changing solvent polarity was reported earlier [148,149]. However, no mechanistic explanation was offered to account for the observed solvent effects. It is rather difficult to rationalize these results based on any of the currently proposed mechanisms of singlet-oxygen ene reactions. However, product distribution depends substantially on solvent polarity and reaction temperature in substrates where both ene and dioxetane products only are produced [150–159].

In the case of $^1O_2$ with α,β-unsaturated esters, there is a small but significant solvent effect on the variation of the ene products [105]. Hydrogen abstraction from the methyl group, which is geminal to the ester functionality in compound **71** [Eq. (17)], producing **71a,** decreases substantially as the solvent polarity increases:

$$\text{(17)}$$

For example, the ratio of ene products **71a/71b** decreases by a factor of 5 in going from carbon tetrachloride to the most polar solvent, dimethyl sulfoxide (DMSO).

The observed solvent effect on the stereoselectivity of singlet oxygen with ester **71** was rationalized by examination of the possible transition states of this reaction (Scheme 28). In transition state TS_2, leading to the minor perepoxide intermediate PE_2 in a limiting step, the oxygen is placed *syn* to the ester group, and the net dipole moment is expected to be larger than that in transition state TS_1, where the oxygen has an *anti* orientation with respect to the ester group. TS_2 is therefore more polar than TS_1 and is expected to be stabilized better by polar solvents than TS_1. Consequently, the ratio **71a/71b** decreases with increase of solvent polarity.

To verify this mechanistic possibility, the solvent dependence of the ene products derived from the photooxygenation of the isomeric α,β-unsaturated esters **105-E** and **105-Z** was examined (Scheme 29). For **105-E** the two ene products are formed from two different perepoxides. When the oxygen atom is placed *syn* to the ester group, **105b** is produced, whereas **105a** is formed from the opposite case. For isomer **105-E,** the expected solvent effect was found (**105a/105b** = 85/15 in CCl$_4$ or benzene; in DMSO, the ratio is 70/30). On the other hand, for the isomer **105-Z,** both products are formed from the same intermediate (oxygen placed *anti* to the ester functionality), and no solvent dependence in the ene products was found (**105a/105b**=95/5 in CCl$_4$ and 93/7 in DMSO).

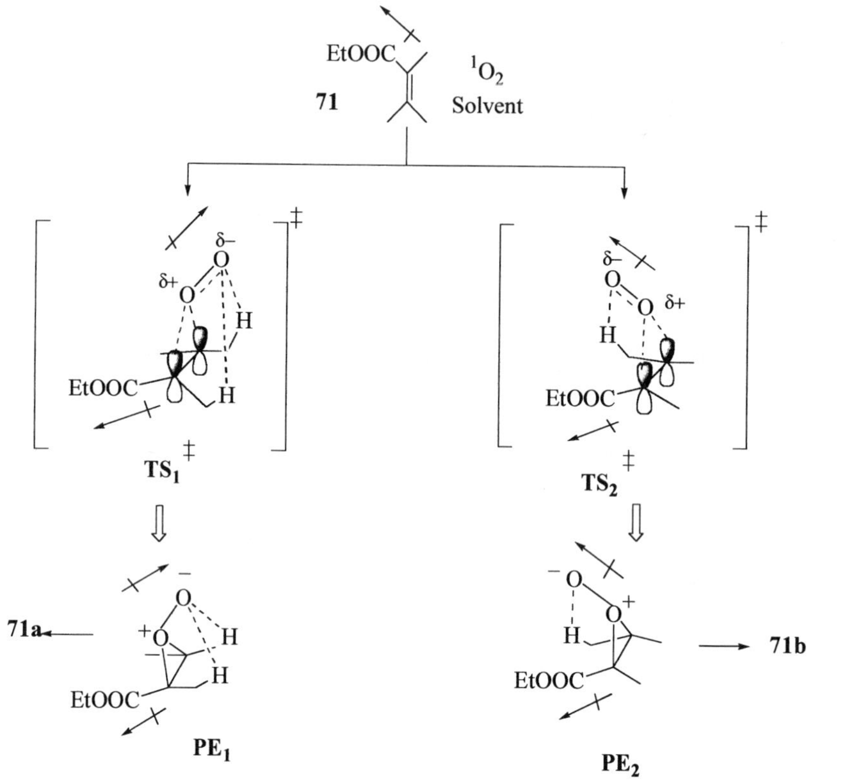

Scheme 28 Proposed mechanisms for the solvent-dependent photo-oxidation of α,β-unsaturated esters.

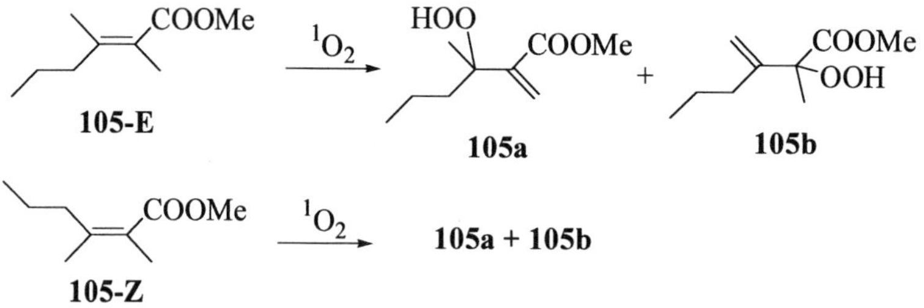

Scheme 29

VI. PERSPECTIVES

Stereo-controlled sensitized photo-oxygenations involving singlet oxygen have been studied the last few years. This useful photochemical oxyfunctionalization of organic substrates will continue to play an important role in chemical and biological systems. The development of new methods for regio-, enantio-, and diastereoselective photo-oxygenations involving singlet oxygen will be of enormous scientific and applied interest. New stereo-controlled processes involving reactive oxygen species in various media such as zeolites surfaces are highly promising.

ACKNOWLEDGMENTS

I would like to thank Professor M. Stratakis and Professor I. Smonou of University of Crete for stimulating discussions and editorial help. The financial support from the Greek General Secretariat of Research and Technology (ΠΕΝΕΔ-1999) is also acknowledged.

REFERENCES

1. Partington, J. R. *A History of Chemistry* London: Macmillan, 1970, Vols. 2–4.
2. Mulliken, R. S. *Nature (London)* **1928,** *122,* 505.
3. Childe, W. H. J.; Mecke, R. *Z. Physik* **1931,** *68,* 344.
4. Herzberg, G. *(Nature, Lond.)* **1934,** *133,* 759.
5. Kautsky, H. *Trans. Faraday Soc.* **1939,** *35,* 216.
6. Kautsky, H.; de Bruijn, II.; *Naturnissenschaften* **1931,** *19,* 1043.
7. Foote, C. S.; Wexler, S. *J. Am. Chem. Soc.* **1964,** *86,* 3879–3881.
8. Corey, E. J.; Taylor, W. C. *J. Am. Chem. Soc.* **1964,** *86,* 3881–3882.
9. Wasserman, H. H.; Murray, R. W. *Singlet Oxygen,* New York: Academic Press, 1979.
10. Ranby, B.; Rabek, J. F. *Photodegradation, Photooxidation and Photostabilization of Polymers,* London: Wiley, 1975.
11. Frimmer, A. A.; Stephenson, L. M. in *Singlet Oxygen. Reactions, Modes and Products;* Frimer, A. A., ed., CRC Press: Boca Raton, FL, 1985.
12. Foote, C. S.; Clennan, E. L. in *Active Oxygen in Chemistry,* Foote, C. S.; Valentine, J. S.; Greenberg, A.; Liebman, J. F., eds., London: Chapman & Hall, 1995, 105–140.
13. Kasha, M.; Khan, A. V. *Ann. NY Acad. Sci.* **1970,** *171,* 5.
14. Kearns, D. R. *Chem. Rev.* **1971,** *71,* 395–427.
15. Khan, A. U.; Kasha, M. *Ann. NY Acad. Sci.* **1970,** *171,* 24.
16. Murray, R. W.; Lin, J. W. P.; Kaplan, M. L. *Ann. NY Acad. Sci.* **1970,** *171,* 121.
17. Foote, C. S. *Science* **1968,** *162,* 963.
18. Pittis, J. N., Jr. *Ann. NY Acad. Sci.* **1970,** *171,* 239.
19. Chan, H. W. S. *J. Chem. Soc., Chem. Commun.* **1970,** 1550–1551.
20. Wasserman, H. H.; Scheffer, J. R.; Cooper, J. L. *J. Am. Chem. Soc.* **1972,** *94,* 4991–4996.

21. Wasserman, H. H.; Ives, J. L. *Tetrahedron* **1981**, *37*, 1825–1852.
22. Schenck, G. O. Patent DE-B 933925, 1943.
23. Schenck, G. O.; Eggert, H.; Denk, W. *Liebigs Ann. Chem.* **1953**, *584*, 177–198.
24. Denny, R. W.; Nickon A. *Org. Reactions* **1973**, *20*, 133.
25. Giese, A. C. in *Photophysiology,* Wilson, T.; Hasting, J.W., eds., New York: Academic Press, 1970, Vol. V, 49.
26. Politzer, J. R.; Griffen, G. W.; Laseter J. L. *Chem. Biol. Interact.* **1971**, *3*, 73.
27. Foote, C. S. in *Free Radicals in Biology,* Pryor, W. A., ed., New York: Academic Press, 1976, Vol. 2, 85–133.
28. Gleason, W. S.; Broadbent, A. D.; Whittle, E.; Pitts, J. N., Jr. *J. Am. Chem. Soc.* **1970**, *92*, 2068–2075.
29. Nickon, A.; Bagli J. E. *J. Am. Chem. Soc.* **1961**, *83*, 1498–1508.
30. Collnick, K.; Schenck, G. O. *Pure Appl. Chem.* **1964**, *9*, 507.
31. Foote, C. S. *Acc. Chem. Res.* **1968**, *1*, 104–109.
32. Jefford, C. W. *Tetrahedron Lett.* **1979**, *20*, 985–988.
33. Harding, L. B.; Goddard, W. A. *J. Am. Chem. Soc.* **1980**, *102*, 439–449.
34. Stephenson, L. M.; Grdina, M. B.; Orfanopoulos, M. *Acc. Chem. Res.* **1980**, *13*, 419–425.
35. Sharp, D. B. "Abstracts of Papers," 138th National Meeting, American Chemical Society, Washington, DC, 1960, 79.
36. Schaap, A. P.; Recher, S. G.; Faler, G. R.; Villasenor, S. R. *J. Am. Chem. Soc.* **1983**, *105*, 1691–1693.
37. Stratakis, M.; Orfanopoulos, M.; Foote, C. S. *Tetrahedron Lett.* **1991**, *32*, 863–866.
38. Clennan, E. L.; Chen, M-F.; Xu, G. *Tetrahedron Lett.* **1996**, *37*, 2911–2914.
39. Inagaki, S.; Fukui, K. *J. Am. Chem. Soc.* **1975**, *97*, 7480–7484.
40. Dewar, M. J. S.; Thiel, W. *J. Am. Chem. Soc.* **1975**, *97*, 3978–3986.
41. Yamaguchi, K.; Yabushita, S.; Fueno, T.; Houk, K. N. *J. Am. Chem. Soc.* **1981**, *103*, 5043–5045.
42. Davies, A. G.; Schiesser, C. H. *Tetrahedron* **1991**, *47*, 1707–1726.
43. Yoshioka, Y.; Yamada, S.; Kawakami, T.; Nishino, M.; Yamaguchi, K.; Saito, I. *Bull. Chem. Soc. Jpn.* **1996**, *69*, 2683–2699.
44. Kearns, D. R. *J. Am. Chem. Soc.* **1969**, *91*, 6554–6569.
45. Kopecky, K. R.; Mumford, C. *Can. J. Chem.* **1969**, *47*, 709.
46. Grdina, M. B.; Orfanopoulos, M.; Stephenson, L. M. *J. Am. Chem. Soc.* **1979**, 101, 3111–3112.
47. Orfanopoulos, M.; Smonou, I.; Foote, C. S. *J. Am. Chem. Soc.* **1990**, *112*, 3607–3610.
48. Clennan, E. L.; Nagraba, K. *J. Am. Chem. Soc.* **1983**, *105*, 5932–5933.
49. Gorman, A. A.; Hamblett, I.; Lambert, C.; Spencer, B.; Standen, M. C. *J. Am. Chem. Soc.* **1988**, *110*, 8053–8054.
50. Aubry, J-M.; Mandard-Cazin, B.; Rougee, M.; Bensasson, R. V. *J. Am. Chem. Soc.* **1995**, *117*, 9159–9164.
51. Baldwin, J. E.; Höfle, G.; Choi, S. C. *J. Am. Chem. Soc.* **1971**, *93*, 2810–2812.
52. Orfanopoulos, M.; Stephenson, L. M. *J. Am. Chem. Soc.* **1980**, *102*, 1417–1418.
53. Schulte-Elte, K. H.; Muller, B. L.; Rautestrauch, V. *Helv. Chim. Acta* **1978**, *61*, 2777.

54. Orfanopoulos, M.; Grdina, M. B.; Stephenson, L. M. *J. Am. Chem. Soc.* **1979,** *101,* 275–276.

55. Hurst, J. R.; McDonald, J. D.; Schuster, G. B. *J. Am. Chem. Soc.* **1982,** *104,* 2065–2067.

56. Hurst, J. R.; Wilson, S. L.; Schuster, G. B. *Tetrahedron* **1985,** *41,* 2191–2195.

57. Gorman, A. A.; Gould, I. R.; Hamblett, I. *J. Am. Chem. Soc.* **1982,** *104,* 7098–7104.

58. Frimer, A. A.; Bartlett, P. D.; Boschung, A. F.; Jewett, J. D. *J. Am. Chem. Soc.* **1977,** *99,* 7977–7986.

59. Bartlett, G. P.; Frimer, A. A. *Heterocycles* **1978,** *11,* 419.

60. Stephenson, L. M. *Tetrahedron Lett.* **1980,** *21,* 1005–1008.

61. Houk, K. N.; Williams, P. A.; Mitchell, P. A.; Yamaguchi, K. *J. Am. Chem. Soc.* **1981,** *103,* 949–950.

62. Orfanopoulos, M.; Stratakis, M.; Elemes, Y.; Jensen, F. *J. Am. Chem. Soc.* **1991,** *113,* 3180–3181.

63. Adam, W.; Nestler, B. *J. Am. Chem. Soc.* **1992,** *114,* 6549–6550.

64. Adam, W.; Nestler, B. *J. Am. Chem. Soc.* **1993,** *115,* 5041–5049.

65. Adam, W.; Brunker, H-G. *J. Am. Chem. Soc.* **1993,** *115,* 3008–3009.

66. Brunke, H-G.; Adam, W. *J. Am. Chem. Soc.* **1995,** *117,* 3976–3982.

67. Stratakis, M.; Orfanopoulos, M.; Foote, C. S. *Tetrahedron Lett.* **1996,** *37,* 7159–7162.

68. Vassilikogiannakis, G.; Stratakis, M.; Orfanopoulos, M.; Foote, C. S. *J. Org. Chem.* **1999,** *64,* 4130–4139.

69. Stratakis, M.; Orfanopoulos, M. *Tetrahedron Lett.* **1995,** *36,* 4875–4879.

70. Gollnick, K.; Schade, G. *Tetrahedron Lett.* **1966,** *7,* 2355.

71. Stratakis, M.; Orfanopoulos, M.; Foote, C. S. *J. Org. Chem.* **1998,** *63,* 1315–1318.

72. Matsumoto, M.; Dobashi, S.; Kuroda, K. *Tetrahedron Lett.* **1977,** *18,* 3361–3364.

73. Matsumoto, M.; Kuroda, K. *Synth. Commun.* **1981,** *11,* 987–992.

74. Rousseau, G.; Leperchec, P.; Conia, J. M. *Tetrahedron Lett.* **1977,** *18,* 2517–2520.

75. Lerdal, D.; Foote, C. S. *Tetrahedron Lett.* **1978,** *19,* 3227–3230.

76. Orfanopoulos, M.; Stratakis, M.; Elemes, Y. *Tetrahedron Lett.* **1989,** *30,* 4755–4758.

77. Clennan, E. L.; Chen, X. *J. Org. Chem.* **1988,** *53,* 3124–3126.

78. Orfanopoulos, M.; Stratakis, M.; Elemes, Y. *J. Am. Chem. Soc.* **1990,** *112,* 6417–6418.

79. Clennan, E. L.; Chen, X.; Koola, J. J. *J. Am. Chem. Soc.* **1990,** *112,* 5193–5199.

80. Stratakis, M.; Orfanopoulos, M. *Synth. Commun.* **1993,** *23,* 425–430.

81. Schulte-Elte, K. H.; Rautestrauch, V. *J. Am. Chem. Soc.* **1980,** *102,* 1738–1740.

82. Dubac, J.; Laporterie, A.; Iloughmane, H.; Pillot, J. P.; Deleris, G.; Dunogues, J. J. *Organomet. Chem.* **1985,** *149,* 281–285.

83. Dubac, J.; Laporterie, A. *Chem. Rev.* **1987,** *87,* 319–334.

84. Adam, W.; Schwarm, M. *J. Org. Chem.* **1988,** *53,* 3129–3130.

85. Laporterie, A.; Manuel, G.; Dubac, J.; Mazerolles, P. *Nouv. J. Chim.* **1982,** *6,* 67.

86. Shimizu, N.; Shibata, F.; Imazu, S.; Tsuno, Y. *Chem. Lett.* **1987,** 1071–1072.

87. Dang, H-S.; Davies, A. G. *Tetrahedron Lett.* **1991,** *32,* 1745–1748.

88. Dang, H-S.; Davies, A. G. *J. Chem. Soc., Perkin 2* **1992,** 2011–2020.

89. Dussault, P. H.; Lee, R. J. *J. Am. Chem. Soc.* **1994,** *116,* 4485–4486.

90. Dussault, P. H.; Zope, U. R. *Tetrahedron Lett.* **1995**, *36*, 2187–2190.
91. Adam, W.; Catalani, L.; Griesbeck, A. *J. Org. Chem.* **1986**, *51*, 5494–5496.
92. Adam, W.; Griesbeck, A. *Angew. Chem. Int. Ed. Engl.* **1985**, *24*, 1070–1071.
93. Orfanopoulos, M.; Foote, C. S. *Tetrahedron Lett.* **1985**, *26*, 5991–5994.
94. Ensley, H. E.; Carr, R. V. C.; Martin, R. S.; Pierce, T. E. *J. Am. Chem. Soc.* **1980**, *102*, 2836–2838.
95. Kwon, B-M.; Kanner, R. C.; Foote, C. S. *Tetrahedron Lett.* **1989**, *30*, 903–906.
96. Akasaka, T.; Kakeushi, T.; Ando, W. *Tetrahedron Lett.* **1987**, *28*, 6633–6636.
97. Akasaka, T.; Misawa, Y.; Goto, M.; Ando, W. *Heterocycles* **1989**, *28*, 445–451.
98. Akasaka, T.; Misawa, Y.; Goto, M.; Ando, W. *Tetrahedron* **1989**, *45*, 6657–6666.
99. Adam, W.; Griesbeck, A. *Synthesis* **1986**, 1050–1052.
100. Adam, W.; Richter, M. J. *Tetrahedron Lett.* **1993**, *34*, 8423–8426.
101. Adam, W.; Bruckner, H-G.; Nestler, B. *Tetrahedron Lett.* **1991**, *32*, 1957–1960.
102. Dussalt, P. H.; Woller, K. R.; Hillier, M. C. *Tetrahedron,* **1994**, *49*, 8929–8940.
103. Adam, W.; Nestler, B. *Liebigs Ann. Chem.* **1990**, 1051–1053.
104. Elemes, Y.; Foote, C. S. *J. Am. Chem. Soc.* **1992**, *114*, 6044–6050.
105. Orfanopoulos, M.; Stratakis, M. *Tetrahedron Lett.* **1991**, *32*, 7321–7324.
106. Stratakis, M.; Orfanopoulos, M. *Tetrahedron Lett.* **1997**, *38*, 1067–1071.
107. Furutachi, N.; Nakadaira, Y.; Nakanishi, K. *J. Chem. Soc., Chem. Commun.* **1968**, 1625–1626.
108. Adam, W.; Griesbeck, A. G.; Wang, X. *Liebigs Ann. Chem.* **1992**, 193–197.
109. Clennan, E. L.; Cheng, X. *J. Am. Chem. Soc.* **1989**, *111*, 8212–8213.
110. Linker, T.; Frohlich, L. *Angew. Chem. Int. Ed. Engl.* **1994**, *33*, 1971–1972.
111. Linker, T.; Frohlich, L. *J. Am. Chem. Soc.* **1995**, *117*, 2694–2697.
112. Adam, W.; Brunker, H-G.; Kumar, A. S.; Peters, E. M.; Peters, K.; Schneider, U.; von Schening, H. G. *J. Am. Chem. Soc.* **1996**, *118*, 1899–1905.
113. Paquette, L. A.; Hertel, L. W.; Gleiter, R.; Bohm, M. C.; Beno, M. A.; Christoph, G. G. *J. Am. Chem. Soc.* **1981**, *103*, 7106–7121.
114. Clennan, E. L.; Koola, J. J. *J. Am. Chem. Soc.* **1993**, *115*, 3802–3803.
115. Clennan, E. L.; Koola, J. J.; Chen, M-F. *Tetrahedron* **1994**, *50*, 8569–8578.
116. Stratakis, Orfanopoulos, M.; Foote, C. S. *Unpublished results.*
117. Isoe, S.; Hyeon, B. S.; Ichikawa, H.; Katsumura, S.; Sakan, T. *Tetrahedron Lett.* **1968**, *9*, 5561.
118. Mousseron-Canet, M.; Dalle, J-P.; Mani, J-C. *Tetrahedron Lett.* **1968**, *9*, 6037.
119. Foote, C. S.; Brenner, M. *Tetrahedron Lett.* **1968**, 6041.
120. Isoe, S.; Katsumura, S.; Hyeon, B. S.; Sakan, T. *Tetrahedron Lett.* **1971**, *12*, 1089.
121. Mori, H.; Ikoma, K.; Masui, Y.; Isoe, S.; Kitaura, K.; Katsumura, S. *Tetrahedron Lett.* **1996**, *43*, 7771–7774.
122. Mori, H.; Ikoma, K.; Katsumura, S. *J. Chem. Soc., Chem. Commun.* **1997**, 2243–2244.
123. Mori, H.; Ikoma, K.; Isoe, S.; Kitaura, K.; Katsumura, S. *J. Org. Chem.* **1998**, *63*, 8704–8718.
124. Arbogast, J. W.; Foote, C. S. *J. Am. Chem. Soc.* **1991**, *113*, 8868–8889.
125. Orfanopoulos, M.; Kambourakis, S. *Tetrahedron Lett.* **1994**, *36*, 1945–1948.
126. Tokyyama, H.; Nakamura, E. *J. Org. Chem.* **1994**, *59*, 1135–1138.

127. Anderson, J. L.; An, Y-Z.; Rubin, Y.; Foote, C. S. *J. Am. Chem. Soc.* **1994,** *116,* 9763–9764.

128. Hamano, T.; Okuda, K.; Tadahiko, M.; Hirobe, M.; Arakane, K.; Ryu, A.; Mashiko, S.; Nagano, T. *J. Chem. Soc., Chem. Commun.* **1997,** 21–22.

129. Zhang, X.; Romero, A.; Foote, C. S. *J. Am. Chem. Soc.* **1993,** *115,* 11,924–11,925.

130. Zhang, X.; Fan, A.; Foote, C. S. *J. Org. Chem.* **1996,** *61,* 5456–5461.

131. Torre-Garcia, G.; Mattay, J. *Tetrahedron* **1996,** *52,* 5421–5426.

132. Vassilikogiannakis, G.; Chronakis, N.; Orfanopoulos, M. *J. Am. Chem. Soc.* **1998,** *120,* 9911–9920.

133. Prein, M.; Adam, W. *Angew. Chem. Int. Ed. Engl.* **1996,** *35,* 477–494.

134. Chronakis, N. *Ph.D. Thesis,* University of Crete, 2000.

135. An, Y-Z.; Viado, A. L.; Arce, M-J.; Rubin, Y. *J. Org. Chem.* **1995,** *60,* 8330–8331.

136. Sen, S. E.; Smith, S. M.; Sullivan, K. A. *Tetrahedron* **1999,** *55,* 12,657–12,698.

137. Li, X.; Ramamurthy, V. *Tetrahedron Lett.* **1996,** *37,* 5235–5238.

138. Li, X.; Ramamurthy, V. *J. Am. Chem. Soc.* **1996,** *118,* 10666–10667.

139. Ramamurthy, V.; Lakshminarasimhan, P.; Grey, C. P.; Johnston, L. J. *J. Chem. Soc., Chem. Commun.* **1998,** 2411–2424.

140. Robbins, R. J.; Ramamurthy, V. *J. Chem. Soc., Chem. Commun.* **1997,** 1071–1072.

141. Clennan, E. L.; Sram, J. P. *Tetrahedron Lett.* **1999,** *40,* 5275–5278.

142. Clennan, E. L.; Sram, J. P. *Tetrahedron* **2000,** *56,* 6945–6950.

143. Stratakis, M.; Froudakis, G. *Org. Lett.* **2000,** *2,* 1369–1372.

144. Joy, A.; Robbins, R. J.; Pitchumani, K.; Ramamurthy, V. *Tetrahedron Lett.* **1997,** *38,* 8825–8828.

145. Joy, A.; Ramamurthy, V. *Chem. Eur. J.* **2000,** *6,* 1287–1293.

146. Foote, C. S.; Denny, R. W. *J. Am. Chem. Soc.* **1971,** *93,* 5168–5171.

147. Gollnick, K.; Griesbeck, A. *Tetrahedron Lett.* **1984,** *25,* 725–728.

148. Manring, L. E.; Foote, C. S. *J. Am. Chem. Soc.* **1983,** *105,* 4710–4717.

149. Rautenstrauch, V.; Thommen, W.; Schulte-Elte, K. H. *Helv. Chim. Acta* **1986,** *69,* 1638–1642.

150. Asveld, E. W. H.; Kellogg, R. M. *J. Am. Chem. Soc.* **1980,** *102,* 3644–3646.

151. Ando, W.; Watanabe, K.; Suzuki, J.; Migita, T. *J. Am. Chem. Soc.* **1974,** *96,* 6766–6768.

152. Bartlett, P. D.; Mendenhall, G. D.; Schaap, A. P. *Ann. NY Acad. Sci.* **1970,** *171,* 79.

153. Hasty, N. M.; Kearns, D. R. *J. Am. Chem. Soc.* **1973,** *95,* 3380–3381.

154. Jefford, C. W.; Kohmoto, S. *Helv. Chim. Acta* **1982,** *65,* 133–136.

155. Chan, Y-Y.; Zhu, C.; Leung, H. K. *Tetrahedron Lett.* **1986,** *27,* 3737–3740.

156. Kwon, B-M.; Foote, C. S. *J. Org. Chem.* **1989,** *54,* 3878–3882.

157. Chan, Y-Y.; Li, X.; Zhu, C.; Zhang, Y.; Leung, H-K. *J. Org. Chem.* **1990,** *55,* 5497–5504.

158. Gollnick, K.; Knutzen-Mies, K. *J. Org. Chem.* **1991,** *56,* 4017–4027.

159. Matsumoto, M.; Suganuma, H. *J. Chem. Soc., Chem. Commun.* **1994,** 2449–2450.

4

Singlet-Oxygen Reactions: Solvent and Compartmentalization Effects

Eduardo A. Lissi
University of Santiago, Santiago, Chile

Else Lemp and Antonio L. Zanocco
University of Chile, Santiago, Chile

I. INTRODUCTION

Singlet molecular oxygen ($^1\Delta_g$) reactions are important in biological systems, where it can play deleterious (damaging valuable biomolecules) and/or beneficial roles [1,2]. A main characteristic of biological systems is their microheterogeneity. Most singlet oxygen reactions are not diffusion-controlled processes. If it is accepted that singlet oxygen is equilibrated between the lipidic regions and the surrounding aqueous media, the extent of damage to a given target will depend both on the steady-state singlet-oxygen concentration in the vicinity of the target and the bimolecular rate constant value in the microenvironment where the target is localized. The damage can be then related to thermodynamic (distribution of the oxygen) and kinetic (specific rate constant) parameters. Furthermore, in situations where diffusion must take place prior to reaction (when singlet oxygen is produced far from the target) or to avoid reaction (if singlet oxygen is produced near the target), other factors, such as diffusion through the different microenvironments and/or exchange rates between microphases, become relevant. The same factors are also relevant in diffusion-controlled processes. This complex situation can be partially mimicked in well-defined systems, such as micellar solutions or liposome suspensions [3–5].

In the present chapter, we shall discuss the main characteristics of singlet-oxygen reactions in three types of compartmentalized systems: micelles, reverse micelles, and unilamellar vesicles. Micelles and reverse micelles are rather small aggregates, where residence times can be considered to be shorter than the life expectancy of singlet-oxygen molecules. The kinetics in these systems can be expected to be very similar to that observed in homogeneous solutions. The main difference between them is that, although in micelles the singlet-oxygen solubility can be expected to be higher than that in the surrounding aqueous solution, the opposite happens in reverse micelles. Liposomes, and particularly large unilamellar vesicles (LUVs), are systems where intraparticle and interparticle distances can be larger than the mean average distance traveled by the singlet oxygen during its lifetime. This renders the kinetic modeling of the system more complex.

The minimum mechanism required to explain singlet-oxygen decay and singlet-oxygen reactions in microheterogeneous systems comprising micelles, reverse micelles, vesicles, and simple biological systems of only one membrane (such as red blood cell ghosts or microsomes) is depicted in Scheme 1 [3–6]. This scheme includes the following as pseudophases:

1.   The dispersing (or external solvent) pseudophase
2.   The micellar pseudophase or the bilayer in vesicles and membranes
3.   The inner pool pseudophase in vesicles and membranes

In more sophisticated models, it can be considered that both the singlet-oxygen generation and/or the interaction with the quencher can take place also at the interfaces. In vesicles (and membranes), this includes the inner and outer interfaces, which can present different characteristics.

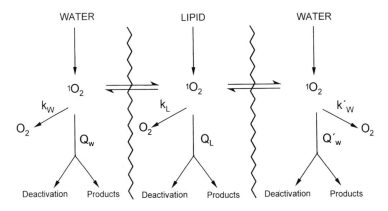

Scheme 1

The study of singlet-oxygen reactions in microheterogeneous systems can be carried out mainly from three experimentally different approaches:

1. In pulsed experiments, following the effect of the additive upon the singlet-oxygen luminescence decay profile measured at 1270 nm.
2. Following the effect of the additive upon the averaged singlet-oxygen steady-state concentration. This can be monitored either from the steady-state intensity of the near-IR (infrared) emission and/or by measuring the effect of the additive upon the rate of consumption of a reference compound. Diphenylisobenzofurane (DPBF) and anthracene derivatives have been extensively employed as reference compounds in this type of studies.
3. Following the rate of consumption of the additive.

The information that can be obtained depends on the experimental approach employed, the characteristics of the reactions involved, and the distribution of the sensitizer that generates the singlet oxygen, and the target molecules. For example, approaches 1 and 2 provide values of the total interaction between the additive and singlet oxygen, whereas approach 3 provides values of the reactive rate constant. Also, the requirements that must be fulfilled by the system depend on the selected experimental approach. In particular, in cases 2 and 3 it is fundamental that the presence of the additive does not alter the quantum yield of singlet-oxygen production. Furthermore, singlet-oxygen reactions must be the only reactions consuming the additive. This is particularly critical in compartmentalized systems when both the sensitizer and the additive are concentrated in the microaggregates, thus favoring the occurrence of Type I photoprocesses and/or secondary reactions that contribute to the additive consumption. On the other hand, the occurrence of these processes is not relevant if the decay kinetics of singlet oxygen is evaluated in time-resolved experiments.

It is interesting to mention what can be obtained for the three above-mentioned experimental approaches in experiments carried out in compartmentalized systems. Time-resolved measurements can, in principle, provide information regarding total quenching rate constants ($k_Q = k_r + k_d$) by the additive in the different microphases, where k_r and k_d are the reactive and physical rate constants, respectively. However, several conditions must be fulfilled to allow these evaluations (*vide infra*). On the other hand, steady-state measurements generally provide information regarding apparent rate constants averaged over all the microenvironments. Values of the quenching rate constants attributable to a given microenvironment can only be obtained when the quencher is localized in a single compartment and/or singlet-oxygen lifetimes (and/or the quenching efficiencies) are widely different in the different media.

It has been already mentioned that important parameters in the kinetics of the singlet-oxygen decay and quenching rates are the intraparticle and interparticle distances and the average distance traveled by a singlet-oxygen molecule during its lifetime. Relevant data are presented in Table 1.

If singlet oxygen, due to its relatively long lifetime even in the presence of reactive targets, can be considered to be equilibrated between both pseudophases, its steady-state concentration in each pseudophase will be proportional to its whole rate of formation and independent of its locus of production (the lipidic or aqueous pseudophase). This situation usually holds in small micelles. In these systems, the small size and the large concentration of aggregates imply that, in most situations, exchange between micelles is fast and singlet oxygen can visit several microaggregates prior to reaction and/or decay. This leads to a situation in which the singlet-oxygen decay becomes independent of its locus of generation [3,4]. Only if the micelles are loaded with very efficient quenchers can it be expected that the fate of the oxygen and its lifetime becomes dependent of the site of generation. On the other hand, and particularly for large liposomes, the distance traveled by the singlet-oxygen molecule during its lifetime is of the order of (or shorter than) the distance between liposomes and/or the diameter of the internal pool. Sin-

Table 1 Intraparticle and Interparticle Distances and Average Lengths Traveled by a Singlet-Oxygen Molecule in Water

Parameter	Length (nm)
Micelle radius	1.5
Distance between micelles (0.1 M)	7
Reverse micelle radius (core, w = 30)	5
Distance between reverse micelles (0.1 M)	11
LUV radius	
Total	500
Aqueous pool	496
Distance between LUVs (2 mM)	1500
SUV radius	
Total	30
Pool	26
Distance between SUVs (2 mM)	200
Bilayer thickness	4
Mean distance traveled by $^1O_2$ in water	
70 μsec	420
5 μsec	100
0.1 μsec	16

Source: Ref. 4.

glet oxygen can barely exchange between liposomes, and the course of the reaction becomes extremely dependent on the site of generation of the singlet oxygen and the reactive substrate localization. Evidently, this situation will also apply in a cellular suspension, where the singlet oxygen produced in a given cell will not be able to diffuse until another cell [5–7]. The decay kinetics in this situation is not a linear combination of the two characteristic decay times in the two pseudophases, but are convoluted with the diffusion rate of singlet oxygen between the lipidic and aqueous compartments [8]. Frequently, in these complex situations, the decay kinetics remains quasimonoexponential, allowing a Stern–Volmer-like treatment of the data. Apparent quenching rate constants are obtained whose values can be influenced by diffusion parameters and the locus of the singlet-oxygen generation.

II. SINGLET-OXYGEN LIFETIMES IN THE ABSENCE OF ADDED QUENCHERS

Scheme 1 predicts a complex decay for the singlet-oxygen phosphorescence. However, there are three simple, limiting situations. For simplicity, let us assume only two pseudophases, the lipidic and the aqueous pseudophases. The three above-mentioned situations are as follows:

1. Very slow interchange. This corresponds to a very unlikely situation in which the interchange between pseudophases is slower than singlet-oxygen decay in both compartments. In this case, we shall observe a bi-exponential decay. The amplitudes of the decays depend on the singlet-oxygen generation locus, whereas the lifetimes are independent of it.

2. Very short lifetime in one of the compartments. This situation can be reached by adding an efficient quencher that is exclusively localized in one pseudophase. It can then be possible to evaluate the lifetime in the other environment. It has to be proved that the added quencher does not interact with the singlet oxygen in the other compartment. Furthermore, it has to be considered that the lifetime measured in this situation corresponds to the sum of the decay rate and the exchange rate, namely

$$\tau^{-1} = k^\circ + k_{exp} \tag{1}$$

   where the rate constants associated with the solvent-promoted decay (k°) and with the exchange between the pseudophases (k_{exp}) correspond to the pseudophase where the singlet-oxygen lifetime is measurable.

3. Fast exchange, maintaining the singlet-oxygen equilibrium distribution between both compartments. The singlet oxygen will follow, in the

absence of quenchers, a first-order decay with a pseudo-first-order rate constant given by

$$k_{exp}^o = fk_s^o + (1 - f)k_m^o \tag{2}$$

where s stands for the external solvent and m for the microphases, and f is the fraction of oxygen present in the external solvent. In this situation, the singlet lifetime becomes independent of its locus of generation.

In the employed pseudophase approximation,

$$f = \frac{V_s/V_m}{K + V_s/V_m} \tag{3}$$

where K is the partition constant of singlet oxygen between the microphases and the external solvent, and V_s and V_m are the volumes of the respective pseudophases. Because in most practical situations employing micelles and, particularly, liposomes, $V_s/V_m > 500$ and the distribution constant of singlet oxygen is <10 (see Secs. V. and VII.), it can be concluded that f is very close to 1. In reverse micelles, the situation is even more unfavorable because it can be expected that $K < 1$. This means that, in micelles and liposomes, k_m^o cannot be accurately determined, even if the value of f is known. This also implies that Eq. (2) can be simplified under most experimental conditions to

$$k_{exp}^o = k_s^o + (1 - f)k_m^o \tag{4}$$

and only if

$$k_m^o \gg k_s^o \tag{5}$$

will the singlet-oxygen lifetime be influenced by the presence of the microaggregates. Expression (2) can be rearranged to

$$(k_{exp}^o - k_s^o)^{-1} = (k_m^o - k_s^o)^{-1} + [K(k_m^o - k_s^o)]^{-1} \frac{V_s}{V_m} \tag{6}$$

A plot of the left-hand side of Eq. (6) against V_s/V_m could allow the evaluation of K and k_m^o. However, due to limitations in the range of V_s/V_m experimentally available, the error in the evaluation of k_{exp}^o is too high and, at most, only the product Kk_m^o can be approximately obtained. This makes the use of this approach for the evaluation of the partition K and/or the singlet lifetime in the dispersed pseudophase difficult.

III. SINGLET-OXYGEN LIFETIMES IN THE PRESENCE OF ADDED QUENCHERS

The steady-state oxygen concentration, in the absence of added quenchers, will be given by

$$[^1O_2] = \frac{R_p}{k_{exp}^\circ} \tag{7}$$

where R_p is the production rate of singlet oxygen. In the presence of a quencher Q, whose average concentration in the dispersed pseudophase is Q_m (given in local concentration), the oxygen decay will take place with a rate constant k_{exp}, given by

$$k_{exp} = k_{exp}^\circ + (1 - f)k_Q Q_m \tag{8}$$

The decay will be monoexponential and will follow a normal Stern–Volmer plot provided that the intraparticle quenching rate is not fast enough to make the entrance to the liposomes the rate-determining step of the quenching process. As far as we know, there are not these types of study. This could be difficult due to the expected very fast rate of oxygen's entrance to the liposomes (as estimated from data obtained employing ground-state oxygen [3,4,9,10]).

Under these conditions, normal Stern–Volmer plots can be expected, both in terms of steady-state singlet-oxygen concentration and decay lifetimes. The slope of the Stern–Volmer plot (K_{SV}) is given by

$$K_{SV} = \frac{(1 - f)k_Q}{k_{exp}^\circ} \tag{9}$$

when the quenching rate constant is expressed in terms of local concentrations. In terms of the singlet-oxygen partition constant, this equation can be expressed as

$$K_{SV} = \frac{Kk_Q(V_m/V_s)}{k_{exp}^\circ} \tag{10}$$

If the quenching is expressed in terms of the analytical quencher concentration, the Stern–Volmer constant becomes

$$K_{SV} = \frac{k_{ap}}{k_{exp}^\circ} \tag{11}$$

with

$$k_{ap} = k_Q K \tag{12}$$

This type of experiment readily provides (Kk_Q) but does not allow an independent evaluation of k_Q and K. Only if K is known (or assumed) can a value of the intravesicular bimolecular quenching rate be obtained. Similarly, if k_Q is known (or assumed), the singlet oxygen partition constant in the system considered can be obtained.

The singlet-oxygen-promoted consumption of the quencher Q will be determined by the reactive quenching constant (k_r). Independent of the quencher localization, its consumption will follow a kinetics that will depend of its concentration and/or k_Q value, ranging from zero order (total trapping of the singlet oxygen) to first order (the presence of the quencher does not significantly modifies the singlet-oxygen lifetime). If the quencher is localized in the microaggregates, its rate of consumption will be given by

$$-\frac{d(Q)}{dt} = k_r(Q)(^1O_2)_m \tag{13}$$

with the steady-state microaggregate singlet-oxygen concentration given by

$$(^1O_2)_m = \frac{R_p K}{k_{exp}} \tag{14}$$

The two limiting situations are

$$-\frac{d(Q)}{dt} = R_p \frac{k_r}{k_Q} \tag{15}$$

and

$$-\frac{d(Q)}{dt} = \left(\frac{R_p K k_r}{k_{exp}^\circ}\right)(Q) \tag{16}$$

where Eq. (15) corresponds to the zero-order limit and Eq. (16) to the first-order limit. These equations show that the measurement of the target consumption only allows an evaluation of k_r/k_Q [either from Eq. (15) or from a comparison of the pseudo-first-order constants obtained employing Eqs. (10) and (16)] and the value of $k_r K$. As in the lifetime and singlet-oxygen steady-state concentration measurements, values of the individual rate constants and partition constants cannot be obtained.

The above discussion shows that, even in conditions where the equilibrium distribution of singlet oxygen is maintained, steady-state measurements, as well as singlet-oxygen lifetime measurements, can provide only values of the product $k_Q K$ averaged over all the pseudophases considered. Furthermore, when liposomes are employed, the derived apparent rate constant does not exactly correspond to the product $k_Q K$ and can be expected to depend on the singlet-oxygen generation locus.

IV. DEPENDENCE OF K_Q IN HOMOGENEOUS SOLUTIONS WITH THE SOLVENT CHARACTERISTICS

The value of k_Q in a dispersed microphase can be expected to depend on the average microproperties of the environment, where the quenching process can take place. In order to predict how the location of the target molecule and the microproperties of the environment will influence the value of k_Q, it is necessary to know the dependence of the quenching rate with those parameters that characterize the properties of the solvent as reaction media. Furthermore, it must be considered that the characteristics of the microphase and the average location of the target molecule also determines the singlet-oxygen steady-state (or equilibrium) concentration at the reaction locus.

The dependence of singlet-oxygen reactions with the solvent has been reviewed by Lissi et al. [3]. Typical results obtained for some compounds whose quenching rate constants have also been measured in microheterogeneous systems are given in Tables 2–4. These data show that quenching rate constant increases with the solvent "polarity" for all the substrates considered. An extensive study of the solvent parameters that can be employed to correlate reactivity and solvent characteristics has been carried out employing 3-methylindole (3-MI) as the substrate [14]. These results are given in Table 3. Figure 1 shows the correlation of the measured quenching constants with the position of 3-MI fluorescence maximum.

The quality of a solvent can be characterized in terms of the molecular descriptors of the solvent characteristic [15]. The contribution of each parameter can be evaluated from multiple regression. For 3-MI [14], the quenching rate can be expressed as

$$\log k_Q = 6.36 + 2.06\pi^* - 0.53\delta \qquad (17)$$

Table 2 Solvent Effect on Singlet-Oxygen Quenching Rate Constants

Quencher	Solvent	$k_Q \times 10^l\ M^{-1}$/Sec	Ref.
DPBF	Hexane	26	11
	Cyclohexane	45	12
	Methylcyclohexane	37	11
	Toluene	67	12
	Acetone	64	12
	Acetonitrile	11, 14.2	11,12
	Methanol	8.1, 7.2	11,12
9, 10-Dimethylanthracene	Benzene	2.5	13
	Ethanol	4.6	13
	Methanol	6.9	13
	Methanol/water (8 : 2)	20	13
	Ethanol/water (1 : 1)	20	13
9-Methylanthracene	Benzene	0.23	13
	Ethanol	0.2	13
	Methanol	0.2	13
	Ethanol/water (1 : 1)	1.7	13
	Water	9.6	13

Table 3 Effect of the Solvent on the Quenching Rate
Constant of 3-Methylindole

Solvent	$k_Q \times 10^7 \ M^{-1}/Sec$
n-Hexane	0.24
n-Heptane	0.12
Benzene	0.69
n-Propanol	1.51
Chloroform	2.05
Iso-Propanol	2.24
n-Pentanol	2.31
Ethyl acetate	2.58
Ethanol	2.78
Methylene chloride	3.71
Methanol	4.32
Acetonitrile	4.56
Tetrahydrofuran	4.71
Dioxane	5.64
Acetone	9.82
Benzonitrile	6.29
Benzyl alcohol	11.5
Methyl formamide	16.5
Dimethyl formamide	17.8
Deuterium oxide, pD = 7.4	32.0

Source: Data from Ref. 14.

The agreement between the experimentally determined values and those calculated with this equation is shown in Fig. 2. The value measured in sodium dodecylsulfate (SDS) solution has been also included.

V. MICELLES

Micelles are small structures in which a higher oxygen solubility than in the surrounding aqueous solution can be expected [3,9,16–18]. The relative singlet-oxygen concentration can be expressed in terms of the partition constant K, defined by

$$K = \frac{[^1O_2]_{mic}}{[^1O_2]_{water}} \tag{18}$$

in terms of the local concentration of singlet oxygen in both pseudophases. Values of K of 2.8 in SDS and 4.0 in cetyltrimethylammonium bromide (CTAB) have been reported [16,17]. These values are very close to that reported for the partition of the ground-state oxygen molecule [18].

Table 4 Experimentally Determined Quenching Rate Constants of DMF, DFTA, and HFUA

Solvent	DMF[a]	DFTA[b]	HFUA[c]
n-Hexane	2.1		
n-Heptane	2.4		
Benzene	5.1	2.8	3.5
Ethylacetate	4.4		
Acetonitrile	3.4	5.4	
Chloroform	3.0	1.9	1.9
Acetone	4.5		
Dichloromethane	4.0		
Methanol	5.3		
Ethanol	3.3	2.7	
n-Propanol	4.2		
Benzyl alcohol	7.5	3.5	
Dimethylformamide	7.9		
Benzonitrile	6.5		
Propylenecarbonate	8.6		
DODAC[d] (LUVs)		1.2	1.2

Note: Values are obtained by time-resolved measurements at 1270 nm (this work). k_Q values are given in $10^8\ M^{-1}$/sec.
[a] DMF = 2,5-dimethyl furane.
[b] DFTA = N,N,N-trimethyl-(5-dodecyl)-2-furyltetramethylenammonium bromide.
[c] HUFA = N,N,N-trimethyl-(5-hexyl)-2-furylundecamethylenammonium bromide.
[d] DODAC = dioctadecyldimethylammonium chloride.

The data of Table 1 show that, in typical micellar systems, the average singlet-oxygen diffusion length is considerably larger than the mean distance between micelles. In the absence of high concentration of scavengers in the aqueous solution or inside the micelles, a singlet-oxygen molecule visits several micelles during its lifetime. For an entrance rate (k_+) equal to that of the ground-state oxygen molecule $(k_+ \geq 1.3 \times 10^{10}\ M^{-1}$/sec [4]), the above-mentioned partition constants imply that the average residence time of a singlet-oxygen molecule inside a micelle (the inverse of k_-) will be ~2.3 and 5.9 nsec in SDS and CTAB micelles, respectively. This simple calculation implies that, irrespective of their locus of generation, singlet-oxygen molecules will escape to the aqueous solution [19–21] and visit several micelles prior to deactivation. The decay kinetics can be expected as monoexponential and well represented by Eq. (1). Furthermore, because the fraction of volume occupied by the micelles is generally low, the decay will be dominated by the decay in the external medium. In agreement with this, singlet-oxygen lifetimes in the presence of micelles have been measured by several groups [16,19,22–25] and they are very similar to those measured in aqueous so-

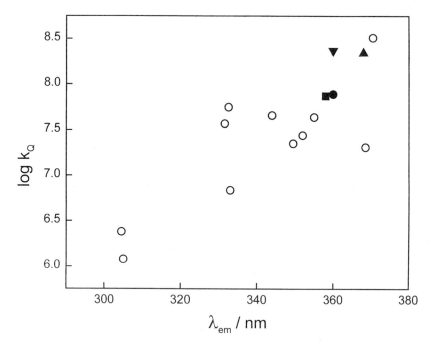

Figure 1 Experimentally determined values of k_Q obtained employing 3-methylindole as the quencher in homogeneous solvents, plotted as a function of the wavelength of maximum fluorescence intensity (data from Ref. 14). k_{ap} values determined in DODAC LUVs (■) and in dipalmitoyl phosphatidylcholine (DPPC) LUVs (▲) have been included. Also are included the experimentally determined value of k_{ap} in sodium dodecyl sulfate micelles (▼) and the value of k_Q estimated from Eq. (21) (●).

lution. The above picture amounts to a pseudophase situation and can also be considered to apply in the presence of water-soluble and/or micelle-solubilized quenchers, provided that the quenching process is not competitive to the entrance and/or exit rates. This implies that in SDS, for water-soluble quenchers,

$$k_Q[Q] < 2 \times 10^7 \ sec^{-1} \tag{19}$$

and for micelle solubilized quenchers,

$$k_Q[Q] < 4 \times 10^8 \ sec \tag{20}$$

where, in this case, the rate constant and the quencher concentration corresponds to that expressed in intramicellar local concentration.

When the above conditions are met, the kinetics of singlet-oxygen decay will be monoexponential and the effect of the quencher upon the steady-state singlet-oxygen phosphorescence will present similar characteristics in the micellar

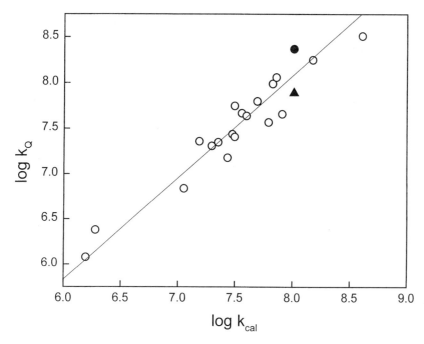

Figure 2 Solvent effect on the quenching rate constant of 3-methylindole. Comparison between experimentally determined k_Q values in homogeneous solvents and those calculated using Eq. (17) (adapted from Ref. 14). The experimentally determined value of k_{ap} in SDS micelles (●) and the value of k_Q estimated from Eq. (21) (▲) have been included.

solution and in homogeneous solvents. Furthermore, the quenching rate constant based on the analytical quencher concentration (k_{ap}) and that defined in terms of intramicellar concentrations (k_Q) are simply related by

$$k_Q = \frac{k_{ap}}{K} \tag{21}$$

when the quencher is only localized in the micellar pseudophase. If the quencher is partitioned between the micelles and the aqueous pseudophases, the apparent quenching rate constant k_{ap} is given by

$$k_{ap} = k_{water} (1 - f_s) + f_s K k_Q \tag{22}$$

where f_s is the fraction of solute incorporated to the micelles.

Because micelles are inhomogeneous microsystems, both the singlet-oxygen local concentration and the average properties where the quenching process takes place will depend on the substrate localization and are modified by the presence of additives or any other factors (such as surfactant concentration and/or the solution ionic strength) that modifies the properties of the micelles and/or the probe localization. An analysis of the reactivity in micelles pointing to the relevance of the probe localization has been carried out in a limited number of systems. Typical results are given in Table 5.

The data in Table 5 show that k_{ap} values for DPBF and 2,3-dimethylindole (DMI) are larger in SDS than in CTAB micelles. This conclusion is further stressed if k_Q values are compared. Because the rate constants of the compounds considered are extremely solvent dependent, increasing with the solvent polarity (see Sec. IV), the differences observed between both surfactants could reflect a higher exposure to water in SDS micelles than in CTAB as a consequence of its smaller, more open structure [27]. Furthermore, because those reactions take place with a negative activation volume [11], the observed differences between both micelles could be due, at least partially, to the smaller Laplace pressure of the larger micelle [28].

The data in Table 5 would indicate that the reactivity in micelles is rather similar to that measure in organic solvents. However, larger differences have been observed in some systems. 1,4-Diazabicyclo[2.2.2]octane (DABCO) reactivity in micelles is considerably lower than in most organic solvents, a result explained in

Table 5   Experimentally Determined Rate Constants ($k$ap) in Micellar Solutions

Quencher	Micelle	$k_{ap} \times 10^8\ M^{-1}$/Sec	Ref.
DPBF	SDS (0.1 M)	11,10	19,26
	CTAB (0.1 M)	6.5, 6.7	19,26
	Igepal (0.1 M)	6.5	19
	Brij 35 (0.1 M)	6.6	19
9-Methylanthracene	SDS (0.1 M)	4.9	13
	SDS (0.6 M)	3.5	13
	SDS + hexane (saturated)	4.9	13
	SDS + 56 mM heptanol	2.2	13
	CTAC (0.1 M)	6.1	13
	CTAC + 56 mM heptanol	4.0	13
	Triton X-100 (0.1 M)	5.4	13
9,10-Dimethylanthracene	SDS (0.1 M)	1.5	13
	CTAC (0.1 M)	2.7	13
	CTAB (0.015 M)	2.0	13
	Triton X-100 (0.1 M)	2.6	13
9-Anthracenecarboxylic acid	CTAB	0.74	4
2,3-Dimethylindole	SDS (0.1 M)	7.0	26
	CTAB (0.1 M)	4.4	26

terms of hydrogen-bonding involving the nitrogen lone pair [26]. On the other hand, an increase between one and two orders of magnitude in the rate constant of singlet-oxygen quenching by poly(chlorophenols) in CTAC micelles has been reported by Bertolotti et al. [29].

The influence of the target localization on k_{ap} has been discussed by Lissi and Rubio [4] employing several anthracene derivatives as substrates and taking advantage of the high sensitivity of their rate constants to the solvent (Tables 2 and 5). Incorporation of 9-methylanthracene (9-MA) into both SDS and CTAB micelles reduces its rate of reaction, relative to that measure in aqueous solution. A similar protection has been reported by Moore and Burt [21] for anthracene. These results, contrary to expectations based on the higher intramicellar singlet-oxygen concentration, reflects the relevance of the intramicellar environment where the reaction takes place. In fact, k_{intra} values are similar to those obtained in ethanol/water (1:1) mixtures. Furthermore, there are noticeable differences among the various compounds considered that can be rationalized in terms of their different intramicellar location. Exposure of the probe to the aqueous pseudophase should probably increase the value of k_{intra} (defined in local concentrations) but decrease the local singlet-oxygen concentration. The interplay of these two factors determines the reaction rate. The fact that 9,10-dimethylanthracene (DMA), the most deeply incorporated substrate, presents the smallest $k_{intra}/k_{ethanol/water}$ ratio would indicate that the effect of the medium polarity is more important than the possible gradient in intramicellar singlet-oxygen concentration [4].

A kinetic study of the photosensitized oxidation of tryptophan-alkyl esters in Triton X-100 micellar solutions has been carried out by Criado et al. [24]. The results obtained are presented in Table 6. These data show an important decrease in the relevance of the photo-oxidative pathway in the esterified compounds in the presence of the micelles. The magnitude of the effect seems to be extremely sensitive to the location of the probe, increasing as the length of the ester hydrocarbon chain increases. These results are interpreted in terms of the competition between the local oxygen concentration and the solvent micropolarity effect that

Table 6 Experimentally Determined Data for Tryptophan (Trp) and Trp Derivatives

Compound[a]	% Incorpor.	$(k_Q)_{water}$	$(k_{ap})_{Triton}$	$(k_r)_{water}$	$(k_r)_{Triton}$
Trp	2.5	0.72	1.93	0.35	0.5
TrpME	14	1.54	3.3	0.22	0.3
TrpBE	28	1.45	2.5	0.25	0.21
TrpOE	52	2.6	0.7	0.22	0.04

Note: Rate constants given in 10^8 M^{-1}/Sec.
[a] trpME: trytophan methyl ester; trpBE: tryptophan butyl ester; trpOE: tryptophan octyl ester.
Source: Data from Refs. 24, 50, and 51.

determines the relevance of the chemical reaction. A surprising feature of these data is that for some of the compounds, the decrease elicited by the surfactant is larger than the fraction incorporated to the micelles. This, at least for tryptophan octyl ester (TrpOE), applies both to the total and reactive quenching constants. In any case, the data show a noticeable protection, both in the total and chemical steps, when the solute is associated to the micelles.

Diphenylisobenzofurane (DPBF) has been the compound most frequently employed as the singlet-oxygen quencher in micellar solutions [16,19,30–35]. However, interpretation of the results, and particularly those obtained in steady-state irradiation experiments, is not straightforward due to the possibility of the occurrence of diffusion-controlled processes. Furthermore, due to the high local concentration of the probe, the occurrence of chain reactions could influence the kinetic data based on the DPBF consumption [36]. Most of the results obtained in these systems employing water-soluble sensitizers have been interpreted in terms of a diffusion-limited process with a very efficient [16] or partially limited [30,32] singlet-oxygen incorporation. However, it has to be considered that the very fast fluorescence quenching of micelle-incorporated probes by ground-state oxygen would imply that the micellar interface should not be a significant barrier to limit singlet-oxygen access to the micellar pseudophase [4,9].

The data obtained employing 3-MI as quencher are particularly suitable for a comparison of results obtained in micellar solutions and in homogeneous solvent [14]. Indole derivatives reactivity toward singlet oxygen are extremely solvent dependent, the rate constant for 3-MI ranging from 0.12×10^7 M^{-1}/sec in n-heptane to 32×10^7 M^{-1}/sec in D_2O (see Table 3). The data can be interpreted in terms of molecular descriptors that characterize the quality of the solvent (Fig. 2). In Fig. 2, the result obtained in SDS has been included. It can be seen than the experimentally determined value is higher than that predicted by Eq. (17) and taking the π^* value of SDS micelles as 0.8 [15]. This difference can be attributed to the higher intramicellar concentration of singlet oxygen. If the experimentally determined value of the apparent rate constant is corrected by the singlet-oxygen distribution [Eq. (12)] to obtain k_Q, the result in SDS agrees with those obtained in homogeneous solution (see Fig. 2). This shows that when the singlet-oxygen distribution is taken into account, the rate in the micellar pseudophase can be predicted from those obtained in homogeneous solutions if the effect of the solvent can be quantitatively expressed in terms of a suitable set of molecular descriptors.

The quenching rate can be modified by the presence of additives that change the micellar capacity to dissolve oxygen and/or the microproperties of the incorporated probe surroundings. The data of Table 5 show that addition of 1-heptanol or increasing the surfactant concentration produces a significant decrease in 9-MA k_{ap} values, as expected from the decrease in the polarity of the probe surroundings [10,37–40]. On the other hand, addition of n-hexane does not change the rate of the process, suggesting a compensation between a less polar environment [10] and a higher intramicellar singlet oxygen concentration.

VI. REVERSE MICELLES

Few singlet-oxygen processes have been studied in reverse micelles [41–45] or water-in-oil microemulsions [46], although these media could facilitate the use of direct singlet-oxygen detection methods employing water-soluble sensitizers. Let us consider AOT sodium 1,4-bis(2-ethylhexyl) sulfosuccinate (0.1 M) in n-hexane ($k° = 3 \times 10^4$ sec^{-1}). If deactivation of the singlet oxygen by the surfactant is disregarded, the total rate of decay in the absence of added quenchers should increase when the amount of water in the system increases. In particular, at high water-to-surfactant (R) ratios (~30), it can be expected that the decay of singlet oxygen in the micellar solution takes place with a rate constant, $k°$, given by

$$k° \text{ (sec}^{-1}) = (1 - 0.05K)(3 \times 10^4) + (0.05) \, 2 \times 10^5 K \tag{23}$$

where K is the singlet-oxygen partition constant between water and the external pseudophase. This value should be considerably smaller than 1 (~0.11) [17]. This implies that the decay rate constant would be indistinguishable from that obtained in the external solvent. Furthermore, given the short residence time expected in the micelles and the small distance between micelles, in these systems the same considerations as those discussed in the normal micelles would apply. The only difference is that the intramicellar singlet-oxygen concentration in reverse micelles is smaller than in the surrounding solvent and, hence, incorporation of singlet-oxygen molecules from the organic solvent into the micelles could be a rather inefficient process [47,48].

Quenching of singlet oxygen in reverse micelles has been carried out employing tryptophan (Trp) [43,45], azide ions [41,43], DPBF [42], and anthracene derivatives [44]. The results reported for azide ions in AOT reverse micelles indicate a noticeable increase in k_{ap} when the water/surfactant ratio (R) increases, from 2.7×10^6 ($R = 1.1$) to 4.2×10^7 M^{-1}/sec ($R = 22$) [43]. This last value is still smaller than the value found in aqueous solution if a K value of 0.11 is taken for singlet oxygen [17]. The extremely low values observed at low R, where a water pool is not present, most probably reflect the localization of azide ions in a small region of high viscosity and reduced singlet-oxygen incorporation. At higher R values, azide ions will be pushed toward the water pool, making the quenching process more efficient. On the other hand, Trp molecules may remain partially bound to the micellar interface, even at rather high R ration [49]. Values of k_{ap} for this compound in AOT reverse micelle range from ~3.2×10^6 M^{-1}/sec at low R, to 4.2×10^6 M^{-1}/sec at $R \cong 20$. When corrected by the singlet-oxygen solubility in the micelle, this value gives a k_{intra} close to that reported in water (see Table 6). Furthermore, it is noticeable that, although in most solvents quenching by azide ions is considerably faster than quenching by Trp, in the reverse micelles at low R values, Trp is more efficient than azide. This can be explained in terms of different localization sites in the relatively "dry" reverse micelle [48].

Rubio and Lissi [44] carried out a study of the reactivity of different anthracene derivatives toward the singlet oxygen in AOT reverse micelles over a wide range of R values and surfactant concentrations. The reactivity of DMA, a compound that can be assumed to remain in the organic solvent, is independent of both AOT and water concentration. For solutes totally incorporated into the micelles {e.g., anthracenecarboxylic acid and ethyldimethyl[3-(9-anthracenenyl)propyl] ammonium bromide}, the consumption rate was independent of AOT concentration but increased when R increases. The dependence with R was larger for the negatively charged substrate, a result explained in terms of a displacement of this compound toward the micellar core. Accordingly, at high R values, the intramicellar reactivity becomes similar to that in bulk water. For hydroxymethylanthracene, partitioned between the solvent and the micelles, the value of k_{ap} changes with AOT concentration and R value. The results are quantitatively described in terms of an equation similar to Eq. (22), rendering an intramicellar bimolecular rate constant of $1.1 \times 10^7 \, M^{-1}/sec$ [3]. This value is larger than that obtained in ethanol but considerably smaller than that measured in bulk water [44], suggesting that the probe remains bound to the interface, even at high R values.

VII. VESICLES

In this section, the term *vesicles* and *liposomes* are applied indistinctly to closed spherical or ellipsoidal bilayers, regardless of their chemical composition. Liposomes have been proposed as suitable models and tools in the understanding of photosensitization mechanisms in biological membranes [52]. There are no evaluations of the singlet-oxygen distribution between vesicles and the external solvent. Estimations of ground-state oxygen incorporation have been carried out by employing a modified Winkler technique [53], electron paramagnetic resonance (EPR) spin labels [54,55], or fluorescence quenching [56–58]. The last two methodologies provide local values of the product solubility × diffusion and show that they depend on the size of the liposome [57] and the fluidity of the membrane [54–58] and that tend to be larger in the central parts of the bilayer [54,55]. The Winkler technique provides an estimate of the average oxygen solubility upon all the bilayer [53]. Measurements carried out in sonicated small unilamellar vesicles (SUVs) of dipalmitoylphosphatidylcholine (DPPC) indicate that oxygen is nearly four times more soluble in the liquid-crystalline bilayer than in the gel-state bilayer. In this state, the oxygen solubility was found to be approximately equal to that in water at the same temperature.

As shown in Table 1, intervesicular and intravesicular distances can be similar or even larger than the average singlet-oxygen diffusion length. This possibility introduces a series of factors (gradient of singlet-oxygen concentration in a given pseudophase, nonequilibrium distribution of oxygen between the aqueous phase and vesicles, multiexponential decay for singlet oxygen) that lead to complex kinetics and to a dependence of the kinetics on the locus of singlet-oxygen generation. The latter point, together with the possibility that the singlet-oxygen generated intravesicularly reacts inside the same vesicle or with target molecules located in the external medium or in other vesicles, has been the central point addressed in most studies in these systems, and the results have frequently been contradictory. Employing SUVs of didodecyldimethylammonium bromide (DDAB), Rodgers and Bates [59] observed normal kinetics for the consumption of DPBF (lipid soluble) and anthracenedipropionic acid (ADPA) (water soluble) in the presence of several quenchers of different hydrophobicity, regardless of the sensitizer location. These results, which imply a fast exchange between vesicles and solvent, can be explained in terms of the rather high surfactant concentration employed and the small size of the vesicles. In agreement with these considerations, singlet-oxygen lifetimes in the absence of quenchers were independent of the generation locus and similar to those reported in aqueous solution [22,59].

Nonell et al. [60] obtained similar singlet-oxygen lifetime values in the presence of SUVs (~26 nm radius, 26 nM in vesicles) of DPPC as in neat D_2O solution, irrespective of the sensitizer. When the sensitizer mesotetrakis(4-sulfonatophenyl) porphine was located in the buffer, singlet-oxygen lifetimes became nearly independent of the DPBF concentration at a high quencher concentration. The measured lifetime (40 μsec) was determined by the decay in the solvent and the irreversible capture of singlet oxygen by the DPBF-loaded vesicles, with $k_+ \cong 2 \times 10^{11}\ M^{-1}$/sec. This is very close to the value estimated by applying the Smoluchowsky formalism ($3.9 \times 10^{11}\ M^{-1}$/s).

Reddi et al. [61] reported that during the photo-oxidation by liposome-bound hematoporphyrin in DPPC SUVs in the presence of 50 μM DPBF, a significant fraction of the singlet oxygen escapes to the water phase, leading to a subsequent, slow, DPBF quenching upon re-entry to the same or to another vesicle. If re-entry takes place mainly to the same vesicle (a process likely to be particularly important in LUVs suspensions), this probability will decrease with the elapsed time and, hence, could be very little dependent on the singlet-oxygen lifetime in the external medium and, hence, to changes from water to D_2O. Similarly, Grossweiner et al. [62–64] concluded that the protection of egg phosphatidylcholine SUVs damage photosensitized by incorporated sensitizers, afforded by water-soluble quenchers, provides evidence that even in vesicles containing reactive lipids, a significant fraction of singlet-oxygen molecules reach the external medium. Lack of "micellelike" kinetics has also been reported by Dearden [65] in a study of the photo-oxidation of unsaturated fatty acids in small egg phosphatidylcholine (eggPC) vesicles, because the rate of the process strongly depends on the sensitizer location. Similar considerations apply to the work of

Hoebeke et al. [66] on the reactions of ADPA (water soluble) and DMA (lipid soluble) in dimyristoylphosphatidylcholine (DMPC) LUVs and to the work of Singh et al. [67] regarding merocyanine 540 photobleaching in dilauroyloylphosphatidylcholine (DLPC) LUVs. From the enhancement effect of D_2O and the partial inhibitory effect of sodium azide, it was concluded that singlet oxygen spends most of its time (87%) in lipidic environments and that ~40% of them remain inside the liposome in which they were originally generated. Despite the fact that these calculations must be considered only as qualitative (because an equilibration treatment is applied to a nonequilibrium situation) and the role of the inner pool is disregarded, these data show that even in small liposomes, when there are high local concentrations of quenchers, equilibrium distributions of the singlet-oxygen molecules are not established. This is a rather general situation, and when LUVs are employed, it holds even in the absence of significant concentrations of quenchers. Rubio and Lissi [4] reported that in dioctadecyldimethylammonium chloride (DODAC) LUVs, there is a smaller steady-state singlet-oxygen concentration inside the intravesicular water pool than in the aqueous solution, when the sensitizer is located in the external solution. This is due to the fact that the average time to diffuse from the external solvent to the inner pool is longer than the singlet-oxygen lifetime. In the presence of reactive substrates, an inhomogeneous distribution of singlet oxygen is evidenced even in systems comprising LUVs. Li et al. [68] have measured the photosensitized oxidation of several olefins and 2,2,6,6-tetramethylpiperidine in vesicles (~ 100 nm radius) of mixed cationic and anionic surfactants. Substrate molecules were solubilized in the bilayer membranes of a set of vesicles, and the sensitizers were incorporated in the bilayers [tetraphenylporphyrin (TPP)] or the aqueous inner pool [methylene blue (MB)] of another set of vesicles. Irradiation was performed after mixing the two sets of vesicles. Product yields were significantly higher in D_2O than in water. The oxidation of the substrate show that singlet oxygen produced in a vesicle is able to react with target molecules associated to other vesicles. The measurements of the quantum yields revealed that only a fraction of the singlet oxygen is able to diffuse from one vesicle to another. Furthermore, this fraction depends on the singlet-oxygen-generation locus: 8% in water and 15% in D_2O when singlet oxygen is generated in the inner aqueous pool and 20% in water and 80% in D_2O when generated in the bilayer. In the photosensitized oxidation of *trans*-stilbene and 1,4-diphenyl-1,2-butadiene, 1,2-cycloaddition products were formed in quantitative yields, which was in sharp contrast to the oxidation in homogeneous solutions, where only 1,4-cycloaddition products were detected. These results would indicate that the organized semirigid environment in vesicles prevents the occurrence of reaction paths involving large conformational changes. Using hypericin as the sensitizer and DPBF as the quencher, Bouirig et al. [69] employed the same experimental approach to show that singlet oxygen produced in a DPPC liposome of 30 nm radius can react with target molecules localized in other liposomes.

The previous considerations imply that only in limiting situations (small vesicles, high surfactant concentration, long singlet-oxygen lifetimes) can it be expected that the systems follows a simple kinetics, rendering monoexponential decays in the presence of quenchers and/or lineal Stern–Volmer plots in steady-state fluorescence measurements. However, pseudo-first-monoexponential decays are even observed in conditions where it can be expected that an equilibrium distribution of singlet oxygen and a homogeneous population of singlet-oxygen molecules cannot be assumed. For example, monoexponential decays and nearly lineal Stern–Volmer plots have been obtained for the singlet-oxygen quenching by 3-MI in DODAC and DPPC LUVs [14]. The apparent quenching rate constants were 7.33×10^7 M^{-1}/sec in DODAC vesicles and 22.0×10^7 M^{-1}/sec in DPPC LUVs. Comparison of these values with those given in Table 3 shows that the rate constant in DODAC suspensions lays between those observed in benzonitrile and acetone, whereas DPPC vesicles show a behavior intermediate between those of dimethylformamide and D_2O, suggesting a more polar environment (and/or a higher oxygen solubility) in the DPPC vesicles. The position of the 3-MI fluorescence spectra is compatible with this interpretation. The wavelength of maximum fluorescence intensity lays at 368 nm in DPPC (very close to that in water) and at 358 nm in DODAC vesicles. In fact, the data given in Fig. 1 show that, regarding the relationship between quenching rate constants and microenvironment of the 3-MI, DODAC and DPPC LUVs behave as homogeneous solvents. This implies that the intravesicular singlet-oxygen concentration is very similar to that present in the external solvent [53]. This comparison is particularly suitable because the "averaged" micropolarity of the target molecule is directly assessed from the position of its fluorescence spectra. The agreement of the k_{intra} measured value in SDS and that expected from the position of 3-MI incorporated to the micelles further support this type of comparison.

The distribution of the singlet oxygen between the external solvent and the liposomes can be further assessed by measuring the quenching rates of substrates whose reaction rate is very little solvent dependent. Alkylfuranes fulfill this requirement. Data obtained for several furanes are included in Table 4.

The data given in Table 4 show that the three furanes present similar quenching rate constants and that they are relatively little dependent on the solvent. Furthermore, it can be seen that both probes in DODAC LUVs appear as less efficient quenchers than in most of the homogeneous solvents. This can be explained in terms of an unfavorable singlet-oxygen distribution, suggesting that the intravesicular singlet-oxygen concentration is approximately two times smaller than in the external D_2O solvent. This is compatible with reported data on the solubility of oxygen in large vesicles in the gel state [53,56,57].

Employing DPPC SUVs, Valduga et al. [70] followed the bleaching of DPBF using the vesicle-bound sensitizer Zn(II) phthalocyanine. A linear relationship between the inverse of the bleaching yield and the inverse DPBF concentration was obtained, leading to a k_{ap} value of ~ 5 × 10^9 M^{-1}/sec. If this value is corrected by the singlet-oxygen distribution ($K = 4$), the value obtained is similar to that reported in acetonitrile. A smaller value (k_{ap} = 6.2 × 10^8 M^{-1}/sec) has been reported in DDAB vesicles [59]. If we consider that k_{intra}, being a non-diffusion-controlled process, cannot be smaller than that measured in alkanes, these results require of a K value smaller than ~ 3 in DDAB vesicles.

Bachowski et al. [71] measured, in LUVs of eggPC and dicetylphosphate (10:1), the protection by azide ions of the lipid peroxidation and lactate dehydrogenase inactivation promoted by chloroaluminum phthalocyanine tetrasulfonate irradiation. Although linear Stern–Volmer plots were obtained, k_{ap} values depended on the substrate considered, implying a strongly inhomogeneous singlet-oxygen distribution. Values of k_{ap} for azide ions in DMPC vesicles were also obtained in protection experiments employing merocyanine and histidine as substrates. The values obtained were also dependent on the substrate considered [67,72]. Fukuzawa et al., in a series of publications [73–75], have presented data on the kinetics of singlet-oxygen scavenging by α-tocopherol, 2,6-ditertbutyl-4-methylphenol (BHT), and several carotenes in phospholipid SUVs and compared the results to those obtained in ethanol. In these works, water solubles [MB and rose bengal (RB)] and a lipid soluble (pyrenedodecanoic acid) were employed as singlet-oxygen sources. A competitive method, employing DPBF as reference, was employed to estimate the additive total quenching constant. However, the fact that singlet-oxygen lifetimes were assumed to be dependent on the employed sensitizer and equal to those in ethanol and *tert*-butanol precludes an evaluation of the quenching rate constants. In spite of this, the results show reactivity ratios widely different in ethanol than in the liposomes and noticeable differences between both sensitizers, suggesting a sensitizer-dependent singlet-oxygen distribution.

Even when the singlet oxygen is produced in the lipid bilayer, most of the molecules will reach the external solvent. Simple calculations based in the bilayer thickness and the intrabilayer diffusion coefficient would indicate that in ~ 10 μsec, more than 100 nm can be traveled (i.e., about 25 times the thickness of the lipid bilayer). To confine the singlet-oxygen molecules to the bilayer would require reducing its lifetime to values considerably shorter than those that can be measured with the presently available equipment. By estimating intravesicular quenching rates, Ehrenberg et al. [8] have estimated lifetimes of 36.4 μsec in DMPC and 12.2 μsec in eggPC in SUVs. The measured lifetimes in a DMPC suspension were 5 μsec in water and 71 μsec in D_2O, as expected if most of the singlet oxygen is present in the aqueous pseudophase. The small fraction remaining in the liposomes does not significantly reduce the measured average lifetime. In DMPC vesicles, nearly lineal Stern–Volmer plots were obtained when D_2O was employed as the external solvent. The apparent quenching rate constant for azide ions was 1.27 × 10^8 M^{-1}/sec, a value that is about 5–10 times lower than that observed in aqueous solution. This difference is larger than that expected from the protection afforded by the liposomes and could be rationalized in terms of an extended protection provided for the inner pool if azide ions are excluded from them. On the other hand, the rather low value measured for the apparent quenching rate constant for DPBF (3.6×10^8 M^{-1}/sec) is readily explainable in terms of a predominant presence of singlet oxygen in the external medium.

Vilensky and Feitelson [76] have discussed the effect of the sensitizer and target molecule location upon the singlet-oxygen reactions in eggPC LUVs (1 mg/mL) in D$_2$O. Meso-tetra(N-methyl-4 pyridyl)porphine tetratosylate (mTPTT), located in the aqueous phase, and hematoporphyrin (HP), which can be considered to be embedded in the membrane, were employed as sensitizers. N-Acetyl tryptophan (NATA), a tryptophan residue with a long aliphatic chain attached, Trp(CH$_2$)$_{16}$, and melittin, a small membrane protein, where employed as target molecules. As in most studies, the presence of the liposomes does not affect the singlet-oxygen lifetime. Apparent reactive quenching constants were obtained from the substrate consumption rates. Total quenching constants were derived from the effect of the quencher upon the rate of singlet-oxygen decay measured in pulsed experiments. The relevant data are summarized in Table 7.

The protection afforded by the membrane can be deduced from a comparison of the rate constants of NATA in D$_2$O and the Trp derivatives in the liposomes when the oxygen is generated in the external medium. This most probably reflects a difference in the environment sensed by the Trp groups in the different compounds. (See Table 3 for the effect of the solvent on the quenching by indole derivatives.) The relevance of the oxygen generation seems to be indicated by the large difference in quenching rate constant observed for the deeply buried Trp moiety in Trp(CH$_2$)$_{16}$ when the lipid-soluble and water-soluble sensitizers are considered. However, it must be mentioned that consumption by Type I processes,

Table 7 Quenching by Trp Residues in eggPC SUVs

Quencher, solvent, and sensitizer[a]	$10^{-7}k_r$	$10^{-7}k_Q$
NATA, D$_2$O	4.9	6.2
NAT, liposomes, mTPTT	5.0	5.4
NATA, liposomes, HP	20	
Melittin, D$_2$O, mTPTT	3.9	5.4
Melittin, liposomes, mTPP	1.2	1.8
Melittin, liposomes, HP	1.7	
Trp(CH$_2$)$_{16}$, liposomes, mTPTT	0.19	
Trp(CH$_2$)$_{16}$, liposomes, HP	2.2	

[a] NATA: N-acetyltryptophan amide; Trp(CH$_2$)$_{16}$: hexadecylamido indole-3-acetate; HP: hematoporphyrin; mTPTT: meso-tetra(N-methyl-4-pyridyl)porphine tetratosylate.
Source: Data from Ref. 76.

favored by the high local concentration of the quencher, cannot be completely disregarded. It is difficult to accept that the Trp derivative traps most of the singlet oxygen prior to its escape to the aqueous pseudophase, as required by the large difference observed when it is generated inside the vesicles. To travel ~4 nm inside the vesicle takes singlet oxygen ~ 5×10^{-9} sec. To avoid escape to the external solvent, the pseudo-first-order intravesicular constant, given by $k_Q(Q)$, must be such that

$$k_Q(Q) \geq 2 \times 10^8 \text{ sec}^{-1} \qquad (24)$$

hence, extremely high local concentrations of very reactive substrates are required. Even if the residence time can be underestimated due to inefficient transfer through the interface [45] as a consequence of the smaller singlet-oxygen solubility in the aqueous pseudophase [17], this simple calculation shows that trapping of singlet oxygen prior to its visit to the aqueous pseudophase is most unlikely. However, intravesicular quenching can take place after re-entry from the external solvent and/or from the inner aqueous pool. It is interesting to note that for melittin, whose Trp groups are more exposed to the solvent than in the Trp derivative (as evidence by the differences in the wavelength of maximum intensity), the apparent quenching constant in the liposomes is very little dependent on the localization of the sensitizer and only three times smaller than that in D_2O.

To liposome data, Fu and Kanofsky [77] applied a one-dimensional model of singlet-oxygen diffusion and quenching previously developed to interpret singlet-oxygen decays in biological membranes [78]. Extruded unilamellar liposomes (100 or 400 nm average diameter) of DMPC labeled with zinc phathalocyanine were employed. The conditions chosen were such that the distance between liposomes was large enough to prevent singlet oxygen generated in one liposome from reaching another liposome. This magnifies the relevance of the inhomogeneous singlet-oxygen distribution and any accurate mathematical analysis must take into account the interaction between singlet-oxygen diffusion and quenching. Results were obtained in the presence of hydrophobic [β-carotene and β-apo-8′-*trans* carotenoate (EAC)] and hydrophilic (histidine and methionine) quenchers. Hydrophobic quenchers principally lowered the initial singlet-oxygen concentration following the excitation pulse and caused only modest changes in the lifetime of the 1270-nm emission. The magnitude of the "initial" decrease in singlet oxygen was related to the size of the liposomes, being larger in the 400-nm liposomes. This was interpreted in terms of a higher re-entry probability to the larger liposomes. In contrast, hydrophilic quenchers mainly decreased the lifetime, the effect being almost independent of the size of the liposomes. The behavior of these quenchers was very close to that obtained in the absence of liposomes. This implies that the quenching takes place after the singlet oxygen has moved far enough from the liposomes to make its probability of re-entry to the original compartment negligible. These results were interpreted in terms of a one-dimensional model of singlet-oxygen diffusion and quenching. The proposed continuity equation, based on Ficks' second law in spherical coordinates, was

$$\frac{\partial[{}^1O_2]}{\partial t} = D\nabla^2\,({}^1O_2) - \frac{[{}^1O_2]}{\tau} + F \tag{25}$$

where the first term of the right-hand side takes into account the diffusion, the second is the oxygen decay, and F is a function that takes into account the singlet-oxygen generation. By taking into account the spherical symmetry, this equation reduces to

$$\frac{\partial[{}^1O_2]_{w'}}{\partial t} = D_{w'}\left(\frac{\partial^2[{}^1O_2]_{w'}}{\partial r^2} + \frac{2}{r}\frac{\partial[{}^1O_2]_{w'}}{\partial t}\right) - \frac{[{}^1O_2]_{w'}}{\tau_{w'}} \tag{26}$$

for the aqueous buffer inside the liposome,

$$\frac{\partial[{}^1O_2]_m}{\partial t} = D_m\left(\frac{\partial^2[{}^1O_2]_m}{\partial r^2} + \frac{2}{r}\frac{\partial[{}^1O_2]_m}{\partial t}\right) - \frac{[{}^1O_2]_m}{\tau_m} + F \tag{27}$$

for the liposome membrane, and

$$\frac{\partial[{}^1O_2]_w}{\partial t} = D_w\left(\frac{\partial^2[{}^1O_2]_w}{\partial r^2} + \frac{2}{r}\frac{\partial[{}^1O_2]_w}{\partial t}\right) - \frac{[{}^1O_2]_w}{\tau_w} \tag{28}$$

for the aqueous buffer outside the liposome. If K is the singlet-oxygen partition coefficient between the membrane and the aqueous phases, r_0 is the radius of the liposome, and d is the liposome membrane thickness, and a rapid equilibrium at the liposome membrane surfaces, is assumed the following boundary conditions apply:

$$D_m\left(\frac{\partial[{}^1O_2]_m}{\partial r}\right) = D_{w'}\left(\frac{\partial[{}^1O_2]_{w'}}{\partial r}\right) \quad \text{at } r = r_0 - d \tag{29}$$

$$[{}^1O_2]_m = K[{}^1O_2]_{w'} \quad \text{at } r = r_0 - d \tag{30}$$

$$[{}^1O_2]_m = K[{}^1O_2]_w \quad \text{at } r = r_0 \tag{31}$$

$$D_w\left(\frac{\partial[{}^1O_2]_w}{\partial r}\right) = D_m\left(\frac{\partial[{}^1O_2]_m}{\partial r}\right) \quad \text{at } r = r_0 \tag{32}$$

and

$$D_{w'}\left(\frac{\partial[{}^1O_2]_{w'}}{\partial r}\right) = 0 \quad \text{at } r = 0 \tag{33}$$

Equations (26)–(28) can then be numerically solved, allowing a detailed description of the singlet-oxygen behavior in this three-pseudophase system. Fitting of the data obtained between 30 and 100 μsec after excitation required apparent quenching constants of 5.0×10^9 M^{-1}/sec for β-carotene and 7.2×10^9 M^{-1}/sec for EAC. These apparent quenching constants correspond to 36% and 74% of the corresponding quenching constants in chloroform solvent.

Most of the above-mentioned complexities can be minimized by employing small concentrations of low reactive targets in fully saturated liposomes and singlet oxygen generated in the aqueous pseudophase. This approach has been employed by Encinas et al. [79] in the evaluation of the reactivity of anthracene derivatives in DODAC vesicles by following the bleaching of the substrate under steady-state conditions. The data are shown in Table 8, together with other related bimolecular rate constants reported in closely related systems.

The values of k_{ap}/k_{water} (Table 8) range from 0.077 to 0.3, implying a protection of the substrates as a consequence of their incorporation to the vesicles. This decrease takes place in spite of the expected favorable distribution of singlet oxygen and reflects the reduced polarity of the medium. The values of k_{ap} for a given compound depend on the size of the vesicle. For those compounds where the aromatic moiety is deeply incorporated into the bilayer [e.g., 9,10-dimethyl anthracene and 3-(9-anthroyl)propionic acid], the values are smaller in the large vesicles, a result that can be explained in terms of a reduced oxygen solubility and/or a less polar environment in the more closely packed vesicles. On the other hand, for those compounds for which the aromatic group located near the inter-

Table 8 Apparent Quenching Rate Constants in Vesicles

Quencher	Vesicle	$k_{ap} \times 10^7$ M^{-1}/Sec	k_{ap}/k_{water}	Ref.
DPBF	DDAB (SUVs)	62		59
	DPPC (SUVs)	88		61
2,3-Dimethylindole	DDAB (SUVs)	4.8		59
9,10-Dimethylanthracene	EPC (SUVs)	32		65
	DODAC (LUVs)	1.8		79
	DODAC (SUVs)	4.1		79
	EPC (SUVs)	78		80
9-Methylanthracene	DODAC (LUVs)	0.84	0.15	79
	DODAC (SUVs)	0.69	0.12	
9-Hydroxymethylanthr acene	DODAC (LUVs)	0.4	0.3	
	DODAC (SUVs)	0.26	0.2	
3-(9-Anthryl)propionic acid	DODAC (LUVs)	0.37	0.077	
	DODAC (SUVs)	0.52	0.11	
9-Antrhacenecarboxylic acid	DODAC (LUVs)	0.087	0.15	
	DODAC (SUVs)	0.094	0.16	

face (e.g., 9-MA), the reactivity is higher in LUVs, suggesting a more exposed location in these structures, similar to that observed for other aromatic compounds [57]. A different location of the aromatics in the vesicles could also explain the small difference in reactivity between 9-MA and DMA (a factor of nearly 2 in LUVs) compared to that observed in homogeneous solvents (e.g., a factor 20 in ethanol).

The characteristics of the vesicles, and hence the singlet-oxygen solubility and the substrate microenvironments, can be drastically altered by incorporation of solutes or changes in temperature. The influence of these factors on singlet-oxygen reactivity with vesicle incorporated solutes has been discussed by Rubio et al. [81]. 1-Octanol addition to LUVs of DODAC lead to a notable decrease in k_{ap} values for DMA and 3-(9-anthryl)propionic acid, in spite of the increased solubility/mobility of oxygen molecules inside the vesicles [57]. These results imply a noticeable reduction in k_{intra} due to changes in the polarity of the microenvironments of the probe and/or its displacement toward less polar regions [57]. The fluidity of DODAC LUVs increases sharply at temperatures above the bilayer phase transition [57]. This should significantly increase the oxygen solubility. In spite of this, the reactivity of anthracene derivatives located inside the vesicles is only slightly dependent on the temperature, increasing by less than a factor of 2 when the temperature changes from 15°C to 42°C [81]. On the other hand, Suwa et al. [82] reported that the yield of cholesterol-derived 5α-OOH hydroperoxide produced in the photosensitized oxygenation of cholesterol in liposomal membranes (DPPC and DMPC) increases sixfold to sevenfold above the transition temperatures. The relatively small effect observed in the anthracene derivatives can then be due to a combination of effects. The increase in singlet-oxygen solubility could be partially offset by a decrease in k_{intra} due to the decreased local polarity at the locus of the anthracene moiety location [81].

A quantitative interpretation of data obtained in more complex systems, such as biological membranes, is even less straightforward [5,83] due to the intrinsic inhomogeneity of the systems and the short lifetimes of the membrane-incorporated singlet-oxygen molecules [7,83,84]. This makes singlet-oxygen monitoring in these systems extremely difficult, due both to the high rate of the process and the low intensities involved [84–86].

ACKNOWLEDGMENTS

This work has been supported by FONDECYT (Project 1000022).

REFERENCES

1. Briviba, K.; Klotz, L-O.; Sies, H. *Biol. Chem.* **1997,** *378,* 1259.
2. Hultén, L. M.; Holmström, M.; Soussi, B. *Free Radical Biol. Med.* **1999,** *27,* 1203.

3. Lissi, E. A.; Encinas, M. V.; Lemp, E.; Rubio, M. A. *Chem. Rev.* **1993**, *93*, 699.

4. Lissi, E. A.; Rubio, M. A. *Pure Appl. Chem.* **1990**, *62*, 1503.

5. Baker, A.; Kanofsky, J. R. *Photochem. Photobiol.* **1993**, *57*, 720.

6. Fu, Y.; Kanofsky, J. R. *Photochem. Photobiol.* **1995**, *62*, 692.

7. Moan, J. *J. Photochem. Photobiol. B: Biol.* **1990**, *6*, 343.

8. Ehrenberg, B.; Anderson, J. L.; Foote, C. S. *Photochem. Photobiol.* **1998**, *68*, 135.

9. Lissi, E. A.; Dattoli, A.; Abuin, E. *Bol. Soc. Chil. Quim.* **1985**, *30*, 37.

10. Abuin, E. B.; Lissi, E. A. *Prog. Reaction Kinetics* **1991**, *16*, 1.

11. Okamoto, M.; Tanaka, F.; Teranishi, H. *J. Phys. Chem.* **1990**, *94*, 669.

12. Gorman, A. A.; Gould, I. R.; Hamblett, I. *J. Am. Chem. Soc.* **1982**, *104*, 7098.

13. Rubio, M. A.; Araya, L.; Abuin, E. B.; Lissi, E. A. *An. Asoc. Quim. Argent.* **1985**, *79*, 301.

14. Lemp, E.; Pizarro, N.; Encinas, M. V.; Zanocco, A. L. *Photochem. Photobiol., submitted.*

15. Abraham, M. H.; Chadha, H. S.; Dixon, J. P.; Rafols, C.; Treiner, C. *J. Chem. Soc. Perkin Trans. 2* **1995**, 348.

16. Matheson, I. B. C.; Massoudi, R. *J. Am. Chem. Soc.* **1980**, *102*, 1942.

17. Lee, P. C.; Rodgers, M. A. J. *J. Phys. Chem.* **1983**, *87*, 4894.

18. Matheson, I. B. C.; King, A. D. *J. Colloid Interf. Sci.* **1978**, *66*, 464.

19. Lindig, B. A.; Rodgers, M. A. J. *J. Phys. Chem.* **1979**, *83*, 1683.

20. Kraljic, I.; Barboy, N.; Leicknam, J. P. *Photochem. Photobiol.* **1979**, *30*, 631.

21. Moore, D. E.; Burt, C. D. *Photochem. Photobiol.* **1981**, *34*, 431.

22. Rodgers, M. A. J. *Photochem. Photobiol.* **1983**, *37*, 99.

23. Gorman, A. A.; Hamblett, I.; Lambert, C.; Prescott, A. L.; Rodgers, M. A. J.; Spence, H. M. *J. Am. Chem. Soc.* **1987**, *109*, 3091.

24. Criado, S.; Bertolotti, S. G.; Soltermann, A. T.; Garcia, N. A. *J. Photochem. Photobiol. B: Biol.* **1997**, *38*, 107.

25. Kraljic, I.; Barboy, N.; Leicknam, J-P. *Photochem. Photobiol.* **1979**, *30*, 631.

26. Lindig, B. A.; Rodgers, M. A. J. *Photochem. Photobiol.* **1981**, *33*, 627.

27. Bonilha, J. B. S.; Zumstein, R. M.; Abuin, E.; Lissi, E. A.; Quina, F. *J. Colloid Interf. Sci.* **1990**, *125*, 238.

28. Menger, F. *J. Phys. Chem.* **1979**, *83*, 893.

29. Bertolotti, S. G.; Gsponer, H. E.; Garcia, N. A. *Toxicol. Environ. Chem.* **1989**, *22*, 229.

30. Miyoshi, M.; Tomita, G. *Z. Naturforsch.* **1978**, *33B*, 622.

31. Usui, Y.; Tsukada, M.; Nakamura, H. *Bull. Chem. Soc. Jpn.* **1978**, *51*, 379.

32. Gorman, A. A.; Lovering, G.; Rodgers, M. A. J. *Photochem. Photobiol.* **1976**, *23*, 399.

33. Miyoshi, N.; Tomita, G. *Photochem. Photobiol.* **1979**, *29*, 527.

34. Chauvet, J. P.; Villain, F.; Viovy, R. *Photochem. Photobiol.* **1981**, *34*, 557.

35. Ko, J. S.; Han, D. S.; Oh, H. S.; Park, V. K.; Kim, C. H.; Oh, S. W. *J. Korean Chem. Soc.* **1991**, *35*, 452.

36. Krieg, M. *J. Biochem. Biophys Methods* **1993**, *27*, 143.

37. Burns, P. A.; Foote, C. S.; Mazur, S. *J. Org. Chem.* **1976**, *41*, 899.

38. González, M.; Vera, J.; Abuin, E. B.; Lissi, E. A. *J. Colloid Interf. Sci.* **1984**, *98*, 152.

39. Croonen, Y.; Gelade, E.; Van der Zegel, M.; Van der Auweraer, M.; Vandendrlessche, H.; De Schryver, F. C.; Almgren, M. *J. Phys. Chem.* **1983,** *87,* 1426.
40. Lianos, P.; Zana, R. *J. Phys. Chem.* **1980,** *84,* 3339.
41. Miyoshi, N.; Tomita, G. *Z. Naturforsch.* **1979,** *34B,* 339.
42. Miyoshi, N.; Tomita, G. *Z. Naturforsch.* **1980,** *35B,* 731.
43. Rodgers, M. A. J.; Lee, P. C. *J. Phys. Chem.* **1984,** *88,* 3480.
44. Rubio, M. A.; Lissi, E. A. *J. Colloid Interf. Sci.* **1989,** *128,* 458.
45. Collins-Gold, L. C.; Barber, D. C.; Hagan, W. J.; Gibson, S. L.; Hilf, R.; Whitten, D. G. *Photochem. Photobiol.* **1988,** *48,* 165.
46. Oliveros, E.; Pheulpin, P.; Braun, A. M. *Tetrahedron* **1987,** *43,* 1713.
47. Wong, M.; Thomas, J. K.; Gratzel, M. *Am. Chem. Soc.* **1976,** *98,* 2391.
48. Sáez, M.; Abuin, E.; Lissi, E. A. *Langmuir* **1989,** *5,* 942.
49. Lissi, E. A.; Encinas, M. V.; Bertolotti, S. G.; Cosa, J. J.; Previtali, C. M. *Photochem. Photobiol.* **1990,** *51,* 53.
50. Reddi, E.; Rodgers, M. A. J.; Spikes, J. D.; Jori, G. *Photochem. Photobiol.* **1984,** *40,* 415.
51. Matheson, I. B. C.; Lee, J. *Photochem. Photobiol.* **1979,** *29,* 879.
52. Hoebeke, M. *J. Photochem. Photobiol. B: Biol.* **1995,** *28,* 189.
53. Smotkin, E. S.; Moy, F. T.; Plachy, W. Z. *Biochim. Biophys Acta* **1991,** *1061,* 33.
54. Subczynski, W. K.; Hyde, J. S. *Biophys. J.* **1983,** *41,* 283.
55. Subczynski, W. K.; Hyde, J. S.; Kusumi, A. *Proc. Natl. Acad. Sci. USA* **1989,** *86,* 4474.
56. Lissi, E. A.; Gallardo, S.; Sepúlveda, P. *J. Colloid Interf. Sci.* **1992,** *152,* 104.
57. Abuin, E.; Lissi, E.; Aravena, D.; Zanocco, A. L.; Macuer, M. *J. Colloid Interf. Sci.* **1988,** *122,* 201.
58. Campos, A. M.; Abuin, E. B.; Lissi, E. A. *Colloids Surf. A: Physicochem. Eng. Aspects* **1995,** *100,* 155.
59. Rodgers, M. A. J.; Bates, A. L. *Photochem. Photobiol.* **1982,** *35,* 473.
60. Nonell, S.; Braslavsky, S. E.; Schaffner, K. *Photochem. Photobiol.* **1990,** *51,* 551.
61. Reddi, E.; Valduga, G.; Rodgers, M. A. J.; Jori, G. *Photochem. Photobiol.* **1991,** *54,* 633.
62. Blum, A.; Grossweiner, L. I. *Photochem. Photobiol.* **1985,** *41,* 27.
63. Grossweiner, L. I.; Patel, A. S.; Grossweiner, J. B. *Photochem. Photobiol.* **1982,** *36,* 159.
64. Goyal, G. C.; Blum, A.; Grossweiner, L. I. *Cancer Res.* **1983,** *43,* 5826.
65. Dearden, S. J. *J. Chem. Soc., Faraday Trans. 1* **1982,** *82,* 1627.
66. Hoebeke, M.; Piette, J.; van de Vorst, A. *J. Photochem. Photobiol. B: Biol.* **1991,** *9,* 281.
67. Singh, R. J.; Feix, J. B.; Kalyanaraman, B. *Photochem. Photobiol.* **1992,** *55,* 483.
68. Li, H-R.; Wu, L-Z.; Tung, C-H. *J. Am. Chem. Soc.* **2000,** *122,* 2446.
69. Bouring, H.; Eloy, D.; Jardon, P. *J. Chim. Phys.* **1992,** *89,* 1391.
70. Valduga, G.; Nonell, S.; Reddi, E.; Jori, G.; Braslavsky, S. E. *Photochem. Photobiol.* **1988,** *48,* 1.
71. Bachowski, G. J.; Ben-Hur, E.; Girotti, A. W. *J. Photochem. Photobiol. B: Biol.* **1991,** *9,* 307.

72. Singh, R. J.; Feix, J. B.; Pintar, T. J.; Girotti, A. W.; Kalyanaraman, B. *Photochem. Photobiol.* **1991,** *53,* 493.

73. Fukuzawa, K.; Inokami, Y.; Tokumura, A.; Terao, J.; Suzuki, A. *Lipids* **1998,** *33,* 751.

74. Fukuzawa, K.; Matsuura, K.; Tokumura, A.; Suzuki, A.; Terao, J. *Free Radical Biol. Med.* **1997,** *22,* 923.

75. Fukuzawa, K.; Inokami, Y.; Tokumura, A.; Terao, J.; Suzuki, A. *BioFactors* **1998,** *7,* 31.

76. Vilensky, A.; Feitelson, J. *Photochem. Photobiol.* **1999,** *70,* 841.

77. Fu, Y.; Kanofsky, J. R. *Photochem. Photobiol.* **1995,** *62,* 692.

78. Baker, A.; Kanofsky, J. R. *Photochem. Photobiol.* **1993,** *57,* 720.

79. Encinas, M. V.; Lemp, E.; Lissi, E. A. *J. Photochem. Photobiol. B: Biol.* **1989,** *3,* 113.

80. Gross, E.; Ehrenbergg, B.; Johnson, F. M. *Photochem. Photobiol.* **1993,** *57,* 808.

81. Rubio, M. A.; Lemp, E.; Encinas, M. V.; Lissi, E. A. *Bol. Soc. Chil. Quim.* **1992,** *37,* 33.

82. Suwa, K.; Kimura, T.; Schaap, P. *Photochem. Photobiol.* **1978,** *28,* 469.

83. Valenzeno, D. P. *Photochem. Photobiol.* **1987,** *46,* 147.

84. Baker, A.; Kanofsky, J. R. *Arch. Biochem. Biophys.* **1991,** *286,* 70.

85. Oelckers, S.; Sczpan, M.; Hanke, T.; Röder, B. *J. Photochem. Photobiol. B: Biol.* **1997,** *39,* 219.

86. Oelckers, S.; Ziegler, T.; Michler, I.; Röder, B. *J. Photochem. Photobiol. B: Biol.* **1999,** *53,* 121.

5

Microreactor-Controlled Product Selectivity in Organic Photochemical Reactions

Chen-Ho Tung, Kai Song, Li-Zhu Wu, Hong-Ru Li, and Li-Ping Zhang
Chinese Academy of Sciences, Beijing, China

I. INTRODUCTION

Selectivity in organic phototransformation continues to be one of the main topics of current interest. Of the various approaches to enhance the selectivity use of microreactors to control the reaction pathways has shown considerable promise [1–11]. Microreactors refer to organized and constrained media which provide cavities and/or surfaces to accommodate the molecules of a substrate and allow chemical reaction(s) to occur. The substrate is usually a small molecule of dimensions of several angstroms, and the microreactor is often tens of angstroms or larger. Thus, microreactors are also known as nanoreactors. A combination of microreactor as a medium and light as a reagent potentially offers a new approach toward "clean chemistry." The implications of this methodology for environmentlly benign synthesis of chemicals are twofold. First, the increased selectivity of the reaction will decrease the production of unwanted side products. Second, in most cases, the use of a microreactor will eliminate the need for expensive and environmentally toxic solvents.

There have been tremendous number of publications on the subject of photophysical and photochemical processes in organized and confined media over the past decades [1–11]. In this chapter, we will not provide a comprehensive survey of the past achievements in the use of microreactors to control the pathways of organic photochemical reactions, as many excellent reviews are already available [3,4,9–11]. Instead, we will summarize contemporaty applications of several types of microreactors to photochemical synthesis of large-ring compounds, photosensitized oxidation of alkenes, photochemical rearrangement, photocycloaddtion, and chiral photochemical synthesis.

II. MICROREACTORS

There exist many classes of microreactor in naturally occurring and synthesized materials. These microreactors can be broadly classified into four categories:

1. *Cavities of macrocyclic hosts.* For example, the cavities of crown ethers, cyclodextrins, calixaranes and other synthesized and naturally occurring macrocycles are several angstroms in size and are capable of recognizing the guest molecules. These microreactors can dramatically influence the rate and product distribution of certain reactions.

2. *Molecular aggregates.* Micelles, microemulsions, vesicles, liquid crystals, dendromers, monolayers, Langmuir-Blodgett (L-B) films, and so forth have dimensions of tens of angstroms or larger, and the photochemistry in this class of microreactor has been extensively investigated.

3. *Cavities and surfaces of microporous and nonporous solids.* Zeolites, silica, aluminia, clays, and other layered solids have large surface areas and/or a great number of cavities. These supports are relatively inert and provide an ordered environment for effecting and controlling photochemical processes more efficiently than can be attained in homogeneous solutions. Thus, these surfaces are known as nonreactive surfaces. In contrast, the surfaces of semiconductor particles such as TiO_2 and CdS directly participate in photochemical reactions by absorbing the incident photon and transferring charge to an adsorbed molecule or by quenching the excited state of the adsorbed molecule. These surfaces are known as reactive surfaces.

4. *New designed and synthesized microreactors.* To meet the needs of enhancing the selectivity of a certain reaction, one can modify the existing microreactors or even design and synthesize new microreactors.

Among the many classes of microreactor which have been used in organic phototransformation, we will limit our discussion only on molecular-sieve zeolites, Nafion membranes, vesicles, and low-density polyethylene films.

A. Zeolites

Molecular-sieve zeolites represent an unique class of material [12–15]. This materials are crystalline aluminosilicates with open-framework structures made up of the primary building blocks $[SiO_4]^{4-}$ and $[AlO_4]^{5-}$ tetrahedra that are linked by all of their corners with no two aluminum atoms sharing the same oxygen. Due to the difference in charge between the $[SiO_4]^{4-}$ and $[AlO_4]^{5-}$ tetrahedra, the net framework charge of an aluminum-containing zeolite is negative and, hence, must be balanced by a cation, typically an alkali or alkaline earth metal cation. The frameworks thus obtained contain pores, channels, and cages of different dimensions and shapes. The pores and cages can accommodate, selectively according to size/shape, a variety of organic molecules of photochemical interest and provide restrictions on the motions of the included guest molecules and reaction intermediates. For example, the internal surface of ZSM-5, a member of the pentasil family, consists of two types of pore system (channels, Fig. 1): one is sinusoidal with a near-circular cross section of ~5.5 Å and the other is straight and perpendicular to the sinusoidal channels. The straight channels are roughly elliptical with dimensions of ~5.2 × 5.8 Å. These channels of ZSM-5 can allow the adsorption of benzene and other molecules of similar molecular size, but prevent molecules which possess a larger size/shape from being sorbed into the internal framework. On the other hand, the structures of X- and Y-type zeolites consist of an intercon-

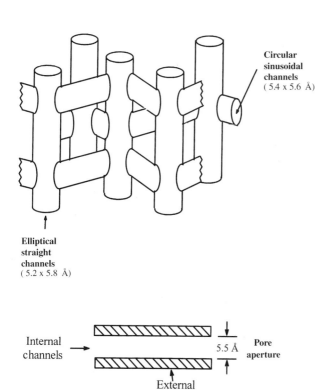

Figure 1   Schematic representation of the void space of the ZSM zeolite.

necting three-dimensional network of relatively large spherical cavities (supercages) with a diameter of about 13 Å (Fig. 2). Each supercage is connected tetrahedrally to four other supercages through 7.4-Å-diameter windows or pores. As an indication of their volume, each supercage can include five molecules of benzene, two molecules of naphthalene, or two molecules of pyrene.

Zeolites can be hydrophilic or hydrophobic due to the different Si/Al ratios within the zeolite framework. Organic molecules rely on H-bonding, electrostatic, and π–cation interactions for effective zeolite absorption, and these interactions will clearly be influenced by the number of cation sites present. As expected, the more Si present, the more hydrophobic the zeolite and, therefore, the greater the ability of these materials to interact with hydrophobic organic molecules or to exclude hydrophilic molecules, such as water. Zeolites X/Y have a Si/Al content at or close to 1 and are highly hydrophilic absorbants. Pentasil zeolite ZSM-5, which

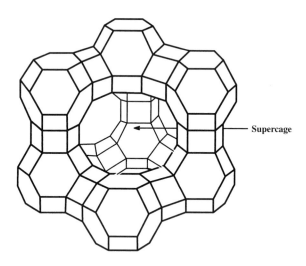

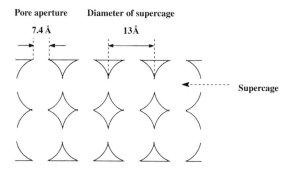

Figure 2 Schematic representation of the supercage of Y-zeolite.

can have a very high Si content, has significant hydrophobic character and has poor affinity for water.

Although zeolites have been known for their adsorption properties for over a century, it was not until 1952, when the first synthetic zeolite was prepared, that their utility in chemical transformations was explored. Since that time, zeolites have been used for a multitude of purposes, and to this day, they are essential catalysts in the petroleum industry, converting large and small hydrocarbons into high-octane compounds. As an outgrowth of this work, zeolites have found utility in industrial fine chemical synthesis for the construction of aromatics, heterocycles, aliphatic amines, and ethers, and the photochemistry within zeolites has already grown out of its infancy.

B. Nafion Membranes

Nafion is a novel and unique family of polymers which consist of a perfluorinated backbone and short pendant chains terminated by sulfonic groups. When swollen in water, the structure of Nafion is believed to resemble that of an inverse micelle (Fig. 3) [16–18]. The hydrated SO_3^- head groups are clustered together in a

$$[(CF_2\!-\!CF_2)_m\text{-}CF\!-\!CF_2]_n$$
$$(O\!-\!CF_2\!-\!CF)_k\!-\!CF_3$$
$$O$$
$$CF_2\!-\!CF_2\!-\!SO_3H$$

$m = 5 - 15$
$n = $ ca. 100
$k = 1, 2, 3$

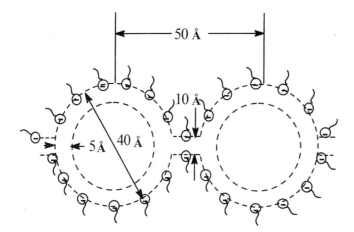

Figure 3 Schematic representation of the two-phase cluster-network model for Nafion membrane.

water-containing pocket of ~40 Å diameter, which are interconnected by short channels (~10 Å in diameter) within the perfluorocarbon matrix. The polymer backbone of Nafion provides exceptional chemical, thermal, and mechanical stability, whereas the sulfonic acid groups provide ion-exchange and swelling abilities. It has been established [19,20] that water-swollen Nafion can incorporate high concentrations of aromatic hydrocarbons and organic dyes, thus raising the possibility of obtaining a high local concentration of organic molecules and inorganic cations. These optically transparent membrane systems are readily amenable to spectroscopic and photochemical investigations. Because of these attractive properties, this polymer has been utilized as a medium for photophysical and photochemical studies in recent years [21–25].

C. Low-Density Polyethylene Films

Low-density polyethylene (LDPE) is a very complex family of materials that consist of ~50% crystalline regions (wherein chains are tightly and regularly packed) and ~50% amorphous or interfacial regions (wherein chains are in somewhat random conformations, Fig. 4) [26–29]. The glass transition of the amorphous part occurs near $-30°C$, and the melting transition is $>100°C$ [30]. A wide variety of organic molecules can be incorporated into LDPE by soaking its films in a swelling solvent containing the guest. It has been established that guest molecules are excluded from the crystalline portions of LDPE at temperatures below the melting transition. Their principal locations are the amorphous parts and the interfacial regions between crystalline and amorphous domains [27,29]. Due to its anisotropic nature, LDPE has been used as a reaction medium to control the

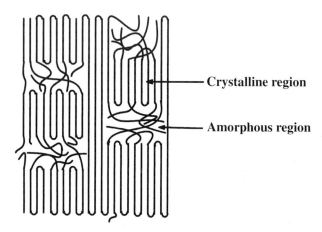

Figure 4   Schematic representation of the structure of a low-density polyethelene fillm.

reaction pathways of a variety of guest molecules [5,6,31,32]. The sites where guest molecules reside need not resemble those in the native polymer. The swelling process by which species are introduced "opens" the network of poly-methylene chains in the amorphous and interfacial regions; when the swelling liq-uid is removed, the nearby chains move into van der Waals contact with the reac-tive guest molecules that have been left behind. In this way, the free volume of the reaction sites can be made sufficiently large [33–35] to accept the guest molecules. These cavities are not well defined in size and shape, and the cavity walls are passive.

D. Vesicles

Vesicle (liposome) is one class of molecular self-assemblies formed by phospho-lipids or synthetic surfactants that form bilayers. It is spherical or ellipsoidal, single-compartment or multicompartment, closed bilayer structures, regardless of its chemical composition (Fig. 5) [36,37]. The aggregates can be defined by three regions: the internal water pool, the hydrocarbon bilayer, and the homogeneous water phases. Vesicles are much more ordered and have longer lifetimes com-pared to other self-assembled structures such as micelles. Fast lateral diffusion within the plane of the bilayer and very slow flip-flop rates for amphiphiles be-tween the two layers of the bilayer are characteristic of liposomes. The size of vesicles can be changed by employing different methods of preparation [38,39]. Small unilamellar vesicles have diameters between 20 nm and 50 nm and large unilamellar vesicles have diameters between 50 nm and 500 nm.

Generally, vesicles are prepared from double-tailed surfactants, and a simple single-tailed surfactant cannot form vesicles due to its relatively large hy-

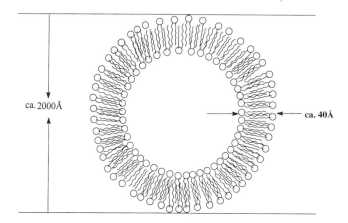

ca. 2000Å

ca. 40Å

Figure 5 Schematic representation of the structure of single-compartment vesicle.

drophilic head effect. However, it was recently established that stable vesicles could be simply produced by mixing of commercially available single-tailed cationic and anionic surfactants [40–45]. This phenomenon arises from the strong electrostatic interaction between the oppositely charged head groups of the components. As a result, the mean effective head-group area decreases considerably, whereas the hydrophobic volume of the tails remains the same. Thus, this dynamic ion pairing yields a pseudo-double-tailed zwitterionic surfactant, which has the preferred geometry of a vesicle-forming surfactant. These vesicles are quite stable when compared with "conventional" vesicles, and their size, charge, or permeability can be readily adjusted by varying the relative amounts and/or chain lengths of the two surfactants.

There has been a great deal of recent interests in vesicles or liposomes as potential reaction media for photochemical reactions due to their ability to selectively solubilize or entrap a wide variety of dissimilar reagents. Much of the interest in such studies originates from possible analogies between these processes and phenomena occurring in biological systems, particularly in membranes and related structures [46–51]. The coexistence of an amphiphilic bilayer and an aqueous compartment in vesicles composes a microheterogeneous system that mimics a specific situation occurring at the cell level. It is already clear that a variety of charged or extremely hydrophilic reagents do not penetrate the vesicle and can be either entrapped within an interior water pool or distributed in the bulk aqueous phase. Other reagents have a certain solubility in the bilayer phase and can penetrate the bilayer structure with reasonable facility and frequency. A number of applications have been proposed, ranging from drug delivery to selective charge separation. However, use of vesicle micro-heterogeneous systems for photochemical reactions is still in its infancy. The reason for this is mostly the high complexity of such systems. Extensive exploration of photochemistry in the vesicle microheterogeneous systems is expected to open a new area of photofunctional science as well as synthetic photochemistry.

III. PHOTOCHEMICAL REACTIONS IN MICROREACTORS

In this section we will discuss photochemical synthesis of macrocycle, photoinitiated oxidation of alkenes, photo-Fries rearrangement, photocycloaddition and chiral photochemical synthesis within a number of microreactors.

A. Zeolites and LDPEs as Hosts for the Preparation of Large-Ring Compounds

The construction of macrocyclic compounds continues to be an important topic of synthetic organic chemistry [52–60]. A bifunctional molecule may undergo either intramolecular or intermolecular reactions. A intramolecular reaction gives macrocyclic ring-closure products, whereas a intermolecular reaction results in dimers, oligomers, and polymers. Thus, the cyclization reaction suffers from the competition of the polymerization reaction. The rates of the latter are dependent on the concentration of the substrate, whereas those of the former are not, because the effective concentration for the reaction is kept constant by the function of a molecular chain linking the two functional groups. Hence, high substrate concentrations favor polymerization, whereas cyclization proceeds in good chemical yields only at low concentrations. Ziegler was the first to apply this principle to the synthesis of large-ring compounds by a high-dilution method [61,62]. In general, the syntheses of many-membered rings are performed at substrate concentrations as low as $1 \times 10^{-5}\,M$. This corresponds in a batch reaction to $1 \times 10^5\,L$ of solvent for each mole of substrate! In order to synthesize large-ring compounds at high substrate concentrations, the proper choice of solvents and using hydrophobic or lipophobic interactions to induce the flexible linking chains to self-coil, thus increasing the probability of the end-to-end encounters, has proved to be effective [60,63–67]. Jiang and coworkers [60] studied the intramolecular photochemical cycloaddition of 1,10-dicinnamoxydecane in aqueous organic binary mixed solvents and obtained 16-membered macrocyclic products with 90% yield. Nishimura et al. [63] photochemically synthesized 29-membered macrorings in high yield using $0.1\,M$ methanol or acetonitrile solutions. Desvergne et al. [64] obtained a 21-membered ring from poly(ethylene glycol) labeled at the chain terminal with 9-anthryls in benzene. Tung and coworkers [65–67] photochemically synthesized a series of large-ring compounds from naphthyl-labeled poly(ethylene glycol) oligomers in nonpolar solvents and from aryl-labeled polymethylene in aqueous organic binary mixed solvents.

Recently, Tung and coworkers reported a new approach to synthesize large-ring compounds in high yields under high substrate concentrations [68,69]. Their approach involves microporous solids or the "cavities" of LDPE films as templates and hosts for the cyclization reactions. The size of the micropore or the cavity has been chosen to permit only one substrate molecule to fit within each. Thus, intermolecular reaction is hindered and cyclization can occur without competition. Because the concentration of the micropores may be very large, they can synthesize large-ring compounds in high yield under conditions of high loadings. They find that the supercages of Y-type zeolite and the "cavities" of LDPE can be used as such microvessels for intramolecular photocycloadditions of diaryl compounds with long flexible chains.

1. Y-Type Zeolites as Microreactors

As mentioned in Section II, each supercage of Y-type zeolite can include five molecules of benzene, two molecules of naphthalene, or two molecules of pyrene [70–75]. Thus, one might expect that both aryl parts of a molecule with α,ω-diaryl groups separated by a flexible chain can be included in one supercage, and the intramolecular reactions between the two aryl groups should be enhanced. On the other hand, according to a Poisson distribution, as long as the loading level is less than 1 guest molecule per 10 supercages, there is less than a 5% probability of finding two substrate molecules in one supercage, and the intermolecular reactions should be inhibited. The unit-cell composition of NaY zeolite crystal is

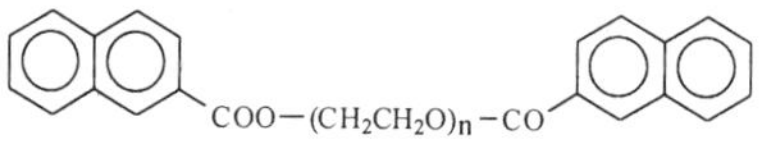

N-P$_n$-N   (n = 4, 5, 10, 12)

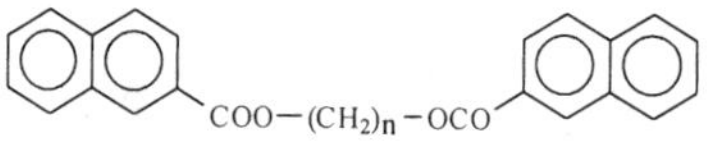

N-M$_n$-N   (n = 3, 5, 8, 10)

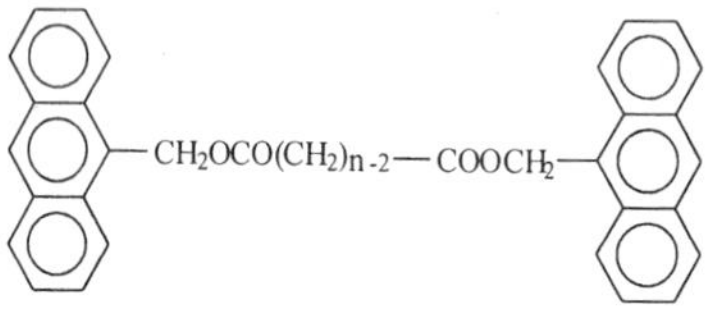

A-M$_n$-A   (n = 6, 10)

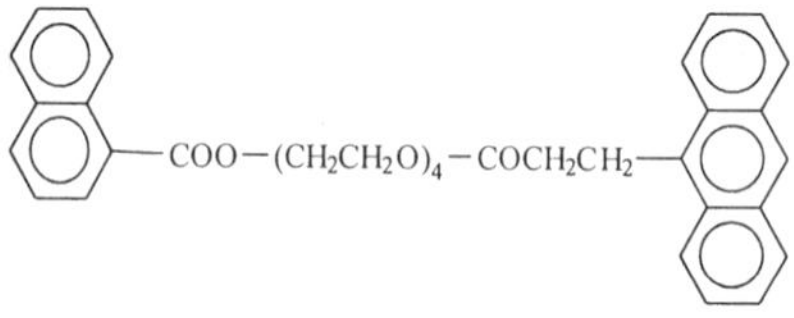

N-P$_4$-A

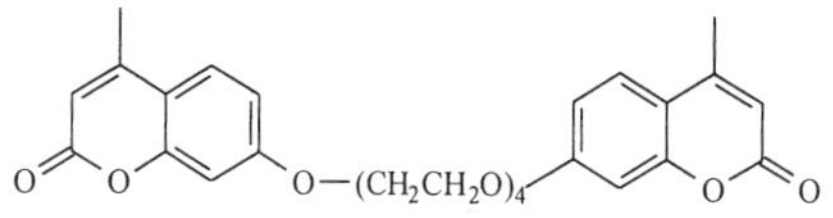

C-P$_4$-C

Figure 6   The diaryl compounds examined for intramolecular photocycloaddition within Y-zeolite and low-density polyethylene.

$Na_{53}[(AlO_2)_{53}(SiO_2)_{139}] \cdot 250H_2O$, and it is cubic with a dimension of ~25 Å [12,76]. Thus, a typical particle 1 μm in diameter contains thousands of interconnected supercages, and the concentration of the supercages is ~500 μmol/g of zeolite [76–79]. Assuming that a loading level of 1 substrate molecule per 10 supercages inhibits intermolecular reaction completely (*vide infra*) while cyclization still proceeds, only 2×10^4 g of zeolite are needed for each mole of substrate! This corresponds to 40 g of NaY zeolite for each gram of substrate whose molecular weight is 500. This amount of zeolite is more than 10^3 times smaller compared with that of the solvent in solution-phase reactions that allow cyclization to dominate.

The intramolecular photocycloadditions of diaryl compounds in Fig. 6 were investigated upon their inclusion in supercages of NaY zeolites [68]. As expected from the above analysis, at loading levels less than 1 molecule of a diaryl compound per 10 supercages of NaY zeolite, only intramolecular photocycloaddition is observed.

Photocycloaddition of N–P_n–N and N–M_n–N. Irradiation of alkyl 2-naphthoates is known to yield a "cubanelike" photocyclomer as the unique product (Fig. 7), although six isomeric cyclomers are formally possible [65–67,80–82]. This selectivity originates from two restrictions. First, the photocycloaddition occurs only between the substituted rings. Second, in the cyclomer the substituents are in a head-to-tail orientation. It has been previously demonstrated that the yield of this

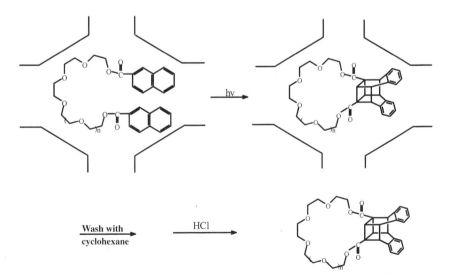

Figure 7 Scheme for intramolecular photocycloaddition of N–P_n–N within the supercage of Y-zeolite.

photocyclomer is dependent on the square of the light intensity, suggesting that the formation of the cubanelike photocyclomer for alkyl 2-naphthoates is a two-photon process [65]. Irradiation of $N-P_n-N$ or $N-M_n-N$ in organic solutions, such as acetonitrile, can lead either to intramolecular or intermolecular cycloadditions. At concentrations higher than 10^{-3} M, the main product is the intermolecular photocyclomer.

By contrast, irradiation of $N-P_n-N$ or $N-M_n-N$ adsorbed on NaY zeolite results in an intramolecular photocyclomer. Inclusion of the diaryl compounds within activated NaY zeolite was achieved using cyclohexane as the solvent. The zeolite with adsorbed substrate was isolated by filtration and dried under reduced pressure. The loading level was kept at ~50 μmol of substrate per gram of zeolite (~1 molecule per 10 supercages). This material was then transferred to a quartz vessel. After evacuation, the vessel was sealed and the sample was irradiated with light $\lambda > 280$ nm. Generally, after 4 hr of irradiation, the conversion was near 100%. Similar conversions were obtained in acetonitrile after about 1 hr irradiation.

Once formed, the photocyclomer is trapped inside the supercage of the zeolite, because the 7.4-Å window is too narrow to allow it to escape. A PC Model 6.00 program calculation showed that the cyclomer of $N-P_4-N$ in its lowest-energy conformation has a size of ~6.6 Å × 8.2 Å × 12.5 Å, suggesting that it is difficult for this molecule to escape through the zeolite window, but it still can be accommodated in the supercage. However, the aluminosilicate framework of Y-zeolite can be dissolved in strongly acidic media so that the photocyclomer is released into solution and can be subsequently isolated [83,84]. Thus, first, the unreacted starting material was extracted with cyclohexane, and then the product was isolated by dissolving the zeolite framework in concentrated HCl followed by extraction with ether. Generally, the products were obtained in high purity by this procedure. *Only the intramolecular ring-closure photocyclomers were obtained and no intermolecular products were detected by high-performance liquid chromatography (HPLC).* The yields of the intramolecular products were close to 100%, based on the consumption of the starting material. The largest rings synthesized with the aid of NaY zeolite contain 42 atoms from $N-P_{12}-N$ and the 17 atoms from $N-M_{10}-N$.

The effect of the loading level of the substrate in NaY zeolite on the photocycloaddition was examined. Between 50 and 5 μmol/g (~40 mg and ~4 mg $N-P_{10}-N$ per gram of zeolite), intramolecular photocyclomers were the only products detected. At higher loading levels, increasing amounts of intermolecular photocyclomers were also obtained. The exclusive formation of intramolecular photocyclomers at loading levels ≥ 50 μmol/g is attributed to the separation of the substrate molecules from each other by the supercages of the zeolite, whereas the terminal aryl groups of one substrate molecule can be compartmentalized in one supercage. In such a case, there is inadequate space in the supercage to include an aryl group of another molecule. Thus, these included substrate molecules will undergo intramolecular photocycloaddition exclusively. Additionally, because the loading level is less than 1 molecule per 10 supercages, it is unlikely that an occupied supercage will have a neighbor with another substrate molecule. Thus, intermolecular photocycloaddition is excluded.

To obtain further information on the compartment experienced by the two terminal groups of a substrate molecule in a supercage, their fluorescence spectra were examined (Fig. 8). In methanol solution at concentrations below 1×10^{-4} M, N–P_n–N show both the monomer (λ_{max} = 360 nm) and intramolecular excimer (λ_{max} = 400 nm) emission. The fluorescence of N–P_n–N differed significantly when included in NaY zeolite compared to those in methanol solution. At the loading levels below 50 μmol/g, the model compound N_p–COO $(CH_2CH_2O)_4COCH_3$ (N–P_4, N_p = 2-naphthyl) only gives monomer emission (λ_{max} = 370 nm), whereas N–P_n–N exclusively exhibit intramolecular excimer fluorescence centered at 400 nm. The excitation spectra of the fluorescence for N–P_n–N and N–P_4 included in NaY zeolite were significantly different, suggesting that there exist interactions between the chromophores in the ground state. In other words, the excimer originates from pair of naphthyl groups which exist prior to excitation. All of the observations indicate that the two terminal aryl groups of a molecule of N–P_n–N are included in one supercage, thus giving intramolecular excimer emission upon excitation.

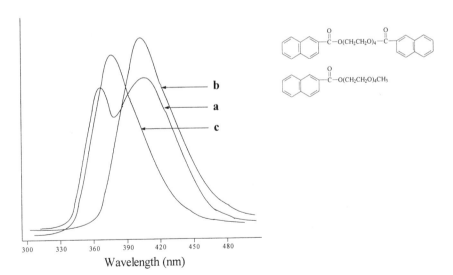

Figure 8 Fluorescence spectra of N–P_n–N in (a) methanol (1×10^{-4} M) and in (b) NaY zeolite (10 μmol/g). The fluorescence spectrum of N–P_4 in (c) NaY zeolite is also given (20 μmol/g). λ_{ex} = 280 nm.

Coadsorbed water in the zeolite significantly affects the behavior of the included N–P$_n$–N and N–M$_n$–N. The zeolite used was activated at 500°C for about 6 hr, and only the sample prepared from well-activated zeolite exhibits excimer fluorescence exclusively. The sample prepared from the zeolite which was not thoroughly activated or the one which after activation was allowed to adsorb water mainly shows monomer emission of the naphthyl chromophore (λ_{max} = 370 nm). It is likely that for the sample having coadsorbed water, the two terminal groups of one N–P$_n$–N or N–M$_n$–N molecule are not accommodated in one supercage. That hydration of NaY zeolite causes separation of dimers of the included pyrene has been established [71]. Accordingly, coadsorbed water also influences the photocycloaddition. For the sample which was not thoroughly activated, the conversion was low even under long period time irradiation.

Photocycloaddition of A–M$_n$–A. In general, photoirradiation of 9-substituted anthracene in organic solution results in [$4\pi + 4\pi$] cycloaddition of the aromatic rings at the 9,10-positions to yield mainly head-to-tail (h–t) rather than head-to-head (h–h) photocyclomers (Fig. 9) [85–91]. This regioselectivity was rationalized in terms of electrostatic and/or steric effects of the substituents [92,93].

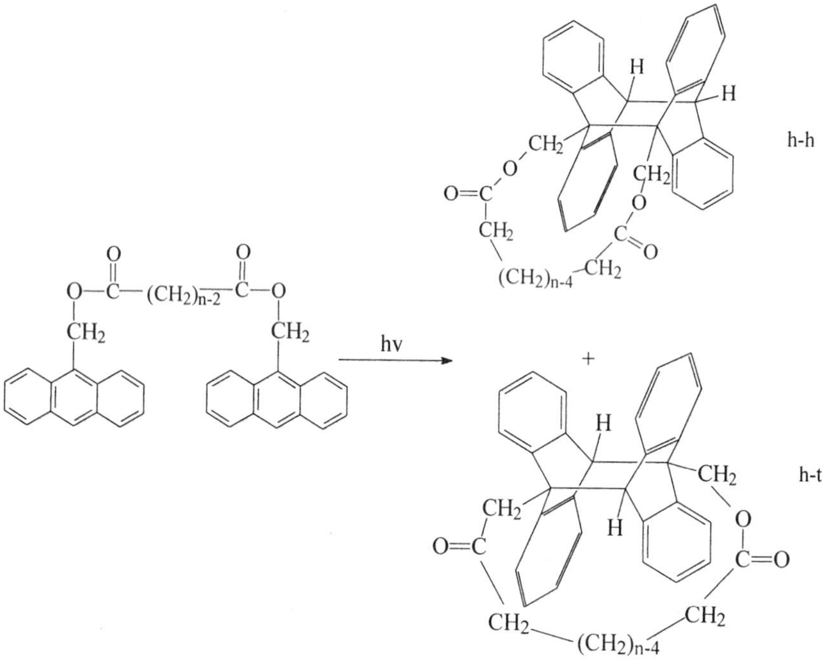

Figure 9 Scheme for intramolecular photocycloaddition of A–M$_n$–A.

For α,ω-di(9-anthryl)-alkanes, it has been established that the compounds may undergo intramolecular $[4\pi + 4\pi]$ cycloaddition, but intermolecular $[4\pi + 4\pi]$ cycloaddition generally dominates when the linking chains are long and the concentration is high [64,92,94–100].

As observed in the case of N–P_n–N and N–M_n–N, the fluorescence study of A–M_n–A included in NaY zeolite with the loading less than 50 μmol/g (~30 mg of A–M_{10}–A per gram of zeolite) suggested that the two terminal aryl groups of one A–M_n–A molecule are included within one supercage of the zeolite. Photoirradiation of the zeolite included A–M_n–A only results in intramolecular $[4\pi + 4\pi]$ cyclomers. As in the case of N–P_n–N, the molecular model calculation indicates that the photocyclomer of A–M_{10}–A in its lowest energy conformation possesses a size of ~8.1 Å × 9.0 Å × 11.3 Å. These photocyclomers are too large to escape form the zeolite supercage through the 7.4-Å window, but still can be accommodated in the supercage. The photocyclomers were isolated by dissolving the zeolite sample with HCl and extracting with ether. The material balance was above 90%. No intermolecular cyclomers were detected by HPLC. At a loading above 100 μmol/g, intermolecular photocyclomer formation was also observed. For example, at a loading of 100 μmol/g, the intermolecular photocyclomer was produced in ~5% of the total products. The assignment of the h–h and h–t photocyclomers relies on the chemical shifts of bridgehead protons in their ^{1}H-NMR (nuclear magnetic resonance) spectra. It has been established [93,100] that the chemical shifts of bridgehead protons for h–h cyclomers appear at lower regions compared with those for the corresponding h–t cyclomers. Photoirradiation of A–M_{10}–A yields both h–h and h–t intramolecular photocyclomers, with the latter as the major product (h–t/h–h = 4). The photocycloaddition of the dianthryl compounds with short linking chains differed from that for A–M_{10}–A. Photoirradiation of A–M_6–A included in NaY zeolite only gives h–h intramolecular photocyclomer.

Intramolecular Photocycloaddition of N–P_4–A. Although the photocycloaddition of anthracene [85–100] and that of naphthalene [65–67,80–82] have been extensively studied, until recently relatively little has been reported on the cross-photocycloaddition between an anthracene and a naphthalene moiety [101–110]. The main reason for this is the large difference in the quantum yields between the photocycloaddition of anthracene and the cross-photocycloaddition of anthracene and naphthalene. Thus, bichromophoric molecules with anthryl as one chromophore and naphthyl as the other generally undergo intermolecular anthryl–anthryl cycloaddition rather than intramolecular cross-cycloaddition when irradiated.

Chandross and Schiebel were the first to study the photochemical processes of 1-(9-anthryl)-3-(1-naphthyl)-propane (ANP) in which the two chromophores are separated by three saturated carbon atoms, thus offering the best geometrical relationship for the observation of excited-state interactions [101,102]. Irradiation of 2×10^{-5} M ANP in organic solvents gave the intramolecular $[4\pi + 4\pi]$ cyclomer, whereas irradiation at concentrations $>10^{-3}$ M resulted in intermolecular $[4\pi + 4\pi]$ cycloadditions involving two anthracene moieties. However, intramolecular cyclomers could not be isolated. Ferguson et al. studied the photochemistry of ANP in more detail [103–105]. They characterized the structure of the intramolecular photocyclomer by NMR spectroscopy and X-ray diffraction. Later, Bouas-Laurent and coworkers studied the intramolecular photocycloaddition of 9-(1-naphthylmethoxymethyl)anthracene and found that the photocycloaddition quantum yield is 10 times higher than that of ANP [106]. Albini and co-workers used the charge-transfer interactions of 9-cyanoanthracene with naphthalene and 2-cyanonaphthalene with anthracene to synthesize the cross-cyclomer [107,108]. Furthermore, the cross-cycloaddition between anthracene and naphthalene moieties in diaza[3,3](1,4)-naphthaleno-(9,10)anthracenes and cyclophanes has been reported [109,110]. In these molecules, the anthracene and naphthalene chromophores are fixed as a sandwich pair.

The difficulty in the synthesis of naphthalene–anthracene cross-photocyclomers can be overcome by irradiation of bichromophoric compounds included in Y-type zeolites. A fluorescence study demonstrated that selective excitation of the naphthyl group in N–P$_4$–A at $\lambda = 280$ nm mainly results in the emission of the anthryl chromophore, although weak fluorescence from the naphthyl is observed (Fig. 10). The excitation spectrum for the anthryl fluorescence corresponds to the ultraviolet (UV) absorption both of the anthryl and the naphthyl chromophores. All of these results suggest that significant energy transfer from naphthyl to anthryl occurs. Furthermore, in addition to the anthryl monomer emission, the fluorescence spectrum of N–P$_4$–A shows exciplex emission between anthryl and naphthyl chromophores. These observations suggest that the two terminal groups of an N–P$_4$–A molecule are included in one supercage of the zeolite, thus enhancing the energy transfer and exciplex formation. Irradiation of 1×10^{-3} M N–P$_4$–A in acetonitrile gives the intermolecular photocyclomer of two anthracene groups. On the other hand, below 1×10^{-3} M, irradiation results in no photochemical reaction, because the lifetime of the singlet excited state of the anthryl (or naphthyl) is not long enough to allow it to encounter the other terminal chromophore of the N–P$_4$–A molecule (intramolecular reaction) or a chromophore of another molecule (intermolecular reaction). However, irradiation of N–P$_4$–A adsorbed on NaY zeolite with loading levels below 50 μmol/g yields the intramolecular $[4\pi + 4\pi]$ photocyclomer exclusively (Fig. 11). A molecular model calculation shows that the photocyclomer of N–P$_4$–A in the conformation with the lowest energy has a size of ~8.3 Å $\times$ 8.8 Å $\times$ 11.3 Å. Thus, this photocyclomer cannot escape from the supercage through the 7.4-Å window. As in the case of N–P$_n$–N, N–M$_n$–N, and A–M$_n$–A, these photocyclomers were easily isolated by dissolving the zeolite framework with concentrated HCl followed by extraction

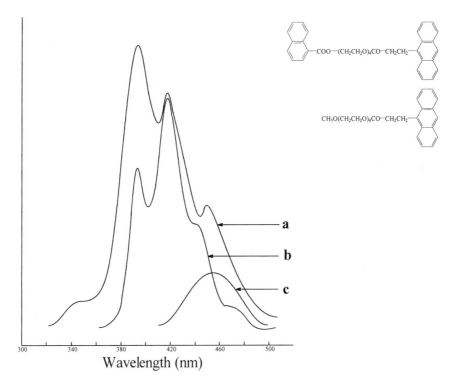

Figure 10 Fluorescence spectra of N–P$_4$–A (curve a, λ_{ex} = 280 nm) and A–P$_4$ (curve b, λ_{ex} = 365 nm) included in NaY zeolite. Exciplex emission spectrum (curve c) is derived by subtraction of the spectrum of A–P$_4$ form that of N–P$_4$–A normalized at 420 nm. The loading level is 10 μmol/g zeolite.

Figure 11 Scheme for intramolecular naphthalene–anthracene cross-photocycloaddition within the supercage of Y-zeolite.

with ether. The mass balance was >90%, suggesting that any unidentified products must be minor.

2. LDPE as a Microreactor

Tung and co-workers [69] reported that the bichromophoric substrates with long flexible chains shown in Fig. 6 can be trapped in the amorphous and interfacial regions of LDPE by soaking its films in a swelling solvent containing the substrates. In this medium, the rates of translational diffusion (which is a prerequisite for intermolecular reactions) of the substrate molecules are attenuated much more than those for rotational diffusion and conformational changes (which is a prerequisite for intramolecular reactions). Consequently, the reaction cavities allow relatively free rotational/conformational changes of the substrate molecules while retarding translational diffusion. Thus, if the loading levels are controlled so that each reaction cavity includes only one substrate molecule, intramolecular reactions can occur, whereas intermolecular reactions will be inhibited because diffusion between cavities is slow on the molecular reaction timescale.

Photocycloaddition of N–P_n–N. As mentioned earlier, the photochemical reaction of alkyl 2-naphthoates yields a "cubane"like photocyclomer as a unique product (Fig. 7), and irradiation of N–P_n–N in organic solvents can lead either to intramolecular or intermolecular cycloadditions. At concentrations higher than 1×10^{-3} *M*, the main product is intermolecular photocyclomers. By contrast, irradiation of N–P_n–N in LDPE films yields only intramolecular photocyclomers; no intermolecular products were detected. The isolated yields of intramolecular products were >90% on the basis of consumption of the starting materials. Due to the immiscibility of the polyether chains of the N–P_n–N with polyethylene, the maximal loading of N–P_n–N in LDPE film was ~1.2×10^{-3} mol/g-film. Because the LDPE films employed are 42% crystalline [111], the actual volume in which the guest molecules reside is only ~60% of the total. As a result, the true concentrations in the amorphous regions are at least 1.7×10^{-3} mol/g-film. These concentrations are much greater compared with those in solution-phase reactions that allow intramolecular reaction to dominate. Despite this, the exclusive formation of intramolecular photocyclomers suggests that each occupied site contains one molecule of N–P_n–N.

The fluorescence spectra of the N–P_n–N were examined in order to obtain further information on the cavities experienced by the N–P_n–N molecules. At a loading level of 1×10^{-3} mol/g-film, the N–P_n–N show monomeric and weak intramolecular excimer emissions. Because the rate of translational diffusion of the naphthoate groups in the LDPE films is slow on the timescale of the decay of the excited singlet states [32,35,112–114], the excimers probably form from chromophores that are at or near excimerlike conformations prior to excitation. On the basis of the relative weakness of the intramolecular excimer emissions, it is proposed that only a small fraction of N–P_n–N molecules have their naphthoyl groups in an excimerlike conformation in the ground state at any given time; the chromophores of most N–P_n–N molecules do not overlap and provide monomer fluorescence upon excitation. However, two naphthoyl groups may diffuse into proximity on larger timescales. When one of them is excited, intramolecular photocycloaddition occurs. In this way, long-term irradiation of N–P_n–N in a film can eventually lead to complete conversion.

The effect of stretching LDPE films doped with N–P_n–N on their photochemical reactivity was also examined. The dimerization efficiency in the stretched films is ~1.4 times greater than in the unstretched ones. This observation has ample precedent [115–118]. The product ratios from competitive intramolecular Paterno–Buchi photocyclization and intermolecular photoreduction of 10-undecenyl benzophenone-4-carboxylate have been employed to probe the free-volume changes that occur at guest sites when LDPE films are stretched [115]. A remarkable increase in the relative yield of the photocyclization product was observed when the films were stretched to ~500% of their original length. Aviv and co-workers studied the photodimerization of tetraphenylbutatriene in LDPE films and compared the reactivity in stretched and unstretched samples [117]. A threefold increase in quantum efficiency for photodimerization of the substrate upon stretching the LDPE films was reported. The interpretation advanced to explain these results is that film stretching decreases the average free volume of guest sites by partially aligning the polymethylene chains that constitute the cavity walls. Therefore, the two photoreactive groups of the substrates will reside, on average, closer to each other in stretched films than in unstretched ones; intramolecular photocycloaddition will be enhanced.

Photocycloaddition of C–P_4–C. The photochemistry of coumarin and its derivatives has been the subject of numerous investigations, mainly as a consequence of its importance in biological systems [119–127]. Irradiation of 7,7′-polymethylene-di(oxycoumarin)s at low concentrations yields only *syn* head-to-head and head-to-tail intramolecular cyclomers. Steric factors introduced in the corresponding molecule by methyl groups at the 4-positions of the coumarins led predominantly the *syn* head-to-tail cyclomer (Fig. 12). At higher concentrations, the amounts of intermolecular photoproducts increase as expected.

C–P_4–C can undergo intra- and intermolecular photoreactions. Irradiation of $<5 \times 10^{-4}$ M C–P_4–C in an organic solvent such as benzene results primarily in intramolecular *syn* head-to-tail cyclomer (Fig. 12). Upon irradiation of more concentrated solutions, a large amount of oligomeric material was formed. For example, at concentration of 1×10^{-3} M, ~20% of oligomers were present in the product mixture. As observed in the case of the N–P_n–N, irradiation of 1×10^{-2} mol/g-film C–P_4–C in LDPE resulted in conversion to intramolecular cyclomer. As in the case of irradiations in homogeneous solutions, only the *syn* head-to-tail

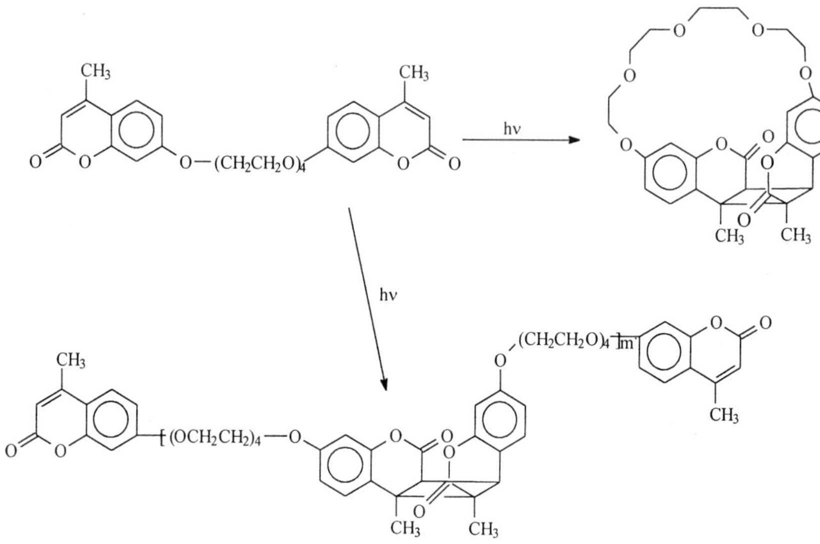

Figure 12 Scheme for photocycloaddition of C–P$_4$–C.

cyclomer was detected. Because the material balance was >90%, any unidentified products must be minor. Thus, at loading levels $<1 \times 10^{-2}$ mol/g-film, each reaction cavity in LDPE must contain no more than one C–P$_4$–C molecule: Intermolecular reaction is inhibited and intramolecular reaction is enhanced.

Further comments on the nature of the reaction cavities are appropriate [33,34]. The intrinsic size of guest sites in LDPE should preclude any of the bichromophoric molecules mentioned earlier from entering them [128]. However, their mode of inclusion, swelling of the amorphous parts of the polymer matrix, opens spaces inside the films large enough for the guests to enter. When the swelling solvent is removed selectively (by its more rapid diffusion to the air), the guest molecules are trapped. Coincidentally, the "walls" of the reaction cavities collapse around the guest molecules in orientations that maximize van der Waals contacts.

It has been suggested that in the LDPE inclusion sample, the groups tethered to the ends of chains experience a more mobile environment than in the bulk of the native LDPE due to the disording effect caused by the guest molecule; large guest molecules can plasticize the chains of polyethylene that constitute the walls of the reaction cages (making them "softer") [6,32,35]. In this way, slow conformational changes bring the two reactive groups at the chain ends into sufficient proximity for them to react within an excited-state lifetime. At any moment, the fraction of molecules in reactive conformations may be small, but eventually all can adopt the necessary shapes. The key is that the rates of these conformational changes are significantly faster than the rates at which two molecules meet in one reaction cavity. It is suspected that polyethylene films will be useful in synthesizing a large variety of macrocycles under concentration conditions that would lead to significant intermolecular (i.e., polymerization) reactions. Because each kind of polyethylene differs from the others in some morphological details, exact reproducibility from laboratory to laboratory should not be expected. However, the gross observation that bichromophoric molecules can be (reactively) isolated from each other at relatively high bulk concentrations should be common to all polyethylenes!

Thus, LDPE complements zeolites, which have been described earlier to serve a similar role in isolating molecules for intramolecular cyclizations [68]. By contrast, the reaction cavities of zeolites have fixed volumes and shapes and a very "hard" wall [12–15]. Each medium has its relative merits; now, it is possible to select the one that is more compatible with a particular reactant and cyclization reaction.

B. Photoinitiated Oxidation of Alkenes Included in Zeolites, Nafion Membranes, and Vesicles

Selective oxidation of small abundant hydrocarbons is the most important type of reaction in organic chemicals production. For example, essentially all building blocks for the manufacture of plastics and synthetic fibers are produced by oxidation of hydrocarbons [129]. Among these, oxidations by molecular oxygen play a particularly important role [130,131]. A key problem is the product specificity in the reactions of hydrocarbon with O_2; here, photoassisted processes hold special promise. Photoinitiated reactions of O_2 furnish access to products that in many cases cannot be obtained by a dark reaction of O_2. Moreover, photochemical reactions can be conducted at or around ambient temperature, thus minimizing the chance for loss of product specificity due to secondary thermal chemistry of the initial products.

There are three well-established types of hydrocarbon photo-oxidation reactions involving O_2 [132]. The most familiar one is dye-sensitized photo-oxidation involving transient singlet oxygen. For example, Diels–Alder reaction of conjugated dienes, "ene" reaction of olefins with allylic hydrogen, and dioxetane reaction of alkenes that do not feature an allylic hydrogen belong to this type [133–135]. Another type of dye-sensitized photo-oxidation involving molecular oxygen that is synthetically useful is sensitized electron-transfer oxygenation of hydrocarbons [135–138]. The third reaction type is hydrocarbon + O_2 photo-chemistry without the need of an added sensitizer. This involves photoexcitation of a hydrocarbon·O_2 charge-transfer state, and reactions so induced have been reported for small olefins as well as arylethenes in solution [139], at surfaces [140], in solid O_2 [141–143], and in zeolite cavitites [132,144–152].

Generally, product selectivity of photo-oxidation of alkenes involving O_2 is still not satisfied. In many cases, several types of photoinitiated oxidations occur simultaneously, and the products are not specific. Furthermore, in the photo-oxidation of hydrocarbon·O_2 charge-transfer state in solution and in solid O_2, the corresponding absorption bands lie in the UV region. As a consequence, photolysis results in extensive fragmentation and/or rearrangement of the hydroncarbon skeleton and poor selectivity. However, many works have demonstrated that one can use microreactors to control reaction pathways and direct the oxidations selectively toward the product(s) desired.

1. Electron-Transfer Versus Energy-Transfer Pathways in Photosensitized Oxidation

As described earlier, among the three well-established types of photoinitiated reactions of alkenes with O_2, two types are dye-sensitized photo-oxidation [135]: energy-transfer oxygenation and electron-transfer oxygenation (Fig. 13). The energy-transfer pathway involves energy transfer from the triplet sensitizer to the ground-state oxygen to generate singlet oxygen ($^1O_2$), then the generated singlet oxygen reacts with the substrate [133–135,153]. In electron-transfer photosensitized oxidation, electron-deficient sensitizers are generally used [135–138]. Electron transfer from alkene to the sensitizer in its excited states results in an alkene radical cation and a sensitizer radical anion, which subsequently reduces O_2 to give superoxide radical anion ($O_2^{-\cdot}$). The generated superoxide radical

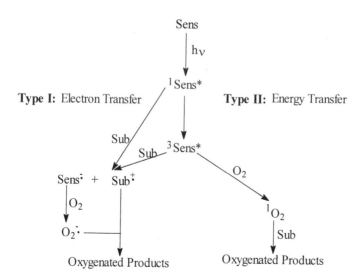

Figure 13   Scheme for Type I and Type II photosensitized oxidation of alkene.

anion reacts with the alkene radical cation to yield the oxidation products. Unfortunately, in many cases, the two types of photo-oxidation occurs simultaneously, and the selectivity of the oxidation reactions is poor.

To gain the selectivity in the photosensitized oxidation of alkenes, various efforts have been made in the past decades, and remarkable control of the reaction pathway has been obtained by use of organized and constrained media [154–159]. Whitten and Horsey [154] reported that the oxidation of protoporphyrin IX in micelles, monolayer films, and organized monolayer assemblies predominately results in the product of 1,2-cycloaddition of $^1O_2$ to the protoporphyrin vinyl groups, whereas the oxidation in solution mainly gives the product of 1,4-cycloaddition of $^1O_2$ to a diene unit in the protoporphyrin consisting of one endocyclic double bond and the vinyl group. Kagiya et al. [155] studied the photosensitized oxidation of methyl 9-*cis*,12-*cis*-octadecadienoate in aqueous emulsion and found that the product distribution is comparable to that in homogeneous solution. Maldotti and co-workers investigated the oxidation of hydrocarbons photocatalyzed by iron porphyrins which are confined inside cross-linked polystyrene [156], bound on semiconducting transition metal oxides such as TiO_2 [157], or included within Nafion clusters [158], and revealed that the heterogeneous media or organized systems affect the chemoselectivity of the oxidation process and increase the photochemical efficiency. Recently, many works have been reported on the photosensitized oxidation of alkenes included within zeolites. Pettit and Fox [159] used $Ru(bpy)_3^{2+}$ exchanged into zeolite Y as the sensitizer to oxidize tetramethylethylene and 1-methylcyclohexene. The photolysis was performed in $Ru(bpy)_3^{2+}$-exchanged zeolite slurry in methanol containing the alkene. They found that the photogenerated singlet oxygen within zeolite cages freely diffuses to the solution, where it reacts with the alkene with normal selectivity.

More recently, Tung [160–164] and Ramamurthy and coworkers [165–171] reported that they are able to direct the photosensitized oxidation of alkenes selectively toward either the singlet-oxygen-mediated or the superoxide-radical-anion-mediated products by controlling the status and location of the substrate and sensitizer molecules in the reaction media. Close inspection of Fig. 13 indicates that in the electron-transfer pathway, the sensitizer and the substrate molecules have to be in close contact for efficient electron transfer, whereas in energy-transfer photosensitized oxidation, such a close contact is not a prerequisite. Thus, isolation of the sensitizer from the substrate will prevent them from undergoing electron transfer, whereas singlet oxygen still can be generated by energy transfer from the triplet-state sensitizer to the ground-state oxygen. The species $^1O_2$ is small and unchanged and has a relatively long lifetime and properties which allow it to diffuse a long distance to meet the substrate molecules. In this case, only the products derived from the energy-transfer pathway can be produced and no superoxide-radical-anion-mediated products would be generated. On the other hand, if the sensitizer and substrate molecules are incorporated in the sample microreactor, the high local concentration of the substrate and the close contact between the sensitizer and the substrate molecules will enhance the electron transfer from the substrate to the singlet excited state of the sensitizer. In this case, the intersystem crossing process from the singlet excited state to the triplet state of the sensitizer will not be able to compete with the electron-transfer quenching. Thus, no triplet-state sensitizer will be generated. This, in turn, results in no singlet-oxygen formation. In this case, only the products derived from electron-transfer pathway will be produced. Zeolite, Nafion, membranes, and vesicles have been used as the microreactors for such photosensitized oxidation.

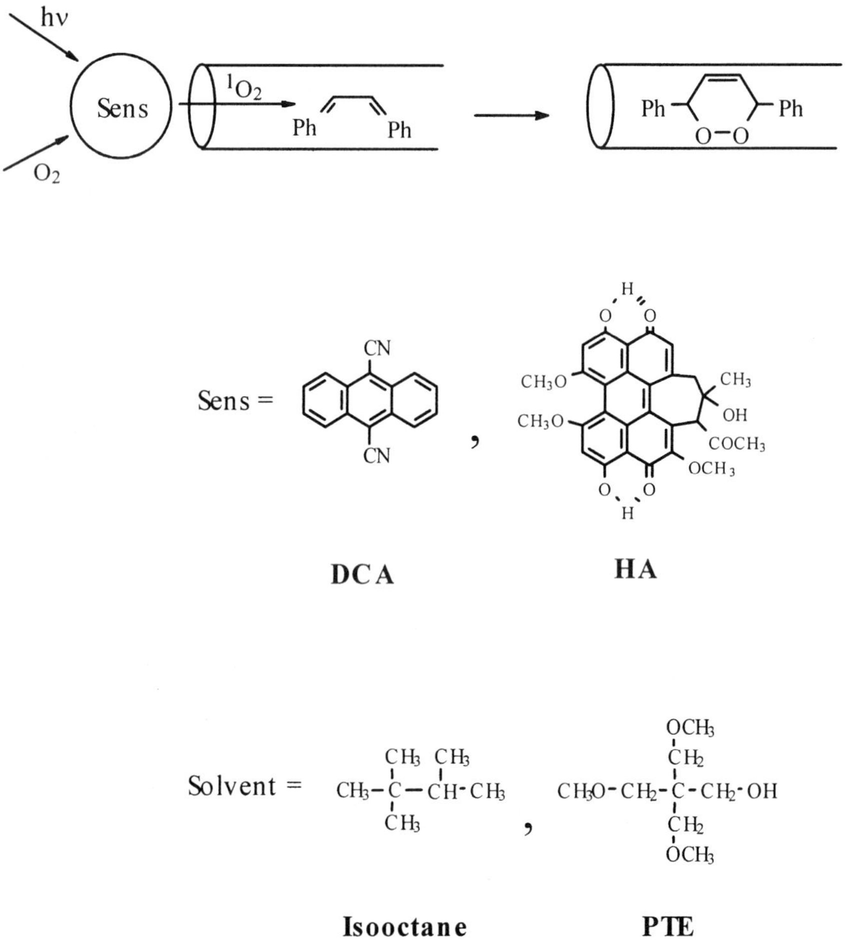

Figure 14 Scheme for DCA- and HA-photosensitized oxidation of alkenes under sensitizer-excluded condition.

Zeolites as the Microreactors. Tung and his coworkers used the channels of ZSM-5 zeolite as the microreactor to conduct the photosensitized oxidation of alkenes [160]. *Trans, trans*-1,4-diphenyl-1,3-butadiene (DPB) and *trans*-stilbene (TS) were selected as representative of alkenes, 9,10-dicyanoanthracene (DCA) and hypocrellin A (HA) were the sensitizers, and iso-octane and pentaerythritol trimethyl ether (PTE) were the solvents. The zeolites used were sodium cation-exchanged forms with different Si/Al ratios (Si/Al = 55 and 25). They trapped the alkenes in the channels of ZSM-5 zeolite, and isolated the photosensitizers in the surrounding solution (Fig. 14). The choice of the solvents and sensitizers was motivated by the desire that they were prevented from being sorbed into the ZSM-5 channel due to their size and shape characteristics. Thus, the internal framework of ZSM-5 is "dry," and the substrate is protected from being extracted to the solution during photolysis. The isolation of the substrate within the zeolite from the sensitizer in the solution outside leads exclusively to the formation of singlet-oxygen products.

Irradiation of oxygen-saturated DPB solution in PTE containing DCA or HA with visible light gave benzaldehyde **1**, cinnamaladhyde **2**, epoxide **3**, ozonide **4**, and endoperoxide **6** (Fig. 15). In addition, a small amount of 1-phenylnaphtha-

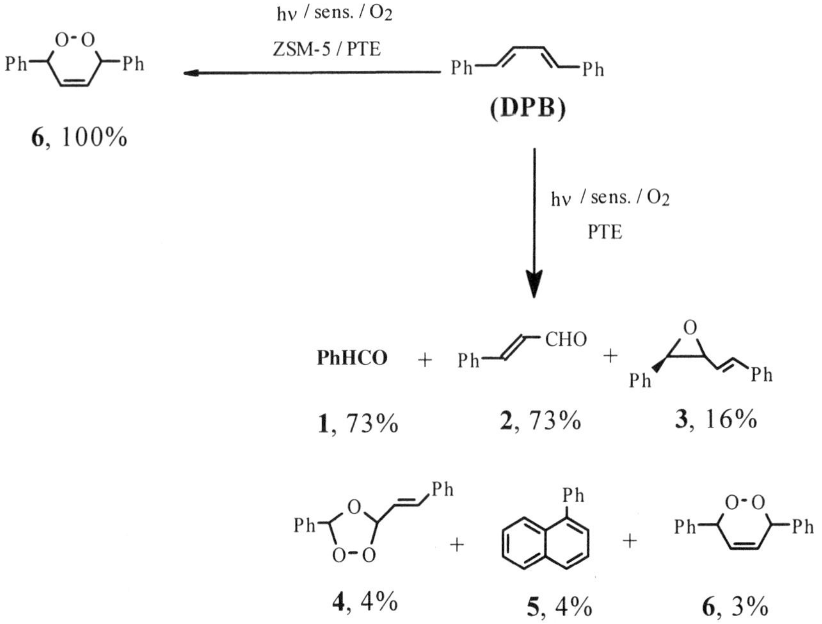

Figure 15 Photosensitized oxidation of *trans,trans*-1,4-diphenyl-1,3-butadiene in solution and within ZSM-5 zeolite. The relative yields of the products are also shown.

lene **5** was detected. The product distribution is slightly dependent on the sensitizers and that for DCA as the sensitizer is shown in Fig. 15. In all cases, the main
products are **1** and **2.** It has been established that DCA [136–138] and HA [172]
can act both as singlet-oxygen sensitizer and electron-transfer sensitizer.
Obviously, **6** is a product of 1,4-cycloaddition of singlet oxygen ($^1O_2$) to DPB.
The other products are presumably derived via the electron-transfer pathway
[136–138]. As Erikson and Foote suggested for the DCA-sensitized oxidation of
stilbene [138], **1** and **2** are most likely derived from an intermediate dioxetane, a
cycloaddition product of DPB$^+$ and superoxide anion (O$^-_2$), which would decompose under reaction conditions. Structure **3** could be formed by the Bartlett
reaction [138]. Structure **4** is possibly a secondary photo-oxygenation product
from **3.** Formation of **5** is probably due to cationic rearrangement of DPB.

In contrast, the DCA- and HA-sensitized photo-oxidation of DPB adsorbed
on the internal surface of ZSM-5 zeolite gave **6** as the unique product (Fig. 15).
The yield of this product was 100% based on the consumption of the starting material. The Al content in the zeolites (Si/Al = 55 or 25) showed no effect on the
oxidation product distribution. Evidently, the isolation of DPB within the zeolite
channels from the sensitizer in the solution outside prevents them from undergoing electron transfer. Thus, no photo-oxidation products derived from the electron-transfer pathway were detected. This proposal was confirmed by the fact that
although the fluorescence of DCA and HA in PTE solution was apparently
quenched by DPB according to an electron-transfer mechanism, the fluorescence
quenching was not observed at all when DPB was included into ZSM-5 zeolite
and DCA or HA was dissolved in PTE. On the other hand, $^1O_2$ can be formed in
the solution by energy transfer from the triplet-state sensitizer to the ground-state
oxygen and it is able to diffuse freely from the surrounding solution to the internal framework of ZSM-5 zeolite. 1,4-Cycloaddition of $^1O_2$ to DPB in the internal
surface of ZSM-5 results in endoperoxide **6.**

As observed in the case of DPB, the photo-oxygenation of TS sensitized by
DCA or HA differed significantly when included in ZSM-5 zeolite compared to
that in homogeneous solution. The oxidation products in homogeneous solution
are remarkably dependent on the sensitizers and conditions used. Erikson and
Foote reported previously [138] that in acetonitrile using DCA as the sensitizer,
the oxidation products were benzaldehyde **1,** *cis*-stilbene **7, trans**-2,3-dipheny-
loxirane **8,** and benzil **9** (Fig. 16). All these products were produced via the electron-transfer pathway. However, Matsumoto et al. reported [173–175] that at
room temperature in CCl$_4$ with tetraphenylporphyrin as the sensitizer, the photo-
oxidation of TS gave diendoperoxide **10** (Fig. 16) in a yield of 16% based on the
reacted starting material. The other major product, benzaldehyde **1,** was produced
in 80% yield. This reaction is proposed to proceed via an energy-transfer pathway.
Tung et al. [160] reported that in PTE using DCA as the sensitizer, the photo-oxidation products are the same as those obtained in the DCA-sensitized reaction in

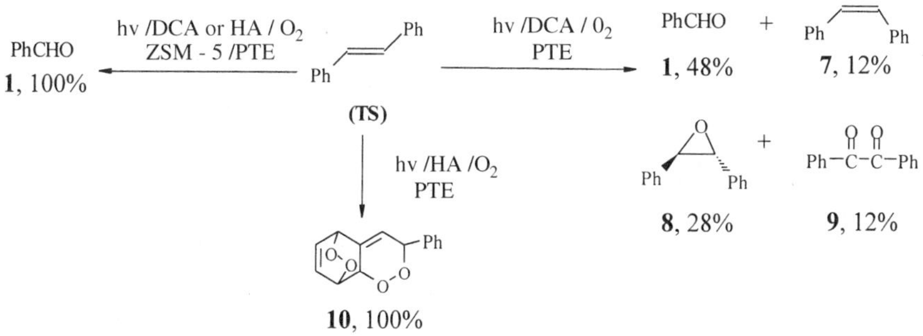

Figure 16 Photosensitized oxidation of *trans*-stilbene in solution and within ZSM-5 zeolite.

acetonitrile, but the product distribution is slightly changed (Fig. 16). On the other hand, in the same solvent with HA as the sensitizer, only product **10** is obtained (Fig. 16). When TS is included within ZSM-5 zeolite (Si/Al = 55 or 25) and the sensitizer is solublized in the surrounding solvent PTE, the photo-oxidation of TS sensitized by DCA or HA yields benzaldehyde **1** as the unique product (Fig. 16). The mass balances of this reaction for the two sensitizers are all close to 100%. As in the case of DPB, the isolation of TS within the zeolite from the sensitizer in the solution outside prevents electron transfer between the substrate and the sensitizer to occur. Thus, no superoxide anion is expected to be produced. Obviously, benzaldehyde **1** is derived via an energy-transfer pathway. The singlet oxygen generated via energy transfer in solution diffuses into the internal framework of ZSM-5 and reacts with TS to form 3,4-diphenyl-1,2-dioxetane, which would decompose to yield benzaldehyde under reaction condition. It is noted that in the HA-sensitized photo-oxidation within ZSM zeolite diendoperoxide **10** was not produced. This observation is contrary to the result of the reaction for TS with singlet oxygen in solution. The absence of **10** in the oxidation products within the zeolite is probably due to the constrained space within ZSM-5 channels, which is not large enough to accommodate the molecule of **10**.

Ramamurthy and Li [166] investigated the photo-oxidation of TS and *trans*-4,4′-dimethoxystilbene (TDHS) sensitized by thionin, methylene blue, or methylene green in X- and Y-zeolites (Fig. 17). This study complements the work described above. Both the sensitizer and substrate were included in X- or Y-zeolite. The zeolite having adsorbed the alkene and the sensitizer was photolyzed in hexane slurry under oxygen atmosphere. The isolated products were benzaldehyde **11**. They found that the oxidation proceeded via an electron-transfer pathway, which is contrasted with that in homogeneous solution (energy-transfer pathway, formation of endoperoxide **12**; Fig. 17). Indeed, the high local concentrations of the alkene in the supercages of the zeolites lead to efficient

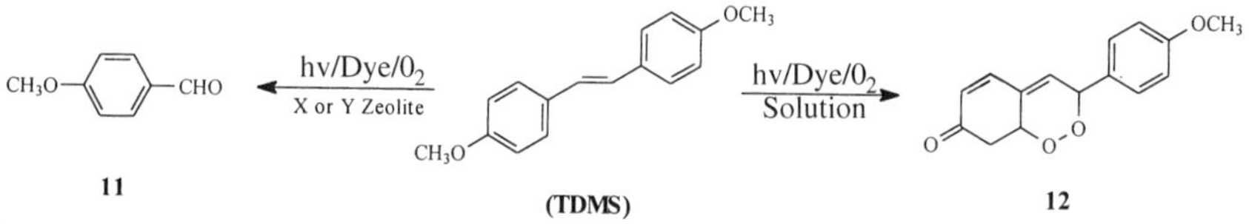

Figure 17 Photosensitized oxidation of *trans*-4,4'-dimethoxystilbene in methylene chloride/methanol solution and within supercages of X- and Y-zeolites.

quenching of the singlet excited state of the sensitizer via the electron-transfer mechanism. As a result, the intersystem crossing from the singlet excited state to the triplet state of the sensitizer cannot compete with the electron-transfer process. Therefore, the subsequent triplet energy transfer to O_2 cannot occur.

Nafion Membranes as Microreactors. Tung and Guan [161] extended the study of photosensitized oxidation of alkenes included in the zeolite to Nafion membranes to establish the scope of the microreactor-controlled selectivity approach. It is well known that water-swollen Nafion can incorporate high concentrations of aromatic hydrocarbons and organic dyes [19,20]. Considering the hydrophobicity and insolubility of the alkenes (DPB and TS) and the sensitizer (DCA) in water, it was proposed that in the water-swollen Nafion, these molecules are primarily solubilized in the hydrophobic perfluorocarbon backbone region. The photosensitized oxidation was performed in two modes. The first one involves irradiation of DCA in dichloromethane solution ($5 \times 10^{-5} M$) in which the water-swollen Nafion sample incorporating the substrate is immersed (substrate in Nafion–DCA in CH_2Cl_2). Because dichloromethane cannot swell Nafion and is insoluble in water, the substrate and water within Nafion cannot be extracted into the solution, and CH_2Cl_2 and DCA in the solution cannot diffuse into the Nafion. Thus, the substrate and the sensitizer are isolated from each other during irradiation. The second mode involves irradiation of the water-swollen Nafion sample which has incorporated both the substrate and the sensitizer (substrate–DCA in Nafion).

As in the case of the ZSM-5 zeolite sample [160], in substrate in Nafion–DCA in CH_2Cl_2, due to the isolation of the sensitizer in solution from the substrate in Nafion, only the products derived from the energy-transfer pathway were detected. The sensitized photo-oxidation of DPB gave **6** as the unique product (Fig. 18), whereas that of TS yielded **10** and **13** (Fig. 19). The oxidation products of TS with $^1O_2$ in this mode are the same as in solution sensitized by HA or tetraphenylporphyrin, but different from that in ZSM-5 zeolite, where benzaldehyde **1** is the unique product. This observation suggests that the constrained space in the water-containing pocket of Nafion membranes can accommodate the molecules of **10** and **13**. In contrast, irradiation of the water-swollen

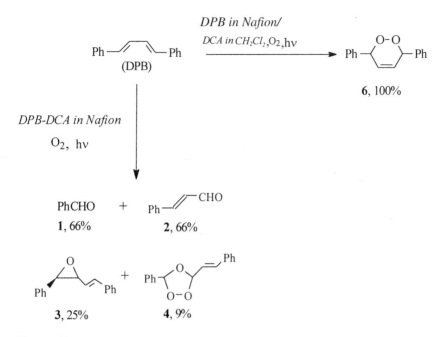

Figure 18 Photosensitized oxidation of *trans,trans*-1,4-diphenyl-1,3-butadiene within Nafion membranes in two modes.

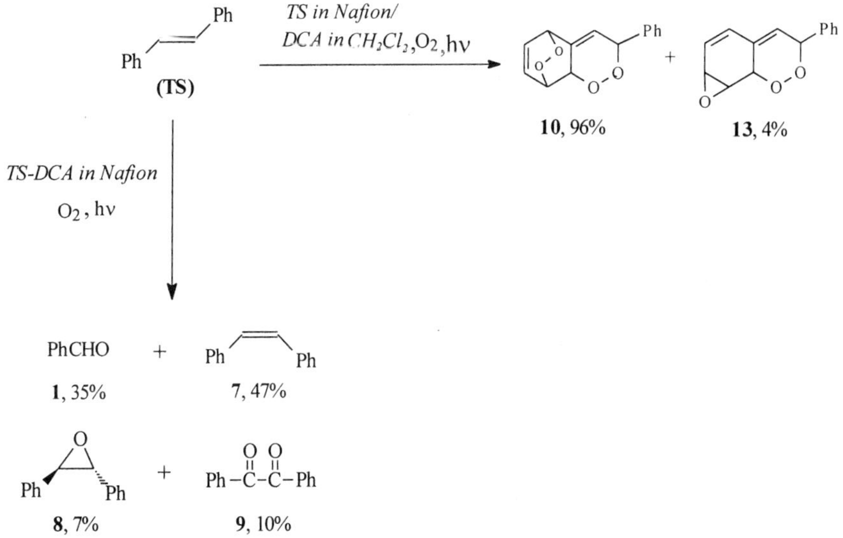

Figure 19 Photosensitized oxidation of *trans*-stilbene within Nafion membranes in two modes.

Nafion sample incorporating both DCA and DPB (DPB–DCA in Nafion mode) only resulted in the electron-transfer-mediated products, **1–4.** No singlet-oxygen product, **6,** was detected (Fig. 18). Material balance was near 100%. Similarly, the photosensitized oxidation of TS in TS–DCA in Nafion mode only produced the electron-transfer-mediated products **1** and **7–9** (Fig. 19).

The loading levels used in the present study were 1 DCA molecule per ~200 water clusters, and 4.2 alkene molecules per water cluster of Nafion. Thus, each DCA molecule is surrounded by a number of the alkene molecules. The high "local concentration" of the alkene and the close contact between DCA and the alkene molecules in the confined cluster of Nafion lead to efficient quenching of the singlet excited state of DCA by the alkene via an electron-transfer process, generating a DCA radical anion and a DPB radical cation. As a result, the intersystem crossing from the singlet excited state to the triplet state of DCA cannot compete with the quenching process by the alkene. Thus, the subsequent triplet energy transfer to O_2 cannot occur and no singlet-oxygen-mediated product was produced. On the other hand, the concentration of oxygen inside Nafion is more than 10 times higher than in organic solvents [176]. The above-generated DCA radical anion would efficiently undergo electron transfer to oxygen to produce a superoxide radical anion, which subsequently reacts with the alkene radical cation to yield electron-transfer-mediated products.

Vesicles as Microreactors. The approach to control the reaction pathways in photosensitized oxidation of alkenes by isolation or close contact of the substrate and sensitizer molecules has also been demonstrated to be effect in vesicle medium [162–164]. Tung and coworkers have shown that singlet oxygen generated in the bilayer or the inner water pool of one vesicle is able to react with the target molecules located in other vesicles [162]. The vesicles were prepared by sonicating the equimolar mixture of a cationic surfactant (octyl-trimethylammonium bromide, 8.2×10^{-2} M) and an anionic surfactant (sodium laurate, 8.2×10^{-2} M) in buffered solution. The substrates were *trans*-1,2-dimethoxystilbene (DMOS) and the amine 2,2,6,6-tetramethylpiperidine (TMP). The sensitizer was either a hydrophobic dye, tetraphenylporphyrin, or a cationic dye, methylene blue. The substrate molecules were solubilized in the bilayer membranes of one set of vesicles, and the sensitizers were incorporated in the bilayers or the aqueous inner compartments of another set of vesicles. The irradiation samples were prepared by mixing the above two sets of vesicle dispersions. Photoirradiation of the oxygen-saturated samples resulted in the oxidation of the substrates, as evidenced by the isolation of the end products in the olefin oxidation and by the detection of the electron spin resonance (ESR) spectrum of the nitroxide radical in the amine oxidation. The quantum yields for the product formation were enhanced two to three times in D_2O dispersion compared with those in H_2O medium. All of these observations suggest that singlet oxygen generated in the bilayer or the inner water pool of one vesicle is able to diffuse out and enter into the bilayer of another vesicle through the aqueous dispersion and react with the target molecules. The measurements of the quantum yields revealed a substantial fraction of the singlet oxygen diffusing from its generated locus to the reaction sites: 8% in H_2O and 15% in D_2O dispersions in the case of singlet oxygen generated in the inner aqueous compartment of the vesicle, and 20% in H_2O and 80% in D_2O dispersions for the singlet oxygen generated in the bilayer of the vesicle. In this context, such easily prepared and inexpensive vesicles were used as the reaction medium to conduct the oxidation of α-pinene (PE) and DPB photosensitized by DCA.

The photosensitized oxidation was performed in two modes (Fig. 20) [163,164]. In the first mode, the sensitizer DCA was incorporated in the bilayer membranes of one set of vesicles, and the substrate was solubilized in another set of vesicles. Equal volumes of the two sets of vesicle dispersions were then mixed to prepare the samples for irradiation. Although sonication was carried out during preparation of the component solutions, the final mixture was not sonicated. In this way, intermixing of solubilizates was prevented. A control experiment was carried out: The mixed solution prepared from DCA-containing vesicles and substrate-containing vesicles was stored in the dark at room temperature for 1 day, and then irradiated. The products and the efficiency of the product formation for the photosensitized oxidation were found to be identical within experimental error limit to those of the sample that was immediately irradiated after the preparation. This observation suggests that the intervesicular exchange both of the substrate and the sensitizer indeed did not occur, and the photosensitized oxidation process in this mode involved the generation of $^1O_2$ in one vesicle and reaction with alkene molecules in other vesicles. In the second mode, both the sensitizer and the substrate were incorporated in the bilayer of the same set of vesicles. Gen-

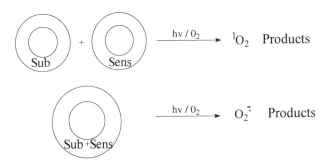

Figure 20 Schematic representation of the modes for photosensitized oxidation of alkenes in vesicles.

erally, the concentration of the olefins was ~1.0×10^{-3} M, corresponding to thousands of substrate molecules in each vesicle, whereas the concentration of the sensitizer was generally ~1.0×10^{-4} M.

The photo-oxidation of PE sensitized by DCA in homogeneous solution followed by reduction of the reaction mixture with sodium sulfite solution gave the ene product pinocarveol **14** and the non-ene products myrtenal **15**, epoxide **16** and aldehyde **17**, as shown in Fig. 21. The ene product and the non-ene products have been proposed to be derived from the energy-transfer and electron-transfer pathways, respectively [177–181]. The product distributions in acetonitrile is given in Fig. 21.

The product distribution of the photosensitized oxidation of PE in vesicles is dramatically altered compared with those in the above homogeneous solutions and is remarkably dependent on the experimental mode. The photosensitized oxidation in the first mode followed by mixing the sample with a sodium sulfite aqueous solution to reduce the reaction mixture exclusively produced the ene product **14** (Fig. 21). No trace of the non-ene products **15–17** was detected. Evidently, the isolation of PE in the bilayer membranes of one set of vesicles from DCA in another set of vesicles prevents them from undergoing electron transfer. On the other hand, singlet oxygen can be generated in the DCA-containing vesicles by the energy transfer. The vesicles used in the study have an aggregation number (number of surfactant molecules per vesicle) in the region of 10^5–10^6, and the average aggregation number is ~7.2×10^5, as estimated from the vesicle size and the volume of the surfactant molecule [162]. Thus, at a surfactant concentration of 8.2×10^{-2} M, the vesicle population is equivalent to a molarity of ~1.0×10^{-7} M, which, in turn, gives the intervesicular distance on average to be ~134 nm [182]. On the other hand, the average diffusion length of the $^1O_2$ molecule in aque-

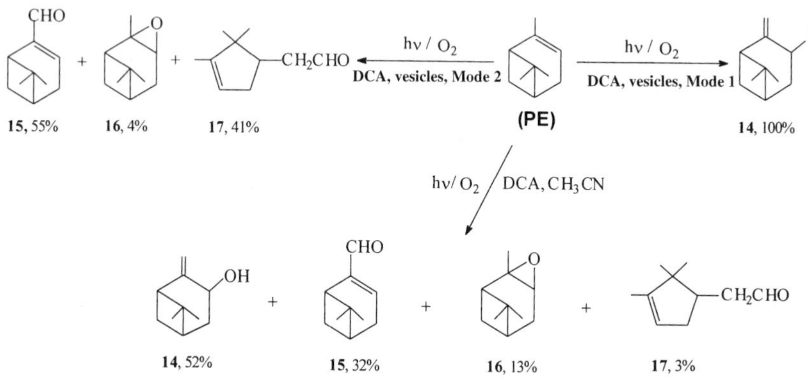

Figure 21   Photosensitized oxidation of α-pinene in CH_3CN and in vesicle dispersions.

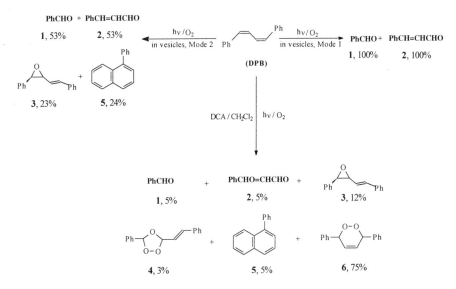

Figure 22 Photosensitized oxidation of *trans,trans*-1,4-diphenyl-1,3-butadiene in CH₂Cl₂ and in vesicle dispersions.

ous solution is estimated to be ~780 nm [46,183,184]. This diffusion length is much longer than the intervesicular distance estimated earlier. Thus, the ¹O₂ generated in DCA-containing vesicles is indeed capable of diffusing into the PE-containing vesicles to react with the alkene molecules to produce the ene product **14.**

In contrast, the photo-oxidation in the second mode followed by mixing the sample with a sodium sulfite aqueous solution to reduce the reaction mixture exclusively gave the non-ene products **15–17** (Fig. 21). No product derived from the energy-transfer pathway was detected. The loading levels used in the study were hundreds of DCA and thousands of PE molecules per vesicle. Thus, each DCA molecule is surrounded by a number of PE molecules. The high "local concentration" of PE in the confined bilayers of vesicles leads to efficient quenching of the singlet excited state of DCA by PE via an electron-transfer process and to inefficient intersystem crossing from the singlet excited state to the triplet state of DCA. Therefore, only the non-ene products **15–17** were produced.

The photosensitized oxidation of DPB in vesicles in the first mode gave **1** and **2** as the unique products (Fig. 22). We believe that these products are derived from the singlet-oxygen pathway. In contrast, the photosensitized oxidation of DPB in vesicles in the second mode only produced the electron-transfer-mediated products **1, 2, 3,** and **5** (Fig. 22). No singlet-oxygen products were detected. These observations demonstrate, once again, that one can control the selectivity in photosensitized oxidation of alkenes by incorporation of the sensitizer and substrate either in different or the same sets of vesicles.

2. Reaction of Alkenes With Singlet Oxygen in Microreactors

Generally, singlet oxygen may undergo three types of reaction with olefins (Fig. 23) [133–135]: (1) [4 + 2] cycloaddition with conjugated dienes to result in endoperoxides; (2) "ene" reaction with olefins bearing allylic hydrogen atoms to form allylic hydroperoxides which can produce ketones after degradation or alcohol after reduction; (3) [2 + 2] cycloaddition with electron-rich olefins to give 1,2-dioxetanes preferably. In many cases, the three types of reaction proceed simultaneously, and the product selectivity is low. However, the above reaction pathways can be controlled by inclusion of the substrates in a microreactor. As described in Section III.B.1, while the reaction of TS with $^1O_2$ in homogeneous solution or in Nafion membranes gives the diendoperoxide **10**, a [4 + 2] product (Figs. 16 and 19) [161], it yields the [2 + 2] product benzaldehyde **1** when TS is included in ZSM-5 zeolite (Fig. 16) [160]. The reason for the exclusive formation of benzaldehyde in the later case is that the channel of ZSM zeolite is too narrow to accommodate the diendoperoxide **10** molecule. Evidently, the size/shape of the microreactor plays an important role in the determination of the reaction pathway.

Another example for microreactor-controlled selectivity in the reaction of olefin with $^1O_2$ can be found in the photosensitized oxidation of DPB in ZSM zeolite [160], Nafion membranes [161], and vesicles [163]. In homogeneous solution, ZSM zeolite, and Nafion membranes, the oxidation of DPB with $^1O_2$ yielded the [2 + 4] reaction product, endoperoxide **6**, as the unique product (Figs. 15 and 18). In sharp contrast, in vesicle medium, the oxidation produced the aldehydes **1** and **2** ([2 + 2] reaction) in quantitative yield as described in Section III.B.1 (Fig.

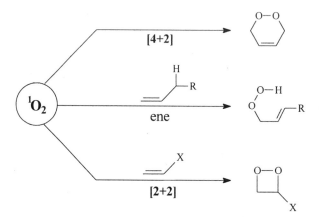

Figure 23 Scheme for reactions of singlet oxygen with alkenes.

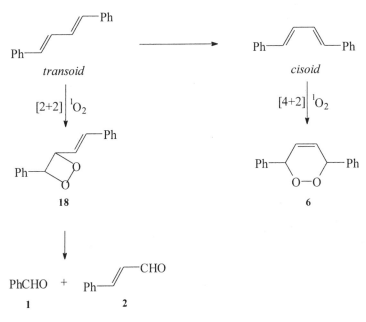

Figure 24 Conformational isomers of *trans,trans*-1,4-diphenyl-1,3-butadiene and their reactions with singlet oxygen.

22). The preferential formation of the product of 1,2-cycloaddition over that of 1,4-cycloaddition in vesicles is probably best explained in terms of the greater difficulty of achieving the necessary geometry for [2 + 4] cycloaddition in this organized medium (Fig. 24). It has been established that DPB in solution exists in two conformational isomers: cisoid and transoid [185,186]. At equilibrium, the main conformer is the transoid (~99%), and the cisoid is presented only in ~1%. The 1,4-cycloaddition of singlet oxygen to 1,3-diene to form endoperoxide is concerted and analogous to the Diels–Alder reaction. This reaction requires a six-membered-ring transition state. Only the cisoid conformer can satisfy such a requirement, and in order to undergo 1,4-cycloaddition with singlet oxygen, the transoid conformer has to be isomerized to the cisoid one first. Due to the kinetic equilibrium between the two conformers in solution, the cycloaddition can proceed until all of the diene is converted to the products. Obviously, in vesicles, the organized semirigid environment prevents DPB molecules from conformational change. Thus, only the 1,2-cycloaddition product dioxetane **18** was obtained, which decomposes to yield **1** and **2**. On the other hand, the exclusive formation of the 1,4-cycloaddition product in the reaction of DPB with $^1O_2$ in Nafion membranes suggests that the water-containing pocket in the membranes allows the

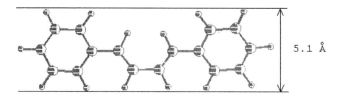

5.1 Å

Cisoid

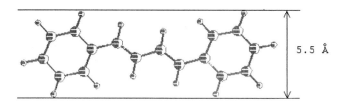

5.5 Å

Transoid

Figure 25   Geometries of *trans,trans*-1,4-diphenyl-1,3-butadiene conformers optimized by the AMF calculation.

conformational change of DPB molecules. However, it is surprising that in the internal surface of ZSM-5, the endoperoxide **6** is quantitatively produced (Fig. 15) and the conversion of the reaction can reach 100%, because on the silica surface, the conformational change is restricted and the DCA-sensitized oxidation of DPB in this medium could not result in the formation of the endoperoxide [187]. This observation is proposed to result from the differences in size and shape between the two conformers, thus showing different adsorption behavior on ZSM-5 zeolite. The AM1 program calculation shows that the molecular width of the cisoid is ~5.1 Å, whereas that for the transoid is ~5.5 Å (Fig. 25). Thus, the transoid possesses shape and size which are relatively large (bulky) to fit into the channels of ZSM-5, whereas the cisoid could readily enter and be accommodated into the channels of ZSM-5. The transoid conformer in solution could be isomerized into the cisoid one first, and then enter into the ZSM-5 channels. As a result, all DPB molecules adsorbed in the channels of ZSM-5 exist in the cisoid conformer. Such remarkable size and shape selectivity displayed by ZSM-5 is not unprecedented [188,189]. For example, *trans*-stilbene can be readily incorporated within the channel systems of this zeolite, whereas, by contrast, the different dimension precludes *cis*-stilbene to penetrate inside the internal voids of the zeolite [83,84].

The "ene" reaction of singlet oxygen with substrates containing several distinct allylic hydrogens can produce several distinct products. Although the peroxidation via singlet oxygen has found much synthetic application, regio- and stereo-control are still difficult issues [190]. Recently, Ramamurthy and coworkers reported that the interior of a zeolite offers one possible solution to this important problem [167,170]. When the sensitizers were thionin, methylene blue, or methylene green, the photosensitized oxidation proceeded via the singlet-oxygen pathway, as evidenced by the observation of singlet-oxygen emission and by quenching studies. A number of alkenes of structure similar to 2-methyl-2-pentene was examined. These alkenes contain two distinct allylic hydrogen atoms and, in an isotropic solution, yield two hydroperoxides with no appreciable selectivity (Fig. 26). Within NaY zeolite, a single hydroperoxide is preferentially obtained [167,170]. Similar selectivity was also observed with related alkenes such as 2-methyl,4-aryl-2-butenes (Fig. 26), and even more impressive results were obtained with 1-methylcycloalkenes (Fig. 27) [167,170]. These alkenes yield three hydroperoxides in solution, and the hydroperoxide resulting from abstraction of the methyl hydrogens is formed in the lowest yield. Surprisingly, the minor isomer in solution was obtained in larger amounts within the zeolite. Thus, the selectivity is a characteristic of hydroperoxidation of alkenes within zeolites. Product hydroperoxides were isolated in ~75% yield. Generally alcohols

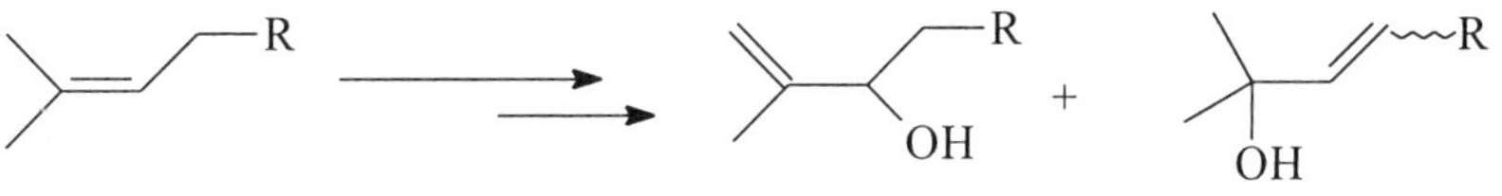

R = CH$_3$	Relative Yield (%)	
Thionin / CH$_3$CN	40	60
NaY / Thionin	100	0

R = CH$_2$CH$_2$CH$_3$ R = CH$_2$(CH$_2$)$_3$CH$_3$

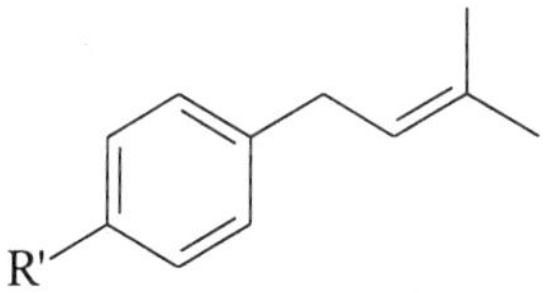

R' = H, CH$_3$, OCH$_3$

Figure 26 Photosensitized oxidation of alkenes with two distinct allylic hydrogen in solution and within NaY zeolite.

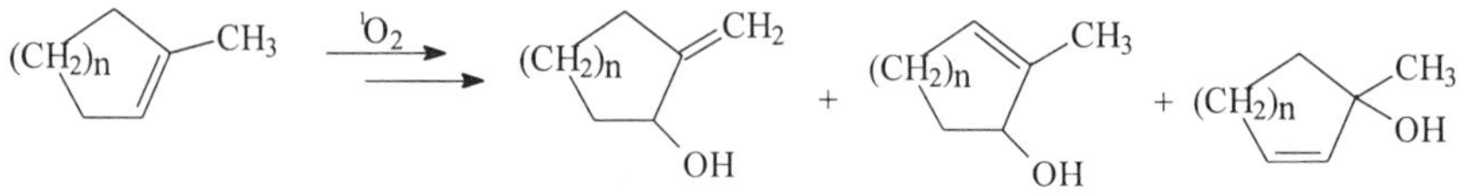

		Relative Yield (%)		
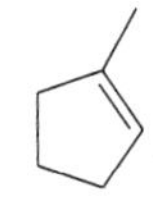	Thionin/CH$_3$CN	6	45	48
	Thionin/NaY/Hexane	100	0	0
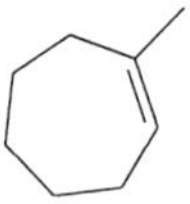	Thionin/CH$_3$CN	10	47	43
	Thionin/NaY/Hexane	100	0	0

Figure 27 Photosensitized oxidation of 1-methylcycloalkenes in solution and within NaY zeolite.

were more easily extracted out of the zeolite than the hydroperoxides. The yields reported in Figs. 26 and 27 are for alcohols.

The above selectivity is rationalized on the basis of two independent models [165,170]. In one, the zeolite is postulated to control the conformation of the reactive alkene; in the other, the zeolite is suggested to polarize the reactive alkene. It is desired to emphasize that the above models are only working hypotheses and further experiments are needed to identify the origin of selectivity. Formation of both hydroperoxides **19** and **20** from 2-methyl-2-pentene in solution has been rationalized on the basis that singlet oxygen attacks the alkene from the top-right side, as shown in Fig. 28. In such an approach, the transition state is stabilized by secondary interactions between the oxygen and the allylic hydrogens which are situated parallel to the π-p orbitals. In this model, the methyl group on the top-left side (Fig. 28) does not participate in the oxidation process. The results within zeolites clearly suggest that the methylene hydrogens H$_a$ of 2-methyl-2-pentene (Fig. 28) are not abstracted by the singlet oxygen. Although the lack of formation of **20** within zeolites is an indication that the methylene hydrogens are excluded from the reaction, selective formation of **19** does not indicate which of the two (or both) methyl groups participates in the oxidation process.

It is proposed that the R group in the alkene (Fig. 28) plays a crucial role in the type of product(s) formed. Although in solution, the most favored conformation places both the methyl and methylene hydrogens in an appropriate geometry for abstraction (Fig. 28), it is quite likely that such a conformation is not favored within a zeolite. In a supramolecular assembly, one must consider the interactions that arise between the adsorbent/guest and the environment. It was speculated that within a zeolite, the alkene will be adsorbed to the surface via cation–π interactions. The rotation of the C3—C4 bond may occur under such conditions to relieve the steric strain that develops between the bulky R group and the surface. Such a rotation will place the methylene hydrogens away from the incoming singlet oxygen (Fig. 28), preventing the formation of the tertiary hydroperoxide. The extent of steric repulsion between the surface and the R

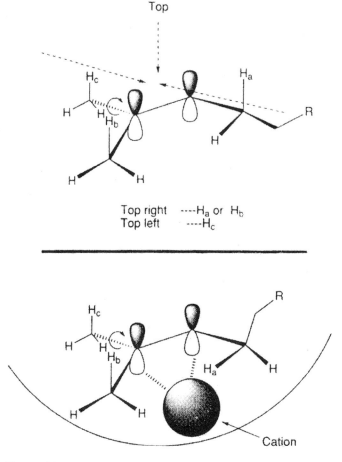

Figure 28 Cation–alkene interaction within a supercage of cation-exchanged Y-zeolite may control the conformations of allylic hydrogens. For allylic hydrogen abstraction, H_a and H_b should be parallel to the π orbital (see top).

group may depend on the distance between the group and the surface, which, in turn, will be controlled by the size and binding strength of the cation. This model predicts that the selectivity should be directly related to the binding energy of the cation with the alkene; reactions involving larger cations such as the Cs^+ ion may be expected to yield lower selectivity than those involving the smaller Na^+ ion.

In the second model, the selectivity is attributed to the polarization of the alkene by the interacting cation. As shown in Fig. 29, when the alkene is asymmetric, the interacting cation will be able to polarize the alkene in such a way that the carbon with the greater number of alkyl substituents will bear a partial positive charge ($\delta+$). The singlet oxygen, being electrophilic, is expected to attack the less substituted electron-rich carbon ($\delta-$) and leads to an ene reaction in which the hydrogen abstraction occurs selectively from the alkyl group connected to the $\delta+$ carbon. The extent of polarizability will depend on the charge density of the cation. Small cations such as Li^+ would be expected to polarize the alkene more effectively than larger cations such as Cs^+. As per this model, selectivity is expected to decrease from Li^+ to Cs^+. Consistent with both the above models, the observed selectivity decreases with the size of the cation ($Li^+ > Na^+ > K^+ > Rb^+ > Cs^+$) [165,170].

Extension of the above oxidation studies to alkenes such as limonene gave a complex mixture of products that resulted from all possible ene reactions to the trisubstituted double bond (Fig. 30) [165]. However, use of NaY zeolite as the microreactor and in the presence of a small amount of pyridine, the photosensitized oxidation of the alkenes is regioselective, yielding only the cis and trans products that result from hydrogen abstraction from the least hindered allylic carbon center. These studies illustrated that a microreactor can provide unprecedented opportunities to conduct selective oxidation of olefins.

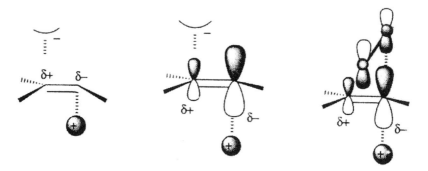

Figure 29 Cation–alkene interaction may polarize the alkene. Polarization is represented in terms of the size of the orbital.

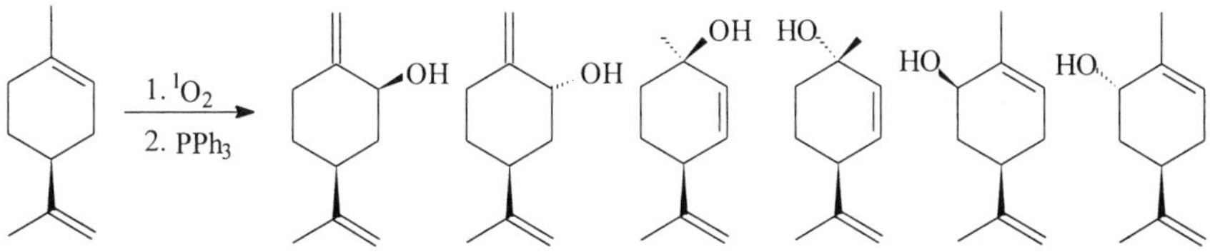

	Relative Yield (%)					
Conditions	24	25	26	27	28	29
Rose Bengal/CH₃CN	20	21	34	10	5	10
Na Y/Thionin	Rearrangement — Complex mixture					
NaY/Thionin/Pyridine	only					

Figure 30 Photosensitized oxidation of limonene in solution and within NaY zeolite in the presence of pyridine.

3. Photochemistry of the Hydrocarbon·O₂ Charge-Transfer State in Zeolites

It is well established that organic molecule·O₂ pairs give rise to contact charge-transfer absorptions and photochemistry [139–152]. Such spectroscopic and chemical observations have been made in oxygen-saturated organic solution [139], solid O₂ [141–143], and within zeolites [132,144–152,191]. In the former two cases, the corresponding absorption bands lie in the UV region, and the photochemical reaction can only be induced by UV light. As a result, the chemistry is much less controlled.

Recently, Frei and coworkers [132,144–151,191] have demonstrated that alkene·O₂ excited charge-transfer states are strongly stablized in zeolite Y matrices. The stabilization is attributed to the very strong electrostatic interactions inside the zeolite cages. The tail of the 2,3-dimethyl-2-butene·O₂ charge-transfer absorption extends to 760 nm when loading the alkene and O₂ into zeolite NaY at −50°C. Thus, hydrocarbons in zeolite Y can be selectively oxidized by visible light with relatively high quantum yields (~0.3) [132,144–151,191]. It has been demonstrated that upon irradiation with visible light, small hydrocarbons such as propylene and 2-butenes loaded in zeolites

BaY and NaY, respectively, can be selectively oxidized to allyl hydroperoxide and 3-hydroperoxy-1-butene by molecular oxygen at low temperatures near $-100°C$ [132,145]. Upon warming to $0°C$, these products further convert to acrolein and methyl vinyl ketone, respectively; water is also a product in the reaction. The mechanism for propylene photooxidation is outlined in Fig. 31. Photolysis of the hydrocarbon·O_2 complex with visible light results in a charge-transfer complex in the zeolite cage [146]. The allyl hydroperoxide is produced from the charge-transfer state, which subsequently goes on to form acrolein. The use of visible rather than ultraviolet irradiation allows access to this low-energy pathway, eliminates many secondary photoprocesses, and leads to product selectivity. The geometric constraints of the zeolite framework are also thought to be important in achieving high reaction selectivity. They have extended their study to toluene·O_2 and cyclohexane·O_2 complexes in alkali and alkaline-earth-exchanged zeolite Y and applied this method to selective oxidation of a tertiary alkane C—H group [146,148,151,191]. From a practical point of view, these reactions need to be done as close to ambient temperature as possible, rather than low temperatures, if they are to be of industrial use.

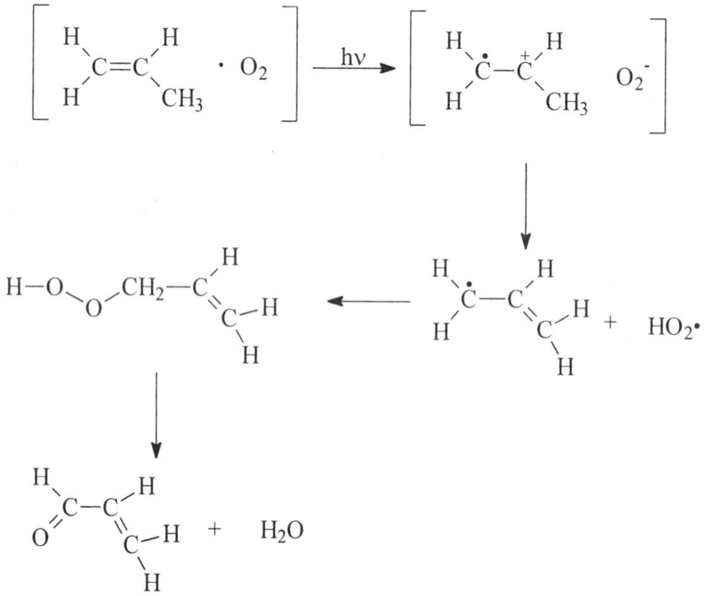

Figure 31 Proposed mechanism of the formation of unsaturated carbonyl compounds from the alkene·O_2 charge-transfer state.

More recently, Grassian and coworkers used *in situ* Fourier transform infrared (FTIR) spectroscopy and *ex situ* NMR spectroscopy to investigate the factors that influence product formation and selectivity in the room-temperature photooxidation of 1-alkenes in zeolites [152]. Upon irradiation with broad-band visible-light propylene, 1-butene and 1-pentene loaded in BaY were photo-oxidized with molecular oxygen to unsaturated aldehydes and ketones through a hydroperoxide intermediate. In addition, epoxide and alcohol products are formed when the hydroperoxide intermediate reacts with an unreacted parent alkene molecule. It is shown that saturated aldehydes and ketones are formed as well through both a thermal ring-opening reaction of the epoxide in BaY and a second photochemical oxidation route involving a dioxetane intermediate. The yield of saturated aldehydes and ketones increased with decreasing wavelength, increasing temperature, and, at a given temperature and wavelength, increasing chain length. Photo-oxidation of propylene in BaX is very similar to that of BaY. In zeolites BaZSM-5 and BaBeta, propylene polymerized upon adsorption. The polypropylene also undergoes photo-oxidation with molecular oxygen to form an oxygenated polymer product. The results of this study show that product formation and selectivity in the photo-oxidation of 1-alkenes in zeolites depend on several factors. These factors include thermal reactions of the reactant and photoproduct molecules in the zeolite at ambient temperatures. It is desired that these photoinitiated oxidations of hydrocarbon will be developed as a useful methodology in environmentally benign synthesis of partial oxidation products.

C. Photochemical Rearrangement in Zeolites, Nafion Membranes, and LDPE Films

The photo-Fries rearrangements of phenyl acetate, phenyl benzoate, phenyl phenylacetates, 1-naphthyl and 2-naphthyl acetates included in zeolites, polyethylene films, and Nafion membranes have been investigated by Ramamurthy [142,143] Weiss [116,193], and Tung [194,195] and their coworkers. In the phenyl phenylacetate photochemistry, Tung and co-workers used esters **31–34** (Fig. 32) to probe the diffusibility and rotational mobility of the radical pair imposed by the surfaces of zeolites and the water-containing pocket of Nafion [144,195]. The photochemistry of esters **31–34** is expected to be analogous to that of phenyl acetate, whose photochemistry in homogeneous solution has been well investigated [196–199]. Figure 33 gives the photochemical reactions of these esters with **31** as the example. Upon photoirradiation, **31** undergoes the C—O bond homolytic cleavage to give two-paired radicals. This geminate radical pair in cage recombines to form the starting ester or *ortho-* and *para*-hydroxyphenones **36** and **37**. The latter reaction is known as the photo-Fries rearrangement. The phenylacetyl radical may undergo decarbonylation in the cage to produce the secondary radical pair, which, in turn, produces phenyl benzyl ether **35**. Thus, the hydrox-

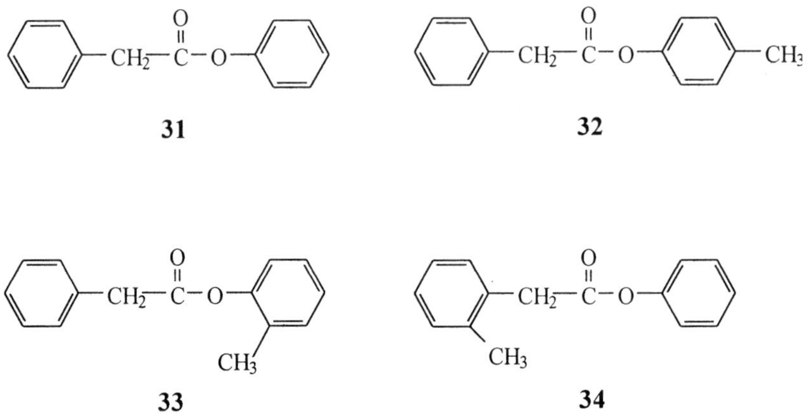

Figure 32 Phenyl phenylacetates examined for photo-Fries rearrangement within zeolites and Nafion membranes.

yphenones and phenyl benzyl ether may be viewed as "cage" products. The radical pair may also undergo diffusive separation to give free phenoxy and phenylacetyl radicals. The active phenoxy radical would easily abstract a hydrogen atom to form phenol **38.** The phenylacetyl radical can be expected to decarbonylate, generating a benzyl radical, which, in turn, couples with each other

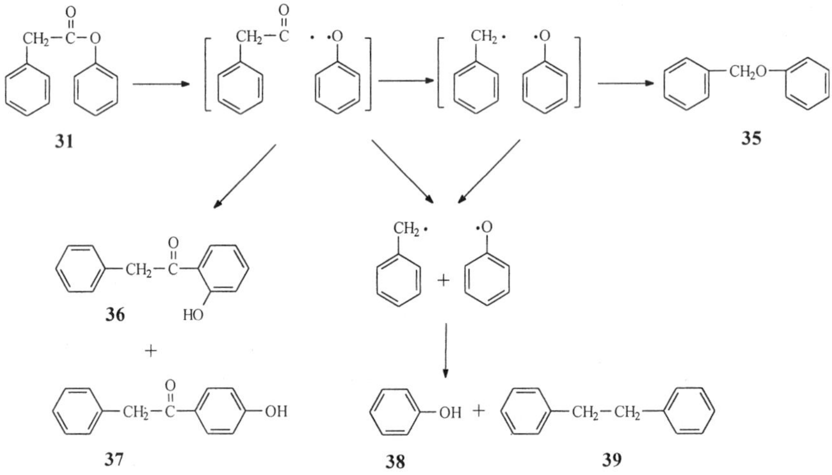

Figure 33 Intermediacy of radical pairs in photochemical reactions of phenyl phenylacetate.

to yield diphenylethane **39.** Thus, phenol and diphenylethane are the "escape" products. The cage effect is defined as a fraction of the yields of hydroxyphenones and phenyl benzyl ether in the total yield of all of the products. Restriction on the diffusion of the radicals would increase the cage effect. Suppression of the rotation of the radicals would decrease the yield of the photo-Fries rearrangement products and enhance the ratio of the *ortho*-hydroxyphenone **36** to the *para*-hydroxyphenone **37,** because formation of a para rearrangement product requires a greater extent of the rotation of the radical than in the formation of an ortho product. Thus, one can use the cage effect and the ratio of *ortho-* to *para*-hydroxyphenone as probes for investigating the diffusibility and rotational mobility of radicals, respectively, in microreactors.

Photolyses of **31–34** in homogeneous solution results in the formation of diphylethanes **39** (5–15%), phenols **38** (5–15%), *ortho*-hydroxyphenone **36** (40–60%), and *para*-hydroxyphenones **37** (20–25%). Small amounts of phenyl benzyl ether **35** (3–8%) were also detected. However, photolyses of all of the four esters on NaY zeolite and Nafion only produce ortho rearrangement products **36.** Molecular models suggest that esters **31–34** can enter into NaY zeolite internal surface and the inverse micelle of Nafion. We believe that the preference for formation of *ortho*-hydroxyphenones **36** in the products is a consequence of the restriction on diffusional and rotational motion of the geminate radical pair.

For samples photolyzed on ZSM-5 zeolite, the product distributions of **31** and **32** are dramatically different from those photolyzed in homogeneous solutions. First, the rearrangement products were totally suppressed. Second, diphenylethane **39** resulted from coupling of benzyl radical was not found. Only phenol **38** and toluene were detected. In contrast, photolyses of **33** and **34** on ZSM-5 follow strikingly different pathways. Both photo-Fries rearrangement **36** and **37** and decarbonylation products **35** and **39** were formed. These results can be understood from consideration of size- and shape-selective sorption combined with restriction on the mobility of the substrates and reaction intermediates imposed by the pentasil pore system.

The structures of **31** and **32** are similar to **p**-xylene and they are expected to diffuse into and be adsorbed within the internal surface of the pentasil. In the pentasil internal channels, **31** and **32** can only adopt a linear conformation to orientate themselves with the long axis of the molecules coinciding with the straight framework channels. The homolytic cleavage of the C—O bond of the excited state of esters results in the production of two paired radicals. The kinetic pore diameter of the ZSM-5 framework channels is about 5.5 Å and the benzene ring has a kinetic diameter of 5.85 Å. Thus, the fit inside the channels would be very tight. Such a tight fit and interaction between the radicals and the surfaces of ZSM-5 would restrict both diffusional and rotational motions of the radicals within the channels of ZSM-5. As observed, restriction on rotational motions of the radicals, which are required for Fries rearrangement, would inhibit formation of *o-* and *p-*

hydroxyphenones **36** and **37.** Abstraction of a hydrogen atom for the phenoxy radical to form phenol **38** is faster than its diffusion to access the benzyl radical and to couple each other, although the two geminate radicals are separated only by a small distance. Because the diffusion of the benzyl radical generated by decarbonylation of the phenylacetyl radical is also seriously inhibited, this radical has enough time to abstract a hydrogen atom before encountering another benzyl radical. Thus, toluene rather than diphenylethane was produced. In this sense, toluene and phenol can be viewed as the "cage" products.

In contrast to **31** and **32,** the structures of **33** and **34** resemble *o*-xylene and possess a size/shape which inhibits them to diffuse into the ZSM-5 channel system. Thus, they are adsorbed on the external surfaces of ZSM-5. Molecules and intermediates on the external surfaces are expected to experience much less restriction to their diffusional and rotational motions. Therefore, both rearrangement and escape products are expected to be produced. Indeed, photolysis of **33** or **34** adsorbed on ZSM-5 zeolite results in both Fries rearrangement and decarbonylation products. However, close inspection of the product distribution reveals that both the cage effect and the ratio of *ortho*-hydroxyphenone **36** to *para*-hydroxyphenone **37** for photolysis of **33** or **34** on ZSM-5 are evidently greater than those for photolysis in homogeneous solution. This observation suggests that the external surface of ZSM-5 significantly restricts the diffusional and rotational motions of the photogenerated radicals.

Ramamurthy and coworkers studied the photo-Fries rearrangement of phenyl acetate and phenyl benzoate and photo-Claisen rearrangement of allyl phenyl ether (Fig. 34) included in two types of zeolite (faujasites X and Y and pentasils ZSM-5 and ZSM-11) [192]. The photolysis was performed with the zeolite slurry in either hexane or iso-octane. One of the most remarkable observations is that the product distribution is altered within zeolites from that in isotropic solvent. Furthermore, while in solution, nearly a 1:1 mixture of ortho and para isomers **40** and **41** (Fig. 34) was obtained, within zeolites one is able to direct the photoreaction selectively toward either the ortho or the para products by conducting the reaction either within faujasites or pentasils, respectively (Fig. 34).

Within Li- and Na-X and -Y zeolites, a small percentage ($<$10%) of the para isomer **41** is formed, whereas in KX and KY zeolites, the ortho isomer **40** is the exclusive product. The above selectivity is not the result of shape exclusion because both the ortho and the para isomers (**40** and **41**) fit well within the supercage. Selectivity results from the restriction imposed on the mobility of the phenoxy and the acyl (benzoyl) fragments by the supercage framework and cations. Several features contribute to the restriction. The cage free volume plays an important role, as evidenced from the increased selectivity with the increase in size of the cation. In addition to the cage free volume, an interaction between the cation and the two reactive fragments must be contributing to the selectivity. This becomes clear when one compares the results of phenyl acetate and allyl phehyl ether. In

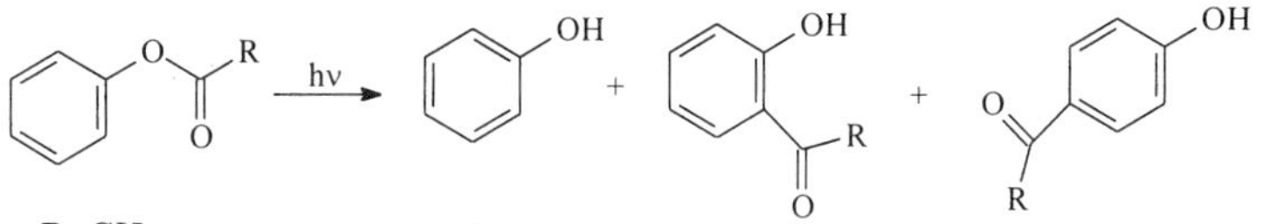

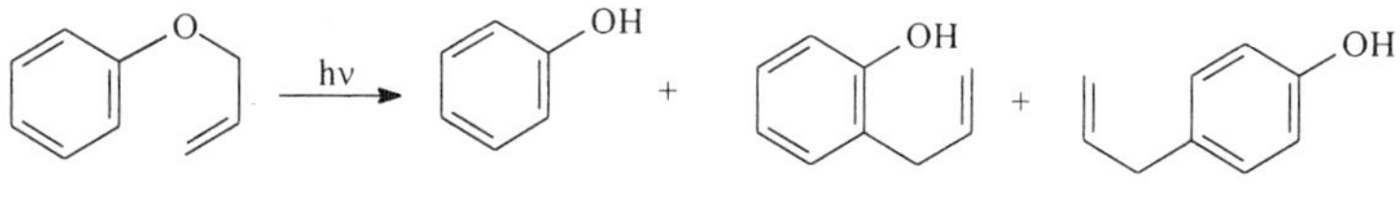

R=CH$_3$
R=Ph
          38   40   41

          38   42   43

Relative Yield (%)

	Phenyl acetate R= CH$_3$			Allyl phenyl ether			Phenyl benzoate R= Ph		
	38	**40**	**41**	**38**	**42**	**43**	**38**	**40**	**41**
Hexane	20	53	27	48	21	31	12	49	37
NaY/hexane	5	91	4	3	79	18	1	93	6
KY/hexane	17	83	0	4	78	18	2	98	0
LiX/hexane	15	77	8	4	81	15	1	96	3
KX/hexane	27	73	0	3	83	14	7	93	0
Na ZSM-5/isooctane	32	5	63	5	5	90	17	38	24
Na ZSM-11/isooctane	18	8	74	7	20	73	8	48	29

Figure 34 Photo-Fries rearrangement of phenyl acetate and phenyl benzoate, and photo-Claisen reaction of allyl phenyl ether in solution and within zeolites.

the latter compound, a small percentage of the para isomer **43** (~15%) always accompanied the major ortho isomer **42**. Whereas the size and shape of acyl and allyl radicals are expected to be similar, the strength of interaction between the cations and these two fragments is expected to differ; the former is expected to bind more strongly than the latter. Such a difference is translated into an increased yield of the para isomer **43** for phenyl allyl ether.

The shape and size of pentasils are such that only the para isomers (**41** and **43,** Fig. 34) fit within the channels of these zeolites. Indeed, a clear preference for the para isomer results (Fig. 34). A small amount of the ortho isomer **40** and **42** that is formed comes from reactions that occur in the solvent portion.

Cui and Weiss reported the photo-Fries rearrangements of 2-naphthyl myristate and 2-naphthyl acetate (Fig. 35) in unstretched and stretched low-density polyethylene films [116]. It was argued that the radical pair generated by irradiation is held in a cavity shaped like the starting ester and whose walls relax more slowly than the radicals recombine. The preferred solution product is too different in shape from the starting ester to be formed.

Recently, Ramamurthy and Weiss and their coworkers reported the photo-Fries rearrangement of three 1-naphthyl phenylacelates (Fig. 36) in cation-exchanged zeolite Y and high-density polyethylene films [193]. When the substrates were irradiated in hexane to ≤30% conversion, the eight photoproducts in Fig. 36 were detected. The photoproduct distributions from polyethylene or a Y-zeolite are drastically different from those in solution. Cage-escape products (**54** and **55**) are absent in both constrained media, and in zeolite Y, only **49** was detected. The

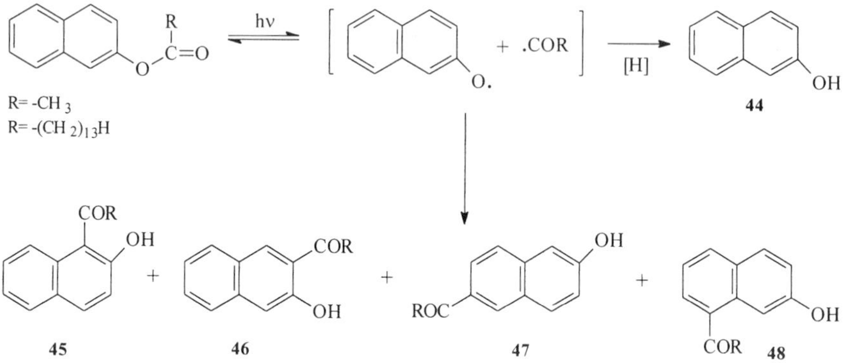

Relative yield (%)

Ester	Solvent	Photoproduct distribution				
		44	45	46	47	48
R= -CH₃	tert-butyl alcohol	32	31	11	17	0
R= -(CH₂)₁₃H	Hexane	85	14	0	1	0
R= -CH₃	LDPE unstretched	21	31	20	28	0
	LDPE stretched	32	0	31	37	0
R= -(CH₂)₁₃H	LDPE unstretched	0	0	75	25	0
	LDPE stretched	0	0	92	8	0

Figure 35 Photo-Fries rearrangement of 2-naphthyl acetates in solution and within low-density polyethylene films.

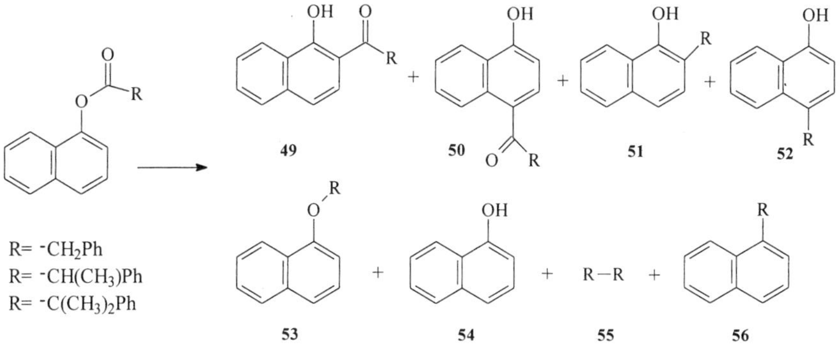

Figure 36 Photo-Fries rearrangement of 1-naphthyl acetates within Y-zeolite and low-density polyethylene films.

selectivity and absence of decarbonylation products in the zeolites is remarkable, and reduction in the number of product from eight to one is rare, if not unprecedented. They suggested that limiting the constrained space of a reaction cavity in an organized medium can be less important than wall–guest interactions in determining the selectivity of guest reactions. Reaction cavities of cation-exchanged Y-zeolites and high-density polyethylene film possess very different properties. The cavity "walls" of zeolites are not "passive." Cations help anchor reactants, intermediates, and products to the surfaces. In addition, the walls are very "hard," so that the shapes and volumes of the cavities do not change as guest molecules react. The cavities of Y-zeolites are much larger than necessary to accommodate a molecule of the 1-naphthyl esters or their rearrangement photoproducts. Supercages are connected directly to each other by 7.4-Å-diameter windows; there are no constricted "tunnels" into which a 1-naphthyl ester or their radical fragments can be placed in their entirety. On the other hand, cavities of polyethylene are less well defined in size and shape. Their free volumes are smaller than the van der Waals volumes of the 1-naphthyl esters. Although the cavity walls of polyethylene are "passive" and "softer" than those of zeolites, they must exert more pressure on guest molecules during reactions. The correspondence between the relaxation times of the polyethylene chains constituting the walls and the lifetimes of the intermediates in the photo-Fries processes determines which products are formed. Both zeolites and polyethylenes can exert significant control over the courses of photo-Fries reactions, but for apparently very different reasons. The more limited control exerted by polyethylene over the reactivity of the radical pairs from the 1-naphthyl esters is due primarily to wall stiffness and limitations on the available space within a cavity. The passive nature of the walls does not allow them to orient radical-pair partners via the strong interactions available in cation-exchanged zeolites. The fact that the space in a zeolite supercage is much larger than the van der Waals volume of a naphthyl ester is of secondary importance in determining medium constraints when there are hard walls with active sites to act as templates for the radical pairs. In topological and energetic terms, surface interactions between radicals and walls of a zeolite cavity are largely "two dimensional" and attractive, whereas escape by radical pairs from polyethylene cavities may be inhibited by "three-dimensional" pressures exerted by the surrounding chains in largely repulsive interactions.

Heavy-atom effects within cation-exchanged zeolites on several photochemical rearrangements where the efficiency of intersystem crossing determinates the types of product formed have been reported by Ramamurthy [200], Ghandi [201], and their coworkers. For example, dibenzobarrelene and benzobarrelene are converted to two different compounds **57** and **58,** and **59** and **60,** respectively, depending on whether the olefins are in the singlet or triplet excited states (Fig. 37) [200]. Using LiX or NaX, a mixture of products forms; however, when a heavy-atom-exchanged zeolite is used, the triplet-derived products **58** and

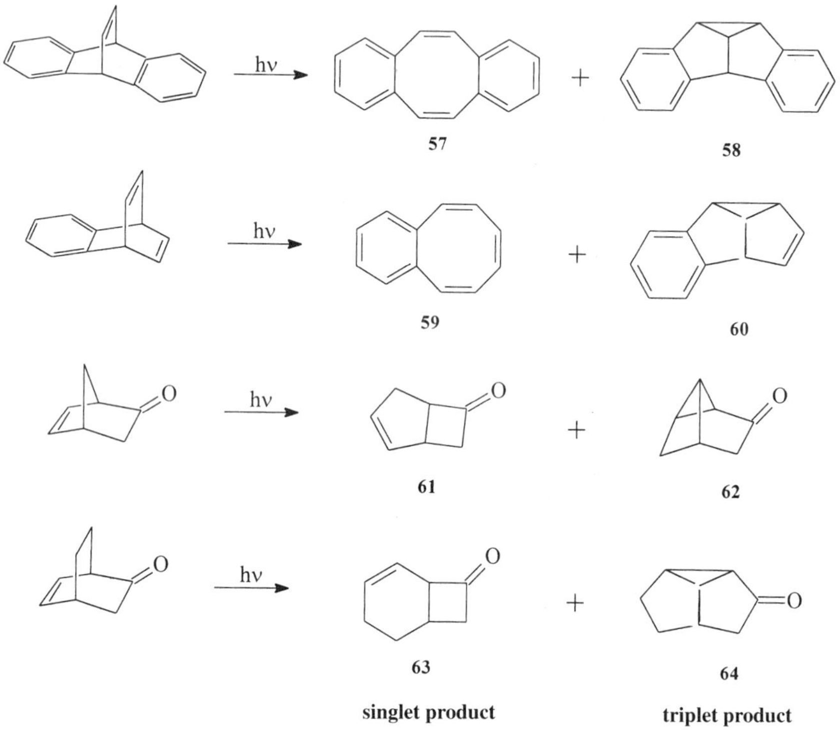

Figure 37 Heavy-atom effect in zeolites: singlet versus triplet photochemistry.

60 are formed predominantly. The photochemical oxa-di-π-methane rearrangement of bicyclic enones has also been examined using heavy-atom-exchanged zeolites (Fig. 37) [201]. In this case, heavy-atom substitution increased the amount of triplet-derived products **62** and **64,** although the zeolite-mediated reactions were not as efficient as the corresponding homogeneous photorearrangement.

D. Photocycloaddition and Photocyclization Within Nafion Membranes, Zeolites, and Vesicles

The zeolite-mediated [2 + 2] cycloaddition of several styrene derivatives has recently been studied [165]. By inclusion of both olefin substrate (*trans*-anethole or 4-vinylanisole) and sensitizer (9-cyanoanthracene) within zeolite NaX, a significant difference was observed for the zeolite-mediated photocyclization compared to the solution reaction (Fig. 38). For example, although both *cis/syn* (**65**) and *trans/anti* (**66**) dimer were formed, the zeolite favors the *cis/syn* product **65** which has a more spherical shape that is similar to the geometry of the supercage. This reflects the fact that the more linear *trans/anti* isomer **66** is best accommodated in two supercages, whereas the *cis/syn* dimer **65** can be formed within a single cage and therefore its formation is the more favorable process.

Tomotaka and coworkers [202] studied the photodimerization of cinnamic acids incorporated in vesicles (Fig. 39). They mixed equimolarly the cinnamic acid with alkyldimethylamine N-oxide (C_nDAO) to produce the ion pair **72** (Fig. 39). These ion pairs form stable vesicular aggregates in water. Whereas photoirradiation of the cinnamic acids in methanol resulted in only the cis-trans isomerization to form **71,** in these vesicle medium, three dimers, **68–70,** were obtained.

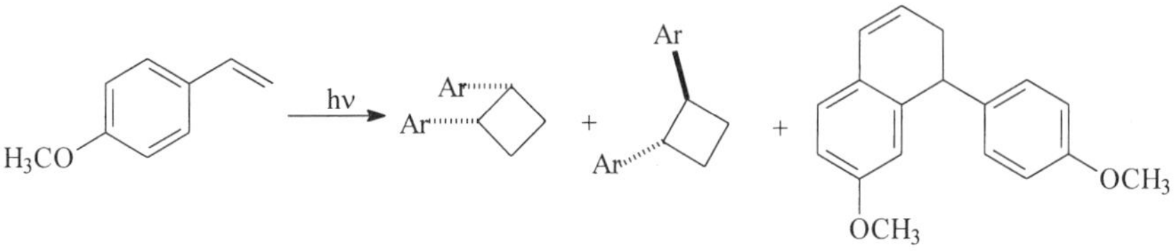

Reaction conditions	Product ratio **65/66/67**
Acetonitrile/PET sensitized	4 / 90 / 6
Na X / 9-cyanoanthracene	66 / 26 /7

Figure 38 [2 + 2] Photocycloaddition with zeolites as hosts.

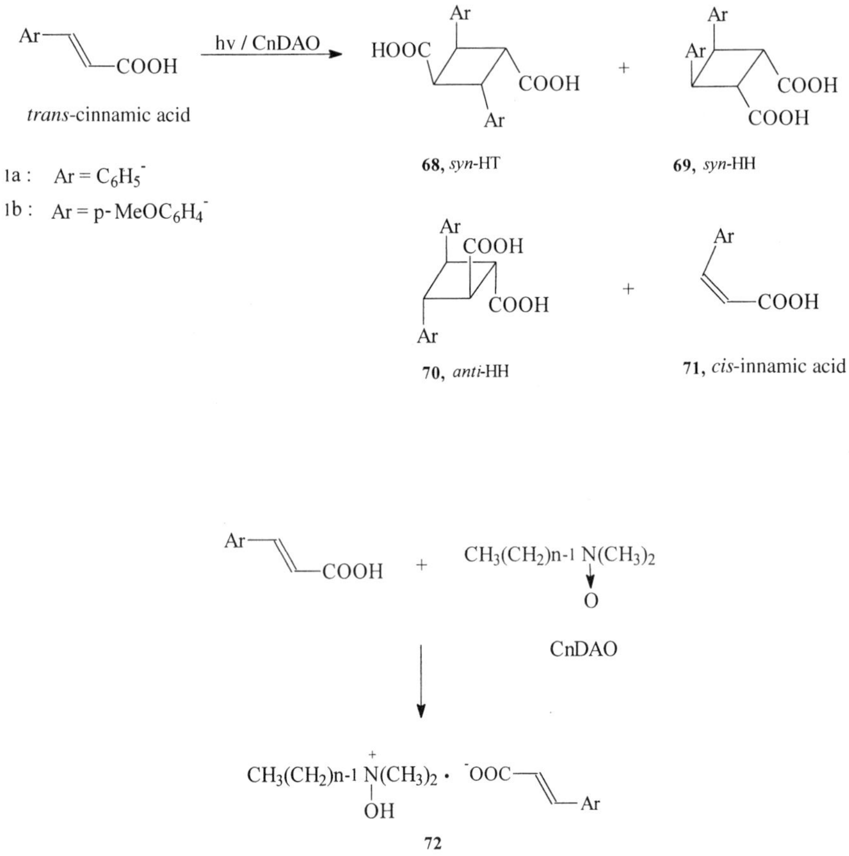

Figure 39 [2 + 2] Photocycloaddition within vesicles.

Photocycloaddition of anthracene and its derivatives continues to be one of the subjects in photochemistry. As mentioned in Section III.A., irradiation of 9-substituted anthracenes in organic solution typically results in cylcoaddition of the aromatic rings at 9,10-positions to yield head-to-tail (h–t) cyclomers, although evidence for the concomitant formation of thermally more labile head-to-head (h–h) cyclomers has been obtained in some instances (Fig. 9) [85–91]. In order to increase the synthetic yields of the h–h photocyclomers, a variety of approaches have been employed [85–99]. Wolff [203], Melo [91] and their coworkers reported that incorporating 9-anthracene carboxylic acid esters into micelles and vesicles of surfactants could induce the anthracene rings in h–h orientation, thus increasing the yield of the h–h photocyclomers. The position of the molecules within the surfactant vesicle membrane is determined by the ester alkyl chain. The orientation of the molecules becomes stricter with increasing chain length of the ester alkyl groups. Because of this preorientation, the planes of the anthracene moieties tend to be perpendicular to the vesicle surface. Consequently, the photodimerization of two neighbored anthracene molecules resulted in the formation of the h–h cyclomer.

Tung and Guan studied the photocycloaddition of anthracene derivatives in Nafion membranes [204]. When anthracene bearing an ionic or a polar substituent is incorporated within Nafion membranes, one might expect that their molecules would be arranged in such a way that the anthracene moiety is directed toward the hydrophobic phase of the polymer backbone, and the polar substituent is directed toward the water-pool interface. This preorientation would favor the formation of a h–h photocyclomer. Irradiation of the solutions of five 9-substituted anthracenes [$AnCH_2N^+(CH_3)_3Br^-$ (**73**), $AnCH_2COO^-Na^+$ (**74**), $AnCH_2OH$ (**75**), $AnCOCH_3$ (**76**), and $AnCH_3$ (**77**), An = 9-anthryl] in organic solvents generally leads to formation of the h–t cyclomers. However, in some instances, the h–h photocyclomers were also obtained as minor products. In contrast, irradiation of **73–76** incorporated within Nafion membranes almost exclusively results in the formation of the h–h cyclomers. On the other hand, under identical conditions, such a regioselectivity for the photocycloaddition of **77** was not observed. The almost exclusive formation of the h–h cyclomers for **73–76** incorporated in Nafion is evidently attributed to preorientation of substrate molecules in the inverse micellelike structure of Nafion. The substrate molecules are located in the fluorocarbon–water interface of the Nafion membranes. Because of the hydrophobicity of the anthracene moiety and the charged or polar nature of the substituent for **73–76,** the anthracene chromophore should reside in the fluorocarbon domain, whereas the substituent is anchored among the sulfonate head groups of Nafion. Thus, the molecular plane would tend to be perpendicular to the interface of the water cluster, and the substituent is directed toward the water pool. Consequently, the photocycloaddition of two neighbored anthracene molecules favors the formation of h–h cyclomers. Because the nonpolar reference substrate **77** lacks such a preorientation in Nafion membranes, the enhancement of h–h cyclomer formation is not observed.

Recently, Tung and Guan studied the photocyclization of azobenzene incorporated in Nafion membranes [205]. Up to now, the majority of the photochemical studies within the water-swollen Nafion have made use of this material just as a heterogeneous and micellelike medium to gain control of the pathways undergone by the substrate excited states. However, this polymer in its protonated form is a superacid [206]. The acid groups can act as active sites and directly intervene in the chemical process occurring in the water-containing pockets of the polymer. The azobenzene system follows two different reaction pathways, depending on the photolysis condition [207]. Thus, upon irradiation in neutral medium, cis-trans isomerization occurs and photostationary mixtures of both stereoisomers are obtained. However, when the photolysis is performed in the presence of acids, azobenzene undergoes the photochemical cyclization to benzo[c]cinnoline **82** and benzidine **84**. In this case, the conjugated acid of azobenzene **78** (not the neutral molecule) is the species that undergoes ring closure. It was demonstrated that by including the azobenzene molecules in solvent-swollen Nafion, one can control the reaction pathways. Photoirradiation of water-swollen Nafion–H$^+$ membrane soaking azobenzene results in the cyclization to give **82**. Variable amounts of **84** is also produced as shown in Fig. 40. The prod-

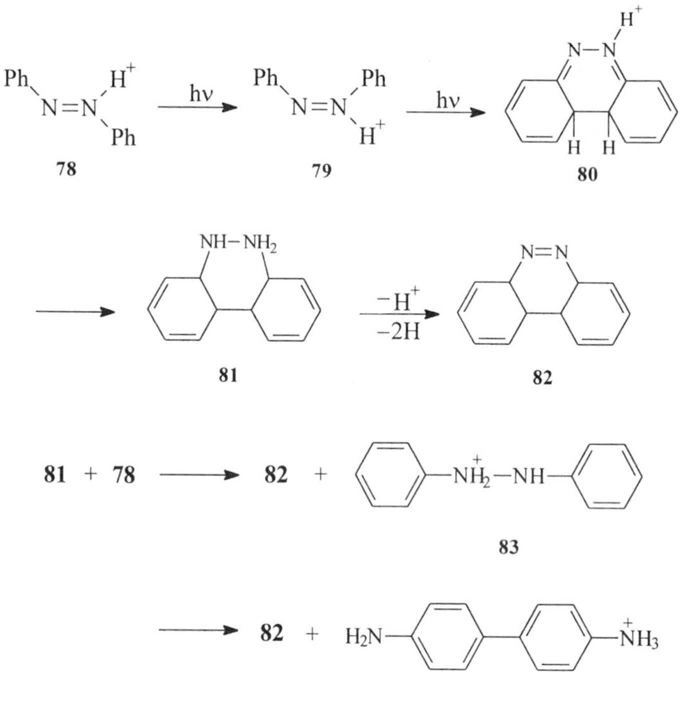

Molar ratio of **84/82** as a function of the occupancy number (n$_{78}$/cluster)

N_{78}/cluster	3.56	2.40	1.82	1.49	118	0.34	0.25	0.016
M_{84}/M_{82}	0.96	0.96	0.77	0.48	0.12	0	0	0

Figure 40 Mechanism of the formation of benzo[c] cinnoline and benzidine.

uct distribution was found to be dependent on the occupancy number of azobenzene in the water cluster of the Nafion sample. Photolysis of the sample with an occupancy number greater than 2 leads to the formation of **84** and **82** with a ratio of ~50:50. However, for the sample with an occupancy number less than 0.4, only the photocyclization product **82** was isolated in a quantitative yield and no **84** was detected. In the photocyclization reaction of azobenzene, the Nafion is more than just a passive medium. The acid sites of the Nafion also play an essential role in the reaction. However, in methanol-swollen Nafion–H^+ membrane, azobenzene molecules are located in the methanol pool. Thus, they are not protonated and the photocyclization cannot occur. Upon irradiation, only trans-cis isomerization could be detected.

Corma [208] and Dutta [209] and their coworkers reported the photocyclization of azobenzene in zeolite HY and microporous aluminophosphate, respectively. They also demonstrated that the acid groups in these media directly intervene in the photochemical reactions of azobenzene.

E. Chiral Photochemical Reaction Within Zeolites

In comparison to the enormous effort expended in developing general methods of asymmetric synthesis for ground-state reactions, relatively little attention has been paid to the subject of *photochemical* asymmetric synthesis; a review article summarizes the field to 1992 [210], and, recently, Ramamurthy and Joy reviewed their work on chiral photochemistry within zeolite [211]. Thus, whereas basic rules of chiral induction of thermal reactions have been reasonably established [212–217], the same is not true of photochemical reactions [210,211,218–221]. Short excited-state lifetime and low activation energies for reactions in the excited state(s) leave very little room for manipulating the diastereomeric transition states. In the past, chiral solvents, circularly polarized light, and chiral sensitizers have been utilized to conduct enantioselective photoreactions [210,218–221]. The highest chiral induction achieved by any of these approaches at ambient temperature and pressure has been about 30% [222–224]. A chiral auxiliaries approach, a popular method in solution thermal chemistry, has also been used with success in photochemistry by Scharf and coworkers [221]. Of the various approaches, the crystalline-state and solid host–guest assemblies have provided encouraging results [225,226].

Recently, Ramamurthy and coworkers reported the most exciting results in chiral photochemistry [168,171,211,227–231]. They used modified zeolites as a microreactor in which a certain proportion of the cages have been rendered chiral by preadsorption of an optically active inductor molecule. The achiral supercages of a zeolite are loaded with an optically active and non-light-absorbing guest molecule by two approaches: (1) by including an optically pure organic molecule as an adsorbent within a zeolite and (2) by exchanging the inorganic cations present in zeolites with chiral ammonium ions. Thus, when a second, photochemically reactive molecule is introduced into the same (or nearby) cage, it senses the asymmetric field due to the first and reacts enantioselectively. The confined space offered by the zeolite supercage forces the reactant and the chiral inductor to interact intimately to yield enantiomerically enriched product. Due to the transitory nature of the reaction cavity in solution, such close interactions are less likely in isotropic solvent media. In principle, this approach can accommodate a wide variety of chiral inductors, photoreactants, and zeolites. They have examined photoelectrocyclization (Fig. 41), the Norrish–Yang reaction (Fig. 42), the Schenk ene reaction (Fig. 43), and the Zimmerman di-π-methane rearrangement within

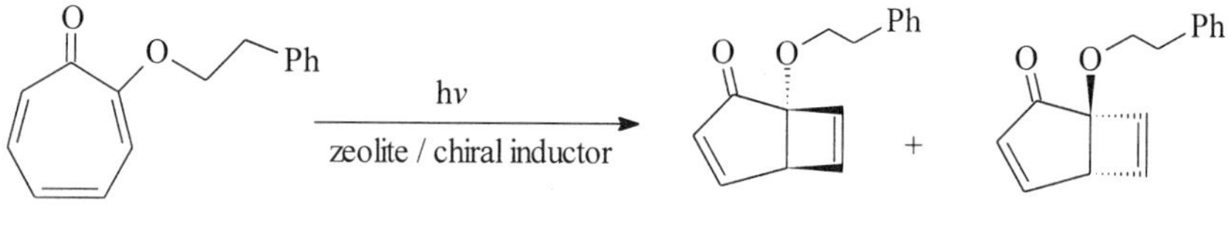

Chiral inductor	Enantiomeric excess (%), the favored enantiomer is shown in the parentheses	
	"wet"	"dry"
Na Y / (+) ephedrine	17 (**85**)	69 (**86**)
Na Y / (−) pseudoephedrine	8 (**86**)	20 (**85**)
Na Y / (−) norephedrine	23 (**86**)	38 (**85**)
Na X / (+) ephedrine	9 (**86**)	
Li Y / (+) ephedrine		22 (**86**)
K Y / (+) ephedrine		11 (**86**)
Rb Y / (+) ephedrine		2 (**86**)

Figure 41 Photocyclization of tropolone alkyl ether within X- and Y-zeolites. The product configuration is arbitrarily assigned based on HPLC retention time. The nature of the favored enantiomer is reversed between "wet" and "dry" zeolites and between NaY and NaX. The extent of *ee* depends on water content, the nature of the cation, and the number of cations (NaY versus NaX).

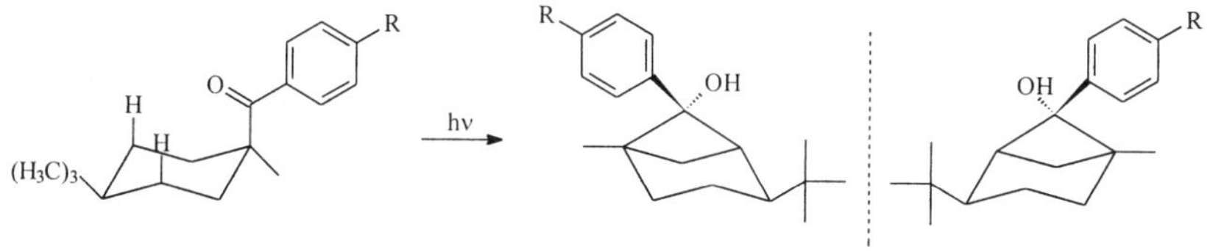

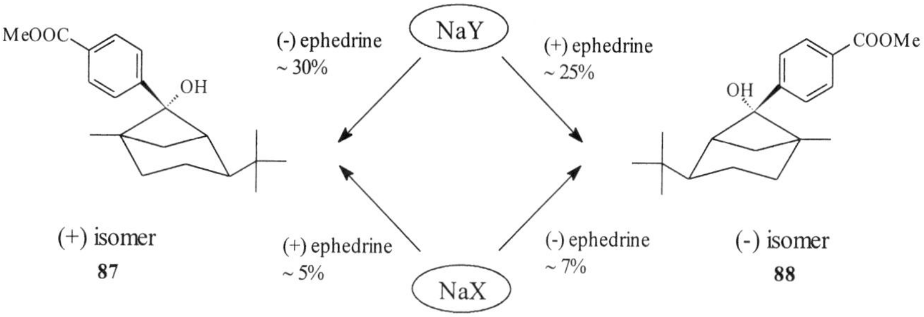

Figure 42 Results on the enantioselective Norrish–Yang photocyclization of *trans*-4-*tert*-butyl-1-methylcyclohexyl aryl ketone. Note that the same isomer of the chiral inductor favors opposite isomers within NaY and NaX zeolites.

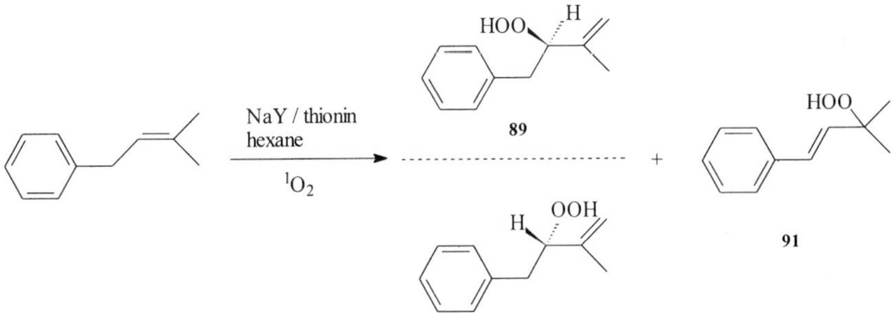

no chiral inductor	0 ~0.05% *ee*
ephidrene hydrochloride	15.5% *ee*

Figure 43 Result on the enantioselective photo-oxidation of an olefin. Products were analyzed by converting the hydroperoxide to alcohol by treatment with triphenylphosphine.

zeolites (Fig. 44) [211]. Most encouraging results have been obtained with electrocyclization of tropolone alkyl ethers (Fig. 41). The high enantiomer excess (*ee*) value of 69% was obtained with tropolone phenylethyl ether.

In addition to the enantiomeric excess of the products, the nature of the enantiomer being enhanced is also of great concern. All the systems studied are well behaved in the sense that, as expected, the optical antipode of the chiral inductor always gives the opposite enantiomer of the product [168,171,211, 227–231]. This is consistent with the expectations. However, the water molecules unexpectedly switch the isomer being enhanced (compare wet versus dry in Fig. 41). For example, in the case of tropolone phenylethyl ether, (+)-ephedrine favored the enantiomer **85** within "wet" NaY and the enantiomer **86** within "dry" NaY. Also, the number of cations makes a difference. For example, in the case of tropolone phenylethyl ether (−)-norephedrine favored the enantiomer **85** in NaY and the enantiomer **86** in NaX; that is, depending on the number of cations present within a supercage, the different enantiomers are favored by the same chiral agent. Even more puzzling is the fact that the nature of the cation itself can alter the isomer being enhanced by the same chiral inductor. A good understanding of these observations is essential for establishing the ground rules for chiral induction within a zeolite.

In the above-discussed examples, a chiral inductor is used to induce chirality on the product of a reaction. In the absence of any specific interaction between a chiral inductor and a reactant, it is unable to force every reactant molecule close to the chiral agent. This is one of the main reasons for poor enantioselectivity. The 69% *ee* obtained in the case of tropolone phenylethyl ether/ephedrine/NaY (dry) despite this limitation is remarkable. In an effort to reduce this limitation and to maintain closeness between the chiral center and the reaction site, Ramamurthy et

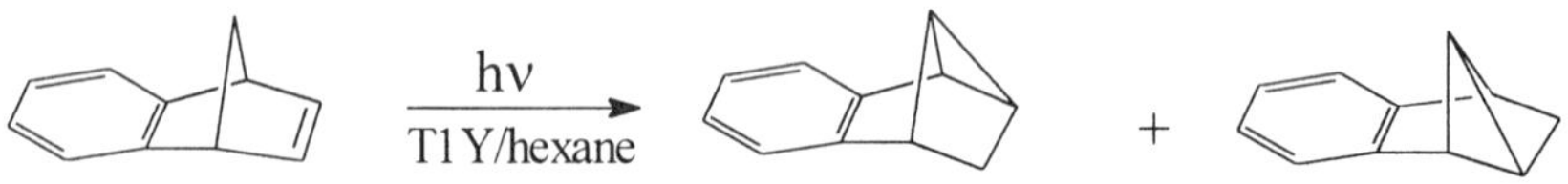

TI Y		*ee* ~0
TI Y/ephedrine·HCl		*ee* ~15%

Figure 44 The enantioselective photoreaction of benzonorbornadiene. The reaction originates from the triplet state. The cation Tl^+ helps to generate the triplet state of the substrate.

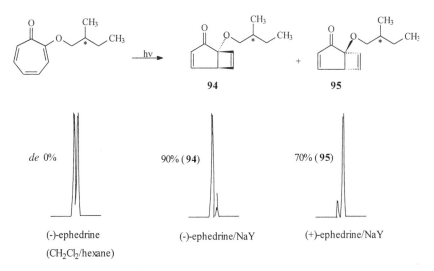

Figure 45 Photocyclization of (*S*)-tropolone 2-methylbutyl ether within chirally modified NaY zeolites. The diastereomeric excess (%) and the isomer enhanced are shown on the HPLC traces. The first eluted isomer on the HPLC column is arbitrarily assigned **93**.

al. have begun investigating examples in which the chiral inductor is covalently linked to the tropolone system (Fig. 45) [211].

Irradiation of the (*S*)-tropolone 2-methylbutyl ether shown in Fig. 45 in solution yields a 1:1 mixture of diastereomeric products. Clearly, in solution, the presence of the chiral auxiliary in proximity to the reactive center has no influence on the product stereochemistry. On the other hand, the same molecule when irradiated within NaY gave the product in about 53% diastereomeric excess. The smaller and controlled space provided by the zeolite supercage apparently has forced the chiral center to establish communication with the reaction site. Considing that the chiral center has only alkyl groups, one would not expect any special interactions to develop between the chiral auxiliary and the reaction centers within a zeolite. It is the small "reaction cavity" that has enforced a certain amount of discipline on the reacting molecule. The most exciting result was obtained when (*S*)-tropolone 2-methylbutyl ether was irradiated within an ephedrine-included NaY (Fig. 45). As mentioned earlier, in the absence of ephedrine, the diastereomer **94** is obtained in 53% diastereomeric excess (*de*). When (−)-ephedrine was used as the chiral inductor, the same isomer was enhanced to the extent of 90%. On the other hand, (+)-ephedrine favors the diastereomer **95** to the tune of 70% diastereomeric excess. These observations are truly remarkable. High *de* values has been achieved through the use of two chiral inducers and a confined space. The importance of this result becomes more apparent when one recognizes that irradiation in solution of the same compound in the presence of ephedrine gave a 1:1 diastereomeric mixture (Fig. 45). Zeolite is essential to achieve the high *de.*

IV. CONCLUDING REMARKS

We have shown that the photochemical reactions of organic compounds in microreactors usually show deviation of product distribution from their molecular photochemical reactions in solution and, in some cases, result in the occurrence of reaction pathways that are not otherwise observed. These effects are attributed to (1) size and shape inclusion selectivities, (2) restriction on rotational and translational motions of the included molecules and intermediates imposed by the microreactor, (3) compartmentalization of the substrate molecules in the microreactors, (4) separation or close contact of the substrate with the sensitizer in photosensitizaition reaction, (5) special interactions between the substrate and the microreactor wall, and (6) intimate interaction between chiral inductor and reactant. The application of microreactor to synthetic organic photochemistry has already grown out of its infancy and now emerges as an interesting field in photochemical research. The challenge for the next decade lies not only in developing new experimental and theoretical approaches to better understand the mechanistic and kinetic details of the photochemical processes occurred in microreactors, but also in extending the scope of this methodology and in designing novel microreactors which enhance the selectivity of phototransformations.

ACKNOWLEDGMENTS

We thank the National Science Foundation of China and the Bureau for Basic Research, Chinese Academy of Sciences for financial support.

REFERENCES

1. Turro, N. J. *Pure Appl. Chem.* **1986,** *58,* 1219.
2. Ramamurthy, V.; Eaton, D. F.; Caspar, J. V. *Acc. Chem. Res.* **1992,** *25,* 299.
3. Ramamurthy, V. in *Photochemistry in Organized and Constrained Media,* Ramamurthy, V., ed., VCH: New York, 1991.
4. Ramamurthy, V. in *Surface Photochemistry,* Anpo, M., ed., Wiley: New York, 1996, 65.
5. Weiss, R. G.; Ramamurthy, V.; Hammond, G. S. *Acc. Chem. Res.* **1993,** *26,* 530.
6. Ramamurthy, V.; Weiss, R. G.; Haammond, G. S. *Adv. Photochem.* **1993,** *18,* 69.
7. Ramamurthy, V.; Turro, N. J. *J. Inclusion Phenom. Mol. Recogn. Chem.* **1995,** *21,* 239.

8. Ito, Y. in *Molecular and Supramolecular Photochemistry, Vol. 3,* Ramamurthy, V.; Schaze, K. S., ed., Marcel Dekker: New York, 1999.
9. Schneider, H. J.; Dürr, H., eds, *Frontiers in Supramolecular Organic Chemistry and Photochemistry,* VCH: Weinheim, 1991.
10. Vaughan, D. E. W., in *Comprehensive Supramolecular Cehmistry,* Pergamon: New York, 1996, 379–388.
11. Balzani, V.; Scandola, F., *Supramolecular Photochemistry,* Prentice-Hall: Englewood Cliffs, NJ, 1991.
12. Breck, D. W. *Zeolite Molecular Sieves: Structure, Chemistry and Use,* Wiley: New York, 1974.
13. Dyer, A. *An Introduction to Zeolite Molecular Sieves,* Wiley: New York, 1988.
14. Van Bekkum, H.; Flanigen, E. M.; Jansen, J. C. *Introduction to Zeolite Science and Practice,* Elsevier: Amsterdam, 1991.
15. Meier, W. M.; Olson, D. H. *Atlas of Zeolite Structure Types,* Butterworth-Heinemann: London, 1992.
16. Komoroski, R. A.; Mauritz, K. A. *J. Am. Chem. Soc.* **1978,** *100,* 7487.
17. Lee, P. C.; Meisel, D. *J. Am. Chem. Soc.* **1980,** *102,* 5477.
18. Sondheimer, S. J.; Bunce, N. J.; Fyfe, C. A. *J. Macromol. Sci., Rev. Macromol. Chem. Phys.* **1986,** *C26,* 353.
19. Lee, P. C.; Meisel, D. *Photochem. Photobiol.* **1985,** *41,* 21.
20. Szentirmay, M. N.; Prieto, N. E.; Martin, C. R. *J. Phys. Chem.* **1985,** *89,* 3017.
21. Niu, E. P.; Ghihhino, K. P.; Smith, T. A.; Mau, A. W. H. *J. Lumin.* **1990,** *46,* 191.
22. Niu, E. P.; Mau, A. W. H. *Aust. J. Chem.* **1991,** *44,* 695.
23. Mohan, H.; Iyer, R. M. *J. Chem. Soc., Faraday Trans.* **1992,** *88,* 41.
24. Mika, A. M.; Lorenz, K.; Azczuek, A. *J. Membrane Sci.* **1989,** *41,* 163.
25. Priydarsini, K. I.; Mohan, H.; Mittall, J. P. *J. Photochem. Photobiol. A: Chem.* **1993,** *69,* 345.
26. Peterlin, A. *Macromolecules* **1980,** *13,* 777.
27. Phillips, P. J. *Chem. Rev.* **1990,** *90,* 425, and references therein.
28. Axelson, D. E.; Levy, G. C.; Mandelkern. L. *Macromolecules* **1979,** *12,* 41.
29. Jang, Y. T.; Phillips, P. J.; Thulstrup, E. W. *Chem. Phys. Lett.* **1982,** *93,* 66.
30. Quirk, R. P.; Alsamarraie, M. A. A. in *Polymer Handbook,* Brandup, J.; Immergut, E. H., eds., Wiley: New York, 1989, V/15.
31. Weiss, R. G. *Spectrum* **1994,** *7(4),* 1.
32. Weiss, R. G.; Ramamurthy, V.; Hammond, G. S. *Acc. Chem. Res.* **1993,** *26,* 530.
33. Jenkins, R. M.; Hammond, G. S.; Weiss, R. G. *J. Phys. Chem.* **1992,** *96,* 496.
34. He, Z.; Hammond, G. S.; Weiss, R. G. *Macromolecules* **1992,** *25,* 1568.
35. Zimerman, O. E.; Weiss, R. G. *J. Phys. Chem.* **1998,** *102A,* 5364.
36. Fendler, J. H., *Membrane Mimetic Chemistry,* Wiley: New York, **1981.**
37. Furhop, J. H.; Mathieu, J. *Angew. Chem. Int. Ed. Engl.* **1984,** *23,* 100.
38. Deamer, D., Bangham, A. D., *Biochim. Biophys. Acta* **1976,** *443,* 629.
39. Carmona, R. A. M., Chaimovich, H., *Biochim. Biophys. Acta* **1983,** *733,* 172.
40. Kaler, E. W.; Murthy, A. K.; Rodriguez, B. E.; Zasadzinski, J. A. *Science* **1989,** *245,* 1371.
41. Oberdisse, J. *Langmuir* **1996,** *12,* 1212.
42. Hoffman, H.; Thunig, C.; Schmiedel, P.; Munkert, U. *Langmuir* **1994,** *10,* 3972.

43. Talhout, R.; Engberts, J. B. F. N. *Langmuir* **1997,** *13,* 5001.
44. Duque, D.; Tarazona, P.; Chacon, E. *Langmuir* **1998,** *14,* 6827.
45. Söderman, O.; Herrington, K. L.; Kaaler, E. W.; Miller D. D. *Langmuir* **1997,** *13,* 5531.
46. Lissi, E. A.; Encinas, M. V.; Lemp, E.; Rubio, M. A. *Chem. Rev.* **1993,** *93,* 699.
47. Song, X. D.; Geiger, C.; Farahat, M.; Peristein, J.; Whitten, D. G. *J. Am. Chem. Soc.* **1997,** *119,* 12481.
48. Khairutdinov, R. F.; Hurst, J. K. *J. Phys. Chem. B* **1999,** *103,* 3682.
49. Nassar, P. M.; Almeida, L. E.; Tabak, M. *Langmuir,* **1998,** *14,* 6811.
50. Nascimento, D. B.; Rapuano, R.; Lessa, M. M.; Carmona-Ribeiro, A. M. *Langmuir* **1998,** *14,* 7387.
51. Jain, A.; Xu, W. Y.; Demas, J. N.; Degraff, B. A. *Inorg. Chem.* **1998,** *37,* 1876.
52. Evans, P. A.; Holmers, A. B. *Tetrahedron,* **1991,** *47,* 9131.
53. Illuminati, G.; Mandolini, L. *Acc. Chem. Res.,* **1981,** *14,* 95.
54. Griesbeck, A. G.; Henz, A.; Hirt, J. *Synthesis,* **1996,** *11,* 1261.
55. Mandolini, L. in *Advanced Physical Organic Chemistry,* Gold, V.; Bethell, D., ed., Academic Press: London, 1986, Vol. 22, 1–112, and references therein.
56. Lehn, J. M. *Supramolecular Chemistry,* VCH: Weinheim, 1995.
57. Vögtle, F. Ed. *Topics Curr. Chem.* **1983,** 113.
58. Weber, E.; Vogtle, F., eds. *Topics Curr. Chem.* **1991,** 161.
59. Proter, N. W. *J. Am. Chem. Soc.* **1986,** *108,* 2787.
60. Jiang, X. K.; Hui Y. Z.; Fei, Z. X. *J. Chem. Soc., Chem. Commun.* **1988,** 689.
61. Ruggli, P., *Liebigs Ann. Chem.* **1912,** *392,* 92; **1913,** *399,* 174; **1916,** *12,* 1.
62. Ziegler, K. in *Methoden der Organischen Chemie (Houben-weyl), Vol. 4/2,* Muller, E., ed., Georg Thieme Verlag: Stuttgart, 1955.
63. Inokuma, S.; Sakai, S.; Nishimura, J. *Topics Curr. Chem.* **1994,** *172,* 87.
64. Desvergne, J. P.; Bouas-Laurent, H. *Isr. J. Chem.* **1979,** *18,* 220.
65. Tung, C. H.; Li, Y.; Yang, Z. Q. *J. Chem. Soc. Faraday Trans.,* **1994,** *90(7),* 947.
66. Tung, C. H.; Wang, Y. M. *J. Am. Chem. Soc.,* **1990,** *112,* 6322.
67. Tung, C. H.; Xu, C. B. in *Photochemistry and Photophysics,* CRC Press: Boca Raton, FL, 1991, Vol. 4, 177.
68. Tung, C. H.; Wu, L. Z.; Yuan, Z. Y.; Su N. *J. Am. Chem. Soc.* **1998,** *120,* 11594.
69. Tung, C. H.; Yuan, Z. Y.; Wu, L. Z.; Weiss, R. G. *J. Org. Chem.* **1999,** *64,* 5156.
70. Suib, S. L. in *Photochemistry and Photophysics,* Rabek, J., ed.; CRC Press: Boca Raton, FL, 1991, Vol. 3.
71. Liu, X.; Iu, K. K.; Thomas, J. K. *J. Phys. Chem.* **1989,** *93,* 4120.
72. Iu, K. K.; Liu X.; Thomas J. K. *Mater. Res. Soc.* **1991,** *223,* 119.
73. Dinesenko, G. I.; Lisovenko, V. A. *Zl. Prikl Spektrosk.* **1971,** *14,* 702.
74. Incavo, J. A.; Dutta, P. K. *J. Phys. Chem.* **1990,** *94,* 3075.
75. Baretz, B. H.; Turro, N. J. *J. Photochem.* **1984,** *24,* 201.
76. Borja, M.; Dutta, P. K. *Nature* **1993,** *362,* 43.
77. Laine, P.; Lanz, M.; Calzaferri, G. *Inorg. Chem.* **1996,** *35,* 3514.
78. Dutta, P. K.; Turbeville, W. *J. Phys. Chem.* **1992,** *96,* 9410.
79. Kim, Y.; Mallouk, T. E. *J. Phys. Chem.* **1992,** *96,* 2879.
80. Collin, P. J.; Roberts, D. B.; Sugowdz, G.; Wells, D.; Sasse, W. H. F. *Tetradron Lett.* **1972,** 321.

81. Kowala, C.; Sugowdz, G.; Sasse, W. H. F.; Wunderlich, J. A. *Tetradron Lett.* **1972,** 4721.
82. Teitei, T.; Wells, D.; Sasse, W. H. F., *Aust. J. Chem.* **1976,** *29,* 1783.
83. Lei, X.-G.; Doubleday Jr. C. E.; Zimmt, M. B.; Turro, N. J. *J. Am. Chem. Soc.* **1986,** *108,* 2444.
84. Maruszewski, K.; Strommen, D. P.; Handrich, K.; Kincaid, J. R. *Inorg. Chem.* **1991,** *30,* 4579.
85. Becker H. D. *Chem. Rev.* **1993,** *93,* 73.
86. Costa, S. M. B.; Melo, E. *J. Chem. Soc., Faraday Trans. 2* **1980,** 76.
87. Cowan, D. O.; Schmiegel, W. W. *J. Am. Chem. Soc.* **1972,** *94,* 6779.
88. Kaupp G. in *Advances in Photochemistry, Vol. 19,* Neckers, D. C.; Volman, D. H.; von Bunau, G., eds., Wiley: New York, 1995, 119.
89. Yamanoto, S.-A.; Grellmann, K.-H.; Weller, A. *Chem. Phys. Lett.* **1980,** *70,* 241.
90. Becker H. D.; Langer V.; Becker H.-C. *J. Org. Chem.* **1993,** *58,* 6394.
91. Moreno, M. J.; Lourtie, I. M. G.; Melo, E. *J. Phys. Chem.* **1996,** *100,* 18, 192.
92. Bouas-Laurent H.; Castellan A.; Desvergne J. P. *Pure Appl. Chem.* **1980,** *52,* 2633.
93. Ito, Y.; Fujita, H. *J. Org. Chem.* **1996,** *61,* 5677.
94. Bouas-Laurent H.; Desvergne, J.-P. in *Photochromism: Molecules and Systems,* Dürr H.; Bouas-Laurent, H., eds., Elsevier: Amsterdam, 1990, 561.
95. Desvergne, J. P.; Fages, F.; Bouas-Laurent, H.; Marsau, P. *Pure Appl. Chem.* **1992,** *64,* 1231.
96. Desvergne, J. P.; Bouas-Laurent, H. *Chem. Commun.* **1978,** 403.
97. Fox, M. A.; Britt, P. F. *Photochem. Photobiol.* **1990,** *51,* 129.
98. Becker, H. D.; Amin, K. A. *J. Org. Chem.* **1989,** *54,* 3182.
99. De Schryver, F. C.; Boens, N.; Huybrechts, J.; Daemen, J.; De Brackeleire, M. *Pure Appl. Chem.* **1977,** *49,* 237.
100. Ito, Y.; Olovsson, G. *J. Chem. Soc., Perkin Trans. 1* **1997,** 127.
101. Chandross, E. A.; Schiebel A. H. *J. Am. Chem. Soc.* **1973,** *95,* 611.
102. Chandross E. A.; Schiebel, A. H. *J. Am. Chem. Soc.* **1973,** *95,* 1671.
103. Ferguson, J.; Mau, A. W.-H.; Puza, M. *Mol. Phys.* **1974,** *27,* 377.
104. Ferguson, J.; Mau, A. W.-H.; Puza, M. *Mol. Phys.* **1974,** *28,* 1457.
105. Ferguson, J.; Mau, A. W.-H.; Whimp, P. O. *J. Am. Chem. Soc.* **1979,** *101,* 2370.
106. Desvergne, J. P.; Bitit, N.; Castellan, A.; Bouas-Laurent, H. *J. Chem. Soc. Perkin. Trans. 2* **1983,** 109.
107. Albini, A.; Fasani, E.; Faiardi, D. *J. Org. Chem.* **1987,** *52,* 155.
108. Albini, A.; Fasani, E. *J. Am. Chem. Soc.* **1988,** *110,* 7760.
109. Tazuke, S.; Watanabe, H. *Tetrahedron Lett.* **1982,** *23,* 197.
110. Ferguson, J.; Puza, M.; Robbins, R. J. *J. Am. Chem. Soc.* **1985,** *107,* 1869.
111. Zimerman, O. E.; Cui, C.; Wang, X.; Atvars, T. D. Z.; Weiss, R. G. *Polymer* **1998,** *39,* 1177.
112. Lu, L.; Weiss, R. G. *Macromolecules* **1994,** *27,* 219.
113. He, Z.; Hammond, G. S.; Weiss, R. G. *Macromolecules* **1992,** *25,* 501.
114. Naciri. J.; Weiss, R. G. *Macromolecules* **1989,** *22,* 3928.
115. Ramesh, V.; Weiss, R. G. *Macromolecules* **1986,** *19,* 1489.
116. Cui, C.; Weiss, R. G. *J. Am. Chem. Soc.* **1993,** *115,* 9820.
117. Aviv, G.; Sagiv, J.; Yogev, A. *Mol. Cryst. Liquid Cryst.* **1976,** *36,* 349.

118. Li, S. k. L.; Guillet, J. E. *Polym. Photochem.* **1984,** *4,* 21.

119. Melo, J. S. S.; Becker, R. S.; Macanita, A. L. *J. Phys. Chem.* **1994,** *98,* 6054.

120. Hallberg, A.; Isaksson, R.; Martin, A. R.; Sandström, J. *J. Am. Chem. Soc.* **1989,** *111,* 4387.

121. Sakellariou-Fargues R.; Maurette, M. T.; Oliveros E.; Riviere M.; Lattes, A. *Tetrahedron* **1984,** *40,* 2381.

122. Muthuramu, K.; Ramnath, N.; Ramamurthy, V. *J. Org. Chem.* **1983,** *48,* 1872.

123. Morrison, H.; Curtis, H.; McDowell T. *J. Am. Chem. Soc.* **1966,** *88,* 5415.

124. Moorthy, J. N.; Venkatesan, K.; Weiss, R. G. *J. Org. Chem.* **1992,** *57,* 3292.

125. Lewis, F. D.; Barancyk, S. V. *J. Am. Chem. Soc.* **1989,** *111,* 8653.

126. Lewis, F. D.; Howard, D. K.; Oxman, J. D. *J. Am. Chem. Soc.* **1983,** *105,* 3344.

127. Muthuramu, K.; Ramamurthy, V. *J. Org. Chem.* **1982,** *47,* 3976.

128. Serna, J.; Abbe, J. Ch.; Duplatre, G. *Phys. Status Solidi A* **1989,** *115,* 389.

129. Haber, J. in *Perspectives in Catalysis;* Thomas, J. M.; Zamaraev, K. I., eds., IUPAC/Blackwell Scientific Publications: London, 1992, 371–385.

130. Parshall, G. W.; Ittel, S. D. *Homogeneous Catalysis,* 2nd ed., Wiley: New York, 1992, 237–268.

131. Sheldon, R. A.; Kochi, J. K. *Metal-Catalyzed Oxidation of Organic Compounds,* Academic Press: New York, 1981.

132. Blatter, F.; Frei, H. *J. Am. Chem. Soc.* **1994,** *116,* 1812.

133. Wasserman, H. H.; Murray, R. W., eds., *Singlet Oxygen,* Academic Press: New York, 1979.

134. Matsumoto, M. In *Singlet O₂,* Frimer, A. A., ed., CRC Press: Boca Raton, FL, 1985, Vol. 2, Part 1, 205–272.

135. Foote, C. S. *Photochem. Photobiol.* **1991,** *54,* 659.

136. Foote, C. S. *Tetrahedron* **1985,** *41,* 2221.

137. Kanner, R. C.; Foote, C. S. *J. Am. Chem. Soc.* **1992,** *114,* 678 and 682.

138. Erikson, J.; Foote, C. S. *J. Am. Chem. Soc.* **1980,** *102,* 6083.

139. Onodera, K.; Furusawa, G.; Kojima, M.; Tsuchiya, M.; Aihara, S.; Akaba, R.; Sakuragi, H.; Tokumaru, K. *Tetrahedron* **1985,** *41,* 2215.

140. Aronovitch, C.; Mazur, Y. *J. Org. Chem.* **1985,** *50,* 149.

141. Hashimoto, S.; Akimoto, H. *J. Phys. Chem.* **1986,** *90,* 529.

142. Hashimoto, S.; Akimoto, H. *J. Phys. Chem.* **1987,** *91,* 1347.

143. Hashimoto, S.; Akimoto, H. *J. Phys. Chem.* **1989,** *93,* 571.

144. Blatter, F.; Frei, H. *J. Am. Chem. Soc.* **1993,** *115,* 7501.

145. Blatter, F.; Moreau, F.; Frei, H. *J. Phys. Chem. Soc.* **1994,** *98,* 13,403.

146. Blatter, F.; Sun, H.; Frei, H. *Catal. Lett.* **1995,** *35,* 1.

147. Frei, H.; Blatter, F.; Sun, H. *CHEMTECH* **1996,** *26,* 24.

148. Sun, H.; Blatter, F.; Frei, H. *J. Am. Chem. Soc.* **1994,** *116,* 7951.

149. Sun, H.; Blatter, F.; Frei, H. *Catal. Lett.* **1997,** *44,* 247.

150. Blatter, F.; Sun, H.; Vasenkov; Frei, H. *Catal. Today* **1998,** *41,* 297.

151. Blatter, F.; Sun, H.; Frei, H. *Chem. Eur. J.* **1996,** *2,* 385.

152. Xiang, Y.; Larsen, S. C.; Grassian, V. H. *J. Am. Chem. Soc.* **1999,** *121,* 5063.

153. Ranby, B.; Rabek, J. F. *Singlet Oxygen Reactions with OrganicCompounds and Polymers,* Wiley: Chichester, 1979.

154. Horsey, B. E.; Whitten, D. G. *J. Am. Chem. Soc.* **1978,** *100,* 1293.

155. Ohtani, B.; Nishida, M.; Nishimoto, S.-I.; Kagiya, T. *Photochem. Photobiol.* **1986,** *44,* 725.
156. Polo, E.; Amadelli, R.; Carassiti, V.; Maldotti, A. *Inorg. Chim. Acta* **1992,** *192,* 1.
157. Amadelli, R.; Bregola, M.; Polo, E.; Crassiti, V.; Maldotti, A. *J. Chem. Soc., Chem. Commun.* **1992,** 1355.
158. Maldotti, A.; Molinari, A.; Andreotti, L.; Fogagnolo, M.; Amadelli, R. *Chem. Commun.* **1998,** 507.
159. Pettit, T. L.; Fox, M. A. *J. Phys. Chem.* **1986,** *90,* 1353.
160. Tung, C.-H.; Wang, H.-W.; Ying, Y.-M. *J. Am. Chem. Soc.* **1998,** *120,* 5179.
161. Tung, C.-H.; Guan, J.-Q. *J. Am. Chem. Soc.* **1998,** *120,* 11874.
162. Li, H. R.; Wu, L. Z.; Tung, C. H. *J. Am. Chem. Soc.* **2000,** *121,* 2446.
163. Li, H. R.; Wu, L. Z.; Tung, C. H. *Chem. Commun.* **2000,** *12,* 1085–1086.
164. Li, H. R.; Wu, L. Z.; Tung, C. H. *Tetrahedron* **2000,** *56,* 7437.
165. Ramamurthy, V.; Lakshminarasimhan, P.; Grey, C. P.; Johnston, L. J. *Chem. Commun.* **1998,** 2411.
166. Li, X.; Ramamurthy, V. *Tetrahedron Lett.* **1996,** *37,* 5235.
167. Li, X.; Ramamurthy, V. *J. Am. Chem. Soc.* **1996,** *118,* 10,666.
168. Leibovitch, M.; Olovsson, G.; Sundarababu, G.; Ramamurthy, V.; Scheffer, J. R.; Trotter, J. *J. Am. Chem. Soc.* **1996,** *118,* 1219.
169. Sundarababu, G.; Leibovitch, M.; Corbin, D. R.; Scheffer, J. R.; Ramamurthy, V. *Chem. Commun.* **1996,** 2049.
170. Robbins, R. J.; Ramamruthy, V. *Chem. Commun.* **1997,** 1071.
171. Joy, A.; Robbins, R. J.; Pitchumani, K.; Ramamurthy, V. *Tetrahedron Lett.* **1997,** *38,* 8825.
172. Hu, Y.-Z.; An, J.-Y.; Jiang, L.-J. *J. Photochem. Photobiol. B. Biol.* **1993,** *17,* 195.
173. Matsumoto, M.; Dobashi, S.; Kondo, K. *Tetrahedron Lett.* **1977,** 2329.
174. Matsumoto, M.; Dobashi, S.; Kuroda, K. *Tetrahedron Lett.* **1977,** 3361.
175. Matsumoto, M.; Kondo, K. *Tetrahedron Lett.* **1975,** 3935.
176. Ogumi, Z.; Kuroe, T.; Takehara, Z. *J. Electrochem. Soc.* **1985,** *132,* 2601.
177. Gollnick, K.; Schenck, G. O. *Pure Appl. Chem.* **1964,** *9,* 507.
178. Foote, C. S.; Wexler, S. S.; Ando, W. *Tetrahedron Lett.* **1965,** 4111.
179. Jefford, C. W.; Boschung, A. F.; Moriarty, R. M., Rimbault, C. G.; Laffer, M. H., *Helv. Chim. Acta* **1973,** *56,* 2649.
180. Gollnick, K.; Kuhn, H. J. in *Singlet Oxygen,* Wasserman, H. H.; Murray, R. W., eds., Academic Press: New York, 1979, 287.
181. Zhang, B. W.; Ming, Y. F.; Cao, Y. *Photochem. Photobiol.* **1984,** *40,* 581.
182. Turro, N. J. in *Modern Molecular Photochemistry, Benjamin/Cummings:* San Francisco, **1978,** 319.
183. Grossweiner, L. I. *Photochem. Photobiol.* **1977,** *26,* 309.
184. Lissi, E. A.; Rubio, M. A. *Pure Appl. Chem.* **1990,** *62,* 1503.
185. Rio, G.; Berthelot, J. *Bull. Soc. Chem. Fr.* **1969,** *5,* 1664.
186. Cao, Y.; Zhang, B. W.; Ming, Y., F.; Chen, J. X. *J. Photochem.* **1987,** *38,* 131.
187. Zhang, B. W.; Chen, J. X.; Cao, Y. *Acta Chim. Sin.* **1989,** *47,* 502.
188. Baldovi, M. V.; Corma, A.; Garcia, H.; Marti, V. *Tetrahedron Lett.* **1994,** *35,* 9447.
189. Gessner, F.; Scaiano, J. C. *J. Photochem. Photobiol. A: Chem.* **1992,** *67,* 91.
190. Prein, M.; Adam, W. *Angew. Chem.* **1996,** *35,* 477.

191. Sun, H; Blatter, F.; Frei, H. *J. Am. Chem. Soc.* **1996,** *118,* 6873.

192. Pitchumani, K.; Warrier, M.; Ramamurthy, V. *J. Am. Chem. Soc.* **1996,** *118,* 9428.

193. Gu, W. Q.; Warrier, M.; Ramamurthy, V.; Weiss, R. G. *J. Am. Chem. Soc.* **1999,** *121,* 9467.

194. Tung, C. H.; Ying, Y. M. *J. Chem. Soc. Perkin Trans. 2* **1997,** 1319.

195. Tung, C. H.; Xu, X. H. *Tetrahrdron Lett.* **1999,** *40,* 127.

196. Lally, J. M.; Spillane, W. J. *J. Chem. Soc., Perkin Trans. 2* **1991,** 803.

197. Kalmus, C. E.; Hercules, K. M. *J. Am. Chem. Soc.* **1974,** *96,* 449.

198. Adam, W. *J. Chem. Soc., Chem. Commun.* **1974,** 289.

199. Meyer, J. W.; Hammond, G. S. *J. Am. Chem. Soc.* **1972,** *94,* 2219.

200. Pitchumani, K.; Warrier, M.; Scheffer, J. R.; Ramamurthy, V. *Chem. Commun.* **1998,** 1197.

201. Sadeghpoor, R.; Ghandi, M.; Mahmoudi N. H.; Farzaneh, F. *Chem. Commun.* **1998,** 329.

202. Tomotake, N.; Katsuhika, T.; Mariko, I.; Kaori, F.; Hiroyuki, K.; Toyoko, I.; Ya-suhiko, S. *J. Chem. Soc., Perkin Trans. 2* **1997,** 2751.

203. Schutz, A.; Wolff, T. *J. Photochem. Photobiol. A, Chem.* **1997,** *109,* 251.

204. Tung, C. H.; Guan, J. Q. *J. Org. Chem.* **1998,** *63,* 5857.

205. Tung, C. H.; Guan, J. Q. *J. Org. Chem.* **1996,** *61,* 9417.

206. Sondheimer, S. T.; Bunce, N. J.; Lemke, M. E.; Fyfe, C. A. *Macromolecules* **1986,** *19,* 339.

207. Griffiths, J. *J. Chem. Soc. Rev.* **1972,** *1,* 481.

208. Corma, A.; Garcia, H.; Ibarra, S.; Marti, V.; Miranda, M. A.; Primo, J. *J. Am. Chem. Soc.* **1993,** *115,* 2177.

209. Lei, Z.; Vaidyalingam, A.; Dutta, P. K. *J. Phys, Chem.* **1998,** *102,* 8557.

210. Inoue, Y. *Chem. Rev.* **1992,** *92,* 741.

211. Joy, A.; Ramamurthy, V. *Chem. Eur. J.* **2000,** *6,* 1287.

212. Stinson, S. C. *Chem. Eng. News* **1998,** *76(38),* 83.

213. Stinson, S. C. *Chem. Eng. News.* **1997,** *75(42),* 38.

214. Avalos, M.; Babiano, R.; Cintas, P.; Jimenez, J. L.; Palacios, J. C. *Chem. Rev.* **1998,** *98,* 2391.

215. Richards, A.; McCague, R. *Chem. Ind.* **1997,** 422.

216. Noyori, R. *Asymmetric Catalysis in Organic Synthesis,* Wiley-Interscience: New York, 1994.

217. Cervinka, O. *Enantioselective Reactions in Organic Chemistry,* Ellis Horwood: London, 1995.

218. Rau, H. *Chem. Rev.* **1983,** *83,* 535.

219. Pete, J. P. *Adv. Photochem.* **1996,** *21,* 135.

220. Everitt, S. R. L.; Inoue Y. in *Molecular and Supramolecular Photochemistry, Vol. 3,* Ramamurthy, V.; Schanze, K. S., eds., Marcel Dekker: New York, 1999, 71ff.

221. Buschmann, H.; Scharf, H. D.; Hoffmann, N.; Esser, P. *Angew. Chem.* **1991,** *103,* 480; *Angew. Chem. Int. Ed. Engl.* **1991,** *30,* 477.

222. Inoue, Y.; Yamasaki, N.; Yokoyama, N.; Arai T. *J. Org. Chem.* **1992,** *57,* 1332.

223. Inoue, Y.; Okano, T.; Yamasaki, N.; Tai, A. *J. Chem. Soc., Chem. Commun.* **1993,** 718.

224. Inoue, Y.; Yamasaki, N.; Yokoyama, N.; Arai T. *J. Org. Chem.* **1992,** *57,* 1011.

225. Ramamurthy, V.; Venkatesan, K. *Chem. Rev.* **1987,** *87,* 433.

226. Vaida, M.; Popovita-Biro, R.; Leserowita, L.; Lahav, M. in *Photochemistry in Organized and Constrained Media,* Ramamurthy, V., ed., VCH: New York, 1991, 247.

227. Sundarababu, G.; Leibovitch, M.; Corbin, D. R.; Scheffer, J. R.; Ramamurthy, V. *Chem. Commun.* **1996,** 2159.

228. Joy, A.; Scheffer, J. R.; Corbin, D. R.; Ramamurthy, V. *Chem. Commun.* **1998,** 1379.

229. Joy, A.; Corbin, D. R.; Ramamurthy, V. in *Proceedings of 12th International Zeolite Conference,* Traecy, M. M. J.; Marcus, B. K.; Bisher, B. E.; Higgins, J. B., eds., Materials Research Society. Warrendale, PA. 1999, 2095.

230. Joy, A.; Scheffer, J. R.; Ramamurthy, V. *Org. Lett.* **2000,** *2,* 119.

231. Joy, A.; Uppiloi, S; Netherton, M. R.; Scheffer, J. R.; Ramamurthy, V. *J. Am. Chem. Soc.* **2000,** *122,* 728.

6

Enantioselective Photoreactions in the Solid State

Fumio Toda
Okayama University of Science, Okayama, Japan

Koichi Tanaka and Hisakazu Miyamoto
Ehime University, Ehime, Japan

I. INTRODUCTION

Because molecules are packed at close positions regularly in crystals, selective photoreactions occur effectively by photoirradiation in the solid state. When enantioselective reactions are required, achiral molecules should be arranged in a chiral form in crystals. In some cases, achiral molecules are arranged in a chiral form in their own crystals. Photoreaction of the chiral crystal gives an optically active reaction product. When an achiral compound is oily material or does not form a chiral crystal, the achiral molecules can be arranged in a chiral form by inclusion complex formation with a chiral host compound. Photoirradiation of the inclusion complex crystals gives optically active product and the chiral host can be used again. These enantioselective photoreactions do not need an organic solvent and a chemical reagent and can be a typical example of a green and sustainable chemical process.

In this chapter, enantioselective photoreactions of achiral molecules in their own chiral crystals and in inclusion crystals with a chiral host compound that have been carried out mainly in our research groups are described.

II. ENANTIOSELECTIVE PHOTOREACTIONS OF ACHIRAL MOLECULES IN THEIR OWN CHIRAL CRYSTALS: GENERATION OF CHIRALITY

In some cases, achiral molecules are arranged in a chiral form in their own crystals. When the chirality can be fixed by photoreaction, this becomes a convenient asymmetric synthesis without using any further chiral source, so it can be described as an absolute asymmetric synthesis. This phenomenon is also important in relation to the mechanism of generation of chirality on Earth. Several such examples of absolute asymmetric synthesis in crystals have been reported so far. In this section, some of these found mainly in the author's laboratory are described.

A. Oxoamides

Recrystallization of the achiral oxo amide N,N-diisopropylphenylglyoxylamide (**1a**) from benzene gave colorless chiral prisms [1]. Each crystal is chiral and shows a CD spectrum in Nujol mulls (Fig. 1). One type of chiral crystal shows a $(+)$-Cotton effect and the other type shows a $(-)$-Cotton effect (Fig. 1). Crystals of $(+)$-**1a** and $(-)$-**1a** can easily be prepared in large quantities by seeding with finely powdered crystals of $(+)$-**1a** or $(-)$-**1a** during recrystallization of **1a** from benzene. Measurement of the CD spectrum of chiral crystals as Nujol mulls is now well established [2].

Irradiation of crystals of $(+)$-**1a** (200 mg) with a high-pressure mercury lamp, by occasional grinding using an agate mortar and pestle at room temperature for 40 h, gave $(+)$-β-lactam 3-hydroxy-1-isopropyl-4,4-dimethyl-3-phenylazetin-2-one (**2a**) of 93% *ee* in 74% yield [1]. Irradiation of $(-)$-**1a** under the same conditions gave $(-)$-**2a** of 93% ee in 75% yield [1]. The reason why achiral **1a** molecules are easily arranged in a chiral from was clarified by x-ray analysis of a crystal [3]. X-ray analysis showed that the two carbonyl groups are twisted

a : R=H	**e** : R=o-Me	**i** : R=p-Cl
b : R=m-Me	**f** : R=o-Cl	**j** : R=p-Br
c : R=m-Cl	**g** : R=o-Br	
d : R=m-Br	**h** : R=p-Me	

Structures 1 and 2

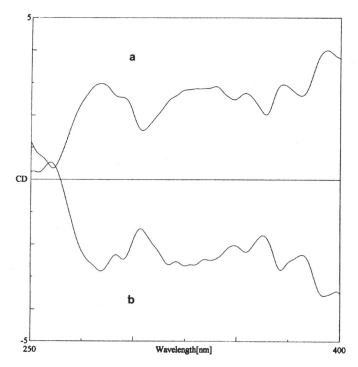

Figure 1 CD spectra of (a) (+)-**1a** and (b) (−)-**1a** chiral crystals in Nujol mulls.

around the single bond connecting these two groups, as schematically depicted in Scheme 1. Photoreaction between the benzoylcarbonyl and the isopropyl group of (+)-**1a** and (−)-**1a** gives (+)-**2a** and (−)-**2a,** respectively (Scheme 1).

In this case, chiral crystals are available in bulk, and mass production of the chiral products is possible. Moreover, the present data may throw some light on the generation of optically active amino acids on Earth [4]. Photocyclization of **1a** proceeds efficiently in sunlight, and hydrolysis of the optically active **2a** gives an optically active β-amino acid.

Of 11 derivatives of **1a, 1b–1j, 3** and **5** the compounds **1b–1e** and **3** also formed chiral crystals, and their photoirradiation in the solid state gave the corresponding chiral β-lactams, **2b, 2c, 2d, 2e,** and **4** in the optical and chemical yields indicated in Table 1 [5].

Although all meta-substituted derivatives **2b–2d** formed chiral crystals, all the para-substituted ones (**1h–1j**) did not. Photoirradiation of **1h–1j** in the solid state gave racemic β-lactams **2h, 2i,** and **2j,** respectively, in the yields indicated (Table 1). In the case of the *o*-substituted derivatives, **1e–1g,** only the *o*-methyl-

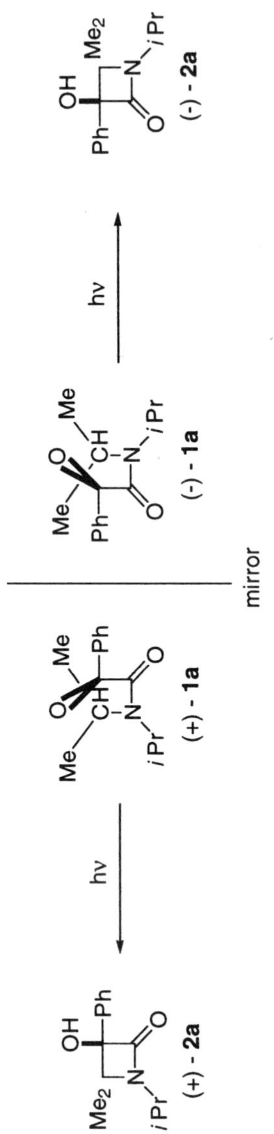

Scheme 1

Structures 3–6

substituted one (**1e**) formed chiral crystals, and its photoirradiation gave optically active **2e** of 92% *ee* in 54% yield. These data suggested that the *m*-substitution pattern is better for forming chiral crystals than the other substitution isomer. However, the *m,m*-dimethyl-substituted derivative **5** did not form chiral crystals and its photoirradiation gave racemic **6**, although the *m,p*-dimethyl derivative **3** formed chiral crystals and its photoirradiation gave optically active **4** of 54% *ee* in 62% yield (Table 1) [5].

Table 1   Photocyclization of Substituted Phenylglyoxylamides in Their Crystals

			Photoproduct	
Phenylglyoxylamide		Irradiation time (hr)	Yield (%)	Optical purity (% *ee*)
a	H	40	75	93
b	*m*-Me	7	63	91
c	*m*-Cl	10	75	100
d	*m*-Br	5	97	96
e	*o*-Me	10	54	92
f	*o*-Cl	24	42	0
g	*o*-Br	24	48	0
h	*p*-Me	5	60	0
i	*p*-Cl	5	50	0
j	*p*-Br	12	65	0
3	3,4-Me$_2$	5	62	54
5	3,5-Me$_2$	5	74	0

By x-ray analysis, the chiral arrangement of **1c** molecules and the achiral arrangement of **1f** and **1i** molecules in their crystals have been proven [6]. The photochemical conversion of **1d** in its chiral crystal to the optically active β-lactam **2d** was monitored by continuous measurements of CD spectra of Nujol mulls (Fig. 2). As the reaction proceeds, the CD spectra of **1d** decrease and new spectra due to **2d** appear.

It has been found that a combination of the two alkyl groups on the nitrogen atom of **1** is important to form a chiral crystal. Of the four derivatives of **1, 7a–7c,** and **9,** only the derivative **7b,** which has *i*-Pr and Et groups, formed chiral crystals. The chirality of **7b** was confirmed by CD spectra measured as Nujol mulls (Fig. 3) [7]. The chiral arrangement of the **7b** molecule within its crystal was proven by x-ray analysis [7]. Photoirradiation of (−)-**7b** and (+)-**7b** crystals in the solid state gave (−)-**8b** and (+)-**8b** of 80% *ee* in 55% yield, respectively. Of course, irradiation of **7a, 7c,** and **9** gave racemic **8a, 8c,** and **10,** respectively.

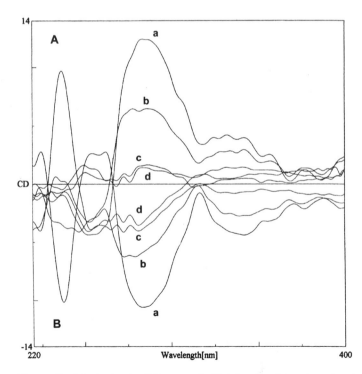

Figure 2   Monitoring by CD spectra in Nujol mulls of the photochemical conversion of (+)-**1d** (A) and (−)-**1d** (B) chiral crystals to (+)-**2d** and (−)-**2d** crystals, respectively; after irradiation for (a) 0 min, (b) 2 min, (c) 4 min, and (d) 6 min.

$$PhCOCON\overset{iPr}{\underset{R}{\diagdown}}$$

7

Ph—OH / Me$_2$ / $\overset{O}{\diagdown}$—N—R

8

a : R=Me
b : R=Et
c : R=nPr

$PhCOCONMe_2$

9

Ph—OH / H$_2$ / $\overset{O}{\diagdown}$—N—Me

10

Structures 7–10

Even for the oxoamides which do not form chiral crystals, enantioselective photoconversion to optically active β-lactams can easily be accomplished by photoirradiation in their inclusion crystals with an optically active host compound. For example, irradiation of a 1:1 inclusion complex crystal of **9** with **11** [8] gave (−)-**10** of 100% *ee* in a quantitative yield [9]. The host compound **11** is recovered and can be used again. The chiral arrangement of **9** molecules in the inclusion complex was studied by x-ray analysis [9]. A schematic stereoview of the inclu-

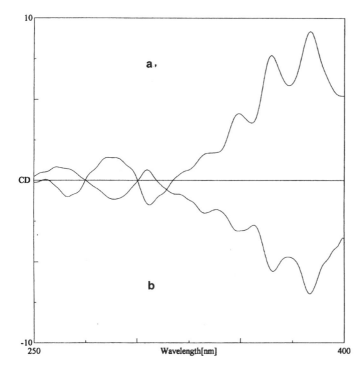

Figure 3 CD spectra of (a) (+)-**7b** and (b) (−)-**7b** chiral crystals in Nujol mulls.

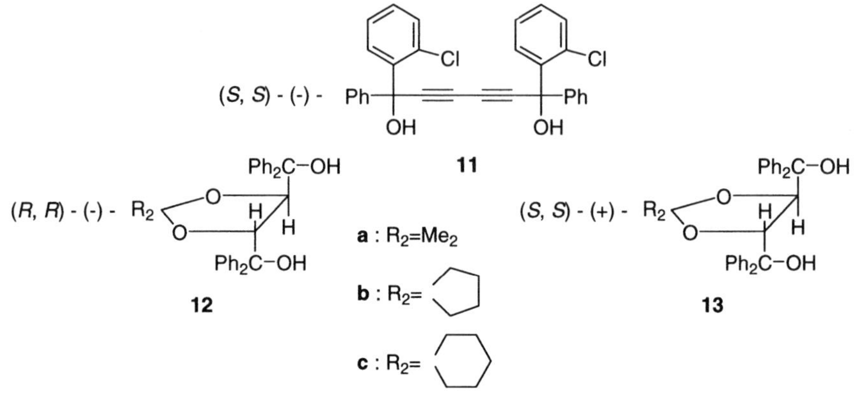

Structures 11–13

sion complex of **9** with **11** and its enantioselective photoreaction to give (−)-**10** of 100% *ee* is shown in Fig. 4.

Optically active hosts **12** [10] and **13** [10] are also useful for an enantioselective photocyclization of **9** [11]. The mechanism of the enantioselective reaction of **9** in an inclusion complex with **12** has also been studied by x-ray analysis [12].

Very interestingly, inclusion complex crystals of oxoamides with **12** can be prepared just by mixing both components in the solid state [13]. Irradiation of the complex formed by such mixing gives the optically active β-lactam. For example, mixing of **12c** and **9** in a 2:1 ratio gave their 2:1 inclusion complex and photoirradiation of the complex gave (+)-**10** of 41% *ee* in 48% yield. However photoirradiation of a 2:1 inclusion complex prepared by recrystallization of **12** and **9** from solvent gave (−)-**10** of 85% *ee* in 39% yield. The chiral arrangement of **9** in these

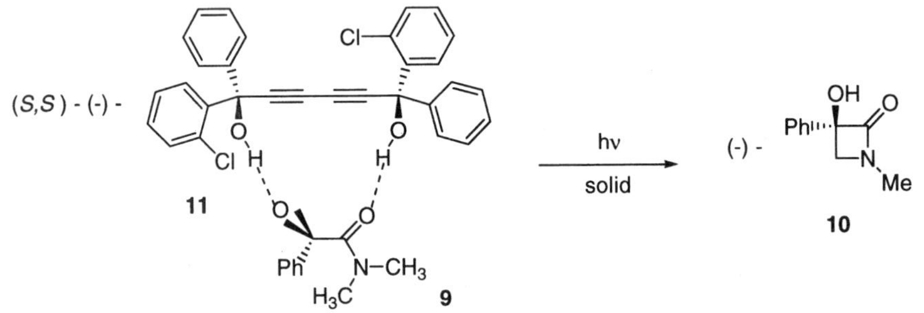

Figure 4 Schematic stereoview of the inclusion complex of **9** with **11** and its photoreaction product (−)-**10.**

two complexes is reversed, although the reason for this drastic change is not clear [13].

When **12b** is used instead of **12c,** irradiation of the two 1:1 inclusion complexes of **12b** with **9,** which are produced by mixing or recrystallization, gave (−)-**10** (29% yield, 82% ee) and (−)-**10** (47% yield, 79% ee), respectively, in the optical and chemical yield indicated [13]. In contrast, the inclusion complex of **12a** with **9,** formed by mixing both components, on its photoirradiation gave (+)-**10** of 61% ee in 70% yield. However, an inclusion complex could not be obtained by recrystallization, despite more than 15 organic solvents being tested. In all cases, **12a** crystallized out separately.

B. *N*-Acyloxoamides and Related Compounds

N-Acyloxoamides such as **14** and **16** also form their own chiral crystals and their photoirradiation at 0°C gave optically active photocyclization products, **15** (84% yield, 35% ee) and **17** (100% yield, 35% ee), respectively, in the chemical and optical yields indicated [14]. However, when the irradiation of **14** and **16** was carried out at −78°C, **15** (87% yield, 99% ee) and **17** (100% yield, 91% ee) were obtained, respectively, in the chemical and optical yields indicated [14].

In the case of **18,** however, photoreaction in its chiral crystals gave **19** (22% yield, 32% ee) in the low chemical and optical yields as indicated, together with the side product **20** (17% yield, 46% ee) [15]. The reason for the low optical yield is not clarified. As the photoreaction in the chiral crystal proceeds, chiral arrangement of the achiral molecules of **18** in their crystal would be disturbed and racemized.

Thiourea derivatives **21a** and **21b** also form chiral crystals, and their photoreactions gave **22a** (16% yield, 20% ee) and **22b** (19% yield, 20% ee), respec-

Structures 14–17

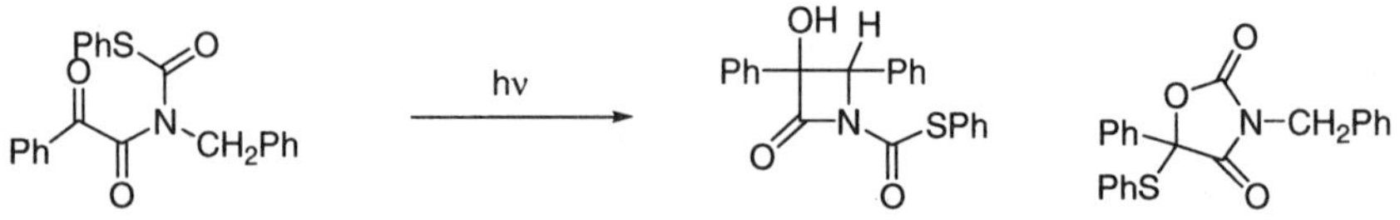

18 **19** **20**

Structures 18–20

tively, in the low chemical and optical yields indicated [16]. Irradiation of a similar thiourea derivative **23** in its chiral crystal at 0°C gave the bicyclic reaction product **24** (75% yield, 10% *ee*) [17]. Although similar photoirradiation at −45°C gave **24** (70% yield at 30% conversion, 40% *ee*) in relatively high chemical and optical yields indicated, no reaction occurred at −78°C [17].

Enantioselectivity of the photochemical conversion of the thioamide **25** in its chiral crystals into **26** (96% yield at 58% conversion, 81–97% *ee*) is quite high, as indicated [18].

C. Tripticenes and Adamantane Derivatives

Recrystallization of **27a** from cyclohexane gave chiral crystals, and their photoirradiation in the solid state afforded optically pure cyclization product **28a** [19]. Similarly, photoirradiation of the ethanol complex crystals of **27b** in which **27b**

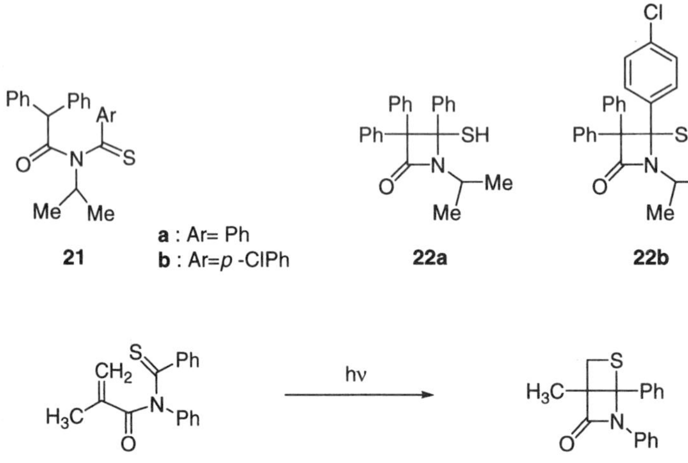

21 **a** : Ar= Ph
b : Ar=*p*-ClPh **22a** **22b**

23 **24**

Structures 21–24

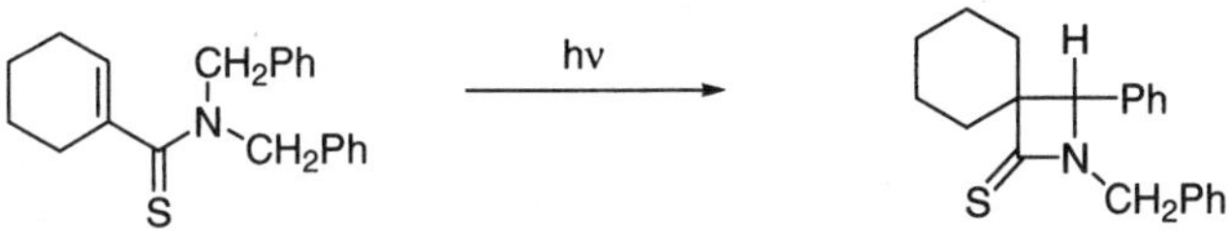

25   **26**

Structures 25 and 26

molecules are arranged in a chiral form gave the optically active compound **28b** in 89% optical yield [19]. Compound **29** also forms chiral crystals, and their photoirradiation at $-20°C$ gave **30** of 70–100% *ee*. Chiral crystals of **31** gave **32** of 80% *ee* in 70% yield on photoirradiation [19].

D. Succinimde Derivatives

The achiral molecule 3,4-bis(diphenylmethylene)succinimide (**33**) was found to arrange itself in a chiral form in the crystalline state [20]. The chirality of **33** was

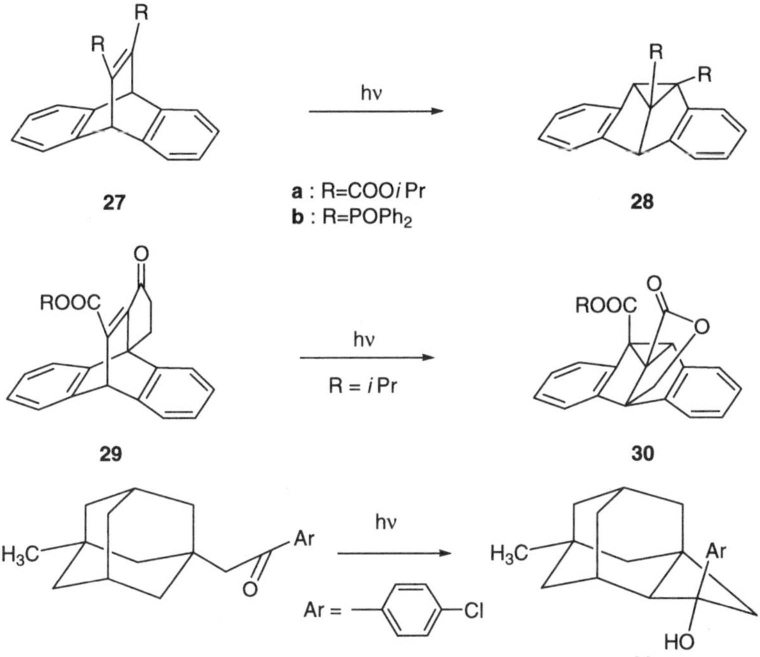

27

a : R=COO*i* Pr
b : R=POPh₂

28

29   **30**

Ar = —⟨benzene⟩—Cl

31   **32**

Structures 27–32

frozen by photoirradiation in the solid state to give the optically active photocyclization product **35** [20]. Recrystallization of **33** from acetone formed chiral crystals as orange hexagonal plates (**A,** which converts to **C** by heating at 260°C) and
two types of racemic crystals as orange rectangular plates [**B,** melting point (mp)
302°C] and yellow rectangular plates (**C,** mp 297°C). The chirality of the crystal
A can be detected easily by measurement of its CD spectra in Nujol mulls. Crystal **A** exhibits strong CD absorptions at around 250 and 330 nm, whereas the crystal types **B** and **C** do not show any CD absorption in these regions (Fig. 5).

Irradiation of powdered (+)-**A** crystals using a 100-W high-pressure Hg
lamp for 50 hr gave (+)-**35** of 64% *ee* in quantitative yield. Similar irradiation of
(−)-**A** crystals gave (−)-**35.** However, photoirradiation of crystals **B** and **C** gave
racemic **35.** This enantioselective photoconversion consists of two steps: the conrotatory ring closure of **33** to the intermediate **34** and then a 1,5-hydrogen shift to
give the product **35** (Scheme 2). X-ray crystal structure analysis disclosed that
crystal **A** consists of **33** molecules in either a purely right-handed or a purely left-
handed helical configuration (Fig. 6) [21]. Crystals **B** and **C** contain equal

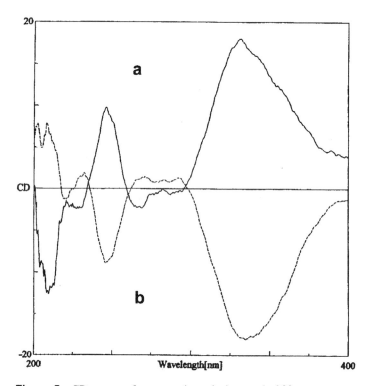

Figure 5   CD spectra of two enantiomeric **A** crystal of **33.**

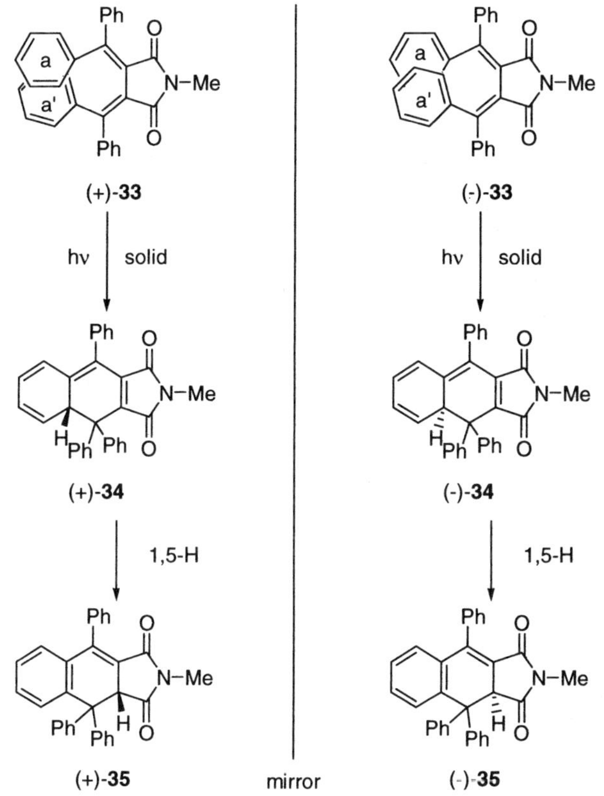

Scheme 2

amounts of the two configurations and, hence, are racemic (Figs. 7 and 8). The phenyl rings a and a' are almost parallel to one another and overlap significantly: The dihedral angles between these rings are within the range 9°–12° (Table 2). The corresponding distances between the unsaturated carbon, which join during the photoreaction, C6···C30 and C14···C22, are 3.38 and 3.27 Å in **A**, 3.35 Å in **B**, and 3.35 and 3.34 Å in **C** (Table 2) [21]. These values match perfectly well the topochemical requirements of the photoreaction in the solid state.

Transformations between the different polymorphs can easily be accomplished both in the solid state and in solution. Addition of one piece of (−)-**A** crystal during the recrystallization of (+)-**A** crystal (50 mg) from acetone gave (−)-**A** (23 mg), **B** (4 mg), and **C** crystals (8 mg). The thermal racemization of **A** to **C** occurs before melting, by heating to 260°C on a hot plate. During this process, the color change from orange to yellow spreads dramatically from one end of the crystal to the other with retention of the crystal morphology.

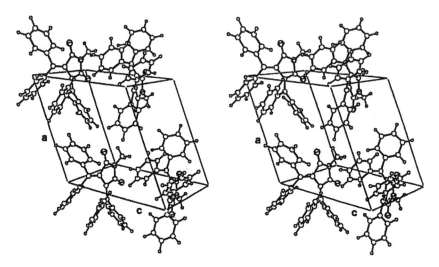

Figure 6 Stereoview of the molecular packing of **33A**.

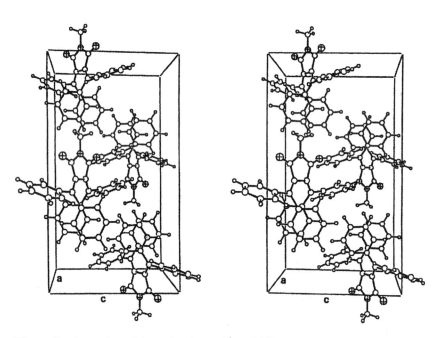

Figure 7 Stereoview of the molecular packing of **33B**.

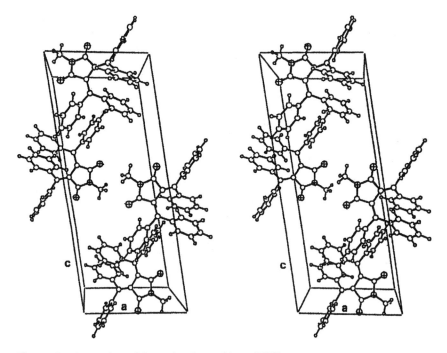

Figure 8 Stereoview of the molecular packing of **33C**.

Table 2 Crystal Data and Selected Structural Parameters of Three Polymorphic Crystals of **33**

Cryst. data	**A**	**B**	**C**
Cryst. system	Monoclinic	Orthorhombic	Monoclinic
Space group	$P2_1$	$Pbcn$	$P2_1/n$
a (Å)	11.640	9.964	9.485
b (Å)	9.257	20.181	11.014
c (Å)	12.103	11.622	22.945
Intramolecular nonbonding distance (Å)			
C6 ••• C30	3.381	3.353	3.348
C14 ••• C22	3.273	3.353	3.338
Torsion angles (deg)			
C6–C5–C21–C22	38.8	34.7	43.2
C4–C3–C5–C6	19.6	13.7	20.2
C20–C19–C21–C22	14.3	13.7	20.9

E. Cinnamic Acid Derivatives

Cinnamic acid derivative **36** crystallizes in the chiral space group $P2_1$ and gives the optically pure dimer **37** upon irradiation in the solid state [22]. Chiral crystals of **38** gave, upon irradiation, the optically active dimer **39** of 90% *ee*, whereas the corresponding methyl ester gave a highly crystalline linear polymer through a typical [2 + 2] topochemical photopolymerization [23].

Optically active [2 + 2] cycloadduct **42** of 95% *ee* was formed upon irradiation of the chiral CT complex crystals of **40** and **41** [24].

Structures 36–42

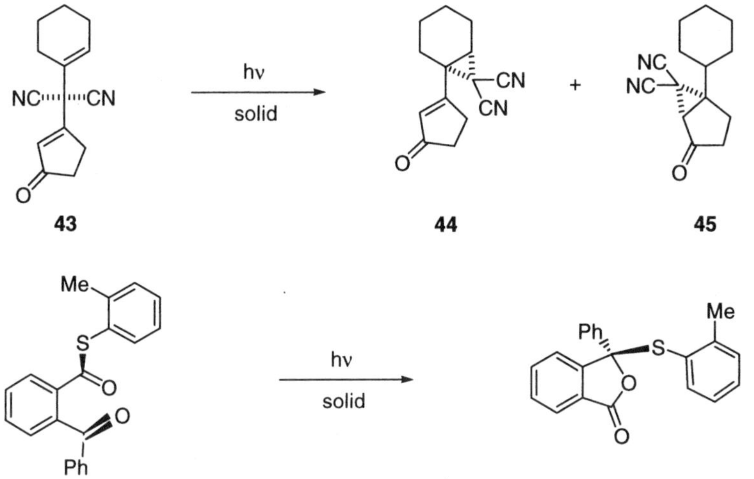

Structures 43–47

F. Some Other Examples

Solid-state photochemical di-π-methane-type rearrangement of chiral crystals of **43** have been found to give a mixture of **44** of 44% *ee* and **45** of 96% *ee* [25]. When chiral crystals of the thioester **46** were irradiated at 0°C for 6 hr, optically active phthalide **47** of 30% *ee* was obtained in 65% yield [26].

Photoirradiation of chiral salt crystals formed from base **48** and acid **49** in the solid state gave the optically active product **50** of about 35% *ee* in 37% yield via photodecarboxylative condensation reaction [27].

III. ENANTIOSELECTIVE PHOTOREACTIONS OF ACHIRAL MOLECULES USING CHIRAL SOURCE

When achiral molecules can be arranged in a chiral form in crystal using a chiral source, its chirality is fixed by photoreaction in the solid state to give an optical

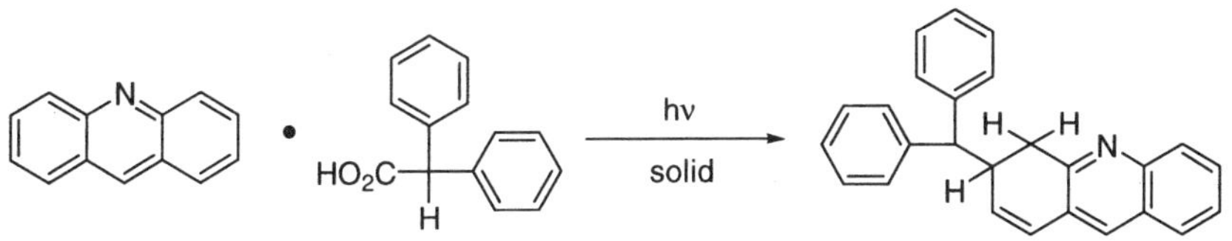

Structures 48–50

active product. As a chiral source, an attached chiral auxiliary and a chiral host compound can be used.

A. Enantioselective Photoreactions of Achiral Molecules Which Have an Attached Chiral Auxiliary

The intramolecular [2 + 2] photocycloaddition of cyclohexadienone derivative **51,** which is substituted with chiral piperidine ring in the solid state, proceeds enantioselectively and optically active products **52** and **53** were obtained [28]. Two crystal modifications of **51,** the α-form (mp 102–104°C) and β-form (mp 127–128°C), gave optically active **52** and **53,** respectively, upon irradiation in the solid state.

A new method for assymmetric induction in the solid state photochemistry of prochiral amine salt with an optically active acid has been developed. For example, (+)-**55** and (−)-**55** were obtained by irradiation of the salts **54a** and **54b,** respectively, at −40°C in the solid state [29].

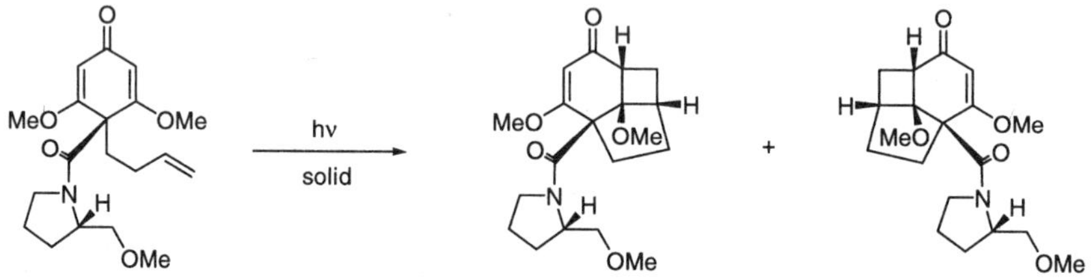

Structures 51–53

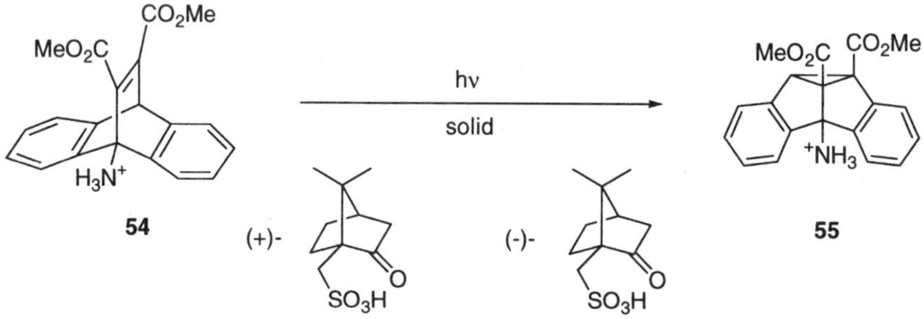

Structures 54–55

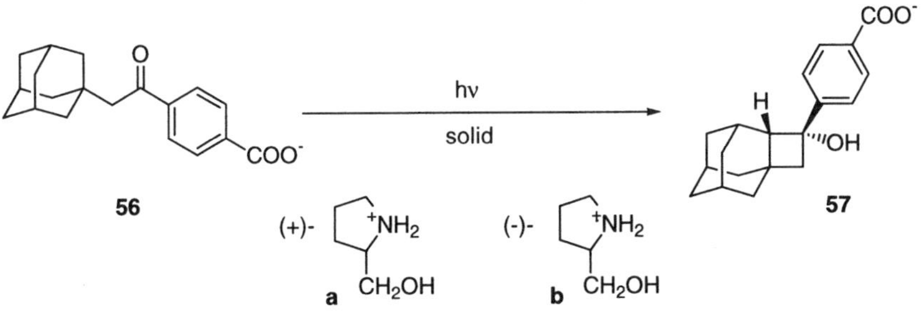

Structures 56 and 57

(+)- or (−)-Prolinol salts of keto acid **56** gave the (+)- or (−)-cyclobutanol **57** of 97% *ee* upon irradiation in the solid state, whereas racemic **57** was obtained on irradiation in solution [30].

Irradiation of the phenethyamine salt **58** in the solid state gave the chiral cyclobutanol derivative **59** of 97% *ee* [31].

Although these enantioselective photoreactions are limited to amide or salt derived from achiral acid and chiral amine, one enantioselective photoisomerization reaction of cobaloxime coordinated with chiral axial ligands such as 1-methylpropylamine, 1-(1-naphthyl)ethylamine, and 2-phenylglycinol has been reported. For example, finely powdered (2-cyanoethyl)cobaloxime (**60**), suspended in liquid paraffin and spread onto a Petri dish, was irradiated to give (*S*)-(−)-**61** of about 80% *ee* after displacement of the chiral auxiliary of the complex with pyridine [32].

B. Enantioselective Intramolecular Photocyclization Reactions of Achiral Molecules in Inclusion Complex With a Chiral Host Compound

1. Tropolone Alkyl Ethers and Pyridones

When the photocyclization reaction of tropolone alkyl ethers **62a** and **62b**, which produces the corresponding cyclization product **63** and its ring-opened derivative

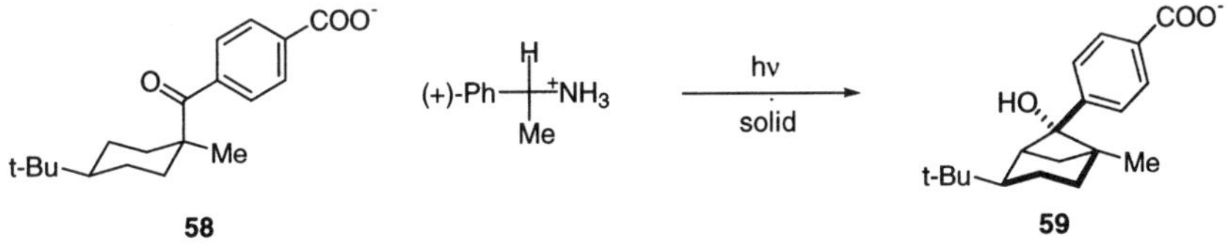

Structures 58 and 59

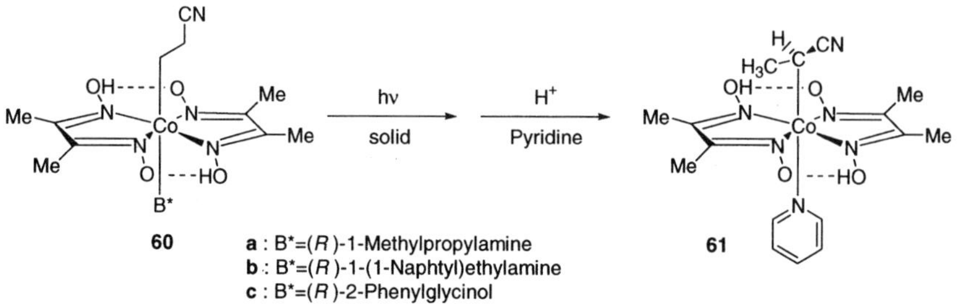

Structures 60 and 61

64 [33], was carried out in a 1:1 inclusion complex with the chiral host **11**, (1*S*, 5*S*)-**63a** (11% yield, 100% *ee*) and (*S*)-**64a** (26% yield, 91% *ee*) and (1*S*, 5*S*)-**63b** (12% yield, 100% *ee*) and (*S*)-**64b** (14% yield, 72% *ee*) were obtained, respectively, in the chemical and optical yields indicated [34]. In the inclusion complex crystal with **11**, a disrotatory [2 + 2] intramolecular photocyclization reaction of **62** would occur in the **A** direction, but not in the **B** direction, due to steric hindrance of the *o*-chlorophenyl group of **11**, hence giving (1*S*,5*R*)-**63** but not (1*R*,5*S*)-**63** [34].

Chirally modified zeolites can also be used as a chiral host for enantioselective photoreaction. For example, a mixture of **62a**, (−)-norephedrine and zeolite (NaY) in CH_2Cl_2–hexane was stirred for 12 hr and filtered to give the zeolite

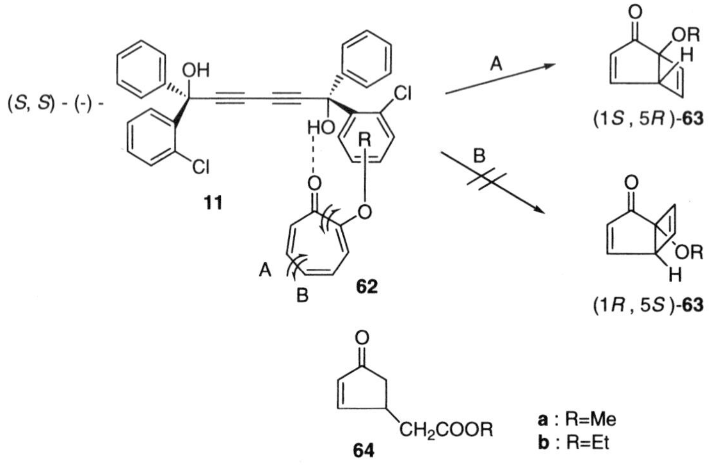

Structures 62–64

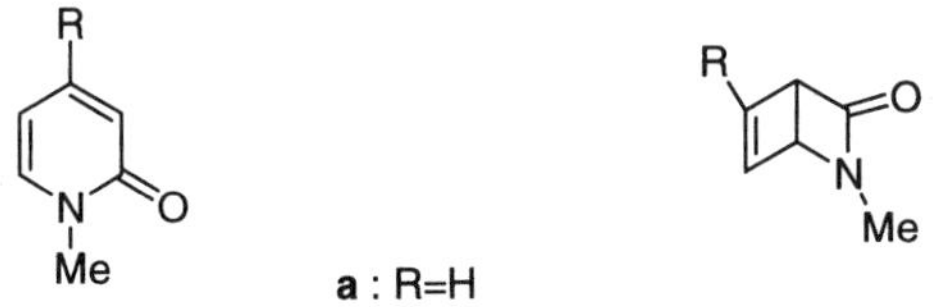

a : R=H
b : R=OMe

65   66

Structures 65 and 66

containing both the reactant and the chiral inductor. Photoirradiation of the zeolite complex as a hexane slurry for 2 hr gave the optically active (1*S*,5*R*)-**63** of up to 50% *ee* [35].

Irradiation of the inclusion complex of *N*-methylpyridone (**65a**) with **12a** in the solid state gave (+)-**66a** of 100% *ee* in 93% yield after 15% conversion. Irradiation of inclusion complexes of **65b** with the chiral host **11** or **12a** gave (−)-**66b** (97% yield, 100% *ee*) or (+)-**66b** (99% yield, 72% *ee*) after 50% and 15% conversions, respectively, in the chemical and optical yields indicated [36]. Photoreaction of 15 other derivatives of **65** as inclusion complexes has also been studied [37].

2. Cyclohexenone Derivatives

Intramolecular [2 + 2] photocyclization reactions of 2-[*N*-(2-propenyl)amino]cyclohex-2-enones (**67**) are also controlled enantioselectively by carrying out irradiation in inclusion complex with a chiral host compound. When inclusion crystals of **67** with **12** are irradiated in the solid state, optically active 9-azatricyclo[5.2.1.0]-decan-2-ones (**68**) were obtained in the chemical and optical yields indicated in Table 3 [30].

Photoirradiation of inclusion crystals of 3-oxo-2-cyclohexanecarboxamide derivatives (**69b–69d**) with the optically active host compound **12b** as a water suspension for 4 hr gave optically almost pure 2-aza-1, 5-dioxaspiro[3,5]nonane derivatives (**70b–70d**) [38]. Optically pure **70b** and **70c** were prepared by the

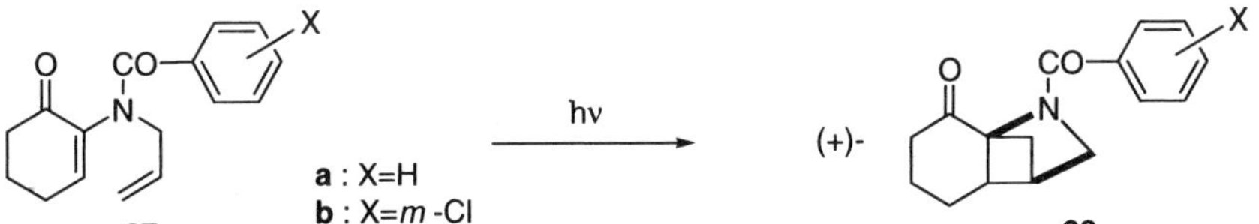

a : X=H
b : X=*m*-Cl
c : X=*p*-Me

67   68

Structures 67 and 68

Table 3 Photocyclization of **67** in a 2:1 Inclusion Compound With **12**

Host	Guest		Yield (%)	Optical purity (% ee)	[α]D (deg) (c in CCl₄)
(−)-**12b**	**67a**	(+)-**68a**	64	99	+70 (0.01)
(−)-**12c**	**67a**	(+)-**68a**	81	99	+90 (1.32)
(−)-**12c**	**67b**	(+)-**68b**	71	99	+77 (0.85)
(−)-**12c**	**67c**	(+)-**68c**	65	99	+61 (0.53)

combination of photoirradiation of **69b** and **69c**, respectively, in their inclusion crystals with **12b** and purification of the reaction product via the corresponding oxime derivatives **71b** and **71c** (Table 4) [38]. However, **69a** did not react in the inclusion crystal.

Photoirradiation of the 1:2 inclusion complexes of **72a** and **72b** and **72d–72g** with (−)-**12c** gave the optically active photocycloaddition products **73a** and **73b** and **73d–73g** in high optical yields (Table 5). However, photoirradiation of the 1:2 inclusion complex of **72c** with (−)-**12c** gave the optically active spiro β-lactam derivative **74c** of 97% ee in 96% yield (Table 5) [39]. This is very interesting result but reason for the difference is not clear.

Enantioselective photocyclization of 2-(arylthio)-3-methylcyclohexan-1-ones (**75**) to dihydrobenzothiophene derivatives **77** was also achieved in inclusion crystals using chiral host molecules. Photoirradiation of the 1:1 inclusion crystals of **75g** with (−)-**12b** as a water suspension gave the corresponding photocyclization product (+)-*cis*-**77g** of 82% ee in 83% yield. Similar photoirradiation of the

a : R¹=R²=H   c : R¹=H; R²=Et

b : R¹=H; R²=Me   d : R¹=R²=Me

Structures 69–71

Table 4 Photocyclization of **69** in a 2:1 Inclusion Compound With **12**

Host	Guest	Product		
			Yield (%)	Optical purity (% ee)
(−)-**12b**	**69a**	(+)-**70a**	a	
(−)-**12b**	**69b**	(+)-**70b**	18	>99.9
(−)-**12b**	**69c**	(+)-**70c**	41	>99.9
(−)-**12b**	**69d**	(+)-**70d**	51	>99.9
(−)-**12c**	**69d**	(+)-**70d**	53	>99.9

ᵃ No reaction occurred.

Table 5 Photocyclization of **72** in a 1:2 Inclusion Compound With (−)-**12c**

Host	Guest	Product		
			Yield (%)	Optical purity (% ee)
(−)-**12c**	**72a**	(+)-**73a**	17	68
(−)-**12c**	**72b**	(+)-**73b**	30	67
(−)-**12c**	**72c**	(+)-**74c**	69	97
(−)-**12c**	**72d**	(+)-**73d**	25	53
(−)-**12c**	**72e**	(+)-**73e**	87	100
(−)-**12c**	**72f**	(+)-**73f**	56	100
(−)-**12c**	**72g**	(+)-**73g**	42	100

72   a : R=Me   e : R=$C_6H_5CH_2$
         b : R=Et   f : R=p-MeC$_6$H$_4$CH$_2$
         c : R=Pr^n g : R=p-ClC$_6$H$_4$CH$_2$
         d : R=Bu^n

Structures 72–74

75	**a** : R=H **e** : R=o-Cl **76**
	b : R=p-Me **f** : R=p-Br
	c : R=o-Me **g** : R=o-Br
	d : R=p-Cl

(+)-cis-**77**

Structures 75–77

inclusion crystals of **75a–75f** with (−)-**12b** afforded (+)-cis-**77a–77f** of the optical purities listed in Table 6. X-ray crystal structural analysis of the 1:1 inclusion complex of **75g** with (−)-**12b** showed the dihedral angle between the two average ring planes of the guest molecule to be about 74°. The photoreactive carbon C12 is 3.7 Å away from the target C3 on one side of the cyclohexenone ring plane, favoring the *R* configuration at C3 (Fig. 9). This assumption was ascertained by x-ray analysis of the photocyclization product (+)-**77g.** The chiral host molecule (−)-**12c** did not show strong absorption nor CD peaks in the 400–250 nm region. In contrast, the inclusion compound of prochiral molecules **75g** exhibited a rather strong CD spectrum. When cocrystallized with the opposite enantiomer, host (+)-**13b**, an almost mirror image solid-state CD spectrum was obtained. Solid-state CD spectra of the reaction product (+)-**77g** and its enantiomer also exhibited mirror image related CD spectra (Fig. 10) [2].

3.   Furancarboxamide

Enantioselective photocyclization of *N*-allylfuran-2-carboxanilide (**78**) in its inclusion crystals with the optically active host compound **12b** was accomplished successfully. More interestingly, (−)-**79** and (+)-**79** were obtained selectivity

Table 6   Photocyclization of **75** in the Inclusion Crystals With **12b**

			Product	
Host	Guest		Yield (%)	Optical purity (% ee)
(−)-**12b**	**75a**	(+)-**77a**	90	32
(−)-**12b**	**75b**	(+)-**77b**	75	60
(−)-**12b**	**75c**	(+)-**77c**	92	62
(−)-**12b**	**75d**	(+)-**77d**	94	65
(−)-**12b**	**75e**	(+)-**77e**	85	77
(−)-**12b**	**75f**	(+)-**77f**	83	59
(−)-**12b**	**75g**	(+)-**77g**	83	82

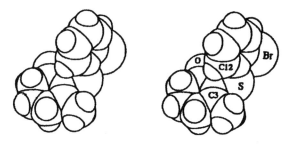

Figure 9 Molecular conformation of **75g** in the inclusion complex with (−)-**12b**.

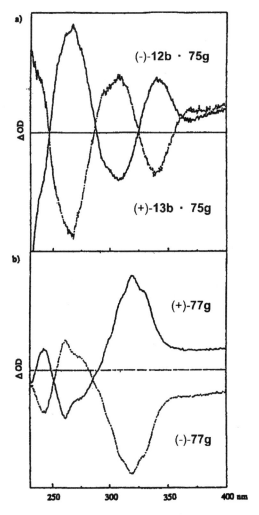

Figure 10 CD spectra in Nujol mulls of inclusion complexes of (a) **75g** with (−)-**12b** and (+)-**13b** and (b) (+)-**77g** and (−)-**75g**.

409

Structures 78 and 79

upon photoirradiation of the 1:1 and 2:1 complexes of **78** with (−)-**12b,** respectively. For example, photoirradiation of the powdered 1:1 complex of **78** with (−)-**12b** gave (−)-**79** of 96% *ee* in 50% yield. On the other hand, similar irradiation of the 2:1 complex of **78** with (−)-**12c** gave the other enantiomer (+)-**79** of 98% *ee* in 86% yield (Table 7) [40]. The chiral arrangement of **78** molecules in the 1:1 and 2:1 complexes with (−)-**12b** were observed by measurement of their CD spectra as Nujol mulls (Figs. 11 and 12). The CD absorptions of the 1:1 complex at 240 nm correspond to the UV absorption of **78** in MeOH at 256 nm. On the other hand, the CD absorptions of the 2:1 complex at around 300 nm correspond to those of their UV absorptions in the solid state at 280 nm. These results indicate that **78** molecules in the 2:1 complex have more planar structure than do **78** molecules in MeOH solution and the 1:1 complex [40].

The 1:1 complexation was achieved by mixing the host and guest in the solid state, and the inclusion complexation was followed by measurement of CD spectra every 30 min in Nujol mulls (Fig. 13). For example, as the complexation between (−)-**12c** and **78** proceeds, the (+)- and (−)-Cotton effects appeared at about 275 and 300 nm, respectively, and these absorptions increased until the complexation was completed after 2 hr, although no absorption was present at the beginning of the mixing process [40].

4. *N*-(Aryloylmethyl)-δ-Valerolactams

Photocyclization reaction of the *N*-(aryloylmethyl)-δ-valerolactams (**80**) occur stereoselectively and enantioselectively in their inclusion crystals with optically

Table 7 Photocyclization of **78** in the Inclusion Crystal With **12**

Host	Host:Guest		Product Yield (%)	Optical purity (% *ee*)
(−)-**12b**	1 : 1	(−)-**79**	50	96
(−)-**12b**	2 : 1	(+)-**79**	86	98
(−)-**12c**	2 : 1	(+)-**79**	77	98

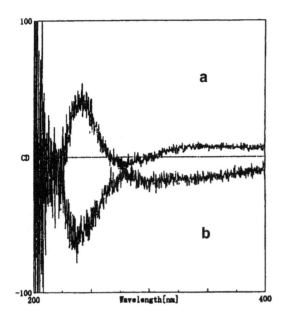

Figure 11 CD spectra of 1:1 complex of **78** with (a) (−)-**12b** and (b) (+)-**13b**.

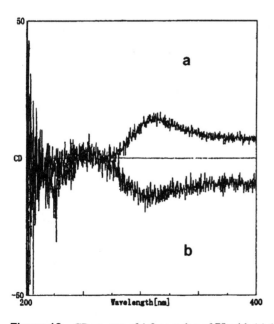

Figure 12 CD spectra of 1:2 complex of **78** with (a) (−)-**12b** and (b) (+)-**13b**.

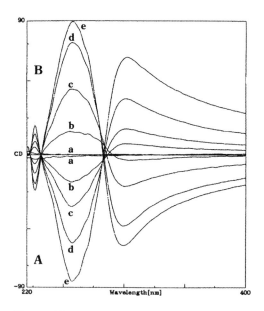

Figure 13   CD spectra of 1:1 mixture of **78** with (−)-**12b** (A) or (+)-**13b** (B) after (a) 0 min, (b) 30 min, (c) 60 min, (d) 90 min, and (e) 120 min in Nujol mulls.

active hosts. For example, irradiation of the powdered inclusion complex of **80a** and the optically active host compound (−)-**12b,** as a suspension in water for 12 hr, gave (+)-**81a** of 98% *ee* in 59% yield. By similar irradiations of the 1:1 inclusion compounds of **80b** and **80c** with (−)-**12c,** (−)-**81b** of 84% *ee* and (−)-**80c** of 98% *ee* were also obtained enantioselectively (Table 8) [41]. Conversely, photoreaction of **80a–80c** in *t*-BuOH afforded a mixture of *rac*-**81a–81c** and *rac*-**82a–82c.** To elucidate the mechanism of the enantioselective photocyclization of **80,** the x-ray structure of the 1:1 complex of **80c** with **12c** was determined (Fig. 14) [42]. The molecules of **80c** are arranged in a chiral form within the crystal so as to give (6*S*, 7*S*)-(−)-**81c** selectively. The distances for C1···C3 and O1···H3B

Structures 80–82

Table 8   Photocyclization of **80** in the Inclusion Crystal and in Solution

80	Host	Irradiation time (hr)	81			82		
				Yield (%)	% ee		Yield (%)	% ee
80a	(−)-**12c**	12	(+)-**81a**	59	98			
80a	(+)-**12c**	12	(−)-**81a**	54	99			
80a	a	27	(±)-**81a**	33	0	(±)-**82a**	50	0
80b	(−)-**12c**	12	(−)-**81b**	45	84			
80b	a	24	(±)-**81b**	34	0	(±)-**82b**	50	0
80c	(−)-**12c**	15	(−)-**81c**	42	98			
80c	a	27	(±)-**81c**	26	0	(±)-**82c**	45	0

[a] Irradiation was carried out in *t*-BuOH.

are 3.24 and 2.82 Å, respectively; hence, the H3B atom is abstracted by O1 to form a hydroxyl group. The C3 atom attacks C1 from the *re* face of the carbonyl group to give (6*S*,7*S*)-(−)-**81c**.

   The chiral arrangement of **80a** molecules in the complex with (−)-**12c** is also easily detected by measurement of CD spectra as Nujol mulls. For example, the inclusion crystal of **80a** with (−)-**12c** showed a (−)-Cotton effect, and that of **80a** with (+)-**12c** showed a (+)-Cotton effect at around 260 and 310 nm, respec-

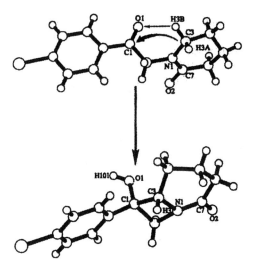

Figure 14   Photocyclization process of **80c** to **81c** in the inclusion crystals with (−)-**12c**.

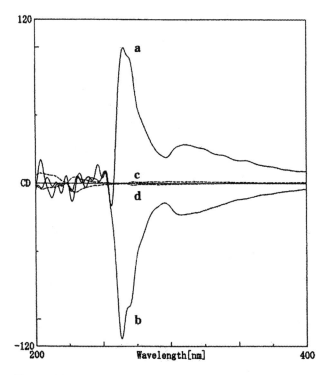

Figure 15 CD spectra of 1:1 inclusion complex of (a) **80a** with (+)-**12c** and (b) **80a** with
(−)-**12c** and of host molecule (c) (+)-**12c** and (d) (−)-**12c** in Nujol mulls.

tively, although the host molecules itself does not show any CD absorption (Fig.
15).

Photoirradiation of inclusion crystals of the 4-(3-butenyl)cyclohexa-2,5-
dien-1-ones (**83**) with the optically active host compounds (**12c**) in the solid state
gave optically active 1-carbomethoxytricyclo[4.3.1.0]dec-2-en-4-ones (**84**). For

83	**a** : R¹=R²OMe	**84**
	b : R¹=OMe; R²=H	

83 a : $R^1=R^2OMe$
         b : $R^1=OMe$; $R^2=H$

84

(+)-

Structures 83 and 84

example, reaction of a powdered 2:1 inclusion crystal of **83a** with (−)-**12b** as a water suspension for 5 hr gave (+)-**84a** of 73% *ee* in 50% yield. In the case of **83b,** enantioselective inclusion complexation occurred to give a 1:1 complex of optically pure (−)-**83b** with (−)-**12b,** and its irradiation in a water suspension gave optically pure (+)-**84b** in 57% yield [43].

5. *N*-Acrylanilides

The photocyclization reaction of acrylanilide to 3,4-dihydroquinolinone was first reported in 1971, and its application to alkaloid synthesis has long been studied [44]. Although stereo- and enantio-control are important in this reaction, no such attempt has been reported, except one enantioselective photocyclization of 1-(methylacryl)-*N*-methylanilide (**85**) in benzene–ether containing (+)-bis(*p*-toluoyl)tartaric acid to give (−)-3-methyl-*N*-methyl-3,4-dihydroquinolinone (**87**) of 12–16% *ee* [45].

Recently, the host–guest inclusion method was found to be useful for a selective photocyclization of acrylanilides. For example, irradiation of the 1:1 inclusion compound of **85** with (−)-**12b** gave (−)-**87** of 98% *ee* in 46% yield. On the other hand, similar irradiation of the 1:1 inclusion compound of **85** with (−)-**12c** gave (+)-**87** of 95% *ee* in 29% yield (Table 9) [46]. It is surprising that the two host compounds (−)-**12b** and (−)-**12c** of so little structural difference (i.e., five-membered ring and six-membered ring) caused such disparate enantioselectivity. The 1,5-hydrogen shift in the photocyclization of **88** in its inclusion crystals with (−)-**12b** or (−)-**12c** is also controlled precisely and finally gives the trans-isomer **89** of high optical purity. When the irradiation of **88** was carried out in solution, a 1:1 mixture of *rac*-**89** and *rac*-**90** was obtained [46].

The selective conversion of **85** to **87** in the inclusion crystal can be interpreted as follows: Of the two possible directions (*S* and *R*) of the conrotatory ring closure of the iminium form (**85′**) of **85,** only the rotation toward *S,* for example, occurs through control by the host **12b** (or **12c**) to give the intermediate **86.** The

Structures 85–87

Table 9 Photocyclization of Anilides in the Inclusion Crystals

Anilides	Host	Product	Yield (%)	% ee
85	(−)-**12b**	(−)-**87**	46	98
85	(−)-**12c**	(+)-**87**	29	95
88a	(−)-**12b**	(−)-**89a**	65	98
88a	(−)-**12c**	(+)-**89a**	44	98
88b	(−)-**12b**	(+)-**89b**	62	70
88b	(−)-**12c**	(−)-**89b**	29	99

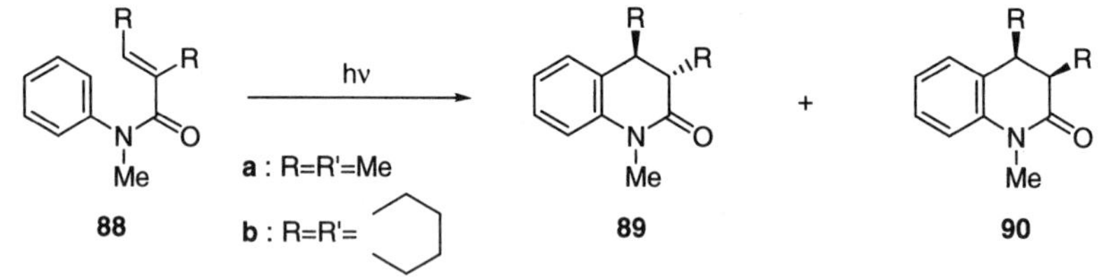

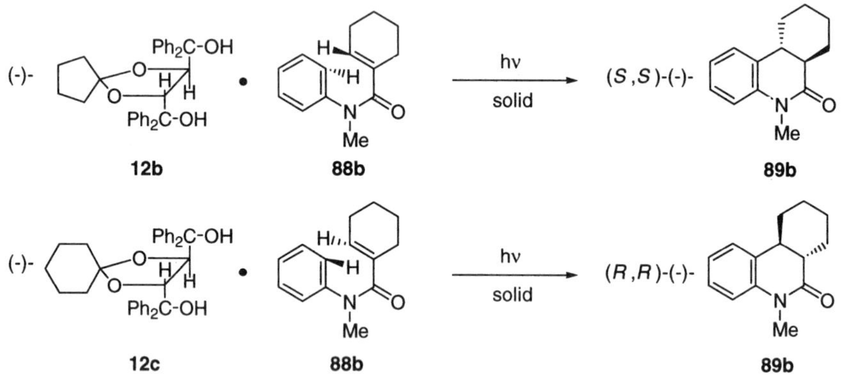

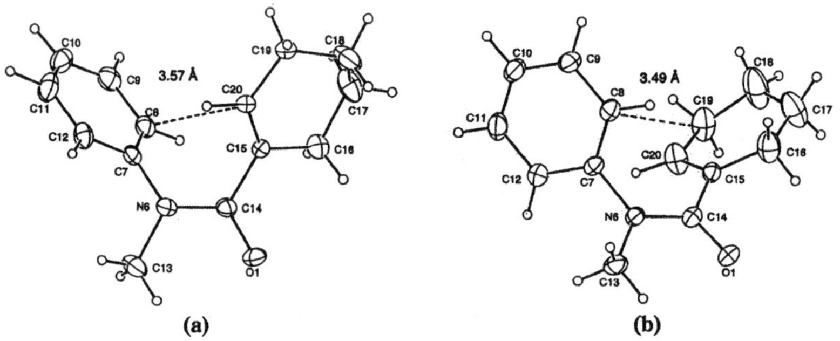

Figure 16 Molecular structures of **88b** in the inclusion crystals with (a) (−)-**12b** and (b)

1,5-hydrogen shift of the intermediate **86** occurs in a suprafacial manner to give the optically active photocyclization product **87**. The chiral conformation of **88** in the inclusion crystals with **12b** and **12c** were determined by x-ray analysis [47]. For example, the molecular structure of **88b** in its inclusion crystals with (−)-**12b** is enantiomorphic in Fig. 16. In the inclusion crystal of **88b** with (−)-**12b,** conrotatory photocyclization of **88b** with a positive torsion angle for C7–N6–C14–C15 will occur in a way that the hydrogens at C8 and C20 do not come into collision. Afterward, a 1,5-hydrogen shift will occur in a suprafacial manner to give (R,R)-(+)-**89b**. (S,S)-(−)-**89b** will be obtained in a similar way upon photoirradiation of **88b** with a negative torsion angle for C7–N6–C14–C15 in its inclusion crystal with (−)-**12c**. These assignments are supported by the known absolute configuration of the photoproducts [47].

Interestingly, the enantioselective conversion of **88a** in the inclusion complex crystal with **12b** into (+)-**89** proceeds by a partial single-crystal to single crystal photocyclization manner [47].

6. Nitrones

Optically active oxazolidinones (**91a–91d**) of 95–100% ee were obtained by irradiation of the 1:1 inclusion complex crystals of nitrones (**90a–90d**) and optically active host compound **11** in the solid state in the chemical and optical yields indicated (Table 10) [48].

C. Enantioselective Intermolecular Photocyclization Reactions of Achiral Molecules in Inclusion Complex With a Chiral Host Compound

1. 2-Cyclohexene and Cycloocta-2,4-Dien-1-One

Photoirradiations of both neat and benzene solution of 2-cyclohexenone (**92**) give a complex mixture of the *syn/trans*-**93** and *anti/trans* dimer, and two other dimers

Structures 90 and 91

Table 10 Yield and Optical Purity (% *ee*) of Oxaziridines **91a–d**

	Ar	Yield (%)	Optical purity (% ee)
a	Ph	41	9.5
b	―⟨◯⟩―Cl	74	30
c	⟨◯⟩ (Cl)	51	100
d	―⟨◯⟩―O―O	52	94

of unknown structure [49]. When a 1:2 inclusion complex of the chiral host compound (−)-1,4-bis[3-(*o*-chlorophenyl)-3-hydroxy-3-phenylprop-1-ynyl]benzene (**94**) with **92** in a water containing a small amount of surfactant was irradiated for 24 h, (−)-**93** of 48% *ee* was obtained in 75% yield [50]. Inclusion complexation of the (−)-**93** of 48% *ee* with **11** gave optically pure (−)-**93.** This photochemical reaction was also found to proceed by a single-crystal to single crystal manner.

Cycloocta-2,4-dien-1-one (**95**) exists as an equilibrium mixture of the conformers **95a** and **95b** in solution, and conversion between these two enantiomeric forms is too fast to allow their isolation at room temperature. Photoreaction of **95** in pentane for 1 hr gives racemic **96** in 10% yield along with polymers. When a solution of (−)-**11** and **95** was kept at room temperature for 12 h, a 3:2 inclusion complex of (−)-**11** with **95a** was obtained as colorless needles. Irradiation of the 3:2 complex of (−)-**11** with **95a** for 48 hr gave (−)-**96** of 78% *ee* in 55% yield. X-ray crystal structure analysis is given in Ref. 51.

(*S,S*)-(-)-**94**   **92**   (-)-   **93**

Structures 92–94

95a   **95b**

96

Structures 95 and 96

2. Coumarin and Thiocoumarin

Enantioselective photodimerization of coumarin (**97**) was accomplished in the inclusion complex (**100**) with **12a.** For example, irradiation of the 1:1 inclusion compound of **97** with **12a** gave the *anti*-head-to-head dimer (−)-**98** of 96% *ee* in 96% yield [52]. This photochemical reaction of **100** to the complex (**101**) of **98** was also found to proceed by a single-crystal to single-crystal manner. However, 1:2 inclusion compound of **11** and **97** gave the *syn*-head-to-head dimer (**99**) upon irradiation in 75% yield [53].

The efficient enantioselective dimerization reaction of **97** in the inclusion complex (**100**) with **12a** suggests that two coumarin molecules are arranged in chiral positions in **100** and their [2 + 2] photodimerization occurs by keeping the chirality in **100**. The chiral arrangement of **97** was detected by CD spectral measurement of the inclusion complex in Nujol mulls. The 1:1 complex of **97** with **12a** and **13a** showed CD spectra with a mirror-imaged relation (Fig. 17). After photoirradiation, the CD absorptions of the complex (**100**) at 225, 275, 300, and 330 nm disappeared, and the new CD absorption due to the inclusion complex (**101**) of **98** at 240 nm appeared. The photodimerization of **97** was also followed by measurement of infrared (IR) spectra as Nujol mulls. Upon photoirradiation, the νCO absorption of **97** in **100** at 1700 cm^{-1} decreased gradually and finally disappeared after 4 hr, and new νCO absorption due to **98** in **101** appeared at 1740 cm^{-1}.

The single-crystal to single-crystal nature and the steric course of the photodimerization of coumarin (**97**) to (−)-*anti*-head-to-head dimer (**98**) in the inclusion complex (**100**) were investigated by x-ray crystallographic analysis and x-ray powder diffraction spectroscopy [54]. X-ray crystallographic analysis showed that two molecules of **97** were arranged by forming a hydrogen bond between the C40═O6 of **97** and the O4—H of **12a** in the direction, which gave the *anti*-head-to-head dimer (**98**) by photodimerization and the molecular aggregation

Structures 97–101

with 3.59 and 3.42 Å were short enough contact, which should react easily and topochemically. After photoirradiation, the bond distances of the cyclobutane ring connecting C38—C38* and C39—C39* are 1.6 and 1.57 Å, respectively (Figs. 18 and 19) [54].

The enantioselective photodimerization of thiocoumarin (**102**) to optically pure (+)-*anti*-head-to-head dimer (**103**) in the 1:1 inclusion complex of **102** with **12b** was also found to proceed in a single-crystal to single-crystal manner. For example, when a mixture of thiocoumarin (**102**) and optically active host compound **12b** in butyl ether was kept at room temperature for 12 hr, a 1:1 inclusion complex of **103** with **12b** was obtained as colorless needles. Photoirradiation of a 1:1 inclusion complex in the solid state (400-W high-pressure Hg lamp, Pyrex filter, room temperature, 2 hr) gave a 2:1 complex of **12b** with **103**,

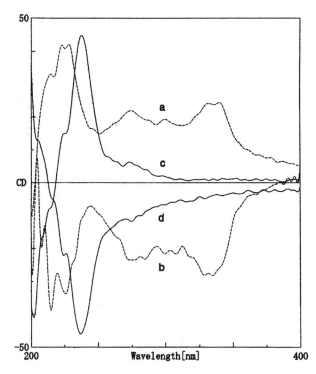

Figure 17 CD spectra in Nujol mulls: (a) a 1:1 complex of **97** with **12a;** (b) a 1:1 complex of **97** with **12b;** (c) a 2:1 complex of **12a** with (+)-**98;** (d) a 2:1 complex of **12b** with (−)-**98.**

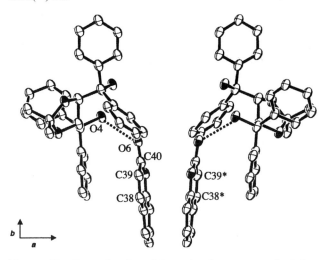

Figure 18 O$_{RTEP}$ drawing of the molecular structure of a 1:1 complex (**100**) of **97** with **12a** viewed along the α axis. All hydrogen atoms are omitted for clarity. The hydrogen-bondings are shown by dotted lines.

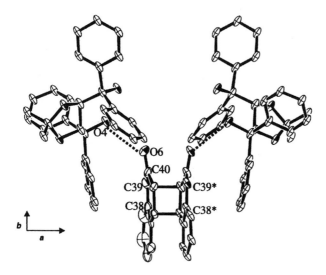

Figure 19 O_{RTEP} drawing of the molecular structure of a 2:1 complex (**101**) of (−)-**98** with **12a** viewed along the α axis. All hydrogen atoms are omitted for clarity. The hydrogen-bondings are shown by dotted lines.

quantitatively; **103** of 100% *ee* was isolated in 73% yield by column chromatographic separation [53].

The single-crystal to single-crystal nature and the steric course of the photodimerization of thiocoumarin (**102**) to (+)-*anti*-head-to-head dimer (**103**) in the inclusion complex were investigated by x-ray crystallographic analysis and x-ray powder diffraction spectroscopy [54].

3. Benzaldehyde

Irradiation of β-cyclodextrin complex (**104**) of benzaldehyde in the solid state gave optically active benzoin (**105**) of 15% *ee* in 56% yield along with the by-product (**106**) [55].

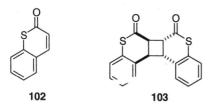

102   **103**

Structures 102 and 103

Structures 104–106

IV. CONCLUSION

Enantioselective photoreactions in the solid state have many advantages as described in Section I. As reviewed in this chapter, highly enantioselective photoreactions can be accomplished in the solid state without using any chiral source. There are hopeful possibilities to find many other examples of similar absolute asymmetric photosynthesis in the solid state. Enantioselective photosynthesis in inclusion complex crystal with a chiral host compound will also be a new general method of asymmetric synthesis in future. By successful designing of many new chiral host compounds, the research field would also be developed widely.

REFERENCES

1. Toda, F.; Yagi, M.; Soda, S. *J. Chem. Soc., Chem. Commun.* **1997,** 413.
2. Toda, F.; Miyamoto, H.; Kikuchi, S.; Kuroda, R.; Nagami, F. *J. Am. Chem. Soc.* **1996,** *118,* 11315.
3. Sekine, A.; Hori, K.; Ohashi, Y.; Yagi, M.; Toda, F. *J. Am. Chem. Soc.* **1989,** *111,* 697.
4. Green, B. S.; Lahav, M.; Rabinovich, D. *Acc. Chem. Res.* **1979,** *12,* 191; Addadi, L.; Lahav, M. *Origins of Optical Activity in Nature,* Elsevier: New York, 1979.
5. Toda, F.; Miyamoto, H. *J. Chem. Soc., Perkin Trans. 1* **1993,** 1129.
6. Hashizume, D.; Koga, H.; Sekine, A.; Ohashi, Y.; Miyamoto, H.; Toda, F. *J. Chem. Soc., Perkin Trans. 2* **1996,** 61.
7. Toda, F.; Miyamoto, H.; Koshima, H. *J. Org. Chem.* **1997,** *62,* 1997.
8. Toda, F.; Tanaka, K.; Omata, T.; Nakamura, K.; Oshima, T. *J. Am. Chem. Soc.* **1983,** *105,* 5151.
9. Kaftory, M.; Yagi, M.; Tanaka, K.; Toda, F. *J. Org. Chem.* **1988,** *53,* 4391.
10. Toda, F.; Tanaka, K. *Tetrahedron Lett.* **1988,** *29,* 551; Seebach, D.; Beck, A. K.; Imwinkelried, R.; Wonnacott, A. *Helv. Chim. Acta* **1987,** *70,* 954.
11. Toda, F.; Miyamoto, H.; Matsukawa, R. *J. Chem. Soc., Perkin Trans. 1* **1992,** 1461.
12. Hashizume, D.; Uekusa, H.; Ohashi, Y.; Matsugawa, R.; Miyamoto, H.; Toda, F. *Bull. Chem. Soc. Jpn.* **1994,** *67,* 985.
13. Toda, F.; Tanaka, K.; Sekikawa, A. *J. Chem. Soc., Chem. Commun.* **1987,** 279; Toda, F.; Miyamoto, H.; Kanemoto, K. *J. Chem. Soc., Chem. Commun.* **1995,** 1719.
14. Sakamoto, M.; Takahashi, M.; Fujita, T.; Watanabe, S.; Iida, I.; Nishio, T.; Aoyama, H. *J. Org. Chem.* **1993,** *58,* 3476; Sakamoto, M.; Takahashi, M.; Fujita, T.; Watanabe, S.; Iida, I.; Nishio, T.; Aoyama, H. *J. Org. Chem.* **1997,** *62,* 6298.

15. Sakamoto, M.; Takahashi, M.; Fujita, T.; Nishio, T.; Iida, I.; Watanabe, S. *J. Org. Chem.* **1995,** *60,* 4682.

16. Sakamoto, M.; Takahashi, M.; Shimizu, M.; Fujita, T.; Nishio, T.; Iida, I.; Yamaguchi, K.; Watanabe, S. *J. Org. Chem.* **1995,** *60,* 7088.

17. Sakamoto, M.; Hokari, N.; Takahashi, M.; Fujita, T.; Watanabe, S.; Iida, I.; Nishio, T. *J. Am. Chem. Soc.* **1993,** *115,* 818.

18. Sakamoto, M.; Takahashi, M.; Kamiya, K.; Yamaguchi, K.; Fujita, T.; Watanabe, S. *J. Am. Chem. Soc.* **1996,** *118,* 10,664; Sakamoto, M.; Takahashi, M.; Arai, W.; Mino, T.; Yamaguchi, K.; Watanabe, S.; Fujita, T. *Tetrahedron* **2000,** *56,* 6795.

19. Evans, S. V.; Garcia-Garibay, M.; Omkaram, N.; Scheffer, J. R.; Trotter, J.; Wireko, F. *J. Am. Chem. Soc.* **1986,** *108,* 5648; Fu, T. Y.; Liu, Z.; Scheffer, J. R.; Trotter, J. *J. Am. Chem. Soc.* **1993,** *115,* 12201; Gudmundsdottir, A. D.; Scheffer, J. R. *Tetrahedron Lett.* **1990,** *31,* 6807.

20. Toda, F.; Tanaka, K. *Supermol. Chem.* **1994,** *3,* 87.

21. Toda, F.; Tanaka, K.; Stein, Z.; Goidberg, I. *Acta Crystallogr.* **1995,** *C51,* 2722.

22. Addadi, L.; Mil, J.; Lahav, M. *J. Am. Chem. Soc.* **1982,** *104,* 3422.

23. Chung, C. M.; Hasegawa, M. *J. Am. Chem. Soc.* **1991,** *113,* 7311.

24. Suzuki, T.; Fukushima, T.; Yamashita, Y.; Miyashi, T. *J. Am. Chem. Soc.* **1994,** *116,* 2793.

25. Roughton, A. L.; Muneer, M.; Demuth, M. *J. Am. Chem. Soc.* **1993,** *115,* 2085.

26. Sakamoto, M.; Takahashi, M.; Moriizumi, S.; Yamaguchi, K.; Fujita, T.; Watanabe, S. *J. Am. Chem. Soc.* **1996,** *118,* 8138; Takahashi, M.; Sekine, N.; Fujita, T.; Watanabe, S.; Yamaguchi, K. *J. Am. Chem. Soc.* **1998,** *120,* 12,770.

27. Koshima, H.; Ding, K.; Chisaka, Y.; Matsuura, T. *J. Am. Chem. Soc.* **1996,** *118,* 12,059.

28. Schultz, A. G.; Taveras, A. G.; Taylor, R. E.; Tham, F. S.; Kulling, R. K. *J. Am. Chem. Soc.* **1992,** *114,* 8725.

29. Gudmundsdottir, A. D.; Scheffer, J. R. *Photochem. Photobiol.* **1991,** *54,* 535.

30. Jones, R.; Scheffer, J. R.; Trotter, J.; Yang, J. *Tetrahedron Lett.* **1992,** *33,* 5481.

31. Leibovitch, M.; Olovsson, G.; Sundarababu, G.; Ramamurthy, V.; Scheffer, J. R.; Trotter, J. *J. Am. Chem. Soc.* **1996,** *118,* 1219; Leibovitch, M.; Olovsson, G.; Scheffer, J. R.; Trotter, J. *J. Am. Chem. Soc.* **1998,** *120,* 12,755.

32. Koura, T.; Ohashi, Y. *Bull. Chem. Soc. Jpn.* **1997,** *70,* 2417; Sekine, A.; Arai, Y.; Ohgo, Y.; Kamiya, N.; Iwasaki, H. *Bull. Chem. Soc. Jpn.* **1995,** *68,* 2517; Ohgo, Y.; Arai, Y.; Hagiwara, M.; Takeuchi, S.; Kogo, H.; Sekine, A.; Uekusa, H.; Ohashi, Y. *Chem. Lett.* **1994,** 715.

33. Daupen, W. G.; Koch, K.; Smith, S. L.; Chapman, O. L. *J. Am. Chem. Soc.* **1963,** *85,* 2616.

34. Toda, F.; Tanaka, K. *J. Chem. Soc., Chem. Commun.* **1986,** 1429.

35. Joy, A.; Scheffer, J. R.; Corbin, D. R.; Ramamurthy, V. *J. Chem. Soc., Chem. Commun.* **1998,** 1379; Koodanjeri, S.; Joy, A.; Ramamurthy, V. *Tetrahedron* **2000,** *56,* 7003.

36. Toda, F.; Tanaka, K. *Tetrahedron Lett.* **1988,** *29,* 4299.

37. Kuzuya, M.; Noguchi, A.; Yokota, N.; Okuda, T.; Toda, F.; Tanaka, K. *Nippon Kagaku Kaishi* **1986,** 1753.

38. Toda, F.; Miyamoto, H.; Takeda, K.; Matsugawa, R.; Maruyama, N. *J. Org. Chem.* **1993,** *58,* 6208.
39. Toda, F.; Miyamoto, H.; Kikuchi, S. *J. Chem. Soc., Chem. Commun.* **1995,** 621.
40. Toda, F.; Miyamoto, H.; Kanemoto, K. *J. Org. Chem.* **1996,** *61,* 6490.
41. Toda, F.; Tanaka, K.; Kakinoki, O.; Kawakami, T. *J. Org. Chem.* **1993,** *58,* 3783.
42. Hashizume, D.; Ohashi, Y.; Tanaka, K.; Toda, F. *Bull. Chem. Soc. Jpn.* **1994,** *67,* 2383.
43. Akutsu, S.; Miyahara, I.; Hirotsu, K.; Miyamoto, H.; Maruyama, N.; Kikuchi, S.; Toda, F. *Mol. Cryst. Liquid Cryst.* **1996,** *277,* 87.
44. Capman, O. L.; Adams, W. R. *J. Am. Chem. Soc.* **1968,** *90,* 2333; Ninomiya, I.; Naito, T. *The Alkaloids,* Academic Press: San Diego, CA, **1983,** Vol. 22, 189–279.
45. Ninomiya, I.; Yamauchi, S.; Kiguchi, T.; Shinohara, A.; Naito, T. *J. Chem. Soc., Perkin Trans 1* **1974,** 1747.
46. Tanaka, K.; Kakinoki, O.; Toda, F. *J. Chem. Soc., Chem. Commun.* **1992,** 1053.
47. Hosomi, H.; Ohba, S.; Tanaka, K.; Toda, F. *J. Am. Chem. Soc.* **2000,** *122,* 1818; Ohba, S.; Hosomi, K.; Tanaka, K.; Miyamoto, H.; Toda, F. *Bull. Chem. Soc. Jpn.* **2000,** *73,* 2075.
48. Toda, F.; Tanaka, K. *Chem. Lett.* **1987,** 2283; Toda, F.; Tanaka, K.; Mak, T. C. W. *Chem. Lett.* **1989,** 1329.
49. Lam, E. Y. Y.; Valentine, D.; Hammond, G. S. *J. Am. Chem. Soc.* **1967,** *89,* 3482.
50. Tanaka, K.; Kakinoki, O.; Toda, F. *J. Chem. Soc., Perkin Trans 1* **1992,** 307.
51. Fujiwara, T.; Nanba, N.; Hamada, K.; Toda, F.; Tanaka, K. *J. Org. Chem.* **1990,** *55,* 4532.
52. Tanaka, K.; Toda, F. *J. Chem. Soc., Perkin Trans 1* **1992,** 943.
53. Moorthy, J. N.; Venkatesan, K. *J. Org. Chem.* **1991,** *56,* 6957.
54. Tanaka, K.; Toda, F.; Mochizuki, E.; Yasui, N.; Kai, Y.; Miyahara, I.; Hirotsu, K. *Angew. Chem., Int. Ed. Engl.* **1999,** *38,* 3523; Tanaka, K.; Mochizuki, E.; Yasui, N.; Kai, N.; Miyahara, I.; Hirotsu, K.; Toda, F. *Tetrahedron* **2000,** *56,* 6853.
55. Rao, V. P.; Turro, N. J. *Tetrahedron Lett.* **1989,** *30,* 4641.

7

Observations on the Photochemical Behavior of Coumarins and Related Systems in the Crystalline State

Kodumuru Vishnumurthy, Tayur N. Guru Row, and Kailasam Venkatesan
Indian Institute of Science, Bangalore, India

I. INTRODUCTION

The occurrence of photoreactivity in crystals was recognized toward the end of the last century. However, vigorous research activity following the first report began only about 30 years ago. The rotational and transitional molecules being arrested in the solid state as against the situations in the gaseous and liquid states of matter are responsible for the regiospecific and stereospecific products obtained from crystals. The generation of photopolymorphs [1,2] and chiral compounds from crystals containing achiral molecules [3–5] has been achieved via photoirradiation of crystal under certain conditions of molecular organization. Progress in the area of solid-state photochemistry depends on the developments in experimental techniques such as automated x-ray diffraction and many very efficient structure-solving packages using the intensity data. This chapter is not a history of solid-state organic photochemistry, as there is a large number of excellent reviews on this subject available [6–8] that the reader would find profitable to read. We discuss here only some recent results obtained during the course of the studies on the photodimerization of coumarins and resulted systems. Also, we touch upon topics such as the effect of temperature and wavelength on the progress of reactions in the crystalline state.

II. TOPOCHEMISTRY

Based on some interesting reactions in certain inorganic crystalline compounds, Kohlschutter [9,10] proposed that the nature and properties of the products obtained take place on the surface or within the solid state. Indeed, he coined the term "topochemistry" for such reactions in the solid state. However, systematic investigations of photoinduced reactions in crystals began from 1964 onward by Schmidt and Cohen [11]. Their studies on the $2\pi + 2\pi$ photoreaction of cinnamic acid derivatives in the crystalline state and correlation with the molecular organization in these crystals led to what are now known as "Topochemical Principles." The most important conclusions reached by them are as follows: (1) The necessary conditions for the reactions to take place are that the reactive double bonds are parallel to one another and the center-to-center distance be within ~4.1 Å; (2) there is one-to-one correspondence between the stereochemistry of the photoproduct and the symmetry relationship between the reactants. The centrosymmetric relationship (called the α-form) leads to centrosymmetric cyclobutane (*anti*-HT), whereas the mirror symmetric arrangements (called the β-form) produce mirror symmetric dimer (*syn*-HH).

According to the topochemical postulates, reaction follows the least-motion path. Also, the parallelism condition ideally dictates that the p_z atomic orbitals of the four reacting centers are in perfect orientation. It is worth mentioning that the series of seminal articles by Schmidt and co-workers form the first structure–reactivity correlations studies. The investigation by these investigators attracted the attention of many chemists and chemical crystallographers and vigorous studies were carried out on many organic systems such as butadiens [12,13], benzylidene cyclopentanones [14], and coumarins [7,8,15].

III. CRYSTAL ENGINEERING

The central problem in organic solid-state photochemistry is the preorganization of molecules satisfying the topochemical postulates. Schmidt coined the term "crystal engineering" for this problem of supramolecular assembly. Indeed, the importance of crystal engineering is fundamental to areas as diverse as nonlinear optics, high-T_c superconductors, and the generation of polymorph forms in pharmaceuticals.

The crystal is a supermolecule which Nature offers to us [16]. Our problem is to synthesize crystals, and to achieve this, we have to know the grammar of crystal packing (i.e., the rules by which the molecules are put together in periodic arrays). It is recognized that hierarchization is one of the basic principles of constructing anything. In crystal syntheses, the hierarchization involves building a complex assembly out of a simpler atoms→molecules→supramolecular assembly. To achieve this, we need to know all about the noncovalent intermolecular interactions—both weak and strong. Our knowledge here is by no means integral. However, analysis of the packing modes in organic crystals available from the Cambridge database has allowed us to develop a few fruitful approaches. These are (1) intramolecular substitutions [11,15,17], (2) mixed-crystal formation [18–20], (3) donor–acceptor strategy [21–23], (4) host–guest complexation [8], (5) generation of new polymorphic forms via controlled impurity addition [8a], (6) molecular shape as design criteria [8a], and (7) molecular recognition through intermolecular hydrogen-bonding [24,25]. In this chapter, we restrict our discussion to strategies 1 and 5–7.

IV. FLUORINE AS A STEERING GROUP

It has been demonstrated that the chloro group is a powerful steering agent for achieving β-packing [15,17]. It has also been found that the bromo group, the acetoxy group, and methylenedioxy substituents, more often than not, favor the β structure. It appears that 7-iodocoumarin takes up β-packing (Table 1). The steering ability of fluorine had not been investigated until recently. However, for example, 6- and 7-fluorocoumarins have been shown to adopt the β structures, giving the *syn*-HH photodimer [26]. Vishnumurthy et al. have examined the effect of fluoro substitution on the photobehavior of 4-styrylcoumarin [27] and 1,4-diphenylbutadiene [28]. Schemes 1 and 2 show the different fluoro derivatives in-

Table 1 Examples of Photodimerization of Halocoumarins and Cinnamic Acids in the Solid State

Compound	Yield (%)
6-Chlorocoumarin	100
7-Chlorocoumarin	80
4-Methyl-7-chlorocoumarin	80
4-Methyl-6-chlorocoumarin	50
6-Iodocoumarin	40
6-Bromocoumarin	90
7-Bromocoumarin	100
2-Chlorocinnamic acid	85
3-Chlorocinnamic acid	70
4-Chlorocinnamic acid	71
2-Bromocinnamic acid	82
3-Bromocinnamic acid	91
4-Bromocinnamic acid	90

Note: For all compounds listed, β-packing is observed and the dimer is *syn*-HH.

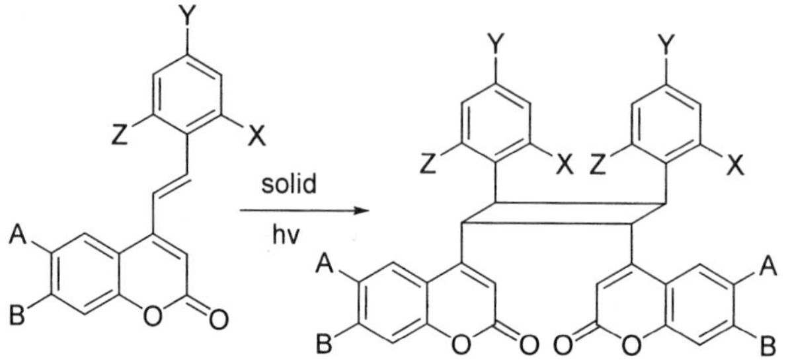

1. X = F; C = Y= Z = A = B = H   2. Y = F; X = C = Z = A = B = H
3. Y = A = F; X = C = Z = B = H   4. X = Z = A = F; C = Y = B = H
5. X = A = F; C = Y = Z = B = H   6. X = Z = B = F; C = Y = A = H
7. X = Z = F; C = Y = A = B = H - No reaction

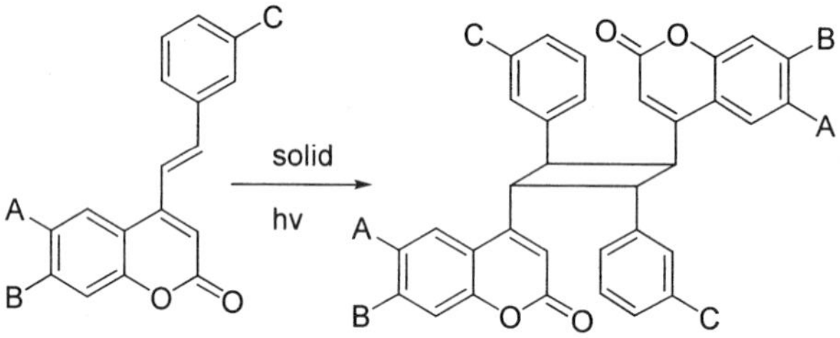

8. C = F; X = Y = Z = A = B = H   9. A = F; X = Y = Z = B = C = H
10. B = F; X = Y = Z = A = C = H - No reaction

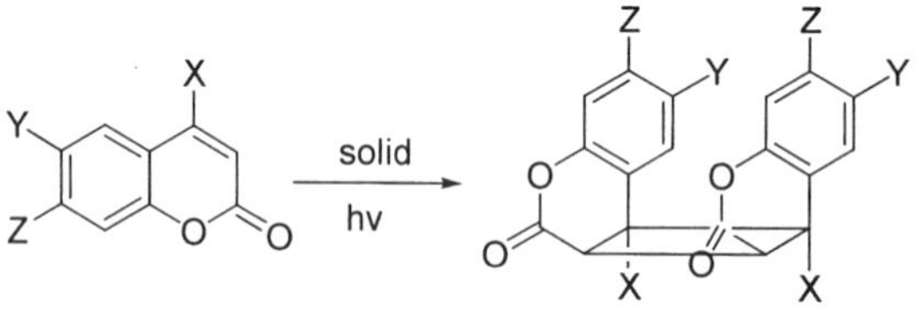

11. Y = F; X = Z = H   12. Z = F; X = Z = H
13. Y = F; X = CH₃; Z = H

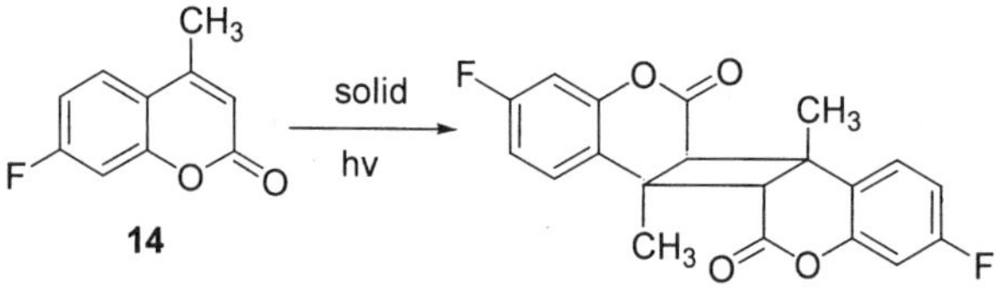

14

Scheme 1

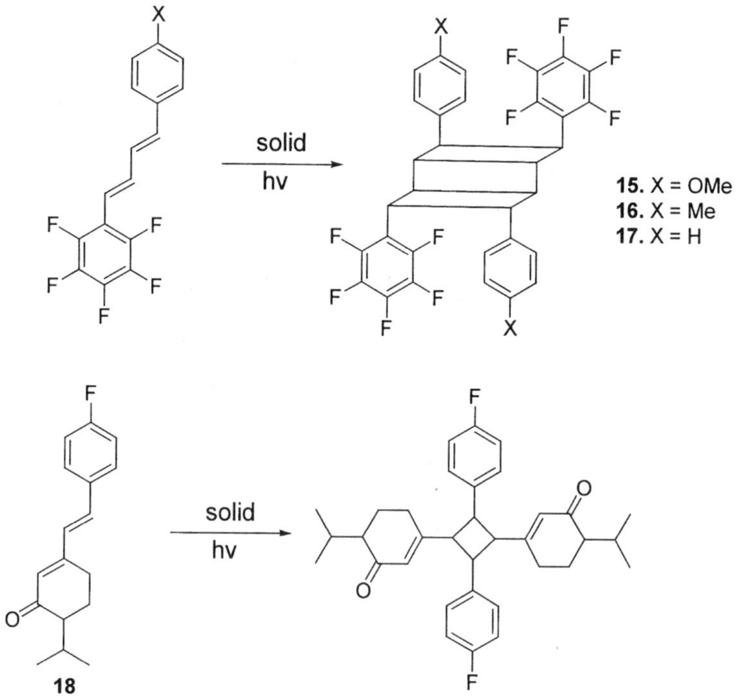

Scheme 2

vestigated and the results are summarized in Table 2. The propensity of fluorine atom to induce β-packing of coumarin and 4-styrylcoumarin molecules is beyond doubt. However, fluoro substitution in benzylidenepiperitone and 1,4-diphenylbutadiene [28] leads to α-packing, giving the *anti*-HT dimer. The reason for the α-packing mode in benzylidene-*dl*-piperitone seems to be due to the nonplanar character of the molecule, which achieves closer crystal packing in the α mode. In the case of butadiene derivatives with one of the phenyl rings highly flourinated, it seems likely there will be strong electrostatic repulsion interaction between the fluorine atoms in the β-packing mode, whereas the attractive interaction between the fluorine and hydrogen atoms favors the experimentally observed α-self-assembly. It is noteworthy that a similar stacking interaction has been observed between phenyl and perfluoro phenyl groups in mono olefinic and di-olefinic systems [29] such that [2 + 2] photocycloaddition takes place.

Inasmuch as monofluoro substitution frequently favors β-packing in molecules that are planar, the prediction of photobehavior in such compounds seems feasible. At this stage, it is worth inquiring into the specific role of the flu-

Table 2 Photobehavior of Fluoro-Substituted Organic Molecules

Compound	Type of packing	Dimer	Yield (%)
4-(2-Fluorostyryl)coumarin (**1**)	β-Packing	syn-HH	80
4-(4-Fluorostyryl)coumarin (**2**)	β-Packing	syn-HH	82
4-(4-Fluorostyryl)-6-fluocoumarin (**3**)	β-Packing	syn-HH	80
4-(2,6-Difluorostyryl)-6-fluocoumarin (**4**)	β-Packing	syn-HH	85
4-(2-Fluorostyryl)-6-fluocoumarin (**5**)	β-Packing	syn-HH	81
4-(2,6-Difluorostyryl)-7-fluocoumarin (**6**)	β-Packing	syn-HH	78
4-(2,6-Difluorostyryl)coumarin (**7**)	γ-Packing	—	—
4-(3-Fluorostyryl)coumarin (**8**)	α-Packing	anti-HT	78
4-(3-Fluorostyryl)coumarin[a] (**8**)	β-Packing	syn-HH	70
4-Styryl-6-fluorocoumarin (**9**)	α-Packing	anti-HT	50[b]
4-Styryl-7-fluorocoumarin (**10**)	γ-Packing	—	—
6-Fluorocoumarin (**11**)	β-Packing	syn-HH	100
7-Fluorocoumarin (**12**)	β-Packing	syn-HT	100
7-Fluoro-4-methylcoumarin (**13**)	α-Packing	anti-HT	25[b]
6-Fluoro-4-methylcoumarin (**14**)	Defect site	syn-HT	30[b]
(1E,3E) 1-(2,3,4,5-Pentafluorophenyl)–4-(4-methoxyphenyl)-1,3-butadiene (**15**)	α-Packing	anti-HT	25[b]
(1E,3E)1-(2,3,4,5-Pentafluorophenyl)-4-(4-methylphenyl)-1,3-butadiene (**16**)	α-Packing	anti-HT	22[b]
(1E,3E)1-(2,3,4,5-pentafluorophenyl)–4-phenyl)-1,3-butadiene (**17**)	α-Packing	anti-HT	23[b]
4-Fluorobenzilidene-dl-piperitone (**18**)	α-Packing	anti-HT	35

[a] Coumarin was added as an additive during crystallization.
[b] Percentage of dimer conversion.

oro group on the molecular organization, such as whether C—H⋯F interactions exist in these crystals.

From the remarks of Pauling [30] it is tempting to conclude that the highly electronegative character of fluorine would be involved in good hydrogen-bonding. The distances are less than the sum of the van der Waals radii (van der Waals radii are F = 1.47, H = 1.20, C = 1.75 Å) [31]. Only those which have a C—H⋯F angle greater than 110° were accepted for the intermolecular contacts. There are a few F⋯F distances less than the sum of the van der Waals radii. Figure 1 depicts the molecular organization in some of the photoreactive 4-styryl-coumarins, viewed down the shortest crystallographic axis in each case. Whereas 4-(2-fluorostyryl)coumarin and 4-(4-fluorostyryl)coumarin crystals have no C—H⋯F interactions, in 4-(4,6-difluorostyryl)coumarin and 4-(2,6-difluorostyryl)-6-fluorocoumarin both C—H⋯O and C—H⋯F interactions are present.

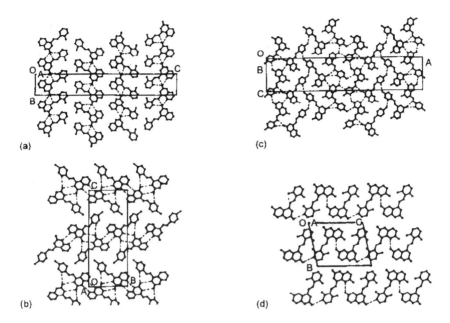

Figure 1 Molecular packing down the shortest crystallographic axis of (a) 4-(2-fluo-rostyryl)coumarin, (b) 4-(4-fluorostyryl)coumarin, (c) 4-(4-fluorostyryl)-6-fluoro-coumarin, and (d) 4-(2,6-difluorostyryl)-6-fluorocoumarin. Interactions of the type C—H···O and C—H···F are shown as dotted lines.

The occurrence of photoreactions in these compounds cannot be attributed to the influence of C—H···F interactions only because compounds **1** and **2,** which are reactive, have neither C—H···F nor F···F interactions. Also, compound **7,** which has a C—H···F interaction is not photoreactive. It may be added that the H···F distance limit of below 2.35 Å as used by Howard et al. [32] in their analysis of hydrogen bonds with F as an acceptor would disqualify the observed H···F lengths as hydrogen bonds. Shimoni and Glusker [33] concluded from their analysis of fluoro-substituted organic compounds that the C—F···H—C interaction is only marginally a hydrogen bond. From a more extensive study of intermolecular interactions in organofluoro compounds, these authors concluded that a C—F group competes unfavorably with a C—O$^-$, C—OH or C=O group to form a hydrogen bond to O—H, N—H, or C—H group [34]. From the very detailed analysis of structural data taken from the CSD, Dunitz and Taylor [35] concluded that F hardly acts as a hydrogen-bond acceptor, and then only in exceptional molecular and crystal environments. It seems that contributions arising from long-range electrostatic attractive interactions in the fluoro-substituted compounds are significant for the stability of the crystals.

To summarize, the steering abilities of both chlorine and fluorine have been investigated extensively and are observed to have remarkable influence on the photobehavior of such compounds. The influence of bromo and iodo substitution has not been examined thoroughly, although indications are that they could be useful as steering agents.

In the chloro- and fluoro-substituted structures, the influence of other weak intermolecular interactions on the supermolecular assembly cannot be over-looked. These include C—H$\cdots\pi$ (phenyl), O—H$\cdots\pi$, $\pi\cdots\pi$, and C=O$\cdots\pi$ interactions, in addition to the contributions from the nondirectional van der Waals interactions. The importance of the C—H$\cdots$O hydrogen bond, which is directional and of common occurrence, has been discussed [38,39]. From the analysis of a large number of crystal structures of esters, Addadi and Lahav [40] have noted that there exists an attractive interaction between the carbonyl and phenyl groups of adjacent molecules; this is attributed to C=O$\cdots\pi$ interaction. The approach geometry of the carbonyl group with respect to the phenyl ring is defined using the parameters α_1, d_1, and d_2 (Fig. 2). Tables 3 and 4 record these geometrical parameters to establish the spatial orientation of carbonyl group with respect to the phenyl group. The arrangement of the C=O groups with respect to the phenyl ring is noteworthy. An electrostatic interaction is presumed to operate between the electron-deficient phenyl group and the electron-rich oxygen atom of the carbonyl group [41].

The fact that chloro and fluoro substitutions change the packing pattern dramatically in the presence of all the other interactions mentioned shows that the contributions to the stability of the crystal structures from the halogen atom interactions (X$\cdots$C, X$\cdots$O, and X$\cdots$H) predominate. The results of the low-temperature single-crystal x-ray analysis of Cl and F are relevant here [42]. In this crystal, no short F$\cdots$F contact was observed (2.92 Å), whereas the interlayer intermolecular Cl$\cdots$Cl contact was shorter (3.07 Å) than the sum of the van der Waals radii, showing that the F$\cdots$F interaction is not significant.

It appears reasonable to assume from the dramatic changes in the supramolecular assembly of fluorine-substituted coumarin, 4-styrylcoumarin, and 1,4-diphenyl-butadiene that significant contributions to the lattice energy arise

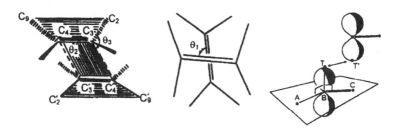

Figure 2   Pictorial representation of geometrical parameters α_1, d_1, and d_2.

Table 3 Geometrical Parameters for Carbonyl Oxygen With Respect to Phenyl Ring

Compound	Molecule[a]	α_1 (deg)	d_1 (Å)	d_2 (Å)
7-Hydroxy-4-styrylcoumarin	A	52.3	4.52	2.93
	B	66.0	4.36	3.42
4-Styrylcoumarin	B	60.8	4.50	1.60
4-(3-Fluorostyryl)coumarin	A	13.98	4.12	1.30
	B	5.74	4.32	2.45
4-Styryl-6-fluocoumarin	A	7.19	4.05	1.67
	B	16.31	7.62	6.21
4-Styryl-7-fluocoumarin		6.20	7.90	5.16
Benzylidene-dl-piperitone		95.9	4.28	2.78
Acetylaminocinnamic acid		60.9	4.04	1.53
2-Benzyl-5-benzylidenecyclopentanone		74.2	5.15	3.00
p-Fluorobenzilidene-dl-piperitone		74.4	4.70	2.90
p-Chlorobenzilidine-dl-piperitone		64.1	4.30	2.63

[a] Indicates molecules **A** and **B** in the asymmetric unit.

from the electrostatic terms. However, it must be stressed that the exact nature of fluorine interaction remains to be understood. It is notable that the molecular packing of the fluoro derivatives are steered in such way that these crystals are photoreactive, whereas the parent crystals of coumarin, benzylidene-dl-piperitone, and 1,4-diphenylbutadiene are photoinert.

Table 4 Geometrical Parameters for Carbonyl Carbon With Respect to Phenyl Ring

Compound	Molecule[a]	α_1 (deg)	d_1 (Å)	d_2 (Å)
7-Hydroxy-4-styrylcoumarin	A	113.37	3.90	1.87
	B	97.85	4.02	2.46
4-Styrylcoumarin	B	104.11	4.04	1.25
4-(3-Fluorostyryl)coumarin	A	52.32	3.73	0.67
	B	114.67	3.83	1.36
4-Styryl-6-fluocoumarin	A	77.44	3.64	0.80
	B	89.66	7.54	6.33
4-Styryl-7-fluocoumarin		151.54	7.93	3.71
Benzylidene-dl-piperitone		68.65	4.57	2.98
Acetylaminocinnamic acid		72.17	3.61	0.76
2-Benzyl-5-benzylidenecyclopentanone		24.67	4.96	2.83
p-Fluorobenzilidene-dl-piperitone		90.70	4.57	2.63
p-Chlorobenzilidene-dl-piperitone		99.59	3.90	1.72

[a] Indicates molecules **A** and **B** in the asymmetric unit.

V. CRYSTAL ENGINEERING VIA INTERMOLECULAR HYDROGEN BONDS

Supramolecular synthesis in the solid state via strong intermolecular hydrogen bonds such as O—H···O and N—H···N is the most reliable route for achieving crystal engineering. Exploitation of such interactions to bring neighboring molecules such that the reactants are favorably juxtaposed to achieve photoreactions between them has been attempted only rarely. Two examples utilizing O—H···H hydrogen bonds are presented here. Feldman and Campbell [24] designed and synthesized a U—shaped diacid that results in the formation of hydrogen bonds between the carbonyl groups (Scheme 3). This brings the reactive double bonds of the alkenes within ≤4.2 Å, resulting in a cyclobutane ring of stereochemistry *anti*-HT and of high yield. Further work following this principle is worthy of consideration. This involves proper design and synthesis of molecular framework, which hopefully would crystallize. One great advantage of this approach is the greater predictability in the preorganization of the molecules in the crystal. The "principle of maximum hydrogen-bonding" almost certainly is satisfied in the building process especially when the strong hydrogen bonds are present.

Formation of inclusion complexes with a variety of small organic molecules in which O—H···O hydrogen bonds play an important role has been discussed by Toda [43]. Toda and Akagi [44] reported that diacetylene diol forms crystalline stoichiometric inclusion complexes with a variety of small molecules. The salient features that assist complex formation are hydrogen-bonding between the poten-

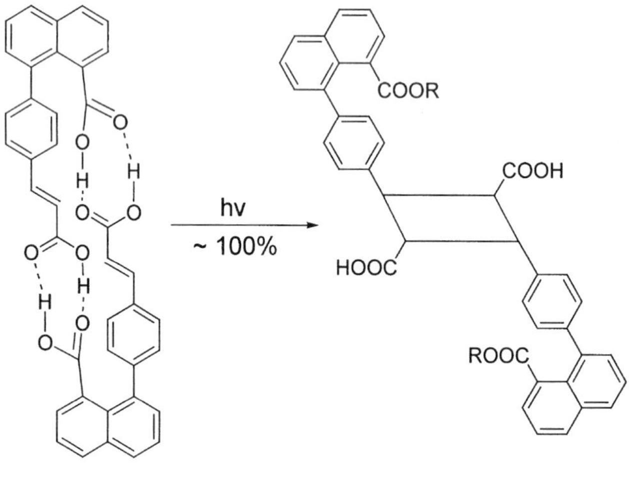

Scheme 3

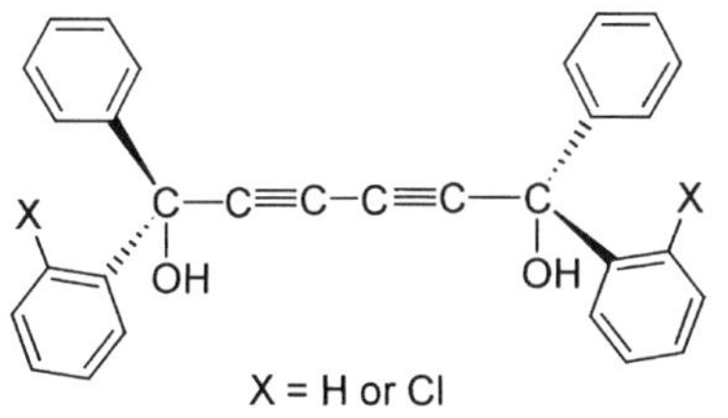

X = H or Cl

Scheme 4

tial hydrogen acceptor of the guest molecules with the two OH groups of the diol and the linear array of alkynic bonds (Scheme 4). The guest molecules will be like pendants attached to the host via O—H···O hydrogen bonds. When the inclusion complexes are irradiated, more often than not photodimerization between the guest molecules takes place. For example, irradiation of benzylidene acetophenone with the achiral host gave a simple product (>80% yield), which has been characterized as a *syn*-HT dimer. In solution, a mixture of cis and trans isomers and polymers is obtained, whereas irradiation of neat crystals gives a mixture of stereoisomeric photodimers in low yield. Photodimerization of a few coumarins with the host (achiral and chiral) has also been examined [8b]. It is observed that inclusion complexes of coumarin, 7-methylcoumarin, and 7-methoxycoumarin with the achiral diol yield the mirror symmetric dimer. As noted earlier, neat crystals of coumarin react nontopochemically as does 7-methylcoumarin. The chiral host–coumarin complex (ratio 1 : 2) crystallizes in the space group $P2_12_12_1$. The two crystallographically independent guest molecules are held in space by O—H···O hydrogen bonds between the carbonyl oxygens of the coumarin and the two hydroxyl groups of the host molecule (Scheme 4). When one of the coumarin molecules is translated along the *a* axis, the reactive bonds of the crystallographically independent coumarins come within a distance of about 3.8 Å.

VI. GENERATION OF POLYMORPHISM VIA ADDITIVES

According to McCrone [46], a polymorph is a solid crystalline phase of a given compound resulting from the possibility of at least two or more different arrangements of the molecules of that compound in the solid state. Polymophs play an important role in a large variety of fields. Lahav and co-workers [47] have carried out very incisive studies on the control of nucleation and crystal polymophism using tailor-made auxiliaries. The importance of polymorphs to an organic solid-state photochemist is obvious.

As mentioned earlier, the 4-styrylcoumarin molecule crystallizes in two different modifications [48]. Controlled addition of coumarin (2.5–20% by molar ra-

Table 5 Crystal Data of the Polymorphic Forms of 4-Styrylcoumarin

Parameters	Plates[a]	Prisms[b]	Needles[b]
Molecular formula	$C_{17}H_{12}O_2$	$C_{17}H_{12}O_2$	$C_{17}H_{12}O_2$
Crystal system	Monoclinic	Triclinic	Monoclinic
Space group	$P2_1/n$	$P-1$	$P2_1/c$
a (Å)	17.057(3)	11.082(2)	13.418(4)
b (Å)	8.229(3)	11.215(3)	5.720(4)
c (Å)	18.261(3)	12.127(3)	17.840(3)
α (deg)	90.00	102.35(2)	90.00
β (deg)	96.33(2)	116.41(2)	110.79(2)
γ (deg)	90.00	98.62(2)	90.00
Z	8	4	4
Volume (Å^3)	2547.5(11)	1265.9(6)	1280.1(10)
D_c (mg/m^3)	1.295	1.303	1.288

[a] Crystals obtained with coumarin as additive.
[b] Crystals obtained without addition of coumarin.

tio) produces a new crystalline form with different morphology—platelike crystals [8]. These crystals differ from the two previously known forms, as seen from the x-ray crystal data (Table 5). The new modification is also photoreactive, the styryl double bond undergoing dimerization giving *anti*-HT isomer.

Similar experiments on 4-(3-fluorostyryl)coumarin produced a modification different from the original one. The new modification yielded a dimer with *syn*-HH stereochemistry, in contrast to the *anti*-HT dimer obtained from the original crystalline form [8a,48]. Accurate single-crystal x-ray analysis of the new forms showed no evidence for the presence of coumarin in the crystals, at least not in a detectable amount (Table 6). It is interesting that the addition of napthalene instead of coumarin did not produce the new forms in both the styrylcoumarins mentioned earlier. Presumably, the interaction between the COO^- units in coumarin and styrylcoumarin may have a role to play in arresting the growth of the original crystalline forms. However, detailed studies are required to understand the exact mechanism.

Of interest in this work is that the polymorphs, generated via the tailor-made impurity, turned out to be photoreactive. It is clear that this approach, despite the uncertainties associated with obtaining a new photoreactive crystalline modification, is worthy of experimentation.

VII. MOLECULAR SHAPE AS DESIGN CRITERION

The ultimate crystal packing observed is a symphony of various weak interactions, such as $\pi\cdots\pi$, $C{=}O\cdots\pi$, $C{-}H\cdots\pi$, and other subtle interactions, although

Table 6 Crystal Data of Dimorphic Forms of 4-(3-Fluoro-styryl)coumarin

Parameters	Form I[a]	Form II[b]
Molecular formula	$C_{17}H_{11}FO_2$	$C_{17}H_{11}FO_2$
Crystal system	Monoclinic	Triclinic
Space group	Pc	$P\text{-}1$
Z	2	4
a (Å)	6.879(3)	7.358(1)
b (Å)	3.929(1)	7.652(1)
c (Å)	23.459(9)	23.829(3)
α (deg)	90.00	97.04(1)
β (deg)	90.35(4)	96.03(1)
γ (deg)	90.00	101.78(1)
Volume (Å^3)	634.0(4)	1291.8(3)
D_c (mg/m^3)	1.395	1.369

[a] Crystals obtained with coumarin as additive.
[b] Crystals obtained without addition of coumarin.

the halogen–halogen interaction plays a major role when a halo group is present in the reactant molecule. The molecules examined in connection with the steering ability of halogens are essentially planar systems. It appears that the influence of the topological features of the molecule, namely the overall three-dimensional shape of the molecule, on the molecular packing, even in the presence of steering groups, cannot be ignored as observed in a few nonplanar organic molecules (Table 7).

Although the reactive double bonds in all these crystals are favorably juxtaposed to give the *anti*-HT dimer, many are photoinert and the reason for this appears to be due to intermolecular steric compression. The number of examples is very limited and requires confirmation based on more data. However, it is strik-

Table 7 Photobehavior of Substituted Nonplanar, Centrosymmetric Organic Molecules

Molecules	Type of dimer	Shortest cell dimension (Å)	Ref.
Benzylidene-*dl*-piperitone	Photoinert	6.116	49
p-Chlorobenzilidene-*dl*-piperitone	*anti*-HT	6.636	50
o-Chlorobenzilidene-*dl*-piperitone	Photoinert	11.154	51
p-Bromobenzilidene-*dl*-piperitone	Photoinert	8.821	52
p-Fluorobenzilidene-*dl*-piperitone	*anti*-HT	6.032	53
Nitrophenylpyrimidine derivative	Photoinert	7.419	54

ing that none of these cases belong to the β-packing mode. Failure to achieve β-packing, irrespective of the presence or absence of the halo substituents, appears to be related to the overall topochemical features of the parent molecule. Unlike cinnamic acids, coumarins, and styrylcoumarins, which are essentially planar molecules, all the benzylidenepiperitones and the pyridine derivative are quite nonplanar. In the piperitone case, the cyclohexanone ring adopts the chair conformation. Further, there is an isopropyl group that is quite anisotropic in shape. It seems reasonable to conclude that the overall nonplanar feature of the molecules would prevent efficient close packing in the β mode with one of the cell dimensions as short as ~4 Å. Figure 3 shows the packing of molecules in benzylidene-*dl*-piperitone, which is typical of the other piperitones in Table 7. When an anisotropically shaped isopropyl, dimethylamino, or isopropenyl group is attached to the parent molecules with a reactive double bond, hopefully a dimer with *anti*-HT stereochemistry can be expected. Further investigations to exploit these

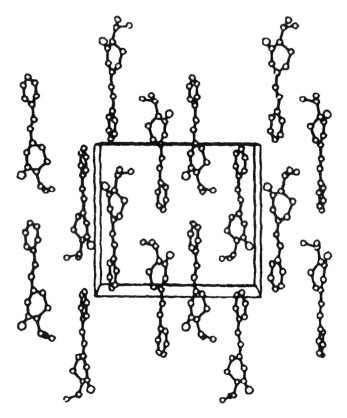

Figure 3   Packing of the molecules in benzylidene-(+)-piperitone crystal.

observations are worth pursuing to generate photoproducts with *anti*-HT stereo-chemistry. Whitesell (personal communication, 1997) has examined the role of the molecular size, shape, and approximate molecular symmetry as a possible route to crystal design, rather than pursuing the use of directional interactions such as hydrogen-bonding.

VIII. NONTOPOCHEMICAL REACTIONS

Several exceptions to Schmidt topochemical least-motion principle has been observed. Although head-to-head photodimerization were expected of several 9-substituted anthracenes [55–57] anti isomers were obtained. It was demonstrated by Craig and Sarti-Fantoni [58] that reactions in these cases take place at defect sites. The results produced by x-ray diffraction being space and time averages over thousands of unit cells would not allow us to gain knowledge of the molecular arrangements in microscopic defects such as microcavities, surfaces, and other irregulations in the macroscopic ordered crystal arrangement. Using electron microscopic techniques, Thomas et al. [59a,b] characterized the nature of defects in which photoreactions originate. Luminescence studies by Ludner [60a,b] as well as by Ebeid and Bridge [67] provided further support for the importance of defects in the reactivity of crystals. An excellent review deals with the different kinds of structural imperfections, their characterization by physical techniques, and their relevance to phototransformations in crystals [62]. Diverse techniques such as electron microscopy, official microscopy, chemical etching, atom–atom potential energy calculations, interference, and fluorescence microscopy techniques have been brought in to the arsenal to characterize the type of defect. It may be mentioned that x-ray diffraction topography remains to be exploited for characterization of defects. The availability of powerful synchrotron x-ray sources can be expected to throw valuable information on the nature of defects.

More recent studies of photodimerization reactions have brought to light that the occurrence of defect-initiated reactions are not uncommon. It was observed that although 4-chloro-, 6-methyl-, and 4-methyl-6-chlorocoumarins [45] and 7-fluoro-4-methylcoumarins [63] are expected to be photoinert based on their molecular packings, all these crystals are photoreactive. It is noteworthy that the defect-initiated cases exhibit significant induction periods [15,63] (Fig. 4). As anticipated, the dimer yields are low (≤25%). The photobehavior of the parent coumarin crystal and 7-methoxy-4-methyl coumarin is particularly interesting. The photoirradiation of crystals of this compound yields two photodimers: a major *syn*-HH product (~80% yield), which is consistent with one of the cell dimensions being as short as 3.98 Å [64], and a mirror *syn*-head–tail product (~20% yield) after 6 hr of irradiation (Fig. 5a). The photobehavior of this crystal seems to be the first instance of the formation of a nontopochemical dimer as dictated by the generation of the topochemical product. Figure 5b shows a steady decrease in

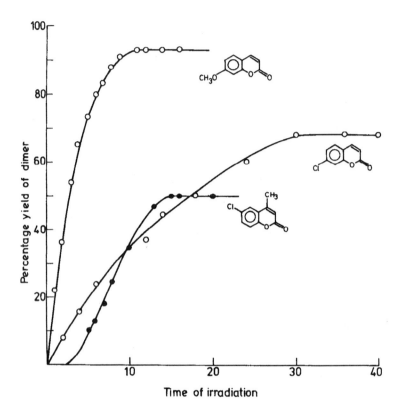

Figure 4 Duration of irradiation versus yield-irradiation time given in hours for 7-methoxycoumarin and 7-chlorocoumarin (topochemical reaction) irradiation in days for 4-methyl-6-chlorocoumarin (defect-initiated reaction)

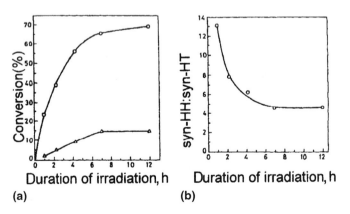

Figure 5 (a) Percentage yield of dimers versus irradiation time: (○) *syn*-HH, (△) *syn*-HT. (b) Ratio of *syn*-HH : *syn*-HT versus irradiation time.

the ratio *syn*-HH : *syn*-HT with time. The formation of the topochemically controlled dimer and the formation of the nontopochemical product occur simultaneously and are coupled. A plausible mechanism is that the site of generations of the *syn*-HH dimer becomes the region of impurity and the reactants proximal to it might be expected to undergo change in the topological relationship in a manner favorable for the generation of the *syn*-HT dimer. In connection with the nontopochemical photoreactivity of a 9-substituted anthracenes. Craig and Sarti-Fantooni have proposed that the excitation energy of molecules in the defect region might be slightly reduced when displaced from its orientation in the bulk of the ordered macroscopic volume of the crystal. This observation seems relevant for the situation in 7-methoxy-4-methylcoumarin. As more and more of the monomers undergo displacement from their original positions upon the formation of the *syn*-HH dimer, one would expect the *syn*-HH : *syn*-HT ratio to decrease, which is indeed the case. It is possible to visualize another mechanism for the observed situation in this crystal.

If we call a molecule related by the translation along the shortest crystallographic axis (3.98 Å) P and assume a few molecules obtained by the twofold axis along the shortest axis Q thus creating a short range order, the arrangement P P P P P P P Q Q Q Q Q P P P could be expected to generate the major *syn*-HH and a minor *syn*-HT dimer. Such an organization of molecules with a long-range order (P) and a short-range order (Q) can be expected to give rise to diffuse x-ray scattering. Detailed experiments using techniques such as electron microscopy and x-ray study of diffuse scattering would be expected to throw light on these possible alternate proposals.

IX. EFFECT OF TEMPERATURE AND WAVELENGTH

Ruzicka opined that "a crystal is a chemical cemetery" (quoted from Ref. [65]). What he meant was that molecules in a crystal are lifeless. However, the reality is that at any given temperature, the atoms and molecules in a crystal are not at rest. In the context of solid-state photochemistry, this immediately suggests that the percentage yield and the rate of reaction would depend on the temperature of irradiation.

A. Effect of Temperature

Hasegawa et al. [66] studied the photodimerization of methylchalcone-4-carboxylate at different temperatures and for different irradiation times. The observed maximum yield occurs at about 258 K, implying that the transport of energy that competes with thermal activation is favorable at this temperature. The photochemical conversion of 1,4-dicinnamoylbenzene (1,4-DCB) at different temperatures shows that the tricyclic dimer yield reaches a maximum at 298 K, the reaction taking place even at 256 K (Fig. 6) [67].

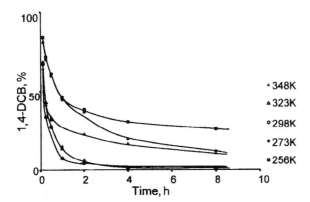

Figure 6 Photochemical conversion of 1,4-dicinnamoylbenzene at different temperatures.

The effect of temperature on the photoproduct obtained is extremely interesting in the case of diethyl-1,4-phenylonediacrylate, which is α-packed [68]. When the irradiation is performed above room temperature, a photodimer which has *anti*-HT stereochemistry is obtained as well as fairly low-molecular-weight oligormers (Scheme 5). However, a linear high-molecular-weight polymer crystal is obtained upon photoirradiation at a temperature below 273 K. The possibility of a phase transition resulting in a different molecular organization when the crystal is cooled cannot be dismissed.

The photobehavior of *n*-propyl-α-cyano-4-[2-(4-pyrimydyl)ethynyl] cinnamate has a valuable message to convey [69]. When the crystals are irradiated at wavelengths above 410 nm at room temperature, a monocyclic dimer, with nearly

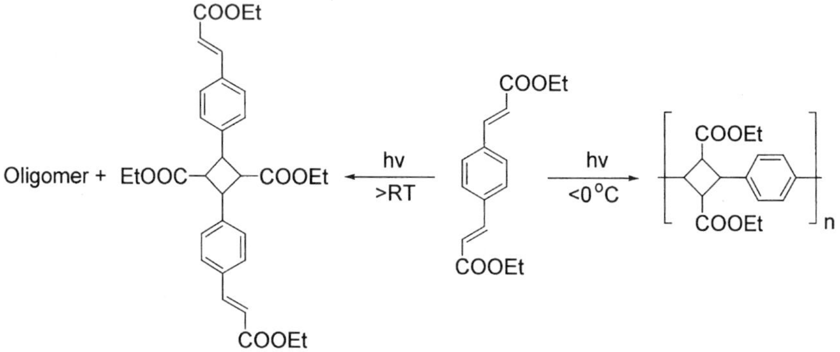

Scheme 5

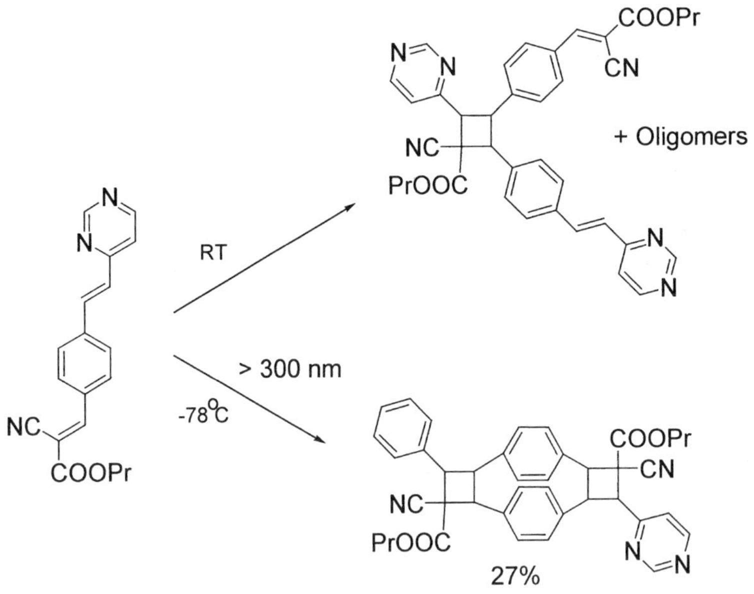

Scheme 6

qualitative yield, is obtained. When the same crystals are irradiated at wavelengths above 300 nm, at lower temperatures, they give the highly strained [2,2]-*para*-cyclophane in yields of 6% and 27% at temperatures of 233 K and 195 K, respectively (Scheme 6).

X-ray crystallographic determination of a crystal structure provides not only the atomic positions but also information regarding their atomic motions. From a very careful analysis of the atomic displacement parameters (ADP), it is possible to derive valuable information regarding the rigid-body motion as well as the internal molecular motion of molecules (i.e., the torsional amplitude of the rigid portion of the molecule around different single bonds) [65]. When the temperature of irradiation is changed, it is obvious that both the rigid-body motions and the internal torsional amplitudes would vary significantly. The change in molecular motion could favor photoreaction in the crystal. In the examples cited earlier (Schemes 5 and 6), rotation about the different single bonds and rigid-body motion of the molecule as a whole might be expected to play a role in the reactivity pattern. Further, the variable-temperature x-ray diffraction results would throw light on the phase changes in the crystal [70]. As mentioned earlier, the possibility of a phase transition when the irradiation temperature is changed must be kept in view.

The case of 7-methoxycoumarin with no associated nonrigid groups was analyzed for librational motion about the molecular axes using the program THMB developed by Trueblood in 1982. The major liberational axes L_1 and L_2 are almost in the plane of the coumarin ring ($L_1 = 34.1°$ and $30.4°$ and $L_2 = 17.8°$ and $14.4°$ for the two independent molecules, respectively) and the minor one L_3 ($10.4°$ and $11.1°$) is perpendicular to the ring [71]. Upon irradiation, the librational motion would be expected to increase, as part of the excitation energy would be utilized to increase the thermal energy of the system, thus increasing the orientational flexibility. It is quite feasible that the two crystallographically independent molecules in 7-methoxycoumarin, having extra freedom of movement, undergo rotation as rigid bodies in the opposite direction. This would bring the reacting bonds into favorable juxtaposition momentarily. The observed variation in the dimer yield with temperature (17.9%, 23.9%, and 17.3% at temperatures of 303, 323 and 343 K, respectively) is expected to be related to the variation in molecular motion with temperature. A thorough analysis of the anisotropic thermal motion parameters obtained from high-precision x-ray diffraction data for the rigid and nonrigid motion of molecules when the irradiation temperature is changed might be expected to help rationalize qualitatively the variation in the product yield. The compound 7-methoxy-4-methylcoumarin undergoes solid-state bimolecular reaction to yield two photodimers: a major *syn*-HH dimer of about 80% yield and a minor *syn*-HT dimer of about 20% yield after 6 hr of irradiation [64] (Fig. 5). The major product is understood in the light of the topochemical postulates, whereas the minor product arises from defect sites and appears to be a direct consequence of the formation of the topochemically controlled dimer. When the crystals were irradiated at 278, 298, and 353 K and the dimer yields estimated, it was found that the ratio of *syn*-HH to *syn*-HT was nearly 10 for the samples irradiated at 278 K and 353 K, whereas the ratio was only 3 at 298 K. The decrease in the minor dimer product (*syn*-HT) at low temperature may be due to smaller rotational and translational motions available for the monomers at that temperature and thus the topochemical product would be dominant. Again, the lower yield of the nontopochemical product at 353 K could be due to increased lattice vibrations competing with energy transport to the disturbed region and due to the excitation process. Thus, we observe that by changing the irradiation temperature, the yields of the product ratio could be changed.

The temperature dependence of photodimer yield has been observed in fluoro-4-methylcoumarin [63]. Upon irradiation of crystalline samples (for 8 hr) at temperatures of 298, 323, and 363 K, the dimer yields are 3%, 10%, 20%, and 26%, respectively. The increase in dimer yield must be due to the greater population of the defect sites generated in the crystal as the temperature is increased.

B. Effect of Wavelength

Dimerization and other topochemical photoreactions in the solid state occur heterogeneously—with the crystal disintegrating as the photoproduct is generated in the parent crystal and phase separation occurring. This is the situation when the radiation used falls in the range of the absorption maximum of the chromophore.

However, certain molecular systems, such as 2-benzyl-5-benzylidene-cyclopentanone, undergo homogeneous single crystal→single crystal transformation upon irradiation [72,73]. Thus, compounds produce 100% yields of the corresponding *anti*-HT dimer (Scheme 7). When the product molecule occupies a similar volume within a crystal and has a molecular topology similar to the reacting partners, single crystal→single crystal chemical transformation can take place. Such a homogeneous photoreaction has also been observed in the crystals of *n*-propyl-α-cyano-4-[2-(4-pyridyl-ethynyl)] cinnamate [74].

An interesting approach for inducing single crystal→single crystal transformations has been reported [75]. When a photoreactive crystal is exposed to irradiation with light for which the substance has a low absorption, the light intensity is uniform throughout the bulk of the crystal. Then, the product distribution within the crystal is homogeneous. This condition can be achieved if the crystal is irradiated with a wavelength corresponding to the tail end of the chromophore's absorption. Homogeneous [2 + 2] photodimerization of styrylpyrilium salt, (E) 2,6-di-*tert*-butyl-4-[2-(4-1-methoxyphenyl)ethylpyrilium]-trifluoro-metha-sulfonate, α-*trans*-cinnamic acid, and distyrylpyrazine have been reported. Figure 6 shows the reaction stages during the production of the dimer in cinnamic acid.

Clearly, homogeneous single crystal→single crystal phototransformation can be achieved only if the reaction cavity free volume is of sufficient size and shape to accommodate the product.

Klaus et al. [76] reported an interesting result concerning the wavelength-dependent photoproduction of 1-thiocoumarin (Scheme 8). When the crystals were irradiated at above 390 nm, dimer **a** was obtained. However, using a shorter wavelength above 340 nm, dimers **a–d** were recorded. However, dimers **b–d** are of nontopochemical origin. This observation is of general application, especially for detecting possible and unanticipated defect-controlled photoproducts.

To summarize, the generations of photoproducts of different stereochemistry through a combination of wavelength and temperature may be looked upon as one more strategy of crystal engineering, with the difference that no chemical modifications are performed on the reactant molecules. However, as pointed out

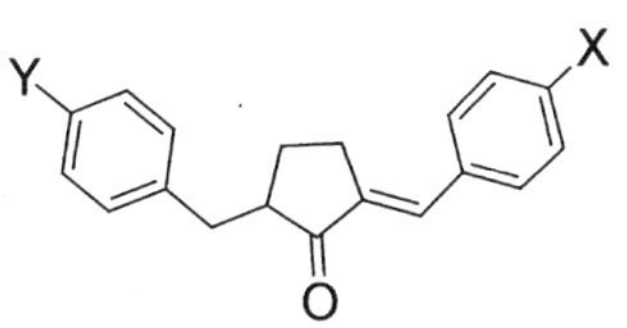

X = Y = H
X = p-Br, Y = H
X = H, Y = p-Cl
X = H, Y = p-Me
X = H, Y = p-Br

Scheme 7

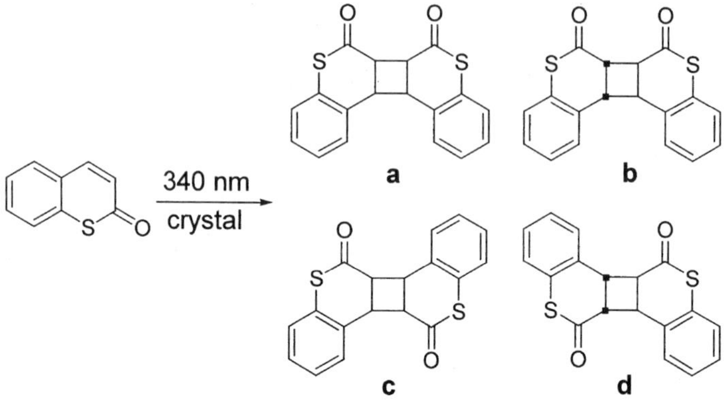

Scheme 8

earlier, the possibility that the parent crystal may undergo a phase change upon heating should not be overlooked.

X. REACTION CAVITY CONCEPT

The topochemical principle has provided a very valuable conceptional basis for the development of organic solid-state chemistry. This principle has led to the formulation of many important ideas, such as the reaction cavity concept [77] and the postulate of steric compression [78]. The reaction cavity is the space of a certain size and shape occupied by the reacting molecules in their ground state, which are surrounded by stationary molecules. Its volume must be able to accommodate changes in the shape of the reacting molecules upon excitation. Changes in the atomic positions will be resisted by the close-packed environment, as a result of which only minimal changes will be energetically allowed (Fig. 7). The concept has been of great help to rationalize, albeit qualitatively, the course of a variety of solid-state reactions. Scheffer and co-workers have used this argument to gain insight into the mechanism of intramolecular reactions [78]. Two different situations encountered in bimolecular photoreactions in crystals may now be discussed: (1) the reaction takes place when the reactants are not ideally juxtaposed in the crystal lattice and (2) a reaction is expected to take place, from topochemical considerations, but the crystal is photoinert. A few representative examples under each category are discussed in the light of the reaction cavity concept.

Ideally, in category 1, the π-orbitals of the reacting partners must overlap, in addition to the distance between them of ~4.2 Å. Figure 8 illustrates the geometrical parameters relevant to [2 + 2] photodimerization. The angle θ_1 corresponds to the rotation of one double bond with respect to the other when projected

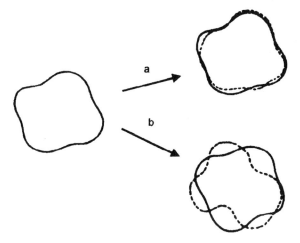

Figure 7 Reaction cavity of (a) a favorable and (b) an unfavorable reaction in the crystalline state.

down the line perpendicular to the plane defined by one of the double bonds and the atoms connected to it. θ_2 corresponds to the acute angle of the parallelogram formed by the double bonds $C_3\!=\!C_4$ and $C_{3'}\!=\!C_{4'}$. θ_3 corresponds to the angle between the least squares plane passing through the atoms C_3, C_4, $C_{3'}$, and $C_{4'}$ and that passing through atoms C_9, C_4, C_3, and C_2. For the best π-orbital overlap, the values of θ_1, θ_2, and θ_3 must be 0°, 90°, and 90°, respectively [79].

Kearsley has used a simple numerical description for the overlap of the reactive orbitals with which to rationalize the reactivity of photoreactive compounds [80]. More often than not, the relative orientation of the reacting partners are such

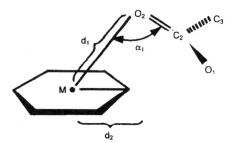

Figure 8 Geometrical parameters describing the relative orientation of reactant double bonds. T and T' are the π-orbitals of the double bonds, and A, B, and C are the point atoms. The distance between T and T' represents the extent of overlap of T and T'.

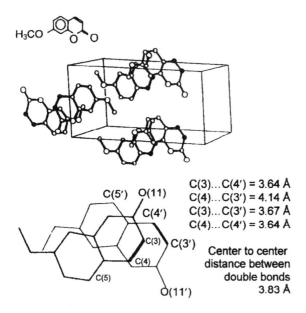

C(3)...C(4') = 3.64 Å
C(4)...C(3') = 4.14 Å
C(3)...C(3') = 3.67 Å
C(4)...C(4') = 3.64 Å

Center to center
distance between
double bonds
3.83 Å

Figure 9 Molecular packing of 7-methoxycoumarin in the unit cell (top) and relative orientation of the reacting parameters.

that the π-orbital overlap is not perfect [i.e., θ_1, θ_2, and θ_3 differ from the ideal values (for the β-packing mode θ_1; eq $0°$, although θ_2 and θ_3 may differ from $90°$)].

Of the coumarins investigated, the most unusual is 7-methoxycoumarin. The crystals belong to the space group *P-1* with two molecules in the asymmetric unit. Figure 9 shows that the molecular packing of the reactive double bonds of the monomer: molecules within the asymmetric unit are rotated by $65°$ with respect to each other, the center-to-center distance being 3.83 Å. Surprisingly, upon irradiation for 10 hr, a *syn*-HT dimer of about 95% yield was obtained. It is significant that the progress of the dimerization with respect to the time of irradiation showed no induction period. It has been observed that in all the cases known to react nontopochemically, there is a significant induction period [15]. Its absence indicates that the reactivity is topochemically controlled. It is worth nothing that the induction period may result from a positive feedback mechanism [81]. The results derived from the lattice energy calculations using the program WMIN [82] show the presence of a certain degree of inherent orientational flexibility of the molecule in the crystal, which would be expected to assist the reacting partners to orient favorably upon excitation [83]. It must be stressed that the lattice energy calculations are crude inasmuch as the dispersion constants used do not correspond to the excited state of the molecule. Craig and co-workers have carried out incisive theoretical investigations of dimer formation, which explain the results in the cases where the reactants are not favorably oriented. According to their analysis, short-term instability caused by photoexitation can have the effect of driving molecules close to a neighbor so as to cause a photochemical reaction [83]. If a reaction proceeds in the short time available before general lattice relaxation to the equilibrium ground state, it can be said to be dynamically preformed.

Other cases have been reported in which the reactive double bonds are skewed to each other and yet react topochemically upon irradiation. In the crystals of 2,5-dibenzylenecyclopentanone [84] and 1,4-dicinnamoylbenzene [67], the reactive bonds are rotated with respect to one another by 56° and 28.5°, respectively, but both are photoreactive. Similarly, the reactive bonds in propyl-4-[2-(4-pyriduyl)] cinnamate [85] are arranged skew to each other with a rotation angle of 74°.

These examples, in which the reacting partners are not ideally oriented and yet are photoreactive, lend support to the idea that the reaction cavity should not be looked upon as having a stiff or hard wall but one that is flexible or soft. An appropriate spatial distribution of the free volume within a cavity must be available for a reaction to occur (Fig. 10) [86].

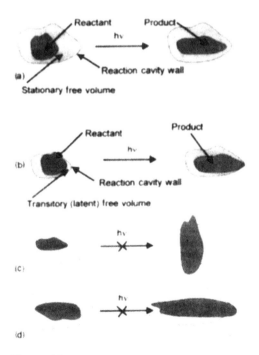

Figure 10   Representation of (a,c,d) a hard and (b) a soft reaction cavity.

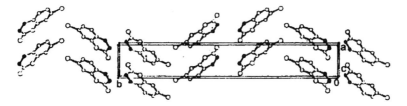

Figure 11   Molecular packing in the unit cell of 7-chlorocoumarin.

The situation with respect to the photobehavior of 7-chlorocoumarin is interesting (Fig. 11). There are two reaction pathways in this crystal: one that favors the formation of the *syn*-HH isomer arising from reaction between the translationally related molecules with a center-to-center distance of 4.54 Å and another that would yield the *anti*-HT dimer, corresponding to the reaction between the centrosymmetrically related molecules, the center-to-center distance being 4.12 Å. Experiment clearly shows that only the *syn*-HH dimer is obtained—not the one that would correspond to the path of least motion. This is supported by the results of the lattice energy calculations. The implication is that the shape of the free volume is anisotropic, with the larger volume or extension in the direction of the translational periodicity of 4.54 Å.

Category 2 concerns examples that satisfy Schmidt's rule but remain photoinactive. Table 8 records a few of such cases in which the distance between the reactive double bonds is less than ~4.2 Å. Scheme 9 shows the molecular structures.

In benzylidene-*dl*-piperitone, the two pairs of reactive double bonds that are related by a center of inversion are properly oriented for [2 + 2] photocycloaddition with $\theta_1 = 0°$, $\theta_2 = 105.2°$, $\theta_3 = 86.4°$, and $d = 3.93$ Å. However, there is no evidence for the presence of a dimer in the irradiated sample. The lattice energy

Table 8   Examples of Photoinert Compounds Satisfying Topochemical Principles

Compound	Double-bond separation (Å)	Ref.
Distyrylpyrazine (**A**)	4.19	71
Enone (**B**)	3.79	71
4-Hydroxy-3-nitrocinnamate (**C**)	3.78	71
Benzylidene-*dl*-piperitone (**D**)	4.00	71
(+)2,5-Dibenzylidene-3-methylcyclopentanone (**E**)	3.87	71
2-Benzylidenecyclopentanone (**F**)	4.14	84
Pyridine-3,5-dicarboxylate (**G**)	3.73	54

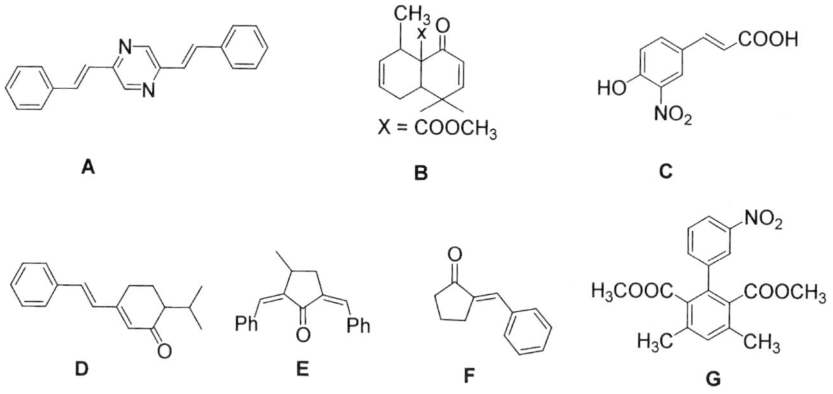

A   B   C

X = COOCH₃

D   E   F   G

Scheme 9

calculations carried out using the program WMIN [82] shows that the increase in energy to achieve the ideal geometry ($\theta_1 = 0°$, $\theta_2 = \theta_3 = 90°$) is found to be as small as 3.35 kj mol, suggesting that the crystal should be photoreactive. However, the results provided by the calculations should be accepted with reservation. First, the dispersion constants used correspond to the ground state. Second, in its reaction pathway involving change in the hybridization of reactive atoms from sp^2 to sp^3 with a lengthy styrene group, it would be expected that upon excitation, the side group would undergo considerable positional changes in the crystal lattice. Thus, the available total free volume as defined by Weiss et al. [86] is insufficient to permit the large molecular topological changes. There is also the possibility that the crystals could undergo phase transition upon irradiation. The photoinertness of compound **G** in Scheme 9, which has bulky substituents, may be rationalized on similar grounds. The photoinertness of the cinnamate derivative (c) is probably due to the extensive network of intermolecular C—H···O and O—H···O interactions which restrict positional changes in atoms demanded for the photoreaction to occur. Also, the reaction cavity volume available might be imposing restrictions on the movements of the reactants. Calculations of the void or the available space around the reaction site using the program derived by Gavezzotti [87] might throw further light on the reasons for the photoinertness of this crystal.

  With the exception of 7-methoxycoumarin, the examples discussed are all with only one molecule in the asymmetric unit in the unit cell, and the reaction takes place between this molecule and the one obtained by a symmetry operation—either a center of inversion or translational symmetry. Table 9 records a few examples where the unit cell has two independent molecules (**A** and **B**) in the asymmetric unit.

  In both 4-(3-fluorostyryl)coumarin and 4-(4-fluorostyryl)coumarin, the two crystallographically independent molecules are favorably juxtaposed for reaction,

Table 9 Photoreactivity of Crystals With Two Molecules in the Asymmetric Unit

Compound	Space group; Z^a	Dimer yield (%)	Ref.
7-Methoxycoumarin	*P-1*; 4	>90	71
4-Styrylcoumarin (prisms)	*P-1*; 4	45	88
4-Styrylcoumarin (plates)	$P2_1/n$; 8	50	88
4-(3-Fluorostyryl)coumarin	*P-1*; 4	80	31
4-(3-Fluorostyryl)-6-flouocoumarin	$Pna2_1$; 8	80	31
o-Chlorobenzylidene-*dl*-piperitone	*P-1*; 4	No reaction	51
4-Styryl-6-fluorocoumarin	*P-1*; 4	50	48
p-Fluobenzylidene-*dl*-piperitone	*P-1*; 4	40	53
Ethyl-4-[2-(4-pyrydy)ethylcinnamate]	$P2_12_12_1$; 8	>90	69
Propyl-4-[2-(4-pyrydyl)ethylcinnamate]	*Pbca*; 8	Two dimers[b]	69

[a] Z = number molecules in the cell.
[b] two types of dimers obtained with different stereochemistry.

and the dimer yield is as large as 80%. In terms of their reactivity, this situation can be considered as two reaction cavities with their free volumes having no interference between these two, and hence the percentage yield is quite high—neglecting other factors such as particle size distribution, local mechanical properties, melting points, and so forth, all of which also would affect the reactivity [89]. The case of 7-methoxycoumarin is unique in that the reaction is between the two crystallographically independent molecules, A and B, rather than between A and A (symmetry related) and between B and B. Therefore, this case does not have two independent reaction cavities to be taken into consideration. However, in the case of *o*-chlorobenzilidene-*dl*-piperitone, there is no reaction, even though molecules A and B are properly oriented. This must be related to the distance between reaction cavities in that there is correlation or interaction between the two free volumes [90].

Where the dimer yields are not very high (≤50%), it is reasonable to conclude that the presence of the unfavorably oriented second molecule (B) does not affect the reaction taking place; in other words, the reaction cavity free volume of the favourably oriented molecules (A) is not interfered by that of B.

To summarize, three different situations occur for crystals with two crystallographically independent molecules (A and B): (1) The independent molecules may react with their symmetry-related partners and undergo bimolecular may react with their symmetry-related partners with high dimer yield of ~75%; (2) it could happen that there is no photoreaction, even when molecules A and B are properly oriented, satisfying Schmidt's criteria; and (3) only one of the molecules (either A or B) undergoes reaction with its symmetry-related partner, but not the second one, as it is not favorably oriented with its symmetry-related partner. It is not surprising that the dimer yield obtained in this case is about 50% or less. A qualitative understanding of all these cases may be gained in the light of the reaction cavity concept and the free volume associated with it.

These observations on the effects of the environment, especially the case with more than one crystallographically independent molecule, may serve as a useful link between the solution and solid-state photochemical behavior [91]. It would be interesting to encounter cases of photoreactive crystals containing more than two crystallographically independent molecules, which would be somewhat akin to, but by no means the same as, the solution–phase reaction.

XI. CONCLUSIONS AND PROSPECTS

From what has been presented in this chapter, it is clear that the problem of crystal engineering, namely synthesis of a crystal—"a supermolecule par excellence" [16]—starting from the chosen synthon remains an immensely challenging task. It is interesting to observe that the great Dutch artist M. C. Escher has achieved supramolecular assembly (Fig. 12), albeit in two dimensions. As it stands now, there is much promise to achieve crystal synthesis (1) through intramolecular substitution and (2) by substitution of functional groups that can participate in strong

Figure 12 M. C. Escher's drawing of black and white lizards: supramolecular organization. (From Ref. 91a.)

intermolecular hydrogen-bonding. Both of these strategies allow us to reduce the number of possible arrangements in the crystal.

It is important that the solid-state chemists keep a watchful eye on the possibility of "disappearing polymorphs" [92] and also of the well-known phenomenon of concomitant polymorphs which has been discussed recently [93].

The generation of new polymophs via the addition of a meaningful choice of impurities in controlled amounts [93,94] is worthy of experimentation.

The formation of photoproducts by the fine-tuning of wavelengths and variation of temperature of the sample being irradiation deserve attention. However, with variable-temperature experiments, the possibility of a phase transition taking place must not be overlooked.

The results of the solid-state photochemistry dealt with the molecules in their ground state and we have no knowledge of the excited-state structures. In this connection, the investigations by Coppens and co-workers [95] on the molecular geometry of long-lived metastable states of transition metal nitrosyl complexes generated upon photoirradiation are extremely interesting and raise our hopes for studying structures on femtosecond timescales or even attosecond time scales (10^{-18} sec). Then, advances in ultrafast time-resolved x-ray diffraction are encouraging [96]. The report on the generation of femtosecond pulses of synchrotron radiation is an important development [97] in connection with studying excited-state structures which would, in turn, permit one to obtain knowledge regarding the reaction mechanism of the photoinduced reactions.

It may be pointed out that the discussions presented in this chapter are concerned with only one aspect, namely the structure features. There are two others, such as the kinetic and thermodynamic factors which influence the reaction rate. A careful investigation of these aspects is also needed [98].

There is still a cultural divide between the synthetic organic chemists and the solid-state photochemists. Hopefully, this situation would change for the benefit of investigations engaged in these two areas of research.

ACKNOWLEDGMENTS

Our grateful thanks go to many of our co-workers for their contributions. The financial support provided by the Council of Scientific and Industrial Research (India) and the Natural Academy of Sciences (India) is gratefully acknowledged.

REFERENCES

1. Hasegawa, M. *Adv. Phys. Org. Chem.* **1995,** *30,* 117–171.
2. Bloor, M.; Chance, R. R.; Nijhoff, M. *Polydiacetylenes,* Reidel: Dordrecht, 1985, 102.
3. Bonner, W. A. *Topics Stereochem.* **1988,** *18,* 1–98.

4. Scheffer, J. R.; Garcia-Garibay, M. A. in *Photochemistry of Solid Surfaces,* Matsuwra, T.; Anpo, M., eds., Elsevier: Amsterdam, 1989.

5. Scheffer, J. R.; Pokkuluri, P. R. in *Photochemistry in Organized and Constrained Media,* Ramamurthy, V., ed., VCH: New York, 1991, 185–246.

6. Thosmas, J. M.; Marsi, S. E.; Desvergne, J. P. *Adv. Phys. Org. Chem.* **1977,** *15,* 63–151.

7. Ramamurthy, V.; Venkatesan, K. *Chem. Rev.* **1987,** *87,* 433–481.

8. (a) Venkatesan, K. in *Reactivity of Molecular Solids,* Boldyrev, E.; Boldyrev, V., eds., Wiley: New York, 1999, 89–131. (b) Narasimha Moorthy, J.; Venkatesan, K.; Weiss, R. G. *J. Org. Chem.* **1992,** *57,* 3292–3297.

9. Kohlschutter, V. *Z. Anorg. Chem.* **1919,** *105,* 1–25.

10. Kohlschutter, V. *Helv. Chim. Acta* **1929,** *12,* 512–529.

11. (a) Cohen, M. D.; Schmidt, G. M. J. *J. Chem. Soc.* **1964,** 1996–2000. (b) Schmidt, G. M. J. *J. Chem. Soc.* **1964,** 2000–2001. (c) Cohen, M. D.; Schmidt, G. M. J.; Sonntag, F. I. *J. Chem. Soc.* **1964,** 2000–2013.

12. Green, B. S.; Lahav, M.; Schmidt, G. M. J. *J. Chem. Soc. B* **1971,** 1552–1564.

13. Lahav, M.; Schmidt, G. M. J. *J. Chem. Soc. B* **1967,** 239–243.

14. Theocharis, C. R.; Jones, W. in *Organic Solid State Chemistry,* Desiraju, G. R., ed., Elsevier: Amsterdam, 1987, 47–67.

15. Gnanaguru, K.; Ramasubbu, N.; Venkatesan, K.; Ramamurthy, V. *J. Org. Chem.* **1985,** *50,* 2337–2346.

16. Dunitz, J. D. in *Perspectives in Supramolecular Chemistry,* Desiraju, G. R., ed., Wiley: Chichester, 1996, 1–30.

17. Desiraju, G. R. *Crystal Engineering; The Design of Organic Solids,* Elsevier: Amsterdam, 1989.

18. Greenwald, J.; Halmann, M. *J. Chem. Soc., Perkin Trans. 2* **1972,** 1095–1101.

19. Cohen, M. D.; Cohen, R. *J. Chem. Soc., Perkin Trans. 2* **1976,** 1731–1735.

20. Sarma, J. A. R. P.; Desiraju, G. R. *J. Am. Chem. Soc.* **1986,** *108,* 2791–2793.

21. Ito, Y. *Synthesis* **1998,** *1,* 1–32.

22. Sarma, J. A. R. P.; Desiraju, G. R. *J. Chem. Soc. Perkin Trans. 2* **1985,** 1905–1912.

23. Desiraju, G. R.; Sharma, C. V. K. M. *J. Chem. Soc., Chem. Commun.* **1991,** 1239–1241.

24. Feldman, K. S.; Campbell, R. F. *J. Org. Chem.* **1995,** *60,* 1924–1925.

25. Whitesides, G. M.; Mathias, J. P.; Seto, C. T. *Science* **1991,** *254,* 1312–1319.

26. Amarendra Kumar, V.; Begum, N. S.; Venkatesan, K. *J. Chem. Soc. Perkin Trans. 2* **1993,** 463–467.

27. (a) Vishnumurthy, K.; Guru Row, T. N.; Venkatesan, K. *J. Chem. Soc. Perkin Trans. 2* **1996,** 1475–1478. (b) Vishnumurthy, K.; Guru Row, T. N.; Venkatesan, K. *J. Chem. Soc. Perkin Trans. 2* **1997,** 615–619.

28. (a) Vishnumurthy, K.; Guru Row, T. N.; Venkatesan, K. Unpublished results. (b) Vishnumurthy, K. *Ph.D. dissertation,* Indian Institute of Science, Bangalore, 1998.

29. Caotes, G. W.; Dunn, A. R.; Henling, L. M.; Ziller, J. W.; Lobkovsky, E. B.; Grubbs, R. H. *J. Am. Chem. Soc.* **1998,** *120,* 3641–3649.

30. Pauling, L. *Nature of the Chemical Bond,* 2nd ed. Cornell University Press: Ithaca, NY, 1940, 286–287.

31. Bondi, A. *J. Phys. Chem.* **1964,** *68,* 441–451.

32. Howard, J. A. K.; Hoy, V. J.; O'Hagan, D.; Smith, G. T. *Tetrahedron* **1996,** *52,* 12,613–12,622.
33. Shimoni, L.; Glusker, J. P. *Struct. Chem.* **1994,** *5,* 383–397.
34. Murray-Rust, P.; Stallings, W. C.; Monti, C. T.; Preston, R. K.; Glusker, J. P. *J. Am. Chem. Soc.* **1983,** *105,* 3206–3214.
35. Dunitz, J. D.; Taylor, R. *Chem. Eur. J.* **1997,** *3,* 383–397.
36. Ponzini, F.; Zagha, R.; Hardcastle, K.; Siegel, J. S. *Angew. Chem. Int. Ed. Engl.* **2000,** *39,* 2323–2325.
37. Robertson, J. M. *Organic Molecular Crystals,* Cornnel University Press: Ithaca, NY, 1953, 239.
38. Desiraju, G. R. *Acc. Chem. Res.* **1991,** *24,* 290–296.
39. Steiner, T. *Cryst. Rev.* **1996,** *6,* 1–57.
40. Addadi, I.; Lahav, M. *Pure Appl. Chem.* **1979,** *39,* 1269–1284.
41. Ueno, K.; Nakanishi, H.; Hasegawa, M.; Sasada, Y. *Acta Crystallogr.* **1978,** *B34,* 2034–2035.
42. Boese, R.; Boese, A. D.; Blaser, D.; Antipin, M. Y.; Eller, A.; Seppelt, K. *Angew. Chem. Int. Ed. Engl.* **1997,** *36,* 1489–1491.
43. Toda, F. *Topics Curr. Chem.* **1988,** *149,* 212–238.
44. Toda, F.; Akagi. K.; *Tetrahedron Lett. 1968,* 3695–3698.
45. Tamaki, T.; Kokubu, T.; Ichimura, K. *Tetrahedron* 1987, 43, 1485–1494.
46. McCrone, E. V. in *Physics and Chemistry of the Organic Solid State,* Fox, D.; Labes, M. M., eds., Interscience: New York, 1995, Vol. 2, 726–767.
47. Weissbuch. I.; Popovitz-Biro. R.; Lahav, M.; Leiserowitz, L. *Acta Crystallogr.* **1995,** *B51,* 115–148.
48. Vishnumurthy, K. *Ph.D. thesis,* Indian Institute of Science, Bangalore, 1998.
49. Kanagapushapam, D.; Ramamurthy, V.; Venkatesan, K.; *Acta Crystallogr.* **1987,** *C43,* 1128–1131.
50. Venugopalan, P.; Venkatesan, K. *Acta. Crystallogr.* **1990,** *B46,* 826–830.
51. Venugopalan, P.; Venkatesan, K. *Acta Crystallogr.* **1992,** *B48,* 532–537.
52. Venugopalan, P.; Bharathi Rao, T.; Venkatesan, K. *J. Chem. Soc. Perkin Trans. 2,* **1991,** 981–987.
53. Amarendra Kumar, V.; Venkatesan, K. *J. Chem. Soc. Perkin Trans 2* **1993,** 2429–2433.
54. Marubayashi, N.; Ogawa, T.; Hamasaki, T.; Hirayama, N. *J. Chem. Soc. Perkin Trans 2* **1997,** 1309–1314.
55. Stevens, B.; Dickinson, T.; Sharpe, R. R. *Nature* **1964,** *204,* 876–877.
56. Cohen, M. D.; Ludmer, Z. *J. Chem. Soc., Chem. Commun.* **1969,** 1172–1173.
57. Heller, E.; Schmidt, G. M. J. *Israel J. Chem.* **1971,** *9,* 449–462.
58. Craig, D. P.; Sarti-Fantoni, P. *J. Chem. Soc., Chem. Commun.* **1966,** 742–743.
59. (a) Cohen, M. D.; Ludmer, M.; Thomas, J. M.; Williams, J. O. *Proc. Roy. Soc.* **1971,** *324A,* 459; (b) Cohen, M. D.; Ludmer, M.; Thomas, J. M.; Williams, J. O. *J. Chem. Soc., Chem. Commun.* **1969,** 1172.
60. (a) Ludmer, Z. *Chem. Phy.* **1977,** *26,* 113–121. (b) Ludmer, Z. *J. Lumin.* **1978,** *17,* 1.
61. Ebeid, E. Z.; Bridge, J. *J. Chem. Soc. Faraday Trans. 1* **1984,** *80,* 1131–1138.
62. Jones, W.; Thomas, W. *Prog. Solid State Chem.* **1979,** *12,* 101–124.

63. (a) Gnanaguru, K.; Ramasubbu, N.; Venkatesan, K.; Ramamurthy, V. *J. Photo. Chem.* **1984,** *27,* 355–362. (b) Vishnumurthy, K.; Guru Row, T. N.; Venkatesan, K. *Tetrahedron* **1998,** *54,* 11,235–11,246.
64. Narasimha Moorthy, J.; Venkatesan, K. *J. Mater. Chem.* **1992,** *2,* 675–676.
65. Dunitz, J. D.; Schomaker, V.; Trueblood, K. N. *J. Phys. Chem.* **1988,** *92,* 856–867.
66. Hasegawa, M.; Arioka, H.; Harashina, H.; Nohara, M.; Kubo, M.; Nishikubo, T. *Israel J. Chem.* **1985,** *25,* 302–305.
67. Hasegawa, M.; Saigo, K.; Mori, T.; Uno, H.; Nohara, M.; Nakanishi, H. *J. Am. Chem. Soc.* **1985,** *107,* 2788–2793.
68. Nahanishi, H.; Nakanishi, F.; Suzuki, Y.; Hasegawa, M. *J. Polym. Sci. Polym. Chem.* **1973,** *11,* 2501–2518.
69. Chung, C. M.; Nakamura, F.; Hashimoto, Y.; Hasegawa, M. *Chem. Lett.* **1991,** 779–782.
70. Yang, Q. C.; Richardson, M. F.; Dunitz, J. D. *J. Am. Chem. Soc.* **1985,** *107,* 5535–5537.
71. Murthy, G. S.; Arjunan, P.; Venkatesan, K.; Ramamurthy, V. *Tetrahedron* **1987,** *43,* 1225–1240.
72. Whiting, D. A. *J. Chem. Soc. C* **1971,** 3396–3398.
73. Nakanishi, H.; Jones, W.; Thomas, J. M.; Hursthouse, M. B.; Motevalli, M. *J. Phys. Chem.* **1981,** *85,* 3636–3642.
74. Hasegawa, M.; Kato, S.; Saigo, K. *J. Photochem. Photobiol.* **1988,** *41A,* 385–394.
75. (a) Enkelmann, V.; Wegner, G.; Novak, K.; Wagener, K. B. *J. Am. Chem. Soc.* **1993,** *115,* 10,390–10,391. (b) Novak, K.; Enkelmann, V.; Wagner, G.; Wagener, K. B. *Angew. Chem. Int. Ed. Engl.* **1993,** *32,* 1614–1616. (c) Enkelmann, V.; Wagner, G.; Novak, K.; Wegener, K. B. *Mol. Cryst. Liquid Cryst.* **1994,** *240,* 121–126.
76. Klaus, C. P.; Thiemann, C.; Kopf, J.; Margaretha, P. *Helv. Chim. Acta* **1995,** *78,* 1079–1082.
77. Cohen, M. D. *Angew. Chem. Int. Ed. Engl.* **1975,** *14,* 386–393.
78. Ariel, S.; Askari, S.; Scheffer, J. R.; Trotter, J.; Walsh, L. *J. Am. Chem. Soc.* **1984,** *106,* 5726–5728.
79. Venkatesan, K.; Ramamurthy, V. in *Photochemistry in Organized and Constrained Media,* Ramamurthy, V., ed., VCH: New York, 1991, 229–249.
80. Kearsley, S. K. in *Organic Solid State Chemistry,* Desiraju, G. R., ed., Elsevier: Amsterdam, 1987, 69–115.
81. Baldyrev, E. V. *J. Therm. Anal.* **1992,** *38,* 89–97.
82. Busing, B. R., WMIN, *A Computing Program to Model Molecules and Crystals in Terms of Potential Energy Function,* Oak Ridge National Lab: Oak Ridge, TN, 1981.
83. (a) Collins, M. A.; Craig, D. P. *Chem. Phys.* **1981,** *54,* 305–321. (b) Craig, D. P.; Mallett, C. P. *Chem. Phys.* **1982,** *65,* 129–142. (c) Craig, D. P.; Lindsay, R. N.; Mallett, C. P. *Chem. Phys.* **1984,** *89,* 187–197.
84. Theocharis, C. R.; Jones, W.; Thomas, J. M.; Motevalli, M.; Hursthouse, M. B. *J. Chem. Soc. Perkin Trans. 2* **1984,** 71–76.
85. Chung, C. M.; Hasegawa, M. *J. Am. Chem. Soc.* **1991,** *113,* 7311–7316.
86. Weiss, R. G.; Ramamurthy, V.; Hammond, G. S. *Acc. Chem. Res.* **1993,** *26,* 530–536.
87. Gavazzotti, A. *J. Am. Chem. Soc.* **1989,** *111,* 1835–1843.

88. Narasimha Moorthy, J.; Venkatesan, K. *Bull. Chem. Soc. Jpn.* **1994,** *67,* 1–6.

89. Singh, N. B.; Singh, R. J.; Singh, N. P. *Tetrahedron* **1994,** *50,* 6441–6493.

90. Boldyrev, E. V. in *Reactivity of Solids, Past, Present, Future,* Boldyrev, V., ed., IU-PAC Series Chemistry for the 21st Century, Blackwell, Oxford, 1996, 141–184.

91. Boldyrev, E. V. *Solid State Ionics* **1997,** *101–103,* 841–849.

91a. McGillavry, C. H., ed. Symmetry Aspects of M. C. Escher Periodic Drawings, International Union of Crystallography, **1965,** *70.*

92. Dunitz, J. D.; Bernstein, J. *Acc. Chem. Res.* **1995,** *28,* 193–200.

93. Bernstein, J.; Davey, R. J.; Henck, J. O. *Angew. Chem. Int. Ed. Engl.* **1999,** *38,* 3440–3461.

94. Weissbuch, I.; Popovitz-Biro, R.; Lahav, M.; Leiserowitz, L. *Acta Crystallogr.* **1995,** *B51,* 115–148.

95. (a) Fomitchev, D. V.; Furlani, T. R.; Coppens, P. *Inorg. Chem.* **1998,** *37,* 1519–1526.
     (b) Coppens, P.; Fomitchev, D. V.; Cardweei, M. D. *J. Chem. Soc. Dalton Trans.* **1998,** 865–872.

96. Wark, J. *Contemp. Phys.* **1996,** *37,* 205–218.

97. Schoenlein, R. W.; Chattopadhyay, S.; Chong, H. H. W.; Glover, T. E.; Helmann, P. A.; Schank, C. V.; Zholents, A. A.; Zolatorov, M. S. *Science* **2000,** *287,* 2237–2240.

98. Boldyrev, V. V. *React. Solids* **1990,** *8,* 231–246.

8

Supramolecular Photochemistry of Cyclodextrin Materials

Akihiko Ueno and Hiroshi Ikeda
Tokyo Institute of Technology, Yokohama, Japan

I. INTRODUCTION

Cyclodextrins (CDs) are cyclic oligosaccharides produced from starch by the action of the amylase of *Bacillus marcerans*. They possess a central cavity capable of accommodating hydrophobic organic molecules in aqueous solution. Although CDs consisting of 9,10,11,12,13,14,15,16, and 17 α-D-glucose members have been found [1–4], the most common CDs are commercially available CDs, α-, β-, and γ-CD, and the inclusion-complex formation of these three distinct CDs have been extensively studied [5–7]. These molecules have different cavity diameters, 4.9, 6.2, and 7.9 Å for α-, β-, and γ-CD, respectively, and are shaped like truncated cones with a smaller and a larger opening at the primary hydroxyl and the secondary hydroxyl faces, respectively. The interior of the cavities are lined with ether oxygens and C_1H and C_4H groups and the wall is relatively hydrophobic. As a consequence of the nature of the cavities, the hydrophobic guests can be stabilized in the cavities primarily by hydrophobic interactions. The stoichiometry of the complex formation is usually 1:1, but in some cases of γ-CD, which have a larger cavity, 1:2 host–guest complexes are formed [8]. In studies of photochemistry, it has been recognized that the restricted shape and size of the cavity geometrically constrain the included guest molecule and thereby stabilize the conformations that are less favored in free solution [9–11]. When a chromophore unit is covalently linked to CD, it is included in the CD cavity, forming a self-inclu-

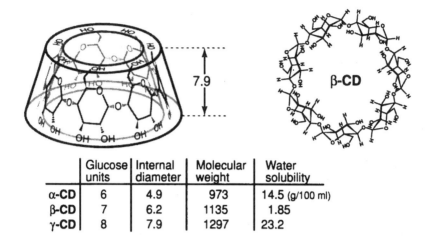

	Glucose units	Internal diameter	Molecular weight	Water solubility
α-CD	6	4.9	973	14.5 (g/100 ml)
β-CD	7	6.2	1135	1.85
γ-CD	8	7.9	1297	23.2

sion complex. The photophysical property of the chromophophore part changes when the part is excluded from hydrophobic cavity to polar bulk water solution upon accommodation of exogenous guest species. This behavior may be used for the construction of molecule-detecting sensors. On the other hand, CDs can be used as scaffolds for constructing multichromophoric systems, in which stereoselective dimerization or effective energy migration may occur. In this chapter, we describe the topics on supramolecular photochemistry of CD-related systems, including sensors [12–14], energy transfer, and unique photoreactions.

II. ROOM-TEMPERATURE PHOSPHORESCENCE

Room-temperature phosphorescence of aromatic compounds, which usually cannot be detected in liquid media, is often observed in aqueous solution containing CD. There are several types of CD systems, which show room-temperature phosphorescence: (1) 1:1 inclusion complexes between a halogenated aromatic guest molecule and CD, (2) 2:1 α-CD-halogenated aromatic guest complexes, (3) complexes that contain CD and plural guest molecules, and (4) complexes of halogenated CD with an aromatic guest molecule. When CD accommodates a triplet guest molecule in its cavity, the collisional quenching by oxygen is reduced. Turro et al. found that room-temperature phosphorescence of 1-bromonaphthalene and 1-chloronaphthalene is readily observable in N_2-purged aqueous solutions containing β-CD [15]. It was reported that the ternary complex of β-CD, 1-bromonaphthalene, and acetonitrile exhibits room-temperature phosphorescence. α-CD has a smaller cavity than β-CD and forms a 2:1 α-CD–halonaphthalene complex. Hamai observed room-temperature phosphorescence from such 2:1

complexes when the halonaphthalene is 6-bromo-2-naphthol [16] or 2-chloron-aphthalene [17]. It is well known that two different kinds of guest molecule are simultaneously accommodated by the same CD cavity to form a ternary inclusion complex. For nonhalogenated phosphorescent guests, the room-temperature phosphorescence may be induced by external heavy-atom effects. Scypinski and Cline Love have reported that when 1,2-dibromoethane is added to an aqueous β-CD solution of phenanthrene, the fluorescence of phenanthrene is significantly reduced, accompanied by the appearance of the intense room-temperature phosphorescence of phenanthrene [18]. Hamai observed the room-temperature phosphorescence of acenaphthene in deaerated aqueous solutions containing β-CD and brominated alcohol (2-bromoethanol or 2,3-dibromo-1-propanol) [19]. Escandar and Pena also observed room-temperature phosphorescence of acenaphthene in aerated solution in the presence of γ-CD and 2-bromoethanol or 2,3-dibromo-1-propanol [20]. On the other hand, upon addition of 1-butanol to the aqueous solution containing β-CD, naphthalene, and 1-bromonaphthalene, the room-temperature phosphorescence of 1-bromonaphthalene appears [21]. The room-temperature phosphorescence is due to a quaternary inclusion complex composed of β-CD, 1-bromonaphthalene, naphthalene, and 1-butanol.

Surfactants were found to induce phosphorescence from 1-bromonaphthalene in aerated aqueous solutions of β-CD [22]. In the case of the 1:1 inclusion-complex formation between γ-cyclodextrin and acenaphthene in aqueous solution, bromoalcohols such as 2,3-dibromopropane-1-ol or 2-bromoethanol induced a decrease of the fluorescence and an enhancement of room-temperature phosphorescence of acenaphthene [23].

In the presence of heptakis(6-bromo-6-deoxy-β-CD) (β-CD7Br), Femia and Cline Love observed the room-temperature phosphorescence of phenanthrene and other polynuclear aromatic hydrocarbons in N_2-purged N,N-dimethylfol-mamide (DMF)–water mixtures [24]. On the other hand, Hamai and Monobe observed room-temperature phosphorescence of 2-chloronaphthalene from a deaerated solution containing a 1:1 complex of 6-iodo-6-deoxy-β-CD (β-CDI) and 2-chloronaphthalene [25]. This result indicates that even only one iodine atom on the β-CD rim can accelerate the intersystem crossing rate of 2-chloronaphthalene included in the CD cavity. The room-temperature phosphorescence of 6-bromo-2-naphthol and 3-bromoquinoline was also observed for the complexes with β-CDI [26].

Mortellaro et al. prepared two β-CD derivatives, which have a bromonaphthalene unit at the primary (**1**) and secondary OH groups (**2**) of β-CD [27]. The phosphorescence of **2** at 530 nm is more than 100 times intense than that observed for an equimolar solution of **1**. These results suggest that the bromonaphthalene moiety of **2** is protected from oxygen quenching, being located inside the CD cavity, whereas the bromonaphthalne moiety of **1** is not protected, being exposed into bulk solution.

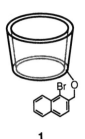

1

Structure 1

De Silva et al. [28] prepared a naphthalene derivative (**3**) with logic functions (Scheme 1). Here, the bromonaphthalene unit exhibits phosphorescence in the presence of both the calcium ion and β-CD [28]. However, without them, oxygen quenches the phosphorescence of 2-bromonaphthalene phosphor because the protection effect of β-CD is absent and photoinduced electron transfer from the tetracarboxylate receptor to the 2-bromonaphthalene phosphor occurs. Thus, phosphorescence output occurs only when the calcium ion and β-CD inputs are active. The operation of these two inputs with a phosphorescence output corresponds to the AND logic function. The input to the NOT gate is oxygen. In the presence of oxygen without either calcium or β-CD, the AND gate is disabled.

III. FLUORESCENT CHEMOSENSORS

A. Naphthalene-Modified CDs

Ueno et al. found that two naphthalene derivatives can be included in the large cavity of γ-CD [29]. This result was obtained by observing enhanced excimer emission in the solution of γ-CD and 2-naphthaleneacetic acid. On this basis, γ-CD with one appending naphthalene moiety was shown to form a complex in

2

Structure 2

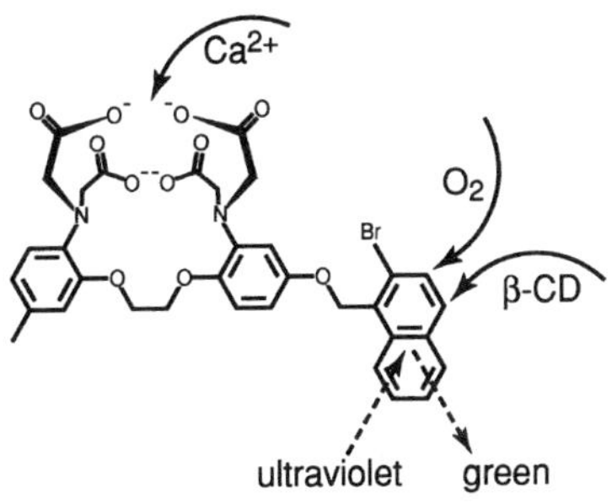

Scheme 1

which the naphthalene moiety and a guest molecule are included together in the γ-CD cavity [30]. Here, the naphthyl moiety was regarded as a spacer that narrows the large cavity so as to include an exogenous guest into the cavity. The next attempt was done with γ-CD with two 2-naphthylacetyl moieties at AE positions (**4**). Compound **4** exhibits a predominant excimer emission that was hardly affected by guest inclusion [31]. Because **4** exhibits marked guest-induced exciton coupling bands in the naphthalene absorption regions, it was concluded that **4** includes two naphthyl moieties in the cavity. The intensities of the circular dichroism bands decreased upon guest addition, suggesting that **4** excludes the two naphthyl moieties from the cavity upon guest accommodation. In this case, the excimer is formed before and after guest addition (Fig. 1, reaction

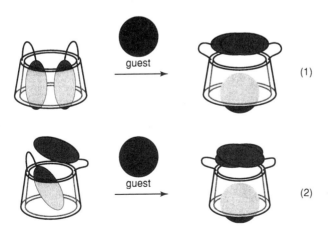

Figure 1 Guest-induced conformational changes of γ-CD (reaction 1) and β-CD (reaction 2) bearing two naphthalene moieties.

1). When 2-naphthalenesulfonyl was attached to γ-CD in place of the 2-naph-thylacetyl moiety as shown by **5–8,** both monomer and excimer emissions were observed for **6**(AC), **7**(AD), and **8**(AE) isomers with comparable intensities [32–35]. The AB isomer (**5**) exhibits predominant monomer emission in spite of the fact that the two naphthyl moieties are located in proximal positions. This observation reflects that the two naphthyl moieties are difficult to take a face-to-face orientation in the AB isomer. The fluorescence behavior may be related to the short and rigid sulfonyl linker existing between the naphthyl moiety and γ-CD. The patterns of the fluorescence spectra of **6**(AC), **7**(AD), and **8**(AE) iso-mers changed significantly upon the addition of guest species, mostly increasing the intensity of the excimer emission. In this system, the naphthyl moieties can form an excimer in the hosts with the two naphthyl moieties inside or outside the γ-CD cavity, so the guest-induced fluorescence variations of the hosts can-not be large (Fig. 1, reaction 1). With this in mind, Ueno et al. prepared corre-sponding β-CD derivatives with two 2-naphthylsulfonyl moieties (**9–11**). Be-cause β-CD can accommodate only one naphthyl unit in the cavity, it is likely that one of two naphthyl moieties is included in the β-CD cavity and another is located outside the cavity [36]. Consequently, the face-to-face interaction is not possible in this system. However, when the included naphthyl moiety is ex-cluded from the cavity upon accommodation of the guest species, the excimer formation based on the face-to-face interaction becomes possible outside the β-CD cavity (Fig. 1, reaction 2). Actually, the guest-induced excimer emission

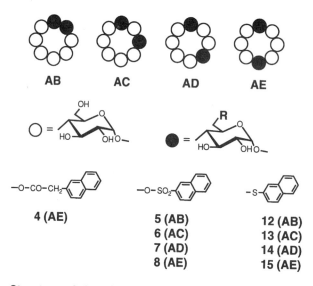

Structures 4–8 and 12–15

AB **AC** **AD**

OH

$O =$ HO— , —OH O—

$\bullet =$ R , HO— —OH O—

—O–SO$_2$—[naphthyl]

9 (AB)
10 (AC)
11 (AD)

Structures 9–11

was observed for **10**(AC) and **11**(AD) of this series. The guest-induced excimer emission was slight for the AB isomer (**9**) for the same reason described for the corresponding AB isomer (**5**) of the γ-CD series.

Ueno et al. also prepared the bis(2-naphthylsulfenyl)-γ-CD series in which the naphthyl moieties are very limited in their movement because the linker between naphthalene and CD is only sulfur [37]. All isomers of **12**(AB), **13**(AC), **14**(AD), and **15**(AE) exhibit only monomer fluorescence due to the rigid linker. It means that the two naphthalene moieties cannot take face-to-face orientation because of the limited flexibility. Although the excimer cannot be used for sensing molecules, Ueno et al. found that the monomer fluorescence intensity increases with increasing guest concentration. Thus, this modified CD series can be used as chemosensors of a different type.

As very different systems, Ueno et al. prepared α-helix peptides bearing γ-cyclodextrin and one (**16,17**) or two (**18**) naphthalene units in their side chains (Fig. 2) [38]. They are alanine-based peptides composed of 17 amino acid residues. The peptides **16** and **17**, which have γ-CD and one naphthalene unit, display a simple monomer emission and the fluorescence intensity slightly decreases with increasing guest concentration. On the other hand, the peptide **18**, which has γ-CD and two naphthalene units on both sides of the γ-CD unit, exhibits considerable excimer emission in addition to the predominant monomer emission and the excimer emission decreases with increasing guest concentration. This indicates that **18** excludes the two naphthalene units from the γ-CD cavity to the opposite directions upon guest accommodation.

A

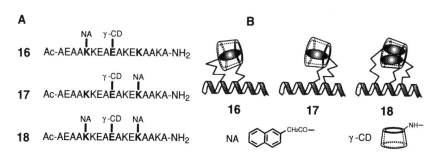

B

16 Ac-AEAAKKEAEAKEKAAKA-NH$_2$

17 Ac-AEAAKKEAEAKEKAAKA-NH$_2$

18 Ac-AEAAKKEAEAKEKAAKA-NH$_2$

Figure 2 Amino acid sequences (A) and plausible structures (B) of γ-CD–peptide hybrids containing one or two naphthalene moieties.

B. Dansyl-Modified CDs

Dansyl is a fluorophore which exhibits strong fluorescence in a hydrophobic environment and weak fluorescence in polar water solution. Ueno et al. prepared a fluorescent CD (**19**), in which the dansylglycine unit is covalently attached to β-CD, and they observed that this fluorescent CD exhibits a fluorescence peak at 535 nm and its fluorescence intensity decreases upon guest addition [39]. This guest responsive fluorescence behavior was explained in terms of guest-induced locational change of the dansyl unit from inside to outside of the CD cavity (Fig. 3); that is, the fluorescent CD forms a self-inclusion type of complex that exhibits strong fluorescence due to the hydrophobic environment of the CD cavity and it excludes the dansyl unit from the CD cavity associated with guest accommodation to bulk water solution, where the dansyl fluorescence is weak. The examined guests were progesterone, corticosterone, cortisone, prednisolone, hydrocortisone, deoxycholic acid, chenodeoxycholic acid, ursodeoxycholic acid, and cholic acid as steroid compounds and (−)-borneol, (−)-menthol, (−)-fenchone, nerol, and cyclohexanol as terpenes or others (Fig. 4). When the fluorescence intensity is abbreviated as I_0 for host alone and I for host in the presence of guest, the $\Delta I/I_0$ value, where ΔI is I_0-I, can be used as a sensitivity factor of the system. The $\Delta I/I_0$ values for the steroids obtained under the conditions of $2.25 \times 10^{-6}M$ for **19** and

Figure 3 Guest-induced conformational change of dansyl-modified CDs.

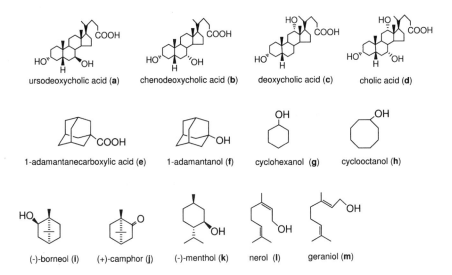

Figure 4 Various guests used for binding measurements of dansylleucine-modified CDs.

0.1 mM for guests were more remarkable for ursodeoxycholic acid (0.63) and chenodeoxycholic acid (0.42) than for other steroids. On the other hand, the $\Delta I/I_0$ values obtained for nonsteroidal compounds under the conditions of 2.25 × $10^{-6}M$ for **19** and 1.0 mM for guests were in the order (−)-borneol>(−)-menthol>(−)-fenchone>nerol>cyclohexanol.

On the other hand, dansylgycine-modified γ-CD (**20**) exhibits a fluorescence peak at 540 nm in 10% dimethyl sulfoxide (DMSO) aqueous solution [40]. The extents of guest-induced fluorescence variations are smaller than those of dansylglycine-modified β-CD, giving $\Delta I/I_0$ values of 0.21 and 0.15 for ursodeoxycholic acid and chenodeoxycholic acid, respectively.

19 n = 7
20 n = 8

Structures 19 and 20

As an extension of this study, the chemosensor abilities of dansylleucine-modified CDs were examined. Leucine is an amino acid which has a hydrophobic side chain, so we can examine how the side chain affects the guest binding. It is also interesting how the leucine chirality affects the guest binding. From this point of view, we prepared dansyl-L-leucine-modified β-CD (**21**), dansyl-D-leucine-modified β-CD (**22**), dansyl-L-leucine-modified γ-CD (**23**), and dansyl-D-leucine-modified γ-CD (**24**) [41]. We compared the binding behaviors of these dansylleucine-modified CDs with those of dansylglycine-modified β-CD (**25**) and γ-CD (**26**). The wavelengths of the fluorescence peaks of **21, 22, 23, 24, 25,** and **26** are 543, 540, 550, 553, 546, and 556 nm, respectively, reflecting the polarity of the environment around the dansyl moiety. From this result, the dansyl moiety is likely to be more deeply involved in the hydrophobic CD cavity in the order **22>21>25>23>24>26.** As indicated by Dunbar and Bright for dansylglycine-modified β-CD [42], the fluorescence lifetime measurements of all these dansyl-modified CDs revealed that they have two lifetimes. The shorter lifetime (5.7–8.2 nsec) and the longer lifetime (13.1–17.8 nsec) correspond to the species with the dansyl chromophore located outside and inside the CD cavity, respectively. The longer-lifetime component is predominant for dansyl-modified β-CDs with the component ratio of 0.67–0.77 whereas the shorter-lifetime component is predominant for dansyl-modified γ-CDs with the component ratio of 0.78–0.79, except for dansyl-L-leucine-modified γ-CD, which gives 0.46 as the shorter-lifetime-component ratio. Upon addition of 1-adamantanol as a guest, the shorter-lifetime component becomes predominant for dansyl-modified β-CDs **21, 22,** and **25,** giving an A_1 value greater than 0.92. These results are consistent with the guest-induced locational change of the dansyl moiety from inside to outside the CD cavity. Similar changes in the values of A_1 were observed upon the addition of (−)-borneol for dansyl-modified γ-CDs **23, 24,** and **26.** However, the ratio of the shorter-lifetime component decreased upon the addition of cyclohexanol for the γ-CD derivatives, suggesting that the dansyl unit is co-included with cyclohexanol in the γ-CD cavity. The co-inclusion behavior is likely to occur for the guests

21 n = 7
23 n = 8

22 n = 7
24 n = 8

25 n = 7
26 n = 8

Structures 21–26

that have smaller sizes like cyclohexanol. This co-inclusion behavior is also suggested by the cyclohexanol-induced enhancement in the fluorescence intensity of the γ-CD derivatives.

The binding constants of **21–26** for various guests were obtained in phosphate buffer (pH 7.0) from the guest-induced variations in the fluorescence intensity. In almost all cases, the order of the binding constants is **21>22>25** for dansyl-modified β-CDs. These results indicate that the leucine-incorporated β-CDs are superior in guest-binding ability to the dansylglycine-incorporated β-CD. This may arise from the increased hydrophobicity around the mouth of the CD cavity due to the presence of the hydrophobic leucine side chain. Another aspect of the results is that L-leucine-incorporated β-CD (**21**) is superior in guest-binding ability to D-leucine-incorporated β-CD (**22**). This difference between **21** and **22** may be related to their structural difference. The detailed nuclear magnetic resonance (NMR) analyses showed that the dansyl moiety of **22** is more deeply included in the CD cavity than that of **21**. This means that the stability of the self-inclusion complex of **22** is higher than that of **21**. As a result, the dansyl moiety of **22** is less easily replaced by the exogenous guest molecules than that of **21,** resulting in larger binding constants of **21**. The case of dansyl leucine-modified γ-CD is different from that of the corresponding β-CD derivatives because D-leucine-incorporated γ-CD (**26**) is superior in guest binding ability to L-leucine-incorporated γ-CD (**25**). The binding constants of **21** and **22** are reflected in the sensitivity parameters ($\Delta I/I_0$) as shown in Fig. 5, in which the $\Delta I/I_0$ values of **22** are always smaller than those of **21,** with large values for ursodeoxycholic acid, chenodeoxycholic acid, 1-adamantanecarboxylic acid, 1-adamantanol, and (−)-borneol. Similarly, the binding constants of **25** and **26** are reflected in the $\Delta I/I_0$ values, as shown in Fig. 5, in which the $\Delta I/I_0$ values of **26** are always larger than those of **25,** with the larger values for ursodeoxycholic acid, chenodeoxycholic acid, and deoxycholic acid.

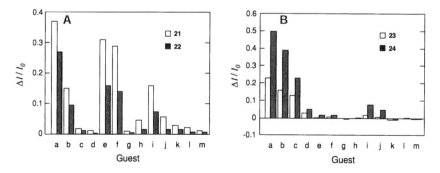

Figure 5 Sensitivity parameters ($\Delta I/I_0$) of dansylleucine-modified CDs for the guests shown in Fig. 4. [**21–24**] = 2 × 10^{-6} M; [guest] = 10^{-5} M (A), 10^{-4} M (B).

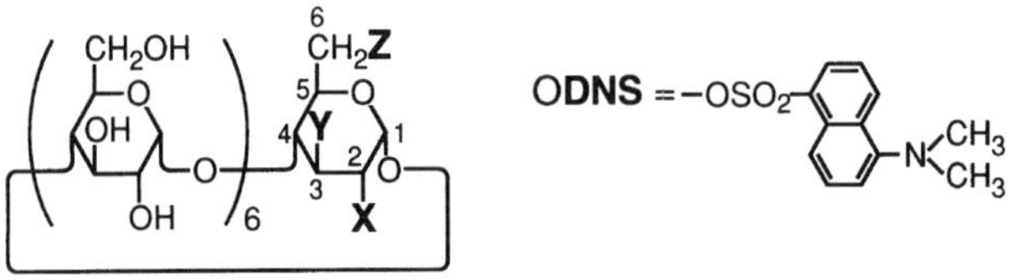

27	X = OH	Y = OH	Z = ODNS
28	X = ODNS	Y = OH	Z = OH
29	X = OH	Y = ODNS	Z = OH

Structures 27–29

Wang et al. reported the sensor abilities of other type of dansyl-modified CDs **27–32** in which **28, 29, 31** and **32** have the dansyl moiety at the secondary side of CD [43,44]. Fluorescence spectra of **27, 28,** and **29** show peaks at 570, 548, and 520 nm, respectively. Remarkably large binding constants were obtained for ursodeoxycholic acid with the values of 1.65×10^6 and $1.05 \times 10^6 \ M^{-1}$ for **27** and **28,** respectively. Fluorescence spectra of γ-CD derivatives **30, 31,** and **32** show peaks at 570, 560, and 520 nm, respectively. They show larger $\Delta I/I_0$ values for ursodeoxycholic acid and chenodeoxycholic acid with the order of the values being **31>30>32.**

The attachment of the dansyl unit to the secondary site of CD can also be accomplished by Hamasaki et al. [45]. They examined the fluorescence properties of **33, 34, 35,** and **36,** in which the dansyl moiety is attached to the secondary site (C3) of CD [43]. Because the preparation of 3-deoxy-3-amino-CDs involves the inversion reaction at C3 that results in the conversion of one glucose member to an altrose residue, the modified CDs derived from 3-deoxy-3-amino-CDs have a distorted cavity with a decreased cavity space. As a result, β-CD derivatives **33** and **34** cannot act as a fluorescence sensor, exhibiting fluorescence peaks at 580

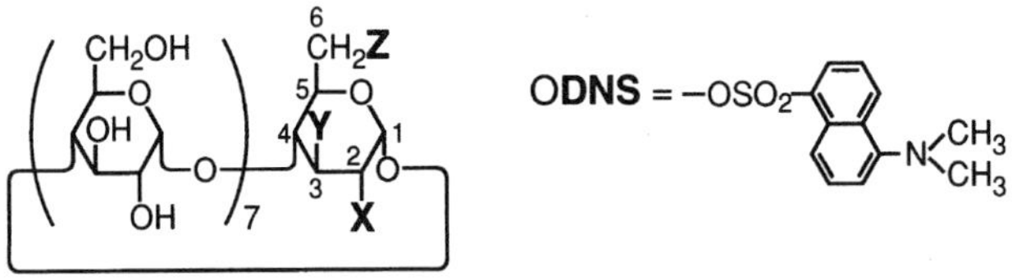

30	X = OH	Y = OH	Z = ODNS
31	X = ODNS	Y = OH	Z = OH
32	X = OH	Y = ODNS	Z = OH

Structures 30–32

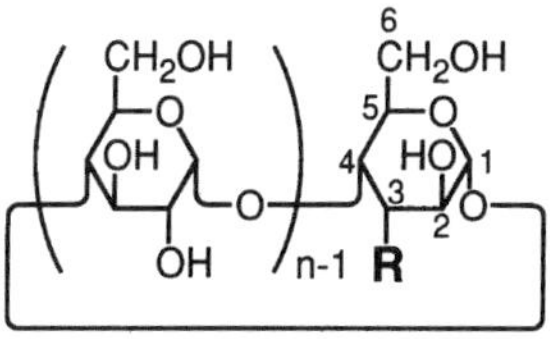

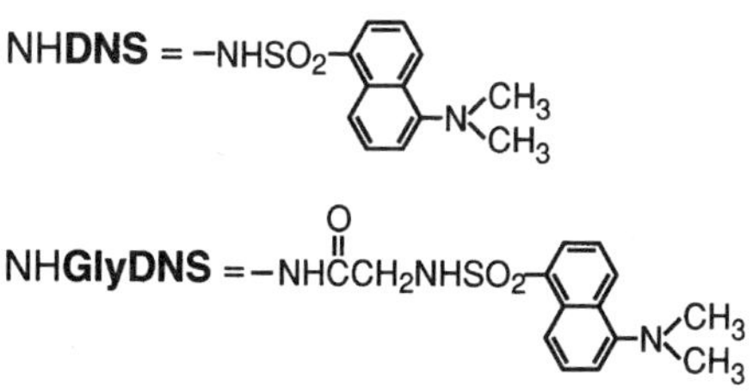

33	n = 7	R = NH**DNS**
34	n = 7	R = NH**GlyDNS**
35	n = 8	R = NH**DNS**
36	n = 8	R = NH**GlyDNS**

Structures 33–36

and 560 nm for **33** and **34,** respectively, which are hardly changeable upon guest addition. On the other hand, the fluorescence spectra of γ-CD derivatives **35** and **36** are slightly guest responsive with very small binding constants.

Corradini et al. reported crystal structure of dansyl-modified β-CD (**37**), in which the dansyl moiety and CD are connected by an ethylenediamine linker [46]. It was shown that the dansyl moiety is fully encapsulated within the CD cavity. The shape of the cavity is considerably flattened, because O(4)—O(4) distances parallel to the naphthalene ring were found to be longer than others. The ^{1}H-NMR data suggested that the orientation of the dansyl moiety in the self-inclusion complex is retained in aqueous solution. Corradini et al. also prepared dansyl-modified β-CD (**38**), in which the dansyl moiety and CD is connected by an diethylenetriamine linker [47]. From the circular dichroism study on **37** and **38,** it was concluded that the orientation of the dansyl moiety is axial for **37** and equatorial for **38,** suggesting that the orientation of the dansyl moiety depends on the length of the linker. Both **37** and **38** showed similar sensing properties, suggesting a similar "in–out" movement of the dansyl moiety due to the competitive inclusion of

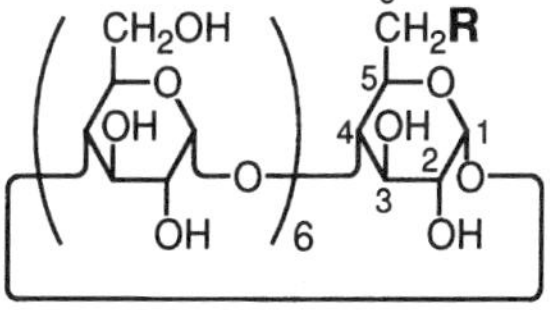

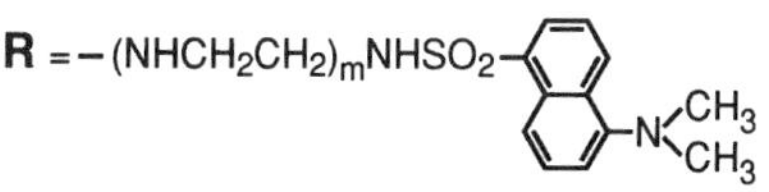

37 m = 1
38 m = 2

Structures 37 and 38

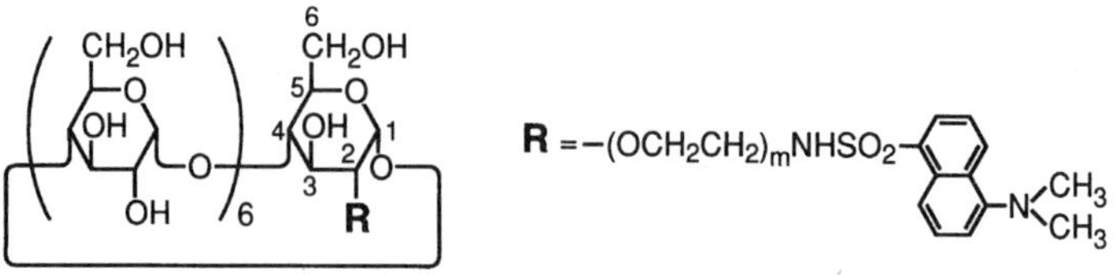

39 m = 1
40 m = 3

Structures 39 and 40

a guest molecule. Unlike **37,** compound **38** was found to decrease the fluorescence intensity by the presence of the Cu(II) ion with a linear response to a 1:1 molar ratio, suggesting that the binding of the metal ion occurs by the amino and sulfonamide groups of **38.** The copper(II) complex of **38** was shown to behave as a chemosensor for bifunctional molecules such as amino acids, increasing the fluorescence intensity upon addition of alanine or tryptophan. The increase in the fluorescence intensity can be attributed to the displacement of the dansyl moiety from the copper ion.

Nelissen et al. prepared the dansyl-appended β-CDs (**39** and **40**), in which the dansyl-appending chain is connected to C2 [48]. At neutral pH, they were not responsive to guest molecules due to the the strong self-inclusion of the dansyl moiety. When the pH is lowered, the self-inclusion becomes less favorable due to the protonation of the dimethylamino group of the dansyl moiety, resulting in the strongly increased response to guests. The guest-binding abilities of **39** and **40** are very large at pH 1.0, as shown by the binding constants 4.5×10^5 and 2.2×10^5 M^{-1} of **39** and **40** for 1-adamantanecarboxylic acid.

As described previously for dansylleucine-modified CDs, the hydrophobic side chain of leucine affected the guest binding of the fluorescent CDs. To examine the presence of hydrophobic units near the dansyl moiety, monensin-incorporated dansyl-L-lysine modified β-CD (**41**) was prepared as an environment-re-

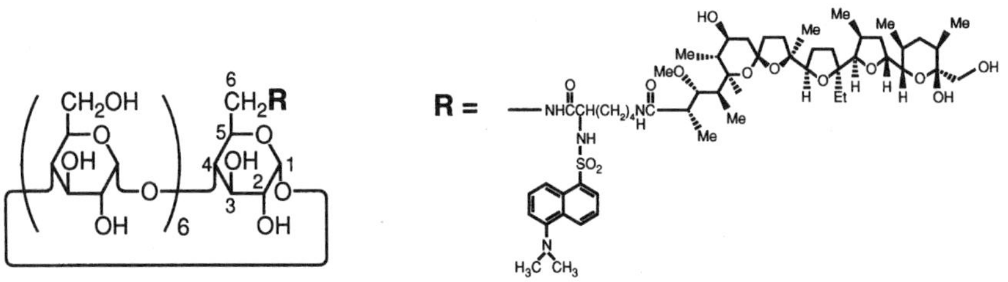

41

Structure 41

sponsive sensor [49]. Monensin is an antibiotic compound, which is a flexible molecule, and can take a circlelike conformation with a sodium ion in the center. On this basis, the effects of sodium cation on $\Delta I/I_0$ and binding constants were examined. The presence of the sodium ion is remarkable, enhancing the $\Delta I/I_0$ value and binding constants for various guests, suggesting that the circlelike monensin acts as a hydrophobic cap of the CD cavity.

Further study to examine the environment was done with biotin-appended dansyl-modified β-CDs (**42** and **43**) [50]. Biotin is well known to bind avidin (protein) tightly; therefore, in this case, the environment is protein. The fluorescence intensities of **42** and **43** are more than three times larger in the presence of avidin. The lifetime measurements indicate that the longer-lifetime component, which newly appears in the presence of avidin, is likely to be the dansyl moiety located in the hydrophobic region of avidin. The addition of hyodeoxycholic acid in the presence of avidin decreased the fluorescence intensities of **42** and **43,** but the degrees of the fluorescence variations induced by hyodeoxycholic acid are limited ($\Delta I/I_0$ <0.16 for **42**, $\Delta I/I_0$ <0.18 for **43**) in comparison to the case in the absence of avidin ($\Delta I/I_0$ <0.29 for **42**, $\Delta I/I_0$ <0.54 for **43**). This may be due to the limited movement of the dansyl moiety in the presence of avidin.

Hamada et al. prepared β-CD and γ-CD derivatives bearing two dansyl moieties (**44–50**) [51,52]. The fluorescence intensities of the β-CD derivatives decrease upon the addition of guests, but those of γ-CD derivatives decrease or increase depending on the guest species. A remarkable increase in the fluorescence

Structures 42 and 43

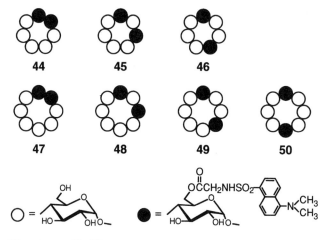

Structures 44–50

intensity was observed for 2,4,6-trichlorophenol as a guest. However, these two-dansyl-modified CDs seem to have smaller binding abilities in comparison to one-dansyl-modified CDs. The guest-binding abilities of the hosts may be lowered due to the steric repulsion between the two dansyl moieties.

On the other hand, β-CD and γ-CD derivatives bearing a dansyl fluorophore and a tosyl unit (**51–57**) were prepared [53]. The fluorescence behaviors of these fluorescent CDs are essentially the same as those of CD derivatives bearing two dansyl moieties.

Although there are many reports on dansyl-modified CDs, almost all are flexible systems in which the distance between the dansyl moiety and CD cannot be controlled. Ueno et al. began the study on CD–peptide hybrids bearing a dansyl fluorophore using the peptide α-helix as a scaffold (Fig. 6) [54]. Prepared samples are EK3 (**58**), EK3R (**59**), EK6 (**60**), EK6R (**61**), and EK (**62**). The last sample, EK, is the reference peptide, which has neither CD nor dansyl fluorophore. All of these are composed of 17-residual alanine (A)-based peptides. The peptides have the β-CD unit in the side chain of glutamic acid (E) and the dansyl unit in the side chain of lysine (K). In the peptides of EK3 and EK3R, β-CD and dansyl units were introduced at an interval of three amino acids and then they were separated by one α-helix turn. On the other hand, in the peptides of EK6 and EK6R, β-CD and dansyl units were introduced at an interval of six amino acids and then they were separated by two α-helix turns. In EK3 and EK6, the β-CD unit is located at the N-terminal side and the dansyl moiety is located at the C-terminal side, whereas in EK3R and EK6R, the arrangement of β-CD unit and the dansyl moiety is reversed. EK3, EK3R, EK6, and EK6R

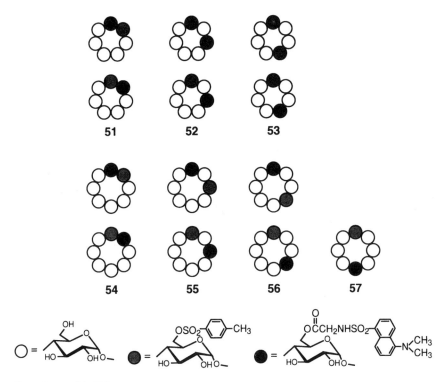

Structures 51–57

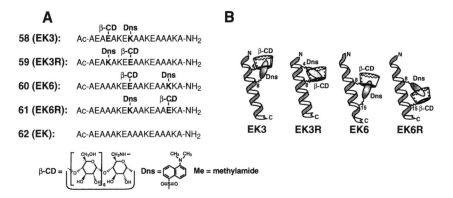

Figure 6 Amino acid sequences (A) and structures (B) of β-CD–peptide hybrids containing a dansyl moiety.

show fluorescence peaks at 534, 524, 540, and 529 nm, respectively. For all peptides, the fluorescence intensity decreases with increasing guest concentration. The binding constants were determined from these fluorescence variations. The binding constants of EK6 and EK6R are much larger than those of EK3 and EK3R. This trend is seen typically for ursodeoxycholic acid, whose binding constants are $2.36 \times 10^5 \, M^{-1}$ and $1.21 \times 10^4 \, M^{-1}$ for EK6 and EK3, respectively. These results suggest that shallow inclusion of the dansyl unit into the CD cavity is better than the deeper inclusion because of the facilitated replacement of the dansyl moiety by guest molecules. It is interesting that the binding constants of EK3 are remarkably larger than EK3R. Similar binding differences were observed between EK6 and EK6R. The result may be related to the orientation of β-CD unit along the α-helix peptide.

Dansyl-modified β-CD dimer (**63**) shows interesting molecular recognition ability for bile acids [55]. It gives very different $\Delta I/I_0$ values, as shown by 0.54, 0.21, 0, and 0.03 for ursodeoxycholic acid, chenodeoxycholic acid, cholic acid, and deoxycholic acid, respectively, under the conditions of 2 μM for the host and 0.3 mM for the guests. The binding constants of 6.9×10^3 and $1.3 \times 10^3 \, M^{-1}$ were obtained for ursodeoxycholic acid and chenodeoxycholic acid, respectively, on the basis of 1:1 stoichiometry.

Reinhout et al. prepared a β-CD–calix[4]arene couple **64** in which a dansyl unit is attached to calix[4]arene and oriented toward the secondary face of β-CD [56]. This compound exhibits a fluorescence peak at 538 nm, and its intensity was hardly influenced by guest molecules. The examination of molecular models suggested a very strong self-inclusion of the dansyl group into the CD cavity.

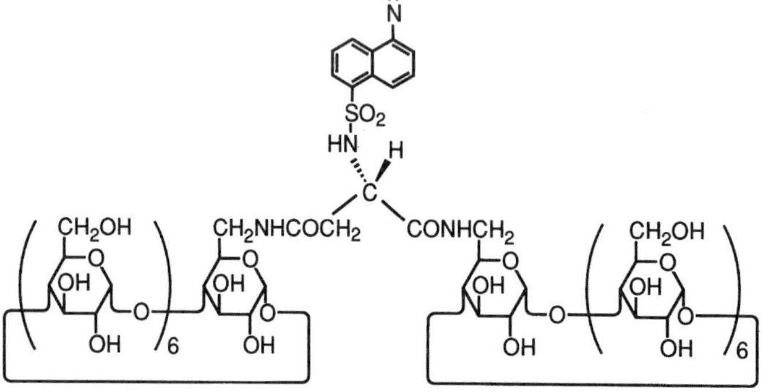

63

Structure 63

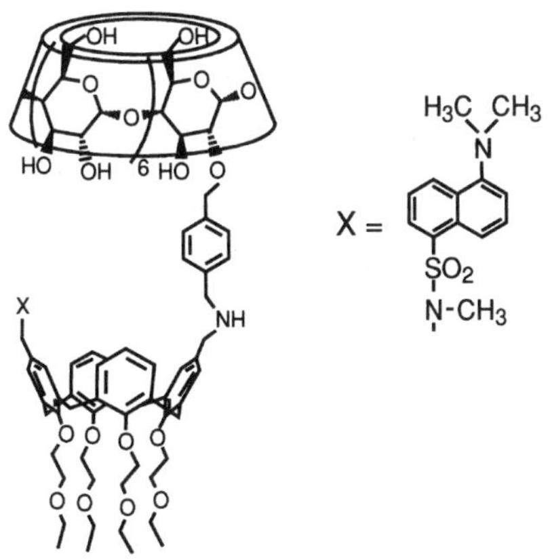

X =

64

Structure 64

C. CD Sensors Using TICT Fluorescence

The effects of CDs on twisted intramolecular charge transfer (TICT) fluorescence of *p-N,N*-dimethylaminobozonitrile in aqueous solution were investigated extensively [57–62]. Hamasaki et al. prepared CDs bearing a *p-N,N*-dimethylaminobenzoyl moiety (**65–67**) and observed dual emissions from normal planar (NP) and TICT excited states. The TICT emission intensity is particular remarkable for **66** and weak for **65** and **67**. β-CD derivative **66** exhibits NP and TICT emissions around 370 and 495 nm, respectively, and the TICT emission intensity decreases upon guest addition [63]. This result indicates that **66** forms a self-inclusion complex by including the *p-N,N*-dimethylaminobenzoyl moiety in the CD cavity and the moiety is excluded from inside to outside the cavity

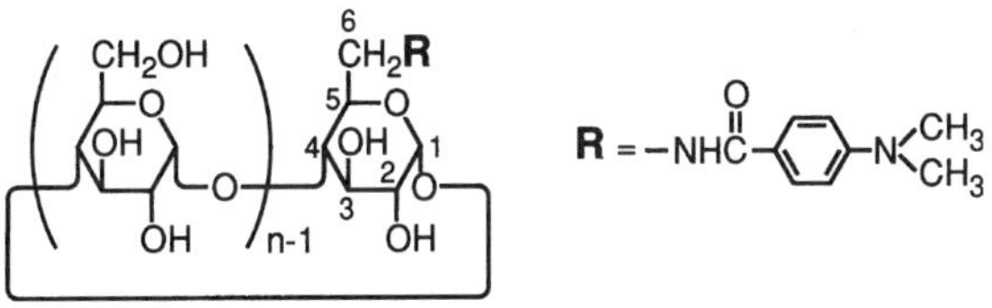

65 n = 6
66 n = 7
67 n = 8

Structures 65–67

upon guest accommodation, resulting in the quenching of the TICT emission in the bulk water solution. Based on this phenomenon, the molecule-sensing ability of **66** was examined using a variety of guest compounds [64]. Recently, it was shown that the self-inclusion form is stabilized mainly by van der Waals interaction [65]. On the other hand, α-CD derivative **65** increases the TICT emission intensity upon the addition of linear alcohols such as *n*-pentanol [66]. In this case, **65** is not sensitive to branched alkanols such as isobutanol and *tert*-butanol, being consistent with the fact that such branched alkanols are hardly included in the small cavity of α-CD.

Wang et al. prepared a β-CD derivative bearing both *p-N,N*-dimethylaminobenzoyl and biotin units (**68**) to examine the effects of avidin binding to the biotin moiety of **68** [67]. Compound **68** exhibits predominant NP fluorescence, and the fluorescence intensity was remarkably enhanced by the presence of avidin. In the absence of avidin, the NP fluorescence intensity usually decreases upon guest addition, but in the presence of avidin, the fluorescence intensity increases with increasing concentration of the guest compounds. These results demonstrate that avidin has remarkable effects on the binding behavior of **68,** probably by acting as a hydrophobic environment around the guest-binding site.

D. Pyrene-Modified CDs

Pyrene is well known to form excimer by excited-state face-to-face interaction. Although pyrene is too large to be accommodated in the cavity of α- and β-CD, it can be included in the cavity of γ-CD. Herkstroeter et al. observed remarkable excimer emission in the solution containing γ-CD and pyrene and proposed the 2:2 host–guest-complex formation in the system [68]. In relation to this proposal, Ueno et al. prepared pyrene-appended γ-CDs **69** and **70** and found that they have a remarkable tendency to form the association dimer, showing strong excimer emission [69]. The excimer emission intensity decreases upon guest addition, indicating that the association dimer is converted into the 1:1 host–guest complexes (Fig. 7). These systems were used as sensory systems for detecting organic compounds [70]. The sensory systems exhibit remarkable molecular recognition

68

Structure 68

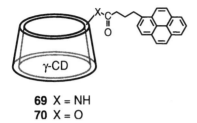

69 X = NH
70 X = O

Structures 69 and 70

abilities and the response of these systems depends on the size, shape, and polarity of guest compounds.

Ueno et al. prepared γ-CD derivatives with two pyrene moieties at AB, AC, AD, and AE glucose units (**71, 72, 73,** and **74,** respectively) [71]. The compounds show predominant excimer emission in aqueous 30% DMSO solution. The excimer intensity slightly increases upon addition of (−)-borneol for all of the hosts, whereas it decreases remarkably upon addition of lithocholic acid for AD and AE isomers.

The β-CD derivative bearing two pyrenes attached at one appending chain (**75**) was also prepared [72]. It exhibits both monomer and excimer emissions even though the excimer emission is remarkable in a 20% DMSO aqueous solution. This system is also responsive to guest compounds, and a remarkable decrease in the excimer emission was observed when lithocholic acid was added.

Suzuki et al. reported that the CD derivative with a pyrene moiety attached to C3 (**76**) is an extremely better host than the one with the moiety attached to C6 (**77**), as shown by the binding constants $1.9 \times 10^5\ M^{-1}$ of **76** in contrast to $2.9 \times 10^2\ M^{-1}$ of **77** for cyclooctanol as a guest [73]. The fact that the capping at the secondary hydroxyl side by a hydrophobic moiety has such a large enhancing effect

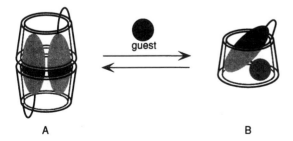

A   B

Figure 7 Guest-induced dissociation of the association dimer of pyrene-modified γ-CDs.

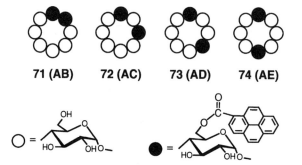

Structures 71–74

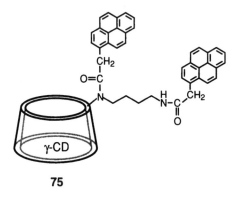

Structure 75

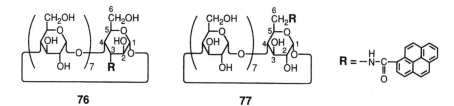

Structures 76 and 77

in guest binding is important for constructing hosts with outstanding binding abilities.

Ueno et al. prepared α-helix peptide bearing γ-CD and pyrene in the side chain (**78**: γ-PR17) where the γ-CD and pyrene units are located at N-terminal and C-terminal sides, respectively and both units are separated by one helix turn [74]. The compound exhibits remarkable concentration dependency in the excimer emission intensity, and the excimer emission intensity decreases upon guest addition. These results suggest that **78** forms an association dimer, which can be dissociated into a 1 : 1 host–guest complex, indicating that the association dimer of the peptide α-helix is dissociated into monomer peptides upon addition of an exogenous guest as an external stimulant molecule. The peptide bearing the pyrene and γ-CD units at opposite sides (**79**: γ-PL17) was also prepared to compare the binding abilities between them [75]. The two peptides exhibit almost pure monomer fluorescence in the very dilute solutions of 5 nM and decrease the fluorescence intensity with increasing guest concentration. From the analysis of the guest-induced fluorescence variations, the binding constants of **78** and **79** were estimated for bile acids. It was shown that the binding constants for bile acids are twofold or threefold larger for **79** than for **78,** indicating that the order of arrangement of pyrene and γ-CD along the peptide chain is important in their binding abilities. The α-helix peptide with one γ-CD and two pyrene units (**80**) was prepared in order to construct the system in which excimer is formed intramolecularly and the excimer is dissociated upon guest addition. Actually, we observed the remarkable excimer fluorescence. However, the excimer formation is also concentration dependent, and the spectrum in a very dilute solution of 5 nM exhibits predominant monomer emission, indicating that the peptide association occurs in this case also. The binding constants of **80** obtained from the guest-induced variations of the monomer fluorescence intensity at 5 nM are larger than those of **79** and comparable to those of **78.**

Native γ-CD was used as a component of a supramolecular system for detecting the potassium ion [76]. The mixture of γ-CD and a crown ether-linked

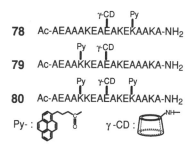

78 Ac-AEAAAKEAEAKEKAAKA-NH$_2$ (γ-CD Py)

79 Ac-AEAAKKEAEAKEAAAKA-NH$_2$ (Py γ-CD)

80 Ac-AEAAKKEAEAKEKAAKA-NH$_2$ (Py γ-CD Py)

Py-: ... γ-CD: ...

Structures 78–80

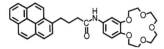

Structure 81

pyrene **81** exhibits only pyrene monomer emission in the absence of the potassium ion, but the excimer emission appears in the presence of the potassium ion. Although the cavity of benzo-15-crown-5 is too small to accommodate a potassium ion, it can form a sandwich-type ccomplex, in which a potassium ion is positioned between the two 15-crown-5 units and, consequently, the excimer is formed between the two pyrene units in the complex $(81)_2 \cdot \gamma$-CD (Fig. 8). Because the excimer emission intensity increases with increasing concentration of potassium ion, the system can be used for detecting the potassium ion.

IV. PHOTODIMERIZATION OF ANTHRACENE-MODIFIED CD

Anthracene is well known to undergo photodimerization. Tamaki investigated photodimerization of anthracene-2-sulfonic acid (2-AS) in aqueous solution [77]. The 2-AS forms a 2:1 complex with γ-CD, but the ratio of the four photodimers was not affected by the presence of γ-CD. In contrast, 2-AS was supposed to form a 2:2 complex and photodimerization gave single photodimer of *anti*-head–tail structure [78].

Ueno et al. prepared γ-CD derivatives bearing one (**82**) or two 9-carbonyl-lanthracene moieties (mainly AE isomer) (**83**). Compound **82** was found to form an association dimer, in which two anthracene units are included in the cavity of the γ-CD dimer and the association dimer of **82** is converted into 1:1 host–guest

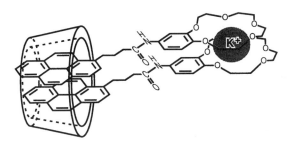

Figure 8 Potassium-ion-induced pyrene excimer formation of **81** in the presence of γ-CD.

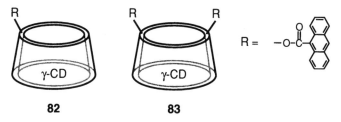

$$R = -O-\overset{O}{\underset{}{C}}-$$

82   **83**

Structures 82 and 83

complexes upon guest addition [79]. On the other hand, **83** forms a 1:1 host–guest complex, changing the location of the anthracene moieties from inside to outside the γ-CD cavity. In the absence of a guest, anthracene photodimerization occurs in the association dimer for **82** whereas the photodimerization proceeds intramolecularly for **83**.

It is also well known that photoirradiation of 9-substituted anthracene gives only the trans photodimer, probably due to the presence of steric repulsion between the substituents for the cis photodimer. With this in mind, Ueno et al. prepared γ-CD derivatives bearing two 9-anthracenecarbony units at AB, AC, AD and AE glucose units [80]. They found that the AB isomer (**84**) undergoes photodimerization and the formed cis photodimer is dissociated into original monomer units with a half-lifetime of 12.5 min at 25°C (Fig. 9). In contrast to the AB isomer, the AE isomer (**87**) gives a stable trans photodimer upon photoirradiation. These results clearly show that the cis photodimer is possible, but it can only exist for a short time. The other isomers of AC (**85**) and AD (**86**) show opposite trends; that is, the photodimer of the AC isomer is unstable with a half-lifetime of 275 min at 25°C, whereas that of the AD isomer is stable. It is noted that these γ-CD derivatives exhibit a remarkable exciton coupling band around 250 nm when they have monomer anthracene units and the shape and the intensity of this band change upon 1:1 host–guest complexation [81].

Photodimerization of 1-substituted anthracene usually give four photodimers. Ueno et al. prepared γ-CD derivatives bearing two 1-carbonylan-

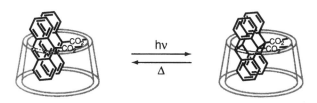

Figure 9 Photodimerization and photochromism behavior of **84** (AB isomer).

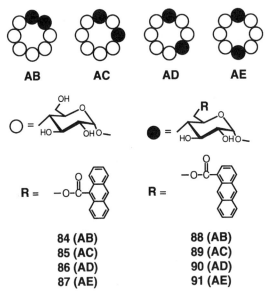

84 (AB)
85 (AC)
86 (AD)
87 (AE)

88 (AB)
89 (AC)
90 (AD)
91 (AE)

Structures 84–91

thracene moieties [AB (**88**), AC (**89**), AD (**90**), and AE (**91**) isomers] to control the stereochemistry of photodimerization [82]. Photodimerization followed by hydrolysis of the ester bonds in these γ-CD derivatives may give the mixture of photodimers *trans*-head–tail (**92**), *trans*-head–head (**93**), *cis*-head–head (**94**), and *cis*-head–tail (**95**). As expected, it was shown that the AB and AC isomers (**88** and **89**) give predominantly **94**, whereas the AD and AE isomers (**90** and **91**) give predominantly **93** (Fig. 10). The product ratio was 48:28:4:20 for **92, 93, 94,** and **95** when photodimerization of 1-anthracenecarboxylic acid was undertaken in methanol, indicating that **94** is very difficult to be produced, probably due to the steric hindrance between the two carboxyl groups. These results demonstrate that this method (the γ-CD template method) is effective in controlling the stereochemistry of the reaction.

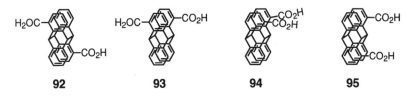

92   **93**   **94**   **95**

Structures 92–95

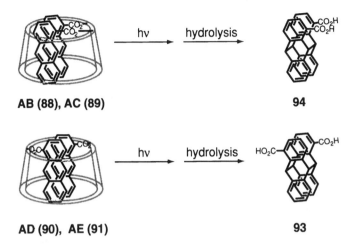

AB (88), AC (89)   **94**

AD (90), AE (91)   **93**

Figure 10 The γ-CD template method for stereoselective preparation of photodimers of 1-anthracenecarboxylic acid.

V. ENERGY TRANSFER IN CHROMOPHORE-MODIFIED CDs

There are several reports on the energy transfer from the appending or capping chromophore of modified CDs to the guest included in the CD cavity. Ueno et al. examined singlet–singlet fluorescence energy transfer from appending the naphthalene moiety of γ-CD (**96**) to a variety of ketone guests by measuring the quenching of the naphthalene fluorescence [83]. The Stern–Volmer constants for the guests are in the order acetone<diethyl ketone<di-*n*-propyl ketone<(−)-fenchone, and this order is the same as the order of the sizes of the guest ketones. The result indicates that efficient energy transfer occurs in the host–guest complexes.

96

Structure 96

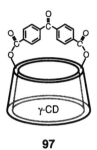

97

Structure 97

Triplet–triplet energy transfer was reported to occur from benzophenone-capped β-CD (**97**) to 1-bromonaphtahlene at 77 K [84]. In this case, the quenching efficiency of phosphorescence of benzophenone donor is 39% and the apparent efficiency of energy transfer was estimated to be 60%.

Nakamura et al. prepared the modified β-cyclodextrin appending a rhenium complex [85,86] and found that triplet–triplet energy transfer occurs from rhenium complex (**98**) to 2-anthracenecarboxylate (2-AS) included in the CD cavity. In this case, the quenching efficiency of the luminescence of the rhenium complex by 2-AS is 0.46.

VI. ENERGY MIGRATION IN MULTICHROMOPHORIC CDs

Gravett and Guillet prepared a β-CD derivative bearing seven naphthalene units (**99**) by the reaction of 6-hydroxy-2-naphthalene sulfonic acid (disodium salt) with heptakis(6-bromo-6-deoxy)-β-CD [87]. Compound **99** shows mainly monomer emission with a small amount of excimer. The depressed excimer formation may be due to the electronic repulsion between the negative charges of the naphthalenesulfonate units. A fluorescence polarization study shows that the fluorescence of **99** is remarkably depolarized, suggesting that energy migration occurs between naphthalene units. Studies using 6-(p-toluidino)-2-naphthalenesulfonic acid (TNS) as a probe showed that **99** has the capability of including a probe

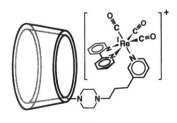

98

Structure 98

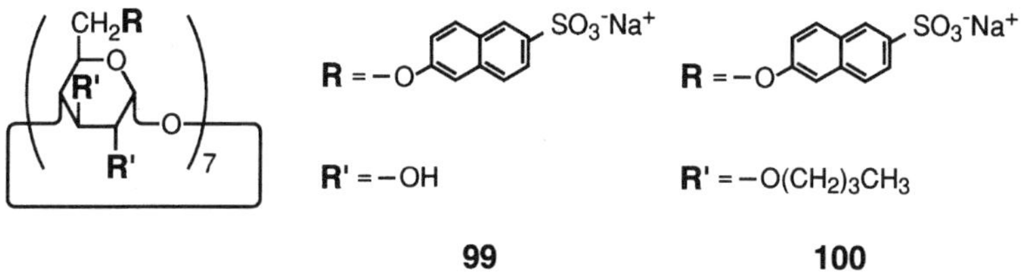

$R = -O$ (naphthalene) $SO_3^-Na^+$

$R' = -OH$

$R = -O$ (naphthalene) $SO_3^-Na^+$

$R' = -O(CH_2)_3CH_3$

99   **100**

Structures 99 and 100

molecule in its cavity and that energy transfer from the naphthalenesulfonate groups to the included molecule is possible.

Compound **99** was made more hydrophobic by butylation of the 2-OH and 3-OH groups [88]. The prepared CD derivative (**100**) solubilized 9-anthrylmethyl pivalate (AP) much more effectively than **99** or β-CD. In the presence of AP, the naphthyl emission ($\lambda_{ex} = 314$ nm) of **100** was efficiently quenched and intense AP emission was observed. Because essentially all of the light is absorbed by the naphthyl chromophores, the strong AP fluorescence demonstrates that energy transfer occurs efficiently. The naphthyl antenna chromophores of **100** were shown to sensitize a selective photoreaction of AP included in the CD cavity (Scheme 2). The photoirradiation of AP in methanol gives 9-neopentylanthracene

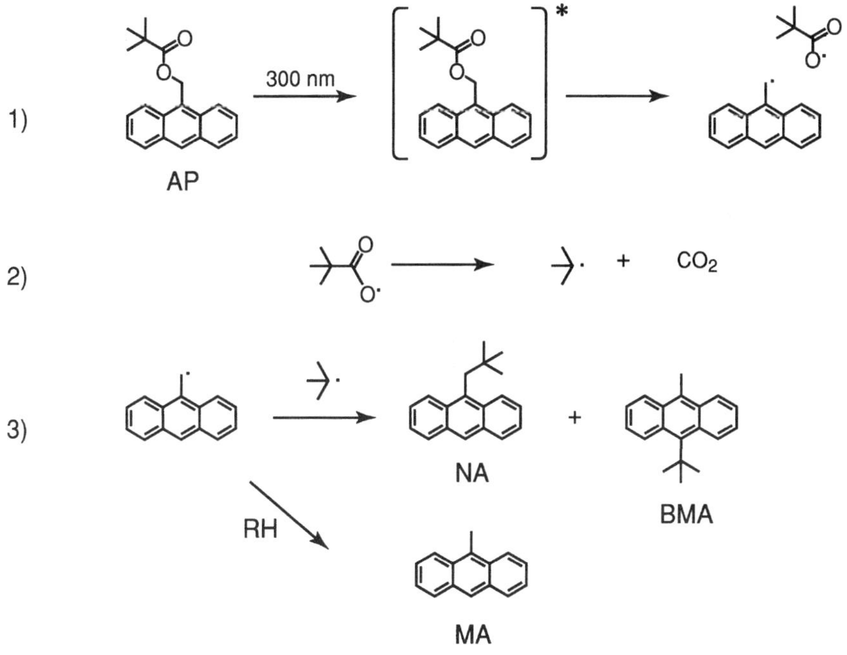

Scheme 2

101   X = −OCOCH$_3$ Y = −OCOCH$_3$ Z = NA
102   X = NA   Y = −OCH$_3$ Z = −CH$_3$
103   X = −OCH$_3$ Y = NA   Z = −OCH$_3$
104   X = NA   Y = −OCH$_3$ Z = NA

Structures 101–104

(NA), 9-*tert*-butyl-10-methylanthracene (BMA), and methylanthracene (MA) with the ratio of 1:1.5:2.7 for MA : BMA : NA, whereas AP in **100** in aqueous solution gives the ratio 1:0:3.4. The result demonstrates that inclusion of AP in **100** modifies the photoreactivity of AP, suppressing the reaction *tert*-butyl radical with the 10-position of the anthryl ring.

On the other hand, the synthesis and photophysical properties of β-CD bearing seven 2-naphthoyloxychromophores on the primary face (**101**) or the secondary face (**102** and **103**) or fourteen 2-naphthoyloxy chromophores, seven on each face (**104**) were reported [89]. The investigation was performed in dichloromethane and in a mixture of ethanol and methanol that can form a glass at low temperature. The absorption spectra show that the interactions between chromophores in the ground state are weak, whereas the fluorescence spectra show the existence of excimers at room temperature but not at low temperature in a rigid glass. Because the excimer acts as an energy trap, the energy hopping process was studied in a glass at low temperature by steady-state and time-resolved fluorescence depolarization techniques. The steady-state anisotropy is found to be one-seventh of the theoretically limited anisotropy (0.4). This result means that the excitation energy hops between chromophores with essentially randomly oriented transition moments at a rate much higher than the chromophore intrinsic decay rate.

Another water-soluble β-CD (**105**) bearing seven naphthoyl chromophores forms very stable 1:1 complexes with a merocyanine laser dye DCM–OH [90]. The energy transfer from the naphthoyl antenna chromophores to the included dye is shown to occur with 100% efficiency (antenna effect) (Scheme 3).

The antenna effect of **105** was also used for reaction of nitrone as a guest [91]. In this case, ismerization of nitrone to the product was promoted by energy transfer from the antenna (Scheme 4).

A Monte Carlo simulation was used to calculate the theoretical anisotropy decays of several β-CDs containing seven naphthyl chromophores [92]. A good

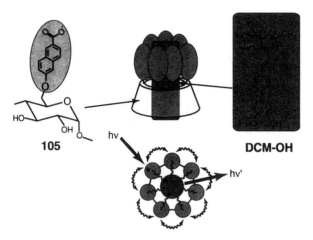

105   **DCM-OH**

Scheme 3

fit to the experimental anisotropy decays was obtained. The nearest-neighbor distances fall in the ranges 5–7 Å in all cases, which is compatible with the nearest-neighbor distances expected from molecular modeling. The observations confirm the validity of the theoretical model.

VII. CD ROTAXANES

Cyclodextrin rings can be threaded by a long chain, and rotaxanes and polyrotaxanes are formed when stopper moieties are connected to the ends of the threading chain [93]. Tamura and Ueno prepared a rotaxane, in which naphthalene-ap-

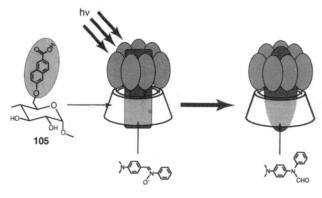

105

Scheme 4

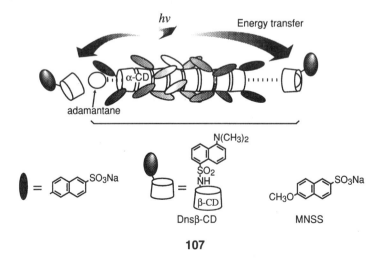

106

Structure 106

pended α-CD is threaded by a poly(ethylene glycol) chain with dansyl moieties at
the ends (**106**) [94]. In this rotaxane, light energy absorbed by the naphthalene
moiety was found to transfer to the terminal dansyl groups.

They also prepared polyrotaxanes composed of about 15 α-CDs, each having roughly 2 naphthalene units and a poly(ethylene glycol) as a threading chain
with adamanatane groups at the ends (**107**) [95]. In this system, energy transfer occurs from the naphthalene units to the dansyl unit of the exogenous host (dansyl-
modified β-CD) that binds to the adamantane ends of the poly(ethylene glycol)
chain (Scheme 5). This energy transfer was reduced by the addition of adaman-
tanol as a competitive binder for the dansyl-modified β-CD.

107

Scheme 5

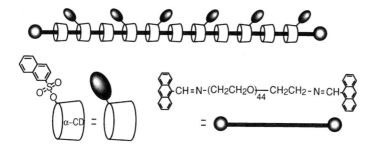

108

Structure 108

Furthermore, they prepared supramolecular antenna systems using polyro-taxanes with naphthalene-modified α-CD as the ring unit and anthracene units at the ends of the threading poly(ethylene glycol) chain (**108**) [96]. In this case, both energy migration between naphthalene units and energy transfer from naphthalene units to anthracene units were observed.

As a different type of photoswitchable rotaxane, Murakami et al. reported on the rotaxane, in which α-CD is threaded by an azobenzene-containing chain with two methylenes as the spacer between azobenzene and viologen units (**109**) [97].

In this rotaxane, α-CD exists at the *trans*-azobenzene part but it moves to the methylene part when the *trans*-azobenzene unit is converted into *cis*-azoben-zene. This light-driven locational change was regarded as a molecular shuttle sys-tem.

VIII.   AZOBENZENE-MODIFIED CDs

Azobebzene undergoes trans–cis photoisomerization and this photoswitchable be-havior can be used for the photoregulation of various molecular phenomena. A simple example is on–off photoregulation of ester hydrolysis by using an azoben-

109

Structure 109

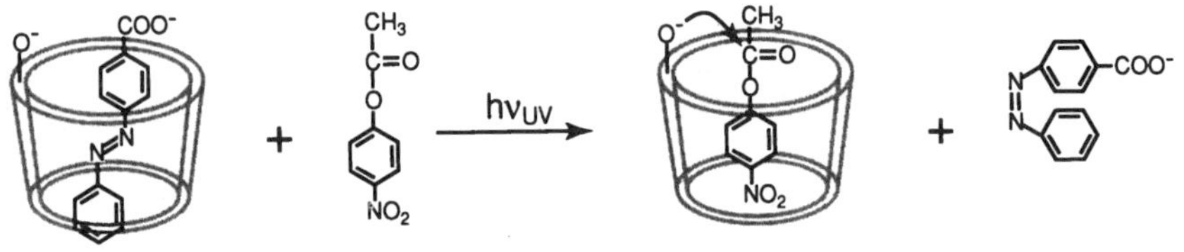

Scheme 6

zene inhibitor. CDs are well known to act as a catalyst for hydrolysis of ester substrates in alkaline solution. The hydrolysis proceeds in the complex of the CD and
substrate, and the alkoxide anion of the secondary hydroxyls of CD attacks the
ester carbonyl in the complex, resulting in the cleavage of the ester bond [96]. In
the system of β-CD, *p*-nitrophenylacetate (substrate) and pottasium *p*-(pheny-
lazo)benzoate in a pH 8.7 Tris buffer solution at 25°C, Ueno et al. observed that
the ester hydrolysis rate is enhanced by photoirradiation. Here, the trans form of
the azo inhibitor forms a complex with β-CD, thus preventing the inclusion of *p*-
nitrophenylacetate. When the azo inhibitor is photoisomerized to the cis form, it
is excluded from the β-CD cavity, resulting in the inclusion of the *p*-nitropheny-
lacetate in place of the azo inhibitor (Scheme 6).

Ueno et al. prepared azobenzene-capped β-CD (**110**) (Scheme 7) and found
that its cis form has a higher binding ability than the trans one by expanding the
hydrophobic environment and then enabling deep inclusion of the guest molecules
[99]. This system was also used to photoregulate the catalytic ability in ester hydrolysis in pH 8.7 Tris buffer solution [98]. The ester hydrolysis rate for *p*-nitro-
phenylacetate was enhanced by photoirradiation even though the intramolecular
reaction rate (k_{cat}) itself is smaller for cis form than for trans form. The enhancement in the overall reaction rate (k_{cat}/K_m) arises from the marked increase in the
binding ability of cis form.

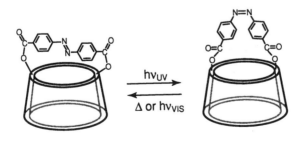

110

Scheme 7

111 CD = β-CD
112 CD = γ-CD

Structure 111–112

Azobenzene-capped β-CD (**111**) and γ-CD (**112**) were also prepared [100,101]. When guest molecules are (−)-borneol, 1-adamantanecarboxylic acid, cyclododecanol, nerol, and (−)-menthol, the trans form of **112** exhibits much larger binding constants than the cis form [102]. This result suggests that the *trans*-azobenzene moiety acts as a hydrophobic cap to enhance guest binding, whereas the cis one is bent inward to the interior of the γ-CD cavity like intramolecular complex formation, thus resulting in the difficulty for the exogenous guests to be included in the cavity.

Several azobenzene-appended CDs (**113–116**) were prepared to examine the effects of photoirradiation on their binding abilities. Azobenzene-appending γ-CD (**113**) shows photoenhanced binding ability, showing binding constants which are approximately two times larger for cis form than for trans form when (+)-fenchone and (−)-borneol are guests [103].

Azobenzene-appending β-CD (**114**) exhibits a positive circular dichroism band around 345 nm associated with the azobenzene π–π* transtion for the trans form, whereas it exhibits strong positive and negative bands at 312 and 425 nm, respectively, after photoirradiation [104]. The circular dichroism intensities of trans and cis forms of **114** decrease upon guest addition, and the analysis

113 CD = γ-CD

Structure 113

114 CD = β-CD

Structure 114

of the circular dichroism intensities indicates that the binding constants of the trans form are much larger than the cis one. In this case, there are two factors that govern the strength of guest binding. One is the stability of the intramolecular complex between CD and azobenzene units. If the intramolecular complex is stable, the binding affinity for the exogenous guest should be reduced. Another is the hydrophobic capping effect, which usually increases the binding affinity for the exogenous guest. From this point of view, the effect of the hydrophobic capping effect of the trans form may be remarkable, being consistent with the fact that the bent shape of the cis form is not appropriate as the cap.

The azobenzene-appending CD (**115**) exhibits a tendency to form association dimers in concentrated solution ($>10 \ M^{-4}$) [105]. This trend has also been shown for **114.** Consequently, in such systems, on–off control of molecular association can be attained.

The azobenzene-appending γ-CD **116** exhibits different circular dichroism bands for trans and cis forms, and **116** was used as a photoswitchable multiresponse sensor on the basis of the guest-induced intensity variations [106].

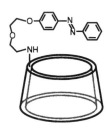

115 CD = β-CD
116 CD = γ-CD

Structures 115 and 116

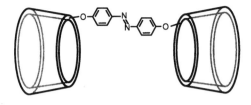

117

Structure 117

Azobenzene with two β-CD units (**117**) was prepared and found that it undergoes trans to cis photoisomerization with 66% cis at the photostationary state [107]. The cis form returns to the original trans form with a half-life of 55 hr at 25°C. The azobenzene derivative was expected to bind one large molecule by the hands of two CD units, but so far no data were reported on this behavior.

IX. STILBENE-CONTAINING CD SYSTEMS

Tabushi et al. reported the photochemistry of stilbene-capped β-CD **118** [108]. The unique property of **118** is that its trans cap is completely converted into a cis cap and the cis cap is further converted into a phenanthrene cap.

Rao et al. observed that photoirradiation of *trans*-stilbene in crystalline γ-cyclodextrin inclusion complexes yields a single isomer of *syn*-tetraphenylcyclobutane (**119**) [109] stereoselectively in high yield (70%). In contrast, the photodimerization of stilbene in solution is very inefficient, and no photodimer was observed even after prolonged irradiation of pure stilbene crystals.

A polyrotaxane that contains stilbene dimers was prepared by Wenz et al. [110]. They once prepared a polypseudorotaxane which has β-CD, γ-CD, *p,p*-dis-

118 CD = β-CD

Structure 118

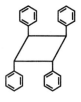

119

Structure 119

ubstituted *trans*-stilbene, and a stibene-containing polymer chain. Although this polypseudorotaxane is dissociable into components, it is converted into stable polyrotaxane after photodimerization of the stilbene units (Scheme 8).

X. CLOSING REMARKS

Cyclodextrins are water-soluble materials and provide a hydrophobic microenvironment in water. On this basis, CDs can accommodate various organic compounds, and unique reactions may proceed in the constrained media. In this chapter, we have described mainly modified CDs with one, two, or more chro-

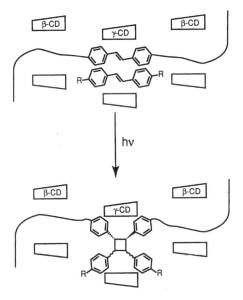

Scheme 8

mophores. Native CDs are spectroscopically inert, but they can be converted into spectroscopically active compound by modification with chromophores. The chromophore-modified CDs usually form self-inclusion complexes and exhibit an induced-fit-type phenomenon in which the chromophore originally included in the CD cavity is excluded upon guest accommodation. This guest-induced locational change results in the perturbation of the spectroscopic property because the inside of the CD cavity is hydrophobic, whereas outside of the CD cavity is the polar water solution. This is why chromophore-modified CDs can act as sensors.

On the other hand, CDs provide a scaffold for arranging two or more chromophores. Stereoselective reactions in the CD templates or energy migration in multichromophoric CD systems are topics in this field. Furthermore, polyrotaxanes, in which many chromophore-modified CDs are threaded by a polymer chain, can be another base for constructing unique supramolecular CD systems. Energy transfer or energy migration may occur in appropriately designed CD polyrotaxane systems. Another aspect of this chemistry is combined systems of CD and peptide, in which CD and a particular chromophore is arranged appropriately in three-dimentional space, by which we can control the distance between a chromophore and a CD unit with desirable orientation. Several examples of CD–peptide conjugates have been shown in connection with an induced-fit type of sensors.

It can be emphasized that, among many hosts, CDs are particularly important because they are the hosts produced in industry from starch. Therefore, if we prepare unique and useful CD derivatives, they may be used on the commercial basis.

The future targets of supramolecular photochemistry in CD chemistry will contain photoresponsive molecular machines, emission-based sensors, and energy transport systems. For construction of such systems, the design of three-dimensionally correct arrangement of component units will become important. The molecular modeling computation approach will be helpful for designing the systems and deeper understanding of structural features of chromophore-modified CDs and their complexes.

REFERENCES

1. Endo, T.; Ueda, H.; Kobayashi, S.; Nagai, T. *Carbohydr. Res.* **1995,** *269,* 369–373.
2. Endo, T.; Nagase, H.; Ueda, H.; Kobayashi, S.; Nagai, T. *Chem. Pharm. Bull.* **1997,** *45,* 532–536.
3. Endo, T.; Nagase, H.; Ueda, H.; Shigihara, A.; Kobayashi, S.; Nai, T. *Chem. Pharm. Bull.* **1997,** *45,* 1856–1859.
4. Harata, K.; Endo, T.; Ueda, H.; Nagai, T. *Supramol. Chem.* **1998,** *9,* 143–150.
5. Bender, M. L.; Komiyama, M. *Cyclodextrin Chemistry;* Springer-Verlag: New York, 1987.

6. Szejtli, J.; Osa, T. in *Comprehensive Supramolecular Chemistry,* Szejtli, J.; Osa, T., eds; Elsevier Science: Oxford, 1996.

7. Wenz, G. *Angew. Chem. Int. Ed. Engl.* **1994,** *33,* 803–822.

8. Ueno, A.; Osa, T. in *Photochemistry in Organized & Constrained Media,* Ramamurthy, V., ed.; VCH: New York, 1991, 739–782.

9. D'Souza, V. T.; Lipkowitz, K. B. *Chem. Rev.* **1998,** *98,* 1741–2076.

10. Ramamurthy, V.; Eaton, D. *Acc. Chem. Res.* **1988,** *21,* 300–306.

11. Bortolus, P.; Monti, S. *Photochemistry in Cyclodextrin Cavities;* Wiley: New York, 1996, 1–133.

12. Czarnik, A. M., ed. *Fluorescent Chemosensors for Ion and Molecule Recognition;* American Chemical Society: Washington, DC, 1992.

13. Ueno, A. in *Chemosensors of Ion and Molecule Recognition,* Desvergne, J. P.; Czarnik, A. W., eds.; Kluwer Academic: Dordrecht, 1997, 105–119.

14. Ueno, A. *Supramol. Sci.* **1996,** *3,* 31–36.

15. Turro, N. J.; Bolt, J. D.; Kuroda, Y.; Tabushi, I. *Photochem. Photobiol.* **1982,** *35,* 69–72.

16. Hamai, S. *J. Phys. Chem.* **1995,** *99,* 12,109–12,114.

17. Hamai, S. *J. Phys. Chem. B* **1997,** *101,* 1707–1727.

18. Scypinski, S.; Cline Love, L. J. *Anal. Chem.* **1984,** *56,* 322–327.

19. Hamai, S. *J. Am. Chem. Soc.* **1989,** *111,* 3954–3959.

20. Escandar, G. M.; Pena, A. M. *Anal. Chim. Acta* **1998,** *370,* 199–205.

21. Du, X.; Zhang, Y.; Huang, X.; Li, Y.; Jiang, Y.; Chen, G. *Spectrochim. Acta* **1996,** *52A,* 1541–1545.

22. Du, X.; Zhang, Y.; Jiang, Y.; Li Lin; Huang, X.; Chen, G. *J. Photochem. Photobiol. A: Chem.* **1998,** *112,* 53–57.

23. Hamai, S. *J. Am. Chem. Soc.* **1989,** *111,* 3954–3957.

24. Femia, R. A.; Cline Love, L. J. *J. Phys. Chem.* **1985,** *89,* 1897–1901.

25. Hamai, S.; Mononobe, M. *J. Photochem. Photobiol. A: Chem.* **1995,** *91,* 217–221.

26. Hamai, S.; Kudou, T. *J. Photochem. Photobiol. A: Chem.* **1998,** *113,* 135–140.

27. Mortellaro, M. A.; Hartmann, W. K.; Nocera, D. G. *Angew. Chem. Int. Ed. Engl.* **1996,** *35,* 1945–1946.

28. de Silva, A. P.; Dixon, T. M.; Gunaratne, H. Q. N.; Gunnlaugsson, T.; Maxwell, P. R. S.; Rice, T. E. *J. Am. Chem. Soc.* **1999,** *121,* 1393–1394.

29. Ueno, A.; Takahashi, K.; Osa, T. *J. Chem. Soc., Chem. Commun.* **1980,** 921–922.

30. Ueno, A.; Tomita, Y.; Osa, T. *J. Chem. Soc., Chem. Commun.* **1983,** 976–977.

31. Ueno, A.; Moriwaki, F.; Osa, T.; Hamada, F.; Murai, K. *Bull. Chem. Soc. Jpn.* **1985,** *59,* 465–470.

32. Minato, S.; Osa, T.; Ueno, A. *J. Chem. Soc., Chem. Commun.* **1991,** 107–108.

33. Minato, S.; Osa, T.; Morita, M.; Nakamura, A.; Ikeda, H.; Toda, F.; Ueno, A. *Photochem. Photobiol.* **1991,** *54,* 593–597.

34. Ueno, A.; Minato, S.; Osa, T. *Anal. Chem.* **1992,** *64,* 2562–2565.

35. Hamada, F.; Minato, S.; Osa, T.; Ueno, A. *Bull. Chem. Soc. Jpn.* **1997,** *70,* 1339–1346.

36. Moriwaki, F.; Kaneko, H.; Ueno, A.; Osa, T.; Hamada, F.; Murai, K. *Bull. Chem. Soc. Jpn.* **1987,** *60,* 3619–3623.

37. Ueno, A.; Minato, S.; Osa, T. *Anal. Chem.* **1992,** *64,* 1154–1157.

38. Toyoda, T.; Matsumura, S.; Mihara, H.; Ueno, A. *Macromol. Rapid Commun.* **2000,** *21,* 485–488.

39. Ueno, A.; Minato, S.; Suzuki, I.; Fukushima, M. *Chem. Lett.* **1990,** 605–608.

40. Hamada, F.; Kondo, Y.; Ito, R. *J. Inclus. Phenom. Mol. Recogn. Chem.* **1993,** *15,* 273–279.

41. Ikeda, H.; Nakamura, M.; Ise, N.; Oguma, N.; Nakamura, A.; Ikeda, T.; Toda, F.; Ueno, A. *J. Am. Chem. Soc.* **1996,** *118,* 10,980–10,988.

42. Dunbar, R. A.; Bright, F. *Supramol. Chem.* **1994,** *3,* 93–99.

43. Wang, Y.; Ikeda, T.; Ikeda, H.; Ueno, A.; Toda, F. *Bull. Chem. Soc. Jpn.* **1994,** *67,* 1598–1607.

44. Wang, Y.; Ikeda, T.; Ueno, A.; Toda, F. *Chem. Lett.* **1992,** 863–866.

45. Hamasaki, K.; Usui, S.; Ikeda, H.; Ikeda, T.; Ueno, A. *Supramol. Chem.* **1997,** *8,* 125–135.

46. Corradini, R.; Dossena, A.; Marchelli, R.; Pamagia, A.; Sator, G.; Saviano, M.; Lombardi, A.; Pavone, V. *Chem. Eur. J.* **1996,** *2,* 373–381.

47. Corradini, R.; Dossena, A.; Galaverna, G.; Marchelli, R.; Panagia, A.; Sartor, G. *J. Org. Chem.* **1997,** *62,* 6283–6289.

48. Nelissen, H. F. M.; Venema, F.; Uittenbogaard, R. M.; Feiters, M. C.; Nolte, R. J. M. *J. Chem. Soc. Perkin Trans. 2* **1997,** 2045–2053.

49. Ueno, A.; Ikeda, A.; Ikeda, H.; Ikeda, T.; Toda, F. *J. Org. Chem.* **1999,** *64,* 382–387.

50. Ikunaga, T.; Ikeda, H.; Ueno, A. *Chem. Eur. J.* **1999,** *5,* 2698–2704.

51. Sato, M.; Narita, M.; Ogawa, N.; Hamada, F. *Anal. Sci.* **1999,** *15,* 1199–1205.

52. Narita, M.; Ogawam, N.; Hamada, F. *Anal. Sci.* **2000,** *16,* 37–43.

53. Narita, M.; Hamada, F. *J. Chem. Soc. Perkin Trans. 2* **2000,** 823–832.

54. Matsumura, S.; Sakamoto, S.; Ueno, A.; Mihara, H. *Chem. Eur. J.* **2000,** *6,* 1781–1788.

55. Nakamura, M.; Ikeda, T.; Nakamura, A.; Ikeda, H.; Ueno, A.; Toda, F. *Chem. Lett.* **1995,** 343–344.

56. Bugler, J.; Engbersen, J. F. J.; Reinhoudt, D. N. *J. Org. Chem.* **1998,** *63,* 5339–5344.

57. Cox, G. S.; Hauptman, P. J.; Turro, N. J. *Photochem. Photobiol.* **1984,** *39,* 597–601.

58. Nag, A.; Bhattacharryya, K. *Chem. Phys. Lett.* **1988,** *151,* 474–476.

59. Nag, A.; Dutta, R.; Chattopadhyay, N.; Bhattacharyya, K. *Chem. Phys. Lett.* **1989,** *157,* 83–86.

60. Nag, A.; Bhattachaayya, K. *Chem. Phys. Lett.* **1990,** *86,* 53–54.

61. Nakamura, A.; Sato, S.; Hamasaki, K.; Ueno, A.; Toda, F. *J. Phys. Chem.* **1995,** *99,* 10,952–10,959.

62. Banu, H. S.; Pitchumani, K.; Shrinivasan, C. *J. Phtochem. Photobiol. A: Chem.* **2000,** *131,* 101–110.

63. Hamasaki, K.; Ikeda, H.; Nakamura, A.; Ueno, A.; Toda, F.; Suzuki, I.; Osa, T. *J. Am. Chem. Soc.* **1993,** *115,* 5035–5040.

64. Hamasaki, K.; Ueno, A.; Toda, F.; Suzuki, I.; Osa, T. *Bull. Chem. Soc. Jpn.* **1994,** *67,* 516–523.

65. Tanabe, T.; Usui, S.; Nakamura, A.; Ueno, A. *J. Inclus. Phenom. Macrocyclic Chem.* **2000,** *36,* 79–93.

66. Hamasaki, K.; Ueno, A.; Toda, F. *J. Chem. Soc., Chem. Commun.* **1993,** 331–333.

67. Wang, J.; Nakamura, A.; Hamasaki, K.; Ikeda, H.; Ikeda, T.; Ueno, A. *Chem. Lett.* **1996**, 303–304.
68. Herkstroeter, W. G.; Mrtic, P. A.; Evans, T. R.; Farid, S. *J. Am. Chem. Soc.* **1986**, *108*, 3275–3280.
69. Ueno, A.; Suzuki, I.; Osa, T. *J. Am. Chem. Soc.* **1989**, *111*, 6391–6397.
70. Ueno, A.; Suzkuki, I.; Osa, T. *Anal. Chem.* **1990**, *62*, 2461–2466.
71. Suzuki, I.; Ohkubo, M.; Ueno, A.; Osa, T. *Chem. Lett.* **1992**, 269–272.
72. Ueno, A.; Takahashi, M.; Nagano, Y.; Shibano, H.; Aoyagi, T.; Ikeda, H. *Macromol. Rapid Commun.* **1998**, *19*, 315–317.
73. Suzuki, I.; Sakurai, Y.; Ohkubo, M.; Ueno, A.; Osa, T. *Chem. Lett.* **1992**, 2005–2008.
74. Hossain, M. A.; Hamasaki, K.; Mihara, H.; Ueno, A. *Chem. Lett.* **2000**, 252–253.
75. Hossain, M. A.; Matsumura, S.; Kanai, T.; Hamasaki, K.; Mihara, H.; Ueno, A. *J. Chem. Soc. Perkin Trans.* **2000**, *2*, 1527–1533.
76. Yamauchi, A.; Hayashita, T.; Nishizawa, S.; Watanabe, M.; Teramae, N. *J. Am. Chem. Soc.* **1999**, *121*, 2319–2320.
77. Tamaki, T. *Chem. Lett.* **1984**, 53–56.
78. Tamaki, T.; Kokubu, T. *J. Inclus. Phenom.* **1984**, *2*, 815–822.
79. Ueno, A.; Moriwaki, F.; Osa, T.; Hamada, F.; Murai, K. *J. Am. Chem. Soc.* **1988**, *110*, 4323–4328.
80. Ueno, A.; Moriwaki, F.; Azuma, A.; Osa, T. *J. Org. Chem.* **1989**, *54*, 295–299.
81. Ueno, A.; Moriwaki, F.; Azuma, A.; Osa, T. *Carbohydr. Res.* **1989**, *192*, 173–180.
82. Ueno, A.; Moriwaki, F.; Iwama, Y.; Suzuki, I.; Osa, T.; Ohta, T.; Nozoe, S. *J. Am. Chem. Soc.* **1991**, *113*, 7034–7036.
83. Ueno, A.; Moriwaki, F.; Tomita, Y.; Osa, T. *Chem. Lett.* **1985**, 493–496.
84. Tabushi, I.; Fujita, K.; Yuan, L. C. *Tetrahedron Lett.* **1999**, 2503–2506.
85. Nakamura, A.; Okutsu, S.; Oda, Y.; Ueno, A.; Toda, F. *Tetrahedron Lett.* **1994**, *35*, 7241–7244.
86. Nakamura, A.; Imai, T.; Okutsu, S.; Oda, Y.; Ueno, A.; Toda, F. *Chem. Lett.* **1995**, 313–314.
87. Gravett, D. M.; Guillet, J. E. *J. Am. Chem. Soc.* **1993**, *115*, 5970–5974.
88. Nowakowska, M.; Loukine, N.; Gravett, D. M.; Burke, N. A. D.; Guillet, J. E. *J. Am. Chem. Soc.* **1997**, *119*, 4364–4368.
89. Berberan-Santos, M. N.; Canceill, J.; Brochon, J.; Jullien, L.; Lehn, J.; Pouget, J.; Tauc, P.; Valeur, B. *J. Am. Chem. Soc.* **1992**, *114*, 6427–6436.
90. Jullien, L.; Canceill, J.; Valeur, B.; Bardez, E.; Lefevre, J.; Lehn, J.; Marchi-Artzner, V.; Pansu, R. *J. Am. Chem. Soc.* **1996**, *118*, 5432–5442.
91. Wang, P. F.; Jullien, L.; Valeur, B.; Filhol, J. S.; Cancell, J.; Lehn, J. M. *New J. Chem.* **1996**, *20*, 895–907.
92. Berberan-Santos, M. N.; Choppinet, P.; Fedorov, A.; Jullien, L.; Valeur, B. *J. Am. Chem. Soc.* **1999**, *121*, 2526–2533.
93. Harada, A.; Kamachi, M. *Nature* **1992**, *356*, 325–326.
94. Tamura, H.; Ueno, A. *Chem. Lett.* **1998**, 369–370.
95. Tamura, M.; Ueno, A. *Bull. Chem. Soc. Jpn.* **2000**, *73*, 147–154.
96. Tamura, M.; Ueno, A. Unpublished data.

97. Murakami, H.; Kawabuchi, A.; Kotoo, K.; Kunitake, M.; Nakashima, N. *J. Am. Chem. Soc.* **1997,** *119,* 7605–7606.

98. Ueno, A.; Takahashi, K.; Osa, T. *J. Chem. Soc., Chem. Commun.* **1980,** 837.

99. Ueno, A.; Yoshimura, H.; Saka, R.; Osa, T. *J. Am. Chem. Soc.* **1979,** *101,* 2779–1780.

100. Ueno, A.; Takahashi, K.; Osa, T. *J. Chem. Soc., Chem. Commun.* **1981,** 94–95.

101. Ueno, A.; Moriwaki, F.; Azuma, A.; Osa, T. *J. Org. Chem.* **1989,** *54,* 295–299.

102. Hamada, F.; Fukushima, M.; Osa, T.; Ikeda, H.; Toda, F.; Osa, T. *Macromol. Chem. Rapid Commun.* **1993,** *14,* 287–291.

103. Ueno, A.; Tomita, Y.; Osa, T. *Tetrahedron Lett.* **1983,** *24,* 5245–5248.

104. Ueno, A.; Fukushima, M.; Osa, T. *J. Chem. Soc. Perkin Trans. 2* **1990,** 1067–1072.

105. Fukushima, M.; Osa, T.; Ueno, A. *J. Chem. Soc., Chem. Commun.* **1991,** 15–17.

106. Fukushima, M.; Osa, T.; Ueno, A. *Chem. Lett.* **1991,** 709–712.

107. Aoyagi, T.; Ueno, A.; Fukushima, M.; Osa, T. *Macromol. Rapid. Commun.* **1998,** *19,* 103–105.

108. Tabushi, I.; Yuan, L. C. *J. Am. Chem. Soc.* **1981,** *103,* 3574–3575.

109. Rao, K. S. S. P.; Hubig, S. M.; Moorthy, J. N.; Kochi, J. K. *J. Org. Chem.* **1999,** *64,* 8098–8104.

110. Herrmann, W.; Schneider, M.; Wenz, G. *Angew. Chem. Int. Ed. Engl.* **1997,** *36,* 2511–2513.

9

Photoactive Layered Materials: Assembly of Ions, Molecules, Metal Complexes, and Proteins

Challa V. Kumar and B. Bangar Raju
University of Connecticut, Storrs, Connecticut

I. INTRODUCTION

This entire series is devoted to the study of supramolecular assemblies and this fact emphasizes the importance of supramolecular chemistry in this new millennium. The transition from the molecular to supramolecular chemistry is due to the inspiration derived from the high-level-molecular architecture observed in biological systems. The details of the intermolecular forces that control this biological organization is not well understood, but efforts in supramolecular chemistry are already helping us understand this organization. Photochemical methods are extremely valuable for exploring these intermolecular interactions that control the molecular organization.

Controlling the photophysical and photochemical behavior of molecules is challenging. Such control over molecular properties is essential in the design of multicomponent systems for solar energy harvesting, photonics, and synthetic applications of photochemistry. Although extensive literature exists on these topics, only a few selected examples are included here to orient the reader regarding the use of layered materials (as solid solvents) in controlling the molecular behavior. In this chapter, our efforts are focused on the use of layered inorganic solids in the construction, characterization, and examination of supramolecular assemblies of metal ions, molecules, metal complexes, and proteins.

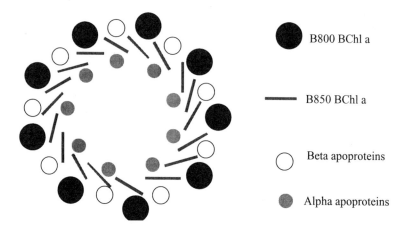

Figure 1 Schematic representation of the bacterial photosynthetic light-harvesting complex. Nine BChl800 pigments on the outside (with their planes perpendicular to that of the membrane) and 18 BChl850 pigments on the inside (with their planes parallel to the membrane) are strongly coupled. The carotenoid pigments cross over from the outer ring to the inner (not shown). The protein α-helices are shown as shaded circles viewed along the helix axis. The two concentric rings of α (inner) and β (outer) apoprotein helices are housed in the two-dimensional structure of the membrane.

A. Supramolecular Assemblies and Layered Materials

Layered materials are used to produce supramolecular assemblies, in which the properties of the assemblies differ from those of the individual members of the assembly, such that the whole is more than the sum of its components. Such a behavior may be used as a criterion to define a supramolecular system, and nature provides numerous examples. The light-harvesting complex of the photosynthetic apparatus is one such example of outstanding organization.

A schematic of the light-harvesting complex, whose structure was determined several years ago [1], is shown in Fig. 1. The structure consists of protein–pigment complexes assembled in the shape of a donut in two dimensions, housed in the lipid bilayer, and the entire protein complex is held together by noncovalent interactions. There are 18 bacteriochlorophyll B850 pigments in the inner ring of protein α-helices, which are strongly coupled with each other. In addition, there are nine bacteriochlorophyll B800 pigments, oriented perpendicular to B850 pigments, placed in the outer ring of α-helices, and these pigments are also strongly coupled with neighbors and with the B850 pigments. Several carotinoids, not shown in Fig. 1, span the two rings and may play an important role in light absorption and energy migration. Excitation of any of the chromophores results in rapid energy migration among the pigments, and for this reason, the system is referred to as a molecular synchrotron. Excitation is also exchanged between adjacent rings that are separated by the protein matrix such that excitation migrates from ring to ring in the light-harvesting complex. Finally, energy is trapped when it reaches one of the rings that houses the reaction center at the center of the earlier described donut of protein–pigment complexes. This is an outstanding example of the noncovalent supramolecular assembly of thousands of chromophores in an ordered fashion, and such control over the organization of the chromophores may, one day, be achieved in the laboratory.

Such a high level of organization of multiple components is still very challenging, but several elegant examples of self-assembly are reported in the literature, and an examination of these systems reveals the underlying principles of supramolecular architecture and assembly. With this goal, we discuss selected examples from our laboratory as well as from literature to examine the role of molecular assemblies in various photoprocesses.

The photophysical and photochemical properties of guests in organized media often differ from those in homogeneous solution and these differences highlight the behavior of assemblies in contrast to that of the individual chromophores [2]. Organized media such as zeolites, layered materials, micelles, deoxyribonucleic acid (DNA), polymer matrices, crystals, and liquid crystals provide a certain degree of control over the excited-state behavior of the guest molecule (Fig. 2). This is because the binding often results in reduced solvent exposure, increased local concentrations, enhanced chromophore ordering, and the host media can impose severe restrictions on guest mobility. Close packing of the guests results in shortened intermolecular distances, altered dielectric constant, desolvation of the guests, and accelerated/decelerated bimolecular events. The environment and the intermolecular interactions that the guest molecules experience in organized media are thus significantly different from those observed in homogeneous solu-

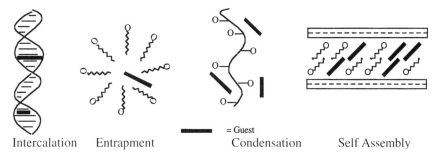

Intercalation   Entrapment   = Guest   Condensation   Self Assembly

Figure 2 Organization of guest molecules by DNA, micelles, polyions, and the layered materials as supramolecular media. A high degree of organization of the guests at the galleries of the layered materials is anticipated from their structural characteristics.

tion. These factors, in turn, influence the photophysical and photochemical processes of the bound guests. Efficient excimer formation, accelerated electron transfer, improved selectivities (chemo, regio, stereo, and chiral), and enhanced energy transfer, for example, are a few processes that the guests may exhibit in organized media with significant improvements over those in solution phase. Guests are organized at DNA, micelles, polyions, zeolites, and layered materials (Fig. 2). Some of these organized assemblies bear close resemblance to the organization found in biological systems and provoke new challenges. Potential areas of applications of immobilized/intercalated/adsorbed chromophores are molecular electronics, nonlinear optics, photocatalysis, artificial photosynthesis, and so forth [3]. Additional examples of organized media include, reverse micelles, cyclodextrins, clays, Langmuir–Blodgett films, clatharates, and so forth and the list is by no means complete but layered materials are special. Layered inorganic materials like group (IV) metal phosphates and phosphonates, transition metal oxides, layered oxides, double hydroxides, and metal sulfides/disulfides can provide novel organized environments for the guest molecules.

One advantage of the lamellar inorganic solids is that the interlamellar gallery can be expanded to accommodate guests of any size from protons, and proteins to protozoa. In recent years, research on intercalation into layered inorganic solids has grown. Metal ions, metal complexes, molecules, cyclodextrins, semiconductor particles, metal clusters, and proteins of various sizes and shapes can be intercalated into the galleries of layered materials. Formation of such intercalates can be verified using absorption, fluorescence, x-ray diffraction (XRD), circular dichroism (CD), and chemical/biological activity methods. The binding affinity and the binding stoichiometry of these guests can be readily estimated from equilibrium centrifugation studies or from spectral studies.

Some of the very interesting applications of these layered intercalates are in material design [3], ion exchange [4], catalysis [5], in the study of quantum-sized semiconductor particles [6], assembly of molecular multilayers at solid–liquid interfaces [7], designer electrode surfaces [8], preparation of low-dimensional conducting polymers [9], and so forth.

The preparation, characterization, and properties of inorganic layered materials is reviewed here first, then we describe the photophysical and photochemical properties of guests intercalated into these hosts; this review is not intended to be exhaustive.

II. LAYERED INORGANIC HOST MATERIALS FOR THE CONSTRUCTION OF MOLECULAR ASSEMBLIES

Group (IV) metal phosphates and phosphonates, transition metal oxides (titanates, silicates, niobates, etc.), layered oxides, and double hydroxides (aluminum, magnesium, iron, etc.) are some of the inorganic compounds used as layered host materials for immobilizing photoactive species. The structural characteristics of these inorganic materials are discussed in the subsections below.

A. Alpha-Zirconium Phosphate

1. Preparation

Layered zirconium phosphate (α-ZrP) ([$Zr(HPO_4)_2 \cdot nH_2O$]) (Fig. 3), has been used for the organization of a number of guests at the galleries and in our laboratory; hence, it will be discussed first [10]. α-ZrP is prepared from an aqueous solution of zirconyl chloride and excess phosphoric acid [11,12]. An amorphous material quickly settles down as a white precipitate, and it is crystallized when refluxed for several days in strong acids (HF or HCl, or $8M$ H_3PO_4). Crystalline layers with extended particle growth, as observed in powder x-ray diffraction experiments, indicate the conversion from the amorphous gel to the crystalline state [13]. The degree of crystallinity of the material varies with refluxing time and concentration of the acid. X-ray diffraction and spectroscopic methods also confirm the physical nature of the material as stacked layers.

Another preparative method in which the rate of precipitation is slow involves slow decomposition of zirconium fluoro complexes [14]. These are first prepared by adding an appropriate amount of hydrofluoric acid (HF) to the zirconyl salt and these complexes are decomposed in the presence of phosphoric acid, with a slow stream of nitrogen or water vapor passing through the system. The rate of precipitation of zirconium phosphate is controlled by the rate of removal of HF from the system, and when this is very slow, a highly crystalline α-ZrP is obtained. The gamma form of the metal phosphate differs significantly from the alpha and current discussion will be concerned with the latter phase.

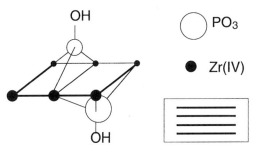

$$ZrOCl_3 + H_3PO_4 \rightarrow Zr(PO_3OH)_2 \cdot nH_2O$$

Figure 3 Partial schematic structure of α-zirconium phosphate lamallae and its preparation.

2. Characterization

The crystal structure of layered α-ZrP [10] shows that a layer of Zr(IV) ions are sandwiched between two layers of phosphate anions, producing a sheet of metal phosphate where the free hydroxyl of the phosphate is oriented perpendicular to the metal phosphate layers [15]. A schematic structure of α-ZrP is shown in Fig. 3. Each phosphate group is coordinated to three zirconium ions and each Zr(IV) is surrounded by six phosphate oxygens in an octahedral geometry. Such an arrangement results in one ionizable hydroxyl group being free in each phosphate, making the α-ZrP acidic, and this ionizable function is responsible for the ion-exchange properties of α-ZrP. As there are no interlayer hydrogen bonds, α-ZrP layers are held together by van der Waals forces. The interlayer distance, as estimated from the powder diffraction patterns, is 7.6 Å.

3. Properties

Alpha-Zirconium phosphate matrices are rigid, robust, chemically inert (at pH $\leq$ 7), and thermally stable. The rigid and inert matrices of zirconium phosphates are useful for the organization of the cationic organic chromophores at the interlayer regions (galleries). The interlayer binding sites can be accessed by the guests via intercalation into the galleries or by exfoliation of the layers followed by binding of the guest and reassembly of the layers into stacks. Some of the attractive features of group (IV) metal phosphates and the corresponding phosphonates (the phosphate OH is replaced by R) are (1) they can be prepared easily from aqueous solutions, (2) the interlamellar gallery can be expanded to accommodate small as well as large guest molecules (from protons to proteins), (3) the matrix does not absorb light in the visible region, albeit they scatter light, (4) they provide surfaces with a high charge density (1 charge per 25 Å^2), and (5) they can impose orientational/mobility restrictions on the bound guests.

4. Tuning the Gallery Spacings

One of the main features of α-ZrP is that the layers can be expanded (Fig. 4) and this facilitates binding of small molecules, ions, and metal complexes at the galleries [16]. Tuning the gallery spacing to match guest size is an important element in the intercalation of large hydrophobic guest molecules into the galleries. When the free energy for binding is greater than the interlayer interaction energy, guests can be exchanged into α-ZrP interlayer regions. However, the small interlayer distance in α-ZrP (7.6 Å) poses a kinetic barrier to open the layers. Short-chain surfactant molecules are introduced into the interlayers as spacers, and guest molecules of size comparable to that of the spacer can be readily exchanged into the expanded galleries [17]. The interlayer distance can also be tuned to accommodate guests of various sizes by varying the surfactant chain length. Intercalation of alkyl amines (RNH_2) into the galleries of α-ZrP is readily accomplished by contacting the amines with suspensions of the metal phosphate (Fig. 4).

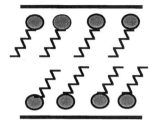

Figure 4 Short-chain surfactant molecules are introduced into the interlayers as spacers. The interlayer distance can be modulated by varying the surfactant chain length.

At low loadings, the alkylamines are found to lie parallel to the α-ZrP layers and the interlayer spacing increases from 7.6 Å to about 10.4 Å, irrespective of the amine chain length [18,19]. At higher loadings, the guests aggregate and begin to incline to the layers at increasing angles. For instance, monoalkyl ammonium ions, such as n-butyl ammonium ions (BA), intercalate into the galleries of α-ZrP to form stable bilayers. The polar ammonium groups are located near the phosphate surface, and the alkyl chains are inclined at 55°–60° to the plane of α-ZrP layers (Fig. 5). Electrostatic attraction between the polar head groups and the negatively charged phosphate surface provides the enthalpic propensity for intercalation, even though the organization of the chains is not entropically favored. Loss of water molecules from the hydrophobic surfaces, when these guests bind in the galleries, contributes to increased entropy of the system. Such a bilayer arrangement of the alkyl amines increases the interlayer distance from 7.6 Å in α-ZrP to 18.6 Å in BAZrP [20].

BAZrP can be used as a precursor to prepare assemblies of several aromatic hydrophobic cations (Fig. 5). In addition to the favorable electrostatic interactions,

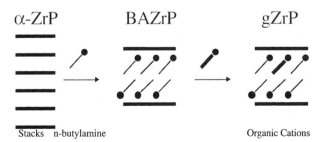

α-ZrP   BAZrP   gZrP

Stacks   n-butylamine   Organic Cations

Figure 5 Intercalation of hydrophobic short-chain amines into α-ZrP to widen the gallery spacings and substitution of these by small organic cations.

the binding in the galleries is also promoted by favorable entropic contributions resulting from the removal of the hydrophobic ions from the aqueous phase into the interior of α-ZrP, as outlined earlier. Hence, aromatic cations of proper size, geometry, and charge density can be readily incorporated in the α-ZrP galleries. Orientation of the guest molecules in the galleries is affected by the interactions among the aromatic cations, the host surface, and the spacer. The proximity of guests, at high loadings, also produces pronounced effects due to strong intermolecular interactions. An ordered orientation of the guests, along with strong intermolecular interactions, favors formation of organized assemblies of the guest molecules in the α-ZrP galleries. Thus, several guest molecules of varying dimensions, such as derivatives of anthracene and pyrene, ethidium bromide, acridinium hydrochloride, rhodamine, fluorescein, chlorophylls, porphyrins, ruthenium complexes, and so forth can be readily intercalated into the galleries of BAZrP, forming monolayers or bilayers (Fig. 6). The variability of the gallery spacing with respect to the guest size is demonstrated in the x-ray diffraction data given in Table 1. More details concerning the characterization of these assemblies are provided later.

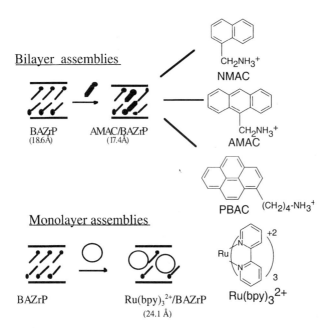

Figure 6 Schematic representation of bilayer (NMAC, AMAC, and PBAC) and monolayer ($Ru(bpy)_3^{2+}$) assemblies of the guest molecules at the galleries of BAZrP.

Table 1 Variability of the Gallery Spacing With Respect to the Guest Size as Observed from the X-ray Diffraction Data

Guest	Interlayer spacing (Å)
α-ZrP	7.6
BAZrP	18.6
NMAC–BAZrP	17.4
AMAC–BAZrP	17.6
Ru(bpy)$_3^{2+}$–BAZrP	24.1

B. Zirconium Phosphonates

Group (IV) metal phosphonates [α-Zr(O$_3$PR)$_2$·H$_2$O where one OH of the phosphate is replaced by R] are another class of useful host materials for organizing molecules in lamellar structures (Fig. 7). The surface functions (R) of these layers can be modified as desired to produce hydrophobic, anionic, cationic, or polar surfaces. An important feature of these layered materials is stability. Like the metal phosphates, the corresponding phosphonates can be used to control the arrangement (orientation) of the intercalated guests. The packing of the R groups in metal phosphonates is similar to that formed by Langmuir–Blodgett (LB) and self-assembly techniques. Metal phosphonates can be prepared as microcrystalline solids, multilayer thin films, and porous solids [21]. Preparation and characteriza-

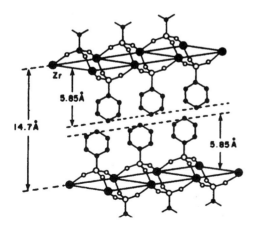

Figure 7 Structural model of zirconium phenylphosphonate. (From Ref. 11g. Copyright 1996 Gordon and Breach Publishers.)

tion of zirconium phosphonates facilitated the development of other layered organic materials for the organization of molecules in a plethora of environments. They are normally prepared from metal ions and the corresponding phosphonic acid [22]. Zirconium phenylphosphonate (Fig. 7) is such an example of an α-type layered compound [11g].

The interlayer spacings of α-Zr(O₃PR)₂·H₂O can be larger (~15 Å) than those for α-ZrP or BAZrP depending on the R group, except that the chains are inclined to the surface for best packing. Synthesis of α-zirconium phosphonates and other group (IV) metal phosphonates is similar to that of α-ZrP and includes extended refluxing of amorphous or semicrystalline precipitate produced initially. The difference is that the phosphoric acid used to prepare α-ZrP is replaced with the corresponding phosphonic acid in the preparation of phosphonates. On the basis of the density and interlayer distances, a structural analogy was predicted between α-ZrP and α-zirconium phosphonates (Fig. 8) [23]. This prediction was later experimentally confirmed to be true [24].

Microcrystalline samples of zirconium phosphonate can also be prepared by adding a strong complexing agent, usually HF, to solubilize the material. As HF is slowly removed from the system, the metal polyfluoride decomposes, leaving a crystalline phosphonate. This method works well for preparing metal phosphonates bearing different functional groups, such as alkyl, aryl, halo, carboxyl, nitro, sulfonato, hydroxy, amino, vinyl, and so forth [11b,25]. They could also be prepared from α-ZrP by an exchange reaction. As in the case of ZrP, α and γ phases of zirconium phosphonate have been identified [21].

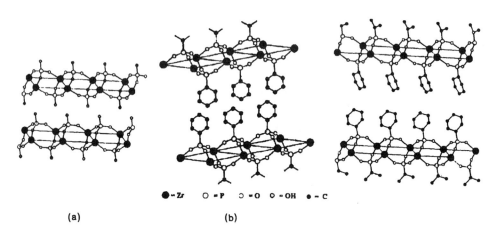

(a)   (b)

Figure 8 Structural models of (a) α-ZrP and (b) zirconium phenylphosphonate. (From Ref. 23. Copyright 1995 Gordon and Breach Science Publishers.)

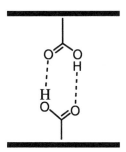

Figure 9 Schematic structure of layered metal phosphonate phosphonates with interlayer hydrogen-bonding.

Spectroscopic and structural determination of zirconium phosphonates revealed that they are layered materials in which the phosphonate oxygens are bound to the metal ions, forming a tightly bound inorganic layer, and the organic groups are between the inorganic lamellae. The phosphonate groups are arranged in a hexagonal pattern above and below the sheets of Zr(IV) ions. The distance between the adjacent alkyl groups was found to be 5.3 Å [26]. In simple alkyl phosphonates, a weak van der Waals bonding prevails between adjacent layers. However, if hydrogen-bonding groups are incorporated at the ends of these alkyl chains, the interlayer hydrogen-bonding becomes strong (Fig. 9). Finally, the adjacent layers can be cross-linked by using biphosphonic acids, instead of monophosphonic acids, resulting in pillared materials (Fig. 10) [26,27].

Amide and ester derivatives of zirconium phosphonate have been prepared via amine and alcohol intercalation and subsequent condensation reactions [28]. X-ray diffraction patterns and spectroscopic measurements show that the inter-

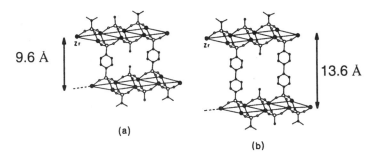

Figure 10 Schematic representation of originally pillared α-ZrP: (a) $Zr(O_3P–C_6H_4PO_3)_{0.5}(O_3PH)$ and (b) $Zr(O_3P–C_6H_4–C_6H_4–PO_3)_{0.5}(O_3PH)$. (From Ref. 16. Copyright 1990 Gordon and Breach Science Publishers.)

layer separation in these materials is directly related to the length of the amide or ester alkyl chains. For amides, analysis of the x-ray diffraction patterns indicated that the amide chains are tilted at 59° to the zirconium layers and 31° to the bilayer normal. Fourier transform infrared (FTIR) spectra showed characteristic peaks for amide I at 1646 and amide II at 1548 cm^{-1}, and N—H stretching at ~3320 and 3080 cm^{-1}, respectively. Thermal and chemical stabilities of the layered amides was found to be greater than analogous compounds prepared by Brönsted acid–base reactions [28,29]. When the two α-ZrRP layers are cross-linked, using either alkyl or aromatic groups to form pillared α-zirconium phosphonates [30], the interlayer spacing could be tailored by varying the height of the pillars (Fig. 10).

In view of this unique property, zirconium and other metal bisphosphonates were termed molecular engineered layered structures (MELS) to emphasize the control over the porosity of these materials [31]. The MELS developed in the earlier years suffered from two drawbacks: low degree of crystallinity and low interlayer porosity. When the cross section of the pillars was made nonuniform and if the pillar cross section at the center was smaller than that at the terminii, pillared α-zirconium bisphosphonates with a high degree of crystallinity and interlayer porosity could be obtained. Such pillared materials with regular interlayer porosity have been prepared by partial replacement of dihydrohgen phosphates with bivalent diphosphonate groups. Computer modeling of the structures of such pillared compounds resulted in a microporous α-zirconium diphosphonate, reminiscent of the architecture of a Greek temple [32]. Pillared materials with well-defined void spaces and, hence, tailored microcomposites could also be engineered by tailoring the length of the organic pillars and their lateral distances within the α-ZrP layers. These materials are likely to find application as molecular sieves and shape-selective catalysts [33].

More recently, crown ether pillared and functionalized α-zirconium phosphonates have been synthesized [34]. Incorporation of crown ethers was achieved by first converting them to their respective phosphonic acids. X-ray diffraction studies showed that the interlamellar spacing was about 15 Å when a biphosphonic acid was used. Both α- and γ-type zirconium phosphonates could be synthesized using this methodology. Preliminary results showed that these layered materials have good selectivity toward binding transition metal ions and that the interlamellar fluoride ions could be replaced by anion exchange.

Sulfonation of zirconium phenylphosphonate with fuming sulfuric acid gave zirconium sulfophenylphosphonates (ZrPS) [35]. The chemical formula of the resultant product is $Zr(HPO_4)(O_3P–C_6H_4SO_3H)\cdot nH_2O$ and has an interlayer spacing of 16.1 Å. The interlayer volume is also relatively large, thus enhancing the rate of diffusion within the lattice. Due to the presence of the sulfonic acid groups, the zirconium phosphophenylsulfonates behave as strong acid ion exchangers (pH range 2–4). The presence of the phosphate groups, on the other hand, enables ion exchange at higher pH values [36].

Metal phosphates and phosphonates do not absorb light in the visible portion of the spectrum, making them ideal for studying the photophysics and photochemistry of organic and inorganic molecules such as rhodamine B, fluorescein, and metalloporphyrins. The white powders do scatter light, and the extent of scattering in a suspension depends on the particle size, wavelength, and concentration.

C. Other Metal Phosphates and Phosphonates

Zinc [37], manganese [38], molybdenum [34], and vanadium [40,41] also form lamellar structures. For example, molybdenyl phenylphosphonate forms a linear structure with double chains in which the molybdenyl oxygens of the adjacent chains point toward each other and the phenyl groups are on the outside [42]. As is the case with zirconium phosphate and phosphonate, the layered nature of the above metal phosphonates is similar to that of the respective phosphates. Among these, vanadium phosphonates have generated greater interest in view of their importance as industrial catalysts.

D. Transition Metal Oxides

Transition metal oxides (TMOs) such as titanates, and molybdates, are also known to form intercalation compounds [42]. A few examples are $KNbO_3$, $KTiNbO_5$, $K_2Ti_4O_9$, and $K_4Nb_6O_{17}\cdot3H_2O$ (Fig. 11). Intercalation of guests into these materials is a slow process and is accomplished by mixing an aqueous solution of guests and hosts. Intercalation at higher temperatures accelerates the process. Chlorophylls, phthalocyanine dyes, and ruthenium complexes are some of the molecules that have successfully been adsorbed on transition metal ox-

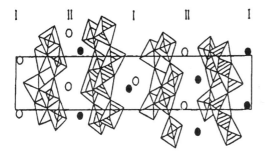

Figure 11 Schematic structure of $K_4Nb_6O_{17}\cdot3H_2O$. The alternating interlayers I and II have different reactivities. (From Ref. 42. Copyright 1995 The American Chemical Society.)

ides. Photocatalysis and photoredox chemistry of the adsorbed molecules are some of the interesting properties studied [43].

E. Layered Double Hydroxides (LDH)

Layered double hydroxides (LDHs) are mixed-metal hydroxides containing a positively charged interlayer that is key to their chemical behavior. The structure of LDH is comprised of brucite–$M(OH)_2$ sheets, where the partial substitution of trivalent cations for divalent cations results in the positive charge for the matrix. The general formula of LDHs is $[M_1^{2+}-xM_x'^{3+}(OH)_2]-[A_{x/n}^{n-}]^{x-}$, where $M^{2+} = Mg^{2+}, Co^{2+}, Ni^{2+}$, and $M^{3+} = Al^{3+}, Cr^{3+}, Fe^{3+}$; A^{n-} is an anion such as CO_3^{2-}, or Cl^- that can be ion-exchanged into the galleries (Fig. 12) [42,43]. A unique aspect of the LDHs is that they consist of positively charged metal hydroxide sheets; hence, organic/inorganic anions or negatively charged proteins or DNA can be readily intercalated into LDHs [44]. Intercalation of anionic species into the interlayers of LDHs can be accomplished by three methods: (1) conventional anion exchange using aqueous solutions of guests, (2) direct synthesis in which a LDH phase precipitates in the presence of guests, thereby encapsulating it [45], and (3) treating a solution of mixed-metal oxide, obtained by heat treatment of LDH–carbonate, with an aqueous solution of guests producing the LDH–guest intercalates [46].

Due to their high charge density [44,47], a high degree of anion exchange (80–100%) can be readily achieved [48]. Like the ZrPs, they are also thermally stable. The intercalation compounds of LDH are found to have potential importance as catalysts, ion exchangers, ceramic precursors, antacid drugs, and as hosts for anionic dyes and metal complexes [49].

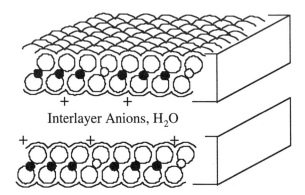

Figure 12   Schematic structure of LDH. Large circles: OH layers, ●: Metal 1 (Zn, Mg, etc.); ○: Metal 2 (usually Al). (From Ref. 42. Copyright 1995 The American Chemical Society.)

F.   Metal Sulfides and Metal Dichalcogenides

Metal disulfides (MS_2) and metal dichalcogenides (MS_4) (metal M = Cd, Mo, Zr, Ti, Nb, Ta, or Sn) are lamellar colloids with strong M—S bonds within the layers and weak van der Waals interactions between the layers. Each metal atom in these lamellae is surrounded by six sulfur anions with an approximate octahedral coordination geometry. Each sulfur anion is, in turn, coordinated to three metal atoms with a trigonal pyramidal geometry [50]. Certain cationic species or polymers are adsorbed onto these surfaces via intercalation or exfoliation [51,52].

III.   PHOTOPHYSICAL AND PHOTOCHEMICAL PROCESSES OF ORGANIC AND INORGANIC GUEST–MOLECULES INTERCALATED INTO α-ZRP

The most important first step preceding the photophysical or photochemical processes is the activation of the molecules from the ground electronic state to the excited electronic state by the absorption of light. Thus, the excited states are rich in energy and they can undergo several spontaneous processes. The time taken for the promotion of a molecule from the ground state to the higher electronic state is much faster than for the vibrational motion; hence, internuclear distances do not change during the absorption process. This forms the basis for the Frank–Condon principle.

A.   Jablonski Diagram

Absorption of light by a polyatomic molecule is followed by the transfer of the molecules from the ground state (S_0) to the vibronic levels of the excited electronic states (Fig. 13). The excited molecules then undergo a rapid internal conversion and relax to the lowest vibronic level of the lowest excited singlet state (S_1). Once in the lowest excited S_1 state and in the absence of an external quencher, the excited molecules may return to the ground state by emitting photons. Other unimolecular processes competing with fluorescence, within the lifetime of the molecule in the excited state, are internal conversion (IC) to the ground state, intersystem crossing (ISC) to the triplet state, or other nonradiative photochemical processes. ISC involves a nonradiative transition accompanied by spin change. A molecule in the triplet state can decay by phosphorescence (radiative process, accompanied by spin change) or by intersystem crossing followed by internal conversion (nonradiative process) to the ground state.

In addition to unimolecular reactions, the excited state may participate in several bimolecular processes. At high concentrations, dimer formation, excimer formation, exciplex formation, solute–solvent complexation, energy transfer, and collosional deactivation may occur. The high-concentration conditions are often experienced when the guest molecules are loaded onto the layered materials with high coverages and specific examples will be provided shortly.

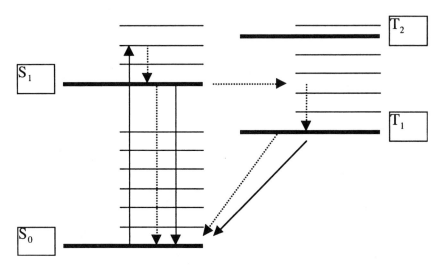

Figure 13 Jablonski diagram showing the energy levels in the ground and excited states and their interconversions involving radiative (solid, down arrows) and nonradiative (dotted arrows) transitions. Photochemical reactions might occur from either the singlet or the triplet manifold.

Spectroscopic properties of organic and inorganic molecules intercalated into the galleries of α-ZrP are significantly different from those in solution. These differences provide an insight into the organization (both orientation and packing) of the guest molecules, the nature of host–guest interactions, and the mobility of the guest molecules.

Depending on the size, shape, hydrophobicity, and the extent of host–guest and guest–guest interactions, monolayer or bilayer assemblies are formed at the galleries. Such assemblies are investigated by using absorption, fluorescence, x-ray diffraction (XRD), and fluorescence polarization techniques. The binding affinity and the binding stoichiometry for various guests can be estimated from equilibrium centrifugation studies (discussed later) when the particles are large enough for rapid sedimentation. Other techniques such as atomic force microscopy (AFM) and transmission electron microscopy (TEM) can also be employed for detailed characterization of these assemblies.

B. Assemblies of Anthryl, Pyrenyl, Ethidium, and Acridine Derivatives

Small hydrophobic cations derived from naphthalene, anthracene, and pyrene and larger molecules such as rhodamine and fluorescein can be readily intercalated into the α-ZrP galleries. Inorganic complexes, ruthenium tris bipyridine derivatives, and others can also be intercalated into the galleries of α-ZrP. This list has been extended recently by the intercalation of globular proteins such as lysozyme, myoglobin, hemoglobin, and glucose oxidase. Enhanced ground-state dimer and excimer formation, excitonic coupling between guests, increased fluorescence polarization, rapid energy transfer suggest ordered assembly of the guests at the galleries.

BAZrP serves as an excellent precursor for the binding and organization of hydrophobic, aromatic cations at the galleries. 9-Anthracene methylamine HCl (AMAC), 4-(1-pyrene) butylamine HCl (PBAC), ethidium bromide (EB), and acridinium HCl (ACR), for example, can be readily exchanged into the galleries of BAZrP. The negatively charged, rigid, inorganic metal phosphate matrix binds guest molecules at the galleries and, to a lesser extent, at the edges of these lamellae. Binding at the galleries also affects the orientation of the chromophores and enhances their local concentrations (concentration per unit area), and at high loadings, ordered self-assembly of the guest molecules can occur. However, first, the binding stoichiometry, binding affinity, and binding mode of the guests with these media must be established. These issues are addressed in the following subsection.

1. Binding Studies

Determining the stoichiometry for binding of the guests into the galleries of BAZrP is useful for estimating the average area occupied by each guest. Binding curves of AMAC, PBAC, EB, and ACR are obtained from centrifugation studies. In this method, the host–guest complexes are spun down in a microcentrifuge (Allied Fischer Scientific, operated at a speed of 10,000 rpm for 5 min) and this method is quite effective micrometer size or larger particles. The concentration of the free probe is determined and the binding curves are constructed from these data. A plot of the percentage of AMAC bound, versus total AMAC concentration, is shown in Fig. 14a. The inflection point at 25 μM of AMAC (for 0.008% BAZrP) is the guest concentration needed to cover all of the available surface area. The area occupied per AMAC is, thus, 53 $Å^2$ of the total surface area (100 m²/g of BAZrP) [17]. Considering a bilayer assembly of AMAC (when the ammonium group faces the ZrP surface, Fig. 14a) and assuming that the anthryl plane is perpendicular to the ZrP planes, the length of the anthryl group was estimated as 15 Å, a value that corresponds to the long axis of the anthryl chromophore.

Binding curves for EB were constructed from the absorption titration experiments as well as from the centrifugation method. A plot of difference in absorbances at 600 (bound form) and 478 (free form) nm versus EB concentration shows a strong transition at 40 μM EB, which agrees with the data from the centrifugation experiments (Fig. 14b). This point corresponds to the saturation of the available binding sites. The area occupied (footprint) by EB estimated from these plots is 33–38 Å square and it is in excellent agreement with molecular modeling studies [17]. The binding plot for Ru(II)trisbipyridyl chloride is also shown in Fig. 14b. In these cases, essentially all of the probe molecules are bound to BAZrP at

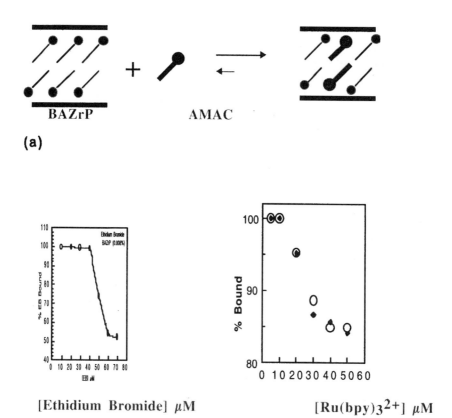

(a)

(b)

Figure 14 (a) Binding plots for the guests with α-ZrP. The binding was monitored in centrifugation experiments. A bilayer arrangement of AMAC at BAZrP galleries. (b) Binding plots for ethidium and the Ru(II)trisbipyridyl chloride with BAZrP. The binding was monitored in centrifugation experiments. (From Ref. 17. Copyright 1995 Overseas Publishers Association.)

low loadings, indicating a strong affinity for these galleries. The stoichiometry of the Ru(II)trisbipyridyl complex indicates a monolayer packing at the galleries, whereas a bilayer is indicated for AMAC and EB. This conclusion is verified from the powder x-ray diffraction data.

2. Enhanced Excimer Formation

Anthracene [19] and pyrene [20] derivatives (AMAC and PBAC, respectively) intercalated into BAZrP galleries show enhanced ground-state dimer and excimer formation. The dimer shows a broad new absorption band, the intensity of which increases quadratically with concentration. In contrast, a 1:1 complex formed between the first electronically excited state and the ground state of the same molecule is called an excimer. Essentially, it is a dimer that is stable only in an electronically excited state. It is represented as M* + M → (MM)*, where M and M* are the ground and the excited-state molecules, respectively. An important aspect of dimer or excimer formation is the indication that the interaction between the two participating molecules is strong and this interaction strongly depends on the distance of separation. The dimer/excimer formation could be due to increased local concentrations or the fact that the intermolecular interactions are strong enough to induce aggregation even at low coverages.

Binding of AMAC in the BAZrP galleries (phosphate concentration of 0.005%), at low loadings, produces uniform distribution of the guest molecules as evidenced by the monomer emission ($<10\ \mu M$, Fig. 15A, curve 1) [19]. At higher

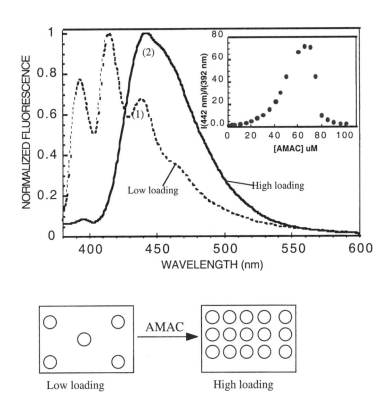

Figure 15 Fluorescence emission spectra of AMAC ($5\ \mu M$) at different loadings of the host (BAZrP): (1) 0.01% phosphate, (2) 0.001% phosphate. Inset: The change in the ratio of the emission intensities at 442 nm to 392 nm with the change in the AMAC concentration at fixed phosphate (0.005%) concentration. (From Ref. 19. Copyright 1993 Elsevier Publications.)

loadings (>10 μM of AMAC), ground-state interactions between the bound an-thryl chromophores are enhanced due to high local concentrations and clustering begins to set in (Fig. 15B). This results in broadening of the absorption spectrum at the red edge, and this is accompanied by a decrease in the monomer emission at 390 nm. The increased local concentrations of AMAC also result in a new broad emission with a maximum at ~440 nm (Fig. 15A, curve 2). This new band was as-signed to the anthryl excimer emission based on literature reports of similar red-shifted emission. The formation of dimers and excimers is dictated by the local concentration of AMAC, as shown in the inset of Fig. 15A. The ratio of excimer to monomer intensities increases until 70 μM AMAC (0.005% BAZrP) and then rapidly decreases. AMAC concentration of ~70 μM corresponds to the maximum loading of AMAC in the phosphate galleries, and any further increase in AMAC concentration results in the increase in the free-probe concentration in the bulk aqueous phase. Thus, the ratio of excimer to monomer emission intensities de-creases after the binding saturation, resulting in a near bell-shaped curve with a rapid fall on the right side (inset, Fig. 15A). Such excimer emission from AMAC or from its sister probe, PBAC, was not observed when these cations bind to DNA or proteins and this illustrates the differences between BAZrP and these other hosts [33]. Thus, in the case of BAZrP, the guest molecules interact with each other strongly, at high loadings, and such a strong interaction suggests the forma-tion of supramolecular assemblies of the guests.

An important requirement for excimer formation in the galleries is the prox-imity of the molecules, enabling a bimolecular encounter within the short lifetime of the singlet excited state (~10 nsec in the case of anthryl chromophores). An or-derly arrangement (parallel geometry) of the participating molecules is necessary for excimer formation and such order is to be expected in these supramolecular as-semblies. Dimer and excimer formation from anthryl chromophores is difficult to observe in the solution phase due to the high concentrations of the guest needed to interact with the short-lived anthryl-singlet excited state. This requirement is met by the formation of supramolecular assemblies of AMAC or PBAC in the gal-leries of BAZrP [14].

Enhanced excimer emission was also observed from PBAC bound to α-ZrP [20]. Excimer formation from pyrene is well known in aqueous solutions [54]. As in the case with AMAC, excimer formation is increased with PBAC concentration (Fig. 16) due to increased local concentrations but with two significant differ-ences. Hydrophobic interactions between the pyrene molecules favor the aggre-gation of PBAC even at moderate coverages and the PBAC singlet excited state is much longer lived (~200 nsec) than that of AMAC (~10 nsec); these factors, in turn, promote excimer formation even at low loadings. The broad, red-shifted flu-orescence band with a peak centered around 470 nm, characteristic of the pyrene excimer emission, is evident in Fig. 16. Rapid formation of the excimer at low coverages is also evident from the plot of the ratio of emission intensities at

A

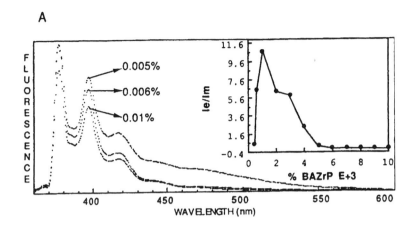

B

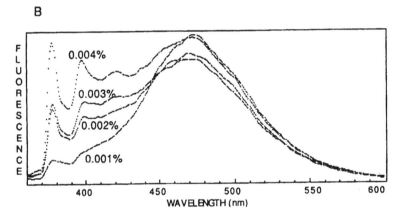

Figure 16 Emission spectra of PBAC (2.6 μ*M*) at (A) low and (B) high concentrations of BAZrP;λ_{ex} = 340 nm. The inset shows the ratio of excimer to monomer emission as a function of phosphate concentration at 3.5 μ*M* PBAC. (From Ref. 20. Copyright 1993 Elsevier Publications.)

470 and 390 nm as a function of BAZrP concentration (inset, Fig. 16A). At higher concentrations of BAZrP, the available surface area for the chromophores is larger and the monomer fluorescence dominates.

The important difference between AMAC and PBAC is that the maximum excimer emission in the latter is observed at much lower loadings than the former. Excimer emission is observed even upon exciting the monomers of PBAC in the BAZrP galleries. The excimer emission from PBAC intercalated into the galleries can arise from two distinct pathways, one being the direct excitation of a ground-state dimer as in the case of AMAC, and the other where excited monomers interact with a ground state. Thus, the excimer formation via the dynamic and static mechanisms distinguishes the two pathways. In supramolecular assemblies, the two pathways compete effectively, whereas in the liquid phase, the static mechanism competes at high concentrations or when the probes aggregate.

3. Binding Inside/Outside

Binding of the guests in the galleries or on the outside of the stacks of lamellae can be distinguished in fluorescence quenching experiments. The binding in the galleries is confirmed from the extensive protection of the probe emission from an aqueous-phase anionic quencher such as iodide. The extent of quenching of excimer emission by iodide is significantly reduced on intercalation into the BAZrP galleries (Fig. 17). This is evident from the dramatic decrease in the Stern–Volmer slope (K_{SV}) when all the probe molecules are bound in the galleries. Thus, K_{SV} decreases from 144 dm^3/mol (in aqueous solution, free probe) to 120 dm^3/mol (at 0.001% BAZrP, outside binding dominates due to limited gallery space), whereas little or no quenching is observed at 0.01% BAZrP (K_{SV} = 5 dm^3/mol) when all the probe molecules are accommodated at the galleries. Thus, the intercalated an-

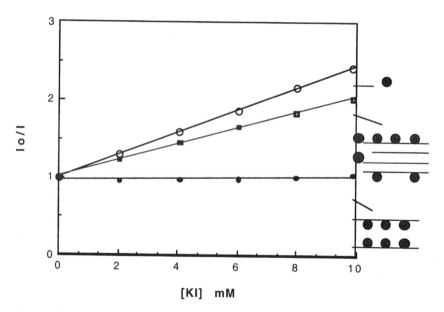

Figure 17 Stern–Volmer plots for the quenching of AMAC excimer emission (excitation at 374 nm and detection at 440 nm) by potassium iodide. ○: No BAZrP; ■: 0.001% BAZrP; ●: 0.01% BAZrP. (From Ref. 19. Copyright 1993 Elsevier Publications.)

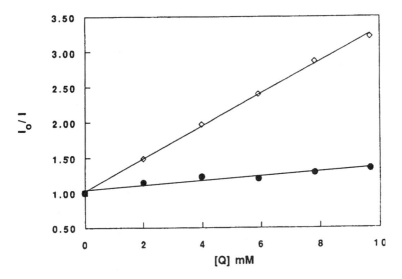

Figure 18 Quenching of PBAC (2.6 μM) monomer emission by KI in the absence ($\square$) and in the presence ($\bullet$) of 0.001% phosphate, monitored at 378 nm (λ_{ex} = 340 nm). (From Ref. 20. Copyright 1993 Elsevier Publications.)

thryl chromophores are not available for the water-bound iodide anions and, hence, the emission is not quenched by iodide to a significant extent.

Quenching of the monomer fluorescence by iodide was used to distinguish the excimers formed at the phosphate surfaces from those in the bulk aqueous phase. Monomer emission from PBAC in the aqueous phase is quenched efficiently by iodide, whereas the negatively charged BAZrP galleries protect the bound PBAC and quenching is inhibited (Fig. 18).

4. Excitonic Interactions

Formation of the supramolecular assemblies of organic chromophores at the galleries of BAZrP is also evidenced from the strong excitonic interactions that the guest molecules experience on binding (Fig. 19); ethidium bromide (EB) serves as an excellent example. Ordered orientation of EB or acridinium hydrochloride (ACR) in these assemblies facilitates excitonic interactions among the guest molecules, as evidenced from new absorption bands [55]. In systems where the molecules are stacked in parallel geometry, such as molecular aggregates, LB films, and polymers, interactions between the transition dipole moments of a given pair of molecules causes splitting (Davydov splitting) of the molecular energy levels (Fig. 20). The strength of this interaction depends on the relative orientations of the two molecules (orientations of the transition dipole moment vec-

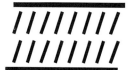 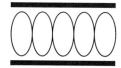

Figure 19 Ordered packing of guests as a bilayer or as a monolayer due to strong intermolecular interactions (shown is an idealized picture) can result in strong excitonic bands in the absorption spectra.

tors) and the distance between them. The intervening medium is also important in this coupling mechanism.

The negatively charged ionic surfaces of BAZrP can reduce the electrostatic repulsion between bound cations and bring them to proximity. This leads to ordered clustering of the guest molecules at the galleries, and ordered clustering of EB is evidenced from the excitonic coupling between EB chromophores. The absorption spectrum of EB ($> 10\ \mu M$) changes dramatically upon binding to BAZrP (0.008%) and shows peaks at 420 and 600 nm (Fig. 21), instead of its normal absorption at 478 nm. The appearance of new bands (420 and 600 nm) at low loadings as well as at high loadings is independent of EB concentration and this independence is ascribed to the clustering of EB due to its high hydrophobicity. This result is surprising because such clustering was not observed for AMAC, comparable in hydrophobicity to EB, at low loadings.

Random orientations of the chromophores in the galleries do not result in such excitonic splitting. The changes produced in the absorption spectrum of EB, upon binding to BAZrP, were used to construct its binding isotherm (Fig. 22) to test it against the centrifugation data presented earlier (Fig. 14). The stoichiometry for binding calculated from these data agrees with that obtained from centrifu-

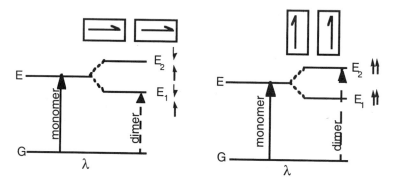

Figure 20 Exciton splitting and geometry (orientation of transition dipoles for absorption) of an ordered array of planar molecules.

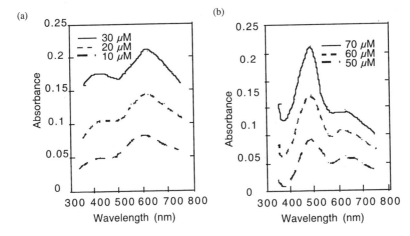

Figure 21 Changes in the absorption spectrum of ethidium bromide, observed at (a) low and (b) high guest loading, upon intercalation into 0.008% BAZrP. The splitting of the absorption spectrum is due to strong excitonic interactions between EB molecules as a result of the self-assembly of the chromophores. (From Ref. 55. Copyright 1996 Elsevier Publications.)

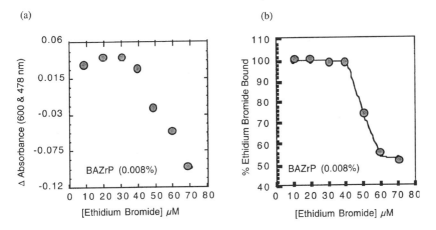

Figure 22 Binding curves for EB with BAZrP (0.008% phosphate). (a) Binding curve obtained from the absorption titration experiments. The difference between absorbance at 478 nm, due to the free form, and 600 nm, due to the bound form, is calculated as delta absorbance (Y axis). (b) Curve constructed from the centrifugation studies in which the amount of bound EB is determined by separating the bound probe from the free probe by centrifugation. The inflection point, which corresponds to the transition from the bound form to the free form, is almost the same in both the cases. (From Ref. 55. Copyright 1996 Elsevier Publications.)

gation studies, suggesting that it is binding that induces the excitonic interactions, even at low loadings. The EB molecules form pools in the galleries and these pools expand as more guests are added.

In contrast to the results obtained with EB, weak excitonic interactions in ACR intercalated at the BAZrP galleries result in broadened absorption spectrum only at high loadings (Fig. 23), suggesting that the intermolecular forces are relatively weak for this chromophore [55]. The fluorescence spectra of ACR intercalated into BAZrP show a large blue shift from 480 nm (free form) to 450 nm (bound form) (Fig. 24), and the fluorescence intensity also decreases monotonously with the concentration of ACR, ascribed to self-quenching. This blue shift of ACR fluorescence is also in contrast to the 20-nm red shift observed for EB bound to BAZrP.

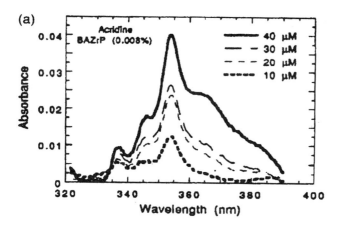

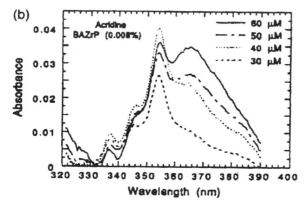

Figure 23   Absorption spectra of ACR in the presence of BAZrP (0.008%) (a) at low guest concentrations and (b) at high guest concentrations. (From Ref. 55. Copyright 1996 Elsevier Publications.)

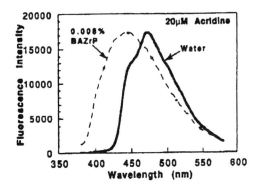

Figure 24 Fluorescence spectra of 20 μM ACR in the absence (——) and presence (----) of 0.008% BAZrP. The spectra of bound ACR intercalated into BAZrP is multiplied by a factor of 13.3. (From Ref. 55. Copyright 1996 Elsevier Publications).

The stronger excitonic interaction in EB assemblies than that of ACR or AMAC is apparently due to a greater hydrophobic surface area of the former, as estimated from computer modeling studies (MacSpartan). Such increased hydrophobic surface is not expected from their structures (three six-membered ring systems); it also results in an enhanced entropic contribution to the binding energy when the probe is transferred from the aqueous phase to the interior of BAZrP, where there is little or no water. Therefore, the formation of these supramolecular assemblies may indeed involve a large entropic component, but this needs to be demonstrated experimentally.

5. Linear Dichroism Studies

Even though the electrostatic host–guest interactions favor the intercalation of aromatic cations into the galleries of BAZrP, the alkyl chains on the guest can influence the binding affinity and orientation of the guest cation in the interlayer region. The linear dichroic (LD) spectra of n-alkylated acridine orange (AO–C_n; n = number of carbon atoms in the alkyl chain) intercalated into γ-ZrP was found to depend strongly on the chain length. A decrease in the absorbance of the probes at 498 nm and a concomitant increase in absorbance at 450 nm was observed upon increasing the concentration of the dye, suggesting the formation of dye aggregates, at the expense of free monomers. The LD of the bound dye decreased with the chain length from 1.3 for $n = 0$ to -0.4 for $n = 9$ (Fig. 25). This strong dependence of LD signal on the chain length clearly demonstrates the packing of the guests with short chains and considerable disorder as the chain length is increased. These differences were suggested to be due to the changes in the orientation of the transition dipoles of the intercalated dyes with the increase in the chain length and loading [56].

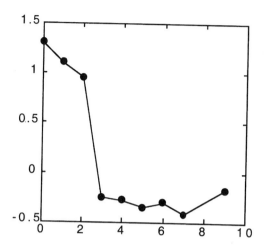

Figure 25 Dependence of the reduced linear dichroism, ρ, on the length of the alkyl chain, *n*. (From Ref. 56. Copyright 1990 The American Chemical Society.)

6. Direct Intercalation

The binding of amphiphilic dyes, such as thionine and methylene blue, into the galleries of prestine α-ZrP results in a monolayer of guests at the galleries [57]. Similar to BAZrP, the α-ZrP interlayer appears to provide an excellent medium for the self-assembly of these dyes. Ordered packing of these dyes in the supramolecular assembly facilitates excitonic coupling between the dye molecules, and the reflectance spectra of thionine and methylene blue intercalated into α-ZrP showed signatures of excitonic coupling between the chromophores [57].

Small chromophores, such as *n*-alkylquinolinium, and isoquinolinium cations can also be intercalated into the galleries of α-ZrP. In most cases, this did not bring about any change in the fluorescence spectra from that observed in solution (acetonitrile). The fluorescence decays of the intercalated compounds were, however, nonexponential and contained a very short-lifetime component. High loadings resulted in self-quenching of the fluorescence and this contributed to the shorter-lived component for the intercalated dyes [58].

Larger molecules, such as methylene blue, rhodamines, and so forth, face a kinetic barrier for intercalation into the galleries of α-ZrP, and intercalation of such chromophores is accelerated in the presence of alkyl amines, as described earlier. The dye-intercalation process involves two steps. First, the amines are intercalated into α-ZrP, increasing the interlayer spacing and then, the intercalation is achieved by substituting the alkylamine (butylamine in the case of methylene blue) with the guest [54]. The flexible chains of the alkylamines are easier to intercalate than the more rigid guests, such as methylene blue.

Intercalation of rhodamine was enhanced by preintercalation of *n*-propylamine into the galleries of α-ZrP [60]. The emission spectrum of the intercalated rhodamine was centered around 675 nm, exhibiting a blue shift in comparison to that of the dye at 690 nm in the solid state. This large blue shift was likely due to the aggregation of the dye molecules in the galleries of α-ZrP, and such aggregation in solution will only occur at high concentrations ($10^{-2}\,M$).

Triphenylmethane dyes such as crystal violet (CV^+), intercalated into the ethanol–α-ZrP complex, showed enhanced dye–dye interactions, ordered in a mutually perpendicular orientation and, hence, a strong bathochromic shift of the absorption band was noted. Two different forms of crystal violet [viz. crystal violet chloride or (CV)Cl and crystal violet acetate or (CV)Ac] were investigated. The former cation was adsorbed on the external surface of the host, whereas the latter is intercalated into the galleries of α-ZrP. X-ray diffraction patterns showed that the basal spacing varied with the increase in dye loading. At low loading, the basal spacing was 11.5 Å, and this increased to ~ 22 Å at high loading. Molecular modeling of dye intercalation was attempted to determine the orientation of the guests intercalated into the α-ZrP galleries, and this showed that at low loadings, the crystal violet cations are likely to be parallel to the α-ZrP layers. At high loadings, the cations are more likely to be mutually perpendicular inside the galleries [61].

The absorption spectra of CV^+ adsorbed on the external surface and that intercalated into the galleries were vastly different from one another. These were also different from that of the cation in water. CV^+ in water showed a band with maximum at ~ 590 nm. At high dye loadings, the absorption band in the red region became partly structured due to a shoulder around 550 nm, and this new band was assigned to the dimer, whereas the band at 590 was assigned to the monomer (Fig. 26). The formation of CV^+ dimers was due to a strong dispersive interaction between the π–electron system and the screening of the electrostatic repulsion between CV^+ ions by the high dielectric constant of the water [62].

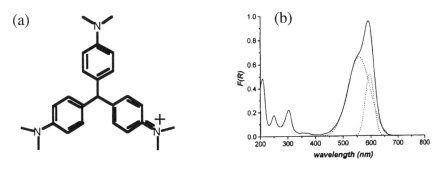

Figure 26 (a) Structure of CV^+. (b) Absorption spectrum of 10 μ*M*/L CV^+ dissolved in water is shown by the solid line. Deconvolution of the spectrum into the monomer (around 590 nm) and the dimer (around 550 nm) bands is shown by the dotted lines. (From Ref. 61. Copyright 1997 The American Chemical Society.)

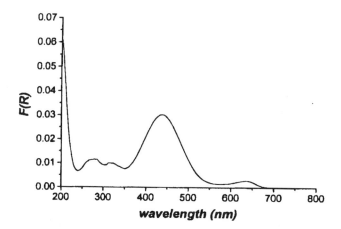

Figure 27 Diffuse reflectance spectra of 1.4 μM crystal violet deposited on the external surface of α-ZrP from aqueous solution. The strong absorption around 430 nm is assigned to CV^{3+}. (From Ref. 61. Copyright 1997 The American Chemical Society.)

Diffuse reflectance spectra (DRS) of CV^+ (1.4×10^{-6} mol/L of guest per gram of α-ZrP) adsorbed on the external surface of α-ZrP differed from that of the cation in water in three ways: (1) The monomer and dimer bands, both, disappeared, (2) two new bands with maxima at 430 nm and 640 nm appeared, and (3) the DRS spectrum in the blue region (< 400 nm) also drastically changed. The appearance of new bands was likely due to the protonation of the dye by the acidic P—OH groups of α-ZrP layers (Fig. 27) and the strong band around 430 nm was assigned to the doubly protonated form (CV^{3+}).

Increase in the loading of CV(Cl) brought about further changes in the DRS spectrum and the 640-nm band broadened, becoming more intense than the 430-nm band (Fig. 28). The 640-nm band could be deconvoluted to three bands: the CV^{2+} band (maximum around 640 nm), CV^+ band (maximum around 590 nm), and the dimer band [$(CV^+)_2$, maximum around 550 nm].

The DRS spectra of CV(Ac) (at low loadings) intercalated into the galleries of α-ZrP had patterns similar to that of CV(Cl) and had two bands in the visible region: one around 430 nm and the other around 640 nm. The increase in dye loading, due to which the CV^+ cations being intercalated in a perpendicular geometry, in the galleries of α-ZrP resulted in a gradual loss in structure and an increase in absorption in the 510–550-nm region (Fig. 29). Because this region corresponded to the absorption of CV^+ dimers in water, this adsorption was associated with higher aggregates [i.e., stacked $(CV^+)_n$ species, or supramolecular assemblies of CV^+, within the galleries of α-ZrP].

Such aggregate formation resulted in a change in the pK_a of the dye from 0 to 1. The absorption spectrum of the protonated form showed maxima around 430

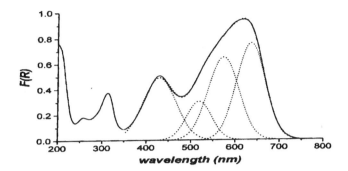

Figure 28 Diffuse reflectance spectra of 70 $\mu M/L$ crystal violet deposited on the external surface of α-ZrP from aqueous solution, exhibiting the deconvolution of CV^{3+} (around 430 nm), CV^{2+} (~ 640 nm), CV^+ (around 590 nm), and $(CV^+)_n$ aggregates (~ 510 nm). (From Ref. 61. Copyright 1997 The American Chemical Society.)

and 640 nm. On the other hand, interaction between the dye molecules perpendicularly oriented to each other resulted in a single absorption band with a maximum centered around 510 nm [57].

Bromides of 3-cyanopyridine, 4-cyanopyridine, and isoquinoline were also intercalated into the galleries of α-ZrP in a similar manner. The diffuse reflectance spectra of the intercalated cations showed a new band signature of charge transfer between the bromide ions and the pyridinium sites in the solid state [63]. Such charge-transfer interactions indicate the coadsorption of counterions into the galleries. These various examples indicate the formation of supramolecular assemblies with significant interactions between the neighbors of the same or opposite charge and such assemblies may be used for solar energy harvesting and conversion.

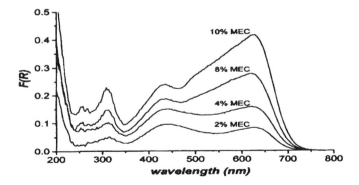

Figure 29 Diffuse reflectance spectra of CV^+ deposited on the external surface of α-ZrP from aqueous solution as a function of loading. (From Ref. 61. Copyright 1997 The American Chemical Society.)

C. Nonlinear Optical Properties of Organic Molecules Intercalated Into α-ZrP

Nonlinear optical (NLO) phenomena imply a nonlinear dependence of the response of the optical system on the intensity of incident light. Two-photon spectroscopy, second and third harmonic generation (SHG and THG), frequency mixing, and so forth are a few examples of such NLO processes. NLO properties are exhibited by noncentrosymmetric molecules that have a large dipole moment in the ground state—a large intramolecular charge-transfer character. These molecules have large hyperpolarizability (β) and observable bulk second-order nonlinearity (χ^2). Such molecules tend to dimerize in a head-to-tail conformation in order to avoid electrostatic repulsion as they form crystals and, hence, exhibit centrosymmetric crystal structures, which diminishes the bulk second-order nonlinearity. Intercalation of such dyes into organic and inorganic host materials could provide an alternate method of organizing the organic chromophores into a noncentrosymmetric arrangement. Such host–guest complexes can then be expected to show useful nonlinear optical characteristics [64]. Extensive research efforts are, hence, garnered toward identifying new (1) nonlinear optical materials and (2) new inert hosts that can efficiently control the molecular orientation of the molecules capable of NLO properties.

Alpha-Zirconium phosphate and its derivatives are some of the established layered materials used as inorganic hosts for such purposes [65]. SHG from a polar azo dye, 4-{4-[N,N-bis(2 hydroxyethyl)amino]phenylazo phenyl-phosph onic acid prepared as a zirconium phosphate–phosphonate multilayer (Fig. 30) is an elegant example of such a design. The second-order nonlinearity of the novel material was found to be of the same order of magnitude as that of $LiNbO_3$, one of the best known inorganic NLO materials. This was made possible because of the high degree of orientation of the chromophore in each layer and, hence, a large β value. In addition to an efficient SHG, this material also exhibits an excellent thermal stability ($\sim$150°C, 3 hr).

Inspired by this success, the effects of molecular rigidity and orientation on the SHG of azo dyes, intercalated into α-ZrP, were also studied. Intercalation of nonpolar azo dyes into the multilayers of α-ZrP had no effect on their SHG properties. Substantially rigid polar azo dyes formed a H aggregate and showed lower SHG efficiency when compared to the flexible analogs. In a multilayer system, the degree of molecular orientation of the dye was also found to depend on the number of layers and the degree of packing of the guest molecules within the layers [66].

Second harmonic generation was observed only from dyes intercalated into γ-ZrP, but not those intercalated into α-ZrP galleries [65c] and this clearly implies that the structure of the host material or its influence on the packing is

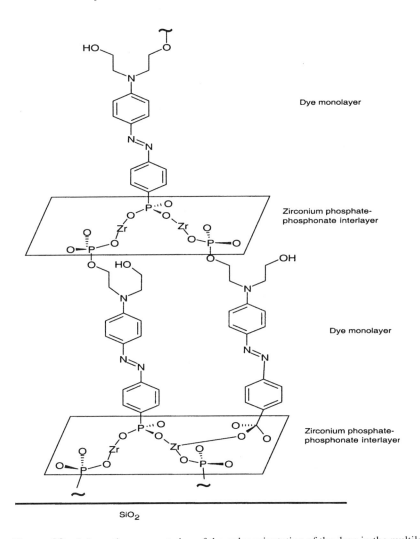

Figure 30 Schematic representation of the polar orientation of the dyes in the multilayers of zirconium phosphate–phosphonate interlayers. This scheme is based on the known structures of Zr(HOPO$_3$)$_2$ and related organic Zr phosphonate salts. (From Ref. 65a. Copyright 1991 The American Association for the Advancement of Science.)

important for the NLO properties. Ion exchange of an organic dye 4-[4-(dimethylamino)-α-styryl]-1-methylpyrydinium (DAMS) with α-ZrP and γ-ZrP gave single-phase materials. SHG efficiency of DAMS-γ-ZrP, at 1.9 μm, was found to be comparable to that of urea, whereas the DAMS-α-ZrP guest–host complex was found to be inactive.

D. Energy Migration/Transfer and Electron Transfer in the Galleries of α-ZrP

1. Use of Supramolecular Assemblies

Efficient singlet–singlet energy transfer and electron transfer can be achieved between chromophores bound to layered materials due to increased local concentrations, preferential orientation of the reactants, and control over the separation of the donors and acceptors. Such a control can also be achieved with crystals, polymers, and Langmuir–Blodgett films where donor–donor energy migration may preceed donor–acceptor energy transfer [2,67]. However, layered materials are easy to prepare, chemically inert, thermally stable, and well suited to design vectoral electron-transfer systems. Efficient energy migration/transfer and electron transfer also have ramifications in efficient degradation of pollutants as well as in supramolecular photochemistry.

In specific cases, the energy migration and transfer rates vastly improved on intercalation into the galleries of α-ZrP. Formation of supramolecular assemblies of the guest chromophores in the inorganic galleries can improve migration of the electronic excitation energy between the donor molecules (antenna effect).

2. Enhanced Excitation Energy Migration and Transfer: Photon Antennas

Ordered assemblies of donors and acceptors at the galleries of BAZrP can give rise to model systems that mimic the light-harvesting complex of natural photosynthetic systems. Efficient singlet–singlet energy transfer and light-induced electron transfer can be achieved between chromophores bound to BAZrP. Singlet–singlet energy transfer between NMAC (donor) and AMAC or PBAC (acceptors) is enhanced when they are intercalated into BAZrP [68]. Fluorescence spectra of NMAC and AMAC with increasing concentrations of the acceptor (AMAC) are shown in Fig. 31. Micromolar concentrations of the acceptor are adequate to quench donor emission. Quenching of the donor fluorescence was accompanied by an increase in acceptor emission (when most of the light is absorbed by the donor), which indicated donor-to-acceptor energy transfer. This was verified by recording the excitation spectrum while monitoring the acceptor emission, and the excitation spectrum clearly indicated peaks that correspond to the donor absorption bands. Therefore, absorption of light by the donor results in emission from the acceptor. Direct light absorption by the acceptor and subsequent emission was minimized by choosing a wavelength at which the acceptor had negligible absorption.

Similar results were obtained with NMAC and PBAC, and in both cases, submicromolar concentrations of the acceptor were sufficient to quench most of the donor emission. In the case of the NMAC–PBAC system, greater than 50%

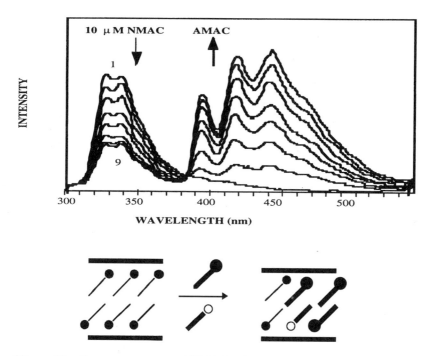

Figure 31 Fluorescence spectra of NMAC and AMAC (0–6.36 μ*M*) in the presence of 0.008% BAZrP. The spectra correspond to 0, 0.99, 1.87, 2.65, 3.35, 3.98, 4.54, 5.5, and 6.36 μM of AMAC. (From Ref. 68. Copyright 1994 The American Chemical Society.)

quenching of the donor fluorescence (50 μ*M*) was achieved by the addition of just 1 μ*M* of acceptor molecules (Fig. 32) with a K_{SV} value of $1.1 \times 10^6 M^{-1}$. Application of the Forster model in two dimentions resulted in Forster radii of the order of 25–30 Å, in reasonable agreement with reported values. The quenching as well as sensitization efficiencies increased with donor concentration.

Donor-to-donor energy transfer in these systems was checked by varying the donor concentration such that even when the donor-to-acceptor ratio was 1000 to 1, more than 50% of the donor emission was quenched by the acceptor. When the donor concentration was 20 μ*M* but at a high loading (0.004% BAZrP), a mere 0.02 μ*M* of PBAC was sufficient to achieve nearly 56% quenching of the donor emission. Due to these large ratios, ground-state complexation between the donors and acceptors cannot account for the large quenching efficiencies. No such rapid/efficient energy transfer was possible in solution at such low concentrations of the donor and acceptor. The strong dependency of quenching on the concentration of the donor (loading) suggests that donor-to-donor energy migration is responsible for the observed enhanced energy transfer.

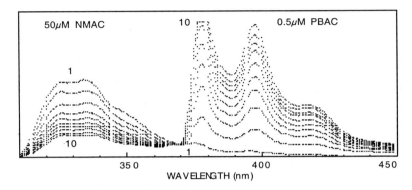

Spectra 1 through 10 contained increasing amounts of BAZrP, at 0.0014, 0.0027, 0.0037
0.0046, 0.0053, 0.006, 0.0066, 0.0071, and 0.0076 by weight.

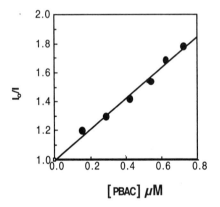

Figure 32   Sensitization of PBAC fluorescence by NMAC (Top). Stern–Volmer plot for
the quenching of NMAC (50 μM) fluorescence intensity by PBAC in the presence of
0.008% BAZrP ($K_{sv} > 1 \times 10^6 M^{-1}$). (From Ref. 68. Copyright 1994 The American Chemical Society.)

     a.   Fluorescein–Rhodamine System.   The method of intercalation dis-
cussed earlier is efficient for cationic, hydrophobic guests but does not work efficiently for guests that are neutral or negatively charged. Several xanthene dyes
cannot be readily intercalated into the galleries of BAZrP because of their hydrophobicity and large size (compared to that of the butylammonium spacer). Intercalation into α-ZrP, as discussed earlier, resulted in dye aggregates on the outside. One way to promote binding of such dyes at the galleries of α-ZrP is to
increase the gallery spacings with long-chain surfactants. This can be accomplished by intercalating surfactants such as cetyltrimethyl ammonium bromide

(CTAB) and dodecyldimethyl ammonium bromide (DDAB) at the ZrP galleries, instead of BA, and use these intercalates for dye binding.

Both CTAB and DDAB readily bind in the galleries of α-ZrP, resulting in the corresponding surfactant–α-ZrP composites CAZrP (CTAB–α-ZrP) and DAZrP (DDAB–α-ZrP) (Fig. 33). The interlayer spacing expands from 7.6 Å in case of α-ZrP to 33 Å for DAZrP and 56 Å for CAZrP, which is large enough to accommodate many organic molecules [64]. These spacings correspond to the formation of a surfactant bilayer in which the surfactant molecules are stacked with an average orientation of $\sim 60°$ with respect to the α-ZrP planes. Large guest molecules, such as fluorescein (Fl) and rhodamine B (RhB), can then be accommodated between the expanded layers. Such incorporation of laser dyes into the galleries of α-ZrP also extends the spectral range for light-harvesting applications

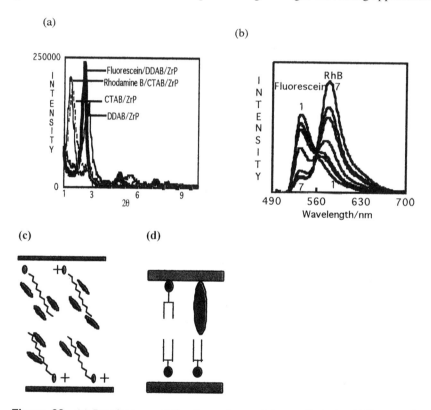

Figure 33 (a) Powder x-ray diffraction patterns of dye–surfactant–α-ZrP composites. (b) Fluorescence spectra of fluorescein–rhodamine B–CTAB–ZrP. The concentrations of fluorescein, CTAB, and ZrP are 0.5 μM, 1 mM, and 1.4 mM, respectively. The mixture is excited at 480 nm. The surfactant–ZrP precursor complexes (SAZrP) are shown in (c) and (d). (From Ref. 69. Copyright 2000 Elsevier Publishers.)

(from that demonstrated with NMAC–AMAC discussed earlier). Measurements of intralayer and interlayer energy transfer between Fl and RhB can also be used as a spectroscopic ruler to probe the intermolecular distances within the supramolecular assemblies [70].

In surfactant solutions, these molecules tend to remain in the hydrophobic domains. Interaction between the dyes and the polar head groups of the surfactants prevents aggregation of these dyes in the α-ZrP galleries. Surfactant–dye and matrix–dye interactions align the chromophores with the charge centers closer to the α-ZrP surface, and the hydrophobic ends are buried in the surfactant hydrocarbon chains. The binding of the dyes to the surfactant–α-ZrP matrix is confirmed from powder x-ray diffraction studies as well as from centrifugation studies [69].

Intercalation of fluorescein and rhodamine B into the surfactant (CTAB and DDAB)-laden α-ZrP galleries results in efficient energy transfer from the Fl donor to the RhB acceptor molecules [69]. Stern–Volmer quenching plots showed that low concentrations (submicromolar–micromolar) of the acceptor were sufficient to obtain 50% quenching of donor fluorescence in α-ZrP (Fig. 34). This is in contrast to the high acceptor concentrations necessary for quenching donor fluorescence in aqueous solution. In the corresponding surfactant media, the quenching was inhibited.

Efficient energy transfer between Fl and RhB, upon binding into the galleries of surfactant-laden α-ZrP, is partly due to the increase in the local concentrations of the acceptor. At high concentrations of the donor, close packing and orderly arrangement of the donor chromophores can bring an acceptor molecule at

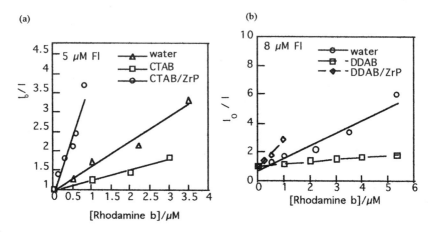

Figure 34 Stern–Volmer plots for the quenching of fluorescein fluorescence by rhodamine B in (a) water, CTAB, DDAB, CTAB–ZrP, and (b) water, DDAB–ZrP. (From Ref. 69. Copyright 2000 Elsevier Publishers.)

a proximal distance to several donors. Under these conditions, excitation energy from multiple donors is accessible to a single acceptor, as in the case of photosynthetic light-harvesting systems.

   b. Uranyl–Europium System.   Photon antennas can also be constructed from metal ions via self-assembly at the galleries of derivatives of α-ZrP. Coordination of the metal ions to functionalities that are directly attached to the metal–phosphonate matrix proved to be effective. An enhancement in energy transfer, as compared to that obtained in solution-phase studies, was observed upon intercalation of uranyl, UO_2^{2+} (donor), and europium, Eu^{3+}, (acceptor), ions by coordination to the carboxyl functions of α-carboxyethyl ZrP (α-ZrCEP) [71]. The efficiency of quenching of the donor fluorescence also increased dramatically with the increase in donor concentration. The Stern–Volmer quenching constant (K_{SV}) estimated for a donor concentration of 2.5 mM was 2627 M^{-1} (Fig. 35). This value is much larger than the value (360 M^{-1}) obtained in solution-phase studies. Because diffusional contribution to the fluorescence quenching of the donor is less probable in the galleries of α-ZrCEP, due to metal coordination, quenching can be attributed to donor–donor energy migration.

   The above studies provide evidence for the donor-to-donor energy migration within the assemblies of the chromophores built from a variety of systems, but they all share a common matrix for assembly. Such systems may be of interest in the construction of artificial light-harvesting antennas [67,72].

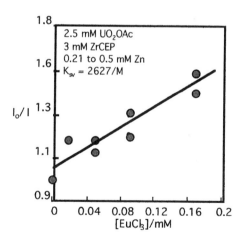

Figure 35   Stern–Volmer quenching of the uranyl donor by europium acceptor when bound to 3 mM ZrCEP. The high value of K_{sv} eliminates the possibility of diffusional contribution to the fluorescence quenching of the donor in the galleries of α-ZrCEP. (From Ref. 71. Copyright 1999 Elsevier Publications.)

E. Preparation and Characterization of Metal Complexes in α-ZrP Galleries

1. Study of Ruthenium(II) Trisbipyridyl Complexes

Due to its stability, long excited-state lifetime, and excellent redox properties, tris(2,2′-bipyridine)ruthenium(II) chloride, $Ru(bpy)_3^{2+}$, is one of the most widely investigated inorganic probes and it is also a good candidate for the construction of artificial light-harvesting systems [73]. $Ru(bpy)_3^{2+}$ absorbs in the visible region; hence, the window for light harvesting can be shifted to visible wavelengths and this metal complex has been extensively investigated for solar energy conversion. Thus, a multitude of energy- and electron-transfer reactions can be carried out using this metal complex [74].

   a. Unusual Photophysical Properties of Ru(bpy)$_3^{2+}$. Upon intercalation into BAZrP galleries, spectroscopic properties of $Ru(bpy)_3^{2+}$ change dramatically from that observed in solution [75,76]. This is in contrast to the great similarities between the absorption and fluorescence spectra of $Ru(bpy)_3^{2+}$ in water and other organized media such as layered oxides [77], zeolites [78], and certain clay materials [79].

   Three types of $Ru(bpy)_3^{2+}$–α-ZrP composites were prepared using different methods [75]. The emission spectrum was strongly dependent on the method of preparation and loading. Incorporation of $Ru(bpy)_3^{2+}$ during the synthesis of α-ZrP in the presence of HF gave a crystalline intercalate. The XRD pattern of the mixture showed a basal spacing of 1.59 nm, indicative of the formation of a monolayer of $Ru(bpy)_3^{2+}$. The emission maxima of $Ru(bpy)_3^{2+}$ for this sample was at 615 nm. On the other hand, adsorption of $Ru(bpy)_3^{2+}$ on the external surface of α-ZrP gave materials with emission maxima at 640–645 nm (Fig. 36) and the emission spectrum of the mixture prepared from either of the above two methods was found to be independent of the excitation wavelength [80].

   Binding of $Ru(bpy)_3^{2+}$ in the BAZrP galleries saturates at a low guest concentration of ~10 μM (Fig. 14B), due to the large footprint of $Ru(bpy)_3^{2+}$. Powder XRD studies confirmed that monolayers of the metal complex were formed at the BAZrP interlayer regions with a spacing of 24.1 Å [81]. Ion exchange of $Ru(bpy)_3^{2+}$ into the galleries of BAZrP dramatically changed its spectroscopic properties [82]. A red shift of the metal-to-ligand charge transfer (MLCT) absorption band from 452 to 481 nm, a blue shift of the emission maximum from 610 nm in aqueous medium to 580 nm, and a fivefold increase in fluorescence intensity were observed on binding of $Ru(bpy)_3^{2+}$ into BAZrP (Fig. 37). The absorption and emission spectral maxima were independent of loading of the metal complex, strongly suggesting the formation of islands or pools of these metal complexes at the interlayer regions of BAZrP, and these pools grow in size, with increased loading.

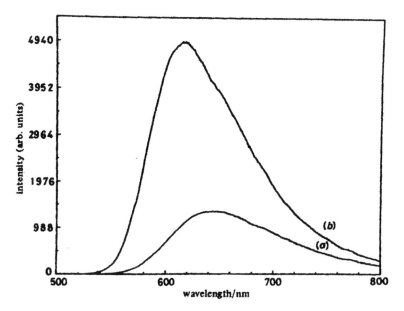

Figure 36 Emission spectrum of Ru(bpy)$_3^{2+}$ (a) intercalated into and (b) adsorbed on the external surface of ZrP. (From Ref. 75a. Copyright 1985 The Royal Society of Chemistry.)

The increase in luminescence intensity and the strong blue shift are suggestive of an interlayer region that is devoid of free water or the water molecules are either frozen or immobile and cannot effectively solvate the initially produced MLCT excited state of the metal complex. The red shift of the absorption spectrum and the

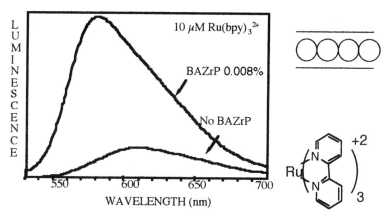

Figure 37 Luminescence spectra of 10 μM Ru(bpy)$_3^{2+}$ with and without 0.008% BAZrP. (From Ref. 76. Copyright 1998 The American Chemical Society.)

blue shift of the emission maximum are in contrast to other studies conducted with this metal complex. Emission intensity increased nearly fivefold and the longest lifetime observed was ~1500 ns when $Ru(bpy)_3^{2+}$ was intercalated into BAZrP [76]. The luminescence intensities and the lifetimes decrease with the loading of the metal complex, indicating self-quenching ($k_q = 7.2 \times 10^9 \ M^{-1}$/sec) (Fig. 38).

This rate constant for self-quenching is larger than expected from the diffusional rate constant suggested for such ions in the galleries of layered materials ($10^{-10} \ cm^2$/sec) [82]. Luminescence quenching of $Ru(bpy)_3^{2+}$ intercalated into the BAZrP galleries by sodium iodide was poor when compared to that in aqueous solutions, confirming that the interlayer regions of BAZrP protect the guests from the quenchers present in the aqueous medium [83]. Interestingly, the addition of potassium iodide or chloride released the metal complex from the galleries, resulting in a decrease in the luminescence from the metal complex due to displacement. Thus, during such quenching experiments, care is to be taken to check if the guest is displaced by the added quenchers.

Refluxing a mixture of $Ru(bpy)_3Cl_2$, zirconium oxychloride, and H_3PO_4 gave a less crystalline material. The emission spectrum of the sample had a maximum at 645 nm, at low loadings. At high loadings of $Ru(bpy)_3^{2+}$, the emission maximum changes as a function of the excitation wavelength, indicating the presence of several forms of the guest. From these changes, two different $Ru(bpy)_3^{2+}$ species were distinguished. Excitation of the mixture at 420 nm gave an emission spectrum with a maximum at 590 nm, whereas excitation at longer wavelengths resulted in a "classical" emission spectrum with a maximum at 615 nm (Fig. 39).

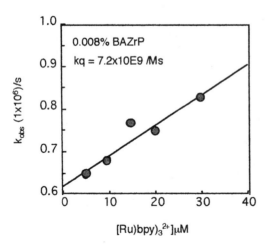

Figure 38 Stern–Volmer plot for self-quenching of the fluorescence lifetime of the intercalated component (0.008% BAZrP) as a function of $Ru(bpy)_3^{2+}$ concentration. k_{obs} is the inverse of the luminescence lifetime. (From Ref. 76. Copyright 1995 The American Chemical Society.)

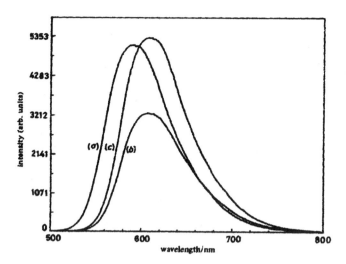

Figure 39 Emission spectrum of the refluxed samples of Ru(bpy)$_3^{2+}$ at different excitation wavelengths: (a) 420, (b) 452, and (c) 480 nm. (From Ref. 75a. Copyright 1985 Royal Society of Chemistry.)

The appearance of the second emission was attributed to (1) strongly bound Ru(bpy)$_3^{2+}$ molecules or (2) a chemical change in Ru(bpy)$_3^{2+}$ during the adsorption process. No significant differences were observed in the diffuse reflectance absorption spectra. A bathochromic shift of the absorption band, a hypsochromic shift of the fluorescence spectrum, and nonexponential luminescence decays are a few other interesting changes observed with these samples.

The absorption and emission spectra of Ru(bpy)$_3^{2+}$ also showed pronounced spectral shifts (compared to that in water) upon intercalating into a sulfonated derivative of layered α-ZrP, zirconium sulfophenylphosphonate, or ZrPS [84]. The XRD and the infrared (IR) spectra of the mixture, on the other hand, indicated that the chemical and structural characteristics of the complex were not significantly affected by the host matrix. The diffuse reflectance spectra, however, revealed three important differences from that observed in aqueous solution. These are that (1) the π–π* band was strongly red-shifted to 317–320 nm from that observed at 285 nm in aqueous solution, (2) the MLCT band also red-shifted, depending on the loading, from 462 to 494 nm, and (c) the ratio of the intensity of the π–π* band to the MLCT band decreased with loading (Fig. 40).

The luminescence spectrum of Ru(bpy)$_3^{2+}$ was also influenced by the microenvironment within ZrPS. The luminescence maximum red-shifted from 604 to 640 nm with loading. The spectral changes have been interpreted to be due to the interactions of the Ru(bpy)$_3^{2+}$ with the phenyl rings of the ZrPS and with bpy rings of the neighboring complexes. The luminescence decay of the metal com-

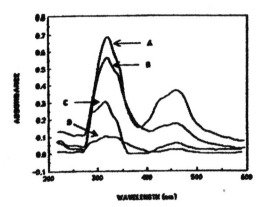

Figure 40   Diffuse reflectance spectra of Ru(bpy)$_3^{2+}$ adsorbed into ZrPS. Percent of sulfonate sites associated with Ru(bpy)$_3^{2+}$ are as follows: (A) 0.43%; (B) 0.22%; (C) 0.11%; (D) 0.02%. (From Ref. 84. Copyright 1990 The American Chemical Society.)

plex in ZrPS was nonexponential and this was analyzed using a distribution of lifetimes assuming a heterogeneous structure with a variety of microenvironments within the ZrPS matrix. The luminescence decay of Ru(bpy)$_3^{2+}$ in ZrPS became faster with loading. The Stern–Volmer plot for quenching of luminescence indicated that although the movement of Ru(bpy)$_3^{2+}$ is restricted in the interlayer spacings of ZrPS, the diffusional quenching mechanism is responsible for the self-quenching process (Fig. 41). The rate constant for self-quenching was found to be $7.0 \times 10^5\ M^{-1}$/sec. The value is four orders of magnitude smaller than the rate constant for self-quenching of luminescence of Ru(bpy)$_3^{2+}$ intercalated into BAZrP (see Fig. 38) [76]. Luminescence quenching by an external quencher, such as methyl viologen (MV^{2+}), showed that the Stern–Volmer plots obtained from the steady-state data and the lifetime data were noncollinear (Fig. 41). A sphere-of-action quenching model, in addition to dynamic quenching, was proposed to explain the quenching of Ru(bpy)$_3^{2+}$ by MV^{2+} in ZrPS [82].

Stable suspensions of α-ZrP and hexylammonium α-ZrP (HexA-ZrP) with Ru(bpy)$_3^{2+}$ are reported [75b]. Powder XRD pattern showed that Ru(bpy)$_3^{2+}$ did not intercalate into α-ZrP and that the cations are located on the external surface. On the other hand, in the case of HexA–ZrP, Ru(bpy)$_3^{2+}$ was adsorbed in the interlamellar regions as well as on the external surface.

The absorption spectrum of Ru(bpy)$_3^{2+}$ in either case showed only minor differences from that in aqueous solution. They are that (1) the maxima of the MLCT band shifted from 452 nm in aqueous solution to 458 nm, (2) the π–π* band also showed a red shift of 3–5 nm as compared to that in aqueous solutions, (3) binding resulted in an increase in the width of the π–π* band by about 10 nm, (4) the ratio of the intensities of the π–π* band to the MLCT band decreased from

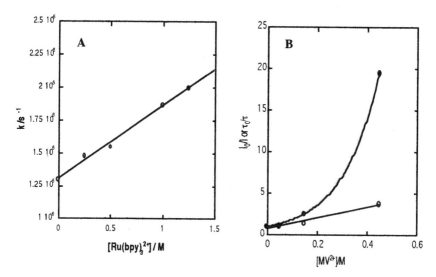

Figure 41 (A) Rate constants for Ru(bpy)$_3^{2+}$ luminescence decay versus concentration of Ru(bpy)$_3^{2+}$ in ZrPS. (B) Stern–Volmer plot for the quenching of Ru(bpy)$_3^{2+}$ by MV^{2+} in ZrPS (I_0/I, open circles; τ_0/τ, closed circles). (From Ref. 82. Copyright 1990 The American Chemical Society.)

2.2 in aqueous solution to 1.7 at high loadings, and (5) the extinction coefficient of the MLCT band at 458 nm decreased drastically with loading.

The emission spectrum of Ru(bpy)$_3^{2+}$ in HZrP and HexA–ZrP had maxima at 620–625 nm. In both cases, quenching by ferricyanide was used to distinguish between the Ru(bpy)$_3^{2+}$ adsorbed on the outer surface and in the interlamellar regions. For HexA–ZrP, the Stern–Volmer plot was nonlinear and this nonlinearity was interpreted as due to the adsorption of Ru(bpy)$_3^{2+}$ at two different places. The Stern–Volmer constant (K_{SV}) for the major component was estimated to be 7200 M^{-1} and this value is of the same order of magnitude as that of Ru(bpy)$_3^{2+}$–kaolin [85], in which Ru(bpy)$_3^{2+}$ is known to bind on the external surface, and both values are much smaller than that observed in aqueous solution (23,000 M^{-1}) [75b].

b.   Electron Transfer in the Galleries of α-ZrP. Intermolecular electron-transfer rates between donors and acceptors can also be increased by orders of magnitude by intercalating these guests at the galleries of α-ZrP. Micromolar concentrations of tris(9,10-phenanthroline)Co(III)$^{3+}$ or Co(Phen)$_3^{3+}$ (an electron acceptor) was found to quench the fluorescence of PBAC (5 μM) intercalated in the galleries of BAZrP (0.008%) (Fig. 42). The corresponding rate constant for quenching by Co(Phen)$_3^{2+}$ is estimated to be ~3 × 10^{12} M^{-1} s while using 200 ns

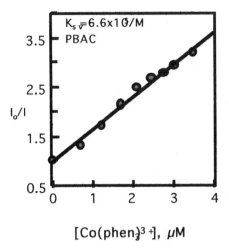

Figure 42 Stern–Volmer plots for fluorescence quenching of PBAC by Co(phen)$_3^{3+}$ in the presence of 0.008% BAZrP. Using 200 nsec as the singlet lifetime of PBAC, the rate constant for quenching is calculated to be ~3 × 10^{12} M^{-1}sec. This is much too fast for a dynamic process and may involve long-range electron transfer. (From Ref. 17. Copyright 1995 Overseas Publishers Association.)

as the singlet excited-state lifetime of PBAC. This is much too fast for dynamic quenching, and the quenching may involve long-range electron transfer [17].

Rapid photoinduced electron transfer was also observed with supramolecular assemblies of Ru(bpy)$_3^{2+}$ (donor) and tris(2,2′-bipyridyl)cobalt(III)$^{3+}$ (Co(bpy)$_3^{3+}$) (acceptor) at the galleries of BAZrP (Fig. 43) [86]. The same donors and acceptors, when immobilized in clays, also showed a rapid intermolecular electron transfer [87]. Luminescence quenching of Ru(bpy)$_3^{2+}$ and Co(bpy)$_3^{3+}$ is diffusion limited in aqueous solutions [88]. The two guests, Ru(bpy)$_3^{2+}$ and Co(bpy)$_3^{3+}$, are similar in size, hydrophobicity, and shape. Hence, they show similar binding characteristics on intercalation into the galleries of BAZrP. The binding of these guests at the galleries of BAZrP results in supramolecular assemblies of donor and acceptor metal complexes. The bimolecular rate constant for the electron transfer, as determined from a time-correlated single-photon-counting method, was found to be nearly independent of solvent viscosity, temperature, and donor loading (Table 2).

The luminescence decay data were analyzed using a number of models, including a long-distance electron-tunneling model, a nonexponential model, and by using a distribution of exponentials. None of these models could satisfactorily fit all the data collected at different loadings, solvent viscosities, and temperatures. However, the biexponential model consistently yielded excellent fits to the data. The rate constants under various conditions are approximately in the range of (1–30) × 10^{11} m^2/mol/s and these are orders of magnitude faster than values an-

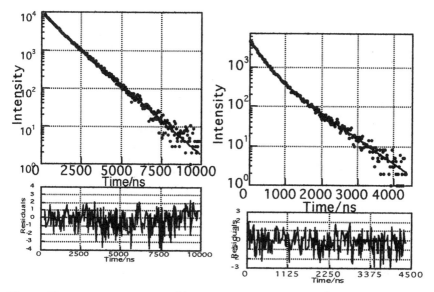

Figure 43 Quenching of Ru(bpy)$_3^{2+}$ (5 μM) emission by Co(bpy)$_3^{3+}$ (1 and 5 μM) while both complexes are intercalated in BAZrP galleries. The luminescence decay data were fitted to a biexponential model over four orders of time and intensity values. Alternative models did not fit the data (see text). (From Ref. 86. Copyright 1998 The American Chemical Society.)

ticipated for diffusional quenching in two dimensions. The activation energy for the above process was also found to be very low and the quenching was independent of temperature up to 77 K. The large value of the rate constant and its invariance to environmental factors suggests a diffusion-independent quenching mechanism. Donor–donor energy migration within the BAZrP matrix followed by electron transfer between the nearest neighbors is a distinct possibility in these

Table 2 Bimolecular Electron-Transfer Rate Constants Measured When the Metal Complexes Were Intercalated in BAZrP (0.008% by Weight)

[Ru(bpy)$_3^{2+}$]	Medium	Temp. (K)	k_q (1) m²/mol/sec	k_q (2) m²/mol/sec
0.5 μM	Aqueous	298	1.6×10^{12}	4.0×10^{11}
5 μM	Aqueous	298	2.1×10^{12}	6.4×10^{11}
10 μM	Aqueous	298	2.2×10^{12}	7.6×10^{11}
5 μM	EG-W (2 : 1)[a]	298	3.0×10^{12}	5.9×10^{11}
5 μM	EG-W (2 : 1)[a]	77	1.8×10^{12}	1.4×10^{11}

Note: The rate constants $k(1)$ and $k(2)$ for the short- and long-lived components, respectively, were calculated as indicated in the text.
[a] EG-W represents a mixture of ethylene glycol and water.

systems [17,86]. This example is also reminiscent of the photoinduced electron-transfer reactions in naturally occurring photosynthetic apparatus.

2. Other Examples With α-ZrP and Its Derivatives

a. Viologens.   Layered zirconium viologen phosphonates (ZrPV) exhibit interesting photoinduced charge-separation reactions [84]. Microcrystalline samples of chloride and bromide derivatives were prepared by the direct reaction of Zr^{4+} with the appropriate biphosphonate salt. The iodide derivative, on the other hand, was prepared by mixing Zr^{4+} with tetraphenylborate salt of the viologen–bisphosphonate in the presence of a large excess of KI. X-ray diffraction patterns indicated that all of these materials are well ordered and the halide is most likely incorporated in the galleries. The proposed structure of these materials is that the viologen groups are arranged between the inorganic zirconium phosphonate layers, with the halide counterions present near the surface of the lamellae (Fig. 44).

The relative rates of photoreduction of ZrPV(X), on photolysis of the samples at 300 nm and monitoring the transmittance at 632 nm, were found to be in

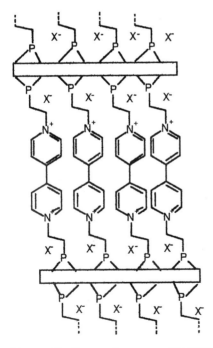

Figure 44   Proposed structure of ZrPV(X). Bars represent the phosphonate oxygen and zirconium atoms. (From Ref. 89b. Copyright 1993 The American Chemical Society.)

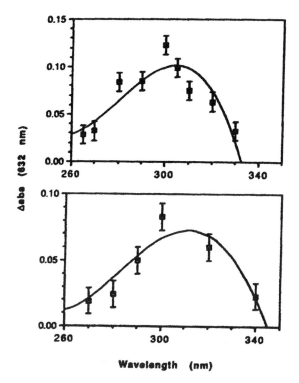

Figure 45 Photoaction spectra for powdered samples of ZrPV(Cl) (top) and ZrPV(Br) (bottom). (From Ref. 89b. Copyright 1993 The American Chemical Society.)

the order Cl > Br > I. The photoaction spectra for ZrPV(Cl) and ZrPV(Br) showed a λ_{max} for the photoreduction of viologen at ~300 nm. The photoaction spectra also shifted as the halide was varied (Fig. 45), indicating that the primary photoprocess involved halide to viologen charge transfer [84b].

ZrPV(X) compounds also exhibited photoluminescence, with a maximum around 340 nm, upon irradiation at 285 nm. The excitation curves and the luminescence lifetimes of all the three materials were similar, and the magnitude of the lifetime (20 ns) suggested that the emission originates from the singlet state. These results, in many aspects, were consistent with that observed with methylviologen intercalated into clays [90].

Similar photochemistry was also observed with thin-film samples, as well as porous versions of these phosphonates [41]. The quantum yield for the formation of charge-separated species in the thin films was reported to be 0.15. The changes observed in the electronic absorption spectra of the samples, with time, were suggested to be due to the formation of both reversible and irreversible charge-separated states. The abstraction of the hydrogen atom by photochemically formed halide radicals, followed by structural rearrangements, resulted in the formation of irreversible products [84b].

b. Substituted Porphyrins Porphyrins are chlorophyll analogs and are of interest due to their catalytic, conductive, and photoactive properties and they have been tested for applications in photodynamic therapy. Hence, intercalation of layered host structures with porphyrins has been widely attempted in an effort to control the attractive properties of these molecules. Direct intercalation of porphyrins into α-ZrP was not successful, but intercalation of aminophenyl (TAPP)- and pyridinium (TMPyP)-substituted porphyrins into α-ZrP was accomplished by exchanging them into the *p*-methoxyaniline (PMA) preintercalated compound $Zr(O_3POH)_2 \cdot 2PMA$ [92].

In the case of materials derived from PMA–ZrP using *p*-H$_2$TAPP, x-ray diffraction patterns of the final product showed an interlayer spacing of 17 Å (including the phosphate layer thickness of 7.6 Å), corresponding to a guest thickness of ~ 9.4 Å. The spacings indicate that the guest molecules are intercalated into the galleries, but these spacings are smaller than the edge-to-edge distance (the distance between the amines on adjacent phenyl groups) of 15 Å in the porphyrin. The observed interlayer spacing of 17 Å then implied either that the porphyrin planes are at ~ 45° angle with respect to the phosphate planes (Fig. 46a) or they form a bilayer in which the prophyrins are roughly parallel to the phosphate layer (Fig. 46b). Intercalation of *p*-H$_2$TAPP into PNA–ZrP by contacting over several weeks indicated an interlayer spacing of 24 Å, and this corresponds to a gallery height of 16.4 Å. Such a large gallery height is ascribed to a bilayer in

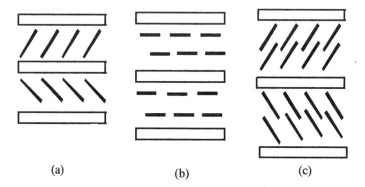

(a)   (b)   (c)

Figure 46 Illustration depicting (a) a canted porphyrin monolayer, (b) a porphyrin bilayer parallel to the matrix, and (c) a canted porphyrin bilayer. (From Ref. 92. Copyright 1993 The American Chemical Society.)

which the porphyrin molecules are canted relative to the host layers, as shown in Fig. 46c. Intercalation of o-H$_2$TAPP into PMA–ZrP on the other hand, resulted in compounds of multiple phases. X-ray diffraction patterns revealed a predominant phase with an interlayer spacing of 19–20 Å, corresponding to a gallery height of ~ 12–13 Å. The arrangement of the porphyrins, in this case, was suggested to be similar to that observed for PMA intercalates (Figs. 46a and 46b) [92].

IV. PROTEIN ASSEMBLY AT α-ZRP AND OTHER LAYERED MATERIALS

So far, intercalation and organization of small molecules/ions has been discussed. We now turn our attention to the intercalation of large, biological molecules such as proteins. The extensive use of proteins in the binding and organization of pigments in the light-harvesting complex has already been discussed. Such protein assemblies, if formed at the galleries of layered materials, can be used in the construction of photon antennas, biosensors, medical implants, enzyme reactors, biomedical devices, and novel biocomposite materials [93].

An appropriate host for proteins should retain the protein in the porous medium, but allow the diffusion of substrates/ligands in and out of the matrix [94]. Proteins have been entrapped at the surfaces of phospholipids, self-assembled monolayers, polymer matrices, sol–gels, and hydrogels [95]. Covalent immobilization of proteins on nylon, alumina, silica gel, cellulose, and controlled pore glass (CPG) has also been possible [96,97]. Layered zirconium phosphates are convenient for the organization of proteins at the interlayer regions for several reasons. However, the small gallery spacings of α-ZrP, BAZrP, or Hex–α-ZrP, with respect to the diameters of even small proteins, present severe kinetic barriers for the intercalation of proteins. Forceful intercalation at high temperatures, long times, and extreme pH's resulted in protein denaturation and partial intercalation.

This problem has been circumvented, recently, by exfoliating the lamellae

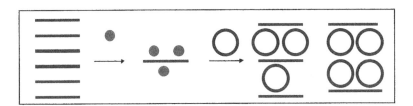

Figure 47 Exfoliation of the ZrP plates with tetrabutyl ammonium ions and subsequent binding of proteins at the α-ZrP galleries. Protein binding results in a monolayer or bilayer packing of the proteins and these can be readily distinguished from XRD data. Binding on the outside of the stacks can also be distinguished from that in the galleries.

and then reassembling the layers, subsequent to binding of the proteins on the α-ZrP surfaces (Fig. 47) [98]. This method can be applied to bind guests of various sizes and shapes, where the size of the guest will no longer be a limiting factor. Armed with this new approach, a number of proteins have been tested for their affinity toward these matrices. Due to the high charge density of α-ZrP (one charge per 5Å × 5Å), electrostatic interactions are expected to dominate the binding of proteins to α-ZrP galleries. A number of proteins are positively charged at pH 7.0 and show a high affinity for these matrices. Whereas α-ZrP is stable at neutral and acidic pH conditions, the corresponding phosphonates are stable even under mildly alkaline conditions. The intercalated proteins are examined using circular dichroism (CD), ultraviolet-visible (UV-Vis), FTIR, and powder XRD techniques. In addition, the biological activities, when appropriate, are also investigated [99].

A. Preparation and Characterization of Protein Assemblies

Proteins such as cytochrome c (Cyt c), met-myoglobn (Mb), met-hemoglobin (Hb), lysozyme (Lys), glucose oxidase (GO), and α-chymotrypsin (CT) bind to the galleries of α-ZrP (at room temperature and pH 7.2) and these proteins also retained their activities and structures upon binding [99].

Spontaneous binding of the above proteins was achieved on exposing the exfoliated α-ZrP platelets to the respective protein solutions. Powder diffraction data of protein–α-ZrP complexes provided a strong evidence for the binding of proteins in the galleries by measuring the interlayer spacings. The spacings expanded from 7.6 Å for α-ZrP to 33, 53, 47, 66, 62, and 108 Å for samples

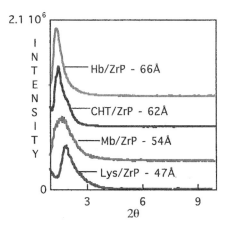

Figure 48 Powder x-ray diffraction patterns of proteins–α-ZrP composites. The interlayer distances increase after protein binding and these correspond to their corresponding hydrodynamic radii. (From Ref. 99a. Copyright 2000 The American Chemical Society.)

Table 3 Interlayer Spacings, Stoichiometries, and Binding Constants Observed for Immobilized Protein–α-ZrP Composites

Protein–α-ZrP	Stoichiometry (μM)	K_b (M^{-1})	d Spacing (Å)	Protein size (Å)
No protein	—	—	7.6	—
Tetrabutylammonium	—	—	18.6	—
Myoglobin	12	2×10^5	54	$30 \times 40 \times 40$
Lysozyme	40	1.33×10^6	47	$32 \times 32 \times 55$
Hemoglobin	14	5.4×10^6	66	$53 \times 54 \times 65$
Chymotrypsin	3	2.5×10^5	62	$40 \times 43 \times 65$
Glucose oxidase	1.1	5.6×10^4	116	$43 \times 51 \times 68$

Note: The protein dimensions are estimates from the known crystal structures and the average of the three dimensions is given in parentheses.

containing Cyt C, Mb, Hb, Lys, CT, and GO, respectively (Fig. 48). The interlayer distances (Table 3) would not have changed if the proteins are binding on the edges or outside of the stacks, and the spacings correlate well with the average diameter of the native proteins reported in the literature [100]. In the case of GO, XRD data revealed bilayer packing of the protein in the galleries. The CD spectra of all the bound proteins are nearly superimposable with those of the native proteins, suggesting that the proteins do not denature upon intercalation into the galleries.

A strong affinity of Cyt c for α-ZrP is evident from its binding constant ($K_b > 10^6\ M^{-1}$) and this is perhaps due to the strong electrostatic interactions between the positively charged lysine side chains of Cyt c and the negatively charged α-ZrP galleries.

B. Redox Activities of Bound Proteins

The redox activity of Cyt c bound to the galleries of α-ZrP was found to be similar to that of the native protein. Dithionite or ascorbate can reduce Fe(IIII)Cyt c to the ferrous form, which can be oxidized by ferricyanide to the Fe(III) form. Retention of this redox property of Cyt c–α-ZrP is indicated by the absorption spectrum of Cyt c–α-ZrP. Although ascorbate and α-ZrP surface are both negatively charged, the addition of ascorbate to Cyt c–α-ZrP rapidly yields the reduced form of Cyt c (Fig. 49A). The absorption at 550 nm, due to Fe(II) form, grows with time and reaches a maximum value after 20 min. The addition of ferricyanide to Fe(II)Cyt c oxidizes it back to the Fe(III) form and restores the original absorption spectrum. These observations highlight the accessibility of bound protein to external small molecules and these immobilized proteins

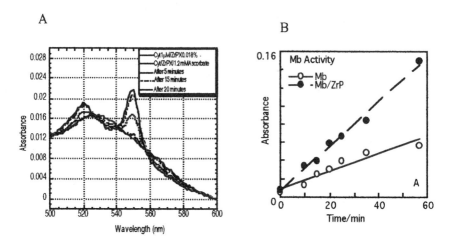

Figure 49 (A) Changes in the absorption spectrum, with time, of 1.1 μ*M* cytochrome *c* bound to α-ZrP (0.018%) on reduction by 1.2 m*M* ascorbate at pH 7.2. (B) Peroxide activity of free Mb and Mb bound to α-ZrP with *p*-methoxyphenol as the substrate. (Part A from Ref. 98. Copyright 1997 The American Chemical Society; Part B from Ref. 99a. Copyright 2000 The American Chemical Society.)

can be used for various applications, including chiral photochemical transformations.

C. Enhanced Catalytic Activities

Retention of the catalytic activities of the immobilized proteins is crucial for their application in biocatalysis and biosensors. The peroxidase activity of Mb intercalated into α-ZrP galleries was probed by the oxidation of 4-methoxyphenol in the presence of hydrogen peroxide. An acceleration of the catalytic activity of immobilized Mb compared to that of free Mb was observed (Fig. 49B). Although the activity of Mb was enhanced, the activities of the immobilized proteins, such as Lys, CT, and GO, are comparable to those in solution (Table 4). These studies clearly establish the utility of layered materials for the binding and self-assembly of large, biologically relevant molecules. However, immobilization of DNA at suitably modified ZrP surfaces is not reported to date. Use of these biological/inorganic composites for phototransformations is yet to be explored, but these materials hold promise to control the chiral selectivities of photoreactions.

D. Enhanced Thermal Stabilities and Protein Refolding

Efficient refolding of met-hemoglobin (Hb) immobilized at the galleries of α-zir-

Table 4 Relative Rates for the Oxidation of Various Substrates With Mb/H$_2$O$_2$ (Bound to α-ZrP Versus Free)

Protein	Amide I/amide II band (cm^{-1})	CD peaks (nm)	Specific activity	K_m	V_{max}
Mb	1651/1530	210/222	3.3×10^{-3} sec^{-1}	1.4 mM	0.08 μM/sec
Mb–α-ZrP	1651/1530	210/222	2.0×10^{-3} sec^{-1}	1.6 mM	0.04 μM/sec
Lys	1650/1521	207/223	11×10^{-3} sec^{-1}	0.5 mM	0.63 μM/sec
Lys–α-ZrP	1650/1521	207/223	9.4×10^{-3} sec^{-1}	0.5 mM	0.63 μM/sec
Hb	1653/1521	210/222	0.013 sec^{-1}	0.103 mM	34 nM/sec
Hb–α-ZrP	1653/1521	210/222	0.038 sec^{-1}	0.112 mM	42 nM/sec
CT	1656/1523	202/231	312 msec^{-1}	—	—
CT–α-ZrP	1658/1523	202/231	400 msec^{-1}	—	—
GO	1645/1538	210/218	2×10^5 msec^{-1}	0.8 mM	0.037 mM/sec
GO–α-ZrP	1645/1538	210/218	2.5×10^5 msec^{-1}	2.5 mM	0.042 mM/sec

Note: Substrate concentrations were the same in all cases (6.6 mM) and the Mb concentration was 5 μM.

conium carboxymethyl–phosphonate (α-Zr(O$_3$PCH$_2$COOH)$_2$·nH$_2$O, α-ZrCMP) was observed recently [99b]. Thermal denaturation of the immobilized protein and subsequent cooling, for example, resulted in the recovery of the secondary and tertiary structures, as well as its activity. The circular dichroism, the infrared absorption spectra, and x-ray diffraction data of the renatured protein indicate that the secondary and tertiary structures of the protein are nearly recovered. The recovery of the peroxidase activity was also established (native Hb–α-ZrCMP, 5×10^{-5} sec^{-1}, renatured Hb–α-ZrCMP, 8×10^{-5} sec^{-1}) to show that the protein undergoes efficient refolding. The extent of recovery to the native structure depended on the nature of support matrix and the recovery was modest for Hb immobilized on α-ZrP [α-Zr(O$_3$POH)$_2$·nH$_2$O] and poor for α-ZrRP (R = CH$_2$CH$_2$COOH). The key role played by the native-like support matrix surface functions in the refolding into a native-like conformation is an interesting result.

Proteins adsorbed at these solids indicate improved thermal stabilities [94c]. Immobilized met-myoglobin (Mb), Hb, and GO, for example, continue to be active after heating at 60°C. Followed by initial thermal denaturation, the adsorbed proteins recovered their activities on continued heating (over days), whereas the free proteins in solution lost activity and did not show any recovery on continued heating. To the best of our knowledge, these are the first examples of activity recovery of immobilized proteins upon continued heating at moderate temperatures. The type of surface functions present on the support matrix also influenced the thermal stabilities. The extent of activity recovery as well as the rate of recovery are in the order α-Zr(CH$_2$CH$_2$COOHPO$_3$)$_2$ > α-Zr(CH$_2$COOHPO$_3$)$_2$

$> \alpha$-Zr(HPO$_4$)$_2$. These studies clearly indicate the important role of surface functions of the support matrix in determining the protein thermal stabilities and their abilities to regain activity subsequent to thermal denaturation.

These aspects of protein adsorption and assembly at layered materials are essential prerequisites for the construction of assemblies consisting of protein–pigment complexes or artificial reaction centers housed in protein assemblies.

V. USE OF OTHER LAYERED MATERIALS FOR ASSEMBLY

A. Layered Double Hydroxides

Photophysical and photochemical properties of porphyrins [101], phthalocyanines [102], and ruthenium complexes [103] intercalated into the galleries of layered double hydroxides (LDHs) have been widely studied. 5,10,15,20-Tetra(4-sulfonatophenyl)porphin (TSPP) was intercalated into the interlayer regions of a Mg–Al LDH by anion exchange. XRD studies showed that the gallery height of LDH expanded from 8.0 to 17.6 Å upon intercalation of TSPP, implying that the porphyrin anions are intercalated with their molecular plane perpendicular to the hydroxide layers (Fig. 50). This is in contrast to the canted geometries of porphyrins intercalated into the layered phosphates described earlier. The absorption spectrum of the LDH–TSPP in the visible region was similar to that of tetrasodium TSPP. The absorbance of LDH–TSPP in the Soret region was, however, different

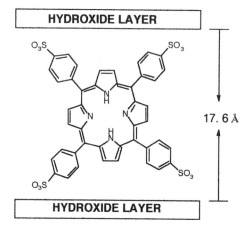

Figure 50   Postulated configuration of TSPP anion upon intercalation into the galleries of LDH. (From Ref. 42. Copyright 1995 The American Chemical Society.)

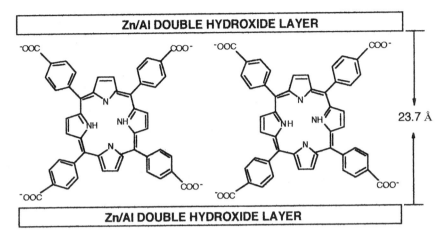

Figure 51 Possible arrangement of TPP-C between the Zn–Al double hydroxide layers. (From Ref. 101b. Copyright 1996 Elsevier Publications.)

from that of the tertasodium TSPP in aqueous solution, suggesting that the TSPP anions are closely packed in the interlayer region.

Intercalation of tetra(4-carboxyphenyl)porphyrin (TPP-C) into the interlayer space of Zn–Al LDH, however, gave a different result [101]. The XRD pattern showed two peaks corresponding to interlayer spacings of 7.8 and 23.7 Å. The 7.6-Å spacing is attributed to the LDH in which the TPP-C molecules were arranged parallel to the plane of the host layers. Expansion of the layer to 23.7 Å indicates that the plane of TPP-C was perpendicular to the plane of the Zn–Al LDH layers (Fig. 51) [101b].

A close packing of the guest molecules was also reflected in the red shift in the absorption spectrum of the TPP-C intercalates. Although the absorption spectrum of TPP-C peaks at 414 nm in water and at 420 nm in dimethyl sulfoxide (DMSO), it shifts to 439 nm upon intercalation into the galleries of LDH (Fig. 52) [101b].

Zinc *meso*-tetrakis (*p*-carboxyphenyl)porphyrin (ZnTPPC), a neutral molecule, can be readily intercalated into the Li–Al LDH–myristate interlayers by replacing the myristate ions [101b]. However, the uptake of the ZnTPPC into this material was minimal. The diffraction pattern showed that, as is the case with other porphyrins, ZnTPPC intercalates with its plane perpendicular to the metal hydroxide layer. The emission spectrum (excitation at 407 nm) of the intercalated guest is similar to that of nonaggregated ZnTPPC in solution (Fig. 53), suggesting that ZnTPPC is solubilized in the LDH in a dispersed form and not as an aggre-

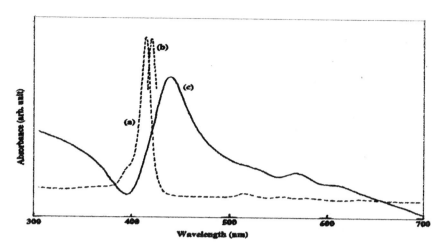

Figure 52   Absorption spectra of TPP-C in (a) water and (b) DMSO and of TPP-C intercalate in ethanol (c). (From Ref. 101b. Copyright 1996 Elsevier Publications.)

gate. Due to increased hydrophobicity of the interlayer space, the intensity of the 604-nm band decreases considerably from that in solution.

The absorption spectrum of ZnTPPC intercalated into LDH shows a small blue shift and a significant broadening of the Soret band compared to that in solution (Fig. 53b). Because the fluorescence spectrum does not give any evidence of aggregates, the changes in the Soret band are attributed to a nonpolar environment surrounding the guest molecule [101d].

Porphyrins are excellent sensitizers for the photochemical reduction of viologen in the presence of sacrificial electron donors. ZnTPPC intercalated into the interlayer spaces of LDH also acts as a sensitizer toward methyl viologen (MV^{2+}) and PVS–EDTA promotes photoreduction of viologen. However, the rate of radical generation for the neutral viologen is higher than that for MV^{2+} (Fig. 54) [101d].

Intercalation of phthalocyanines into the galleries of LDH also occurs via ion exchange [104]. Like the prophyrins, the phthalocyanines are also oriented perpendicular to the cationic LDH. This is evident from the XRD pattern, which shows an increase in the basal spacing in LDH from 4.8 to about 18–25 Å for LDH–phthalocyanine intercalates.

Photophysical properties of ruthenium tris(4,7-dipheny-1,10-phenanthrolinedisulfonate) or $Ru(BPS)_3^{4-}$ intercalated into Mg–Al LDH have also been studied [103]. The absorption spectrum of the intercalate was found to be similar to that of $Ru(BPS)_3^{4-}$ in water; a decrease in the absorption at 285 nm and a small shoulder around 310 nm are the only changes in the absorption spectrum of the in-

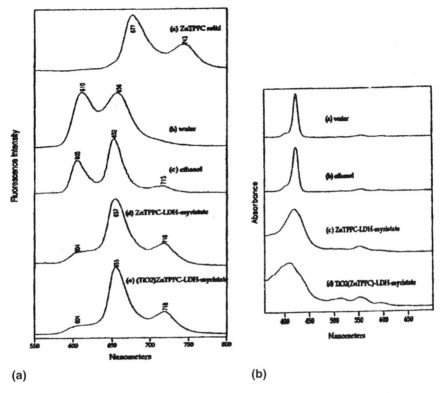

(a)   (b)

Figure 53 (a) Emission spectra of ZnTPPC in various environments (excitation wavelength ~407 nm); (b) electronic spectra of ZnTPPC in various environments. (From Ref. 101d. Copyright 1996 The American Chemical Society.)

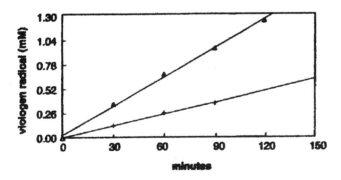

Figure 54 Comparison of the yields of the viologen radical as a function of photolysis time for ZnTPPC in Li–Al LDH with 0.01 M EDTA; (Δ) 0.01 M PSV and (+) 0.01 M HV^{2+}. (From Ref. 101d. Copyright 1996 The American Chemical Society.)

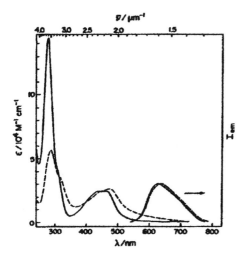

Figure 55   Absorption and emission spectra (excitation = 437 nm) of Ru(BPS)$_3^{4-}$ in water (——) and in LDH (----). (From Ref. 103. Copyright 1987 The American Chemical Society.)

tercalated Ru(BPS)$_3^{4-}$ (Fig. 55). The luminescence spectrum also did not reveal any differences. Luminescence decays of Ru(BPS)$_3^{4-}$ in LDH were, however, multiexponential with a long lifetime of 2.6 μsec and a short lifetime of 0.3 μsec, and this has been attributed to the excited-state self-quenching processes. Self-quenching of Ru(II) emission has been discussed in previous sections.

B. Oxides

1. Photophysical Studies

This general area will be covered only briefly, as many reviews have already appeared recently. The photophysics and photochemistry of organic molecules (aromatics, aza aromatics, diphenylpolyenes, azobenzene, triphenylmethane dyes, thioindigo) adsorbed on metal oxides (alumina, silica, thoria, titania) were investigated by steady-state and time-resolved diffuse reflectance and luminescence methods. In some of these cases, a small red-shift of the fluorescence spectrum was observed (e.g., 1,6-diphenyl-hexatriene) [105]. The fluorescence intensity of these organic molecules was found to be strongly dependent on the energy gap between the electronic energy levels of the adsorbents and the adsorbates. For instance, fluorescence of rhodamine B, fluorescein, pyrene, and anthracene adsorbed onto titania was rapidly quenched. The fluorescence quenching was suggested to be due to energy or (reversible) charge transfer between the guest and the hosts.

Incorporation of chlorophyll and phthalocyanine molecules into the layers of transition metal oxides (V_2O_5, WO_3, MoO_3, etc.) produced photosensitive materials [106]. Being large in size with a long hydrophobic tail, these dyes could not be directly intercalated into the galleries. Instead, the transition metal oxide galleries were pretreated with a surfactant, hexadecyltrimethyl ammonium chloride (C16-TMA). Chlorophyll and phthalocyanine molecules were then embeddd into the hydrophobic galleries. This enabled the guests to self-organize into a periodic array. The phthalocyanine–WO_3 mesoporous material also showed photocatalytic properties [107]. Chlorophyll-related pigments adsorbed onto porous TiO_2 nanospheres were used in the preparation of nanostructured electroceramic materials for solar cells and batteries [108].

Intercalation of tris(2,2′-bipyridine)ruthenium(II) [$Ru(bpy)_3^{2+}$, bpy = 2,2′-

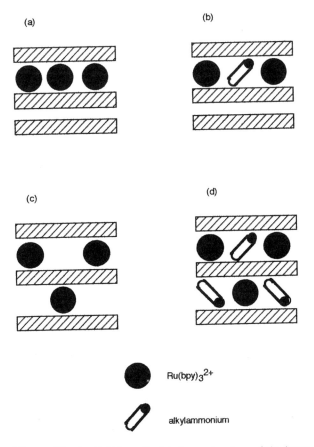

Figure 56 (a–d) Schematic interlayer structure of the intercalation compounds of the layered niobates and titanates with $Ru(bpy)_3^{2+}$ ions. See text for more details. (From Ref. 109b. Copyright 1995 The American Chemical Society.)

bipyridine] ions into layered niobates and titanates was achieved either by direct reaction or by displacement of preintercalated alkylammonium ions [104]. The latter process involved a stepwise intercalation of alkylammonium ions, and intercalation of $Ru(bpy)_3^{2+}$ was then achieved by replacing the alkylammonium ions. X-ray diffraction and elemental analyses showed that four different types of intercalation compounds were formed. These were (Fig. 56) (1) $Ru(bpy)_3^{2+}$ ions directly intercalated into every other layer of the niobate host, (2) $Ru(bpy)_3^{2+}$ ions intercalated along with the alkylammonium ions in every other layer of the host, (3) $Ru(bpy)_3^{2+}$ ions intercalated alone in all interlayer spaces, and (4) $Ru(bpy)_3^{2+}$ ions intercalated in all interlayer spaces with alkylammonium ions.

Such seggregation of the bound guest molecules is an important aspect in these systems. Photoluminescence spectra of intercalated $Ru(bpy)_3^{2+}$ showed a broad, partly structured emission. The luminescence maximum was at 591 nm for the direct intercalated ions (Fig. 57a) and around 597 nm for the guest ions intercalated via preintercalation of alkylammonium ions (Fig. 57b). In all the cases, the

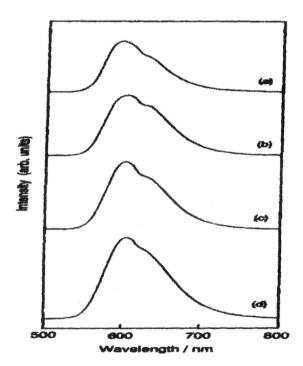

Figure 57 Luminescence spectra of (a) $Ru(bpy)_3^{2+}$–$K_4Nb_6O_{17}$, (b) $Ru(bpy)_3^{2+}$–BA^+–$K_4Nb_6O_{17}$, (c) $Ru(bpy)_3^{2+}$–HA^+–$K_4Nb_6O_{17}$, and (d) an aqueous solution of $Ru(bpy)_3^{2+}$. (From Ref. 109b. Copyright 1995 The American Chemical Society.)

luminescence maximum is blue-shifted from its value in aqueous solution (605 nm) (Fig. 57d). This was in contrast to the red shift of the MLCT bands in the absorption spectra. The blue shift of the luminescence maxima is indicative of a hydrophobic and/or rigid microenvironment of Ru(bpy)$_3^{2+}$ ions, as reported in the case of BAZrP. For the preintercalated compounds, the extent of blue shift of the luminescence maximum and the contribution of the shoulder at 625–630 nm increased with the increase in basal spacing of the samples. Excitation spectra of the samples recorded at two different luminescence maxima were essentially the same and indicated that the two luminescence peaks are due to contributions from the same excited state of the Ru(bpy)$_3^{2+}$ ions.

Luminescence lifetimes of all the four kinds of intercalated Ru(bpy)$_3^{2+}$ ions were biexponential. The lifetime of the short-lived component was 100–300 ns and this is shorter than that in an aqueous solution. Ru(bpy)$_3^{2+}$ intercalated into the interlayers of niobates, therefore, experienced self-quenching, and this was attributed to ground-state aggregation. The luminescence lifetimes of the compounds preintercalated with alkylammonium ions were longer than that of the directly intercalated Ru(bpy)$_3^{2+}$ compounds. Thus, the alkylammonium ions acted as spacers between the host and the intercalated guest ions and these ions are successful in reducing the luminescence quenching. However, host–guest interactions were also operative when the interlayer spacings decreased to distort the intercalated Ru(bpy)$_3^{2+}$ ions and, thereby, quench the luminescence [112].

The fluorescence intensity of crystal violet adsorbed onto alumina was found to be almost independent of temperature [434]. This is in contrast to the

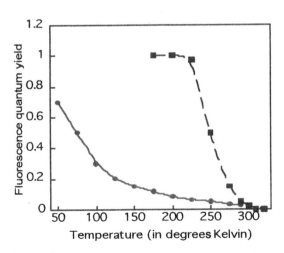

Figure 58 Fluorescence quantum yield of crystal violet as a function of temperature adsorbed on alumina (——) and dissolved in glycerol (- - - -). (From Ref. 43g. Copyright 1986 IUPAC.)

rapid quenching of fluorescence of crystal violet in solution with the increase in temperature (Fig. 58) even though the yield was low for the immobilized sample. The latter phenomenon has been explained on the basis of the twisted intramolecular change-transfer hypothesis [110] and involved a low barrier to the torsional twisting of the phenyl rings in the solution phase. In the case of the intercalated guest, the twisting was inhibited by the host structure.

2. Photoisomerization

Cis–trans photoisomerization of thioindigo and stilbene was significantly reduced on adsorption to alumina, silica, and titania. *Trans*-thioindigo, contrary to its behavior in solution, did not undergo photoisomerization on being adsorbed to these surfaces. The luminescence spectrum of *trans*-thioindigo also showed a nearly 2000-cm^{-1} red shift from that in solution, indicating a strong binding of the guest molecules to the surface. This strong binding acted as an additional activation barrier for the internal mobility (twist around the central double bond) of the photoexcited thioindigo molecules [111]. The low photoisomerization efficiency of *trans*-stillbene adsorbed onto alumina was also explained based on guest–guest interactions. The probability of photoisomerization at room temperature was found to be very small (0.12) when compared to the value in a low-viscosity solvent (0.55).

3. Electron/Energy Transfer and Other Photoreactions

Intercalation of three chromophores (viz., coumarin, fluorescein, and metalloporphyrin) into the galleries of $HTiNbO_5$ yielded well-organized three-component light-harvesting assemblies [112]. The three components were coumarin–poly(allylamine) hydrochloride or Coum-PAH polymer, fluorescein–poly(allylamine) hydrochloride or Fl-PAH polymer, and palladium(II)tetrakis(4-*N*,*N*,*N*-trimethylanilinium)porphyrin, chloride $(PDTAPP^{4+})$ or palladium(II)tetrakis(4-sulfonatophenyl)porphyrin $(PDTSPP^{4-})$. Excitation of Coum-PAH at its absorption maximum (about 418 nm) produced emission from $PDTAPP^{4+}$, with a maximum around 720 nm. The spectral range for light harvesting was, thus, spread over the visible range. The efficiency of the coumarin $\rightarrow$ fluorescein $\rightarrow$ porphyrin energy transfer within the two inorganic layers was about 80% (Fig. 59).

Considerable overlap of the donor fluorescence and acceptor absorbance was the reason for the high quantum yield of energy transfer within the galleries. Interlayer electron transfer to methyl viologen quenched the emission from the $PDTAPP^{4+}$ molecule. Transient diffuse reflectance spectra of the four-component system, acquired 200 μsec after a 532-nm laser excitation, showed maxima at 380 and 600 nm, characteristic of reduced viologen. The four-component energy/electron transfer had an overall quantum yield of 47%. Thus, the intercalated guests formed light-harvesting assemblies that participated in energy transfer and eventual electron transfer across the inorganic spacer layers.

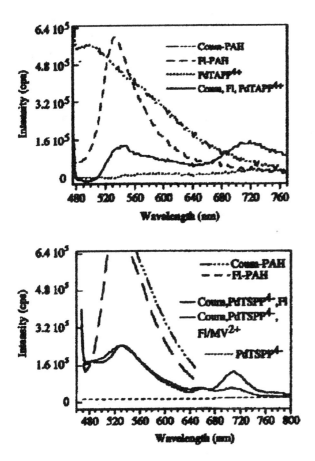

Figure 59 (Top) Steady-state emission spectra (λ_{exc} = 450 nm) of adsorbed monolayers of Coum-PAH and Fl-PAH polycations, PdTAPP$^+$, and the coadsorbed coumarin–fluorescein–porphyrin triad. (Bottom) Emission spectra of a similar triad containing PdT-SPP^{4-}, with and without an added viologen electron acceptor layer. (From Ref. 43a. Copyright 1999 Elsevier Publications.)

H-Aggregates are formed on the adsorption of Cresyl violet (CV$^+$), on the negatively charged SiO$_2$ and SnO$_2$ colloids [113]. These aggregates exhibited broad absorbance around 520 nm. Triplet-state energy transfer from the H-aggregates to a sensitizer, Ru(bpy)$_3^{2+}$, coadsorbed onto the surface was completed within 20 ps. Electron transfer from the sensitizer to the H-aggregates was found to be promoted only when coadsorbed onto the SnO$_2$ layer and not the SiO$_2$ layer.

4. Nonlinear Optical Properties

Intercalation of an organic dye, 4-[4-(dimethylamino)-α-styryl]-1-methylpyry-dinium (DAMS$^+$), into the lamellar MnS$_3$ or CdS$_3$ gave materials that showed excellent SHG efficiencies and these efficiencies are several hundred times that of urea [114]. XRD patterns of the thin films of these material showed an increase in basal spacing of the MnS$_3$ from 6.5 to 12.6 Å, confirming the successful binding of the guest in the galleries. In the case of CdS$_3$, the increase in the basal spacing was found to be smaller than expected.

The absorption spectrum of DAMS$^+$ intercalated into CdS$_3$ showed features that are vastly different from that of pure dye and these are characteristic of the J-aggregates: a strong absorption at 600 nm and a broad shoulder around 530 nm. The SHG efficiency of the corresponding intercalation compound, measured on a picosecond Nd:YAG pulsed laser operating in the fundamental mode (1064 nm), was found to be negligible.

VI. CONCLUSIONS AND FUTURE WORK

The supramolecular organization of molecules at layered materials provides a powerful tool for the photochemist to control molecular behavior. The organization is best achieved when the intermolecular interactions are designed such that the assemblies are produced in a predetermined manner with expected properties. The molecular design will have to take into account the nature of the organizing matrix, the expected orientation of the guests within the matrix, and various non-covalent interactions between the guests. Our understanding of these interactions as a function of molecular structure is growing at a tremendous pace and such understanding is not only central to the current topic but also important for biomolecular recognition, molecular medicine, drug design, intelligent molecular devices, smart materials, and environmental applications. Photosciences will continue to provide molecular details of these assemblies and usher in a new era of molecular design for controlling the structure and properties of supramolecular assemblies. Although such a control over the structure and properties of the molecular assemblies is challenging, this field will be one of the most exciting areas of chemistry in this new century.

As is evident from the various examples discussed here, use of proteins in the orientation, organization, and self-assembly of organic pigments, although spectacularly used by Mother Nature (briefly described in the Introduction), is yet to be realized. Current effort by several groups worldwide indicate that we are ready to take on this new challenge. Exciting times are ahead in this field.

ACKNOWLEDGMENTS

The authors thank the National Science Foundation (DMR-9729178) and the donors of the Petroleum Research Fund (PRF#33821-AC4) for financial support.

REFERENCES

1. McDermott, G.; Prince, S. M.; Freer, A. A.; Hawthornthwaite-Lawless, A. M.; Papiz, M. Z.; Cogdell, R. J.; Isaacs, N. W. *Nature (London)* **1995,** *374* (6522), 517–521.

2. (a) Thomas, J. K. *The Chemistry of Excitation at Interfaces;* American Chemical Society: Washington, DC, 1984. (b) Ramamurthy, V., ed. *Photochemistry in Organized & Constrained Media;* VCH: New York, 1991. (c) Fendler, J. H. *Membrane Mimetic Chemistry;* Wiley: New York, 1982. (d) Kalyanasundaram, K. *Photochemistry in Microheterogeneous System;* Academic Press: New York, 1987. (e) Mallouk, T. E., Gavin, J. A. *Acc. Chem. Res.* **1998,** *31,* 209.

3. (a) Clearfield, A. in *Design of New Materials;* Cocke, D. L.; Clearfield, A., eds.; Plenum Press: New York, 1988. (b) Giannelis, E. P.; Mehrotra, V.; Russell, M. W. *Mater. Res. Soc. Symp. Proc.* **1990,** *180,* 685–696.

4. (a) Alberti, G.; Constantino, U. *Intercalation Chem.* **1982,** 147–180. (b) Alberti, G.; Constantino, U.; Marmottini, F. *Recent Dev. Ion Exch., [Pap. Int. Conf. Ion Exch. Processes]* **1987,** 249–256. (c) Clearfield, A. *Chem. Rev.* **1988,** *88,* 125–148.

5. Alberti, G.; Constantino, U. *J. Mol. Catal.* **1984,** *27,* 235–250.

6. Cao, G.; Rabenberg, L. K.; Nunn, C. M.; Mallouk, T. E. *Chem. Mater.* **1991,** *3,* 149–156.

7. Lee, H.; Kepley, L. J.; Hong, H. G.; Mallouk, T. E. *J. Am. Chem. Soc.* **1988,** *110,* 618–620.

8. (a) Li, Z.; Lai, C.; Mallouk, T. E. *Inorg. Chem.* **1989,** *28,* 178–182. (b) Rong, D.; Kim, Y., II; Mallouk, T. E. *Inorg. Chem.* **1990,** *29,* 1531–1535.

9. (a) Kanatzidis, M. G.; Wu, C. G.; Marcy, H. O.; DeGroot, D. C.; Kannewurf, C. R. *Chem. Mater.* **1990,** *2,* 222–224. (b) Kanatzidis, M. G.; Hubbard, M.; Tonge, L. M.; Marks, T. J.; Marcy, H. O.; Kannewurf, C. R. *Synth. Met.* **1989,** *28,* C89–C95.

10. (a) Clearfield, A.; Smith, G. D. *Inorg. Chem.* **1969,** *8,* 431–436. (b) Alberti, G.; Constantino, U.; Allulli, S.; Tomassini, N. *J. Inorg. Nucl. Chem.* **1978,** *40,* 1113–1117. (c) Clearfield, A. *J. Mol. Catal.* **1984,** *27,* 251–262. (d) Alberti, G.; Casciola, M.; Constantino, U. *J. Colloid Interf. Sci.* **1985,** *107,* 256–263.

11. (a) Alberti, G. *Acc. Chem. Res.* **1974,** *11,* 163. (b) Vliers, D. P.; Schoonheydt, R. A.; De Schrijver, F. C. *J. Chem. Soc., Faraday Trans. 1* **1985,** *81,* 2009–2019. (c) Vliers, D. P.; Collin, D.; Schoonheydt, R. A.; De Schryver, F. C. *Langmuir* **1986,** *2,* 165–169. (d) Taniguchi, M.; Yamagishi, A.; Iwamoto, T. *J. Phys. Chem.* **1990,** *94,* 2534–2537. (e) Lee, C. F.; Thompson, M. E. *Inorg. Chem.* **1991,** *30,* 4–5. (f) Kim, R. M.; Pillion, J. E.; Burwell, D. A.; Groves, J. T.; Thompson, M. E. *Inorg. Chem.* **1993,** *32,* 4509–4516. (g) Snover, J. L.; Thompson, M. E. *J. Am. Chem. Soc.* **1994,** *116,* 765–766. (h) Alberti, G.; Dionigi, C.; Murcia-Mascaros, S.; Vivani, R. in *Crys-*

tallography of Supramolecular Compounds; Tsoucaris, G.; et al., ed.; Kluwer Academic Publishers: Amsterdam, 1996. (i) Alberti, G. *Acc. Chem. Res.* **1978,** *11,* 163–170.

12. (a) Clearfield, A.; Oskarsson, A.; Oskarsson, C. *Ion Exch. Membr.* **1972,** *1,* 91–107. (b) Amphlett, C. B. *Inorganic Ion Exchangers;* Elsevier: Amsterdam, 1964.

13. (a) Clearfield, A.; Kullberg, L. H. *J. Phys. Chem.* **1974,** *78,* 1812–1817. (b) Horsley, S. E.; Nowell, D. V. *J. Appl. Chem. Biotechnol.* **1973,** *23,* 215–224. (c) Alberti, G.; Constantino, U.; Giulietti, R. *J. Inorg. Nucl. Chem.* **1980,** *42,* 1062–1063. (d) Troup, J. M.; Clearfield, A. *Inorg. Chem.* **1969,** *8,* 431.

14. Alberti, G.; Torracca, E. *J. Inorg. Nucl. Chem.* **1968,** *30,* 317–318.

15. Michel, E.; Weiss, A. Z. *Naturforsch.* **1965,** *B20,* 1307.

16. Clearfield, A. *Comments Inorg. Chem.* **1990,** *10,* 89–128.

17. Kumar, C. V.; Williams, Z. J. *Spectrum* **1995,** June, 17.

18. Clearfield, A.; Tindwa, R. M. *J. Inorg. Nucl. Chem.* **1979,** *41,* 871–878.

19. Kumar, C. V.; Asuncion, E. H.; Rosenthal, G. *Micropor. Mater.* **1993,** *1,* 123–129.

20. Kumar, C. V.; Asuncion, E. H.; Rosenthal, G. *Micropor. Mater.* **1993,** *1,* 299–308.

21. Thompson, M. E. *Chem. Mater.* **1994,** *6,* 1168–1175.

22. Cao, G.; Hong, H. G.; Mallouk, T. E. *Acc. Chem. Res.* **1992,** *25,* 420–427.

23. Alberti, G.; Constantino, U.; Dionigi, C.; Murcia-Mascaros, S.; Vivani, R. *Supramol. Chem.* **1995,** *6,* 29–40.

24. Poojary, M. D.; Hu, H. L.; Campbell, F. L., III; Clearfield, A. *Acta Crystallogr. Sect. B: Struct. Sci.* **1993,** *B49,* 996–1001.

25. (a) Dines, M. B.; Griffith, P. C. *Inorg. Chem.* **1983,** *22,* 567–569. (b) DiGiacomo, P. M.; Dines, M. B. *Polyhedron* **1982,** *1,* 61–68. (c) Dines, M. B.; Griffith, P. C. *Polyhedron* **1983,** *2,* 607–611. (d) Alberti, G.; Constantino, U. in *Inclusion Compounds, Inorganic and Physical Aspects of Inclusion;* Atwood, J. L.; Davies, J. E. D.; MacNico, D. D., eds.; Oxford University Press: Oxford, 1991.

26. Troup, J. M.; Clearfield, A. *Inorg. Chem.* **1977,** *16,* 3311–3314.

27. Clearfield, A. *J. Mol. Catal.* **1984,** *27,* 251–262.

28. Burwell, D. A.; Thompson, M. E. *Chem. Mater.* **1991,** *3,* 730–737.

29. Ortiz-Avila, C. Y.; Bhardwaj, C.; Clearfield, A. *Inorg. Chem.* **1994,** *33,* 2499–2500.

30. (a) Dines, M. B.; DiGiacomo, P. M. *Inorg. Chem.* **1981,** *20,* 92–97. (b) Dines, M. B.; DiGiacomo, P. M.; Callahan, K. P.; Griffith, P. C.; Lane, R. H.; Cooksey, R. E. in *Chemically Modified Surfaces in Catalysis and Electrocatalysis;* Miller, J. S., ed.; American Chemical Society: Washington, DC, 1982.

31. Alper, J. *Chem Ind* **1986,** 335.

32. Alberti, G.; Costantino, U.; Marmottini, F.; Vivani, R.; Zappelli, P. *Angew. Chem.* **1993,** *105,* 1396–1398 [see also *Angew. Chem. Int. Ed. Engl.,* **1993,** *1332* (1399), 1357–1399].

33. Alberti, G.; Dionigi, C.; Murcia-Mascaros, S.; Vivani, R. in *Crystallography of Supromolecular Compounds;* Tsoucaris, G.; et al., eds.; Kluwer Academic Publishers: Amsterdam, 1996, 143–157.

34. Zhang, B.; Clearfield, A. *J. Am. Chem. Soc.* **1997,** *119,* 2751–2752.

35. Yang, C. Y.; Clearfield, A. *React. Polym. Ion Exch., Sorbents* **1987,** *5,* 13–21.

36. Kullberg, L. H.; Clearfield, A. *Solvent Extr. Ion Exch.* **1990,** *8,* 187–197.

37. Martin, K. J.; Squattrito, P. J.; Clearfield, A. *Inorg. Chim. Acta* **1989,** *155,* 7–9.

38. Cao, G.; Lee, H.; Lynch, V. M.; Mallouk, T. E. *Inorg. Chem.* **1988,** *27,* 2781–2785.
39. Poojary, D. M.; Zhang, Y.; Zhang, B.; Clearfield, A. *Chem. Mater.* **1995,** *7,* 822–827.
40. Huan, G.; Jacobson, A. J.; Johnson, J. W.; Goshorn, D. P. *Chem. Mater.* **1992,** *4,* 661–665.
41. Guliants, V. V.; Benziger, J. B.; Sundaresan, S.; Wachs, I. E.; Jehng, J. M. *Chem. Mater.* **1995,** *7,* 1493–1498.
42. Ogawa, M.; Kuroda, K. *Chem. Rev.* **1995,** *95,* 399–438.
43. (a) Kaschak, D. M.; Johnson, S. A.; Waraksa, C. C.; Pogue, J.; Mallouk, T. E. *Coordin. Chem. Rev.* **1999,** *185–186,* 403–416. (b) Zhou, H. S.; Honma, I. *Adv. Mater.* **1999,** *11,* 683. (c) Schoonman, J. *Chem. Mag. (Den Haag)* **1998,** 354–357. (d) Levy, B. *J. Electroceram.* **1997,** *1,* 239–272. (e) Al-Shamery, K. *Appl. Phys. A: Mater. Sci. Process.* **1996,** *A63,* 509–521. (f) Vogler, A.; Kunkely, H. *Topics Curr. Chem.* **1990,** *158,* 1–30. (g) Oelkrug, D.; Flemming, W.; Fuellemann, R.; Gunther, R.; Honnen, W.; Krabichler, G.; Schaefer, M.; Uhl, S. *Pure Appl. Chem.* **1986,** *58,* 1207–1218. (h) Cowdery-Corvan, R.; Spooner, S. P.; McLendon, G. L.; Whitten, D. G. *Adv. Chem. Ser.* **1993,** *238,* 261–279.
44. (a) Kanezaki, E.; Kinugawa, K.; Ishikawa, Y. *Chem. Phys. Lett.* **1994,** *226,* 325–330. (b) Carlino, S.; Hudson, M. J. *J. Mater. Chem.* **1995,** *5,* 1433–1442. (c) Tagaya, H.; Ogata, A.; Kuwahara, T.; Ogata, S.; Karasu, M.; Kadokawa, J.-I.; Chiba, K. *Micropor. Mater.* **1996,** *7,* 151–158. (d) Chibwe, K.; Jones, W. *J. Chem. Soc., Chem. Commun.* **1989,** 926–927. (e) Kuk, W.-K.; Huh, Y.-D. *J. Mater. Chem.* **1997,** *7,* 1933–1936. (f) Kuk, W.-K.; Huh, Y.-D. *Bull. Korean Chem. Soc.* **1998,** *19,* 1032–1036. (g) Cavani, F.; Trifiro, F.; Vaccari, A. *Catal. Today* **1991,** *11,* 173–301.
45. Park, I. Y.; Kuroda, K.; Kato, C. *J. Chem. Soc., Dalton Trans.* **1990,** 3071–3074.
46. Chibwe, M.; Pinnavaia, T. J. *J. Chem. Soc., Chem. Commun.* **1993,** 278–280.
47. Lagaly, G.; Beneke, K. *Colloid Polym. Sci.* **1991,** *269,* 1198–1211.
48. Meyn, M.; Beneke, K.; Lagaly, G. *Inorg. Chem.* **1990,** *29,* 5201–5207.
49. (a) Cavani, F.; Trifiro, F.; Vaccari, A. *Catal. Today* **1991,** *11,* 173–301. (b) Reichle, W. T. *Chemtech* **1986,** *16,* 58–63. (c) Reichle, W. T. *J. Catal.* **1985,** *94,* 547–557. (d) Suzuki, E.; Okamoto, M.; Ono, Y. *Chem. Lett.* **1989,** 1485–1486. (e) Suzuki, E.; Ono, Y. *Bull. Chem. Soc. Jpn.* **1988,** *61,* 1008–1010. (f) Twu, J.; Dutta, P. K. *J. Phys. Chem.* **1989,** *93,* 7863–7868. (g) Miyata, S. *Clays Clay Miner.* **1980,** *28,* 50–56. (h) Corma, A.; Fornes, V.; Rey, F. *J. Catal.* **1994,** *148,* 205–212. (i) Corma, A.; Fornes, V.; Rey, F.; Cervilla, A.; Llopis, E.; Ribera, A. *J. Catal.* **1995,** *152,* 237–242. (j) Constantino, V. R. L.; Pinnavaia, T. J. *Inorg. Chem.* **1995,** *34,* 883–892.
50. Ibarz, A.; Ruiz, E.; Alvarez, S. *Chem. Mater.* **1998,** *10,* 3422–3428.
51. Lagadic, I.; Lacroix, P. G.; Clement, R. *Chem. Mater.* **1997,** *9,* 2004–2012.
52. Lacroix, P. G.; Veret Lemarinier, A. V.; Clement, R.; Nakatani, K.; Delaire, J. A. *J. Mater. Chem.* **1993,** *3,* 499–503.
53. Kumar, C. V.; Asuncion, E. H. *J. Chem. Soc., Chem. Commun.* **1992,** 470–472.
54. Birks, J. B. *Photophysics of Aromatic Molecules;* Wiley: New York, 1970.
55. Kumar, C. V.; Williams, Z. J.; Kher, F. *Micropor. Mater.* **1996,** *7,* 161–171.
56. Taniguchi, M.; Yamagishi, A.; Iwamoto, T. *J. Phys. Chem.* **1990,** *94,* 2534–2537.
57. Souto, F.; Rodriguez, E.; Siegel, G.; Jimenezz, A.; Rodriguez, L.; Olivera, P.; Meri-

daa, J.; Perez, F.; Alcantara, M. *Mol. Cryst. Liquid Cryst. Sci. Technol., Sect. A* **1998**, *311*, 817–824.

58. Okuno, S.; Matsubayashi, G. *Inorg. Chim. Acta* **1996**, *245*, 101–104.

59. Danjo, M.; Tsuhako, M.; Nakayama, H.; Eguchi, T.; Nakamura, N.; Yamaguchi, S.; Nariai, H.; Motooka, I. *Bull. Chem. Soc. Jpn.* **1997**, *70*, 1053–1060.

60. Aloisi, G. G.; Costantino, U.; Elisei, F.; Nocchetti, M.; Sulli, C. *Mol. Cryst. Liquid Cryst. Sci. Technol., Sect. A* **1998**, *311*, 653–658.

61. Hoppe, R.; Alberti, G.; Costantino, U.; Dionigi, C.; Schulz-Ekloff, G.; Vivani, R. *Langmuir* **1997**, *13*, 7252–7257.

62. Rabinowitch, E.; Epstein, L. *J. Am. Chem. Soc.* **1941**, *63*, 69.

63. Okuno, S.; Matsubayashi, G. *Inorg. Chim. Acta* **1995**, *233*, 173–177.

64. (a) Prasad, P. N.; Williams, D. J. *Introduction to Nonlinear Optical Effects in Molecules and Polymers;* Wiley–Interscience: New York, 1991. (b) Munn, R. W.; Ironside, C. W., eds. *Principles and Applications of Nonlinear Optical Materials;* CRC: London, 1993.

65. (a) Katz, H. E.; Scheller, G.; Putvinski, T. M.; Schilling, M. L.; Wilson, W. L.; Chidsey, C. E. D. *Science (Washington, D. C.)* **1991**, *254*, 1485–1487. (b) Katz, H. E.; Wilson, W. L.; Scheller, G. *J. Am. Chem. Soc.* **1994**, *116*, 6636–6640. (c) Coradin, T.; Backov, R.; Jones, D. J.; Roziere, J.; Clement, R. *Mol. Cryst. Liquid Cryst. Sci. Technol., Sect. A* **1998**, *311*, 683–688.

66. Doughty, S. K.; Simpson, G. J.; Rowlen, K. L. *J. Am. Chem. Soc.* **1998**, *120*, 7997–7998.

67. (a) Whitten, D. G.; Spooner, S. P.; Hsu, Y.; Penner, T. L. *React. Polym.* **1991**, *15*, 37–48. (b) Webber, S. E. *Chem. Rev.* **1990**, *90*, 1469–1482. (c) Mooney, W. F.; Whitten, D. G. *J. Am. Chem. Soc.* **1986**, *108*, 5712–5719.

68. Kumar, C. V.; Chaudhari, A.; Rosenthal, G. L. *J. Am. Chem. Soc.* **1994**, *116*, 403–404.

69. Kumar, C. V.; Chaudhari, A. Unpublished work.

70. (a) Haugland, R. P. in *Molecular Probes Handbook of Fluorescent Probes and Research Chemicals;* Molecular Probes Inc.: Eugene, OR, 1992–1994; 20. (b) Kaschak, D. M.; Mallouk, T. E. *J. Am. Chem. Soc.* **1996**, *118*, 4222–4223.

71. Kumar, C. V.; Chaudhary, A. *Micropor. Mesopor. Mater.* **1999**, *32*, 75–79.

72. Gust, D.; Moore, T. A.; Moore, A. L. *Acc. Chem. Res.* **1993**, *26*, 198–205.

73. (a) Lin, C.-T.; Sutin, N. *J. Phys. Chem.* **1976**, *80*, 97–105. (b) Creutz, C.; Sutin, N. *Inorg. Chem.* **1976**, *15*, 496–499. (c) Purugganan, M. D.; Kumar, C. V.; Turro, N. J.; Barton, J. K. *Science (Washington, D. C.)* **1988**, *241*, 1645–1649.

74. (a) Kalyanasundaram, K. *Coordin. Chem. Rev.* **1982**, *46*, 159–244. (b) Juris, A.; Balzani, V.; Barigelletti, F.; Campagna, S.; Belser, P.; Von Zelewsky, A. *Coordin. Chem. Rev.* **1988**, *84*, 85–277.

75. (a) Vliers, D. P.; Schoonheydt, R. A.; De Schrijver, F. C. *J. Chem. Soc., Faraday Trans. 1* **1985**, *81*, 2009–2019. (b) Vliers, D. P.; Collin, D.; Schoonheydt, R. A.; De Schryver, F. C. *Langmuir* **1986**, *2*, 165–169.

76. Kumar, C. V.; Williams, Z. J. *J. Phys. Chem.* **1995**, *99*, 17,632–17,639.

77. (a) Wheeler, J.; Thomas, J. K. *J. Phys. Chem.* **1982**, *86*, 4540–4544. (b) Kajiwara, T.; Hashimoto, K.; Kawai, T.; Sakata, T. *J. Phys. Chem.* **1982**, *86*, 4516–4522. (c) Willner, I.; Yang, J.-M.; Laane, C.; Otvos, J. W.; Calvin, M. *J. Phys. Chem.* **1981**,

85, 3277–3282. (d) Memming, R. *Surf. Sci.* **1980,** *101,* 551–563. (e) Perry, J. W.; McQuillan, A. J.; Anson, F. C.; Zewail, A. H. *J. Phys. Chem.* **1983,** *87,* 1480–1483.
78. (a) DeWilde, W.; Peeters, G.; Lunsford, J. H. *J. Phys. Chem.* **1980,** *84,* 2306–2310. (b) Quayle, W. H.; Lunsford, J. H. *Inorg. Chem.* **1982,** *21,* 97–103.
79. (a) Krenske, D.; Abdo, S.; Van Damme, H.; Cruz, M.; Fripiat, J. J. *J. Phys. Chem.* **1980,** *84,* 2447–2457. (b) Abdo, S.; Canesson, P.; Cruz, M.; Fripiat, J. J.; Van Damme, H. *J. Phys. Chem.* **1981,** *85,* 797–809. (c) Nijs, H.; Cruz, M.; Fripiat, J.; Van Damme, H. *J. Chem. Soc., Chem. Commun.* **1981,** 1026–1027. (d) Nijs, H.; Fripiat, J. J.; Van Damme, H. *J. Phys. Chem.* **1983,** *87,* 1279–1282. (e) Schoonheydt, R. A.; Pelgrims, J.; Heroes, Y.; Uytterhoeven, J. B. *Clay Miner.* **1978,** *13,* 435–438.
80. Schoonheydt, R. A.; De Pauw, P.; Vliers, D.; De Schrijver, F. C. *J. Phys. Chem.* **1984,** *88,* 5113–5118.
81. Kumar, C. V.; et al. unpublished results.
82. Colon, J. L.; Yang, C. Y.; Clearfield, A.; Martin, C. R. *J. Phys. Chem.* **1990,** *94,* 874–882.
83. Kumar, C. V. Unpublished results.
84. Colon, J. L.; Yang, C. Y.; Clearfield, A.; Martin, C. R. *J. Phys. Chem.* **1988,** *92,* 5777–5781.
85. DellaGuardia, R. A.; Thomas, J. K. *J. Phys. Chem.* **1983,** *87,* 990–998.
86. Kumar, C. V.; Williams, Z. J.; Turner, R. S. *J. Phys. Chem. A* **1998,** *102,* 5562–5568.
87. Awaluddin, A.; DeGuzman, R. N.; Kumar, C. V.; Suib, S. L.; Burkett, S. L.; Davis, M. E. *J. Phys. Chem.* **1995,** *99,* 9886–9892.
88. Ungashe, S. B.; Wilson, W. L.; Katz, H. E.; Scheller, G. R.; Putvinski, T. M. *J. Am. Chem. Soc.* **1992,** *114,* 8717–8719.
89. (a) Vermeulen, L. A.; Thompson, M. E. *Nature (London)* **1992,** *358,* 656–658. (b) Vermeulen, L. A.; Snover, J. L.; Sapochak, L. S.; Thompson, M. E. *J. Am. Chem. Soc.* **1993,** *115,* 11,767–11,774.
90. Villemure, G.; Detellier, C.; Szabo, A. G. *J. Am. Chem. Soc.* **1986,** *108,* 4658–4659.
91. Vermeulen, L. A.; Thompson, M. E. *Chem. Mater.* **1994,** *6,* 77–81.
92. Kim, R. M.; Pillion, J. E.; Burwell, D. A.; Groves, J. T.; Thompson, M. E. *Inorg. Chem.* **1993,** *32,* 4509–4516.
93. Braun, S.; Rappoport, S.; Zusman, R.; Shtelzer, S; Druckman, S.; Avnir, D.; Ottolenghi, M. in *Biotechnology: Bridging Research and Applications;* Kamely, D.; Chakrabarty, A.; Kornguth, S. E., eds.; Kluwer Academic: Amsterdam, 1991, 205–218.
94. Avnir, D.; Levy, D.; Reisfeld, R. *J. Phys. Chem.* **1984,** *88,* 5956–5959.
95. (a) Hamachi, I.; Fujita, A.; Kunitake, T. *J. Am. Chem. Soc.* **1994,** *116,* 8811–8812. (b) Fujita, A.; Senzu, H.; Kunitake, T.; Hamachi, I. *Chem. Lett.* **1994,** 1219–1222. (c) Fang, J.; Knobler, C. M. *Langmuir* **1996,** *12,* 1368–1374. (d) Mrksich, M.; Sigal, G. B.; Whitesides, G. M. *Langmuir* **1995,** *11,* 4383–4385. (e) Miksa, B.; Slomkowski, S. *Colloid Polym. Sci.* **1995,** *273,* 47–52. (f) Yoshinaga, K.; Kondo, K.; Kondo, A. *Polym. J. (Tokyo)* **1995,** *27,* 98–100. (g) Tiberg, F.; Brink, C.; Hellsten, M.; Holmberg, K. *Colloid Polym. Sci.* **1992,** *270,* 1188–1193. (h) Yoshinaga, K.; Kito, T.; Yamaye, M. *J. Appl. Polym. Sci.* **1990,** *41,* 1443–1450.

96.   Shabat, D.; Grynszpan, F.; Saphier, S.; Turniansky, A.; Avnir, D.; Keinan, E. *Chem. Mater.* **1997,** *9,* 2258–2260.
97.   (a) Kennedy, J. F.; White, C. A. in *Handbook of Enzyme Biotechnology;* Weisman, A., ed.; Ellis Horwood Ltd.: Chichester, 1985; 147. (b) Gubitz, G.; Kunssberg, E.; van Zoonen, P.; Jansen, H.; Gooijer, C.; Velthorst, N. H.; Fei, R. W. in *Chemically Modified Surfaces Vol. 2.;* Leyden, D.; Collins, W. T., eds.; Gordon and Breach: London, 1988, 129. (c) Gorton, L.; Marko-Varga, G.; Dominguez, E.; Emneus, J. in *Analytical Applications of Immobilized Enzyme Reactors;* Lam, S.; Malikin, G., eds.; Blackie Academic & Professional: New York, 1994, 1.
98.   Kumar, C. V.; McLendon, G. L. *Chem. Mater.* **1997,** *9,* 863.
99.   Kumar, C. V.; Chaudhari, A. *J. Am. Chem. Soc.,* **2000,** *122,* 830. (b) Kumar, C. V., Chaudhari, A. Submitted (2000). 180. Kumar, C. V.; Jagannadham, V. Submitted (2000).
100. (a) Takano, T. *J. Mol. Biol.* **1993,** *229,* 12. (b) Imoto, T.; Johnson, L. N.; North, A.; Phillips, D. C.; Rupley, J. A. in *The Enzymes;* Boyer, P. D., ed.; Academic Press: New York, 1972, 665–808. (c) Hecht, H. J.; Kalisz, H. M.; Hendle, J.; Schmid, R. D.; Schomburg, D. *J. Mol. Biol.* **1993,** *229,* 153–172.
101. (a) Park, I. Y.; Kuroda, K.; Kato, C. *Chem. Lett.* **1989,** 2057–2058. (b) Tagaya, H.; Ogata, A.; Kuwahara, T.; Ogata, S.; Karasu, M.; Kadokawa, J. I.; Chiba, K. *Micropor. Mater.* **1996,** *7,* 151–158. (c) Bonnet, S.; Forano, C.; de Roy, A.; Besse, J. P.; Maillard, P.; Momenteau, M. *Chem. Mater.* **1996,** *8,* 1962–1968. (d) Robins, D. S.; Dutta, P. K. *Langmuir* **1996,** *12,* 402–408.
102. (a) Carrado, K. A.; Forman, J. E.; Botto, R. E.; Winans, R. E. *Chem. Mater.* **1993,** *5,* 472–478. (b) Tagaya, H.; Ogata, A.; Kuwahara, T.; Ogata, S.; Karasu, M.; Kadokawa, J.; Chiba, K. *Micropor. Mater.* **1996,** *7,* 151–158.
103. Giannelis, E. P.; Nocera, D. G.; Pinnavaia, T. J. *Inorg. Chem.* **1987,** *26,* 203–205.
104. Darwent, J. R.; Douglas, P.; Harriman, A.; Porter, G.; Richoux, M. C. *Coordin. Chem. Rev.* **1982,** *44,* 83–126.
105. Oelkrug, D.; Flemming, W.; Füllemann, R.; Günther, R.; Honnen, W.; Krabichler, G.; Schäfer, M.; Uhl, S. *Pure. Appl. Chem.* **1986,** *58,* 1207–1218.
106. Zhou, H. S.; Honma, I. *Adv. Mater.* **1999,** *11*(8), 683–685.
107. Honma, I.; Zhou, H. S.; Moon, S. C., Mater. Res. Soc. Symp. Proc. (1999), 549 (Advanced Catalytic Materials—1998), 131–136.
108. Schoonman, J. *Chem. Mag. (Den Haag)* **1998,** *10,* 354–357.
109. (a) Nakato, T.; Sakamoto, D.; Kuroda, K.; Kato, C. *Bull. Chem. Soc. Jpn.* **1992,** *65,* 322–328. (b) Nakato, T.; Kusunoki, K.; Yoshizawa, K.; Kuroda, K.; Kaneko, M. *J. Phys. Chem.* **1995,** *99,* 17,896–17,905.
110. (a) Ben-Amotz, D.; Harris, C. B. *Chem. Phys. Lett.* **1985,** *119,* 305–311. (b) Vogel, M.; Rettig, W. *Ber. Bunsen-Ges. Phys. Chem.* **1985,** *89,* 962–968.
111. Breuer, H. D.; Jacob, H. *Chem. Phys. Lett.* **1980,** *73,* 172–174.
112. Kaschak, D. M.; Lean, J. T.; Waraksa, C. C.; Saupe, G. B.; Usami, H.; Mallouk, T. E. *J. Am. Chem. Soc.* **1999,** *121,* 3435–3445.
113. Liu, D.; Hug, G. L.; Kamat, P. V. *J. Phys. Chem.* **1995,** *99,* 16,768–16,775.
114. Lacroix, P. G.; Veret Lemarinier, A. V.; Clement, R.; Nakatani, K.; Delaire, J. A. *J. Mater. Chem.* **1993,** *3,* 499–503.

10

Fluorescence of Excited Singlet-State Acids in Certain Organized Media: Applications as Molecular Probes

Ashok Kumar Mishra

Indian Institute of Technology Madras, Chennai, India

I. INTRODUCTION

In a review of the theories of excited-state proton-transfer (ESPT) reactions, Arnaut and Förmosinho [1] observed that excited-state prototropism as a subject of research has been much less popular, even in the realm of photochemistry, despite its unquestionable importance in fundamental and applied photochemistry. The pace of research in this area has more or less remained the same during all the years since then.

Excited-state proton transfer relates to a class of molecules with one or more ionizable proton, whose proton-transfer efficiency is different in the ground and excited states. The works of Förster [2–4] and Weller [5–7] laid the foundation for this area on which much of the subsequent work was based. Förster's work led to the understanding of the thermodynamics of ESPT. He constructed a thermodynamic cycle (Förster cycle) which, under certain acceptable approximations, provides the excited-state proton-transfer equilibrium constant (pK_a^*) from the corresponding ground-state value (pK_a) and electronic transition energies of the acid (protonated) and base (deprotonated) forms of the ESPT molecule:

$$pK_a - pK_a^* = \left(\frac{N_A hc}{2.303RT} \right) (\bar{\nu}_{\text{acid}} - \bar{\nu}_{\text{base}})$$

where N_A is Avogadro's number, h is Planck's constant, c is the velocity of light, R is a gas constant, T is the absolute temperature, $\bar{\nu}_{acid}$ and $\bar{\nu}_{base}$ are the frequencies (in the cm^{-1} scale) of the longest-wavelength transitions of the protonated (P) and deprotonated (DP) forms, respectively.

According to the Förster cycle, if the longest wavelength electronic transition of the deprotonated form is of lower energy compared to that of the protonated form (red-shifted electronic absorption or emission spectrum of the deprotonated form with reference to the protonated-form spectrum), the molecule has enhanced excited-state acidity (i.e., the pK_a^* of the molecule is lower than pK_a). Equation (1) provides a quick and effective method for evaluating a molecule for its ESPT behavior.

Weller's work [5–7] on the kinetics of ESPT brought out the importance of competition between the rates of deactivation of the excited states and the rates of proton transfer. In cases where the deactivation rates are slow enough for a complete establishment of excited-state equilibrium, fluorimetric titrations provide a method for experimental determinations of $pK^*{}_a$. However, it has been realized that for a fairly large number of ESPT molecules, there is a frequent mismatch of pK_a^* values obtained from Förster cycle and fluorimetric titrations methods. There are also examples of extended fluorimetric titration curves resulting from low proton availability in the mid-pH region (4–10). Various modifications of the Förster cycle and extensions of Weller's original kinetic considerations have been made from time to time and have been reviewed periodically. Some of the earlier important ones include those by Weller [7] in 1961, Vander Donckt [8] in 1970, Schulman [9] in 1974, and Klöpffer [10] in 1977. The review by Ireland and Wyatt [11] contains extensive references of experimental results available in the literature until 1976.

The 1977 review of Martynov et al. [12] discusses existing mechanisms of ESPT, excited-state intramolecular proton transfer (ESIPT) and excited-state double-proton transfer (ESDPT). Various models that have been proposed to account for the kinetics of proton-transfer reactions in general. They include that of association–proton-transfer–dissociation model of Eigen [13], Marcus' adaptation of electron-transfer theory [14], and the intersecting state model by Varandas and Förmosinho [15,16]. Gutman and Nachliel's [17] review in 1990 offers a framework of general conclusions about the mechanism and dynamics of proton-transfer processes.

The advent of nanosecond time-resolved fluorimetry in the late 1970s and early 1980s made it possible to obtain fluorescence lifetimes of the prototropic forms as a function of proton concentration. Dynamic analysis at nanosecond resolution enabled a more accurate estimation of pK_a^* values. Phenomenon like proton-induced fluorescence quenching in naphthylamines could then be given a satisfactory explanation. Shizuka's review [18] in 1985 summarizes the dynamic analysis techniques that were employed. Certain intermolecular proton-transfer reactions (as in naphthols) and intramolecular proton-transfer reactions are known to occur in a picosecond time domain. Introduction of picosecond spectroscopic technique made it possible to measure proton-transfer rates in real time. Consequently, many ESPT molecules were and are being reinvestigated in this time domain. Initial work in this area has been reviewed by Kosower and Huppert [19]. A brief review by Dogra [20] discussed ESPT and biprotonic phototautomerism in amino- and hydroxyphenyl-substituted benzimidazole, benzoxazole, and benzthiazole. In the context of photochemical reactions involving oxygen and carbon acids and carbon bases, the use of excited-state acid–base behavior was reviewed by Wan and Shukla [21] in 1993. The two reviews by Arnaut and Förmosinho [1] and Förmosinho and Arnaut [22] cover progress in the areas of intermolecular and intramolecular ESPT, respectively until 1993. Gas-phase studies on ESPT molecules with solvent base clusters in the picosecond timescale revealed the importance of solvation [23–27]. Some of the earliest studies in this time domain was excited-state proton transfer in phenol–ammonia [23,24] and naphthol–ammonia clusters [26]. Studies in the femtosecond time domain are especially important for molecules with ESIPT and ESDPT, where the proton-transfer rate is in the subpicosecond time range. A 1993 review by Elsaesser [28] gives early work in this time domain. Research interest in this area is growing rapidly and more and more ESPT work is being reported with ever-increasing frequency. The topic awaits a comprehensive review.

A. Photon-Initiated Acids

It has been well known for a long time [1] that for aromatic hydroxy compounds and protonated aromatic amines, the lowest singlet excited state is much more acidic than the ground state by a factor in the range 10^5–10^9. This has been explained as a result of increased charge flow from the electron-donating substituents to the aromatic ring. Conversely, aromatic acids and the conjugate acids of certain heterocyclic amines become more basic in the excited singlet state. The pK_a^* of excited triplet states are known to be intermediate between the ground-state and excited-singlet-state values, being close to the respective ground-state values [29]. A much larger number of ESPT molecules studied so far are excited-singlet-state acids. This review concentrates on only these molecules.

B. Aromatic Hydroxy Compounds

Aromatic hydroxy compounds are known to show remarkable enhancement in their acidity in the excited singlet state and often serve as prototypes in this area of research. They are some of the earliest ESPT molecules studied by Förster [2–4]. Their ESPT behavior has been extensively studied over the past three decades [1,7–12,17–19]. The ultrafast nature of ESPT rate constants was realized as early as 1979 [30]. Picosecond time-resolved measurements by Webb et al.

[31,32] for 1-naphthol in water gave a value of the excited-state deprotonation rate of 2.5×10^{10} sec^{-1} and reprotonation rate of 6.8×10^{10} M^{-1}/sec. This gives a pK_a^* value [32] of 0.4, which is taken as the most accurate value. During the 1980s, ultrafast kinetic studies in the picosecond timescale in pure and mixed aqueous solutions and in aqueous electrolytic solutions led to the emergence of two different explanations of ESPT dynamics. Lee and co-workers [33–37] suggested that a cluster of 4 ± 1 water molecules acts as a base responsible for accepting the dissociated proton. The deprotonation rate is thus limited by the time water takes to wrap itself around the proton. The overall rate constants were computed using the Markov random walk theory. Studies in aqueous electrolytic solutions [36–38] also supported the cluster-base proposition. The ESPT of 1-naphthol to water was studied [39] as a function of solvent system size—from supersonically cooled neutral clusters 1-naphthol·(H$_2$O)n, $n = 1$–50, to bulk ice and water. It was found that ESPT is completely absent in ice, which is evidence of the orientational relaxation of water molecules during proton transfer. Under the experimental conditions, a minimum cluster size of 1-naphthol·(H$_2$O)$_{30}$ was seen to show ESPT. This essentially implies that the 1-naphthol·(H$_2$O)$_4$ acid–base pair in The Lee model undergoes ESPT only when solvated by bulk water.

Studying the ESPT of hydroxy aromatic sulfonates, Huppert and co-workers [40–44] suggested an alternative model based on the geminate proton–anion recombination, governed by diffusive motion. The analysis was carried out by using Debye–Smoluchowskii-type diffusion equations. Their ESPT studies in water–methanol mixtures showed that solvent effects in the dissociation rate coefficient are equal to the effects in the dissociation equilibrium constant [45]. 4-Hydroxy-1-naphthalenesulphonate in a water–propanol mixture as the solvent system has been found to behave somewhat differently than water–methanol or water–ethanol media, with a possible role of a water dimer [46,47].

Naphthols with electron-withdrawing groups such as cyano sulfonyl and methanesulfonyl at C-5 and C-8 exhibit greatly enhanced photoacidity [48]. 5,8-dicyano-1-naphthol shows remarkable photoacidity with a pK_a of 7.8 and a Förster cycle pK_a^* of -4.5. The kinetics and mechanism of ESPT of these superphotoacid molecules have been subjected to extensive investigations in recent times [49–53].

C. Excited-State Intramolecular Proton Transfer

If a proton-accepting basic center is present within an excited-state acid molecule and the basic center is close to the photodissociable proton capable of forming ground-state intramolecular hydrogen-bonding (e.g., methylsalisylate [7,54–56]), there is the possibility of a proton being transferred to this center, forming a tautomer T^* in the excited state, whose fluorescence is usually red-shifted compared to the normal emission. This is commonly referred to as ESIPT. As is expected, this monoprotonic process is extremely fast [55]. Numerous example of such molecules exist; some of them are hydroxyalkylnaphthols [57], hydroxyphenyl- and aminophenyl-benzazoles [20,58–60], 2-(2′-acetamidophenyl)benzimidazole [61], 2-hydroxyphenyl-lapazole [62], hydroxyflavones [63], aromatic α-hydroxy azo aldehydes and their derivatives [64], esters of *o*-hydroxynaphthoic acid [65], 2-hydroxybenzoyl compounds [66], and hypericin [67].

D. Excited-State Double Proton Transfer

If the basic center in the ESPT molecule is not suitably disposed to form a direct hydrogen bond with the dissociable proton, involvement of protons from the solvent medium becomes necessary and the processes is biprotonic (e.g., 5- and 6-aminoindazole [68,69], 7-azaindole monomer [70]). This type of ESPT has recently been called ESDPT. The dimer of 7-aminoindazole shows ESDPT between the two monomeric units that are suitably hydrogen-bonded [71]. Femtosecond studies by Douhal et al. [72] has suggested a mechanism in which the two protons are transferred in sequence. Some other molecules in which ESDPT has been observed are 1-H-pyrrolo[3,2-*h*]quinoline [73], [2,2′-bipyridyl]-3,3′-diol [74,75].

E. ESPT and Organized Media

The importance of the immediate environment of a molecule on its ESPT behavior is well recognized [1,60,76]. A recent review on different classes of photophysical probes for organized assemblies by Bhattacharyya [60] brings this out clearly. Both in homogeneous liquid mixtures as well as in the water-deficient environment of a variety of organized media, the rates of forward and reverse ESPT are affected [1]. These changes in ESPT rates are often reflected in the steady-state fluorescence spectra of ESPT molecules, as well as their fluorescence lifetimes. Thus, monitoring fluorescence spectral parameters of ESPT molecules in an organized microheterogeneous medium, in principle, provides a method of probing the microenvironment of the medium around the fluorophore molecule. However, having mentioned this, it is relevant to observe that not many ESPT probes exist that have been generally accepted by nonphotochemists as routine probes. Only a few examples of such well-accepted probes where excited-state proton transfer is involved [77] include pyranine (as a probe useful at near-neutral pH's and as a probe for endocytosis), hypericin (as a probe for protein kinases, protein phosphatases and cyclic nucleotides), and ethidium bromide (the popular DNA-binding probe in which ESPT is implicated as the major pathway for nonradiative decay of its excited state [78]). A fluorescence probe is a molecular probe capable of reflecting environment properties of a few cubic nanometers volume around itself. This has positioned fluorescence spectroscopy uniquely in the study of microheterogeneous media [79,80]. Some of the impor-

tant characteristics any ideal fluorescence probe for organized microheterogeneous media should meet are as follows:

1. The molecule should have high extinction coefficient and high fluorescence quantum yield.
2. There should be sensitive and large changes in one or more fluorescence parameters: λ_{max}^{fl}, intensity, lifetime and anisotropy, as the probe
   partitions to the organized media.
3. The absorption and fluorescence spectra of the host media should not
   be interfering with the corresponding spectra of the probe.
4. The probe should efficiently partition to the microheterogeneous
   medium.
5. The size of the probe molecule should be small enough so as to minimally perturb the guest physical chemical properties.
6. It should have good photostability.

The ESPT probes are essentially two-state probes, the acid form and the
base form having fluorescence spectra distinct from each other. The general
characteristics of reversible two-state reactions in the excited state have been
discussed by Lakowicz [81]. If the pK_a and the pK_a^* are conveniently disposed
with respect to the pH of the medium, it is possible to excite the molecule in exclusively one form in the ground state and obtain fluorescence from both the
acid and base forms in the excited singlet state. In the context of excited-singlet-
state acids with $pK_a^* < pK_a$, it implies exciting the neutral form in the ground
state and the possibility of observing fluorescence of either or both of the two
forms. The more separated is the difference in fluorescence maxima of the two
forms, more convenient the measurements are expected to be. It is interesting to
note that in Förster cycle terms, this essentially implies a larger ΔpK_a (pK_a
$\sim pK_a^*$).

F. Objective and Organization

This review focuses on the *ESPT behavior of excited-singlet-state acid* molecules
in the following organized media: *liposomes, proteins, cyclodextrins, polymers,
sol–gel glasses, monolayers, LB films,* and *solids.* Applicability of these
molecules as molecular probes for organized media have also been discussed for
systems where such possibilities exist. ESPT in the relatively more dynamic organized systems like micelles and reverse micelles have not been included in this
review. The need of frequent reference to the prototropic forms in ground and excited states necessitated the use of abbreviations for them. In the review, P and DP
refer to the protonated and deprotonated forms, respectively, with respect to the
prototropic reaction of interest for a molecule in the ground state, and P* and DP*
refer to the corresponding forms in the excited singlet state.

There are three sections in the review. This section is a brief introduction to the subject. Because aromatic hydroxy compounds are taken as prototype excited-state acids, they have been discussed in a separate subsection. Section II has subsections for each medium and each subsection is further branched according to the classes of ESPT molecules studied in the medium. Section III gives conclusions.

II. ORGANIZED MEDIA

A. Liposomes and Natural Lipid Bilayers

Liposomes are aqueous dispersions of phospholipids forming closed structures enclosing a volume of water. The membrane usually forms a bilayer structure, which is relatively impermeable to ions. Liposomes, or lipid vesicles as they are otherwise known, can be composed of natural or synthetic lipids. The phospholipids are usually of two main classes: glycerophospholipids and sphingophospholipids. Glycerophospholipids are the predominant phospholipids in biological membranes. These phosphoglyceric esters of higher fatty acids can contain a head group such as be choline, serine, ethanolamine, acid, and so forth, which determines the charge on the lipid. Both multilamellar vesicles (MLV) and unilamellar vesicles (UV) can be prepared. The size of MLV's range is 100–1000 nm and those of unilamellar vesicles range from ~25 nm (small unilamellar vesicles, SUV) to ~1000 nm (large unilamellar vesicles, LUV). The functional properties of lipid bilayers like fluidity, permeability, dynamics and order depend on the dynamic properties of the constituent lipids. At different temperatures, lipid bilayer membranes exist in different phases. Below the thermotropic phase transition temperature T_c, they are in a tightly ordered solid gel phase, and above T_c, they are in a fluid liquid-crystalline phase. The membrane properties show marked changes with phase transition. A more detailed description of the system, methods of preparation, and so forth can be found elsewhere [82].

1. Aromatic Hydroxy Compounds

a. Naphthols and Derivatives. The ESPT dynamics of 1-naphthol (1NpOH) (1) and 2-naphthol (2NpOH), the prototype ESPT molecules, and some of their substituents have been studied by Kuzmin and co-workers [83–85].

The quantum yield of photodissociation in neutral aqueous suspension of vesicles like egg lecithin (EL) and dipalmitoylphosphatidylcholine (DPPC) are significantly smaller than in the aqueous solution [83]. 1NpOH ($pK_a = 9.2$, $pK_a^* = 0.4$) emits only from its DP* form (2) in water. On incorporation into liposome membrane, a substantial increase of the P* form (1) fluorescence is seen with concomitant decrease of DP* form fluorescence. A similar effect is seen for 2NpOH with membrane incorporation. The biexponential fluorescence decay of the P* form in fully incorporated naphthol suggests the presence of two localization sites

OH
$(\lambda_{ex} = 290nm)$

$pK_a = 9.2$

$pK_a^* = 0.4$

O⁻

P

DP*

$\lambda_{em} = 360nm$

$\lambda_{em} = 470nm$

1   **2**

1-naphthol (1-NpOH)

Structures 1 and 2

in liposome membranes [85]. Naphthols in a water-inaccessible inner site in the membrane does not undergo excited ESPT, where as naphthol present in a relatively more water-accessible surface site shows excited-state deprotonation. The ESPT kinetics of 1NpOH and 2NpOH, their hydroxy and chloro substitutions in the DPPC membrane studied as a function of temperature showed that the temperature dependence of the rate constants of photodissociation in water, micelles, and bilayer membranes of egg lecithin can be described by the Arrhenius equation [84]. The increase in activation energy observed in surfactant assemblies as compared with aqueous solutions can be explained by an increase in the enthalpy and entropy of the protolytic equilibrium for weak acid dissociation of less polar media. An abrupt change in the dissociation rate constant was observed near the main phase transition temperature in DPPC liposomes. The ESPT dynamics of 1NpOH in the bilayer membrane of cationic vesicles of didodecyldimethylammonium bromide (DDAB) and dioctadecyldimethylammonium bromide (DOAB) have been studied in comparison with the same reaction in vesicles of zwitterionic DPPC [85]. The rate constants of ESPT for both the forms are much higher in bilayer membranes of cationic surfactants than for zwitterionic lipids (DPPC and egg lecithin). The activation enthalpy of ESPT for both fractions of NpOH in the membrane of DDAB is ~40 kJ/mol, which is much higher than in homogenous solutions and zwitterionic surfactants. Fluorescence kinetic data in the nanosecond timescale for DOAB vesicles did not allow reliable conclusions to be drawn from the temperature dependence of excited-state protolytic dissociation rate constants of these vesicles, as the reaction rates were faster. No significant decrease in the excited-state proton-transfer rate constants at the membrane phase-transition temperature of vesicles of cationic surfactants was observed, in contrast to the zwitterionic lipids. Taking into account the fluorescence lifetimes of the two-prototropic forms of 1NpOH, Sujatha and Mishra [86] have derived a modified expression for obtaining the true partition coefficient (K_p) of 1NpOH distribution to the dimyristoylphosphatidylcholine (DMPC) membrane. K_p was found to be fairly large both for the gel phase (2.5 × $10^6 \ M^{-1}$) and the liquid-crystalline phase (5.3 × $10^6 \ M^{-1}$). This high K_p ensures that a fairly low lipid/fluorophore ratio is required for complete partitioning of 1-naphthol to membrane, thus satisfying an important probe criterion.

The understanding gained by the work of Kuzmin and co-workers on ESPT kinetics of naphthols in lipid bilayers has been used by Mishra and co-workers [87–92] for developing a novel steady-state fluorimetric method to study certain membrane properties. It was observed that the two-state distribution (Fig. 1) of 1NpOH and 2NpOH between the hydrophobic inner site and water-accessible surface site is a sensitive function of the membrane physical state [87].

As per the two-state distribution model, the kinetic scheme for prototropism of 1NpOH incorporated in liposome membrane is given in Scheme 1.

The change in the two-state distribution is easily monitored by a convenient one-wavelength measurement of the neutral form fluorescence, and this can be used for probing the membrane. The fairly large differences in wavelengths of excitation (300 nm), fluorescence of the neutral form (360 nm), and fluorescence of the anion form (480 nm) makes the fluorescence free from spectral interference. The variation of the P* form fluorescence intensity with temperature showed a maximum at phase-transition temperatures (T_c) for both DMPC (23°C) (Fig. 2) and DPPC (42°C) membranes (Fig. 3). Figures 2 and 3 show a very nice correspondence of this variation with DPH fluorescence polarization and self-diffusion rate [93] of $^{22}Na^+$. The coexistence of solid gel and fluid liquid-crystalline phases at T_c and the consequent imperfection of the membrane [93] result in a redistribution of

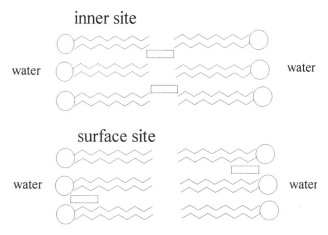

Figure 1   Two-state distribution of 1-naphthol in lipid bilayers. (Adapted from Ref. 93a. Copyright 1998 American Chemical Society.)

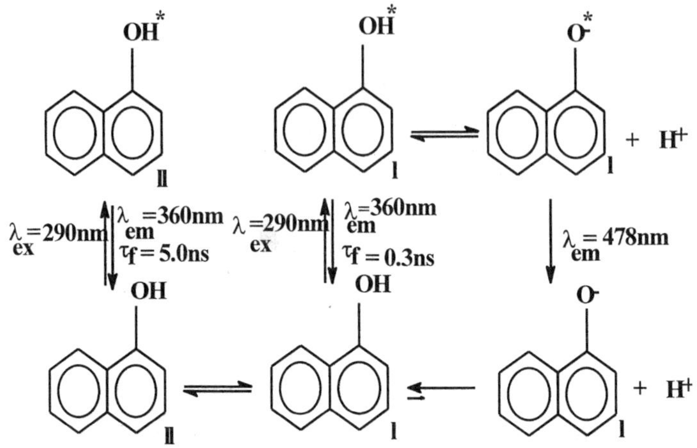

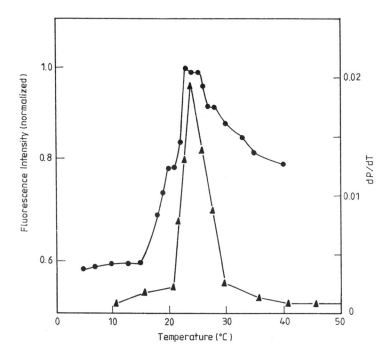

Scheme 1 Kinetic scheme for prototropism of 1-naphthol incorporated in liposome membrane.

Figure 2 Variation of P* form fluorescence intensity of 1-naphthol (•–•–•) and polarization (dP/dT) of DPH (▲–▲–▲) with temperature in DMPC liposome membranes. (From Ref. 93a. Copyright 1998 American Chemical Society.)

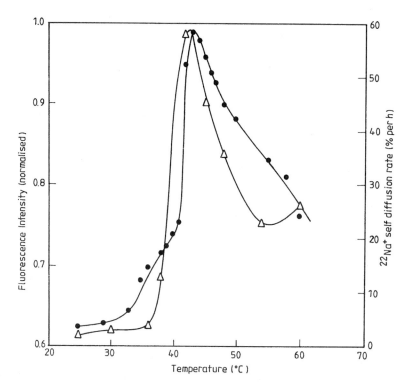

Figure 3 Variation of P* form fluorescence intensity of 1-naphthol (•–•–•) and ^{22}Na$^+$ self-diffusion rate (△–△–△) with temperature in DPPC liposome membrane. (The curve for ^{22}Na$^+$ self-diffusion rate has been adapted from Ref. 93 with permission from Elsevier Science. Figure reprinted with permission from Ref. 93a. Copyright 1998 American Chemical Society.)

the 1NpOH population in these two states. At T_c, there is an increase in population of the inner site at the expense of the surface site population, which is reflected in the enhancement of P* form fluorescence. Fluorescence quenching studies by a hydrophilic quencher I$^-$, showed that for the 1-naphthol-incorporated DMPC membrane, the water-accessible fraction of 1NpOH is 66% in the gel state (10°C), 55% at the phase transition temperature (23°C), and 64% in the liquid crystalline state (35°C). These, as well as the variation of amplitudes of biexponential fluorescence decay of membrane incorporated 1NpOH, were taken as supportive evidences for the model. This method of monitoring P* fluorescence to get the T_c is sensitive enough to show pretransition temperatures. It was found that 1NpOH shows more sensitivity than 2NpOH in this regard. The method was also found to be successful in predicting the T_c of membranes constituted from natural lipids: human erythrocyte membranes [88]. The fraction of surface site occupation in erythrocyte membranes was 83% in the liquid-crystalline state as compared to 64%

for the same state in DPPC membranes. Thus, the lipid bilayer membrane reconstituted from erythrocyte lipids seems to be more impermeable to 1NpOH.

The response of 1NpOH fluorescence to the presence of certain membrane property modifiers has been studied. The known effect of cholesterol in blurring the T_c of the membrane and making it more compact could be explained satisfactorily by this model [87]. For DPPC and DMPC mixed liposomes, the progressive shift of T_c with the increase in one component as well as the corresponding phase behavior was also reflected in the neutral form fluorescence [87]. The effect of Triton X-100, sodium dodecyl sulfate (SDS), and cetyl trimethylammonium bromide (CTAB) on the physical properties of DMPC liposome was studied [89] in the sublytic concentration range of the detergent using 1NpOH and conceptually different probes like anilino-1-naphthalene sulfonate (ANS) and 1,6-diphenyl-hexatriene (DPH). Depending on the nature and concentration of the detergent and the position of the hydrophilic and hydrophobic parts of different detergents in liposomes, the fluidity and permeability of the bilayers were shown to be affected. The fluidity was found to decrease at very low concentrations of detergent because of an improvement in the packing density of phospholipid molecules. The extents of changes in these properties were greater in the solid-gel phase than in the fluid liquid-crystalline phase. The thermotropic phase transitions of phospholipid bilayers are affected moderately by neutral surfactants and substantially by charged surfactants because of an alteration in the water of hydration of the lipid molecules. The effect of tripeptide leucinyl-phenylalanyl-valine (LFV) on DMPC liposome membrane has been studied [90] by 1NpOH in addition to some other fluorescent probes and it has been found that LFV increases the compactness of the membrane.

The steady-state fluorescence response of a series of substituted naphthols (4-chloro-1-naphthol, 4-methoxy-1-naphthol, 5-amino-1-naphthol, 7-methoxy-2-naphthol, 3-amino-2naphthol, 1-bromo-2-naphthol, and 6-bromo-2-naphthol) was studied [91] in the DMPC membrane. It was seen that 4-chloro-1-naphthol is almost as sensitive as 1NpOH. 7-methoxy-2-naphthol and 4-methoxy-1-naphthol were also found to be useful. However, the amino and bromo naphthols did not show significant probe possibilities. Although the ESPT equilibrium in the membrane is much different from that in bulk water, a relatively large magnitude of ΔpK_a ($pK_a \sim pK_a^*$) was suggested to serve as a rough indicator of the applicability of an ESPT molecule as a probe. N,N-dimethylaminomethyl-2-naphthol and morpholinomethyl-2-naphthol, which are essentially excited-state intramolecular proton-transfer systems, were seen to show a less sensitive response to T_c-related changes [92].

The ESPT equilibria for d-equilenin [d-3-hydroxyestra-1,3,-5(10),6,8-pentaen-17-one] (3) and dihydroequilenin (4), two fluorescent estrogens, are found to depend on both pH and on proton-acceptor concentration in dimyristoyllecithin (DML) vesicles [94]. These molecules basically include a 2NpOH moiety, which

$\lambda_{ex} = 330\text{nm}$

$\lambda_{em}(P^*) = 350\text{nm}$

$\lambda_{em}(DP^*) = 415\text{nm}$

3

d-equilenin

$\lambda_{ex} = 337\text{nm}$

$\lambda_{em}(P^*) = 360\text{nm}$

$\lambda_{em}(DP^*) = 440\text{nm}$

4

dihydroequilenin

Structures 3 and 4

is responsible for the ESPT behavior. Both the P* and DP* forms of the excited molecules are fluorescent. The excited-state pK_a^* value of this molecule is substantially lower than the ground-state pK_a. As expected, both of these molecules are found to partition to DML vesicles. The ESPT rates are greatly reduced when dihydroequilenin partitions to vesicles. The steady-state emission anisotropy of the P* form of equilenin incorporated into DML vesicles with temperature shows a decrease near the phase-transition temperature. It is suggested that dihydrocqui lenin, which retains the steroidal character, preferentially partitions into cholesterol-rich regions, where the 3'-hydroxyl group may be exposed to external ionic species. The accessibility of the bound probe to acetate as a proton acceptor depends on the cholesterol content of the vesicles. Cholesterol disrupts the overall bilayer architecture, resulting in "looser packing" of phospholipid molecules and increased vesicle diameter [95,96].

Neyroz et al. [97] have covalently linked 2NpOH to phosphatidylethanolamine moiety by the Schiff-base formation between the NH_2 of the phospholipid and the aldehyde moiety of 2-hydroxy-1-naphthaldehyde, followed by selective reduction of the imine to obtain a stable secondary amine. This fluorescent phospholipid easily incorporates into DML vesicle membrane and exhibits the typical behavior of ESPT probes. The emission spectrum of this probe inserted in the liposome is similar to that in ethanol medium and is affected by acetate used as a proton acceptor.

b. Pyranine. Pyranine (8-hydroxy-1,3,6-pyrenetrisulfonate) (**5**) has been one of the most extensively studied ESPT probe. This was introduced as a

$\lambda_{ex} = 403$ nm

$pK_a = 7.22$

$pK_a^* = 0.4$

$\lambda_{em} = 440$ nm $\lambda_{em} = 510$ nm

5   6

pyranine

Structures 5 and 6

sensitive pH probe for liposomes as early as 1978 [98]. It is a strong excited-singlet-state acid with a pK_a of 7.22 and a pK_a^* of 1.38. It has high fluorescence quantum yield, large Stokes shift, and the neutral (440 nm) and anionic (520 nm) form well separated. Due to its negative charge, it is well shielded from the phospholipid head groups by water molecules in the interior of anionic liposomes, but it is electrostatically bound to the surface of the cationic liposomes. Using pyranine fluorescence, it was observed that the liposome surface pH is equal to, greater than, and less than bulk pH for neutral, cationic, and anionic liposomes, respectively [98]. The probe was shown to work in the pH range 4–10 [99] and permitted surface potential measurement for cationic liposomes as well as pH gradients across membranes. The polyanionic character of pyranine keeps it away from anionic liposome membranes, and once entrapped inside a vesicle, pyranine does not leak out easily. The entrapped pyranine can then be used as a reliable reporter of aqueous pH changes within anionic vesicles. It was observed that when HCl is rapidly added to the suspension of unilamellar soybean phospholipid (asolectin) vesicles preincubated at alkaline pH, there is a biphasic decrease in the pH of the vesicle inner aqueous compartment. An initial, very rapid, and electrically uncompensated H^+ influx ($t_{1/2} < 1$ sec) results in the generation of a transmembrane electric potential opposing further H^+ influx. This leads to the development of much slower ($t_{1/2} \approx 5$ min), valinomycin-sensitive, proton–counterion exchange which continues until the proton concentration gradient is eliminated. Similar results were obtained in asolectin vesicles prepared by detergent dilution, in sonicated egg phosphatidylcholine vesicles, and in multilamellar asolectin liposomes [99].

Morphological consequences of the photopolymerization of vesicles prepared from $(C_{18}H_{37})_2N^+(CH_3)CH_2C_6H_4$-$p$-CH$=CH_2Cl^-$ were investigated by using pyranine [100]. It was seen to bind appreciably to the surface of the vesicle. Changes in the relative intensities of neutral and anionic form fluorescence were interpreted in terms of the gradual penetration of the probe into unpolymerized vesicles and the absence of such fluorescence changes in terms of long-term stabilization of the probe in the clefts formed on the vesicle surface upon pulling the surfactant head groups together by photopolymerization. Fluorescence polarization studies and ESPT rate constants obtained by nanosecond fluorimetry supported this rationalization.

Pyranine has been used to study the proton dissociation and diffusion dynamics in the aqueous layer of multilamellar phospholipid vesicles [101]. There are 3–10 water layers interspacing between the phospholipid membranes of a multilamellar vesicle, and their width gets adjusted by osmotic pressure [102]. Pyranine dissolved in these thin layers of DPPC and DPPC+cholesterol multilamellar vesicles were used as a probe for the study. Before the photoreleased proton escapes from the coulombic cage, the probability of a proton excited-anion recombination was found to be higher than in bulk. This was attributed to the diminished water activity in the thin layer. It was found that the effect of local forces on proton diffusion at the timescale of physiological processes is negligible.

The binding of pyranine to phosphatidylcholine (lecithin) vesicles as a function of the probe and electrolyte concentration has been investigated [103]. The binding of the probe to the internal leaflet of lecithin small unilamellar vesicles (SUVs) was found to be larger than that to the external leaflet. The addition of salt up to 2 M did not prevent binding, even at low probe concentrations. The ground-state reprotonation rate constant was found to depend on the probe content per vesicle.

c. Hypericin. There is a growing awareness in the studies on photodynamic therapy on the role of photogenerated protons. Changes in pH have been implicated in inhibition of virus replication [104], antitumor activity [105,106], and apoptosis [107–109]. Hypericin (**7**) is a naturally occurring polycyclic quinone possessing light-induced antiviral activity against the human immunodeficiency virus (HIV) and other closely related enveloped lentiviruses such as equine infectious anemia virus (EIAV) [110]. Fast intramolecular proton transfer is known to occur in this molecule [111–113] producing the excited state tautomer (**8**).

Steady-state illumination of hypericin effects a pH drop in its surroundings [110]. When hypericin and an indicator dye, 3-hexadecanoyl-7-hydroxycoumarin, are both embedded in the DPPC vesicle, steady-state illumination causes hypericin to transfer a proton to the indicator within a time commensurate to its triplet-state lifetime. Triplet hypericin is seen to be a source, but not necessarily the only source of protons.

2. pH Jump

The phenomenal increase in excited-singlet-state acidity of aromatic hydroxy compounds makes them good excited-state proton emitters, with deprotonation

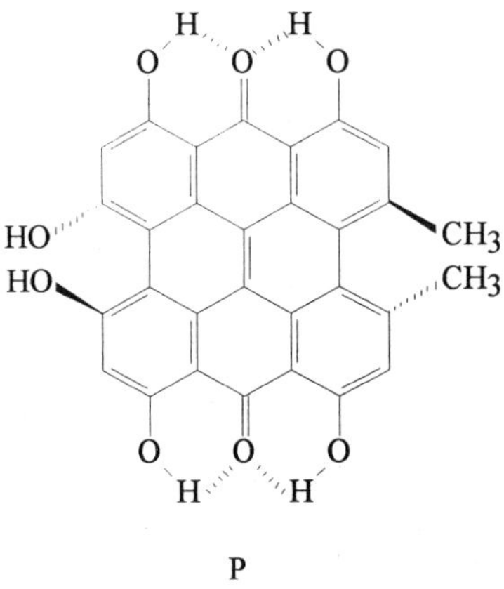

P

7

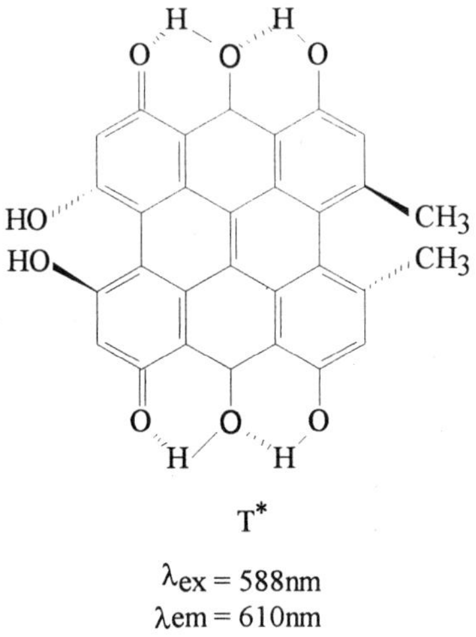

T*

$\lambda_{ex} = 588nm$
$\lambda_{em} = 610nm$

8

Structures 7 and 8

rate constants among the fastest in chemistry. When they absorb picosecond pulses of light, they release a transient pulse of protons, causing a pH jump. Because the lifetime of the excited state is very short (a few nanoseconds), the measured reaction is that which takes place in a volume corresponding to the diffusion distance of the proton during the lifetime of the excited state. This pulse of proton can then be used to perturb the proton distribution in the microenvironment of the emitter and the consequent relaxation dynamics are studied. The dynamics of proton dissociation from an acidic moiety and its subsequent dispersion in the bulk is regulated by the physical and chemical properties of the solvent environment. The solvent has to provide a potential well to accommodate the discharged protons, screen it from the negative charge of the conjugated base, and provide an efficient mode for the diffusion of the proton to the bulk. On measuring the dynamics of proton dissociation in the time-resolved domain, the kinetic analysis of the reaction can quantify the properties of the immediate environment. The pioneering and exhaustive work of Gutman and co-workers in this field has resulted in the development of pH jump methods to monitor proton diffusion dynamics in the microenvironment of the proton emitter [114–116]. The details of this technique, along with the instrumentation, list of proton emitters, and their efficiency, analytical procedures, and their possible application to complex systems like proteins and membranes are described in Gutman's review of 1986 [117]. The formation and recombination dynamics of the proton pulse, which is in a timescale comparable to or faster than many physiological processes involving bilayer membranes and proteins, make them particularly suitable for probing the microenvironments. The propagation of protons at the water–membrane interface was discussed in a review by Gutman et al. [118]. The symmetry of the diffusion space, the physical properties of the solvent, the chemical interaction between the proton and the surface, and the exchange of proton between surface groups was discussed. Results concerning surfaces insulated from bulk water and macroscopic measurements of proton diffusion in the system exposed to bulk water are presented. The heterogeneity of surface composition and dielectric discontinuity normal to the surface markedly alters the proton diffusion dynamics [119]. A more recent review by Gutman and Nachliel [120] describes the kinetic analysis for evaluating the properties of small cavities of proteins and the diffusion of protons within narrow channels. The clustering of proton binding sites on a surface can endow the surface with an enhanced capacity to attract protons and to funnel them toward a specific site. Pyranine and naphthol sulfonates have been molecules of choice to serve as proton emitters.

Using 2NpOH and 2-naphthol-3,6-disulfonate [121] as excited-state proton emitters, a transient high proton concentration is achieved on the membrane surface. With bromocresol green dye adsorbed on the membrane serving as a pH indicator, it has been found that the protons first react with the acidic ionized moieties on the surface and then reach the strongest base on the surface by rapid exchange.

The diffusion of protons at the immediate vicinity of (less than 10 Å from) a phospholipid membrane has been studied by proton pulse techniques using pyranine as the proton emitter trapped exclusively in the hydration layers of multilamellar vesicles made of egg phosphatidylcholine [122]. The recombination of the proton with the pyranine anion was monitored by time-resolved spectroscopy and analyzed by a diffusion-controlled formalism. The measured diffusion coefficient was found to be only slightly smaller that the diffusion coefficient of protons in bulk water. Modulating the width of the hydration layer by external pressure had a direct effect on the diffusibility of the proton: The narrower the hydration layer, the slower the diffusion of protons. Using the same technique, the exchange of protons between the bulk and the mitochondrial membranes was measured in the time-resolved domain with a submicrosecond resolution [123]. The protons were discharged by photoexcitation of pyranine by a short laser pulse, and the reprotonation of the pyranine anion was monitored at 458 nm. In parallel, the protonation of the membrane was followed at 496.5 nm, looking at the transient absorbance of fluorescein, covalently attached to the M side of the membrane. The analysis of the relaxation dynamics was carried out by a simulation procedure that reconstructed the observed dynamics of the two chromophores. The analysis revealed the presence of two membrane-indigenous buffering moieties: one at a low pK (4.1) and the other at a pK of 6.9. Both types of indigenous buffers reacted with the protons and pyranine anion in unhindered diffusion-controlled reactions. On the other hand, the exchange of protons between the indigenous buffer moieties was rather slow. No evidence was found for the presence of sites capable of retaining a proton, secluded from the bulk, for a time frame longer than 100 μsec required by the models of localized proton gradient. Studies have been carried out on the dynamics of proton transfer at the C side of the mitochondrial membrane also by the same method [124]. Upon initiation of respiration, pyranine in the inner aqueous space of inside-out submitochondrial particles (SMP) recorded acidification of this space. Incorporation of the high concentration of the probe (100 nmol/mg of protein) had no effect on the respiratory functions of the vesicles nor on their capacity to execute $\Delta\mu H^+$ coupled reverse electron transfer. The respiratory control ratio (RCR) remained as high as >4. The release of the proton and its reaction with the matrix of the inner space of SMP were monitored at two time intervals: Nanosecond fluorimetry measured the dissociation of the proton from the excited dye molecule, and microsecond spectroscopy followed the reaction between the proton and the ground-state anion. The apparent dielectric constant of that space is rather low ($\varepsilon = 20$). The diffusion coefficient of the proton is 2.3×10^{-5} cm^2/sec, and the activity of water is $a_{H2O} = 0.87$. All of these values imply that a large fraction of the intervesicular aqueous phase is taken up by the hydration layer of the lipids and proteins of the C side of the membrane. The microsecond dynamics measurements indicated that the rates of proton binding to the membrane surface components reach equilibrium within 60 μsec. On the basis of these figures, it was concluded that, under physiological conditions, the inner space of submitochondrial vesicles is in a homogenous state of protonation.

The dynamics of proton binding to the extra cellular and the cytoplasmic surfaces of the purple membranes were measured by the pH jump methods [125]. The purple membranes selectively labeled by fluorescein Lys-129 of bacteriorhodopsin were pulsed by protons released in the aqueous bulk from excited pyranine and the reaction of the protons with the indicators was measured. Kinetic analysis of the data implied that the two faces of the membrane differ in their buffer capacities and in their rates of interaction with bulk protons. The extracellular surfaces of the purple membrane contains one anionic proton binding site per protein molecule with p$K = 5.1$. This site is within a Coulomb cage radius from Lys-129. The cytoplasmic surface of the purple membrane bears four to five protonable moieties that, due to close proximity, function as a common proton binding site. The reaction of the proton with this cluster is at a very fast rate (3×10^{10} M^{-1}sec^{-1}). The proximity between the elements is sufficiently high that even in 100 mM NaCl, they still function as a cluster. Extraction of the chromophore retinal from the protein has a marked effect on the carboxylates of the cytoplasmic surface, and two to three of them assume positions that almost bar their reaction with bulk protons. Quantitative evaluation of the dynamics of proton transfer from photoactivated bacteriorhodopsin to the bulk has been done by using numerical simulation of chemical reaction dynamics for analyzing the delayed appearance of protons in the bulk. This indicates that the low-pK surface groups of the membrane, which form an undilutable concentrated matrix of proton binding sites, retain the protons in this space according to the mass action law. The main sites for proton accumulation are the clusters of carboxylates on the cytoplasmic side of the membrane. The protonation of an indicator in the bulk does not proceed by its reaction with the free proton, but rather through self-diffusion of the indicator in the membrane and abstraction of a proton from the protonated surface groups [126]. Monensin (Mo) is an ionophore that supports an electroneutral ion exchange across the lipid bilayer. Because of this, under steady-state conditions, no electrical signals accompany its reaction. Using the laser-induced proton pulse as a synchronizing event [127], selective acidification of one face of a black lipid membrane impregnated by monensin has been done. The short perturbation temporarily upsets the acid–base equilibrium on one face of the membrane, causing a transient cycle of ion exchange. Under such conditions, the molecular events could be discerned as a transient electric polarization of the membrane lasting approximately 200 μsec. The proton-driven chemical reactions that lead to the electrical signals was reconstructed by numeric integration and differential rate equations describing the multiequilibrium nature of the system. The analysis of the reaction reveals that the ionic selectivity of the monensin ($H^+ > Na^+ > K^+$) is due to more than one term. In addition to the well-established different affinity for the various cations, the selectivity is also derived from a large difference in the rates of cross-membrane diffusivities (MoH > MoNa > MoK) which have never been detected previously. Quantitative analysis of the membranes' crossing rates of the three neutral complexes reveals a major role of the membranal dipolar field in regulating ion transport.

3. Carbazoles

Carbazole (CBZ) (9) has an ionizable proton at the nitrogen atom whose ESPT properties have been studied in detail. It has a pK_a value of 21.1 and a pK_a^* value of 11.0 [128]. Its absorption maximum is at 325 nm, with a moderately high extinction coefficient of 4200 M^{-1}/cm. The P* form (9) and the DP* form (10) emit at 362 and 417 nm, respectively. The fluorescence quantum yield in nonpolar medium is 0.41 [129]. All these make it a possible ESPT probe for microheterogeneous media, the ΔpK_a suggesting a possible use in alkaline pH ranges.

This possibility has been explored by Pappayee and Mishra [130]. It was found that in DMPC and DPPC liposomes, simultaneous presence of both P* and DP* form fluorescence is observed in a pH interval of 11–13. Within this pH interval, the maximal fluorescence sensitivity (largest changes in neutral to anionic form fluorescence ratio) was observed at pH 12. At this pH, CBZ partitions well into the liposome membrane with a large K_p value of 2×10^6 M^{-1} for DMPC and 3×10^6 M^{-1} for DPPC liposomes. Fluorescence quenching studies with the hy-

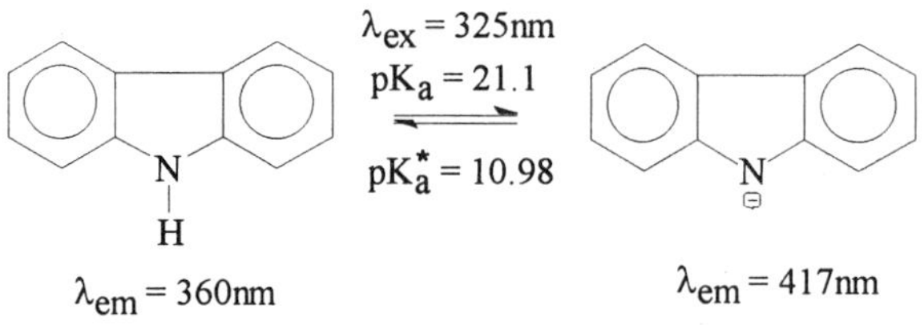

$\lambda_{ex} = 325$nm

$pK_a = 21.1$

$pK_a^* = 10.98$

$\lambda_{em} = 360$nm

$\lambda_{em} = 417$nm

9

10

carbazole(CBZ)

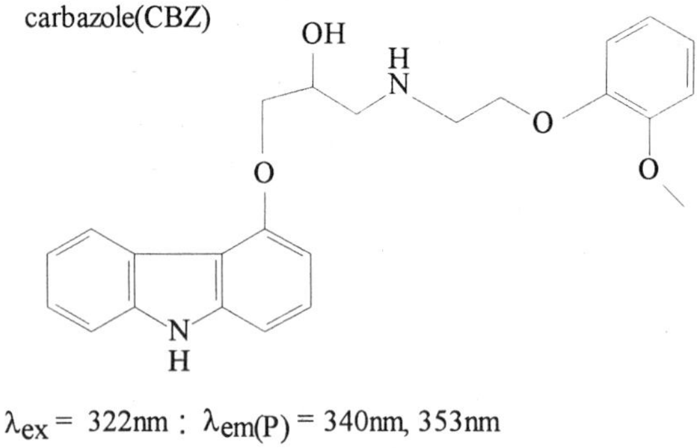

$\lambda_{ex} = 322$nm : $\lambda_{em(P)} = 340$nm, 353nm

11

carvedilol

Structures 9–11

drophilic I$^-$ quencher showed that this also has a two-site distribution in the membrane bilayer similar to naphthols [87]: a surface-water-accessible site and an inner hydrophobic site (Fig. 1). The fraction of surface site occupation was seen to be high at ~75%. The P* form fluorescence shows a sensitive response to thermotropic gel–liquid-crystalline-phase changes, which can be explained by a redistribution of CBZ in the two sites, in a manner analogous to that of 1NpOH. The cholesterol-induced changes in the membrane state was also sensed by it. The steady-state fluorescence anisotropy values were low even after incorporation into the membrane, which could be due to the preference of CBZ to surface sites. Carbazole as a probe in alkaline pHs thus becomes useful, as naphthol (with $pK_a = 9.2$) cannot be used at these pHs.

Carvedilol (**11**), which contains a CBZ moiety, is a multiple-action antihypertensive drug that has been shown to protect cell membranes from lipid peroxidative damages. Cheng et al. have studied [131] carvedilol, CBZ and 4-hydroxycarbazole in a 9 : 1 DMPC : DMPG membrane. Fluorescence anisotropy measurement showed that carvedilol is relatively mobile and does not have a rigidly defined molecular orientation in the membrane. The fluorescence spectra and fluorescence quenching studies with a nitrate quencher suggested a depth of 11 Å for the moiety of the drug. The drug was found to be an effective membrane fluidizer, as it dose-dependently lowers the thermotropic phase-transition temperature and broadens the endothermic transition.

4. ESIPT

3-hydroxyflavone (3HF) (**12**) and its analogs are known to exhibit intramolecular excited-state proton transfer (ESIPT). ESIPT leads to a tautomeric form (**13**) in the excited state (T^*) whose fluorescence shows a large red shift compared to the fluorescence of the normal form. The interaction of 3-HF [132] and 4-N,N'-dimethylamino-3HF [133], with synthetic vesicle membranes have been studied by Sengupta and co-workers. Both the molecules showed a conspicuous T^* fluorescence. Upon incorporation in the membrane, they show a substantially large increase in fluorescence anisotropy values compared to that in a homogeneous aqueous medium. A sigmoidal decrease of this anisotropy with increasing temperature permits monitoring the thermotropic phase-transition temperature of the membrane. A comparison of the intensity ratio of the tautomeric form and the normal form suggested that the molecule resides in a polar, aprotic environment in the vesicle membrane, presumably proximal to the carbonyl groups of the acyl chains in the interfacial region. Although a substantial red-edge-excitation-spectral (REES) effect was observed, indicating heterogeneity of probe distribution, similar excitation profiles of fluorescence of both forms suggest that the gross environment around the probe is similar.

2-(2'-hydroxyphenyl)imidzo[1,2-a]pyridine (HPIP) (**14**) also shows ESIPT. Rotation around the C(2)—C(1') single bond makes it possible to have four

λex = 340nm
λem = 410nm

12

λex = 340nm
λem = 533nm

13

3-hydroxyflavone (3-HF)

Structures 12 and 13

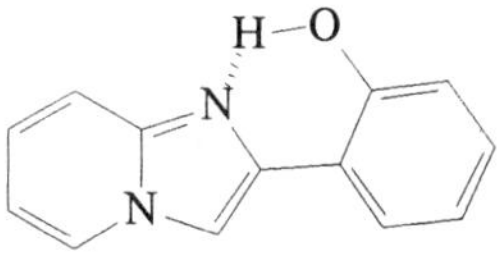

Structures 14 and 15

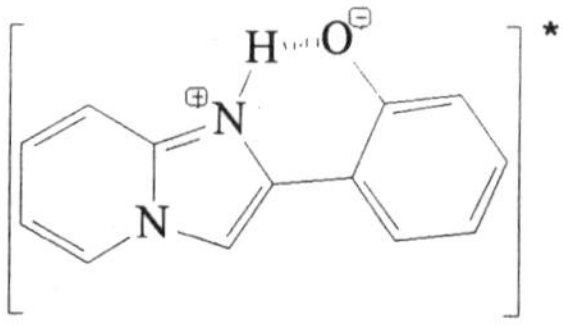

different structures in the ground and the excited singlet states. This has been introduced by Mateo and Douhal [134] as a new and sensitive probe for structural changes in lipid bilayers. In cyclohexane, HPIP has two absorption bands at 340 ($\varepsilon = 22,000$ M^{-1}/cm) and 357 nm assigned to (**14**), and a structureless emission band with an exceptional large shift at 590 nm, assigned to the rotamer (**16**) after ESPT [135]. Intramolecular ESPT is prevented in aqueous solution, and both absorption (max. 320 nm) and emission (max. 390 nm) spectra arise from (**17**). Upon the addition of DMPC vesicles, the ESIPT-related fluorescence band in the range 560–590 nm appears at the expense of 390-nm fluorescence. The K_p of the probe for DMPC is 4×10^4 at 30°C. Relative quenching of ESIPT fluorescence by lipophilic spin probes 5-doxyl-stearic acid and 16-doxyl-stearic acid suggested deep penetration of the probe. The spectral shift of the ESIPT-related fluorescence band with temperature, from 558 nm in the gel state to 590 nm in the liquid-crystalline state was sigmoidal in nature, giving the phase-transition temperature for both DMPC and DPPC. The probe has a reasonably large fluorescence anisotropy

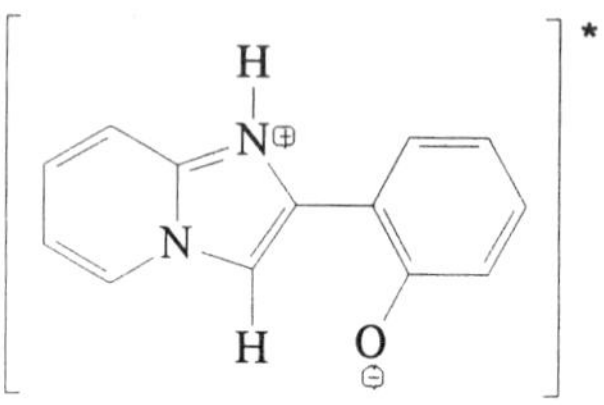

$\lambda_{ex} = 340$nm $\lambda_{ex} = 340$nm

$\lambda_{em} = 590$nm $\lambda_{em} = 390$nm

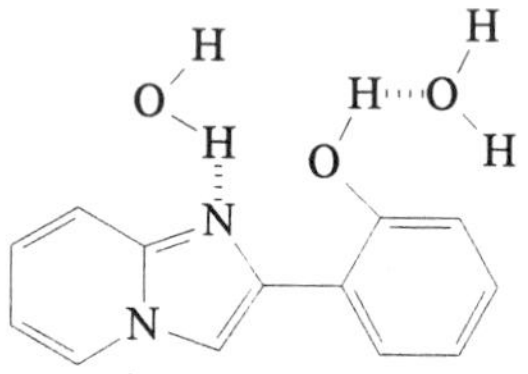

16   **17**

2-(2'-hydroxyphenyl)imidazo[1,2-a]pyridine (HPIP)

Structures 16 and 17

(~0.25 in gel and ~0.13 in liquid-crystalline states), which also responds to phase transition. The probe responds to cholesterol-induced membrane changes too.

B. Proteins

Proteins are made of amino acids linked by peptide bonds. The sequence within the polypeptide chains determines the primary structure of a protein. These chains fold together to form a helix or a β-sheet structure, which is known as the protein's secondary structure. The tertiary structure in proteins results from additional folding due to interactions between groups that are distant in the primary sequence. In an aqueous environment, the tertiary structure often brings the nonpolar hydrophobic side chains together, thus forming hydrophobic cavities. The dynamics of the protein structure plays an important role in its activity. The ability of protein to bind to a variety of small molecules and the possibility of alteration in fluorescence properties of protein-bound ESPT fluorophores has attracted a lot of attention in such studies. ESPT equilibria being sensitive to the microenvironment provides certain vital and unique information about the environment of the binding site.

1. Aromatic Hydroxy Compounds

a.   Naphthols and Derivatives.   The interaction of naphthols with a variety of proteins has been studied through ESPT. The absorption spectra of protein-intrinsic fluorophores tryptophan and tyrosine overlap with those of 1-naphthols (~290 nm), which makes it difficult to selectively excite the fluorophore. Therefore, most of such studies have been carried out with 2NpOH (~320 nm) and its sulfonates. In one of the very early uses of ESPT as a biological probe, Loken et al. [136] used 2-naphthol-6-sulfonate (NSOH) noncovalently bound to bovine serum albumin (BSA). The binding resulted in a sharp increase of P* form fluorescence at the expense of DP* form. A 1 : 1 molar complex is formed with a high value of binding constant ($>10^6 M^{-1}$). Nanosecond fluorimetry was used to monitor the substantial suppression of the proton-dissociation rate in the protein.

The lactone (**18**) of 2-hydroxyl-1-naphthaleneacetic acid can be used as a reagent for a novel highly efficient, covalent attachment of an ESPT fluorescence probe (i.e., 2NpOH) to the protein amino group [137]. The reaction of lactone with BSA was found to be faster than its reaction with corresponding concentrations of small organic amines, which suggested that the lactone first adsorbs to BSA and subsequently reacts covalently to an amino group at the site of adsorption (**19**). Equilibrium studies including instantaneous and time-dependent fluorescence changes indicate a 1 : 1 lactone : BSA labeling ratio at neutral pH. This particular method of covalent incorporation of an ESPT probe is significant for two reasons: (1) The lactone form is nonfluorescent; fluorescence appears only after binding of the probe to protein and (2) usually finding a substitution on naphthol for covalent

$\lambda_{ex} = 325\text{nm}$

$\lambda_{em} = 440\text{nm}$

18   **19**

Structures 18 and 19

linking to protein is difficult and the —OH group itself is reactive; this problem is neatly solved in the suggested method, as the —OH group is blocked in the lactone and the covalent reaction is also a deblocking reaction. A different method of covalently linking NSOH was developed by Jankowski and co-workers [138–141]. 2-acetyloxynaphthalene-6-sulfonic acid was used in a procedure [138] to covalently link NSOH to BSA on an active surface and inner sites and to pappain. The water accessibility of the bound probe was checked by fluorescence quenching methods using I^- and acrylamide. The equilibrium constants in the ground and excited states and rate constants of ESPT for the introduced fluorophoric groups were estimated by steady-state fluorescence spectroscopy and compared with the data for low-molecular-weight analogs of 2NpOH. The pK_a and pK_a^* values of bound NSOH were slightly lower than NSOH itself due to the surface charge of protein. It was observed that the fluorophores bound inside macromolecules are, in most cases, characterized by a higher rate of ESPT than groups situated on the surface. Proton transfer through intramolecular hydrogen bonds of the protein was suggested as a reason for this observation [139]. When the amine and the carboxyl groups in BSA were modified by neutral groups by following the method of Means and Feeney [142], the variation in both pK_a and pK_a^* values of the NSOH groups bound to the protein was induced. The dependence of $\log k_{pt}$ on $\log K_a^*$ for BSA–NSOH conjugates was fitted to a modified Brønsted relation. This plot was treated as a reference curve for comparison with analogous data for NSOH groups bound to other protein samples. Deviations from this plot were explained by kinetic isotope effects. When the fluorophore is bound preferentially to the surface of the protein globule by a spacer (approximately 2.5Å), the diffusion of water as a proton acceptor is the rate-limiting step in ESPT. When the same fluorophore is bound to the surface of the macromolecule by a sulfonamide bond anchored directly to an amino group of the protein, partition of a protein is the rate-determining step, leading to a "negative" isotope effect and a low rate of ESPT. When the fluorophore is bound inside the protein molecule, proton transfer through a chain of preformed hydrogen bonds is responsible for a much higher rate of ESPT than in the other samples [140]. Centers of gravity (CGs) of the fluorescence bands of P* and DP* forms of NSOH bound to proteins were determined at various experimental conditions and were correlated with the solvent relaxation time around fluorophores. The variation in the rate of ESPT for NSOH groups bound to proteins could be explained in terms of the solvent polarizability and the solvent relaxation rate with respect to a moving proton [141]. The ESPT dynamics of 2-naphthol 6,8,-disulfonate (NDSOH) noncovalently bound to BSA has been studied by Das et al. [143] using nanosecond and picosecond time-resolved fluorimetry. A 3 : 1 NDSOH : BSA stoichiometry was found to form which changed to a 1 : 1 complex in the presence of urea. In a concentrated salt solution, the excited-state proton dissociation is slowed down as the exponential function of activity of water in the solution. This kinetic parameter was used to probe the microenvironment of the binding site of BSA at which NDSOH is bound. The dissociation of the proton in water is a very fast reaction, with a rate constant of 7.2 $\times$ 10^9 sec^{-1}; however, upon binding to BSA, the rate of proton dissociation is slowed down significantly to 2.0×10^9 sec^{-1}. This slow dissociation rate constant suggests a strong interaction of the water molecules with the inner walls of the cavity. On the addition of urea, it increases to 2.5×10^9 sec^{-1} because of the increased availability of water molecules to hydrate the dissociated proton. The ability of the water molecules to hydrate the dissociated proton in the site is equivalent to a homogeneous salt solution with the activity of water at ~0.67, but in the presence of urea, the activity of the water molecules in the binding site is 0.78.

The estrogen d-3-hydroxy-1,3,5(10),6,8-estrapentaen-17-one (equilenin) (**3**) contains 2NpOH moiety. The interaction of equilenin with the human and rabbit sex-steroid-binding proteins (hSBP and rSBP, respectively) was investigated [144]. Equilenin was found to compete for the binding of 5α-dihydro-testosterone. The binding constant of equilenin for rSBP was calculated to be 1.9×10^7 M^{-1} at 4°C, comparable with the binding constant of 5.7×10^7 M^{-1} reported for hSBP [145,146]. The results of fluorescence quenching experiments for the anionic form fluorescence with the collisional quenchers I$^-$ and acrylamide indicated that the bound steroid has limited accessibility to the bulk solvent and that there are no anionic surface groups near the steroid-binding site. The fluorescence excitation spectra of SBP–equilenin complexes are similar to the absorption spectra of equilenin in low-dielectric solvents. The fluorescence emission of the complexes, however, were found to exhibit a red shift opposite to those of the steroids in low-dielectric solvents or complexed with β-cyclodextrin (blue shift) but similar to the red shift produced by the addition of the proton-acceptor triethylamine to equilenin in cyclohexane. These data indicated that the steroid-binding site of hSBP and rSBP is a nonpolar cavity containing a proton acceptor that participates in a specific interaction, possibly a hydrogen bond, with the 3′-hydroxyl group of the bound steroid.

   b.   Pyranine.   Compared to naphthols, the absorption (405 nm) and fluorescence (440 nm for the P* form, and 510 nm for the DP* form) maxima of pyranine are more red-shifted with little overlap with the corresponding spectra of protein-intrinsic fluorophores. The microenvironment of the hydrophobic site of BSA and apomyoglobin heme-binding site has been studied for pyranine as the probe by Gutman et al. [147,148]. The 1 : 1 complexes were seen to form. A two-component decay of P* form fluorescence was observed in both. The rapid component of decay represented proton dissociation in the binding site accompanied by a rapid recombination reaction. During the slow phase, the excited DP* form and the proton are in equilibrium, but the escape of the proton from the cavity shifts the P* form population toward the DP* form. The binding affinity in apomyoglobin was pH dependent. Two protonatable groups with $pK = 6.5$ participate in the stabilization of the negatively charged ligand in the binding site of apomyoglobin. The displacement of pyranine by hemin suggested that pyranine is bound to the heme-binding site of apomyoglobin. The rate of proton dissociation is slowed to 7% of the rate measured for the free ligand, which implies the water activity in the site at 0.67. On adsorption of pyranine to lysozyme, a fluorophoric site with a total charge of -3 is created on the positively charged protein. ESPT dynamics in this system has been studied [148]. The probe reactions were studied with a series of dye–enzyme complexes where the number of free carboxylate was reduced by amidation, increasing the total charge of the complex from $+5$ to $+12.6$. Microsecond absorption measurements were used to monitor bulk proton penetration in the Coulomb cage of the bound dye and time-resolved fluorescence was used to monitor the escaping of proton out of the Coulomb cage.

2.   pH Jump

The principles of proton pulse or pH jump technique developed by Gutman has been discussed in Section II.A.2. Such techniques have also been used for probing protein microenvironments. Using this technique, the fluorescein–isothiocyanate adduct of BSA has been used as a model for studying the kinetic of protonation of a distinguishable site on a protein [150]. Fast perturbation of the bulk pH, by the laser-induced proton pulse obtained from 2-naphthol-3,6-disulfonate, led to transient protonation of the dye as monitored at 441 nm. The dynamics were measured over a wide range of prepulse pH values (7–8.5) and analyzed by a numerical solution of four coupled differential equations which define the kinetics of the following reactants: the proton emitter, the covalently bound dye, and the protonable groups of the proteins, carboxyls, and amines. Both carboxyls and lysines of the proteins were found to be involved in the protonation of the covalently bound dye. The carboxyls act as efficient primary acceptors of the bulk protons, followed by proton transfer to more basic moieties. Within a microsecond, the proton distribution between bulk and protein is in equilibrium, whereas the proton distribution between surface groups is over in less than 5 μsec.

The pH jump technique with picosecond fluorescence measurements were used to study apomyoglobin and the anionic specific channel, porin of *E. coli* [151]. The results indicated that the water in the sites deviates markedly from the liquid state in the bulk, having a lower dielectric constant and smaller diffusivity of protons.

The interaction of bulk protons with the cytoplasmic surface (proton-input side) of cytochrome *c* oxidase has been studied by Yael et al. [152] by labeling the cytoplasmic surface of the enzyme with fluorescein covalently bound on the surface. Protonation of flurescein was observed to proceed through multiple pathways involving diffusion-controlled reactions and proton exchange among surface groups. The surface of the protein carries an efficient system made of carboxylate and histidine moieties that are sufficiently close to each other so as to form a proton-collecting antenna. It is the passage of protons among these sites that endows cytochrome *c* oxidase with the capacity to pick up protons from the buffered cytoplasmic matrix within a time frame compatible with the physiological turnover of the enzyme.

3. ESPT in Certain Biomolecules

The presence of ESPT has been implicated in explaining the fluorescence behavior of some biomolecules.

    a. *Luciferin.* Dehydroluciferin (**20**) possesses a phenolic proton which shows ESPT behavior upon excitation [153] ($pK_a = 8.7$, $pK_a^* = -1.0$). Fluorescence lifetime measurements have been used by DeLuca et al. [154] to obtain

$\lambda_{ex} = 338nm : \lambda_{em}(P^*) = 435nm : \lambda_{em}(DP^*) = 530nm$

$pK_a = 8.7 : pK_a^* = -1.0$

20

dehydroluciferin

Structure 20

nanosecond time-resolved emission spectra of the molecule in various solvents and when bound to luciferase.

The blue fluorescence of the P* species is at 435 nm and the green emission of the DP* form is at 530 nm. This large shift permits accurate measurement of the time course of excited-state ionization. The rate of proton transfer is very fast in aqueous solution but slower in 80% ethanol. The addition of imidazole increases the rate of proton transfer. When bound to luciferase, dehydroluciferin has a slow rate of proton transfer, which is shown by an increase of relative fluorescence intensity of the DP* form at longer times (11 nsec) after pulsing. This suggests that the binding site is hydrophobic. Dehydroluciferol, a fluorescent substrate analog of firefly luciferin, has ESPT properties similar to dehydroluciferin with the necessary spectral characteristics to be a good active-site probe [155]. The binding of the activator, adenosine monophosphate (AMP), was seen to cause changes in the fluorescence emission of bound dehydroluciferol, which indicated that the microenvironment becomes more hydrophobic with binding. The effect of AMP may have significance in normal firefly bioluminescence because it facilitates this ionization.

b. Green Fluorescent Protein. The green fluorescent protein (GFP) of the jellyfish *Aequorea victoria* has attracted widespread interest since the discovery that its chromophore is generated by the autocatalytic, posttranslational cyclization, and oxidation of a hexapeptide unit [156]. This permits fusion of the DNA sequence of GFP with that of any protein whose expression or transport can then be readily monitored by sensitive fluorescence methods without the need to add exogenous fluorescent dyes. The excited-state dynamics of GFP has been studied following photoexcitation of each of it two strong absorption bands in the visible using fluorescence upconversion spectroscopy (about 100 fsec time resolution) [157]. The absorption bands are suggested to originate from two conformations that interconvert slowly in the ground state but rapidly in the excited state. The observed isotope effect suggested that the initial excited-state process involved a proton-transfer reaction that is followed by additional structural changes.

c. Hypericin. The photodynamic drug hypericin (**7**) is known to acidify its surroundings upon light absorption. Ultrafast time-resolved fluorescence microscopy of hypericin interaction with fetal rat neurons showed a nonexponential decay of fluorescence which were attributed to excited-state electron transfer, excited-state proton transfer, or both [158]. Hypericin associates with membranes throughout the cell. Ultrafast time-resolved fluorescence and absorption measurements have been performed on hypericin complexed with human serum albumin (HSA) (1 : 4, and ~5 : 1 hypericin : HSA complexes) [159]. Detailed comparisons with hypocerllin A (an analog of hypericin)–HSA complexes (1 : 4 and 1 : 1) were made. The results were consistent with the conclusions of earlier studies [160], indicating that hypericin binds to HSA by means of a specific hydrogen-bonded interactions between its carbonyl oxygen and the B1-H of the tryptophan residue in the IIA subdomain of HSA. A single-exponential rotational diffusion time of 31 nsec is measured for hypericin bound to HSA, indicating that it is very rigidly held. In contrast, hypocrellin binds largely to the HSA surface in a nonspecific manner. Although it is tightly bound to HSA, hypericin is still capable of executing excited-state intramolecular proton (or hydrogen atom) transfer in the ~5 : 1 complex, albeit to a lesser extent than when it is free in solution. The possible role of ESIPT in light-induced antiviral and antitumor activity has been discussed.

   d.   ESPT and Intrinsic Tyrosine Fluorescence in Proteins.   A discussion on tyrosine fluorescence in proteins is relevant, as the phenolic proton in tyrosine has a pK_a near 10, whereas the first excited singlet state, according to Förster cycle calculation has a pK_a^* between 4 and 5 [161]. Thus, at a physiological pH, one would expect observation of emission from the anionic form. There is, however, no experimental evidence for ESPT by tyrosine to water [162]. Tyrosine fluorescence has a constant quantum yield as a function of pH through the region of the pK_a^*. The excited-state deprotonation rate of tyrosine is nearly two orders of magnitude smaller than that of 2NpOH [30,163]. The fluorescence lifetime of tyrosine being around 3 nsec, the deprotonation rate is too slow compared to deactivation rate of the excited state. The review by Ross et al. [162] discusses the connection between ESPT and tyrosine fluorescence in detail.

4.   ESIPT

Only a few works have appeared in literature on the interaction of ESIPT molecules with proteins. In a comparative study on fluorescence of the three hydroxyflavones 3-hydroxyflavone (3HF) (**12**), 3,3',4',7-tetrahydroxyflavone (fisetin) (**21**), and 4'-diethylamino-3-hydroxyflavone (DHF) (**23**), Sytnik et al.

$\lambda_{ex} = 365nm$ $\lambda_{ex} = 365nm$
$\lambda_{em} = 463nm$ $\lambda_{em} = 540nm$

21   **22**

Structures 21 and 22

$N(C_2H_5)_2$ $N(C_2H_5)_2$

$\lambda_{ex} = 413nm$ $\lambda_{ex} = 413nm$

$\lambda_{em} = 516nm$ $\lambda_{em} = 570nm$

23   **24**

4'-diethylamino-3-hydroxyflavone (DHF)

Structures 23 and 24

[164] have seen that for all three flavonols, the P* fluorescence in the 400-nm region is largely replaced by the T* fluorescence in the 550-nm region in aprotic solvents at room temperature.

For DHF in polar solvents, the normal fluorescence becomes a charge-transfer fluorescence (460–500 nm) which competes strongly with the still dominant proton-transfer fluorescence (at 570 nm). In protic solvents and at 77K, the interference with intramolecular hydrogen-bonding gives rise to greatly enhanced normal fluorescence, lowering the quantum yield of proton-transfer fluorescence. The utility of DHF as a discriminating fluorescence probe for protein binding sites in rat serum albumin is suggested by the strong dependence of the charge-transfer fluorescence on polarity of the environment and by various static and dynamic parameters of the charge-transfer and proton-transfer fluorescence which can be determined.

4-hydroxy-5-azaphenanthrene (HAP) (**25**) has a nitrogen center conveniently located near the photodissociable proton. The ESIPT fluorescence [from (**26**)] of this molecule has been proposed as a probe for protein conformation and binding-site monitoring [165]. A typical grossly wavelength-shifted PT fluorescence for HAP is observed in the 600-nm region for this UV-absorbing molecule (absorption onset, 400 nm), for which ESPT occurs as a protein-binding-site static-polarity calibrator, shifting from a λ_{max} of 612 nm in cyclohexane to 585 nm in ethanol at 298 K, contrary to the usual dispersion red shift. A small mechanical solvent-cage effect is noted in ethanol at 77 K, but solvent dielectric relaxation is not apparent from the fluorescence spectrum. Thus, HAP serves to distinguish static solvent-cage polarity from dynamical solvent dielectric relaxation and other

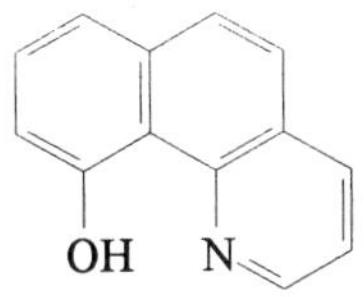

$\lambda_{ex} = 381$nm

$\lambda_{em} = 585$ nm

25

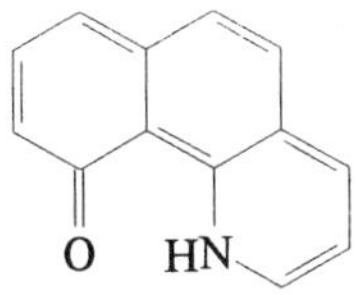

$\lambda_{ex} = 381$nm

$\lambda_{em} = 612$nm

26

4-hydroxy-5-azaphenanthrene (HAP)

Structures 25 and 26

solvent-cage effects (mechanical restriction of molecular conformation). HAP as an ESIPT fluorescence probe has been applied to HSA and beaver-apomyoglobin. Preliminary studies on temperature dependence of phototautomer fluorescence indicate changes in micropolarity of protein binding sites.

The ESIPT of 2-(2'-hydroxyphenyl)-4-methyloxazole (HPMO) (**27**) has been explored by Douhal and co-workers [166] for its probe characteristics in a variety of organized media which include cyclodextrin, calixarene, micelle, and HSA. The incorporation of HPMO into hydrophobic cavities in an aqueous medium involves the rupture of its intermolecular hydrogen bond to water and formation of an intramolecular hydrogen bond in the sequestered molecule. Upon excitation (280–330 nm) of this entity, a fast intramolecular proton-transfer reaction of the excited state produces a phototautomer (**28**), the fluorescence of which ($\lambda_{max} = 450$–470 nm) shows a largely Stokes-shifted band. Because of the existence of a twisting motion around the C_2—C_1' bond of this phototautomer, the absorption and emission properties of the probe depend on the size of the host cav-

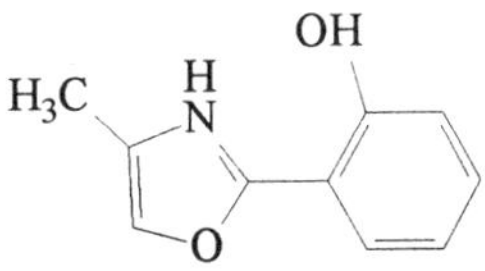

$\lambda_{ex} = 320$nm

$\lambda_{em} = 350$nm

27

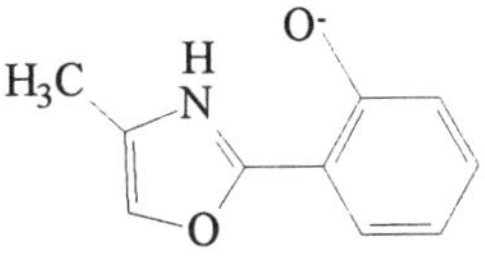

$\lambda_{ex} = 320$nm

$\lambda_{em} = 465$nm

28

Structures 27 and 28

ity. Incorporation to the host also results in a large increase of fluorescence anisotropy. The emission of the sequestered phototautomer of HPMO is shown to be a simple and efficient tool for detecting and exploring the size of hydrophobic nanocavities.

C. Cyclodextrins

Cyclodextrins (CD) are α-(1,4)-linked glucopyranose rings forming truncated-cone-shaped molecules. The interior of the cavity is hydrophobic and accommodates a wide variety of organic molecules. α-, β-, and γ-CDs are made of 6, 7, and 8 D(+)-glucopyranose units, respectively, and the width of the cavity increases correspondingly. The outer side of the CD is hydrophilic, with primary hydroxyl groups on the narrower rim of the truncated cone and secondary hydroxyl groups on the wider rim.

Unlike lipid bilayer membranes and proteins in which the ESPT of aromatic hydroxy compounds have been extensively used for probe purposes, a diverse variety of ESPT molecules have been studied as inclusion complexes in cyclodextrin cavities. The well-defined CD cavities often accommodate molecules in distinctly different but well-defined orientations. This is reflected in their ESPT behavior. For the conventional range of ESPT molecules, the size of β-CD seem to be more appropriate and a fairly large number of studies are reported on it; α-CD appears too small and γ-CD too large for forming suitable inclusion complexes. The effects of local polarity and water accessibility in ESIPT have been topics of active interest recently. These aspects are discussed next.

1. Aromatic Hydroxy Compounds

a. Naphthols and Derivatives.   Structural differences among inclusion complexes of α-, β-, and γ-CDs with 2NpOH have been studies by means of optical absorption, fluorescence, and circular dichroism spectrometry [167]. The rate of proton dissociation is markedly depressed by inclusion. The hydroxyl group of 2NpOH is hidden in the α-CD cavity, whereas the naphthyl group is largely exposed to a water phase. The fluorescence study revealed that 2NpOH is very loosely packed in a γ-CD cavity, whereas it is tightly packed in a β-CD cavity. It has been suggested from fluorescence lifetime studies that there is a complete suppression of excited-state deprotonation in the 2NpOH–β-CD complex, as the basicity of the hydroxyl groups on the sugar subunit in the cavity is too low to accept the proton [168].

The pH-dependent fluorescence of 1,6-naphthalenediol (ND) (**29**), 1NpOH, and 2NpOH has been studies in the presence of α-CD and β-CD [168]. Inclusion complexation of ND inside the CD cavity greatly reduced proton transfer in photoexcited ND. This led to the interesting observation of four pH-dependent bands, attributed to four different forms of ND: molecular (**29**), β-ionic (**30**), α-ionic (**31**), and diionic (**32**), having fluorescence maxima at 365, 400, 470, and 440 nm, respectively. Ground-and excited-state ionization constants of 2NpOH in inclusion

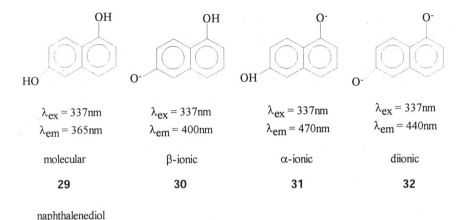

$\lambda_{ex} = 337nm$	$\lambda_{ex} = 337nm$	$\lambda_{ex} = 337nm$	$\lambda_{ex} = 337nm$
$\lambda_{em} = 365nm$	$\lambda_{em} = 400nm$	$\lambda_{em} = 470nm$	$\lambda_{em} = 440nm$
molecular	β-ionic	α-ionic	diionic
29	**30**	**31**	**32**

naphthalenediol

Structures 29–32

complexes with α-CD, β-CD, and some substituted β-CDs have been determined using steady-state and time-resolved optical spectroscopy techniques [170]. These pK_a's relate to the complexation constants of the two prototropic forms. As the standard free energy of inclusion is larger for the molecular form than for the ion, this aromatic acids becomes less acidic upon complexation. In the case of α-CD as host, 1:2 guest:host complexation suppresses excite-state deprotonation. Scheme 2 summarizes the energetics of the 2NpOH–cyclodextrin system.

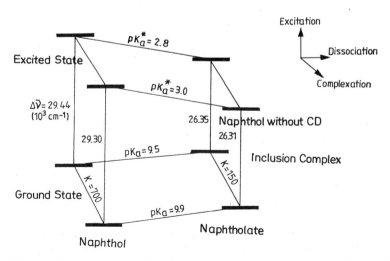

Scheme 2 Thermodynamic and spectroscopic quantities relating to the 2-naphthol–cyclodextrin complex. (From Ref. 170a. Copyright 1994 American Chemical Society.)

The effect of third-party interaction with the 1:1 complex of 2NpOH–β-CD has been studied using fluorescence, the third party being a series of short-chain linear alcohols (methanol–pentanol) [171]. The apparent complexation constant for 2NpOH decreases linearly with an increasing number of carbon atoms in the chain of the alcohol which is attributed to a competition between 2NpOH and alcohol for the β-CD cavity. Fluorescence studies confirm the redistribution of 2NpOH from the CD environment to the aqueous phase when alcohols are present. 2NpOH fluorescence is quenched by iodide in all the systems studied. At 2 mM β-CD, alcohols increase the Stern–Volmer constants above the value found in the absence of alcohols. These results suggest that alcohols occupy space within the β-CD cavity, with the result that the aqueous 2NpOH concentration is increased.

b. *Hypericin.* The ESIPT of hypericin (**7**), the antiviral drug, has been studied in presence of β-CD [172]. Hypericin is insoluble in water at pH 4, but goes into solution with β-CD because of the formation of a complex. The absorption and fluorescence spectra of the complex are very similar to those of hypericin in less polar solvents. From the nature of the spectra, it has been suggested that the corner of hypericin bearing the carbonyl group and the β-hydroxyl group adjacent to it are accommodated in the β-CD cavity.

2. Nitrogen Heterocycles and Aromatic Amino Compounds

Fluorimetric investigations of the ESPT reactions of carbazole (CBZ) (**9**) and 2-naphthylamine in the presence of β-CD show that the prototropic reaction depends not only on the microenvironment of the molecular imposed by CD but also on the nature of the molecule itself [173]. Thus, in comparison with the bare chromophore, the deprotonation rate of the CD inclusion complex is enhanced when the guest molecules is CBZ, whereas it is decreased for guests like naphthylamine or naphthols. CBZ includes in the cavity with its >NH proton hydrogen bonded to the outer rim hydroxyl groups. Invoking a bidirectional flipflop hydrogen-bonding mechanism [175] Chattopadhayay has suggested a cooperative proton-transfer scheme involving participation of the rim hydroxyl groups to explain the ESPT to OH⁻. The addition of urea caused the expulsion of the probe from the CD environment to the bulk aqueous phase, as was reflected in the DP* form fluorescence intensity, prototropic rates, and fluorescence quenching studies [176]. A brief review by Chattopadhyay et al. [177] discusses the effects of environmental perturbations by crown ether, a micelle-forming surfactant, and CD on the kinetics of ESPT reactions.

When 1-aminopyrene (**33**) is complexed with β-CD, the dissociation rate of its P* form in the excited state increases by a factor of 2–3, compared to pure water [178]. This is in contrast to the well-known decrease of excited-state deprotonation of 1NpOH. It was suggested that the hydroxyl rims of β-CD have a similar effect on their immediate water environment as do simple alcohols in bi-

λ_{ex} = 295nm

pK_a = ~3.0

pK_a^* = -1.2

λ_{em} = 374nm λ_{em} = 440nm

33   **34**

1-aminopyrene

Structures 33 and 34

nary aqueous mixtures. Two distinct binding orientations for 1-aminopyrine in β-CD were observed with the amino group exposed to water. The immediate environment around the nitrogen center of 1-aminopyrene resembles a 75% by volume ethanol–water mixture, whereas that near 1NpOH is similar to 80% ethanol–water.

2-(2′-pyridyl)benzimidazole (2-PyBI) (**35**) shows considerable changes in fluorescence spectra and decay behavior when complexed with α-, and β- CD, and do not form complexes with γ-CD [179]. The red shift of the tautomer form of the excited-state cation essentially implies that the molecule is an excited-state base in Förster cycle terms. Steady-state and time-resolved fluorescence measurements at different pHs, semiempirical calculations of structures, and nuclear magnetic resonance (NMR) measurement revealed different structures of inclusion complexes and the effect of this on ESPT. In α-CD, the pyridyl nitrogen is shielded by the cavity, whereas in β-CD, the benzimidazole part resides in the cavity. Emission at 500 nm attributable to the T^* state (**36**) is absent in α-CD complexes and is nonexponential in β-CD complexes, which have been explained through the structures of the complexes.

3. ESIPT

10-hydroxybenzo[*h*]quinoline [alternatively named 4-hydroxy-5-azaphenanthrene (HAP)] (**25** and **26**) has a structure in which the phenolic hydrogen and the basic nitrogen are nicely disposed to form a strong intramolecular hydrogen bond. Photoexcitation leads to the tautomeric structure (**26**), which emits with one of the largest Stoke's shift (absorption max. 370 nm, fluorescence max. 600 nm) in ESPT literature [181]. HAP forms 1:1 inclusion complexes with β- and γ-CDs with reasonably large complexation constants (513 for β-CD and 263 for γ-CD) [182]. There is a substantial enhancement of the T^* fluorescence with complexation due to the reduction of nonradiative deactivation rates.

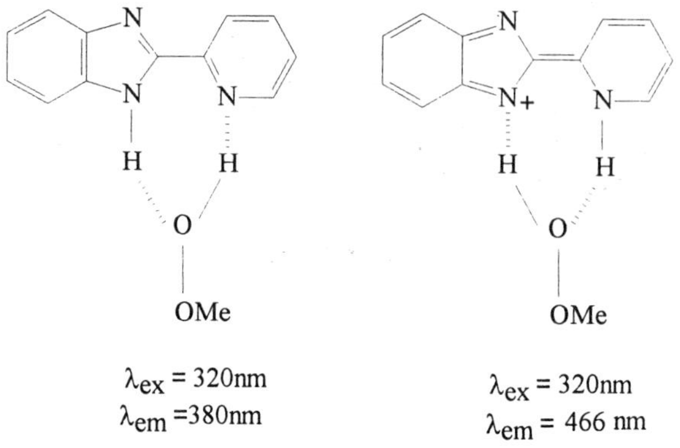

λ_{ex} = 320nm
λ_{em} =380nm

35

λ_{ex} = 320nm
λ_{em} = 466 nm

36

2-(2'-pyridyl)benzimidazole (2-PyBI)

Structures 35 and 36

2-(2'-aminophenyl) benzimidazole (2-APBI) (**37**) forms a 1 : 1 inclusion complex with β-CD which results in red-shifted absorption and fluorescence spectra [183]. In the complex, the aminophenyl moiety is toward the wider rim and the benzimidazole is toward the narrower rim. Both of the ground-state rotational isomers (**37**) and (**39**) are present. Fluorescence originates from the phototautomer (**38**) as a shoulder at ~500 nm and from (**39**) at ~440 nm. The decrease in the fluorescence quantum yield of the large red-shifted fluorescence band is due to back-proton-transfer from the β-CDx hydroxyl groups to the imine group to yield the amine group in the S_1 state. In contrast, a 1:1 inclusion complex forms between the neutral form of 2-(4'-N,N-dimethylaminophenyl)benzimidazole (DMAPBI) and β-CD in aqueous media with only one orientation, with the —NMe$_2$ group close to the narrower rim and the nitrogen atoms of the benzimidazole moiety near the wider rim [180]. Fluorescence spectra, fluorescence excitation spectra, different values of binding constants observed at 440 nm and 520 nm, and time-resolved fluorescence studies suggest that the monocation species (**40** and **41**) can form two kinds of inclusion complex.

The effects of α-, β-, γ-, and 2,6-di-o-methyl-β-CDs on the ground- and excited-state properties of 2-(2'-hydroxyphenyl)benzoxazole (HBO), 2-(2'-hydroxyphenyl)benzothiazole (HBT), and 2-(2'-hydroxyphenyl)benzimidazole (HBI) (**42**) in aqueous media have been investigated using fluorescence [184]. The molecules form 1:1 complexes with a reasonably high association constant, and they enter the cavity axially from the wider-rim side of β-CD with the benzazole

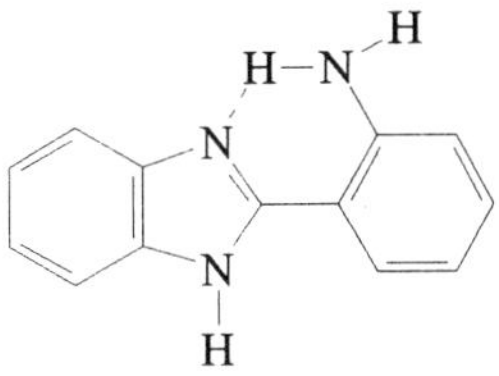

$\lambda_{ex} = 325nm$
$\lambda_{em} = 412nm$

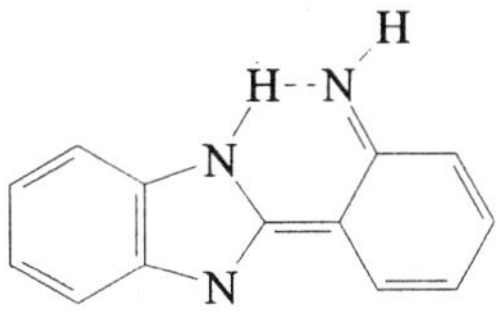

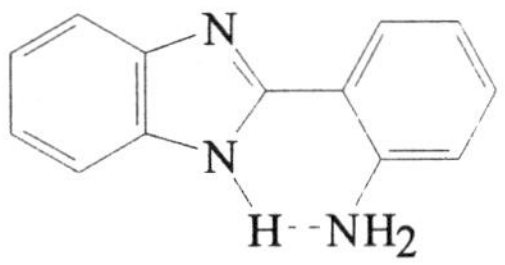

$\lambda_{ex} = 325nm$
$\lambda_{em} = 520nm$

P

T*

37

38

39

2-(2'-aminophenyl)benzimidazole (2-APBI)

Structures 37–39

moiety forward. Measurement of acidity constants and the data from induced circular dichroism indicate increased ground- and excited-state acidities of the phenolic protons of HBT and HBI with inclusion, but for HBO, there is an increase in ground-state acidity and decrease in excited-state acidity. The data further suggest a planar structure for HBO and a twisted confirmation for both HBT and HBI. It was suggested that the HBA molecule exists as a zwitterionic tautomer in the presence of CDs. The intramolecular interactions of HBI with CD and various solvents appear to weaken intramolecular hydrogen-bonding in HBI and facilitate the formation of strong intermolecular hydrogen bonds with the various CD and solvent molecules [185]. 2-(2'-hydroxyphenyl)-4-methyloxazole (HPMO) (**27**) is com-

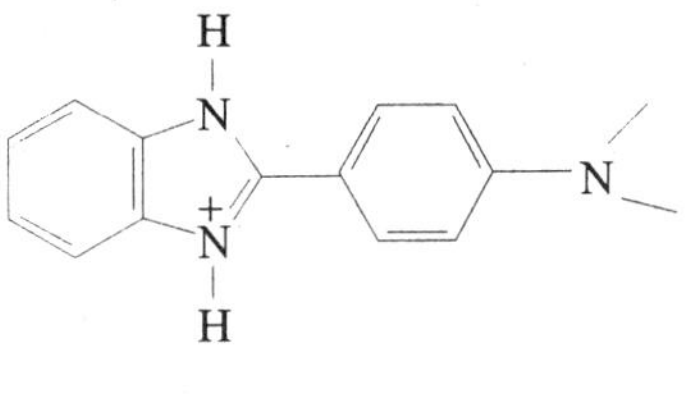

$\lambda_{ex} = 328nm$
$\lambda_{em} = 396nm$

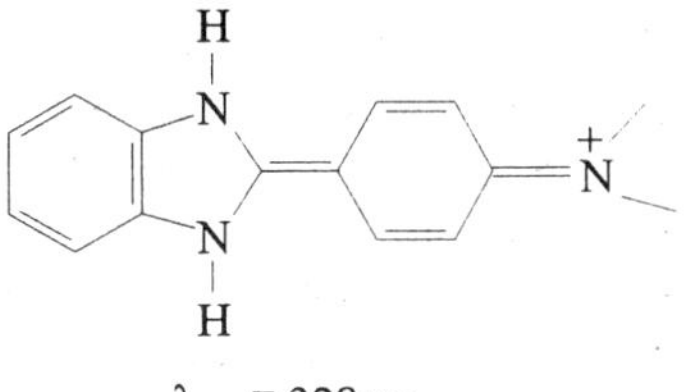

$\lambda_{ex} = 328nm$
$\lambda_{em} = 586nm$

40

41

2-(4'-N,N-dimethylaminophenyl)benzimidazole (DMAPBI)

Structures 40 and 41

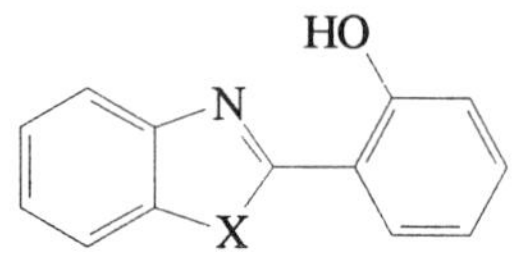

X		λ_{ex}	$\lambda_{em}(P^*)$	$\lambda_{em}(DP^*)$
O	HBO	320nm	440nm	478nm
NH	HBI	350nm	350nm	427nm
S	HBT	340nm		460nm

42

2-(2'-hydroxyphenyl)benzazoles

Structure 42

plexed by β-CD [166] with a large complexation constant of 1400 M^{-1}. The oxazole moiety was found to be sequestered by the CD cavity. The red shift of the absorption maximum of the inclusion complex when the diameter of the CD cavity was increased (from α-CD to β-CD to γ-CD), is related to the possibility of C2—C1' bond rotation.

2-(2'-hydroxyphenyl)imidazo[1,2-a]pyridine (HPIP) (**14**) is encapsulated by all three CD cavities [186]. As the tautomer fluorescence maxima (**16**) are related to a single bond rotation around the C2—C1' bond axis [187,188], the red shift observed with increasing cavity size (500 nm for α-CD, 550 nm for β-CD, and 570 nm for γ-CD) directly relates to the freedom offered by the cavity for the single bond rotation. This red shift in fluorescence of HPIP as a probe is thus proposed to provide a unique, simple, and direct measure of the diameter of the encapsulating cavity.

The molecule 4-methyl-2,6-dicarbomethoxyphenol (CMOH) (**43**), which undergoes ultrafast ESIPT, has been employed as a guest to probe ESIPT within the interior of the host α-CD and β-CD in water and highly polar nonaqueous media [184]. It absorbs at 322 nm and emits from the T^* form (**45**) at 455 nm. Enhancement in the tautomer fluorescence upon inclusion can be used to obtain association constants. The ESIPT reaction of CMOH is favored in its microencapsulated form in both aqueous and nonaqueous media. As a consequence, resultant enhancement of tautomer emission, following ESIPT, relative to that in homogenous bulk media is observed. CMOH forms 1:2 and 1:1 complexes

P   P*   T*

43   **44**   **45**

4-methyl-2,6-dicarbomethoxyphenol (CMOH)

Structures 43–45

with α-CD and β-CD, respectively, in both water and highly polar nonaqueous media. However, tautomer emission due to ESIPT is significantly enhanced in water compared to nonaqueous solvents, despite the formation of stronger guest–host complexes with a much higher binding constant in the nonaqueous media. The results have been interpreted as the formation of ternary complexes comprising CMOH, second guest, and CD in nonaqueous media, where the nonaqueous solvent acts as the second guest molecule.

D. Crown Ethers

The ESPT of naphthylammonium [190] and phenanthrylammonium [191] ions in their 18-crown-6 ether complexes in MeOH-H_2O (9:1) solvent shows that the excited-state proton-transfer rate decreases markedly on complexing. The back-protonation rate in the excited state is negligibly small compared with those of the other decay processes, which essentially means that there is no excited-state protropic equilibrium in the crown complexes. The one-way proton-transfer reaction is elucidated by the presence of the excited neutral amine–crown complex (RNH_2–crown) produced by deprotonation of (RN^+H_3–crown). There is a large steric effect on protonation to the amino group of the excited neutral complex.

E. Monolayers and Langmuir–Blodgett Films

The monolayer formed by 6-dodecyl-2-naphthol [192] on neutral water shows no fluorescence due to the DP* form at 420 nm. The absence of a water–cluster–base [33–37], and a reduction of apparent surface dielectric constant have been offered as explanations of this observation. A similar observation of the absence of excited DP* form fluorescence was made in Langmuir–Blodgett (LB) films of

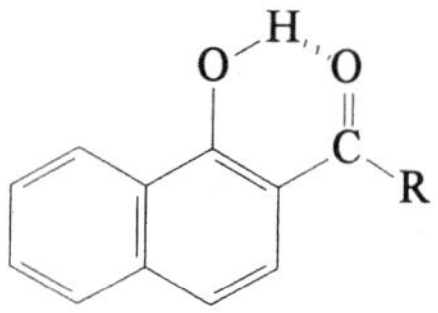

P

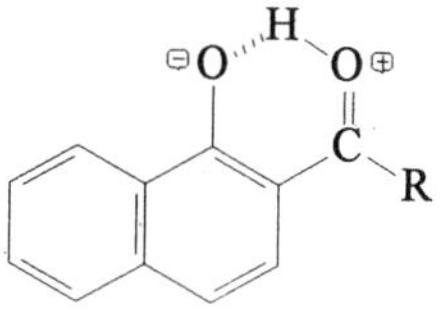

T*

46

47

2-acyl-1-naphthol

Structures 46 and 47

stearic acid mixed with long-chain-substituted naphthols (2-octadecyl-1-naph-
thol, 1-octadecyl-2-naphthol) [193]. Possible reasons for stearate as the base for
not accepting the proton from excited-state naphthol could be the low polarity of
the film interior and/or insufficient availability of protons. In LB films made of
stearic acid and 2-stearoyl-1-naphthol (**46**), however, intramolecular ESPT [47]
was found to proceed with a high rate constant ($>10^{10}$ sec^{-1}), similar to that in
homogeneous media.

F. Polymers and Sol–Gel Glasses

6-(2-hydroxy-5-methylphenyl)-s-triazines (**48**)–(**50**) are strongly intramolecular
hydrogen-bonded closed-ring-structure molecules. The ESPT of the triazines give
fluorescence of the phototautomer with a large Stokes shift (~10^4 cm^{-1}) in a
poly(methyl methacrylate) (PMMA) matrix [194]. When the torsional motions are
restricted and the volume changes due to temperature are extremely small, the rate

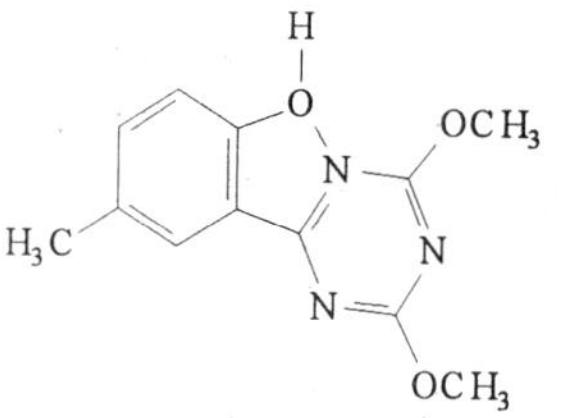

48

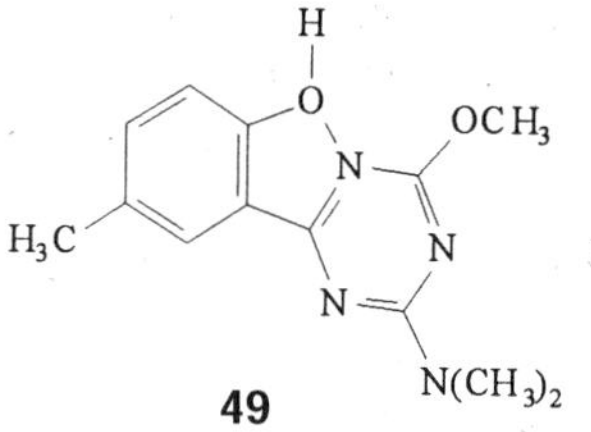

49

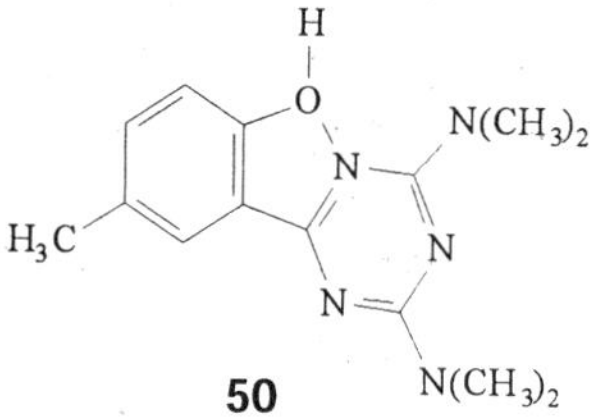

50

triazines

Structures 48–50

constant of ESPT is very fast ($>2 \times 10^{11}$ sec^{-1}) at 77 K as well as 300 K with no potential barrier.

The excited-state dynamics of the 2-hydroxypheylbenzotriazole (HPB) photostabilizer copolymerized with polystyrene (**51**) are reported in Ref. 195. The HPB fluorescence from these copolymer films is observed at ~630 nm, characteristics of the proton-transferred excited state of HPB, and it has a rise time of <10 psec and a decay time of 28 ± 4 psec at room temperature. Measurement of the relative fluorescence quantum yield as a function of temperature gives the activation energy for nonradiative decay of this state as 259 ± 25 cm^{-1}.

The ESIPT fluorescence of 3-hydroxyflavone (3HF) (**12**) and 2,5-bis-(2-benzoxzoyl)-4-methoxyphenol (BBMP) (**52**) were studied in polyethylene films [146]. It was found that perturbations due to H-bonded impurities are largely suppressed in the polymer matrix as compared with other media. Thus, the fluorophores behaved almost as unperturbed molecules, with spectral quality similar to those obtained from extensively purified solvents or embedded in rare matrices, even if they are introduced into the film from unpurified and undried chloroform.

Solid-state laser oscillation has been induced in 2-(2'-hydroxy-5'-fluorophenyl) benzimidazole (**53**) dissolved in PMMA, based on a four-level ESIPT mechanism [197]. The energy conversion efficiency at 493 nm was ~1% for N$_2$ laser pumping with pulses of 2 mJ at 337 nm.

The spectral behavior and quantum yields of an ESDPT molecule [2,2'-bipyridyl]-3,3'-diol [BP(OH)$_2$] (**54**) in sol–gel glasses modified by organic side groups were studied with the properties of BP(OH)$_2$ in solutions, in a plain sol–gel glass and in PMMA [148]. The fluorescence intensity of the dye in a fresh sol–gel glass is similar to that in water, whereas the corresponding blue shift is smaller and

$\lambda_{ex} = 355$nm: λ_{em}(P*) = 390nm: λ_{em}(T*) = 580nm

51

2-hydroxyphenylbenzotriazole (HPB)

Structure 51

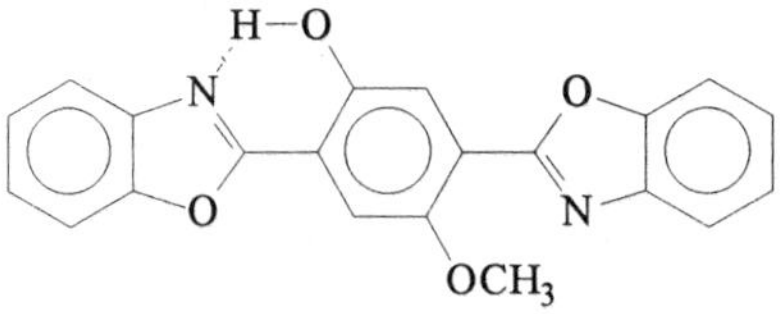

52

2,5-bis-(2-benzoxzoyl)-4-methoxyphenol (BBMP

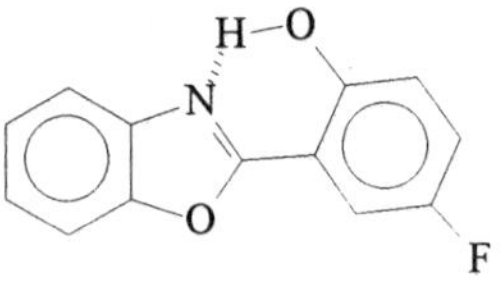

53

2-(2'-hydroxy-5'-fluorophenyl)benzimidazole (FBIM

Structures 52 and 53

hints at the cage effect. The overall spectral behavior of the dye in organically modified sol–gel glasses is similar to that observed in some polar solvents and in PMMA, whereas the fluorescence intensity is higher and the lifetimes are longer. This increase is ascribed to the rigidity of the silica–polymer network in these glasses. The double-proton-transfer dynamics of BP(OH)$_2$ has been studied in sol–gel glasses of tetramethylorthosilicate (TMOS) using the femtosecond fluorescence upconversion technique [199]. The double-proton transfer was found to be a two-step process involving the reaction from the excited dienol-to-monoketo tautomer (<100 fsec), followed by the monoketo-to-diketo step (with time constant of a few picoseconds).

Fluorescence in 1NpOH–aliphatic amine hydrogen-bonded systems in nonpolar rigid matrices like polyethylene films at 77 K is dual in nature [200–202]. The ESPT in the hydrogen-bonded complex results in two types of ion pairs, one with in-plane orientation between excited naphtholate and ammonium ions (fluorescence maximum at 395 nm with a lifetime of 5.3 nsec) and the other with an

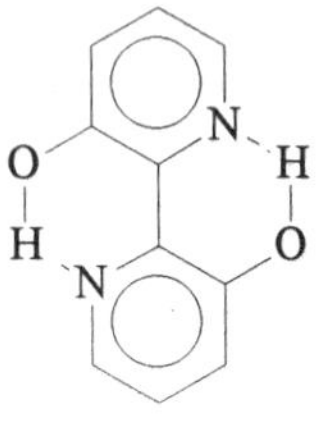

P

54

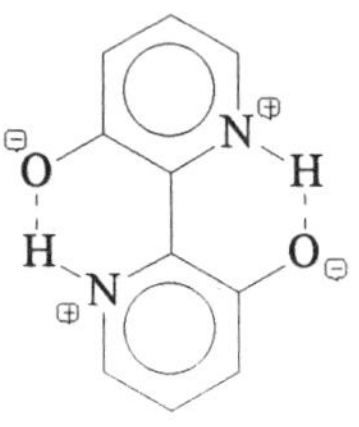

T*

55

[2,2'-bipyridyl]-3,3'-diol (BP(OH)$_2$)

Structures 54 and 55

out-of-plane orientation (fluorescence maximum at 420 nm with a lifetime of 13.8 nsec) [201]. The temperature dependence of fluorescence decay showed that at higher temperatures, two different quenching processes are operative [202]. A rapid quenching competing with ESPT results in a considerable loss of fluorescence quantum yield, with no significant change in fluorescence lifetimes. This could be as a result of a fast internal conversion due to the out-of-plane bending motion of O—H···N bond. The other quenching process is a dynamic one involving charge-transfer interaction.

The effect of phase transitions on ESPT of 2NpOH in poly (N-isopropylacryamide) (PNIPAM) covalently labeled with 2NpOH has been studied [203]. The 2NpOH undergoes ESPT at a rate that is dependent on water concentration in aqueous solvent systems. Upon conversion to the phase-separated state in naphthol-labeled PNIPAM, the efficiency of ESPT diminished, consistent with the formation of a globular phase in which exposure of the probe molecules to the water phase was reduced.

The effect of molecular size, conjugation length, and competition with excimer formation on ESIPT in conjugated polymers, in which the intramolecularly hydrogen-bonded moieties reside in the main domain, has been examined [204]. The polymers were poly(5'-(2'-hydroxy)phenylenebenzobisoxazole (**56**) and poly(4'-(2'-hydroxy)phenylenebenzobisoxazole (**57**). A large extent of π–electron delocalization is found to inhibit the ESIPT process. Excimer formation is found to be competitive with ESIPT in the polymers studied, whereas molecular size does not inhibit ESIPT.

The fluorescence spectra of 9-anthrol (**58**) were studied in the sol–gel–xerogel transitions of Si Al glasses of various composition prepared from tetraethyl orthosilicate and diisobutoxyaluminium triethylsilane [205]. A biprotonic ESIPT between the anthrone and anthrol tautomers was observed in which a proton of the surface Brønsted acid site on the —O—Si—O—Al—O— network was implicated.

56

poly(5'-(2'-hydroxy)phenylene benzobisoaxazole

57

poly(4'-(2'-hydroxy)phenylene benzobisoaxazole

Structures 56 and 57

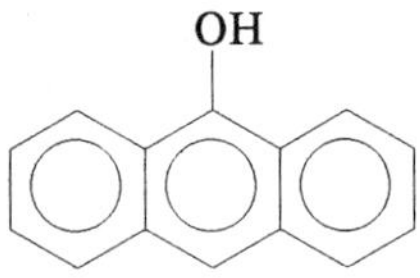

58

9-anthrol

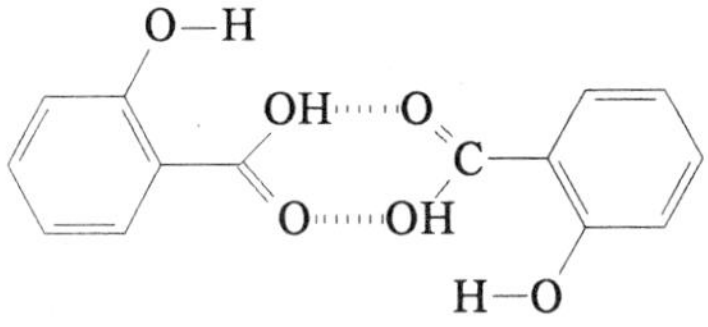

59

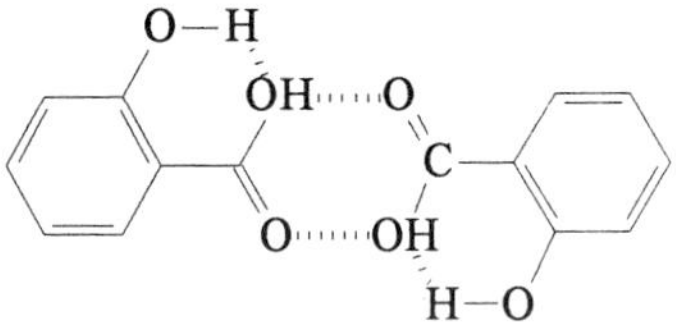

60

salicylicacid dimer

Structures 58–60

G.   Solids

Picosecond time-resolved total internal reflection fluorescence spectroscopy was applied to analyze the proton-transfer reaction of 1NpOH in water–sapphire interface layers [206]. The rate constant of the proton-transfer reaction from excited neutral species became slow in the interface layer as compared with that in the bulk aqueous solution and decreased smoothly with increasing penetration depth in the interfacial layer up to 100 nm. The anomaly was interpreted in terms of rotational fluctuations of water aggregates in the interface layer.

The photophysics of solid salicylic acid (SA) has been studied by using steady-state and time-resolved spectroscopic techniques [207,208]. Dimers of SA form in two possible structures (**59** and **60**) due to fast ground-state double proton transfer. Dual fluorescence is observed at 380 nm and 440 nm, which are ascribable to the excited-state double proton transfer between different dimeric structures of SA. The enol form is more stable in the ground state. However, in the excited singlet state, the keto form has a lower potential energy [207]. This excited enol–keto tautomerism has a barrier height of ~1250 cm^{-1}, as is calculated from the dependence of dual fluorescence on excitation wavelength in the temperature range 10–293 K. The observation that the fluorescence lifetimes of both forms become the same (3.0 nsec) at a higher-energy excitation and at a higher temperature indicates the establishment of equilibrium between the two forms [208].

Photochromism in crystalline 2-(2,4-dinitrobenzyl) pyridine and its derivatives has been shown to be due to ESPT [204]. Different photoinduced and thermal processes were found to have a relatively strong primary isotope effect, indicating that the rate-determining steps are proton-transfer processes. The supramolecular effects of the crystalline matrix control the relative rates and efficiencies of the observed proton-transfer process in both the ground and excited states, yielding, at room temperature, photoproducts having lifetimes ranging between hours and weeks. At least two proton-accepting groups may be active in the abstraction of the proton from its relatively stable benzylic position. 7-Azaindole has only a weak normal fluorescence at 350 nm in the solid state originating from the monomeric form, but the 3-iodo-7-azaindole dimer in single crystal exhibits a large Stokes-shifted fluorescence at ~500 nm in the temperature range 10–298 K [210]. Phosphorescence is also observed at 600 nm. These bands originate from excited-state double proton transfer in the dimer, as is clear from the absence of large Stokes-shifted band in the N-methylated counterpart.

H. Summary Table

Tables 1 and 2 give lists of all the ESPT molecules reviewed in this chapter with a summary of ESPT investigations.

III. CONCLUSIONS

A perusal of ESPT behavior of excited-singlet-state acids in various organized media makes it amply clear that the ESPT equilibrium and its consequent effects on fluorescence properties are sensitively dependent on the local properties of the microenvironment like polarity, steric, and solvation effects. This difference of ESPT behavior in organized media and homogeneous media like water makes ESPT an attractive probe concept. As the previous discussions show, the usefulness of quite a few ESPT fluorescent probes has been well researched. However, so far, ESPT probes have not been able to reach the status of certain other classes of fluorescent probes which are well accepted and routinely used by nonphotochemist researchers in other disciplines [77,80]. This review on ESPT in organized media is aimed at providing a perspective which, it is hoped, will enable workers looking for newer molecular probes to investigate a particular organized medium of interest.

Table 1 ESPT Molecules Reviewed: Aromatic Hydroxy and Amino Compounds

ESPT molecules	Spectral characteristics in polar solvent (Ref.)						Photophysical studies (Ref.)							Section in text
				Emission wavelength (nm)			k_d (Deprotonation rate constant) (sec^{-1})	k_p (Protonation rate constant) (M^{-1}/sec)	τ_p^*	τ_{DP}^*	Organized media			
	pK_a	pK_a^*	λ_{ex} (nm)	λ_p^*	λ_{DP}^*	λ_T^*					Medium	Kind of study	Ref.	
1NpOH	9.2 (1)	0.4 (1)	290 (86)	360 (86)	470 (86)		2.5×10^{10} (1)	6.8×10^{10} (1)	33 psec (31)	8.0 nsec (31)	Liposomes	Photo dissociation	83,85	II.A.1.a
												Temperature studies	84	
												Partition coefficient	86	
												Phase transition and cholesterol	87,92	
												Effect of cholesterol	89	
												Effect of surfactants		
												Effect of peptide	90	
												Phase transition	88	
											Natural lipids	ESPT behavior	200–202	II.F
											Polymer Liquid–solid interface	ESPT behavior	206	II.G
4-Chloro-1NpOH	8.75 (91)	1.73 (91)	300 (91)	378 (91)	460 (91)				0.32 nsec (91)	9.43 nsec (83)	Liposomes	Partition coefficient, Phase transition, cholesterol	91	II.A.1.a

Compound	pK	pK*	λ	λ	λ	k	k	τ	τ	Medium	Property	Ref.	Section
2NpOH	9.3 (1)	2.7 (1)	320 (33)	360 (33)	420 (33)	1.0×10^8 (1)	5.8×10^8 (1)	4.8 nsec (30)	9.5 nsec (30)	Liposomes	Dynamics	121	II.A.1.a
										Protein	Rate constants	136	II.B.1.a
											Dynamics	121	
										Cyclodextrin	ESPT behavior	167,168	II.C.1.a
1-Chloro-2-NpOH										Polymer	Effect of alcohols	170,171	II.F
										Liposomes	Phase transition	203	II.A.1.a
								0.79 nsec (83)	8.77 nsec (83)		ESPT behavior	83	
2-NpOH-6-sulfonate	9.12 (1)	1.95 (1)	325 (136)	360 (136)	445 (136)	1.0×10^9 (1)	9.0×10^{10} (1)			Protein	ESPT behavior	136	II.A.1.a
										Protein	Binding sites	139–141	II.B.1.a
2 NpOH-6,8 disulfonate	9.3 (1)	0.13 (1)	330 (143)	385 (143)	460 (143)	1.7×10^{10} (1)	2.3×10^{10} (1)	0.13 nsec (143)	4.0 nsec (143)	Protein	Binding site and effect of urea	143	II.B.1.a
2NpOH-3,6 disulfonate		0.71 (1)				5.8×10^8 (1)	3.0×10^9 (1)			Liposomes	Dynamics	121	II.A.1.a
										Proteins	Dynamics	121	II.B.1.a
											Kinetics	150	
1,6-Naphth-alenediol	~9 (169)	~3 (169)	337 (169)	$\lambda_{molecular} = 365$	$\lambda_{\beta ionic} = 400$ (169), $\lambda_{\alpha ionic} = 470$ (169), $\lambda_{dianionic} = 440$ (169)			2.1 nsec (169)	8.4 nsec (169)	Cyclodextrin	Kinetics	169	II.C.1.a
6-Dodecyl-2 NpOH			290 (192)	350 (192)	420 (192)					Air–water interface	ESPT behavior	192	II.E
2-Stearoyl-1 NpOH			360 (193)		460 (193)	$>10^{10}$ (193)				LB films	ESPT behavior	193	II.E

Table 1 Continued

ESPT molecules	Spectral characteristics in polar solvent (Ref.)						Photophysical studies (Ref.)				Organized media			
				Emission wavelength (nm)			k_d (Deprotonation rate constant) (sec^{-1})	k_p (Protonation rate constant) (M^{-1}/sec)	$\tau_p{}^*$	$\tau_{DP}{}^*$	Medium	Kind of study	Ref.	Section in text
	pK_a	$pK_a{}^*$	λ_{ex} (nm)	$\lambda_p{}^*$	$\lambda_{DP}{}^*$	$\lambda_T{}^*$								
N-[1-(2NpOH)]-Phosphatidyl ethanolamine	8.42 (97)	1.81 (97)	337 (97)	360 (97)	430 (97)				2.60 and 5.75 nsec (97)	6.35 nsec (97)	Liposomes	ESPT behavior	97	II.A.1.a
Pyranine	7.22 (98)	0.4 (114)	403 (103)	440 (103)	510 (103)		3.2×10^{10} (114)		4.6 nsec (103)	5.5 nsec (103)	Liposomes	pH gradients	98, 99	II.A.1.b
												Proton diffusion dynamics	101	
												Polymerization	100	
												ESPT behavior	103	
												Dynamics	122, 123	
											Purple membrane	Binding dynamics	125, 126	II.A.2
											Black lipid membrane	pH gradient	127	II.A.2
											Proteins	Binding site	124, 147, 148	II.A.2
											Sol–gel glass	ESPT dynamics	149	II.B.1.b
Anthrol			368 (205)	454 (205)	550 (205)							ESPT behavior	205	II.F
Equilenin			330 (144)	350 (144)	415 (144)				6.0 nsec (144)	1.2 nsec (144)	Liposomes	ESPT behavior	94	II.A.1.a
											Proteins	Binding site	144–146	II.B.1.a
Dihydroequilenin			337 (94)	360 (94)	440 (94)				2.7 nsec (94)	6.5 nsec (94)	Liposome	ESPT behavior	94	II.A.1.a

Aminopyrene	~3.0 (178)	−1.2 (1)	295 (178)	374 (178)	440 (178)		2.2×10^9 (1)	1.3×10^8 (1)	4.97 nsec (178)	5.18 nsec (178)	Cyclodextrin	ESPT behavior	178	II.C.2
Hypericin (1 : 1 EtOH : MeOH)			588 (113)		~610 (113)				5.6 psec (113)		Membranes	Antiviral studies	100, 111, 112	II.A.1.b
												Temperature studies	113	II.A.1.c
												Photodynamic action		II.A.1.c
											Neurons	Photodynamic action	158	II.B.3.c
											Proteins	Photophysics	159	II.B.3.c
											Cyclodextrins	ESIPT behavior	172	II.C.1.b
GFP			398 (157)	460 (157)	508 (157)				3.6, 12.0 120.0 psec (157)		Proteins	Dynamics	157	II.B.3.b
HNAA			325 (137)	360 (137)	440 (137)				3.15 nsec (137)	10.2 nsec (137)	Protein	ESPT behavior	137	II.A.1.a
2-Naphthylamine	4.1 (1)	−0.8 (1)					1×10^9 (1)	1.5×10^8 (1)			Cyclodextrin	ESPT behavior	173	II.C.2
CMOH			322 (189)			455 (189)					Cyclodextrin	ESIPT behavior	189	II.C.3
Naphthylammonium ion 18-crown-6			278 (190)	333 (190)	401 (190)		~10^9 (190)				Crown ether	ESPT behavior	190	II.D
Phenanthrylammonium ion 18-crown-6			296 (191)	367 (191)	447 (191)				44 nsec (191)	7 nsec (191)	Crown ether	ESPT behavior	191	II.D
Salicylic acid			350 (207)		$\lambda_D = 370$ (207)	450 (207)			$\tau_D = 3.11$ nsec (207)	$\tau_T = 2.92$ and 5.41 nsec (207)	Solid sate	ESPT behavior	207	II.G

Table 2 ESPT Molecules Reviewed: Heterocyclic Compounds

ESPT molecules	Spectral characteristics in water (Ref.)						Photophysical studies (Ref.)				Organized media			Section in text
	pK_a	pK_a^*	λ_{ex} (nm)	λ_p	λ_{DP}	λ_T	k_d (Deprotonation rate constant) (sec^{-1})	k_p (Protonation rate constant) (M^{-1}/sec)	τ_p^*	τ_{DP}^*	Medium	Kind of study	Ref.	
Carbazole	21.1 (1)	10.98 (1)	325 (130)	360 (130)	417 (130)		9.0×10^9 (1)	8.5×10^6 (1)			Liposomes	Phase transition, cholesterol	130	II.A.3
											Cyclodextrin	ESPT behavior	173	II.C.2
												Effect of urea	176	II.C.2
3-Iodo-7-aza-indazole			~300 (210)	350 (210)	$\lambda_D =$ 500 (210)		1.7×10^9 (210)		0.6 nsec (210)		Single crystal	ESDPT	210	II.G
Carvedilol			322 (131)	340, 353 (131)					8.2 nsec (131)		Liposomes	Phase transition Drug loading	131	II.A.3
3HF			340 (132)	410 (132)		533 (132)					Liposomes	Physical properties	132	II.A.4
											Protein	Binding sites	164	II.B.4
											Polymer	ESPT behavior	196	II.F
DMA3HF			410 (133)	510 (133)		562 (133)					Liposomes	ESPT behavior	133	II.A.4
BBMP			355 (196)	445 (196)		590 (196)					Polymer	ESPT behavior	196	II.F
DHF			413 (164)	516 (164)		570 (164)					Proteins	Binding site	164	II.B.4

Compound	pKa	λ (nm)	λ (nm)	λ (nm)	λ (nm)	Lifetime	System	Property	Ref.	Section
Fisetin		365 (164)	463 (164)		540 (164)		Proteins	Binding site	164	II.B.4
Dehydro-luciferin	8.7 (154), −1.0 (154)	338 (155)	435 (155)	530 (155)		3.7–4.0 nsec (155)	Luciferase	Binding studies	153–155	II.B.3.a
FBIM		337 (197)			~490 (197)		Solid organic	ESPT behavior	197	II.F
HPIP		340 (134)	390 (134)		590 (134)	1.8 nsec (cyclo-hexane) (134)	Liposomes	ESPT behavior	134	II.A.4
							Cyclodextrin	ESPT probing	186	II.C.3
HPB		355 (195)	390 (195)		580, 600 (195)		Polymer	ESPT behavior	195	II.E
HAP		380 (165)			585 (165)		Protein	ESPT probe	165	II.B.4
							Cyclodextrin	ESPT behavior	181,182	II.C.3
2PyBI	4.41 (179)	320 (179)	380 (cationic) 466 (tautomer) (179)	~380 (179)		1.8 nsec (179), 0.84 nsec (179)	Cyclodextrin	ESPT behavior	179	II.C.2
HBO	10.48 (184)	320 (184)	440 (184)	478 (184)			Cyclodextrin	ESPT behavior	184	II.C.3
HBI	9.98 (184)	330 (184)	350 (184)	427 (184)		2.2 and 0.7 nsec (185)	Cyclodextrin	ESPT behavior	184,185	II.C.3

Table 2 Continued

ESPT molecules	Spectral characteristics in water (Ref.)						Photophysical studies (Ref.)				Organized media			
	pK_a	pK_a	λ_{ex} (nm)	Emission wavelength (nm)			k_d (Deprotonation rate constant) (sec^{-1})	k_p (Protonation rate constant) (M^{-1}/sec)	τ_p^*	τ_{DP}^*	Medium	Kind of study	Ref.	Section in text
				λ_p	λ_{DP}	λ_T								
HBT	10.16 (184)		340 (184)	.	460 (184)						Cyclodextrin	ESPT behavior	184	II.C.3
2APBI			325 (58)		412 (58)	520 (58)			0.85 and 5.92 nsec (58)		Cyclodextrin	ESPT behavior	183	II.C.3
HPMO			320 (166)	350 (166)		465 (166)					Proteins, cyclodextrin	ESPT behavior	166	II.B.4
Triazines														
OO			343 (194)	470 (194)	530 (194)						Polymer	ESPT behavior	194	II.F
ON			336 (194)	490 (194)	510 (194)									
NN			328 (194)	494 (194)	497 (194)									
BP(OH)$_2$			365 (198)	445 (198)						1.0 nsec (198)	Sol–gel glass Polymer	ESPT behavior	198,199	II.F
2-(2,4-dinitrobenzyl) pyridine											Solids	ESPT-related photochromism	209	II.G

ACKNOWLEDGMENTS

Ms. N. Pappayee helped a great deal at every stage of writing this review, starting with literature collection. Ms. J. Shobini's assistance in manuscript typing and correction and structure drawing made my job easier. I greatly appreciate their help and thank them sincerely. I thank Professor S. K. Dogra, I.I.T. Kanpur, for his encouragement and help. Thanks are due to Dr. P. B. Bisht and Dr. P. B. Sunilkumar, I.I.T. Madras, and Professor S. Tobita, Gunma University, Japan, for their useful suggestions.

REFERENCES

1. Arnaut, L. G.; Förmosinho, S. J. *J Photochem. Photobiol. A: Chem.* **1993**, *75*, 1.
2. Förster, T. *Naturwissenschaften.* **1949**, *36*, 186.
3. Förster, T. *Z. Elektrochemie* **1950**, *54*, 43.
4. Förster, T. *Pure Appl. Chem. Int. Ed. Engl.* **1964**, *3*, 1.
5. Weller, A. *Naturwissenschaften.* **1955**, *42*, 175.
6. Weller, A. *Z. Electrochemistry* **1952**, *56*, 662.
7. Weller, A. *Prog. React. Kinet.* **1961**, *1*, 187.
8. Vander Donckt, E. *Prog. React. Kinet.* **1970**, *5*, 273.
9. Schulman, S. G. in *Physical Methods in Heterocyclic Chemistry;* Katritzky, A. R., ed.; Academic Press: New York, 1974, Vol. VI, 147.
10. Klopffer, W. *Adv. Photochem.* **1977**, *10*, 311.
11. Ireland, J. F.; Wyatt, P. A. H. *Adv. Phys. Org. Chem.* **1976**, *12*, 131.
12. Martynov, I. Yu.; Demyashkevich, A. B.; Uzhinov, B. M.; Kuzmin, M. G. *Russ. Chem. Rev.* **1977**, *46*, 3.
13. Eigen, W. *Angew. Chem. Int. Ed. Engl.* **1964**, *3*, 1.
14. Marcus, R. A. *Faraday Symp. Chem. Soc.* **1975**, *10*, 60.
15. Varandas, A. J. C.; Formosinho, S. J. *J. Chem. Soc., Chem. Commun.* **1986**, 163.
16. Varandas, A. J. C.; Formosinho, S. J. *J. Chem. Soc. Faraday Trans. 2* **1986**, *82*, 953.
17. Gutman, M.; Nachliel, E. *Biochim. Biophys. Acta* **1990**, *1015*, 391.
18. Shizuka, H. *Acc. Chem. Res.* **1985**, *18*, 141.
19. Kosower, E. M.; Huppert, D. *Annu Rev. Phys. Chem.* **1986**, *37*, 127.
20. Dogra, S. K. *Proc. Indian. Acad. Sci.* **1992**, *104*, 635.
21. Wan, P.; Shukla, D. *Chem. Rev.* **1993**, *93*, 571.
22. Förmosinho, S. J.; Arnaut, L. G. *J Photochem. Photobiol. A: Chem.* **1993**, *75*, 28.
23. Steadman, J.; Syage, J. A. *J. Chem. Phys.* **1990**, *92*, 4630.
24. Steadman, J.; Syage, J. A. *J. Am. Chem. Soc.* **1991**, *113*, 6786.
25. Steadman, J.; Fournier, E. W.; Syage, J. A. *Appl. Opt.* **1990**, *29*, 4962.
26. Breen, J. J.; Peng, L. W.; Willberg, D. M.; Heikal, A.; Cong, P.; Zewail, A. H. *J. Chem. Phys.* **1990**, *92*, 805.
27. Syage, J. A.; Steadman, J. *J. Chem. Phys.* **1991**, *95*, 2497.
28. Elsaesser, T. in *Femtosecond Chemistry;* Manz, J.; Woste, L., eds.; VCH: Weinheim, 1995, Chap. 18.
29. Jackson, G.; Porter, G. *Proc. Roy. Soc. (London)* **1961**, *A260*, 13.

30. Laws, W. R.; Brand, L. *J. Phys. Chem.* **1979,** *83,* 795.
31. Webb, S. P.; Yeh, S. W.; Philips, L. A.; Tolbert, L. M.; Clark, J. H. *J. Am. Chem. Soc.* **1984,** *106,* 7286.
32. Webb, S. P.; Philips, L. A.; Yeh, S. W.; Tolbert, L. M.; Clark, J. H. *J. Phys. Chem.* **1986,** *90,* 5154.
33. Lee, J.; Griffin, R. D.; Robinson, G. W. *J. Chem. Phys.* **1985,** *82,* 4920.
34. Robinson, G. W.; Thistlethwaite, P. J.; Lee, J. *J. Phys. Chem.* **1986,** *90,* 4224.
35. Lee, J.; Robinson, G. W.; Webb, S. P.; Philips, L. A.; Clark, J. H. *J. Am. Chem. Soc.* **1986,** *108,* 6538.
36. Lee, J. *J. Am. Chem. Soc.* **1989,** *111,* 427.
37. Lee, J. *J. Phys. Chem.* **1990,** *94,* 258.
38. Shizuka, H.; Ogiwara, T.; Narita, A.; Sumitani, M.; Yoshihara, K. *J. Phys. Chem.* **1986,** *90,* 6708.
39. Knochenmuss, R.; Leutwyler, S. *J. Chem. Phys.* **1989,** *91,* 1268.
40. Pines, E.; Huppert, D. *J. Chem. Phys.* **1986,** *84,* 3576.
41. Pines, E.; Huppert, D.; Agmon, N. *J. Chem. Phys.* **1988,** *88,* 5620.
42. Agmon, N.; Pines, E.; Huppert, D. *J. Chem. Phys.* **1988,** *88,* 5631.
43. Agmon, N. *J. Chem. Phys.* **1988,** *88,* 5639.
44. Agmon, N. *J. Chem. Phys.* **1988,** *89,* 1524.
45. Agmon, N.; Huppert, D.; Masad, A.; Pines, E. *J. Phys. Chem.* **1991,** *95,* 10,407.
46. Htun, M. T.; Suwaiyan, A.; Klein, U. K. A. *Chem. Phys. Lett.* **1995,** *243,* 506.
47. Htun, M. T.; Suwaiyan, A.; Klein, U. K. A. *Chem. Phys. Lett.* **1995,** *243,* 512.
48. Tolbert, L. M.; Haubrich, J. E. *J. Am. Chem. Soc.* **1994,** *116,* 10,593.
49. Huppert, D.; Tolbert, L. M.; Samaniego, S. L. *J. Phys. Chem. A* **1997,** *101,* 4602.
50. Gopich, I. V.; Solntsev, K. M.; Agmon, N. *J. Chem. Phys.* **1999,** *110,* 2164.
51. Solntsev, K. M.; Huppert, D.; Agmon, N. *J. Phys. Chem. A* **1999,** *103,* 6984.
52. Solntsev, K. M.; Huppert, D.; Agmon, N.; Tolbert, L. M. *J. Phys. Chem. A* **2000,** *104,* 4658.
53. Cohen, B.; Huppert, D. *J. Phys. Chem A* **2000,** *104,* 2663.
54. Klopffer, W.; Naundorf, G. *J. Lumin.* **1974,** *8,* 457.
55. Herek, J. L.; Pedersen, S.; Banares, L.; Zewail, A. H. *J. Chem. Phys.* **1992,** *97,* 9046.
56. Felker, P. M.; Lambert, W. R.; Zewail, A. H. *J. Chem. Phys.* **1982,** *87,* 1603.
57. Tolbert, L. M.; Harvey, L. C.; Lum, R. C. *J. Phys. Chem.* **1993,** *97,* 13,335.
58. Santra, S.; Dogra, S. K. *Chem. Phys.* **1998,** *226,* 285.
59. Nagaoka, S.; Kusunoki, J.; Fujibuchi, T.; Hatakenaka, S.; Mukai, K.; Nagashima, U. *J. Photochem. Photobiol A: Chem.* **1999,** *122,* 151.
60. Bhatacharyya, K. in *Organic Molecular Photochemistry;* Ramamurthy, V.; Schanze, K., eds.; Marcel Dekker: New York, 1999, 283.
61. Santra, S.; Krishnamoorthy, G.; Dogra, S. K. *J. Phys. Chem. A* **2000,** *104,* 476.
62. Carvalho, C. E. M.; Brinn, I. M.; Pinto, A. V.; Pinto, M. C. F. R. *J. Photochem. Photobiol A: Chem.* **1999,** *123,* 61.
63. Brucker, G. A.: Swinney, T. C.; Kelly, D. F. *J. Phys. Chem.* **1991,** *95,* 3190.
64. Ledesma, G. N.; Ibañez, G. A.; Escandar, G. M.; Oliveri, A. C. *J. Mol. Struct. (Theochem.)* **1997,** *415,* 115.
65. Catalán, J.; de Valle, J. C.; Palomar, J.; Diaz, C.; de Paz, J. L. G. *J. Phys. Chem. A* **1999,** *103,* 10,921.

66. Catalán, J.; Palomar, J.; de Paz, J. L. G. *J. Phys. Chem. A* **1997,** *101,* 7914.
67. English, D. S.; Zhang, W.; Kraus, G. A.; Petrich, J. W. *J. Am. Chem. Soc.* **1997,** *119,* 2980.
68. Swaminathan, M.; Dogra, S. K. *J. Am. Chem. Soc.* **1983,** *105,* 6263.
69. Mishra, A. K.; Dogra, S. K. *Indian J. Chem,* **1985,** *24A,* 285.
70. Chou, P. T.; Martinez, M. L.; Cooper, W. C.; McMorrow, D.; Collins, S. T.; Kasha, M. *J. Phys. Chem.* **1992,** *96,* 5203.
71. Taylor, C. A.; El-Bayomi, M. A.; Kasha, M. *Proc. Natl. Acad. Sci. USA* **1969,** *63,* 253.
72. Douhal, A.; Kim, S. K.; Zewail, A. H. *Nature,* **1995,** *378,* 260.
73. del Valle, J. C.; Dominguez, E.; Kasha, M. *J. Phys. Chem. A* **1999,** *103,* 2467.
74. Marks, D.; Zhang, H.; Glasbeek, M.; Borowicz, P.; Grabowska, A. *Chem. Phys. Lett.* **1997,** *275,* 370.
75. Zhang, H.; van der Meulen, P.; Galsbeek, M. *Chem. Phys. Lett.* **1996,** *253,* 97.
76. Kalayanasundaram, K., in *Photochemistry in Organized and Constrained Media;* Ramamurthy, V., ed.; VCH: New York, 1991, 39.
77. Haugland, R. P. *Handbook of Fluorescent Probes and Research Chemicals;* Molecular Probes Inc., 1996
78. Ohmsted, J. III; Kearns, D. R. *Biochemistry* **1977,** *16,* 3647.
79. Lakowicz, J. R. *Topics in Fluorescence Spectroscopy: Probe Design and Chemical Sensing;* Plenum Publishing: New York, 1994, Vol. 4.
80. Lakowicz, J. R. *Principles of Fluorescence Spectroscopy,* 2nd ed.; Plenum Press: New York, 1999.
81. Lakowicz, J. R. *Principles of Fluorescence Spectroscopy;* Plenum Press: New York, 1999, 385.
82. New, R. R. C. *Liposomes, a Practical Approach;* Oxford University Press: New York, 1991.
83. Il'ichev, Yu. V.; Demyashkevich, A. B.; Kuzmin, M. G. *J. Phys. Chem.* **1991,** *95,* 3438.
84. Il'ichev, Yu. V.; Demyashkevich, A. B.; Kuzmin, M. G.; Lemmetyinen, H. *J. Photochem. Photobiol. A: Chem.* **1993,** *74,* 51.
85. Il'ichev, Yu. V.; Solntsev, K. M. S.; Kuzmin, M. G.; Lemmetyinen, H. *J. Chem. Soc. Faraday. Trans.* **1994,** *90,* 2717.
86. Sujatha, J; Mishra, A. K. *J. Photochem. Photobiol A: Chem.* **1996,** *101,* 215.
87. Sujatha, J; Mishra, A. K. *Langmuir,* **1998,** *14,* 2256.
88. Pappayee, N.; Mishra, A. K. *Spectrochim. Acta A* **2000,** *56,* 2249.
89. Sujatha, J; Mishra, A. K. *J. Photochem. Photobiol. A: Chem.* **1997,** *104,* 173.
90. Shobini, J.; Mishra, A. K. *Spectrochim. Acta. A* **2000,** *56,* 2239.
91. Pappayee, N.; Mishra, A. K. *Indian. J. Chem. A* **2000,** *39A,* 964.
92. Sujatha, J.; Mishra, A. K. *J. Fluoresc.* **1997,** *7,* 165(s).
93. Papahadjopoulos, D.; Jacobson, K.; Nir, S.; Isac, T. *Biochim. Biophys. Acta* **1973,** *311,* 330.
93a. Sujatha, J.; Mishra, A. K. *Langmuir* **1998,** *14,* 2256–2262.
94. Davenport, L.; Knutson, J. R.; Brand, L. *Biochemistry* **1986,** *25,* 1186.
95. Huang, C. H. *Biochemistry* **1969,** *8,* 344.
96. Newman, G. C.; Huang, C. H. *Biochemistry* **1975,** *14,* 3363.

97. Neyroz, P.; Franzoni, L.; Spisni, A.; Masotti, L.; Brand, L. *Chem. Phys. Lipids* **1992,** *61,* 255.
98. Kano, K.; Fendler, J. H. *Biochim. Biophys. Acta* **1978,** *509,* 289.
99. Clement, N. R.; Gould, J. M. *Biochemistry* **1981,** *20,* 1534.
100. Nome, F.; Reed, W.; Politi, M.; Tundo, P.; Fendler, J. H. *J. Am. Chem. Soc.* **1984,** *106,* 8086.
101. Rochel, S.; Nachliel, E.; Huppert, D.; Gutman, M. *J. Membr. Biol.* **1990,** *118,* 225.
102. Parsegian, V. A.; Rand, R. P.; Fuller, N. L.; Race, D. C. *Methods Enzymol.* **1986,** *127,* 400.
103. Fernandez, C.; Politi, M. J. *J. Photochem. Photobiol. A: Chem.* **1997,** *104,* 165.
104. Pinto, L. H.; Holsinger, L. J.; Lamb, R. A. *Cell,* **1992,** *69,* 517.
105. Newell, K. J.; Tannock, I. F. *Cancer Res.* **1989,** *49,* 4447.
106. Newell, K. J.; Wood, P. Stratford, I.; Tannock, I. *Br. J. Cancer* **1992,** *66,* 311.
107. Barry, M. A.; Reynold, J. E.; Eastman, A. *Cancer Res.* **1993,** *53,* 2349.
108. Gottlieb, R. A.; Nordberg, J.; Skowronski, E.; Babior, B. M. *Proc. Natl. Acad. Sci. USA* **1996,** *93,* 654.
109. Li, J.; Eastman, A. *J. Biol. Chem.* **1995,** *270,* 3203.
110. Fehr, M. J.; McCloskey, M. A.; Petrich, J. W. *J. Am. Chem. Soc.* **1995,** *117,* 1833.
111. Das, K.; English, D. S.; Fehr, M. J.; Smirnov, A. V.; Petrich, J. W. *J. Phys. Chem.* **1996,** *100,* 18,275.
112. English, D. S.; Das, K.; Ashby, K. D.; Park, J.; Petrich, J. W.; Castner, E. W., Jr. *J. Am. Chem. Soc.* **1997,** *119,* 11,585.
113. Das, K.; Ashby, K. D.; Wen, J.; Petrich, J. W. *J. Phys. Chem. B* **1999,** *103,* 1581.
114. Smith, K. K.; Kaufmann, K. J.; Huppert, D.; Gutman, M. *Chem. Phys. Lett.* **1979,** *64,* 522.
115. Gutman, M.; Huppert, D.; Pines, E. *J. Am. Chem. Soc.* **1981,** *103,* 3709.
116. Gutman, M. *Methods Biochem. Anal.* **1984,** *30,* 1.
117. Gutman, M. *Methods Enzymol.* **1986,** *127,* 522.
118. Gutman, M.; Nachliel, E.; Tsfadia, Y. *CRC Rev., Biochem.* **1995,** *25,* 9.
119. Gutman, M., in *Biomembrane and Electrochemistry.* Blank, M.; Vodyanoy, I., eds. *Advances in Chemistry Series No 235*, American Chemical Society, **1994,** 27.
120. Gutman, M.; Nachliel, E. *Annu. Rev. Phys. Chem.* **1997,** *48,* 329.
121. Nachliel, E.; Gutman, M. *J. Am. Chem. Soc.* **1988,** *110,* 2629.
122. Gutman, M.; Nachliel, E.; Moshiach, S. *Biochemistry* **1989,** *28,* 2936.
123. Gutman, M.; Kotlyar, A. B.; Borovok, N.; Nachliel, E. *Biochemistry* **1993,** *32,* 2942.
124. Kotlyar, A. B.; Borovok, N.; Kiryati, S.; Nachliel, E.; Gutman, M. *Biochemistry* **1994,** *33,* 873.
125. Nachliel, E.; Gutman, M.; Kiryati, S.; Dencher, N. A. *Proc. Natl. Acad. Sci. USA* **1996,** *93,* 10,747.
126. Nachliel, E.; Gutman, M. *Proc. Natl. Acad. Sci. USA* **1996,** *93,* 221.
127. Nachliel, E.; Finkelstein, Y; Gutman, M. *Biochim. Biophys. Acta* **1996,** *1285,* 131.
128. Samanta, A.; Chattopadhyay, N.; Nath, D.; Kundu, T.; Choowdhury, M. *Chem. Phys. Lett.* **1985,** *121,* 507.
129. Chattophadhyay, N.; Dutta, R.; Chowdhury, M. *J. Photochem. Photobiol. A: Chem.* **1989,** *47,* 249.

130. Pappayee, N.; Mishra, A. K. *Spectochim. Acta A* **2000,** *56,* 1027.
131. Cheng, H. Y.; Randall, C. S.; Holl, W. W.; Constantinides, P. P.; Yue, T. L.; Feuerstein, G. Z. *Biochim. Biophys. Acta* **1996,** *1284,* 20.
132. Guharay, J.; Chaudhuri, R.; Chakrabarti, A.; Sengupta, P. K. *Spectrochim. Acta A* **1997,** *53,* 457.
133. Dennison, S. M.; Guharay, J.; Sengupta, P. K. *Spectrochim. Acta A* **1999,** *55,* 1127.
134. Mateo, C. R.; Douhal, A. *Proc. Natl. Acad. Sci. USA* **1998,** *95,* 7245.
135. Douhal, A.; Amat-Guerri, F.; Acuna, A. U. *J. Phys. Chem.* **1995,** *99,* 76.
136. Loken, M. R.; Hayes, J. W.; Gohlke, J. R.; Brand, L. *Biochemistry* **1972,** *11,* 4779.
137. Laws, W. R.; Posner, G. H.; Brand, L. *Arch. Biochem. Biophys.* **1979,** *193,* 88.
138. Jankowski, A.; Stefanowicz, P. *Studies Biophys.* **1989,** *131,* 55.
139. Jankowski, A.; Stefanowicz, P.; Dobryszycki, P. *J. Photochem. Photobiol. A: Chem.* **1992,** *69,* 57.
140. Jankowski, A.; Stefanowicz, P. *J. Photochem. Photobiol. A: Chem.* **1994,** *84,* 143.
141. Jankowski, A.; Wiczk, W.; Janiak, T. *J. Photochem. Photobiol. A: Chem.* **1995,** *85,* 69.
142. Means, G.; Feeney, R. in *Chemical Modification of Proteins;* Holden-Day: San Francisco, 1971, 214, 223.
143. Das, R.; Mitra, S.; Nath, D.; Mukherjee, S. *J. Phys. Chem.* **1996,** *100,* 14,514.
144. Örstan, A.; Lulka, M. F.; Eide, B.; Petra, P. H.; Ross, J. B. A. *Biochemistry* **1986,** *25,* 2686.
145. Ross, J. B. A.; Torres, R.; Petra, P. H. *FEBS Lett.* **1982,** *149,* 240.
146. Petra, P. H.; Stanczyk, F. Z.; Senear, D. F.; Namkung, P. C.; Novy, M. J.; Ross, J. B. A.; Turner, E.; Brown, J. A. *J. Steroid Biochem.* **1983,** *19,* 699.
147. Gutman, M.; Nachliel, E.; Huppert, D. *Eur. J. Biochem.* **1982,** *125,* 175.
148. Gutman, M.; Huppert, D.; Nachliel, E. *Eur. J. Biochem.* **1982,** *121,* 637.
149. Yam, R.; Nachliel, E.; Smader, K.; Gutman, M. *Biophys. J.* **1991,** *59,* 4.
150. Yam, R.; Nachliel, E.; Gutman, M. *J. Am. Chem. Soc.* **1988,** *110,* 2636.
151. Gutman, M.; Nachliel, E. *Solid State Ionics* **1993,** *61(1–3),* 1.
152. Yael, M.; Nachliel, E.; Aagaard, A.; Brzezinski, P.; Gutman, M. *Proc. Natl. Acad. Sci. USA* **1998,** *95,* 8590.
153. Morton, R. A.; Hopkins, T. A.; Seliger, H. H. *Biochemistry* **1969,** *8,* 1598.
154. De Luca, M.; Brand, L.; Cebula, T. A.; Seliger, H. H.; Makula, A. F. *J. Biol. Chem.* **1971,** *246,* 6702.
155. Bowie, L. J.; Irwin, R.; Loken, M.; De Luca, M.; Brand, L. *Biochemistry* **1973,** *12,* 1852.
156. Heim, R.; Prassher, D. C.; Tsien, R. Y. *Proc. Natl. Acad. Sci. USA* **1994,** *91,* 12,501.
157. Chattoraj, M.; King, B. A.; Bublitz, G. U.; Boxer, S. G. *Proc. Natl. Acad. Sci. USA* **1996,** *93,* 8362.
158. English, D. S.; Doyle, R. T.; Petrich, J. W.; Haydon, P. G. *Photochem. Photobiol.* **1999,** *69,* 301.
159. Das, K.; Smirnov, A. V.; Wen, J.; Miskovsky, P.; Petrich, J. W. *Photochem. Photobiol.* **1999,** *69,* 633.
160. Miskovsky, P.; Jancura, D.; Sánchez-Cortés, S.; Kocisova, E.; Jancura, D.; Chinsky, L. *J. Am. Chem. Soc.* **1998,** *120,* 6374.
161. Rayner, D. M.; Krajcarski, D. T.; Szabo, A. G. *Can. J. Chem.* **1978,** *56,* 1238.

162. Ross, J. B. A.; Laws, W. R.; Rousslang, K. W.; Wyssbrod, W. R., in *Topics in Fluorescence Spectroscopy;* Lakowicz, J. R. ed. Plenum Press: New York, 1992, Vol. 3, Chap. 1.
163. Harris, C. M.; Selinger, B. K. *J. Phys. Chem.* **1980,** *84,* 891.
164. Sytnik, A.; Gormin, D.; Kasha, M. *Proc. Natl. Acad. Sci. USA* **1994,** *91,* 11,968.
165. Sytnik, A.; Kasha, M. *Proc. Natl. Acad. Sci. USA.* **1994,** *91,* 8627.
166. Garcia-Ochoa, I.; López, M. A. D.; Viñas, M. H.; Santos, L.; Atáz, E. M.; Amat-Guerri, F.; Douhal, A. *Chem. Eur. J.* **1999,** *5,* 897.
167. Yorozu, T.; Hoshino, M.; Imamura, M.; Shizuka, H. *J. Phys. Chem.* **1982,** *86,* 4422.
168. Eaton, D. F. *Tetrahedron,* **1987,** *43,* 1551.
169. Agbaria, R. A.; Uzan, B.; Gill, D. *J. Phys. Chem.* **1989,** *93,* 3855.
170. Park, H. R.; Mayer, B.; Wolschann, P.; Kohler, G. *J. Phys. Chem.* **1994,** 98, 6158.
170a. Park, H. R.; et al. *J. Phys. Chem.* **1994,** *98,* 6158.
171. Stam, J. V.; De Feyter, S.; De Schryver, F. C.; Evans, C. H. *J. Phys. Chem.* **1996,** *100,* 19,959.
172. Gai, F.; Fehr, M. J.; Petrich, J. W. *J. Phys. Chem.* **1994,** *98,* 5784.
173. Chattopadhyay, N. *J. Photochem. Photobiol. A: Chem.* **1991,** *58,* 31.
174. Saenger, W.; Betzel, C.; Hingerty, B.; Brown, G. M. *Nature (London)* **1982,** *296,* 581.
175. Saenger, W. in *Inclusion Compounds;* Academic Press: London, 1984, Vol. 2, 231.
176. Kundu, S.; Chattopadhyay, N. *J. Mol. Liquids* **1997,** *71,* 49.
177. Chattopadhyay, N.; Dutta, R. Chowdhury, M. *Indian J. Chem.* **1992,** *31A,* 512.
178. Hansen, J. E.; Pines, E.; Fleming, G. R. *J. Phys. Chem.* **1992,** *96,* 6904.
179. Rath, M. C.; Palit, D. K.; Mukherjee, T. *J. Chem. Soc. Faraday Trans.* **1998,** *94,* 1189.
180. Krishnamoorthy, G.; Dogra, S. K. *J. Photochem. Photobiol. A: Chem.* **1999,** *123,* 109.
181. Martinez, M. L.; Cooper, W. C.; Chou, P. T. *Chem. Phys. Lett.* **1992,** *193,* 151.
182. Roberts, E. L.; Chou, P. T.; Alexander, T. A.; Agbaria, R. A.; Warner, I. M. *J. Phys. Chem.* **1995,** *99,* 5431.
183. Santra, S.; Dogra, S. K. *J. Photochem. Photobiol. A: Chem.* **1996,** *101,* 221.
184. Roberts, E. L.; Dey, J.; Warner, I. M. *J. Phys. Chem.* **1996,** *100,* 19681.
185. Roberts, E. L.; Dey, J.; Warner, I. M. *J. Phys. Chem. A.* **1997,** *101,* 5296.
186. Douhal, A.; Amat-Guerri, F.; Acuña, A. U. *Angew. Chem. Int. Ed. Engl.* **1997,** *36,* 1514.
187. Douhal, A.; Amat-Guerri, F.; Lillo, M. P.; Acuna, A. U. *J. Photochem. Photobiol. A* **1994,** *78,* 127.
188. Guallar, V.; Moreno, M.; Lluch, J. M.; Amat-Guerri, F.; Douhal, A. *J. Phys. Chem.* **1996,** *100,* 19,789.
189. Mitra, S.; Das, R.; Mukherjee, S. *J. Phys. Chem. B.* **1998,** *102,* 3730.
190. Shizuka, H.; Kameta, K.; Shinozaki, T. *J. Am. Chem. Soc.* **1985,** *107,* 3956.
191. Shizuka, H.; Serizawa, M. *J. Phys. Chem.* **1986,** *90,* 4573.
192. Grieser, F.; Thistlethwaite, P.; Triandos, P. *J. Am. Chem. Soc.* **1986,** *108,* 3844.
193. Il'ichev, Yu. V.; Solntsev, K. M.; Demyashkevich, A. B.; Kuzmin, M. G.; Lemmetyinen, H.; Vuorimaa, E. *Chem. Phys. Lett.* **1992,** *193,* 128.

194. Shizuka, H.; Machii, M.; Higaki, Y.; Tanaka, M.; Tanaka, I. *J. Phys. Chem.* **1985,** *89,* 320.
195. O'Connor, D. B.; Scott, G. W.; Coulter, D. R.; Gupta, A.; Webb, S. P.; Yeh, S. W.; Clark, J. H. *Chem. Phys. Lett.* **1985,** *121,* 417.
196. Mordzinski, A. *Chem. Phys. Lett.* **1988,** *150,* 254.
197. Acuña, A. U.; Amat-Guerri, F.; Costela, A.; Douhal, A.; Figuera, J. M.; Florida, F; Sastre, R. *Chem. Phys. Lett.* **1991,** *187,* 98.
198. Eyal, M.; Reisfeld, R.; Chernyak, V.; Kaczmarek, L.; Grabowska, A. *Chem. Phys. Lett.* **1991,** *176,* 531.
199. Prosposito, P.; Marks, D.; Zhang, H.; Glasbeek, M. *J. Phys. Chem. A.* **1998,** *102,* 8894.
200. Mishra, A. K.; Shizuka, H. *Chem. Phys. Lett.* **1988,** *151,* 379.
201. Mishra, A. K.; Sato, M.; Hiratsuka, H.; Shizuka, H. *J. Chem. Soc. Faraday Trans.* **1991,** *87,* 1311.
202. Mishra, A. K.; Shizuka, H. *J. Chem. Soc. Faraday Trans.* **1996,** *92,* 1481.
203. Linares-Samaniego, S.; Tolbert, L. M. *J. Am. Chem. Soc.* **1996,** *118,* 9974.
204. Tarkka, R. M.; Jenekhe, S. A. *Chem. Phys. Lett.* **1996,** *260,* 533.
205. Fujii, T.; Kodaira, K.; Kawauchi, O.; Tanaka, N.; Yamashita, H.; Anpo, M. *J. Phys. Chem. B* **1997,** *101,* 10631.
206. Yanagimachi, M.; Tamai, N.; Masuhara, H. *Chem. Phys. Lett.* **1993,** *201,* 115.
207. Pant, D. D.; Joshi, H. C.; Bisht, P. B.; Tripathi, H. B. *Chem. Phys.* **1994,** *185,* 137.
208. Bisht, P. B.; Tripathi, H. B.; Pant, D. D. *J. Photochem. Photobiol. A: Chem.* **1995,** *90,* 103.
209. Scherl, M.; Haarer, D.; Fischer, J.; DeCian, A.; Lehn, J. M.; Eichen, Y. *J. Phys. Chem.* **1996,** *100,* 16,175.
210. Chou, P. T.; Liao, J. H.; Wei, C. Y.; Yang, C. Y.; Yu, W. S.; Chou, Y. H. *J. Am. Chem. Soc.* **2000,** *122,* 986.

11

Photophysics and Photochemistry of Fullerenes and Fullerene Derivatives

Jochen Mattay, Lars Ulmer, and Andreas Sotzmann
University of Bielefeld, Bielefeld, Germany

I. INTRODUCTION

In 1985, the existence of a form of elementary carbon with a definite number of carbon atoms, 60 carbon atoms in the shape of a soccer ball (Fig. 1), became certain with the discovery of the fullerenes by Kroto and coworkers [1]. Five years later, Krätschmer et al. [2] developed a simple procedure for macroscopic generation of fullerenes. This produced a flood of investigations aimed at new molecules and materials with interesting behavior like superconductivity [3], ferromagnetism [4], biological activity [5–7], nonlinear optical properties [8–10], etc. As a result, great numbers of reports and reviews have been published describing the full scope of fullerene chemistry, including functionalization and physical properties [11–18]. In 1994, photochemistry was the topic of a review by Foote [19]. Since then, intensive research concerning all aspects of fullerene chemistry has been carried out. It is the aim of this chapter to close the gap between 1994 and the present, but because literature concerning fullerene chemistry is widely spread across the disciplines of material science, physics, and chemistry we do not make the claim of being comprehensive, especially in the coverage of conference-related literature.

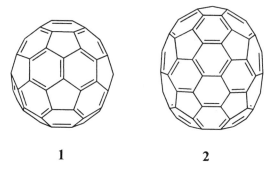

1   **2**

Figure 1 C_{60} (1) and C_{70} (2).

II. PHOTOPHYSICAL PROPERTIES

A. C_{60} and C_{70}

In 1991, Foote et al. [20,21] carried out the first investigations of basic photo-physical properties of pure fullerenes. Since then research has lead to extensive knowledge about the photophysical behavior of the fullerenes in general [22–48]. Scheme 1 shows the main photophysical processes. The most relevant photo phys-ical properties are summarized in Table 1.

1.   Ultraviolet (UV)/Vis–Absorption and Fluorescence

The electronic absorption spectra of both C_{60} and C_{70} show strong absorption bands in the ultraviolet (UV) region due to allowed $^1T_{1u}$–^1A_g transitions (ε_{max} $\sim 10^5$ M^{-1} cm^{-1}; λ_{max} (C_{60}) = 211, 256, 328 nm; λ_{max} (C_{70}) = 215, 236, 313)

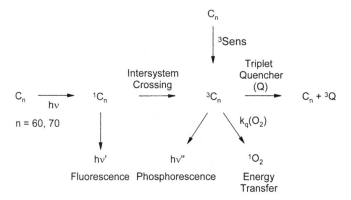

Scheme 1   Photophysical processes in C_{60} and C_{70}.

Table 1 Photophysical Properties of C_{60} and C_{70}

Property	C_{60}[a]	C_{70}[a]
E_s	46.1 kcal/mol[b]	44.1 kcal/mol[b]
E_T	37.5 ± 4.5 kcal/mol[c]	33.0 kcal/mol $\leq E_T <$ 42 kcal/mol[c]
ε_T (480 nm)	$(2.8 \pm 0.2) \times 10^3\ M^{-1}\,s^{-1}$	$(1.4 \pm 0.1) \times 10^4\ M^{-1}\,s^{-1}$
τ_T	$40 \pm 4\ \mu s^d$	$130 \pm 10\ \mu s^d$
k_q (O_2)	$(1.9 \pm 0.2) \times 10^9\ M^{-1}\,s^{-1}$	$(9.4 \pm 1.5) \times 10^8\ M^{-1}\,s^{-1}$
$\Phi\ 1_{O_2}$ (355 nm)	0.76 ± 0.05^e	0.81 ± 0.15^e
$\Phi\ 1_{O_2}$ (532 nm)	0.96 ± 0.04^f	0.81 ± 0.15^f
k_q ($^1O_2$)	$(5 \pm 2) \times 10^5\ M^{-1}\,s^{-1g}$	not measurable[g]

[a] All measurements in C_6H_6 or C_6D_6 were taken at room temperature, with concentrations of $C_{60} \approx 3 \times 10^{-5}$ and 3×10^{-4} M (OD $\approx$ 0.5) at λ_{exc} = 355 and 532 nm; $C_{70} \approx 1 \times 10^{-5}$ and 4×10^{-5} M (OD $\approx$ 0.3) at λ_{exc} = 355 and 532 nm, respectively, unless otherwise noted.
[b] Calculated from the presumed 0–0 band at 620 nm for C_{60} and at 638 nm for C_{70}.
[c] Average of the triplet energy of TPP (33.0 kcal/mol) and anthracene (42 kcal/mol); however, the low quenching rate with TPP suggests the actual energy is closer to the level of TPP.
[d] In argon-saturated solution.
[e] C_{60}: avarage of the five determinations with TPP (Φ_Δ = 0.62) and acridine (Φ_Δ = 0.84); C_{70}: average of five determinations at varied laser energies (1–3 mJ/puls), OD = 0.3–0.5, with TPP (Φ_Δ = 0.62).
[f] C_{60}: avarage of three determinations with TPP (Φ_Δ = 0.62); C_{70}: avarage of seven determinations, same conditons as for Φ^1O_2 (355 nm).
[g] C_{60}: total of physical and chemical quenching of $^1O_2$ by C_{60} (TPP, 532 nm). The maximum concentration of C_{60} was limited by saturation; C_{70}: No change in $^1O_2$ lifetime on adding C_{70} up to 4.9×10^{-4} M.
Source: Refs. 20–22.

[22,23]. The absorption in the visible region due to the low-lying electronic transitions is only weakly allowed because of the high degree of symmetry in the closed shell electronic configuration [26,65] (λ_{max} 540 nm, ε_{max} ~710 $M^{-1}\,cm^{-1}$ for C_{60} and λ_{max} 468 nm, ε_{max} ~15 000 M^{-1} for C_{70} [22,23]. The longest wavelength absorption of C_{60} attributed to the $S_0 \rightarrow S_1$ transition appears at 620 nm [22]. The lower symmetry of C_{70} leads to stronger absorption in the visible region due to the fact that the longest wavelength transition is less forbidden [21].

C_{60} shows very weak fluorescence with low quantum yield (λ_{max} 720 nm, quantum yield Φ_F ~$10^{-5} - 10^{-4}$) [24–29], while fluorescence of C_{70} is somewhat stronger (Φ_F ~8.5×10^{-4}) and 77 K structured fluorescence occurs with λ_{max} 682 nm [21,30,32,33]. It has been reported that fluorescence of C_{70} at 77 K strongly depends on the solvent matrix. This is most likely associated with deformations of the solute molecules [31]. The S_1–T_1 splitting of both C_{60} and C_{70} is small (~10 kcal/mol in C_{60} [48,50]; ~7 kcal/mol in C_{70} [29,31]). Surprisingly, the S_1–S_2 energy difference of C_{70} was found to be quite large, almost equal to the S_0–S_1 energy difference [31]. The low fluorescence of both C_{60} and C_{70} most likely results from the short lifetime of the singlet excited state and from the symmetry forbiddeness of the lowest-energy transition. The decay rates (k_F) were determined to be $1.5 \times 10^9 \text{ s}^{-1}$ for $^1C_{60}^*$ and $1.61 \times 10^9 \text{ s}^{-1}$ for $^1C_{70}^*$ [33,41]. Both fullerenes display wavelength independent photophysical properties as well as absorption and emission and thus obey Kasha's rule [84].

2. Excited-State Spectroscopy and Phosphorescence

The singlet excited states (S_1) of C_{60} and C_{70} are characterized by transient absorption studies. The estimated lifetimes are for $^1C_{60}^*$ 1.2 ns and for $^1C_{70}^*$ 0.7 ns. The absorption maxima λ_{max} appear for C_{60} at 513, 759, and 885 nm and for C_{70} at 675 and 840 nm [34–38]. The reason for the very short singlet lifetimes of both C_{60} and C_{70} is related to the high intersystem crossing rate to the triplet-excited state with triplet quantum yields close to 100% [35,40–43]. The intersystem crossing rates (k_{ISC}) for C_{60} and C_{70} are in good agreement with the values for the fluorescence decay rates (k_{ISC} [C_{60}]: $1.5 \times 10^9 \text{ s}^{-1}$; k_{ISC} [C_{70}]: $1.25 \times 10^9 \text{ s}^{-1}$) [33]. These values were determined by spectroscopic and photothermal methods and agree with lower limits determined from singlet oxygen quantum yields [20,21].

Triplet–triplet absorption spectra and extinction coefficients (C_{60}: $\lambda_{max} = 330, 750$ nm, ε_{max} 40 000, 20 000 $\text{M}^{-1} \text{ cm}^{-1}$; C_{70} $\lambda_{max} = 335, 480, 690, 970$ nm, ε_{max} 35 000, 12 000, 2250, 3800 $\text{M}^{-1} \text{ cm}^{-1}$) were measured by several groups with varying details of the spectra due to difficulties of accurating the spectra for ground-state depletion. The triplet lifetimes of both fullerenes in solution are quite short (50–100 μs) due to triplet–triplet annihilation and ground-state quenching [20,21,33,36,38,42–46]. However, Fraelich et al. [39] reported significantly longer T_1 lifetimes for both fullerenes. For C_{60} they determined a triplet lifetime of 133 μs and for C_{70} 2.2 ms. In mixed solutions of C_{60} and C_{70} they observed intermolecular energy transfer between nearly isoenergetic triplet states. The T_1–S_0 gap of C_{60} was estimated to be $102 \pm 30 \text{ cm}^{-1}$ greater than that of C_{70}.

The triplet energy levels (E_T) of C_{60} and C_{70} were estimated to be close to 35 kcal/mol by triplet–triplet energy transfer [20,21]. Both compounds quench the triplet states of sensitizers with $E_T \geq 42$ kcal/mol with essential diffusion-controlled rate constants. Quenching of the tetraphenylporphyrine (TPP) triplet ($E_T = 33$ kcal/mol) [47] is significantly slower ($\sim 2 - 4 \times 10^7 \text{ M}^{-1} \text{ s}^{-1}$) mostly likely because energy transfer with this sensitizer is slightly endothermic. Energy transfer from C_{60} and C_{70} triplets to acceptors with $E_T \leq 28$ kcal/mol is diffusion controlled. Despite the high quantum yield of triplet formation there is no phosphorescence of C_{60} observed and C_{70} shows only weak phosphorescence at 77 K, which also confirmed the triplet energy of C_{70} [37]. Heavy-atom-induced phosphorescence can be observed and gives a triplet energy of 36.3 kcal/mol [48]. These values are in agreement with those determined by photothermal [42,43,49] and spectroscopic methods [23,36].

EPR studies of the triplet states of the fullerenes give much insight into the rotational dynamics of the triplet. It can be shown that for both fullerenes two triplet states exist over a temperature range from $T = 8$ K to $T = 213$ K. The difference is their relaxation rates, one type having a faster decay of magnetization (by one order of magnitude) than the other type. Additionally, there is a small difference in the resonance frequency (small shift in the g-value). The faster decaying triplet is found to have a stronger intensity than the other one. Also, triplet–triplet annihilation and quenching by nitroxide radicals was measured by EPR technique [37,50–58].

B. Fullerene Derivatives

In comparison with pristine C_{60} and C_{70}, the fullerene derivatives (Fig. 2) show partly different photophysical properties due to the pertubation of the fullerene's π-electron system. The degree of variation is dependent on (1) the electronic structure of the functionalizing group, (2) the number of addends, and (3) in the case of multiple adducts on the addition pattern at the fullerene core [59–112].

1. UV/Vis–Absorption

The functionalization of the fullerenes leads to differences in the UV–Vis spectra compared to the pure fullerenes. The UV region is relatively less affected by addends as well as for [6,6] ring-closed [62,73,75] (Fig. 2) and [6,5] ring-opened [59,96,99] fullerene adducts [61]. All these monoadducts exhibit strong absorption bands that are very similar to the features observed for C_{60} in this spectral region at nearly the same wavelengths and with similar absorption coefficients (210 nm, 255–260 nm with ε ~100 000–160 000 M^{-1} cm^{-1} and 325–330 nm with ε ~35 000–50 000 M^{-1} cm^{-1}) [61,65,73,88].

The visible region, however, shows significant changes. Some new bands appear whereas others are obscured or disappear. A characteristic feature of C_{60} is the sharp absorption band at 407 nm [88]. This band appears at ca. 420–430 nm for all C_{60} derivatives functionalized at one [6,6] double bond by addition of separate groups, cycloaddition or additions involving triangular bridging (Fig. 3) [59–102].

The bisamino C_{60} derivative **3** (Fig. 4) is an exception since this band does not appear, possibly because of the direct connection of two nitrogen atoms to the fullerene cage [65]. Methanofullerenes, for example $C_{61}H_2$ [99,100] and other fullerene derivatives where the adduct is linked to the fullerene core by a Δ bridge (e.g., $C_{60}O$ [98] and $C_{60}NR$ [101]), exhibit a broad band at around 500 nm (Fig. 3) [59–61,85]. This seems to be characteristic for three-ring fused fullerene derivatives, because the other classes of C_{60} derivatives do not show this broad absorption [62,65,69,70,72,75,79,84,85].

The longest wavelength absorption appears again for all [6,6]-functionalized fullerene monoadducts at around 700 nm corresponding to the 0→0 transition [61,65,73,75,79,81,85,88–93]. Like the absorption at 420 nm the absorption

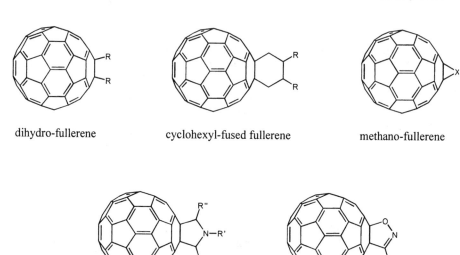

dihydro-fullerene   cyclohexyl-fused fullerene   methano-fullerene

pyrrolidino-fullerene   isoxazolino-fullerene

Figure 2   Types of 1,2-monofunctionalized fullerene derivatives

at 700 nm is also significantly red-shifted in comparison with the corresponding absorption band in the C_{60} spectra (620 nm) [22,79,89]. The prominent broad bands at 550 and 600 nm in the C_{60} spectra do not appear in the spectra of the fullerene derivatives [62,72].

The spectral features of the [6,5] ring-opened fulleroids **4** (Fig. 5) in the visible region are similar to those of C_{60} [94,97]. It is not only the spectra shape that is similar to C_{60}, but also the absorption coefficients are the same of pristine C_{60} [67]. However, the absorption band at 420 nm is missing in the spectra of the methano- and azafulleroids because of the undisturbed π-type electron system of the fullerene core [59,67,94–96]. This specific feature of fulleroides derivatives have often been used to differentiate between ring-opened and ring-closed C_{60} derivatives. These values suggest that substituents in the case of a [6,6] junction in the derivatives have bathochromic effects in the visible region, implying that the electronic forbidden transitions of C_{60} are loosened owing to a reduction in the molecular symmetry. In comparison, the allowed transitions in the UV region are relatively less affected [88]. The electronic properties of the addends hardly influence the absorption band positions, however. Luo et al. [88] investigated a series of aryl substituted pyrrolidinofullerenes and found that the influence of electron-withdrawing groups like -CHO and electron-donating groups like -OMe to the absorption band at around 700 nm are in the order of only a few nanometers

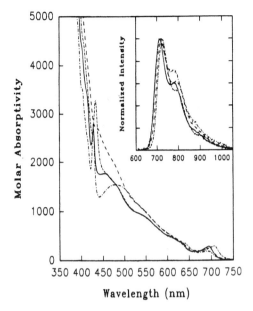

Figure 3 Comparison of absorption and fluorescence (inset) spectra of different classes of C$_{60}$ derivatives: the methano-C$_{60}$ derivative 1′-carboxy-1,2-methano[60]fullerene (-•-•-), the pyrrolidino-C$_{60}$ derivative *N*-ethyl-*trans*-2′, 5′-dimethylpyrrolidino[3′,4′:1,2][60] fullerene 33 (Fig. 24) (-••-••-),[B14a] the C$_{60}$ derivative 2-hydroxy-tetrahydrofuran[4′,5′:1,2] [60]fullerene (—), and the amino-C$_{60}$ derivative 3 (Fig. 4) *N,N*′-dimethyl-piperazine [2′,3′:1,2][60]fullerene (- - -). (From Ref. 65.)

[88]. This indicates that there is neither a strong through-band electronic interaction nor a through-space interaction between the fullerene moiety and the attached group [75,81,110]. An exception to this is the fullerene derivatives **5** (Fig. 6) in which the UV spectrum shows two new bands at 455 and 500 nm that are not observed in the pyrene or in the pure fullerene spectra [75]. Moreover, the solvent

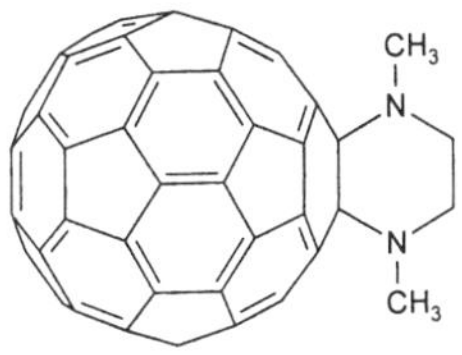

3

Figure 4 Bisamino-C$_{60}$-derivative 3. (From Ref. 65.)

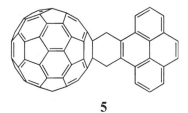

X = C, N, O

4

Figure 5 C$_{60}$ fulleroid structure.

influences the electronic absorption spectra as well. The spectra resolution increases in saturated hydrocarbons like hexane or methylcyclohexane showing more detailed vibronic structures in the 700 nm region. However, the band position and the absorption coefficients are not influenced by different solvents. In more polar solvents like benzonitrile the fine structures of the absorption bands mainly in the 600–700 nm region become indistinct. By using DMF as solvent bands become more indistinct and new broad bands extend to 800–1000 nm, suggesting interactions like charge transfer between the adduct and the solvent in the ground state [88]. However, the UV–Vis spectra from C$_{60}$ derivatives as well as from C$_{60}$ differ in the 420 nm and 407 nm absorption region, respectively, by using a solvent series of benzenes with various numbers of methyl substituents, for example toluene, o-xylene, 1,2,4-trimethylbenzene, and 1,2,3,5-tetramethylbenzene. There is a significant progressive red-shift observed for the band at 430 nm with solvents from toluene to 1,2,3,5-tetramethylbenzene, whereas all other absorption bands in the visible are not effected [26,85,86]. From the investigations of C$_{70}$-derivatives it is known that the absorptions in the UV region are broadened compared to C$_{70}$. The absorption maxima in the visible region is slightly red-shifted compared to the absorption maxima of pristine C$_{70}$ (e.g., 380 nm for C$_{70}$ and 400 nm for cyclohexylfused derivatives and for C$_{70}$H$_2$). The absorption coefficients are similar to those of C$_{70}$ [82,83,92].

5

Figure 6 Fullerene derivative with pyrene unit as addent. (From Ref. 75.)

In the case of multiple functionalized fullerene derivatives the situation is quite different. By increasing the number of addends, either cycloaddition type addends like methano- and pyrrolidinofullerenes or dihydro-type addends like hydrogen, halogen and phenyl groups, the molar extinction coefficients decrease (e.g., $C_{60}H_4$, $C_{60}H_{18}$ and $C_{60}H_{36}$). The absorption spectra of $C_{60}H_4$ shows the typical absorption of C_{60} in the UV region, a sharp peak at 433 nm and a continuous unstructured drop of the absorption without a peak at 700 nm [72]. $C_{60}H_{18}$ shows a strong absorption at 210 nm, two peaks at 260 and 340 nm, and at least two shoulders at 370 and 420 nm while in the visible region there is no significant absorption as in the spectra of the monoadducts. The absorption coefficients are much weaker than those of the corresponding absorption bands of pure fullerene [104]. Earlier investigations showed slightly different absorption maxima but the trend to weaker absorption coefficients compared with pure C_{60} is similar [103]. The spectrum of $C_{60}H_{36}$ shows a strong absorption at 210 nm, a weak absorption at 329 nm, and then only very weak absorption without further maximum at longer wavelengths. The molar absorption coefficients could not be determined due to the very low solubility in cyclohexane [104]. As in the case of $C_{60}H_{18}$, an earlier investigation revealed slight differences in absorption bands position which may be caused by an anthracene impurity [103]. In any case, both groups reported continuous decrease of the absorption coefficients in the UV–Vis region with increasing functionalization [103,104]. Similar results were reported for four $C_{70}H_{10}$ isomers. Compared to pure C_{70} all four isomers show broader absorption in the UV region with less resolution. The visible region is dominated by a continuous drop of the absorption without further maxima. The absorption onset occurs at around 600 nm [105]. Similar results were obtained for $C_{60}Cl_6$, $C_{60}(C_6H_5)_5Cl$ and $C_{60}(C_6H_5)_5H$. In the UV region three strong absorptions appear with significantly weaker absorption coefficients compared to C_{60}. In the visible region absorption bands appear for all three compounds between 390 and 470 nm. Derivatives with phenyl groups show additional absorptions in the 300–450 nm region [106,107]. It is remarkable that the absorption at ~470 nm is not due to the $S_0 \rightarrow S_1$ transition as can be seen by the fluorescence spectra [107].

The spectral features are also influenced by the addition pattern as can be seen by the different absorption spectra of the multiple malonic-ester C_{60} derivatives (Fig. 7). The spectra of one bis- **(6)** and one tris-isomer **(7)** are quite different from the spectra of the corresponding monomer. The absorptivity is partly increased in the 450–600 nm region, while the sharp band at 420 nm is missing [67]. The hexakis adducts **(8, 9)** exhibit no spectral feature in the visible region and only slight resolution in the UV region, but the spectra of both compounds are not similar [67]. Nakamura et al. [110] investigated [60]fullerene o-quinodimethane bisadducts. They found a strong dependence of the absorption maxima on the addition pattern, whereas different substituents do not effect the spectroscopic features [110].

2. Fluorescence

The increasing pertubation of the fullerene π-type electron system leads to a reduction of the symmetry, e.g., from I_h to C_{2v} or C_S, and converts some vibronic-forbidden states to allowed states. Besides the absorptivity in the UV–Vis, the fluorescence properties of the adducts are also influenced, as well as the fluorescence lifetime, the emission band positions, and especially the quantum yield, which increases compared with that of C_{60} [61,65–67,75,88–90,92].

One characteristic feature for all derivatives is that $0 \rightarrow 0$ bands in the fluorescence spectra correspond well to the $0 \rightarrow 0$ bands in the absorption spectra indicating that the distribution of vibrational levels in the first excited state resembles those in the ground state (Fig. 8) [65,70,71,75–91,108]. The mirror—image relationship between the $0 \rightarrow 0$ absorption and the fluorescence bands suggest that the lowest-energy absorption and emission are associated with the same excited singlet state for the C_{60} derivatives in room temperature solution [65].

The main features observed in the fluorescence emission spectra of C_{60} are multiple peaks in the 686–762 nm region and three spectral shoulders between 620–677 nm. In comparison, the fluorescence spectra of the monoadducts generally show a main emission peak around 690 nm which is slightly blue-shifted (C_{60}: 720 nm). Three shoulders in the 620–677 nm region are still present in the spec-

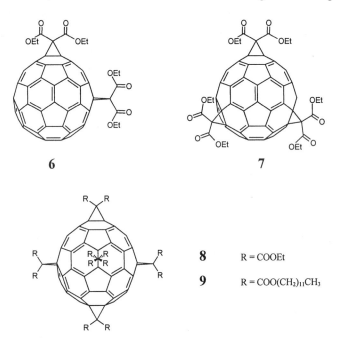

Figure 7   Multiple malonic ester C_{60} derivatives.

tra of the derivatives but are obscured and, in addition, there are two shoulders at longer wavelengths (~780 and 890 nm) [61,85,89,90,92]. The main emission peaks of the adducts range from 680–710 nm [67,71,75,80,89,90] depending on the type of connection between addend and fullerene core (e.g., methano-, pyrrolidino-, or dihydrofullerenes) but independent of the excitation wavelength [65,75,90]. Changing the solvent from methylcyclohexane to dichloromethane or benzene, for example, causes a broadening of the emission spectra, a noticeable red-shift of the *0 → 0 transition bands, and a significant loss of resolution [61,108]. Nevertheless, no influence of the solvent has been reported by Ma et al. [85].

In C_{70} monoadducts (Fig. 9) the emission maxima appear around 690–710 nm and a shoulder at 750 nm, which is slightly red-shifted compared to C_{70} [82,83,92]. The exact emission band position depends strongly on the regiochemistry of the monoadduct. The two possible 1,2-isomers (**10, 10a**) with an addend attached near to the pole show similar emission spectra, while in the case of the 5,6-isomer (**11**) the emission maxima is bathochromic shifted about 8 nm [92]. The fluorescence lifetime was determined to be in the range of 1 ns, somewhat higher than that of C_{70} with 0.65 ns, while the fluorescence quantum yield is 7×10^{-4} and nearly identical to that of C_{70} [82].

The emission band position is also influenced by the electron-withdrawing and electron-donating effect of the addend, respectively, and by the type of chemical functionalization (e.g., various size of the rings fused to the fullerene core). For example, it has been reported for three different fullerene derivatives that an enhanced electron-withdrawing effect causes an increasing red-shift of the emission band [108]. Another example is a C_{60} derivative **12** (Fig. 10). The emission maxima appears at 705 nm, but if there are methoxy groups additionally attached to the anthryl moiety the emission maxima is shifted to 712 nm [92].

Since fluorescence intensity is known to depend on the molar extinction coefficient of the ground-states and C_{60} and fullerene derivatives do display different molar extinction coefficients the relative emission yields, i.e., intensities are quite different. For most fullerene derivatives, independent of the type of functionalization, the quantum yield is increased by a factor of 2 or even 3 due to the reduction of the structural symmetry of the fullerenes (Φ_F of C_{60} is 0.32×10^{-3}) [65–67,71,90]. The Φ_F values are independent of the excitation wavelength [61,66,71,80–88,90,92]. In the case of fulleroid derivatives, the fluorescence properties (emission band maxima, fluorescence quantum yields) are similar to that of pristine C_{60} [67]. However, in many examples observed there is a strong solvent dependence of the quantum yields. Derivatives with strong electron-donating groups attached to the addend-like dimethyl aniline moiety [66,80, 84,88,110] or methoxy substituted aromatic addends [92] show a significant decrease of the fluorescence quantum yield from 6×10^{-4} to 0.2×10^{-4} with increasing polarity of the solvent (e.g., from methylcyclohexane to benzonitrile).

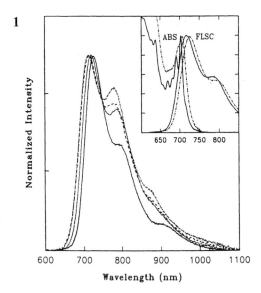

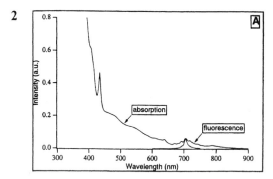

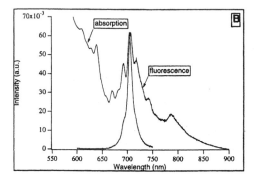

Figure 8 Fluorescence spectra of different type of C_{60} derivatives and mirror image; (1): methano-C_{60} derivatives (•••), (---), and (-••-), and the pyrrolidino-C_{60} derivative *N*-ethyl-*trans*-2′, 5′-dimethylpyrrolidino [3′,4′:1,2][60]fullerene 33 (Fig. 24) (—) in room temperature toluene (6×10^{-5} M). Inset: comparison of absorption and fluorescence spectra of the methano-C_{60} derivative in hexane (—) and toluene (-•-). (2): Absorption and fluorescence spectra of C_{60}[11]DMA 16 (Fig. 15) in methylcyclohexane at room temperature (A, full range; B, expansion above 550 nm). (From Refs. 80, 85.)

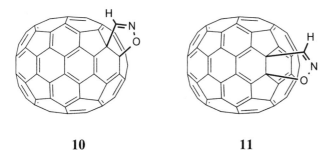

10 **11**

Figure 9 Isomeric C_{70} derivatives with 1,2- (10) and 5,6-addition pattern (11). (From Ref. 92.)

The fluorescence intensity increases again by adding trifluoro acetic acid. As a result the emission and the absorption maxima are slightly hypsochromic shifted [84,88]. Presumably in polar solvents another deactivation pathway competes with the radiative path. However, the decreasing triplet state formation in polar solvents implies that the low fluorescence quantum yield cannot be caused by an efficient intersystem crossing. The extra deactivation pathway can be assigned to an intramolecular electron transfer from the NMe_2 moiety or the methoxy-substituted aromatic addends to the excited singlet state of the C_{60} or C_{70} moiety [84,88,92,110]. Verhoeven et al. investigated five pyrrolidino- and cyclohexyl-fused fullerene derivatives. For one of the C_{60} pyrrolidino derivatives (**13,** Fig. 11) they observed only a very weak fluorescence in polar as well as in nonpolar solvents ($\Phi_F = 2 \times 10^{-5}$). However, after addition of acid to the solutions fluorescence could be observed as in other pyrrolidino monoadducts [80].

The lifetimes of the singlet excited states are only partly influenced; the values range from 1.1 to 1.8 ns [61,66,84,85,89–92], and for C_{60} a value of about ~1.3 ns was estimated. As for the fluorescence quantum yields, there is also a solvent dependence of the fluorescence lifetimes for fullerene derivatives with electron-

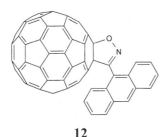

12

Figure 10 Isoxazoline-C_{60}-derivative bearing an anthryl moiety. (From Ref. 92.)

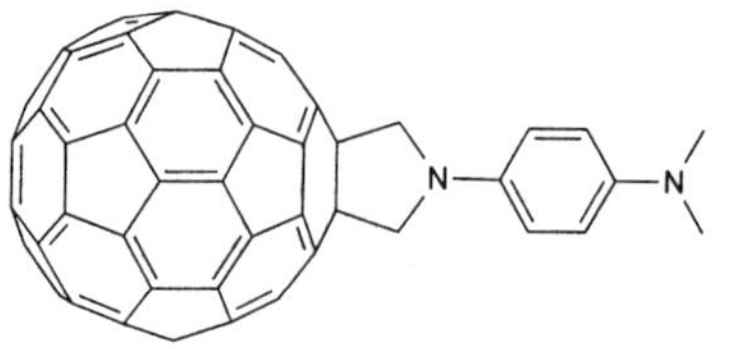

13

Figure 11 Pyrrolidino-C_{60}-derivative bearing a dimethylaniline moiety at the pyrrolidi-nonitrogen. (From Ref. 80.)

donating substituents. For some of the fullerene derivatives described previously a dramatic decrease in fluorescence lifetimes from ~1.5 ns to 0.06 ns in benzonitrile was observed [80,84,88,92].

In the case of the monoadducts, the fluorescence compared with that of C_{60} and C_{70} is effected for most adducts in the same direction blue-shift of the main emission maxima, increased fluorescence quantum yield (Table 2), the influence on the fluorescence properties in multiple functionalized fullerenes strongly depends on the type of addend. The hydrogenated adducts $C_{60}H_{18}$ and $C_{60}H_{36}$ show very similar spectra with emission maxima at 480 and 520 nm, strongly blue-shifted compared with the maxima of C_{60} at 720 nm [103,104]. The emission band of $C_{60}H_{18}$ is much broader than that of $C_{60}H_{36}$ and the fluorescence quantum yield of $C_{60}H_{18}$ is with $\Phi_f = 0.06$ about 6 times smaller than that of $C_{60}H_{36}$ with $\Phi_f = 0.37$ ($\Phi_f(C_{60}) = 10^{-4}$). The fluorescence lifetimes of both $C_{60}H_{18}$ and $C_{60}H_{36}$ are remarkably increased compared to the fluorescence lifetime of C_{60}. The estimated values are 2.8 ns for $C_{60}H_{18}$ and 27 ns for $C_{60}H_{36}$ [104].

Adducts of the same type of addition but with a smaller number of addends like $C_{60}Cl_6$, $C_{60}(C_6H_5)_5Cl$ and $C_{60}(C_6H_5)_5H$ show similar emission spectra, with the difference being that the emission maxima are less blue-shifted. The maxima corresponding to the $S_1 \rightarrow S_0$ transition appear approximately at 620 nm [107]. The singlet lifetime of the adducts $C_{60}(C_6H_5)_5Cl$ and $C_{60}(C_6H_5)_5H$ is with ~1.2 ns in the same range of C_{60} [106,107]. However, the singlet lifetime of $C_{60}Cl_6$ is with 0.5 ns partly decreased caused by the heavy atom effect on the nonradiative transfer rate from the S_1 state to the lowest triplet-excited state T_1 [107].

Upon further addition of addends of the malonic ester type (Fig. 12), e.g., 2, 3, or 6 addends, the emission spectra are partly effected. However, the direction of change in the spectral features seems to correlate with the symmetry of the adducts. Addition of 2 or 3 malonic ester units lead to a small red-shift of the main emission band, while the emission band in the spectra of the hexakis adducts is barely blue-shifted compared to the corresponding monoadduct [108]. Changing the solvent from methylcyclohexane to the more polar benzene or toluene the main emission

Table 2   Fluorescence Properties of Fullerene Derivatives Compared to Those of C_{60}/C_{70}

	C_{60} [24–29,34,36]	C_{60}-monoadducts [65–67,71,80–92]	C_{70} [21,30,32,33,36]	C_{70}-monoadducts [82,83,92]
Emission bands λ_{max}	720 nm	690–710 nm	682 nm	690–710 nm
Fluorescence-quantum yield Φ_F	0.3×10^{-3}	$0.5–1.0 \times 10^{-3}$	0.85×10^{-3}	$\sim0.7 \times 10^{-3}$
Singlet-lifetime τ_F	1.1 ns	1.1–1.8 ns	0.6 ns	~1 ns

is (1) red-shifted, (2) remarkably broadened, and (3) a significant loss of resolution is observed [108]. In methylcyclohexane at 77 K the monobis(ethoxycarbonyl)-methylene adduct shows the lowest emission band at 690 nm. By adding a second addend the emission band is shifted to 699–702 nm depending on the addition pattern. Adding a third bis(ethoxycarbonyl)methylene group the emission band appears at 702 nm [108]. Presumably, one reason is the change in adduct symmetry. In comparison with C_{60}, the emission bands are still blue-shifted [108]. Sun et al. observed a drastically red-shift of the emission band for the e-bis-adduct (780 nm) and the e,e,e-tris-adduct (723 nm) if fluorescence measurements are carried out at room temperature in toulene [67].

In the case of the hexaadduct where the symmetry is increased to T_h or D_3 compared to C_{2v} for the corresponding monoadduct, the emission band is blue-shifted to 660–690 nm depending on the solvent used [67,111,112]. The fluorescence spectra of o-quinodimethane bis-adducts also strongly depend on the addition pattern. The spectra for three different regioisomers (cis-2, cis-3, e) are all remarkably different from one another, as in the case of the absorption spectra.

All fluorescence spectra exhibit mirror-image of the longest wavelength absorption. The quite low fluorescence for two isomers (cis-2, e) is probably ascribed to the fact that no absorption maxima in the visible region could be observed [109]. The same tendencies in the fluorescence spectra are also observed for the corresponding regioisomeric derivatives with different substituents. Therefore, the differences in the fluorescence spectra among the regioisomers are apparently brought about by the differences in the addition pattern [110]. However, beside the symmetry the electron-withdrawing effect of the addends also influence the emission band position. By substituting the bis(ethoxycarbonyl)methylene groups in the hexakis adduct with pyrrolidino addends the emission band is blue-shifted. This is caused by the higher electron-withdrawing properties due to the lower degree of delocalization by homoconjugation [111]. For both isomers of the hexapyrrolidino adducts fluorescence appears at $\lambda_{max} = 550$ nm [111,112].

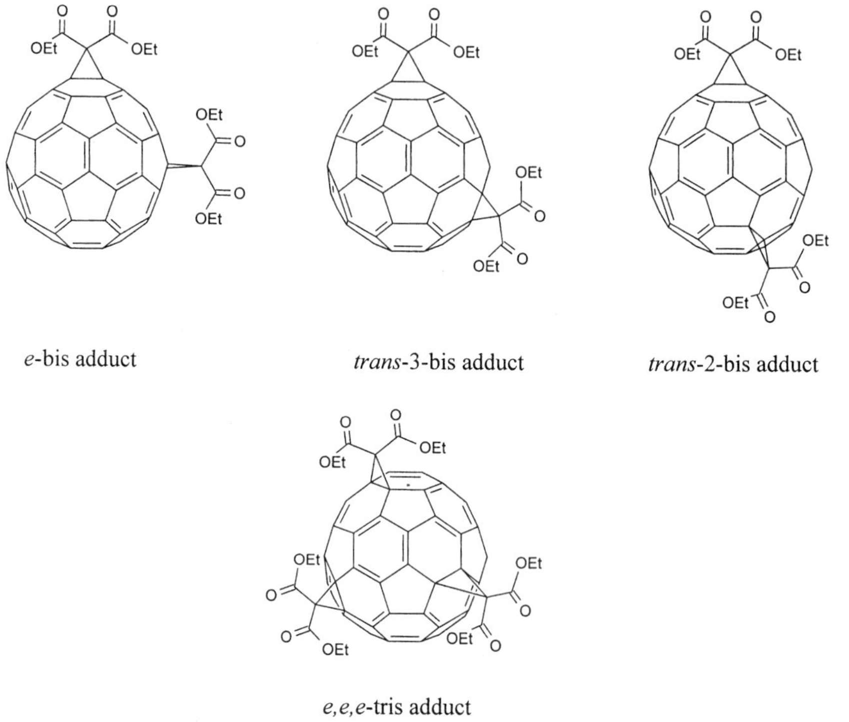

e-bis adduct   *trans*-3-bis adduct   *trans*-2-bis adduct

e,e,e-tris adduct

Figure 12   Multiple malonic ester type adducts.

Unlike the reversed shift of the emission band compared to the dihydrogen addend type, the singlet lifetime in the bis- and tris(bis-(ethoxycarbonyl)-
methylene) derivatives is increased comparable to the former multiple adducts.
The values range from 1.7 to 3.1 ns (tris-adduct), depending on the distorted
π-electron system of the fullerene core [67,108]. In comparison to C_{60}, the
fluorescence quantum yield for the malonic ester hexaadduct is increased by
the factor 10 (30×10^{-4}) [67,111]. In the case of both pyrrolidino hexaadducts (T_h **14** and D_3 **15,** Fig. 13), the effect is remarkably higher. The fluorescence quantum yields are increased about 100-fold (~0.02) compared to
C_{60}. On the other hand, the singlet lifetime is only partly increased with ~3.5
ns [111,112].

The photophysical properties of the pyrrolidino-T_h-hexaadduct **14** are significantly affected upon adding acid to the solutions while pyrrolidino-D_3-hex-
aadduct **15** does not show change at any acid concentration either in organic or
aqueous media. The UV–vis absorption spectrum is only minimally affected, the

absorption bands are slightly shifted, and the absorption coefficients are partly increased. However, the fluorescence properties are considerably influenced. At fixed excitation wavelength (300 nm) the emission band is significantly red-shifted and the quantum yields increase from 0.024 to 0.043 most likely due to the longer singlet lifetime of the protonated species in comparison to the neutral form. Also, the change of the excitation wavelength has a pronounced effect on the emission spectra. Changing to longer excitation wavelength (300 → 370) the weak emission at about 540 nm grows stronger and gets a small blue-shift [112].

3. Excited-State Spectroscopy and Phosphorescence

As a consequence of the inefficient fluorescence mainly caused by the high inter-system crossing rate, the quantum yield of triplet formation is near unity for the fullerene monoadducts as in the case of the pure fullerenes [62,64,65,71–73]. The values were determined by photothermal and spectroscopic methods and agree with lower limits determined by singlet oxygen production [62,65,72]. In comparison with pure C_{60}, the derivatives exhibit deviations in the triplet–triplet absorption spectra (Table 3). All adducts show nearly identical spectra regardless of the solvents or the excitation wavelength [75,88]. The absorption maxima are somewhat broader and slightly blue-shifted compared to that of C_{60} with estimated maxima at 330 and 750 nm and extinction coefficients of $\epsilon = 40000$ and $20200 \ M^{-1} \ cm^{-1}$. The absorption is due to the $T_1 \rightarrow T_n$ transition of the C_{60} moiety in the derivatives (Fig. 14) [75]. For methano adducts the maxima in the visible are about 30 nm blue-shifted to ~720 nm and the extinction coefficients are

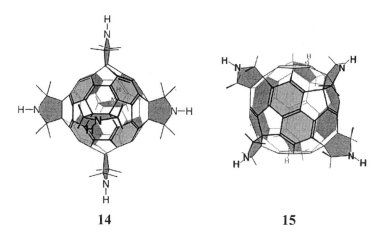

14   **15**

Figure 13 T_h (14) and D_3 (15) hexapyrrolidino C_{60} derivatives. (From Ref. 112.)

about 25 percent lower than that of C_{60} (~14000M^{-1} cm^{-1}) [62,63]. Additionally, the maxima shoulders at 820 nm appear in these spectra [62,84,91]. In the case of cyclohexylfused fullerene derivatives, pyrrolidinofullerenes, and dihydro-fullerenes the absorption maxima in the visible are about 20 nm more hyp-sochromic shifted and appear around 700 nm [71–75,84,87,88,91]. The absorption maxima in the 300–400 nm region are blue-shifted as well and appear near 400 nm, the extinction coefficients are considerably lower than those of C_{60}, the values range from 4000 to 17000 M^{-1} cm^{-1} [71,72,75,84,91].

As mentioned earlier, fullerene derivatives with addends bearing strong electron donating groups can undergo photoinduced intramolecular electron transfer in polar solvents. This leads to a large decrease of singlet → triplet intersystem crossing and therefore to a decrease of the triplet quantum yield [62,84,88]. Williams et al. reported a cyclohexylfused C_{60}-derivative with an dimethylaniline group attached to the addend (16, Fig. 15) which shows triplet absorption spectra with a significant solvent dependence. By changing from nonpolar solvents to benzonitrile the spectrum changes drastically. A new and characteristic absorption around 460 nm with a lifetime of 0.25 μs appears, which is attributed to the radical cation of the dimethylaniline chromophore. Furthermore, two absorptions with the same lifetime at 750 and 590 nm can be assigned to the radical anion of the fullerene moiety. This indicates an intramolecular charge separation that quenches the locally excited state [80].

The triplet lifetimes of the adducts are ~25 μs apparently shorter than that of C_{60} (Table 3) [71,75]. Exceptions are adducts that show strong solvent dependence of the triplet quantum yield and the triplet absorption spectra. These adducts have significantly shorter lifetimes in polar solvents [88]. The triplet states of the adducts can be quenched by oxygen as in the case of C_{60}. In solutions containing oxygen the quenching rate is 100 times faster than in argon bubbled solutions. The quenching rates for the adducts are somewhat faster than that of the pure fullerenes [88]. Despite the high intersystem crossing rate of the adducts there is only less phosphorescence implying that the radiative transitions between the dif-

Table 3 Triplet Properties of Fullerene Derivatives and C_{60}

	Triplet-triplet absorption maximum λ_{max}/nm	Extinction coefficient ε_T/mol^{-1} cm^{-1}	Triplet-lifetime τ/μs	Triplet quantum yield Φ_T
C_{60}	747 [34,36]	~15000 [34,36,38,42,43–45]	50–100 [20,34,39,42]	~1 [20,35,40–43]
C_{60}-monoadducts	700–720[a]	4000–17000	~25	0.88–0.95

[a] For methanofullerenes, the triplet-triplet absorption appears arround 720 nm, for cyclohexylfused derivatives, pyrrolidino- and dihydro-fullerene, the absorption appears arround 700 nm [62,71,72].

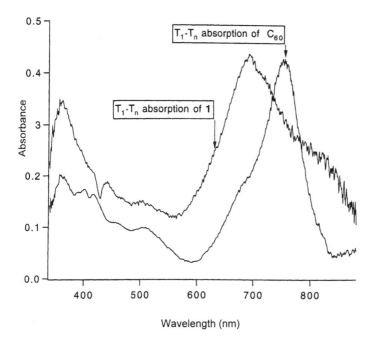

Figure 14 Comparison of scaled triplet–triplet absorption spectra of the *N,N*-dimethylaniline substituted pyrrolidino C$_{60}$ derivative 22 (Fig. 20) and C$_{60}$ in methylcyclohexane. (From Ref. 84.)

ferent multiplicity states are principally forbidden [88]. However, adding iodo ethane as heavy atom donor fluorescence is quenched and a new weak band appears at 825 nm that is attributed to phosphorescence. As for the other photophysical properties, there exists an influence of the substituent on the phosphorescence. Electron-donating groups cause a blue-shift of the band [88]. The triplet energy is estimated to E_T^* 1.50 eV [88] and EPR studies have shown that the triplet state is located on the fullerene moiety [78].

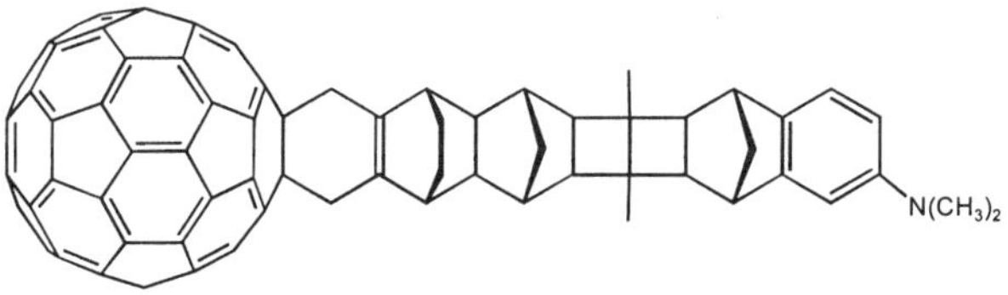

16

Figure 15 C$_{60}$ derivative of cyclohexylfused type bearing a *N,N*-dimethylaniline moiety (16).

The phosphorescence of C_{70} derivatives is noticeably red-shifted, like the absorption and the fluorescence. The maxima appear at ca. 850 nm for cyclohexylfused derivatives as well as for $C_{70}H_2$. However, the phosphorescence lifetime was determined to be 6.2 ms indicating it to be somewhat smaller than C_{70} under similar conditions [82,83].

Due to the fact that further functionalization of fullerene monoadducts leads to an exponential increase of possible adducts compared with the high number of isomers of each multiple adduct it is very difficult to elaborate the tendencies in the excited state properties upon continuous functionalization. Upon excitation of the derivatives $C_{60}H_{18}$ and $C_{60}H_{36}$ by laser flash photolysis after 35 ps the growth of a transient at 480 nm was observed for both adducts. This absorption is due to the $S_1 \rightarrow S_n$ absorption, the absorption coefficient of $C_{60}H_{18}$ with 5700 dm^3 mol^{-1} cm^{-1} is somewhat higher than for $C_{60}H_{36}$ with 4500 dm^3 mol^{-1} cm^{-1}. In both S $\rightarrow$ S absorption spectra a very weak absorption appears at 880 nm that is suppressed compared to that of C_{60}. The singlet-excited lifetimes perfectly match with the values obtained by fluorescence measurements ($C_{60}H_{18}$: 2.5 ns, $C_{60}H_{36}$: no decay of the singlet absorption up to 6 ns, the estimated fluorescence lifetime is 27 ns). In the case of $C_{60}H_{18}$, the singlet absorption disappears and a new absorption with a major peak at 640 nm is observed in addition to a smaller one at 500 nm [103,104] pointing to markable blue-shifts for both singlet and triplet absorption. In addition, the 880 nm absorption is drastically decreased compared to C_{60}. $C_{60}(C_6H_5)_5Cl$ belongs to the same type of multiple addition products, however, with less addends. This hexaadduct shows some differences in the excited state properties. Excitation by laser flash photolysis leads to a growth of a band at 750–920 nm after 35 ps due to the $S_1 \rightarrow S_n$ absorption, comparable with the excited singlet state of C_{60}. The molar absorbtivity of $^1C_{60}(C_6H_5)_5Cl^*$ is with 9000 dm^3 mol^{-1} cm^{-1}, about 30% higher than that of $^1C_{60}^*$ (6300 dm^3 mol^{-1} cm^{-1}). After 6 ns a decay of the absorption at 750–920 nm is observed and a transient at 670 nm grows up. This transient is assigned to the excited triplet state and is about 70 nm blue-shifted compared to that of C_{60}. The molar absorptivity is much higher than that of pure C_{60} (25 000 vs 12 000 dm^3 mol^{-1} cm^{-1}). On the other hand, the triplet quantum yield is only ~0.3 and the triplet lifetime is much higher than that of C_{60} (100 μs) [106].

As shown for multiple methano adducts, the increasing distortion of the π-electron system of the fullerene core plays an important role in affecting the excited state properties. While the mono-bis(ethoxycarbonyl)methylene adduct shows similar singlet and triplet absorption spectra compared to C_{60}, the further functionalization with bis(ethoxycarbonyl)methylene groups lead to continuous blue-shift of the excited absorption bands. Also, the triplet quantum yield is strongly decreased by higher functionalization, e.g., for the hexakis adducts a value of 0.18 was determined. The triplet lifetime with 50 μs is only less affected [73]. An additional blue-shift can be observed by substituting the methano groups by more electron withdrawing addends like cyclohexano or pyrrolidino groups. The singlet lifetimes are partly increased as well from 1.9–3.2 ns compared to 1.2 ns for $^1C_{60}^*$ [108]. However, the phosphorescence of the multiple adducts is not much effected. Upon cooling the solutions to 77 K, two bright orange, well resolved, and intense phosphorescence emissions (Fig. 16) with lifetimes of ~4 s appear between 615 and 800 nm. This high phosphorescence intensity is primarily due to the 10^6 times slower thermal decay of their triplet states than that of C_{60} [111,112].

Triplet absorption spectra of *o*-quinodimethane bis-adducts exhibit some differences to the corresponding monoadduct. The absorption band positions are only less affected, whereas the lifetime strongly differs between the regioisomers. Also, the initial absorbance differs noticeably, which may be associated with the quantum yield of intersystem crossing from S_1 to T_1 [110].

III. PHOTOCHEMICAL PRIMARY PROCESSES

Since the absorption features of fullerenes extend over the UV and visible region, light of different wavelengths can be used for excitation. Depending on the frequency, various vibrational levels of singlet excited states are populated followed by relaxation to the lowest vibrational level of S_1. This process is completed be-

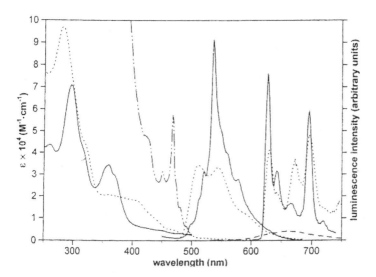

Figure 16 Electronic absorption (CH₂Cl₂, 25°C), fluorescence (25°C) and phosphorescence (77K) spectra of the T_h (—) and D_3 (•••) hexaadducts 14 and 15 (Fig. 13). Excitation spectra of 14 (-••-••-) at 77 K (λ_{em} = 540 nm) and fluorescence of the Bingel-type T_h-(dicarboethoxycyclopropyl)hexaaduct 8 (Fig. 7). (From Ref. 111.)

fore efficient intersystem crossing to the triplet-excited state occurs [19,35]. Usually, the triplet-excited state is then the starting point for a slow regeneration to the singlet ground-state [19]. Excitation of molecules to their excited states makes them stronger reductants and oxidants in comparison to their ground-states (Table 4) [19,113–115,120–123,133].

In principal, electron transfer reactions with fullerenes could occur via both the singlet- and triplet-excited state. However, due to the short singlet lifetime and the efficient intersystem crossing, intermolecular electron transfer reactions usually occur with the much longer lived triplet-excited state. The result of the electron transfer is a radical ion pair of fullerene and electron donor or acceptor.

A. Photochemical Processes of Excited C_{60} and C_{70}

1. Reduction

It is known from electrochemical studies that fullerenes are easily reduced. Up to 6 electrons can be added reversibly [19], and, as mentioned earlier, the excited states are even more easily reduced. A large number of electron donors were investigated including aromatic and alkyl amines [29,43,79,119–140,152,161], nitroxide radicals [57,117], suspensions of TiO_2 [118], polyaromatic compounds, [19,127] organo-silicon compounds, [133,158] phenothiazine, [133] acridine [145,154], β-carotene [141], tetrathiafulvalenes [146], tetraethoxyethene [147], phthalocyanines [148], porphyrines [151,153], NADH and analogues [150,154, 155], borates [156,159], and naphtoles [23] to name a few representative cases.

Quenching the triplet-excited states of the fullerenes by the electron donors occurs efficiently in polar solvents like benzonitrile. The mechanism is a primary electron transfer [Eq. (1) [120,125,133,139,148,154–162].

$$C_{60/70} \xrightarrow{h\nu} {}^1C_{60/70}{}^* \xrightarrow{ISC} {}^3C_{60/70}{}^*$$
$$^3C_{60/70}{}^* + Donor \xrightarrow{ET} C_{60/70}{}^{\bullet-} + D^{\bullet+}$$

(1)

This is shown by the formation of transient absorptions of both the donor radical cations and the C_{60} or C_{70} radical anion [120,125,127,133,139,141,

Table 4 Reduction and Oxidation Potentials of Ground-State and Excited C_{60}

Reduction potential/electron affinity in benzonitrile	Oxidation potential in benzonitrile
C_{60} −0.43 V vs SCE	+1.76 V vs SCE
$^1C_{60}^*$ +1.56 V vs SCE	−0.23 V vs SCE
$^3C_{60}^*$ +1.13 V vs SCE	+0.19 V vs SCE

Source: Ref. 116.

146,148,152–162]. The transient absorption of the fullerene radical anions appear in the infrared (for $C_{60}^{\bullet -}$ at 1070 nm with shoulders in the region between 900 and 1000 nm; for $C_{70}^{\bullet -}$ at 1380 nm, Fig. 17) [120,125,127,139,146–148, 152–162]. Time profiles of laser flash photolysis experiments show mirror—image of increasing $C_{60}^{\bullet -}/C_{70}^{\bullet -}$ and decreasing $^3C_{60}^*/^3C_{70}^*$. The maximum of the radical anion absorption is reached after 150 ns for both the fullerenes, the decay occurs within a timescale of 10 μs. The triplet states of both C_{60} and C_{70} do not decay within 1 μs in the absence of an electron donor. Addition of donors, e.g., nitroxide radicals, result in progressive decrease of triplet lifetime while no decrease of the triplet yield is observed. This is in agreement with the reported intersystem crossing yield (ΦISC), which is near unity [117,147]. These observations clearly indicate that the electron transfer occurs from the donors to the triplet-excited states of the fullerenes [120,125,133,146,148,152–162].

However, the initial step of the electron transfer reaction strongly depends on the solvent polarity. By changing the solvent to less polar or nonpolar solvents like benzene or nonaromatic hydrocarbons the transient absorptions of $^3C_{60}^*$, $C_{60}^{\bullet -}$ and donor radical cation appear immediately after the laser pulse. The decay of all the absorptions is also completed at the same time. The fast appearance and the fast decay of the $C_{60}^{\bullet -}$ and donor radical cation absorption suggest that there is an interaction between fullerene and donor in less polar and nonpolar solvents before laser irradiation [120,125,133–139].

Time profiles of the formation of fullerene radical anions in polar solvents as well as the decay of $^3C_{60}^*$ obey pseudo first-order kinetics due to high concentrations of the donor molecule [120,125,127,146,159]. By changing to nonpolar solvents the rise kinetics of $C_{60}^{\bullet -}$ changes to second-order as well as the decay kinetics for $^3C_{60}^*$ [120,125,133,148]. The analysis of the decay kinetics of the fullerene radical anions confirm this suggestion as well. In the case of polar solvents, the decay of the radical ion absorptions obey second-order kinetics, while changing to nonpolar solvents the decay obey first-order kinetics [120,125, 127,133,147]. This can be explained by radical ion pairs of the $C_{60}^{\bullet -}$ and the donor radical cation in less polar and nonpolar solvents, which do not dissociate. The back-electron transfer takes place within the ion pair. This is also the reason for the fast back-electron transfer in comparison to the slower back-electron transfer in polar solvents, where the radical ions are solvated as free ions or solvent-separated ion pairs [120,125,147]. However, back-electron transfer is suppressed when using mixtures of fullerene and borates as donors in o-dichlorobenzene (less polar solvent), since the borate radicals immediately dissociate into Ph_3B and $Bu^{\bullet}/Ph^{\bullet}$ [Eq. (2)][156].

$$C_{60} \xrightarrow{h\nu} {}^1C_{60}^* \xrightarrow{ISC} {}^3C_{60}^*$$

$$^3C_{60}^* + {}^-BPh_3R \xrightarrow{k_{et}} C_{60}^{\bullet -} + {}^\bullet BPh_3R \longrightarrow C_{60}^{\bullet -} + BPh_3 + R^{\bullet} \qquad (2)$$

$$(R = Bu \text{ or } Ph)$$

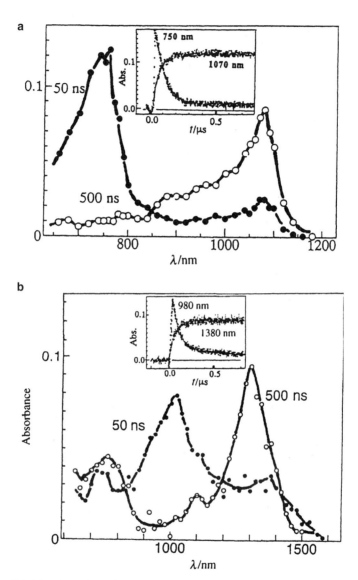

Figure 17 Transient absorption spectra obtained by 532 nm laser flash photolysis of (a) C_{60} and (b) C_{70} (both 2.5×10^{-4} mol dm^{-3}) in the presence of tetraethoxyethene (1.0×10^{-2} mol dm^{-3}) in deareated benzonitrile. (●) 50 ns and (○) 500 ns. (a) Time profiles at 750 nm and 1070 nm; (b) time profiles at 980 nm and 1380 nm. (From Ref. 147.)

In solvent mixtures both pathways (electron transfer via free $^3C_{60/70}^*$ or ground-state complex) occur and the proportion is dependent on the ratio of polar to nonpolar solvent. The same holds for back-electron transfer [125,133,141]. In general, the efficiency of electron transfer via $^3C_{60}^*/^3C_{70}^*$ increases with increasing solvent polarity [156,158].

The first-order rate constant can be evaluated from the decay curves of $^3C_{60}^*$ and the rise curves of $C_{60}^{\bullet-}$ and the donor radical cation [125,154]. The observed electron transfer rate constants for C_{60} are usually in the order of 10^9-10^{10} dm^3 mol^{-1} s^{-1} and thus near the diffusion controlled limit which depends on the solvent (e.g., diffusion controlled limit in benzonitrile ~5.6 $\times$ 10^9 M^{-1} s^{-1}) [120,125,127,141,154–156].

An exception are naphthols as donors (e.g., 2-naphthol [2-NOH], 2,3- and 2,7-dihydroxynaphthalene, 1,1'-bi-2-naphthol [BN(OH)$_2$]), which quenches the fullerene triplet with rate constants of only ~10^6 M^{-1} s^{-1}. The addition of pyridine increases the quenching rate to about 2–3 orders of magnitude [162]. This observation can be explained by a reduced oxidation potential of the naphthols by 20 kcal/mol on addition of pyridine. This indicates that pyridine attracts proton from the naphtholes resulting in the increase of quenching rates [Eq. (3) [162].

$$^3C_{60/70}^* + NOH - Py \xrightarrow{k_{et}} C_{60/70} + NO^{\bullet} - PyH^+ \qquad (3)$$

The quantum yields decreases by changing from planar naphtholes to the perpendicular binaphthole independent on the absence or presence of pyridine [162]. Also, in the case of optical active 1,1'-binaphthyl-2,2'-diamine (BNA) compared with N-phenyl-β-naphthylamine (PNA), the quantum yield of electron transfer is for BNA by a factor of 0.5 lower due to the perpendicular structure [161]. A possible explanation is that the approach of the perpendicular donors BN(OH)$_2$ and BNA to the almost spherical fullerenes is hindered, while the planar 2-NOH and PNA more easily contact with the fullerenes [161,162]. The quantum yields of C_{70} in the absence of pyridine are slightly higher than those of C_{60} suggesting a slightly stronger acceptor ability of C_{70} [162].

By using cyclic silicon and germanium compounds as donors the quenching rate constants depend on the ring size of the donor. There, a significant decrease of the quenching rate constant with increasing ring size is observed. The quantum yields of electron transfer range from 0.50 for three-membered rings to 0.10 for five-membered rings. The differences in the quenching rate constants and the quantum yields can be interpreted on the basis of the oxidation potentials of the cyclic silicon and germanium compounds. Also, the ring size effects the electron transfer quantum yields whereas different substituents cause no influence [158].

Ito et al. investigated the influence of O$_2$ in the quenching process of $^3C_{60}^*$

by tetrathiafulvalenes as donors. They observed a considerable decrease in the yield of $C_{60}^{\bullet-}$ with an increase of O_2 concentration (fig. 18). In aerated solutions two-thirds of $C_{60}^{\bullet-}$ disappears, while in O_2-saturated solutions 90% of $C_{60}^{\bullet-}$ disappears [146,159]. This also confirms that $C_{60}^{\bullet-}$ is mainly formed via $^3C_{60}^*$, as shown by other investigations. But it also confirms that about 10% of $C_{60}^{\bullet-}$ may be produced via $^1C_{60}^*$ or an exciplex route. The rate constant for electron transfer via $^1C_{60}^*$ is estimated to be $k_{et}^S = 1.1 \times 10^{11}\,M^{-1}\,s^{-1}$. This value is about 20 times greater than the diffusion controlled limit ($5.2 \times 10^9\,M^{-1}\,s^{-1}$ in benzonitrile), which suggests that electron transfer through the $^1C^*_{60}$ channel occurs via an exciplex with the tetrathiafulvalene [146].

These observations confirm that electron transfer via the singlet excited state represents an alternative route to electron transfer via the triplet. It has been reported for several aromatic amines such as N,N-dimethylaniline, N,N,-diethylaniline, diphenylamines, triphenylamine, triethylamine, and phenothiazines, [29,134–137,139] as well as for a number of substituted naphthalenes [160] to form such charge transfer complexes in the ground-state (Scheme 2). The absorption spectra of C_{60} and phenothiazine, for example, show a progressive increase of the optical absorption of the solution on the continuous addition of phenothiazine combined with a continuous blue-shift of the absorption at 542 nm to 520 nm [133,138]. Irradiation into this absorption band leads to the formation of contact ion pairs [120,125,133,135,138,141]. However, the charge transfer complexes are weak (in neat N,N-diethylaniline only 53% of C_{60} exists in the form of CT-complex) and therefore high excess of donor is required to study the characteristics of fullerene containing CT-complexes [130,131]. Mittal et al. reported that CT-complex formation between fullereness and some aromatic amines is inhibited in benzene solutions due to the π-π interaction between fullerenes and solvents in the ground-state. In nonaromatic solvents like methylcyclohexane they observed interactions between fullerene and amine [135,139].

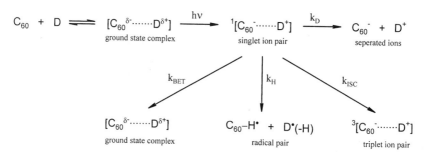

Scheme 2 Possible processes following the singlet ion pair formation. (From Ref. 134.)

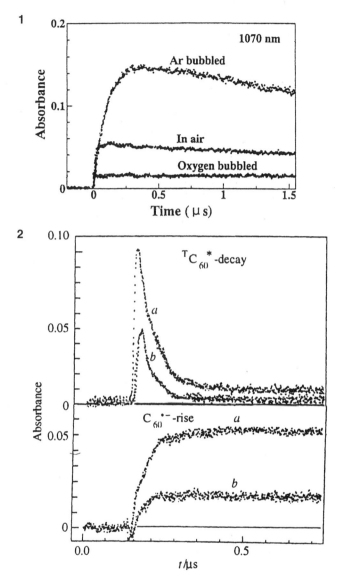

Figure 18 O_2-effect on time profiles. (1) rise and decay curves of $C_{60}^{\bullet-}$ in Ar-, air- and O_2-saturated solutions of tetrathiafulvalene (1.0 mM). (2) time profiles of $^3C_{60}^*$-decay and $C_6^{\bullet-}$0-rise: (a) in Ar-saturated (2.5×10^{-4} mol dm^{-3} of C_{60} and 1.0×10^{-2} mol dem^{-3} of TEOE) and (b) in O_2-saturated solution. (From Refs. 146, 147.)

The electron transfer processes involving singlet excited states of the fullerenes were investigated by picosecond time resolved transient spectroscopy. Depending on the equilibrium of the CT-complex, the excitation of the fullerene donor solution leads first to a singlet ion pair or to singlet excited fullerene. The latter can undergo intersystem crossing to the triplet or could be quenched by the donor. However, if the ground-state complex is excited followed by the formation of a singlet ion pair four processes are possible: (1) dissociation to the charge seperated radical ions k_D, (2) intersystem crossing to the triplet ion pair k_{ISC}, (3) back-electron transfer k_{BET}, or (4) hydrogen transfer leading to radical k_H. This depends on the donor and the solvent [133,134,138]. The radical cations of some amine donors can transfer a proton to the fullerene radical anion under fomation of radical pairs [134,140]. When proton transfer is not possible further processes depending on the solvent polarity are taking place. In nonpolar solvents, intersystem crossing to the triplet ion state followed by charge recombination to the fullerene triplet becomes predominant [134]. In more polar solvents (benzonitrile and solvent mixtures with benzonitrile), ion dissociation dominates leading to more solvent separated ions [134].

Kim et al. observed a very fast ion pair formation (below their detection limit of about 1 ps) from transient absorption spectra of fullerenes in the presence of aromatic amines such as N,N-dimethyl- or N,N-diethyl-aniline, corresponding to a rate $> 1 \times 10^{12}$ M^{-1} s^{-1}. An explanation for such extremly fast electron transfer is most likely a ground-state complex of fullerene and amine. Excitation leads to the neutral amine/$^1C_{60}^*$ contact pair followed by electron transfer. The decay of the both transient absorption from $^1C_{60}^*$ and $C_{60}^{\bullet-}$/amine$^+$ occurs with the same rate suggesting that charge recombination is the major nonradiative relaxation channel [138].

As an equivalent analysis method for time resolved absorption spectra EPR measurements indicate comparable results with regard to quenching rates and quantum yields. The EPR of the photochemically generated fullerene radical anions have been reported. Several groups agree on characteristic g-values in the region of 2.000. On the other hand, the peak-to-peak width values range from 0.01 mT to ~3 mT and are widely discussed. Staško et al. explained the apparent conflicting results by the generation of $C_{60}^{\bullet-}$ on various time scales using different reduction systems. Thus, immediately after electron transfer $C_{60}^{\bullet-}$ is characterized by an extremely narrow line (pp = 0.01 mT). In polar solvents, solvation processes lead to an increase of the line width to 0.1 mT. This species may aggregate to form radical products with relatively wide lines (pp ~3 mT) [57,124,127,151, 154,157].

Investigations of the photoinduced electron transfer between fullerenes and porphyrines show that electron transfer occur from the porphyrin to $^3C_{60}^*/^3C_{70}^*$. Additionally, due to the absorption of the porphyrines at the excitation wavelength, electron transfer from triplet-excited porphyrine to fullerene may occur. Both pathways lead to the radical ion pair, the ratio depends on the ratio of concentrations of both compounds. In the case of an excess of fullerene, electron transfer occurs via $^3C_{60}^*/^3C_{70}^*$, in the case of an excess of porphyrin electron-transfer runs via triplet-excited porphyrine. Besides the electron transfer from triplet-excited porphyrin to the fullerene, triplet energy transfer to the fullerenes was observed [151,153].

2. Energy Transfer

In several cases, dependent on the donor, the electron transfer triplet energy transfer from the triplet state of the fullerenes to the donor was observed. For example, excitation of C_{60}/perylene (Pe) mixtures leads to ^{3}Pe and C_{60} in a fast reaction $((1.4 \pm 0.1) \times 10^9 \ M^{-1} \ s^{-1})$. The electron transfer from Pe to $^3C_{60}^*$ occurs with a rate one-third of triplet energy transfer [127]. Ito et al. investigated the photoexcitation of mixed system of C_{60} and β-carotene [141]. They observed triplet energy transfer from $^3C_{60}^*$ to β-carotene in polar as well as in nonpolar solvents besides electron transfer from β-carotene to $^3C_{60}^*$. However, the electron transfer rate constant increases with solvent polarity while the energy transfer is only less effected by the change of solvent polarity (Table 5).

The energy gap between the two levels (^{3}β-carotene* + C_{60} and β-carotene$^{\bullet+}$ + $C_{60}^{\bullet-}$) is also dependent on the solvent polarity, becoming smaller in polar solvents (Scheme 3) [141]. The decay of the β-carotene$^{\bullet+}$-absorption follows second-order kinetics suggesting that back-electron transfer occurs from $C_{60}^{\bullet-}$ [141]. Similar results were observed for mixtures of fullerenes with phthalocyanines as donors [148]. In nonpolar solvents triplet energy transfer from the triplet-excited fullerenes to the phthalocyanine becomes prominent [148].

The most investigated system for energy transfer from the excited triplet states of fullerenes to an acceptor is the formation of singlet oxygen. Both the fullerenes, C_{60} and C_{70}, in their triplet states efficiently transfer energy to $^3O_2$ to

Table 5 Quantum Yields and Rate Constants for Electron and Energy Transfer Between C_{60} and β-carotene

BN:BZ	Φ_{ET}	Φ_{ENT}	k_{ET} $M^{-1} \ s^{-1}$	k_{ENT} $M^{-1} \ s^{-1}$
10:0	0.16	0.84	8.4×10^8	4.5×10^9
7:3	0.15	0.85	5.9×10^8	3.3×10^9
5:5	0.09	0.91	3.5×10^8	3.5×10^9
3:7	0.07	0.93	7.7×10^8	10.0×10^9
0:10	0.00	1.00	0.0×10^8	7.8×10^9

BN, benzonitrile; BZ, benzene.
For evaluation of electron- (Φ_{ET}) and energy- (Φ_{ENT}) transfer quantum yields, ε for β-carotene$^{\bullet+}$ = 1 $\times 10^5 \ M^{-1} \ s^{-1}$ and ε for $^3C_{60}^* = 1.6 \times 10^4 \ M^{-1} \ s^{-1}$ are used.
Source: Refs. 20, 141–144.

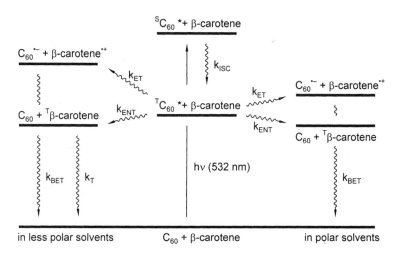

Scheme 3   Schematic illustration of energy diagram. (From Ref. 141.)

give $^1O_2$ [Eq. (4)]. The yield of $^1O_2$ is measured by the intensity of its luminescence at 1270 nm giving quantum yields Φ ($^1O_2$) near unity for C_{60} as well as for
C_{70}. The values differ slightly, maybe because of the relatively large experimental errors in measuring quantum yields. The values for C_{60} range from 0.76 to 0.96,
for C_{70} a value of ~0.8 was estimated. These values represent lower limits of the
quantum yields of triplet formation [20,21]. The rate of quenching for $^3C_{60}^*$ by
oxygen is 1.4×10^9 M^{-1} s^{-1} [62]. The reversed process is possible too, i.e., $^1O_2$
is quenched by the fullerenes with rate constants of $(5 \pm 2) \times 10^5$ M^{-1} s^{-1} for C_{60}
and $(2.8 \pm 0.7) \times 10^6$ M^{-1} s^{-1} for C_{70} [20,149].

$$^3C_{60/70}^* + {}^3O_2 \rightarrow C_{60/70} + {}^1O_2 \qquad (4)$$

Other groups also investigated the influence of O_2 in the photochemical pro-
cesses, but they focused on the reaction of $C_{60}^{\bullet-}$ with O_2 leading to $O_2^{\bullet-}$ [150,59].
With DMPO as spin trapping agent, $O_2^{\bullet-}$ and the following $^{\bullet}OH$ were detected by
means of EPR [150]. In the presence of molecular oxygen in solutions of C_{60} and
$^-BPh_4$ as donor electron transfer from $^-BPh_4$ to C_{60} is followed by a second electron transfer step from the $C_{60}^{\bullet-}$ to O_2 under formation of $O_2^{\bullet-}$. However, this process is overlapped by energy transfer from $^3C_{60}^*$ to molecular oxygen leading to
the formation of $^1(O_2)^*$ [159].

3. Oxidation

In contrast to the reduction, fullerenes are difficult to oxidize [163–165]. First, the
energy levels of the HOMO and LUMO orbitals are very low lying [122]. Second,
the structural demand of the cation: cations are normally planar, therefore oxida-

tion does not lead to a decrease of the tension. On the other hand, anions prefer a pyramidal structure, therefore reduction reduces the tension.

From electrochemical studies it is known that an irreversible oxidation of C_{60} occurs at $+1.76$ V vs SCE in benzonitrile [165]. One of the first methods to generate the radical cation by radiation was performed by γ-irradiation of C_{60} in glass matrix at 77 K [166]. A new absorption in the near IR at 980 nm was assigned to the $C_{60}^{\cdot+}$ (Fig. 19) [19]. Despite the sufficiently high reduction potential of $+2.0$ V of dicyanoanthracene, first attempts to generate the radical cation by photoinduced electron transfer were unsuccessful [Eq. (5)] [19]. One reason may be a fast back-electron transfer that competes with ionic dissociation in benzonitrile [19].

$$^1DCA + C_{60} \longrightarrow [DCA^{\cdot-} C_{60}^{\cdot+}] \longrightarrow DCA + C_{60} \tag{5}$$

The first really successful experiments to generate the fullerene radical cation by photoinduced electron transfer were carried out by Foote and coworkers (Fig. 19) [167]. They used singlet excited N-methylacridinium hexafluorophosphate (MA^+) as an electron acceptor which has a reduction potential of $+2.31$ V vs SCE, enough to oxidize C_{60} [Eq. (6)] [19].

$$^1(MA^+)^* + C_{60} \longrightarrow MA^{\cdot} + C_{60}^{\cdot+} \tag{6}$$

The singlet lifetime is somewhat longer than that of $^1DCA^*$. Additionally, the reduced form of the oxidant is a neutral radical exerting less coulombic at-

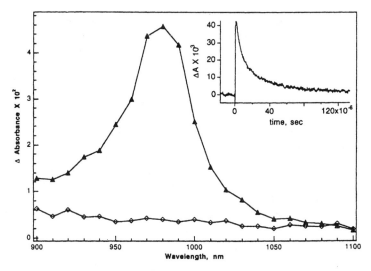

Figure 19 Transient absorption spectra from MA^+ and C_{60} (2×10^{-4} M, $\lambda_{exc} = 420$ nm) containing biphenyl (0.2 M) in CH_2Cl_2 (▲); C_{60} absent (◊). Inset: transient decay (λ_{obs} = 980 nm, average of 10 shots). (From Ref. 167.)

traction to the $C_{60}^{\bullet +}$ and, therefore, facilitates dissociation [167]. The transient absorption spectra shows a weak absorption at 980 nm similar to the spectra of the radical cation reported from pulse radiolysis and in argon matrix [19]. The electron transfer rate constant was estimated to be about $2 \times 10^{10} \, M^{-1} \, s^{-1}$, essentially diffusion controlled [167]. To increase the low quantum yield of $C_{60}^{\bullet +}$ they used cosensitization with biphenyl (BP, E (BP$^+$/BP) = 1.96 V vs SCE). The first step of this process is electron transfer from BP to MA$^+$ in its singlet excited state giving BP$^{\bullet +}$. The lifetime of this species is much longer than that of ^{1}(MA$^+$)* and has a sufficiently high oxidation potential to oxidize C_{60}. The transient absorption spectra shows the well characterized absorption at 980 nm with a ten-fold increase in intensity compared to that without cosensitization. The electron transfer from C_{60} to BP$^{\bullet +}$ is also diffusion controlled [167].

Mattay and coworkers extended the investigations of photoinduced electron tansfer from C_{60} to excited sensitizers and cosensitizers (Scheme 4) [173–175]. They used dicyanoanthracene (DCA), dicyanonaphthaline (DCN), N-methylacridinium hexafluorophosphate (MA$^+$), and triphenylpyrylium tetrafluoroborate (TPP$^+$) as sensitizers. In the case of DCA, DCN, and MA$^+$, the addition of a cosensitizer (biphenyl) was necessary to produce the fullerene radical cation in sufficiently high yields [175]. Otherwise, fast back-electron transfer seems to be predominant. However, by using TPP$^+$ the formation of $C_{60}^{\bullet +}$ could be detected by EPR measurements even in the absence of a cosensitizer. This can be explained by (1) the high reduction potential ($E_{red}^* = 2.53$ V vs SCE) and (2) the neutral form of the reduced sensitizer (electron shift) [174–176]. Nevertheless, no influence of the cosensitizer on the EPR signal was observed under irradiation [175].

It should be stressed that irradiation of a solution containing C_{60} and sensitizers such as N-methylacridinium hexafluorophosphate or dicyanoanthracene mainly leads to excited sensitizers, while only less than 10% of the absorbed photons were absorbed by C_{60} [167].

Excitation of C_{60} in the presence of electron acceptors like tetracyanoquinodimethane (TCNQ) or tetracyanoethylene (TCNE) leads first to $^3C_{60}^*$ via intersystem crossing [168]. The triplet-excited state is now quenched depending on the donor and the solvent.

In the case of TCNQ, transient time-resolved spectra do not exhibit neither the prominant peak at 980 nm as a characteristic absorption feature of $C_{60}^{\bullet +}$ nor the absorption at 850 nm as the characteristic feature of TCNQ$^{\bullet -}$. However, a new transient absorption band is observed in the region $450 < \lambda < 850$ nm in toluene as well as in benzonitrile. The transient product decays in both solvents with nearly similar rate contants and is quenched by oxygen with a rate constant of $4 \times 10^9 \, M^{-1} \, s^{-1}$. In addition, the more conjugative π-system of TCNQ with respect to TCNE implies low lying energetic levels probably determining the exciplex formation with negligible electron transfer and indicating triplet energy transfer $^3C_{60}^*$ to TCNQ. These observations together with the assumption of triplet energy

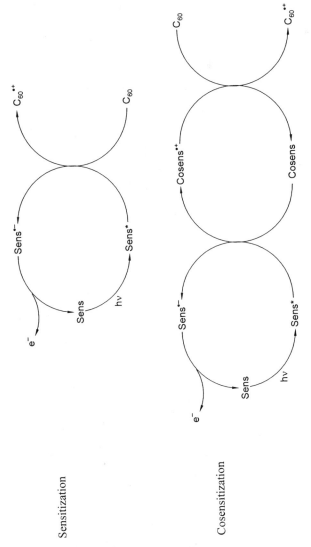

Scheme 4 General mechanism of photosensitized formation of the C_{60} radical cation. Sens = sensitizer, e.g., MA^+ or TPP; Cosens = cosensitizer biphenyl; TPP can be applied without a cosensitizer. (From Ref. 175.)

transfer explain the observed independence of the quenching process from solvent polarity [168].

On the other hand, in polar solvents like benzonitrile TCNE quenches $^3C_{60}^*$ by electron transfer most likely by an exciplex. This was confirmed by the characteristic transient absorptions of $C_{60}^{•+}$ at 980 nm [167,168] and at 425 nm of TCNE$^-$ [168]. The decay of the radical ion absorptions follow first-order kinetics. In nonpolar solvents like toluene the quenching process does not lead to radical ions, which is also reflected in lower rates of electron transfer (4.2×10^8 M^{-1} s^{-1} in benzonitrile; 7.9×10^5 M^{-1} s^{-1} in toluene) [168]. EPR measurements of a benzonitrile solution of C_{60} and TCNE show the quenching of $^3C_{60}^*$ by TCNE and the formation of TCNE$^{•-}$ [169]. The detection of the $C_{60}^{•+}$ signal failed, presumably due to inhomogenous line broadening and short spin relaxation [169]. The authors also observed an interaction between $^3C_{60}^*$ and TCNE$^{•-}$, which yielded a chemically induced dynamic electron polarization of TCNE$^{•-}$. The mechanism is a triplet-doublet quenching process [169]. CCl_4, $CHCl_3$, CH_2Cl_2, $C_2H_4Cl_2$, or CCl_3CF_3 can also be used as electron acceptors [170–172]. The mechanism of formation of $C_{60}^{•+}$ depends on the excitation method. In the case of laser flash photolysis, first $^3C_{60}^*$ is produced. After formation of an exciplex electron transfer to the chlorinated hydrocarbons occurs. As determined by transient absorption spectroscopy the generation of $[C_{60}CCl_3]^•$, $[C_{60}CHCl_2]^•$, or $[C_{60}CH_2Cl]^•$ competes with electron transfer [170,171]. If pulse radiolysis is used $C_{60}^{•+}$ is also produced, however, the mechanism is hole transfer, e.g., from $CCl_4^{•+}$ to C_{60}. The formation of $[C_{60}CCl_3]^•$ could also be observed by this method, but the reactions occurs via radical addition of $^•CCl_3$ to C_{60} [170–172].

4. Radical Addition

Irradiation of phenyliodonium salts lead to the formation of phenyl radicals. In the presence of C_{60} these radicals are efficiently trapped under formation of phenylated C_{60} derivatives, mainly the monoadduct. In reaction mixtures of C_{60}, phenyliodonium salts and spin traps like nitroso-*tert*-butane (^tBuNO) or nitroso-durene (ND) no phenyl adducts with the spin traps could be observed after irradiation. This suggests that C_{60} is a more efficient scavenger for phenyl radicals than the spin traps [177]. Other investigations yielded similar results, e.g., the photolysis of organomercury compounds in the presence of fullerenes leads to fullerene-derived radical adducts. These radical adducts can combine to form dimers that are thermally stable and accumulate in the samples [Eq. (7)] [178].

$$[(CH_3)_2CH]_2Hg \xrightarrow{h\nu} 2(CH_3)_2CH^• + Hg$$
$$C_{60} + (CH_3)_2CH^• \longrightarrow (CH_3)_2CH - C_{60}^•$$
$$2(CH_3)_2CH - C_{60}^• \longrightarrow [(CH_3)_2CH - C_{60}]_2 \qquad (7)$$

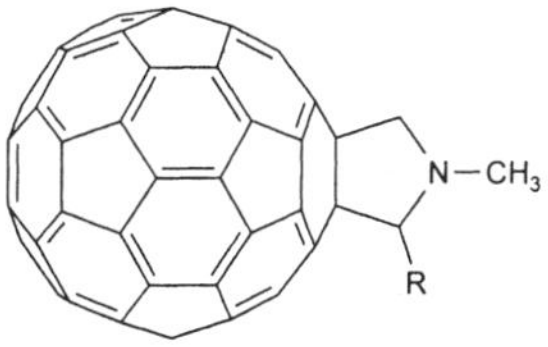

R = H (17), $C_6H_4NO_2$ (18), C_6H_4CHO (19), Ph (20), C_6H_4OMe (21), $C_6H_4NMe_2$ (22)

Figure 20 Substituted pyrrolidino-C_{60} derivatives. (From Ref. 179.)

B. Photochemical Processes of Excited Fullerene Adducts

Considering the large variety of fullerene adducts, only a few investigations on photochemical processes of fullerene adducts have been reported. These studies deal mainly with reductive processes due to the easy reduction of fullerenes and its adducts. Studies on electron transfer processes of several pyrrolidino fullerenes (Fig. 20) with donors such as dimethylaniline (DMA) or tetrakis(dimethylamino)ethylene (TDAE) show that the electron transfer rate constants decrease compared to that of C_{60} ($3.5 \times 10^9 \, M^{-1} \, s^{-1}$ for C_{60}, ~$0.5 \times 10^9 \, M^{-1} \, s^{-1}$ for the derivatives (Table 6, Fig. 21) [179]. This can be interpreted in terms of decreasing π-conjugation of the C_{60} moiety, which causes an increase in LUMO energy level and a decrease in the lowest triplet energy. Therefore, the substituents influe-

Table 6 Electrontransfer Rate Constants and Back-Electron Transfer Rate Constants of $^3C_{60}$ and $^3C_{60}$-Pyrrolidino Derivatives With DMA in Benzonitrile and Benzonitrile/Toluene (1:1)

	C_{60}	H (17)	$C_6H_4NO_2$ (18)	C_6H_4CHO (19)	Ph (20)	C_6H_4OMe (21)	$C_6H_4NMe_2$ (22)
			In benzonitrile				
Φ_{ET}	1.00	1.00	1.00	1.00	1.00	1.00	1.00
k_{ET} ($\times 10^9 \, M^{-1} \, s^{-1}$)	3.5	0.57	0.95	0.73	0.44	0.42	0.57
k_{BET} ($\times 10^9 \, M^{-1} \, s^{-1}$)	19	13	6.2	8.7	8.4	7.1	7.9
			In benzonitrile/toluene 1:1				
$\Phi_{ET}^{1:1 \, a}$	0.52	0.58	0.95	0.80	0.50	0.61	0.06
k_{ET} ($\times 10^9 \, M^{-1} \, s^{-1}$)	2.1	0.026	0.12	0.078	0.025	0.013	0.017
k_{BET}^a ($\times 10^9 \, M^{-1} \, s^{-1}$)	41	20	16	18	12	16	25

[a] ε_a values were assumed to be the same to those in benzonitrile.
Source: Ref. 179.

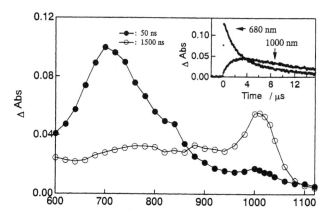

Figure 21 Transient absorption spectra of 18 (0.1 mM) with DMA in BN/Tol: (a) [DMA] = 0; (b) [DMA] = 16 mM. Inset: decay at 700 nm and rise at 1000 nm in the presence of DMA. (From Ref. 179.)

ence the electron accepting ability of the fullerene, but electron transfer still occurs to the fullerene core. Time resolved transient spectroscopy indicates that electron transfer occurs via the triplet-excited states of the fullerene derivatives [Eq. (8)] [179].

$$C_{60}R \xrightarrow{h\nu} {}^1C_{60}R^* \xrightarrow{k_{ISC}} {}^3C_{60}R^*$$
$${}^3C_{60}R^* + DMA \xrightarrow{k_{ET}} C_{60}R^{\bullet-} + DMA^{\bullet+} \tag{8}$$

Among the pyrrolidinofullerenes electron-withdrawing groups increase the electron transfer rates, while electron-donating groups slightly decelerate the electron transfer reaction. Changing to less polar solvents, like mixtures of toluene and benzonitrile, the substituent effect becomes more prominent. While the electron transfer rate for pristine C_{60} is only less reduced, the electron transfer rate for the derivatives are decreased about 1 order of magnitude. The decrease of the quenching rate for pyrrolidinofullerenes bearing electron-donating groups like $-OMe$ or $-NMe_2$ is stronger than for those with electron-withdrawing groups [179].

Changing the solvent from polar to less polar solvents effects not only the electron transfer but also the back-electron transfer. Back-electron transfer rate constants are in less polar solvents larger than those in polar solvents, which can reasonably be interpreted in terms of desolvation process and loose in ion pair formation. The transient absorptions of the pyrrolidino fullerene radical anions are slightly blue-shifted compared to that of $C_{60}^{\bullet-}$ ($C_{60}^{\bullet-}$: 1076 nm, derivatives radical anions: 991–1002 nm) [179].

Mittal et al. investigated the electron transfer from diphenylamine (DPA) or triphenylamine (TPA) to two pentaphenylated fullerene derivatives (PPF). The studies, carried out in benzonitrile with concentrations of fullerene ~1 × 10^{-4} mol dm^{-3} and a great excess of amine (0.3 mol dm^{-3}), indicate the formation of a 1:1 ground state CT-complex due to the ratios of concentration. From time resolved transient absorption spectra in the ps-time domain they showed that electron transfer occurs to the singlet-excited state of the fullerene derivatives in the CT-complexes with quenching rates near the diffusion controlled limit [180]. At lower amine concentrations (~1 × 10^{-3} mol dm^{-3}) in benzonitrile no CT-complex could be detected, however, following the kinetics the formation of the amine radical cation must occur via the singlet-excited state of the fullerene derivative. The pseudo first order rate of electron transfer between $^{1}PPF^{*}$ and DPA ($k_{ET} = \tau_s^{-1} = 2.4 \times 10^9 \text{ s}^{-1}$) is approximately greater than five times faster than that of the intersystem crossing process (k_{ISC} ~5.3 × 10^8 s^{-1}) [180].

Contrary to the electron transfer in organic solvents, the reduction process of functionalized fullerenes in aqueous solutions is very complex. Although some adducts are sufficiently water-soluble (Fig. 22) no reduction could be observed [181,183,187]. This is attributed to the irreversible formation of fullerene clusters in aqueous media, which seem to prevent electron transfer. Consequently, efficient triplet-triplet annihilation within the clusters is observed resulting in short triplet lifetimes (< 0.1 μs compared to microseconds for their monomeric analogue) [182,187].

Therefore, inhibiting cluster formation is a possible way to avoid the problem of suppressed electron transfer. There are several methods to reduce cluster formation, e.g., by capping the surface with surfactants like lauryl-sulfate or cetyltrimethylammonium chloride, or by incorporating the fullerene derivatives into the cavity of γ-cyclodextrines [182–185,187]. Transient absorption spectroscopy show that excitation to the singlet-excited state and intersystem crossing to the triplet are not effected by surfacting or incorporating fullerene derivatives

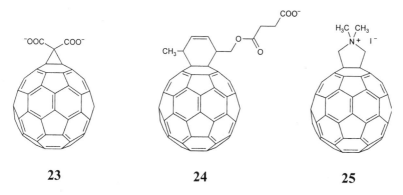

23   24   25

Figure 22 Water soluble fullerene derivatives. (From Ref. 182.)

compared to the fullerene clusters. However, the decay rates of the triplet-excited states of the fullerene clusters are accelerated by 2 orders of magnitude faster than those for the fullerene monomers [182,187].

It was clearly shown by EPR measurements that irradiation of aqueous suspensions of TiO_2 and surfaced or encapsulated fullerene derivatives leads to the formation of fullerene radical anions by electron transfer from excited TiO_2. The reduction of the embedded derivative compared to the surfaced one is more efficient [183]. However, in comparison to organic solvents, the reduction yield of the fullerenes radical anions in aqeous media is lower. This may be explained by the influence of the aquatic environment and the stability of the radical anions formed [183,184]. Nevertheless, electron transfer still occurs from the donor to the fullerenes triplet-excited state [182,185,186].

In the case of bis- and tris-adducts (e.g., **26, 27, 28;** Fig. 23), micellar aggregation in aqueous solution plays only a minor role, as can be seen by triplet lifetimes [187]. While the triplet lifetime for $\{(^3C_{60})C(COO^-)_2\}_n$ clusters is $\tau = 0.4$ μs, the lifetime for the bis- and tris-adducts is in the same range as for truly monomeric fullerene solutions [185–187]. Quenching the triplet-excited state of the derivatives by 1,4-diazadicyclo[2.2.2]octane (DABCO) as donor occurs with somewhat smaller rate constants than for pristine C_{60} [182,185,186]. In the sequence of increasing functionalization the quenching rate decreases continuously under identical conditions [185]. The quenching rates for surfaced fullerenes are somewhat lower than those of the fullerenes incorporated in γ-cyclodextrin. These results demonstrate that pristine C_{60} and $C_{60}C(COO^-)_2$ (**23**), either surfaced with Triton X-100 or γ-CD incorporated, are still better electron acceptor units than the malonic multiple adducts despite their heterogeneous nature (Table 7) [185].

However, replacement of the malonic bis-carboxylate units by pyrrolidinium ion moieties leads to slightly increased quenching rates, for mono- as well as for bis-functionalized derivatives (Table 8) [186,187]. Among the different monofunctionalized fullerene derivatives, the triplet-excited state of the γ-CD

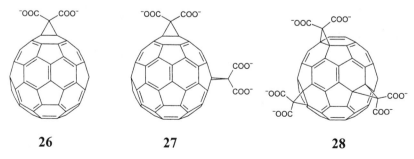

Figure 23 $C_{60}C(COO^-)_2$, e-$C_{60}[C(COO^-)_2]_2$, e,e,e- $C_{60}[C(COO^-)_2]_3$.

Table 7 Photophysical Data of Monomeric Fullerene/γ-CD and Fullerene/Surfactant Complexes in Aqueous Media

Compound	λ_{max} ($^3C_{60}R$) (nm)	λ_{max}^a ($C_{60}R^{\bullet-}$) (nm)	$k_{quenching}$ (DABCO) ($M^{-1} s^{-1}$)
C_{60}/λ-CD	750		1.6×10^8
$[C_{60}]_{surfactant}$	750		7.4×10^7
$C_{60}C(COO^-)_2/\gamma$-CD	720		6.9×10^6
$[C_{60}C(COO^-)_2]_{surfactant}$	700	1015	4.8×10^6
$C_{60}(C_9H_{11}0_2)(COO^-)/\gamma$-CD	710		6.8×10^6
$[C_{60}(C_9H_{11}O_2)(C00^-)]_{surfactant}$	690	1010	4.9×10^6
$C_{60}(C_4H_{10}N^+)/\gamma$-CD	700		2.7×10^7
$[C_{60}(C_4H_{10}N^+)]_{surfactant}$	690	1015	7.8×10^6

[a] absorption maxima for radiolytic reduction of monomeric fullerene/surfactant complexes in aqueous media.

Source: Ref. 182.

complex with the monopyrrolidinium salt $C_{60}(C_4H_{10}N^+)_2$ (**25**) is subject to the most efficient reductive quenching by DABCO with a rate constant of 2.7×10^7 $M^{-1} s^{-1}$ [186,187]. The enhanced quenching process can be explained by the lower electron-withdrawing effect of the quartery ammonium cation in $C_{60}(C_4H_{10}N^+)$ (**25**) compared to the functional groups of the other derivatives (see also Fig. 21) [182,185,186].

The photochemical reaction of C_{60} with triethylamine leads in deoxygenated toluene to the formation of **29** (Scheme 5). The reaction is initiated by photoinduced electron transfer between $^1C_{60}^*$ and the amine (ground-state CT-complex). As a model compound for intramolecular electron transfer compound

Table 8 Photophysical and Spectroscopic Data for the Excitation of Functionalized-Fullerene Derivatives in Aqueous Media

Compound	λ_{max} ($^3C_{60}R$) (nm)	τ Triplet (s^{-1})	$k_{quenching}$ (DABCO) ($M^{-1} s^{-1}$)
e-$C_{60}[C(COO^-)_2]_2$	690	1.7×10^4	7.2×10^5
$trans$-3-$C_{60}[C(COO^-)_2]_2$	670	1.4×10^4	5.1×10^5
$trans$-2-$C_{60}[C(COO^-)_2]_2$	670	1.7×10^4	8.3×10^5
e,e,e-$C_{60}[C(COO^-)_2]_3$	650	1.2×10^4	$< 1.0 \times 10^5$
e-$C_{60}[(C_4H_{10}N^+)]_2$	690	3.1×10^4	2.5×10^6
$trans$-3-$C_{60}[C_4H_{10}N^+)]_2$	670	2.9×10^4	3.7×10^6
$trans$-2-$C_{60}[(C_4H_{10}N^+)]_2$	660	2.8×10^4	4.7×10^6

Source: Refs. 185, 186.

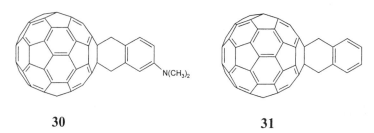

Scheme 5 Photochemical reaction of C_{60} with triethylamine in (a) deoxygenated toluene and (b) air-saturated toluene. (From Ref. 79.)

29 was investigated. Fluorescence studies of **29** in different solvents clearly show that there is a substantial quenching of the fullerene singlet-excited state by intramolecular electron transfer from the tertiary amino group. However, electron transfer could only be observed in polar or polarizable solvents. The electron transfer rate constants k_{ET} thus obtained are ~2-5 × $10^9 \, M^{-1} \, s^{-1}$ depending on the solvent [79].

Nishimura et al. reported the comparison between the two C_{60}-o-quinodimethane adducts **30** and **31** (Fig. 24). The absorption spectra of **31** shows with increasing the concentration of dimethylaniline (DMA) a linear increase of the absorption in the 400–650 nm region indicating the formation of a ground-state CT-complex [188].

In the case of compound **32,** no change in the absorption spectrum compared with that of **33** could be observed, neither in cyclohexane nor in benzonitrile. This indicates that intramolecular CT interaction is negligible in the ground-state. Nevertheless, fluorescence and transient absorption spectroscopy show that processes after excitation strongly depends on solvent polarity. In cyclohexane the S_1 state

Figure 24 C_{60}-o-quinodimethane adducts 30 and 31. (From Ref. 188.)

of the fullerene is only less quenched by the DMA moiety. After excitation to the singlet-excited state intersystem crossing to the triplet occurs it is followed by radiation less conversion to the singlet ground-state. Changing to polar solvents like benzonitrile the fluorescence is quenched with a rate constant of 1.5×10^{10} M^{-1} s^{-1}. Due to the calculated free enthalpies of electron transfer ($\Delta G = -0.36$ eV in benzonitrile, $\Delta G = +0.29$ eV in cyclohexane) the quenching in benzonitrile is reasonably attributed to the intramolecular electron transfer. It could be shown that electron transfer occurs via through-bond interaction and not via through-space interaction. In case of compound **30,** no solvent dependence of the fluorescence quenching could be observed [188].

Bunker et al. also reported approximately three aminofullerene derivatives (**3, 32, 33,** Fig. 25) that show intramolecular electron transfer from the amino group to the singlet excited state of the fullerene core [189]. Although the intramolecular electron transfer shares some characteristics with the classical twisted intramolecular charge transfer in molecules represented by *p-N,N-*dimethylaminobenzonitrile the amino-C_{60} derivatives are in fact better classified as redoxdyads [189].

With the synthesis of donor-linked fullerene derivatives investigations concerning the intramolecular electron transfer were greatly extended. The variety of donor-attached derivatives range from simple systems like **13** (see Fig. 11) and **33** (Fig. 25) to ferrocene- [64] or tetrathiafulvalene-linked [190–192] fullerenes and to very complex systems with linkage between fullerenes and chromophoric units like ruthenium(II)polypyridyl [193,194] complexes or metalloporphyrines [195,196]. In the simple systems, electron transfer occurs from the donor to the singlet excited state of the fullerenes while in the case of the more complex systems there is a drastic change in the electron transfer process. Electron or energy transfer now occurs from the photoexcited chromophore to the fullerene core [116]. One reason for the synthesis and the investigations of photophysical and

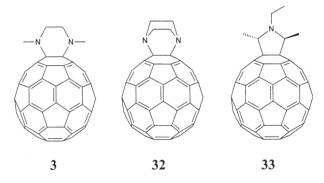

3   **32**   **33**

Figure 25   Different amino fullerene derivatives. (From Refs. 65, 189.)

photochemical properties of these donor-linked fullerene systems is to evaluate electron transfer in relation to the primary processes of photosynthesis [197–199]. Because of the fact that many reports and reviews on this subject of fullerene chemistry exist we only want to refer to this literature [74,116,190,200–204].

The efficient production of singlet oxygen is one of the characteristics of C_{60} and C_{70}. Since it is known that functionalization leads to changes in photophysical and photochemical properties it is of great interest whether functionalization causes noticeable changes in singlet oxygen sensitization or not. The investigated monoadducts show nearly no influence of the addend to the singlet oxygen production. The quenching rates are in the order of $\sim 1.5 \times 10^9$ $M^{-1}s^{-1}$ and the quantum yields are similar to those of C_{60} [62,64,71,72,82,88, 110,159,184]. This seems to be generally applicable since these values are determined for different types of monoadducts like cyclohexyl-fused derivatives, methanofullerenes, pyrrolidinofullerenes, or fullerene derivatives with different addends.

In the case of multiple adducts the influence of the addends to the singlet oxygen sensitization is somewhat larger. For example, in the case of bisadducts of cyclohexyl-fused type (**34, 35, 36,** Fig. 26) the $^1O_2$ quantum yield is only the half of that of the corresponding monoadduct (Table 9) [110]. Similar quantum yields were estimated for bis-malonic derivatives. Further functionalization leads to further decrease of the quantum yield [64]. Additionally, the $^1O_2$ generation ability depends on the addition pattern [64,110]. However, epoxide derivatives show a dramatic decrease of $^1O_2$ generation by addition of a second addend. For $C_{60}O$ the quantum yield is in the range of other monoadducts, while the quantum yield for $C_{60}O_2$ is only 10% of that of pure C_{60} [64]. Higher epoxides like $C_{60}O_4$ and $C_{60}O_5$ do not show significant singlet oxygen generation.

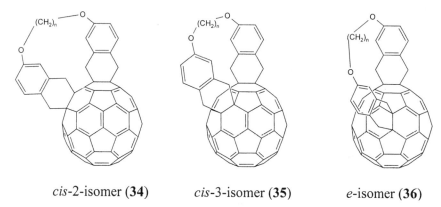

      cis-2-isomer (**34**)   *cis*-3-isomer (**35**)   *e*-isomer (**36**)

Figure 26 Cyclohexylfused bis–adducts with isomeric addition pattern. (From Ref. 110.)

Table 9 Quantum Yield of $^1O_2$ Generation ($\Phi(^1O_2)$) and Intersystem Crossing (Φ_{ISC}) by Isomeric Bis-Adducts (34,35,36)

Compounds	$\Phi(^1O_2)$ in benzene	$\Phi(^1O_2)$ in benzonitrile	Φ_{ISC} in benzene
cis-2 (n = 2, 3) **(34)**	0.54	0.64	0.55
cis-3 (n = 2, 3) **(35)**	0.35	0.27	0.54
e (n = 5) **(36)**	0.56	0.52	0.57

Source: Ref. 110.

IV. PHOTOCHEMICAL REACTIONS OF FULLERENES AND ITS DERIVATIVES

In the early 1990s, the derivatization of fullerenes was based on thermal reactions (e.g., the Diels-Alder reaction of C_{60} with substituted dienes leading to cyclohexyl-fused fullerene derivatives), and only few examples were known where a fullerene adduct was obtained by photochemical reactions. For example, irradiation of C_{60} with UV or visible light led to the polymerized C_{60}. Mechanistically, this reaction proceeds via a photochemical [2 + 2] cycloaddition [205]. In 1993, Wilson et al. reported the [2 + 2] photocycloaddition of enones to C_{60} [206]. It was known that the formation of fullerene epoxides can be obtained by photooxidation [98]. Today, more and more fullerene derivatives are synthesized by photochemical reaction pathways. Photoreactions are based on the processes described in Chapter 3. Photoreactions via an excited partner, which may involve any type of intermediate formed from this partner in its singlet or triplet state are discussed in Chapter 4. Because much less photoreactions of C_{70} are known these reactions will be described in context with the corresponding reactions of C_{60}.

The photoinduced functionalization of fullerenes can be achieved either by electron transfer activation leading to radical ions or by energy transfer processes, either by direct excitation of the fullerenes or the reaction partner [Eq. (9)]. In the latter case, both singlet and triplet species are involved whereas most of the reactions of electronically excited fullerenes proceed via the triplet states due to their efficient intersystem crossing.

Electron transfer activation: Photochemical Activation:

$$C_{60} \xrightarrow{+e^-} C_{60}^{\bullet-} \qquad\qquad C_{60} \xrightarrow{h\nu} C_{60}^* \xrightarrow{A} C_{60}A$$
$$C_{60} \xrightarrow{-e^-} C_{60}^{\bullet+} \qquad\qquad A \xrightarrow{h\nu} A^* \xrightarrow{C_{60}} C_{60}A \qquad\qquad (9)$$

A. Principles of Fullerene Chemistry

The chemical behavior of C_{60} is largely governed by three properties due to the structure of C_{60} [12,15,236]. First, the bonds at the junctions of two hexagons

([6,6] bonds) are shorter than the bonds at the junctions of a hexagon and a pentagon ([5,6] bonds, see Fig. 27). Second, the highly pyramidalized sp^2 C-atoms in spherical C_{60} cause a large amount of strain energy within the molecule. Third, C_{60} is an electronegative molecule that can be easily reduced, but difficult to oxidize.

The following rules of reactivity that are based on these properties can be deduced from the results obtained by the chemical transformation of fullerenes.

> The reactivity of C_{60} is that of a fairly localized electron deficient polyolefin, thus, typical reactions are additions at a [6,6] double bond
> The driving force is the relief of strain in the fullerene cage
> The regiochemistry of the addition reaction is governed by the minimization of [5,6] double bonds within the fullerene framework

The chemical reactivity of C_{70} is very similar to that of C_{60}. While all positions in C_{60} are identical and two different types of C–C bonds exist, five sets of carbon atoms and eight distinct types of C–C bonds are available in C_{70} (Fig. 27). It can be anticipated that the different bonds will display different reactivity [207].

Since there are up to 30 reactive double bonds in the C_{60} molecule, a great number of polyadducts are possible. For a description of the spatial arrangement of the addends the C_{60} molecule is divided into three sections with regard to the location of the second attack [208]. The second attack can take place on the same hemisphere (*cis*), at the equator (*e*), or on the opposite hemisphere (*trans*). For example, for the addition of an identical second addend eight regioisomeric bisadducts are possible as shown in Fig. 28.

To identify modified fullerenes a numbering system for the carbon atoms is necessary. According to the IUPAC nomenclature [209], based on the proposal of Taylor in 1993 [210], the carbon atoms are numbered as outlined for C_{60} and C_{70} in Fig. 29.

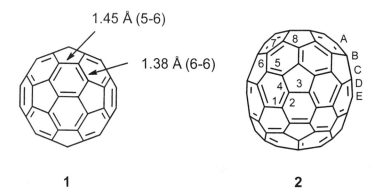

1.45 Å (5-6)

1.38 Å (6-6)

1   **2**

Figure 27   Different bond lengths in C_{60} and different bond types in C_{70}.

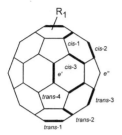

Figure 28 Positions of the ligand-carrying bonds in the eight possible regioisomers.

B. Photochemistry of C_{60}/C_{70}

1. Synthesis of Substituted 1,2-Dihydrofullerenes by Photoinduced Electron Transfer

As a first example, the photochemical synthesis of substituted 1,2-dihydro-[60]fullerenes will be discussed. These compounds can be synthesized by various photochemical reaction pathways. In the first one the radical cation $C_{60}^{\bullet+}$ is involved in the reaction. In 1995, Schuster et al. reported the formation of C_{60} radical cations by photosensitized electron transfer that were trapped by alcohols and hydrocarbons to yield alkoxy or alkyl substituted fullerene monoadducts as major products [211]. Whereas Foote et al. used N-methylacridinium hexafluorophosphate NMA^+ as a sensitizer and biphenyl as a cosensitizer [167], Schuster et al. used 1,4-dicyanoanthracene (DCA) as a sensitizer for the generation of $C_{60}^{\bullet+}$. The

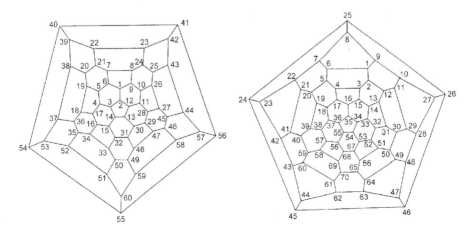

Figure 29 Schlegel diagrams with carbon atom numbering for C_{60} and C_{70}.

proposed mechanism involving various reaction paths of the photosensitized addition is outlined in Scheme 6. It involves (1) initial excitation of DCA, (2) formation of an encounter complex between excited sensitizer and ground state **1** (= C_{60}), (3) electron transfer to form a radical ion pair, and (4) solvent separation of the fullerene radical cation and the DCA radical. To form alkoxy-substituted fullerenes attack of an alcohol on $C_{60}^{\bullet+}$ generates the fullerene alkoxy radical (5a) that is followed by back electron transfer (6a) from $DCA^{\bullet-}$ to the alkoxyfullerene radical to give the alkoxyfullerene anion and neutral sensitizer. Finally, protonation of the anion by the alcohol (7a) to yield the neutral alkoxyfullerene. The formation of the fullerene adducts of toluene, cyclohexane, and cyclohexene, when **1** and DCA are irradiated at 425 nm in the absence of a nucleophile, indicates that under these conditions $C_{60}^{\bullet+}$ can behave as a typical free radical and abstract hydrogen from a suitable hydrogen donor in contrast to reaction at the cationic center when alcohol is present. This process becomes competitive with nucleophilic attack on the radical cation in the case of benzyl alcohol as evidenced by the formation of the benzylated as well as alkoxylated fullerene. Therefore, the proposed mechanism for the alkylation of **1** is as follows: after formation and solvent separation of the radical ion pair $C_{60}^{\bullet+}$ abstracts a hydrogen from an appropriate donor to generate the hydrogenated fullerene cation (5b). Back-electron transfer from $DCA^{\bullet-}$ to the fullerene cation regenerates the fullerene radical and the neutral sensitizer DCA (6b). Coupling of the fullerene radical and the alkyl radical finally results in the formation of the alkylated fullerene (7b).

Mattay et al. also generated $C_{60}^{\bullet+}$ under PET conditions by the method reported by Schuster et al. [173,211]. Other well known PET sensitizers such as

(1) $\text{DCA} \xrightarrow[425]{h\nu} \text{DCA}^{\bullet}$

(2) $\text{DCA}^{\bullet} + \mathbf{1} \longrightarrow \text{DCA}^{\bullet}\text{---}\mathbf{1}$

(3) $\text{DCA}^{\bullet}\text{---}\mathbf{1} \longrightarrow [\mathbf{1}]^{\bullet+}\text{---}[\text{DCA}]^{\bullet-}$

(4) $[\mathbf{1}]^{\bullet+}\text{---}[\text{DCA}]^{\bullet-} \xrightarrow{CH_3CN} [\mathbf{1}]^{\bullet+}_{solv} + [\text{DCA}]^{\bullet-}_{solv}$

(5a) $[\mathbf{1}]^{\bullet+}_{solv} + \text{ROH} \longrightarrow [\mathbf{1}\text{-OR}]^{\bullet} + H^+$

(5b) $[\mathbf{1}]^{\bullet+}_{solv} + \text{RH} \longrightarrow [\mathbf{1}\text{-H}]^{+}_{solv} + R^{\bullet}$

(6a) $[\mathbf{1}\text{-OR}]^{\bullet} + [\text{DCA}]^{\bullet-}_{solv} \longrightarrow [\mathbf{1}\text{-OR}]^{(-)}_{solv} + \text{DCA}$

(6b) $[\mathbf{1}\text{-H}]^{+}_{solv} + [\text{DCA}]^{\bullet-}_{solv} \longrightarrow [\mathbf{1}\text{-H}]^{\bullet} + \text{DCA}$

(7a) $[\mathbf{1}\text{-OR}]^{(-)}_{solv} + \text{ROH} \longrightarrow \text{H-1-OR} + RO^-$

(7b) $[\mathbf{1}\text{-H}]^{\bullet} + R^{\bullet} \longrightarrow \text{H-1-R}$

Scheme 6   Mechanism for the formation of alkoxylated and alkylated fullerenes.

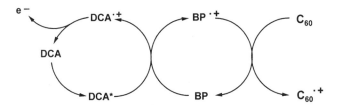

Scheme 7 Generation of $C_{60}^{\cdot+}$ by cosensitization (hv: 419 nm, DCA, BP: biphenyl).

(NMA^+) and 2,4,6-triphenyl-pyrylium tetrafluoroborate (TPP^+) in the presence of biphenyl as cosensitizer were suitable for this reaction [174]. The assumed mechanism of formation of $C_{60}^{\cdot+}$ by this cosensitization is shown in Scheme 7. Reaction of $C_{60}^{\cdot+}$ with H-donors such as *tert*-butylmethylether, propionaldehyde and alcohols results in the formation of 1:1 adducts, the 1-substituted 1,2-dihydro-[60]fullerenes. Product structure support a H-abstraction process [212,213] rather than nucleophilic addition. In Scheme 8, the general formation of 1-substituted 1,2-dihydro-[60]fullerenes is shown. Selected examples of the products obtained by this method are summarized in Table 10.

Time resolved laser flash photolysis and electric spin resonance (ESR) spectroscopic investigations were used to get further insight to the reaction mechanism. Both methods demonstrate the formation of $C_{60}^{\cdot+}$ using PET conditions [175,214,215]. Upon addition of H donors the signal of $C_{60}^{\cdot+}$ is quenched [214]. The oxidation of C_{60} is followed by H abstraction from the H donor as shown in Scheme 9. Nucleophilic addition can be excluded because no alkoxyfullerenes were detected at all [173]. After reduction of $H-C_{60}^+$, e.g., by electron transfer from the reduced sensitizer molecule $H-C_{60}^{\cdot}$ might recombine with $R^{\cdot}$ to the final product. Decay experiments of $C_{60}^{\cdot+}$ by the addition of alcohols support the proposed mechanism of H abstraction as a first step. The involved radical products reveal $C_{60}^{\cdot+}$ as an electrophilic radical.

The authors reported the formation of some unexpected products in the above mentioned reaction. For example, when using aldehydes as H donor reactants the products depend on the reactivity of the intermediately formed radi-

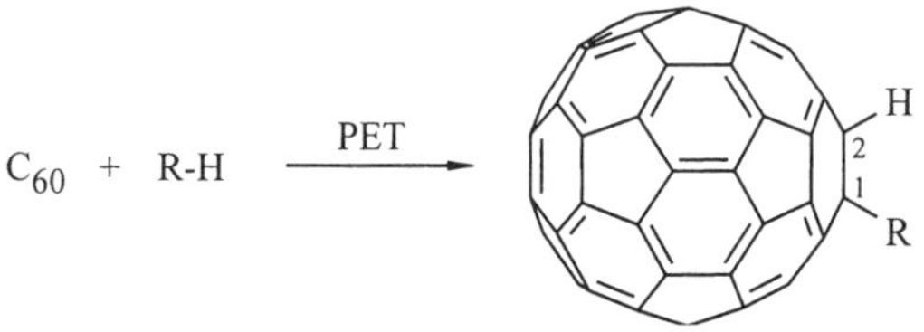

$$C_{60} \;+\; R\text{-}H \;\xrightarrow{\text{PET}}\;$$

Scheme 8 Formation of dihydro[60]fullerene under PET conditions (TPP^+ or DCA/BP, 419nm).

Table 10 Selected Examples of 1-Substituted 1,2-Dihydro[60]-Fullerenes

H-donor: fullerene product	H-donor: fullerene product

HOCH$_2$CH$_2$OH

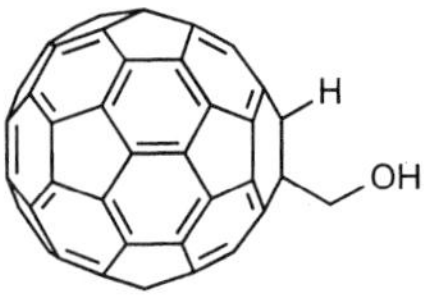

37

CD$_3$OD

38

t-BuOH

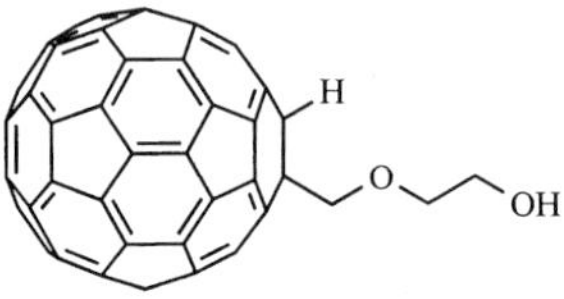

39

+

40

1,3-Dioxolane

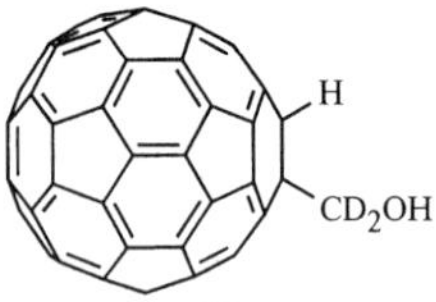

41

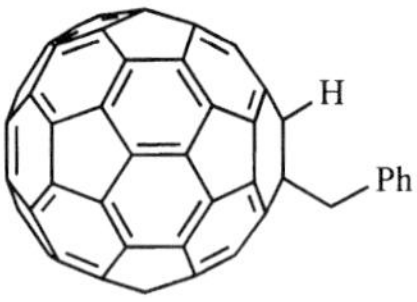

C$_{60}$H$_2$

CH$_3$OCH$_2$CH$_2$OH

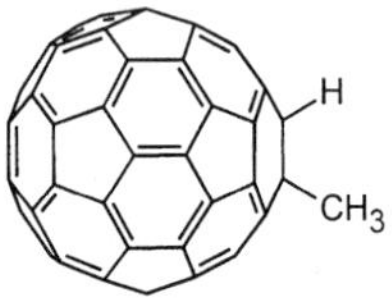

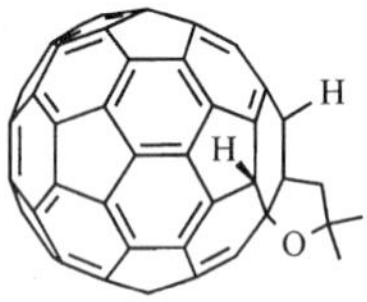

42

PhCH$_2$CHO

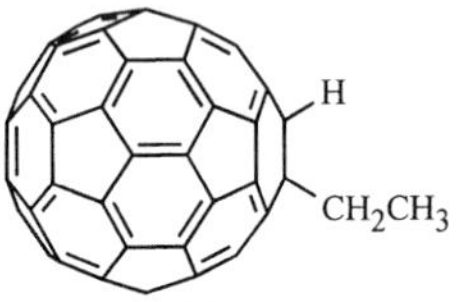

43

CH$_3$CH$_2$COOH

44

(CH$_3$)$_2$NCHO

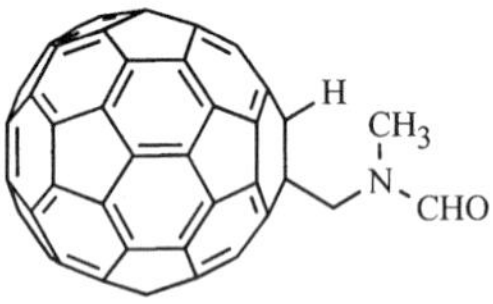

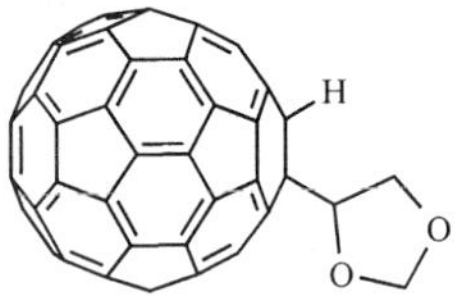

45

HCOOCH$_3$

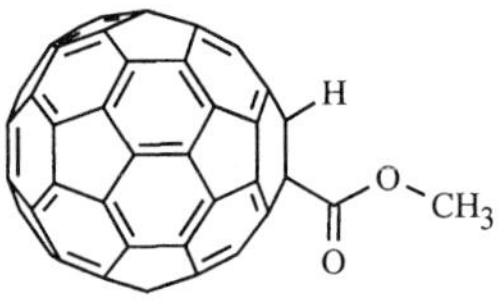

46

Source: Ref. 174.

$$C_{60} \xrightarrow{\text{PET}} C_{60}^{\bullet+}$$

$$C_{60}^{\bullet+} + RH \longrightarrow H-C_{60}^{+} + \bullet R$$

$$H-C_{60}^{+} + \bullet R + e^{-} \rightleftharpoons H-C_{60}-R$$

Scheme 9   Proposed mechanism of the addition of H donors to C_{60}.

cal. Propionaldehyde as starting material did not lead to decarbonylation because a primary alkyl radical is less stable than a benzyl radical. Consequently, the decarbonylation during the addition of phenylacetaldehyde leading to **43** strongly suggests that this reaction involves an acyl radical. In addition, the reaction of $C_{60}^{\bullet+}$ with glycol, propionic acid, and *tert*-butanol could only be explained by loss of the hydrogen from a O–H bond, either from the neutral or from the oxidized substrate. For example, the reaction of *tert*-butanol with C_{60} leading to the dihydrofullerene **39** (path A in Scheme 10) and the tetrahydro[60]fullerene **40** (path B in Scheme 10) can be easily explained by homolytic bond breaking at the O–H position in the former case and abstraction at the C–H position in the latter. Upon reaction of alcohols and C_{60} under PET conditions hydroxylated dihydro [60]fullerenes are isolated. It was first considered that the α-centered hydroxyalkyl radicals were formed by direct abstraction of a hydrogen from the C–H bond. Instead, it was shown that such radicals can be formed in a bimolecular process from O-centered radicals as a second step in a pseudo-[1,2]-shift manner. Analogously, the product of the reaction with methoxyethanol can be explained. The primarily generated O-centered radical is transformed into the C-centered radical via a [1,5]-H-shift in a six-membered transition state. Whether simple alcohols react with $C_{60}^{\bullet+}$ by direct H abstraction from the α-position or via O-centered radicals in a two-step mechanism cannot be determined as both reaction paths may operate.

   In all examples discussed up to now the radical cation of C_{60} is involved in the reaction mechanism. However, due to the electronic features reduction of the fullerenes leading to radical anions should be much easier performed. For example, a useful method to synthesize 1-substituted 1,2-dihydro-[60]fullerenes is the irradiation of C_{60} with ketene silyl acetals (KAs) first reported by Nakamura et al. [216]. Interestingly, when unstrained KAs are used, this reaction did not yield the expected [2 + 2]-cycloaddition product either by the thermal, as observed by the use of highly strained ketene silyl acetals [217], or by the photochemical pathway. In a typical reaction C_{60} was irradiated for 10 h at 5°C with a high pressure mercury lamp (Pyrex filter) in a degassed toluene solution with an excess amount of the KA in the presence of water (Scheme 11). Some examples of the addition of KAs are summarized in Table 11.

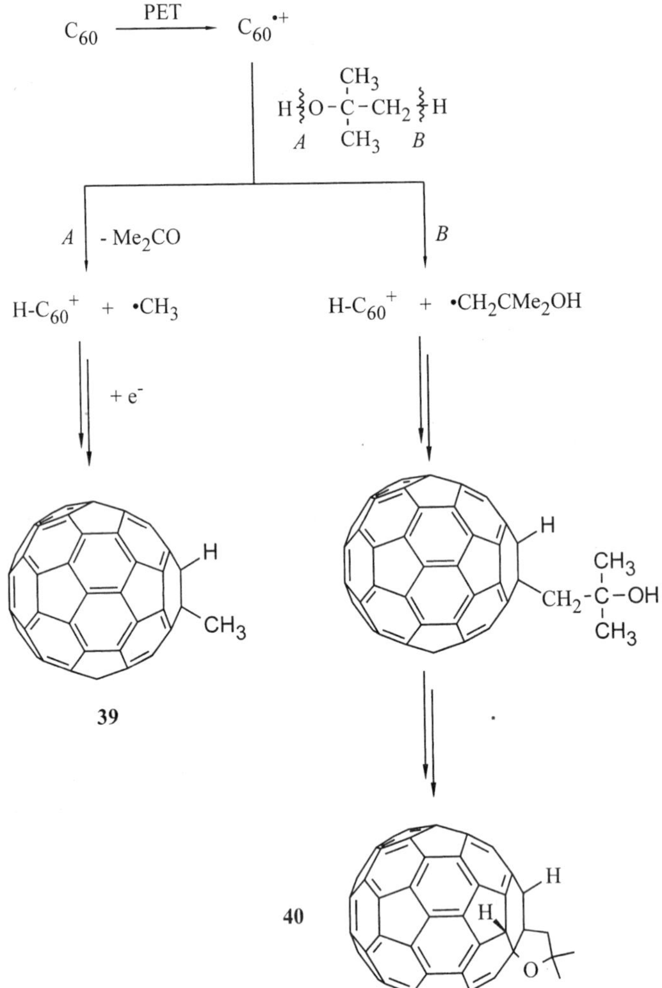

Scheme 10   Reaction of *tert*-butanol and C_{60} under PET conditions.

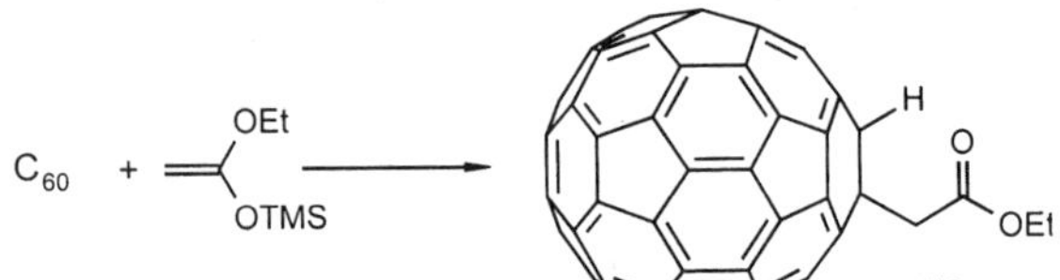

Scheme 11   Reaction of C_{60} with 47a in benzene.

Table 11 Light-Induced Carbon–Carbon Bond Formation of C_{60} With Ketene Silyl Acetals

SI–NU	Equiv.	Time (min)	Product	Yield of monoadduct (%)
OEt / OSiMe$_3$ **47A**	20	90	**48A**	71 (100)
OEt / OSiMe$_2$t-Bu **47B**	10	470	**48A**	63 (100)
OMe / OSiMe$_3$ **47C (E)**	10	90	**48B**	37 (51)
OMe / OSiMe$_3$ **47C (Z)**	10	90	**48B**	36 (58)
OMe / OSiMe$_3$ **47D**	10	30	**48C**	16 (24)
OMe / OSiMe$_2$t-Bu **47E**	10	30	**48C**	28 (46)
OSiMe$_2$t-Bu **47F**	10	470	—	0
SiMe$_3$ **47G**	10	60	—	0

Source: Ref. 218.

It is observed that the reactivity of the KA increases as the number of electron-donating substituents on the olefin increases. Steric hindrance leads to a decreasing reactivity. The authors rationalized the formation of the photoadduct as follows (see Scheme 12). First, the radical ion pair **49** is formed by single electron transfer between photoexcited $^3C_{60}$ and the electron rich ketene silyl acetal fol-

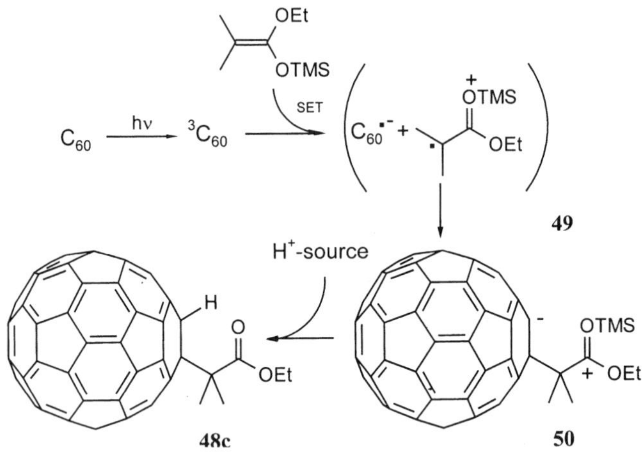

Scheme 12 Reaction mechanism for the photoinduced C–C bond formation of C_{60} with ketene silyl acetals.

lowed by coupling to a zwitterionic species **50** which finally leads to product **48c** [218]. The active role of water in the photoaddition in relation to the TMS/TB-DMS-dependent reactivity difference indicates that the oxonium intermediate **50** is the direct precursor of the photoproduct **48c**, which is formed by the water as-sisted loss of the silyl group from **50**. The radical ion pair **49** is the most plausible precursor to **50** in light of the high electron affinity of excited [60]fullerene [20] and the excellent single-electron donating ability of silyl KA [219]. No free radi-cal species are involved in the reaction since addition of a radical trap (6 equiv. TEMPO) has no effect on the photoreaction. At least the formation of the [2 + 2] cycloadduct as an intermediate, which undergoes photoassisted heterolytic C-C bond cleavage via **50,** cannot be excluded by the authors.

Mikami et al. also investigated the addition of ketene silyl acetals. They found that addition of the silyl enol ether of acetone and allylic silanes did not re-sult in the synthesis of substituted 1,2-dihydro[60]fullerenes [218a,220]. In 1997, Mikami et al. [221] reported the photoaddition of allylic stannanes that leads to monoallylation of C_{60} (Scheme 13).

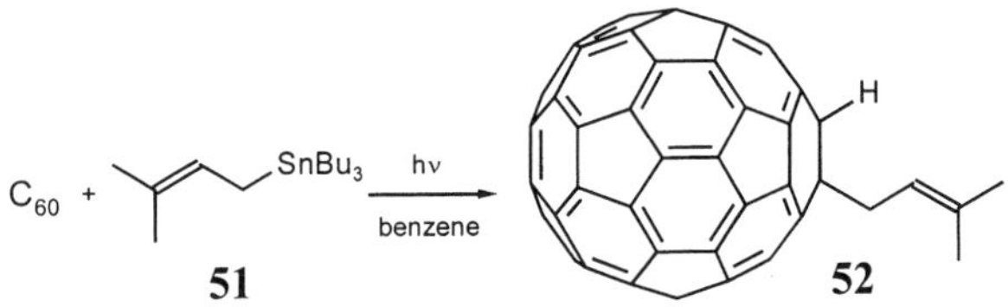

Scheme 13 Reaction of prenyltributyltin 51 with C_{60} upon irradiation.

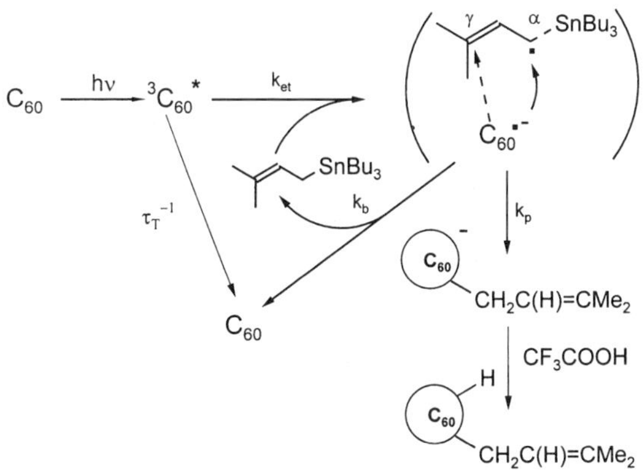

Scheme 14 Reaction mechanism for the reaction of prenyltributyltin 51 with C_{60}.

The C–C bond formation of C_{60} with group 14 organometallic compounds is attained through photoinduced electron transfer from group 14 organometallic compounds acting as electron donors to the triplet-excited state of C_{60}. When an unsymmetric allylic stannane such as γ,γ-dimethyl substituted allyl stannane **51** [222,223] (Scheme 14) is employed, the allylic group is introduced selectively at the α-position to yield C_{60}–1,2–$CH_2CH=CMe_2$. Since the allylic group is introduced selectively at the α-position and no γ-adduct has been formed, the Sn–C bond of the allylstannane radical cation formed in the photoinduced electron transfer may be nearly cleaved prior to the C–C bond formation with $C_{60}^{\bullet+}$ in the radical ion pair [223].

The radical anion $C_{60}^{\bullet-}$ can also be easily obtained by photoinduced electron transfer from various strong electron donors such as tertiary amines, ferrocenes, tetrathiafulvalenes, thiophenes, etc. In homogeneous systems back-electron transfer to the reactant pair plays a dominant role resulting in a extremely short lifetime of $C_{60}^{\bullet-}$. In these cases no net formation of $C_{60}^{\bullet-}$ is observed. These problems were circumvented by Fukuzumi et al. by using NADH analogues as electron donors [154,155]. In these cases selective one-electron reduction of C_{60} to $C_{60}^{\bullet-}$ takes place by the irradiation of C_{60} with a Xe lamp ($\lambda > 540$ nm) in a deaerated benzonitrile solution upon the addition of 1-benzyl-1,4-dihydronicotinamide (BNAH) or the corresponding dimer [$(BNA)_2$] (Scheme 15) [154]. The formation of $C_{60}^{\bullet-}$ is confirmed by the observation of the absorption band at 1080 nm in the near infrared (NIR) spectrum assigned to the fullerene radical cation.

As shown in Scheme 16 the triplet-excited state $^3C_{60}^*$ is quenched by electron transfer from [$(BNA)_2$] generating the radical ion pair [$(BNA)_2^{\bullet+}$ and $C_{60}^{\bullet-}$] in competition with the decay to the ground-state. However, back-electron transfer is reduced by the fast cleavage of the C–C bond in the dimer [$(BNA)_2^{\bullet+}$]. Finally, a second electron transfer from BNA$^\bullet$ occurs leading to two molecules of $C_{60}^{\bullet-}$. The

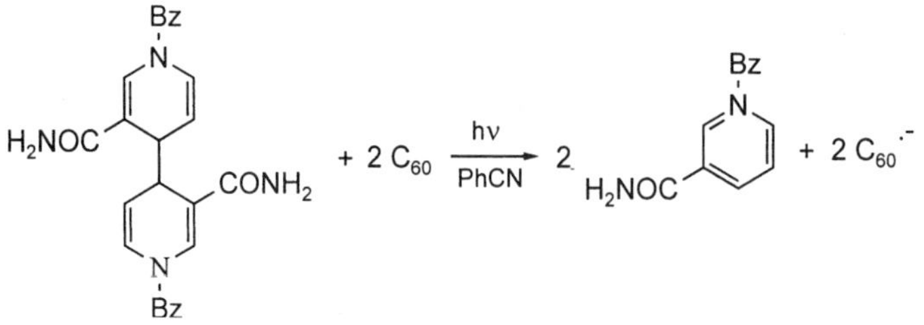

Scheme 15 Generation of $C_{60}^{\cdot-}$ by selective one-electron reduction with [(BNA)$_2$] in deaerated benzonitrile at $\lambda > 540$ nm.

photoreduction of C_{60} with BNAH also proceeds via photoinduced electron transfer. In this case C_{60} is also reduced to the monoanion by BNA$^{\cdot}$, which is formed through deprotonation of BNAH$^{\cdot+}$. Formation of $C_{70}^{\cdot-}$ is also attained through photoinduced electron transfer from BNAH and [(BNA)$_2$] to the triplet-excited state of C_{70} [155].

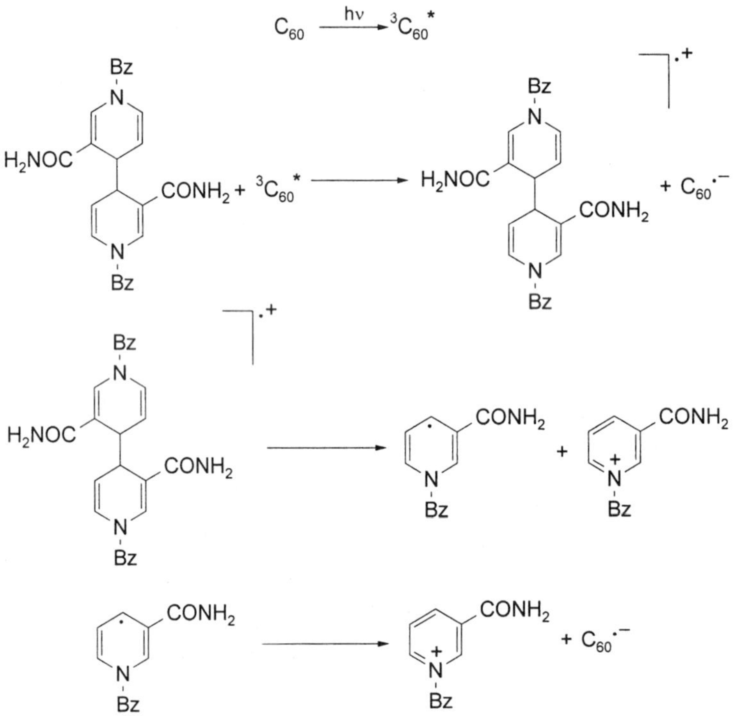

Scheme 16 Mechanism for the generation of $C_{60}^{\cdot-}$ by selective one-electron reduction with [(BNA)$_2$].

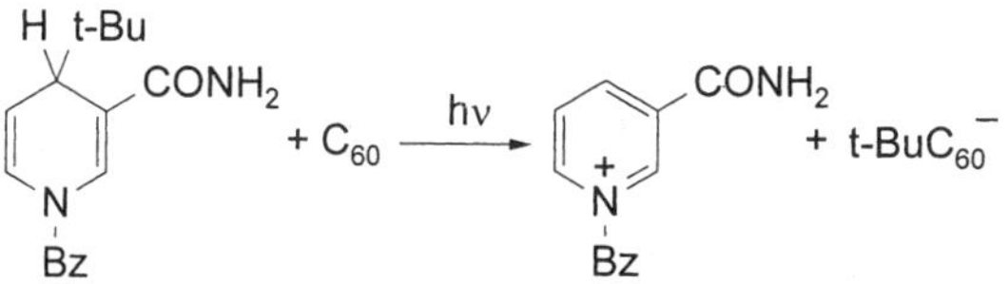

Scheme 17 Selective two-electron reduction of C_{60} to $t\text{-BuC}_{60}^-$.

In addition, Fukuzumi et al. observed selective two electron reduction of C_{60} when BNAH is replaced by the 4-*tert*-butylated BNAH (*t*-BuBNAH) in the above reaction leading to the formation of $t\text{-BuC}_{60}^-$ and the two electron oxidation of BNAH to BNA^+ (Scheme 17). Subsequent trapping of initially formed $t\text{-BuC}_{60}^-$ by CF_3COOH and $PhCH_2Br$ results in formation of the corresponding 1,2-substituted dihydrofullerene for the former and 1,4-substituted adduct for the latter [224]. The initial product of $1,4\text{-}t\text{-BuC}_{60}H$ is rearranged to the 1,2-isomer during the isolation procedure to give the 1,2-adduct exclusively [225].

Again, photoinduced electron transfer to the triplet excited state $^3C_{60}^*$ is the initial step to yield $t\text{-BuBNAH}^{\bullet+}$ and $C_{60}^{\bullet-}$. Upon C(4)–C bond cleavage of $t\text{-BuB-NAH}^{\bullet+}$ the *tert*-butyl radical is formed which readily adds to the fullerene radical anion yielding $t\text{-BuC}_{60}^-$. In competition back-electron transfer takes place. Similar results are obtained in the photochemical reaction of C_{70} with the NADH analogues mentioned previously. The reaction mechanism remains the same in the case of *t*-BuBNAH. However, there are some differences in the reaction mechanism for $[(BNA)_2]$ and BNAH. After the photoinduced electron transfer reaction from $[(BNA)_2]$ to C_{70} results in the formation of the radical ion pair, C–C bond cleavage of the dimer leads to C_{70}^{2-} which is followed by facile electron transfer from C_{70}^{2-} to C_{70} to give two equivalents of $C_{70}^{\bullet-}$. In the case of BNAH, C–H bond cleavage instead of C–C bond cleavage in the case of $t\text{-BuBNAH}^{\bullet+}$ and $C_{70}^{\bullet-}$ occurs to give HC_{70}^- and BNA^+. Subsequent electron transfer from C_{70}^{2-} being in equilibrium with HC_{70}^- to C_{70} leads to the formation of two equivalents of $C_{70}^{\bullet-}$. Using 10-methyl-9,10-dihydroacridine ($AcrH_2$) in the photochemical reaction with C_{60} in the presence of CF_3COOH $1,2\text{-}C_{60}H_2$ is formed exclusively (Scheme 18) [154]. The reaction mechanism is shown in Scheme 19. Proton transfer from $AcrH_2^{\bullet+}$ to $C_{60}^{\bullet-}$ is significantly exergonic and occurs efficiently in the radical ion pair to give $C_{60}H^{\bullet}$. The product $1,2\text{-}C_{60}H_2$ is obtained by fast electron transfer from $AcrH^{\bullet}$ in the pres-

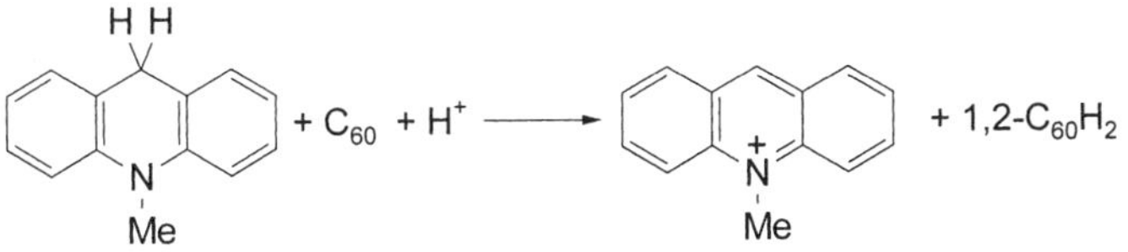

Scheme 18 Selective two-electron reduction of C_{60} to $1,2\text{-}C_{60}H_2$.

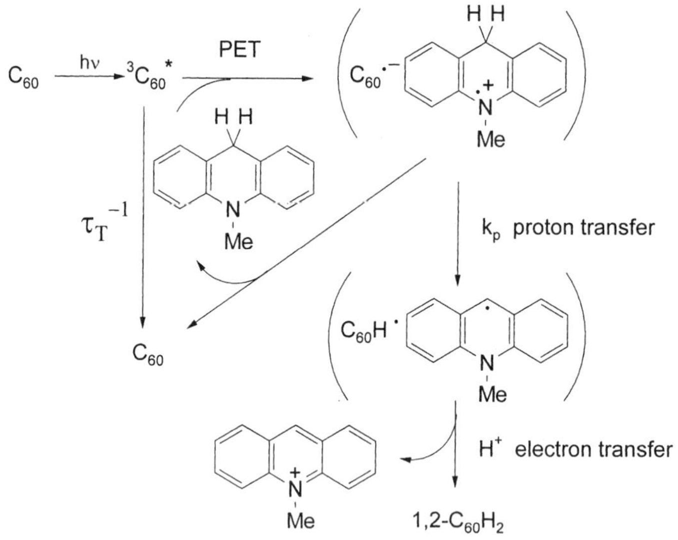

Scheme 19 Mechanism for the selective two-electron reduction of C_{60} to $1,2\text{-}C_{60}H_2$.

ence of CF_3COOH. Protonation of $C_{60}^{\bullet-}$ by the acid and subsequent hydrogen transfer from $AcrH_2^{\bullet+}$ to $C_{60}H^{\bullet}$ yielding the final products is also possible.

The reaction of fullerenes with amines has been the subject of considerable interest. Whereas primary and secondary amines readily add to C_{60} to give 1-H-2-$(NR'R'')C_{60}$ [226], tertiary amines such as triethylamine, N,N-diethylaniline and N,N-dimethylaniline do not undergo the same type of addition due to the absence of an N–H bond. Instead, weak charge transfer complexes with C_{60} are formed in the ground state [135,137,227]. Photoexcitation of amine-[60]fullerene solutions leads to electron transfer from amines to the fullerene and the formation of $C_{60}^{\bullet-}$ amine$^+$ ion pairs [20,29,84,131,228]. For amines possessing α hydrogens, proton transfer from amines to C_{60} is possible to give photochemical products [134,229]. In most cases the photoreaction of amines with C_{60} leads to pyrrolidine cycloadducts. These reactions will be discussed later. A few examples are known where this reaction does not lead to pyrrolidine-fused fullerenes. Liou et al. reported the photochemical reaction of triethylamine and N,N-dimethylaniline with C_{60} [230]. Irradiation of C_{60} at 350 nm in the presence of an excess amine **53a–b** in toluene slowly led to the isolation of **54a–b** (Scheme 20). No cycloadduct similar to the pyrrolidinofullerenes obtained from the irradiation of triethylamine and C_{60} was detected. In the case of **54b,** further irradiation led to the pyrrolidinofullerenes **55** and a dihydrogen derivative, $1,2\text{-}H_2C_{60}$, suggesting that **54b** is an intermediate for **55.**

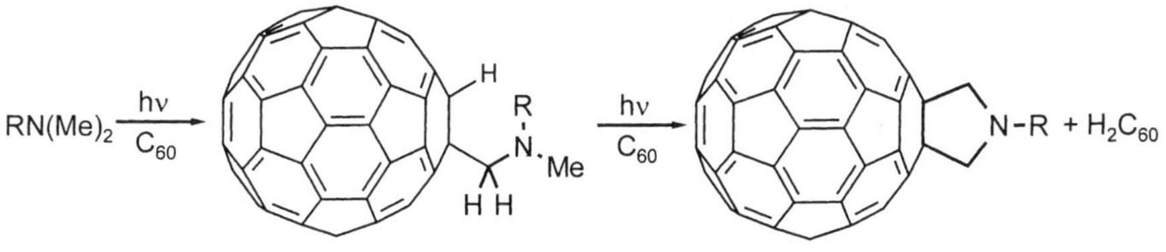

Scheme 20 Reaction of C_{60} and tertiary amines $RN(Me)_2$ (R = Me, Ph) in toluene at 350 nm.

     Photoinduced electron transfer from the amine to C_{60} to yield a radical ion pair is suggested to be the initial step for the formation of **54a–b.** This is followed by deprotonation of the amine cation by the fullerene anion to give an α-aminoalkyl and HC_{60} radical pain [134]. Subsequent combination of the radical pair leads to the final product. Formation of **55** is likely to be initiated by PET from **54b** to C_{60}. This is then followed by successive intermolecular proton transfer, hydrogen abstraction, and ring closure to give 1,2-H_2C_{60} and **55** (Scheme 21).

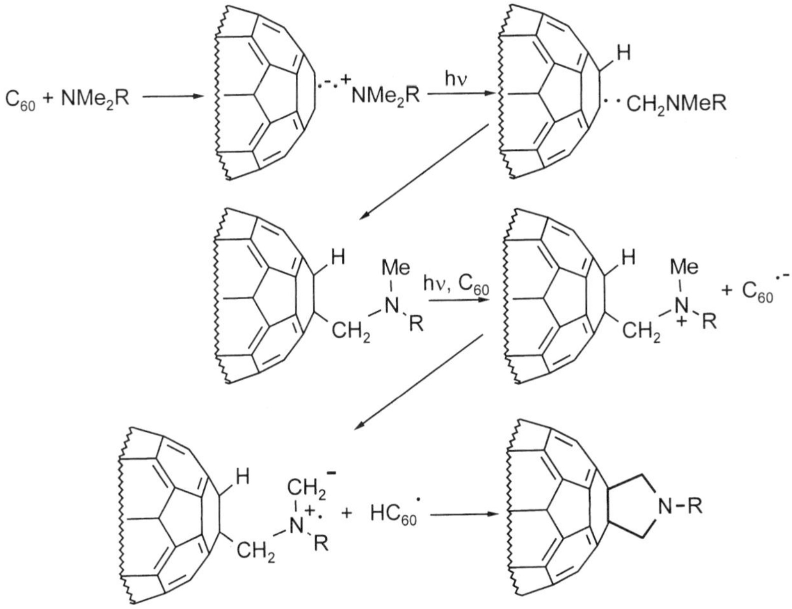

Scheme 21 Mechanism for the reaction of C_{60} and tertiary amines $RN(Me)_2$.

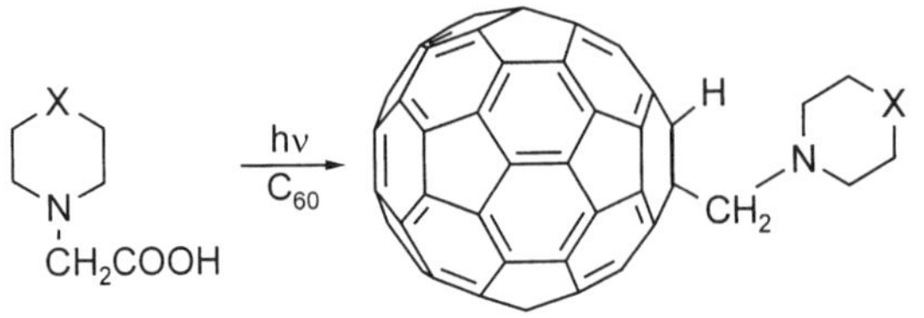

Scheme 22 Formation of 1,2-dihydro[60]fullerenes 57a–b upon irradiation of C_{60} and cyclic amino acids 56a–b.

Gan et al. used amino acids such as piperidino acetic acid or morpholino acetic acid in photochemical reactions with C_{60} [231,232]. Whereas the reaction of the ester derivatives of the amino acids result in the formation of pyrrolidine-fused C_{60} derivatives, the reaction of the free amino acids give the 1-substituted 1,2-dihydro[60]fullerenes **57a–b** (Scheme 22). In a typical experiment a methanol solution of the amino acid **56a–b** was added to a toluene solution of C_{60} and irradiated with a 250 W overhead projector lightbulb. Prolonged irradiation led to multiadducts. To avoid hydroxylation of C_{60} the pH of the amino acid solution should be no more than 9.0.

The first step in the formation of the dihydrofullerenes is electron transfer and proton loss. The CO_2 loss from carboxyl radicals is well known [233]. The so-formed aminomethyl radical then adds to C_{60}. The final step is the abstraction of hydrogen from the environment (Scheme 23).

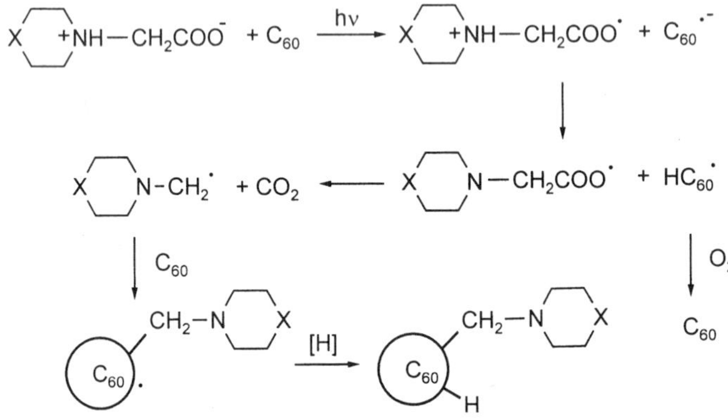

Scheme 23 Mechanism for the ormation of 1,2-dihydro[60]fullerenes 57a–b upon irradiation of C_{60} and cyclic amino acids.

C. Photocycloaddition

Among the reactions applied in the synthesis of fullerene derivatives cycloaddition reactions such as [2 + 1]-, [2 + 2]-, [3 + 2] and [4 + 2] cycloadditions play a dominant role. In these reactions ring-fused fullerene derivatives are obtained, at least with incorporation of heteroatoms such as oxygen, nitrogen, or silicon. In this section photochemical reactions leading to cycloalkyl ring-fused fullerene adducts will be presented. Photocycloaddition reactions leading to C_{60}-fused heterocycles will be discussed later.

1. [2 + 1] Photocycloaddition

C_{60} reacts with diazomethane to yield fulleroids [97,99,100,234]. Carbene generated from the thermolysis of precursors such as diazirines, sodium trichloroacetate, cyclopropene, oxadiazole, and tosylhydrazone [60,235] adds onto C_{60} leading to methanofullerenes [12,15,236]. Recently, Akasaka et al. described the photochemical reaction of diazirine with C_{60} [237]. Irradiation of a benzene solution of 2-adamantane-2,3'-[3H]-diazirine **58** and C_{60} with a high pressure mercury lamp (cutoff <300 nm) at 15°C in a Pyrex tube resulted in the formation of mixture of the isomers **59a** and **59b** in a ratio of 51/49 (Scheme 24).

UV–vis spectroscopy and NMR investigations clearly demonstrate that isomer **59b** is the 6,6-adduct of C_{60} with C_{2v} symmetry, namely the methano fullerene. There is crucial evidence supported by ^1H and ^{13}C NMR for the isomer **59a** being the 5,6-adduct of C_{60} with C_s symmetry. No isomerization of the pure fulleroid to the methanofullerene or vice versa upon photolysis is observed as reported by Wudl et al. [238]. Thus, the isomeric ratio of the two isomers reveals the formation ratio of carbene and diazo compound during the reaction.

2. [2 + 2] Photocycloaddition

A very versatile method of formation of fullerene adducts is the [2 + 2] photocycloaddition. The preparation and isolation of well characterized [2 + 2]

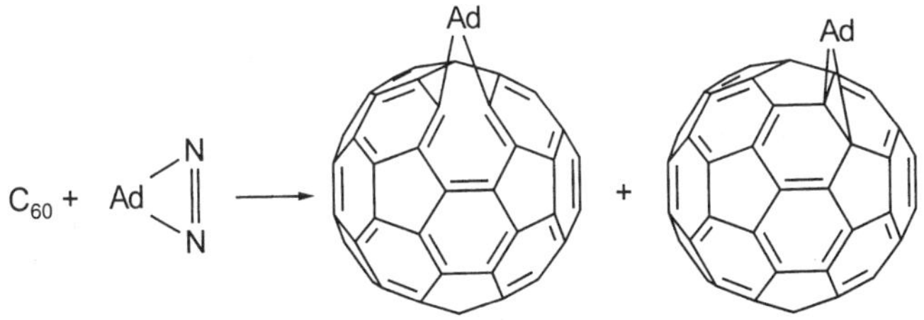

Scheme 24 Photolysis of 2-adamantane-2,3'-[3H]-diazirine and C_{60} with a high pressure mercury lamp (cutoff < 300 nm).

monoadducts of C_{60} with alkynes [239], arylalkenes [240], dienes [241], cy-cloenones [242], acyclic enones [243], and diones [244] has been reported. The pioneering work on this useful synthetic pathway was executed by Schuster et al. in 1993. In their first attempt a crown-ether tagged fulleroid and various cyclo-hexen-2-one derivatives were used in the cycloaddition reaction [206,245]. Once the reaction was successfully performed Schuster and coworkers synthesized cy-clobutane-fullerene adducts by the irradiation of C_{60} using both the broad spec-trum of a Hanovia high pressure mercury lamp and the emission of a XeCl ex-cimer laser at 308 nm (Scheme 25) [206]. Up to seven enone units are incorporated on longer irradiation time. As shown by ^{1}H-NMR, IR, and HPLC analysis the isolated monoadducts consist of a mixture of two stereoisomers, the *cis-* and *trans-*isomers arising from the [2 + 2] cycloaddition across the [6,6]-ring junction in C_{60}, the major product normally being *trans*. When 3-methyl-2-cyclo-hexen-1-one is used in the reaction, the obtained fullerene derivatives are chiral. Semipreparative HPLC analysis with a chiral stationary phase (S,S)-Whelk-O HPLC column results in the separation of the four stereoisomers [244]. No prod-uct formation was observed upon irradiation at 532 nm where the fullerene is the only light absorbing component. Thus, fullerene triplets do not undergo addition to ground-state enones. Instead, the reaction proceeds by the same mechanism op-erating in the photoaddition of enones to ordinary alkenes [246], namely stepwise addition of enone triplet excited states to the fullerene via an intermediate triplet 1,4-biradical.

The photochemical addition of cyclic 1,3-diones such as dimedone, 1,3-cylohexandione **62,** or their respective silyl enol ethers leads to the formation of two fused furanylfullerenes, (1) achiral **63** and (2) chiral **64** [244]. The latter having an unusual bis-[6,5] closed structure. In the initial step of this reaction, [2 + 2] photocycloaddition across a [6,6] bond to form cyclobutanols or the cor-responding TMS ethers is involved (Scheme 26). Oxidation with $^1O_2$ yields in the formation of the radical **65a**. Cleavage to **66a** followed by cyclization gives furanyl radical **67a**. H abstraction by $^1O_2$ or a peroxy radical finally leads to product **63**. In competition, formation of fullerene triplets by absorption of a

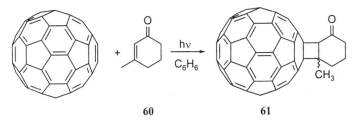

60   61

Scheme 25   Reaction of C_{60} with cyclic enone 60.

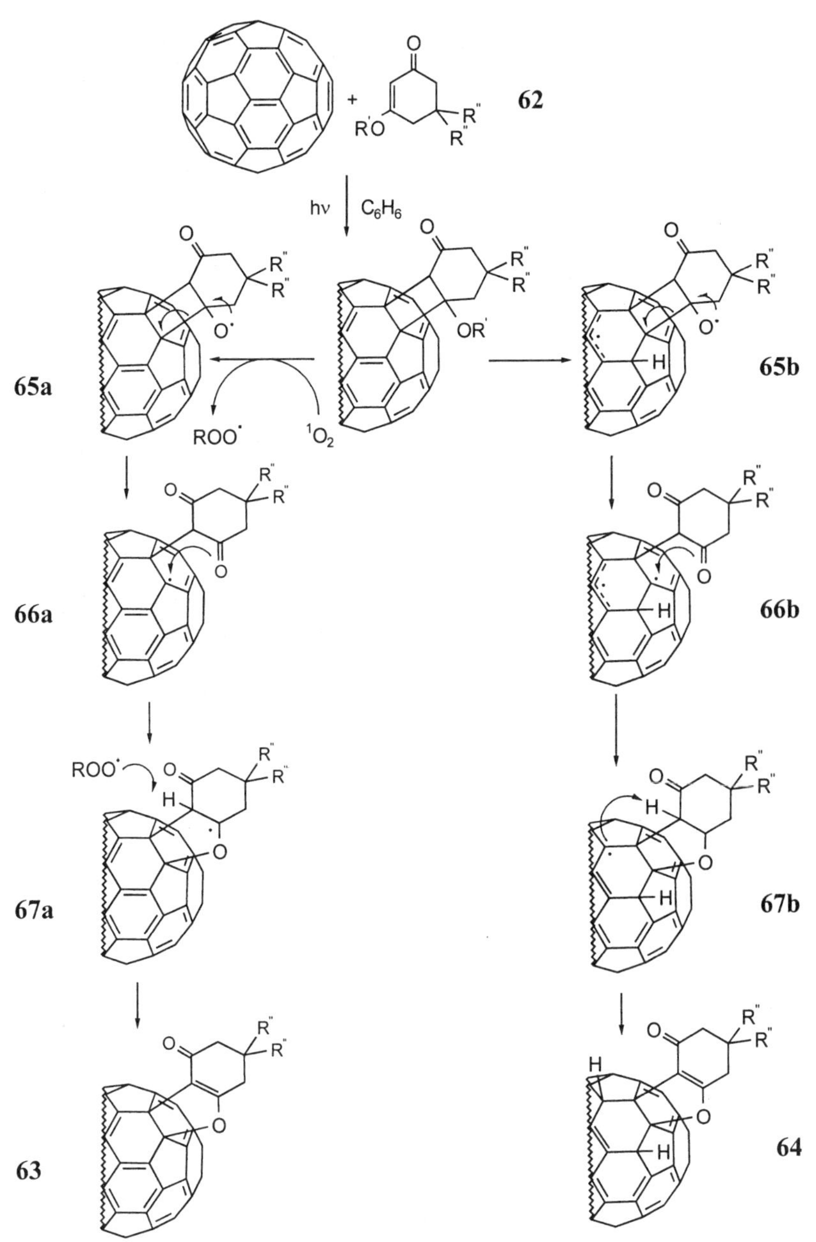

Scheme 26 Proposed mechanism for the [2 + 2] photocycloaddition of C_{60} and cyclic 1,3-dione 62.

photon by the cycloaddition products followed by intramolecular H abstraction results in the production of radical **66a.** Cleavage and cyclization as mentioned earlier yield the furanyl radical **67b.** Product **64** is obtained via a thermodynamically driven 1,3-H atom shift. Note that compound **64** is a unique type of fullerene adduct in which four adjacent stereogenic centers are formed on the fullerene sphere. Only the (S,S,R,R)/(R,R,S,S) enantiomeric pair is formed due to the nature of the reaction.

Orfanopoulos et al. studied the photochemical reaction of alkenes, arylalkenes, dienes dienones, and acyclic enones with [60]fullerene to obtain various substituted cyclobutylfullerenes [240,241,243,247]. For example, the photocycloaddition of *cis*- and *trans*-1-(p-methoxyphenyl)-1-propene **68** to C_{60} gives only the *trans* [2 + 2] adducts (Scheme 27), thus the reaction is stereospecific for the most thermodynamically stable cycloadduct. A possible mechanism includes the formation of a common dipolar or biradical intermediate between $^3C_{60}^*$ and the arylalkene. Subsequent fast rotation of the aryl moiety around the former double bond leads exclusively to the *trans*-**69** [2 + 2] adduct. Irradiation of this product, yielded 90% *trans*-**68,** 10% *cis*-**68** and cycloreversion products. Thus, a concerted mechanism can be excluded because the photocycloreversion is expected to give the *trans*-**68** as the only product. These results can be explained by the formation of a common dipolar or diradical intermediate. Similarly, cycloreversion products from C_{60} and tetraalkoxyethylene

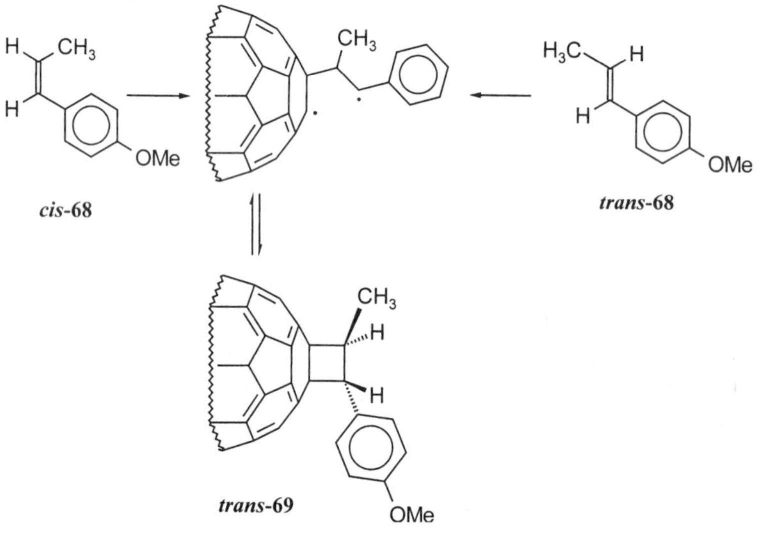

Scheme 27 Reaction of C_{60} and *cis*-68/*trans*-68 at $\lambda > 530$ nm and proposed mechanism for this reaction.

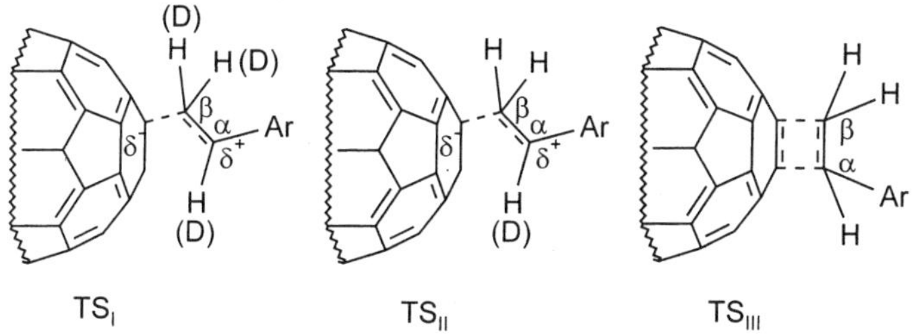

Figure 30 Possible transition states in the reaction of C_{60} with arylalkene 70.

adducts were also observed by Foote et al. [239b] As reported by this research group, the cycloreversion proceeds via the triplet excited state of the adduct because rubrene and oxygen efficiently inhibit photocycloreversion. However, these results do not exclude an electron transfer or charge transfer mechanism with loss of stereochemical integrity.

Studies of the secondary isotope effect also confirmed the two-step mechanism, involving the formation of a dipolar or diradical intermediate in the rate determining step. The observed isotope effects in the reaction of 1-(p-methoxyphenyl)- ethylene **70** and its deuterated analogues 1-(p-methoxyphenyl)-ethylene-d_1 **70-d_1** and 1-(p-methoxyphenyl)-ethylene-d_3 **70-d_3** with C_{60}, when taken in conjunction, exclude the formation of transition state TS_{III} (concerted mechanism, see Fig. 30) because in that case substitution at either C_α or C_β would have given an inverse isotope effect.

The addition reactions of alkyl-substituted 1,3-butadienes to C_{60} [241] supported the suggested two-step mechanism via a dipolar or diradical intermediate also. The formation of this intermediate is responsible for the observed regioselectivity and the secondary isotope effects. The second step (collapse of the initial ion pair) dictates the diastereoselectivity that favors the more thermodynamically stable *trans*-cyclobutane products. The lifetime of the biradical intermediate is long enough to allow rotation around the C_2–C_3 bond in the 1,3-butadienes leading to loss of stereochemical integrity of the cycloadducts. However, the stereochemistry of the unreacted double bond of the dienes is retained, and this result is consistent with the known propensity of allylic radicals to resist rotation around the partial double bond. The regio- and stereoselectivity can also be observed in the reaction of acyclic enones to C_{60} [243]. Foote et al. used electron-rich alkynes such as ynamines in the [2 + 2] cycloaddition reaction with C_{60} leading to C_{60}-fused cyclobutenanimes (Scheme 28) [239b,248]. The authors proposed that the addition of ynamines to fullerenes proceed via electron or at least charge transfer from the electron-rich ynamines followed by rapid collapse of the initial ion pair or charge-transfer complex to the covalent adducts.

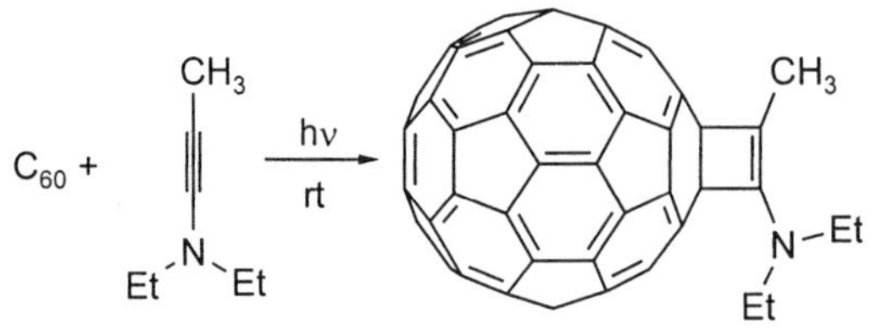

Scheme 28 Photolysis of C_{60} with ynamine 71 at $\lambda > 500$ nm.

3. [2 + 4] Cycloaddition

The thermal Diels-Alder reaction ([4 + 2] cycloaddition) is widespread in the synthesis of fullerene derivatives. In contrast, only a few examples of the photochemical Diels-Alder reaction in solution or in the solid state are known. The first example is described by Tomioka and coworkers [249]. Irradiation of ketone **73** and C_{60} at 10°C with a high pressure mercury lamp through a Pyrex filter led to the formation of 61-hydroxy-61-phenyl-1,9-(methano[1,2]benzenomethano) fullerene **75** (Scheme 29). This compound is unusually unstable and yields the monoalkyl-1,2-dihydrofullerene **76** either by silica gel chromatography or upon heating.

Upon photoexcitation **73** undergoes intramolecular H-abstraction to generate 7-hydroxy-7-phenyl-o-quinodimethane **74,** which is trapped by the dienophile C_{60}. As a control experiment the reaction was carried out with 3-phenylbenzocy-

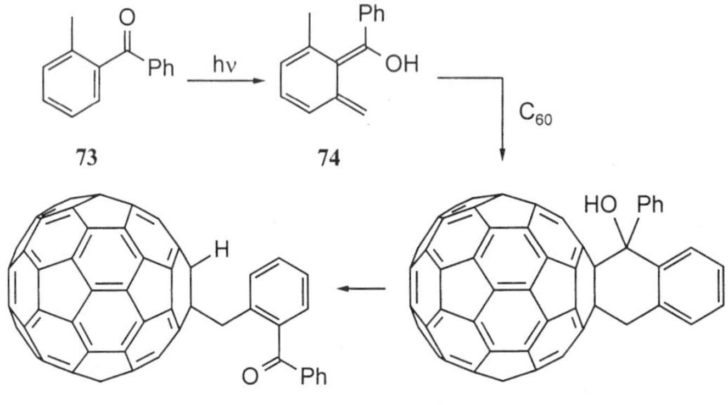

Scheme 29 Irradiation of ketone 73 and C_{60} with a high pressure mercury lamp (Pyrex filter).

clobutenol, where the above o-quinodimethane is known to be generated, leading again to product **76**. The reaction must involve cleavage of the 60–61 σ bond of **75** directly connected to the fullerene core. Hydrolytic cleavage of the bond in fullerene adducts is an example for producing alkyl fullerenes [217,248,250]. Obviously, the 61-hydroxy group must play an important role in this transformation because cleavage to the dihydrofullerene is not observed when the methyl ether of compound **75** is subjected to the same reaction conditions. Proton transfer from the hydroxyl group to the adjacent C_{60} will be favored in this system, but intermolecular protonation can not be excluded. Substitution of the 61-phenyl with n-butyl leads to complete quenching of the reaction, therefore stabilization of positive charge developing at the 61 carbon in the transition state is crucial for this bond cleavage reaction. This can be interpreted as implying that the σ bond cleavage is involved at the early stage of the transition state.

Mikami et al. investigated the photochemical Diels-Alder reaction of anthracenes with C_{60} in the solid state [251]. Irradiation of a mixture of C_{60} and 9-methylanthracene with a high pressure mercury lamp through a Pyrex vessel and stirring by using a cooling jacket to keep the reaction at 28°C led to the formation of mono- and bisadducts. No formation of 9-methylanthracene dimers as in solution was observed. The reaction did not work with less bulky anthracenes. The reaction rate depends on the ionization potential of the anthracenes. With decreasing ionization potential the Diels-Alder reaction of C_{60} proceeds much easier. Therefore, the reaction may proceed via photoinduced electron transfer from anthracenes to the triplet excited state of C_{60}. The energetics for the photoinduced electron transfer in the solid state is significantly different from that in solution where the solvation plays an important role. Such a difference leads to the different reactivity of the anthracene derivatives in the solid state as compared to that in solution. The [2 + 2] photocycloaddition of ketene silyl acetals to C_{60} described previously was successfully tested in a solvent free system.

Recently, Fukuzumi et al. reported the photochemical Diels-Alder reaction with Danishefsky's diene [252]. A mixture of C_{60} and the stereochemically defined (1E,3Z)-1,4 disubstituted Danishefsky's diene **77** was irradiated for 9h at -30°C to avoid the thermal reaction (Scheme 30). A high pressure mercury lamp

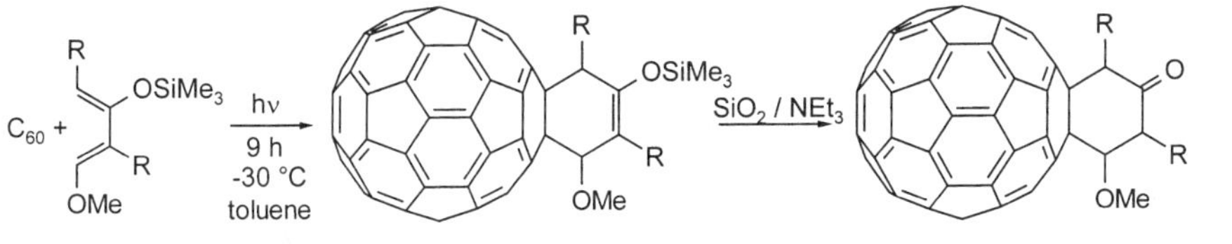

Scheme 30 Irradiation of C_{60} and Danishefsky's diene 77.

was used as a light source. After work-up two desilylated adducts **79** were obtained, the major being the *trans*-product the minor the *cis*-product. The same *trans* product **79** is obtained as a single isomer after a short irradiation time indicating the *cis*-isomer may be formed by isomerization of the *trans*-adduct after a longer irradiation period. Such stereochemistry indicates that the Diels-Alder reaction proceeds by a stepwise mechanism rather than a concerted mechanism. In the initial step of the photochemical Diels-Alder reaction the triplet radical ion pair (**A**) is formed by photoinduced electron transfer from **77** to triplet-excited C_{60} (k_{et}) in competition with the decay to the ground state (Scheme 31). The triplet

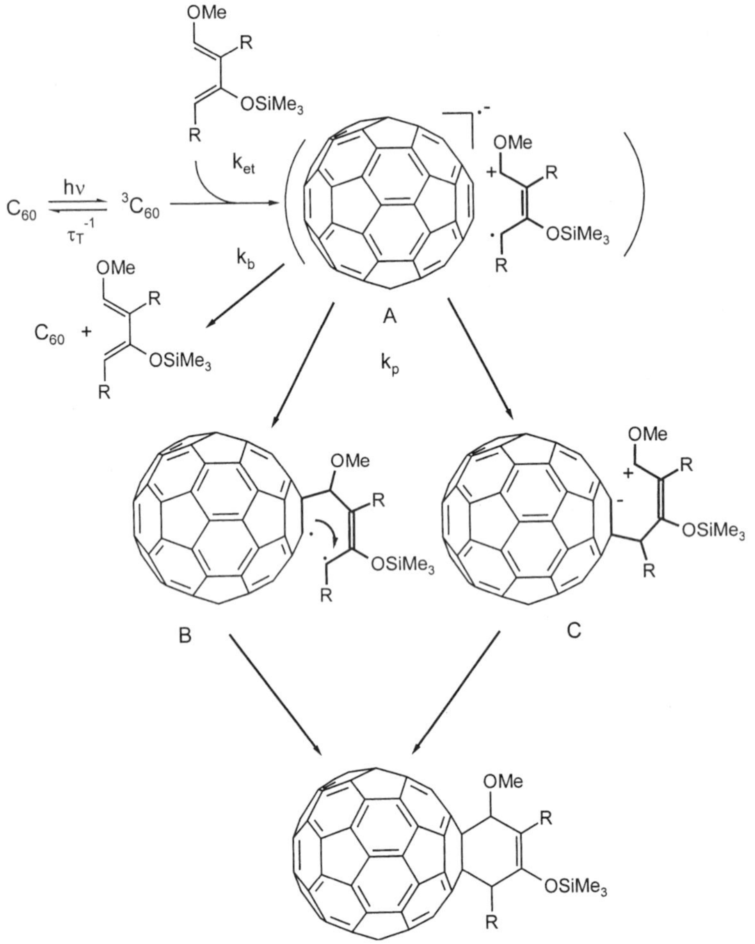

Scheme 31   Proposed mechanism for the irradiation of C_{60} and Danishefsky's diene **77**.

radical ion pair then provides the single radical ion pair leading to a diradical intermediate (**B**) or a zwitterionic intermediate (**C**) in competition with back electron transfer to the reactant pair (k_b). In case of the diradical pathway the initial C–C bond formation should occur between C-1 carbon of **77**$^{\bullet+}$ and the $C_{60}^{\bullet-}$, while it occurs between C-4 of carbon of **77**$^{\bullet+}$ and the $C_{60}^{\bullet-}$ in the case of the zwitterionic pathway. There are difficulties in distinguishing between the two possibilities but in any case the formation occurs stepwise, and thus there is no symmetry restriction for bond formation. Both *trans*-adduct and *cis*-adduct **79** are obtained as the final products.

D. Photocycloaddition Reactions Leading to C₆₀-Fused Heterocycles

As mentioned previously, photocycloaddition reactions are a useful method for obtaining fullerene derivatives fused directly to heterocycles. Such compounds have attracted much interest because they might have interesting properties, e.g., amino acid fullerene derivatives as biologically active compounds or pyrrolidinofullerenes as key precursors for a great number of fullerenes donor-acceptor bridged dyads and triads. Fullerenes functionalized with silicon compounds are of great interest in materials science. The synthesis of these compounds will be described in this section.

1. Heterocycles Containing Oxygen

Since the synthesis and characterization of $C_{60}O$ in 1992 by Smith et al. [98] several studies on $C_{60}O$ have appeared, showing that the fullerene epoxide can be generated through both thermal and photochemical pathways [98]. Smith et al. obtained $C_{60}O$ by the extended irradiation of C_{60} in an oxygen atmosphere. In 1998, Schuster et al. reported the photochemical synthesis of $C_{60}O$ **81** by irradiation of C_{60} with endoperoxides **80** (Scheme 32) [253]. The light source was a Hanovia 450 W medium pressure Hg arc lamp cooled by a quartz jacket fitted with a Pyrex filter between the lamp and the inner wall of the jacket.

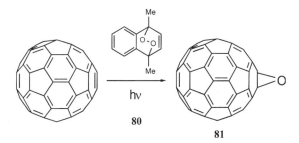

Scheme 32 Irradiation of C₆₀ and endoperoxide 80.

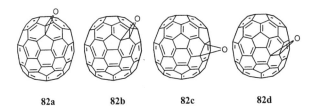

      82a   **82b**   **82c**   **82d**

Figure 31 Possible $C_{70}O$ isomers.

After irradiation $C_{60}O$ is obtained in a higher yield than that reported from irradiation of C_{60} in an oxygen atmosphere under analogous conditions. Singlet oxygen $^1O_2$ released from the endoperoxides is the key species in the mechanism for fullerene oxide formation under photochemical reaction conditions, and suggests that $C_{60}O$ results from reaction of $^1O_2$ with C_{60} triplet excited states as primarily suggested by Juha et al. [254].

Heymann et al. [255] and Smith et al. [256] reported the synthesis of $C_{70}O$ **82** by photooxygenation. Irradiation of an oxygenated toluene solution of C_{70} and rubrene as a sensitizer resulted in the formation of a nearly 1:1 mixture of two $C_{70}O$ isomers **82a** and **82b** that can not be separated (Fig. 31). The structures of **82a** and **82b** were elucidated by UV–vis spectroscopy, ^{13}C-NMR studies as well as ^3He NMR studies. No evidence for the formation of isomers **82c** and **82d** was found.

2. Heterocycles Containing Nitrogen

The thermal reaction of diazomethane followed by N_2 extrusion and of nitrenes leads to the formation of the nitrogen containing analogues of the fullerene epoxides, the aziridinofullerenes, and azafulleroids [257]. The photochemical synthesis of aziridino-fullerenes was first reported by Keana et al. in 1994 [258]. When a solution of C_{60} and N-succinimidyl 4-azido-2,3,5,6-tetrafluorobenzoate **83** in dry chlorobenzene was purged with Ar and irradiated at 300 nm for 5h the aziridinofullerene **84** was obtained in 10% yield (Scheme 33). The photochemical reaction may proceed via the addition of a highly reactive nitrene intermediate generated by photolysis of **83** to a [6,6] double bond of C_{60} to give the aziridine **84**. Since perfluorophenyl azides have been reported to react with aromatic molecules such as benzene or toluene upon photolysis [259], reaction of the nitrene derived from **83** with the solvent chlorobenzene may contribute to the low yield of **84**.

Averdung et al. also obtained several aziridinofullerenes by the reaction of C_{60} with acylnitrenes, generated by photolysis of aroylazides **85a–d,** leading to the fullerene adducts **86a–d** (Scheme 34) [260]. In a typical experiment a solution of C_{60} and a five-fold excess of azide **85a–d** in oxygen free 1,1,2,2-tetrachloroethane was irradiated for 60 min in Pyrex tubes using a RPR 100 Rayonet

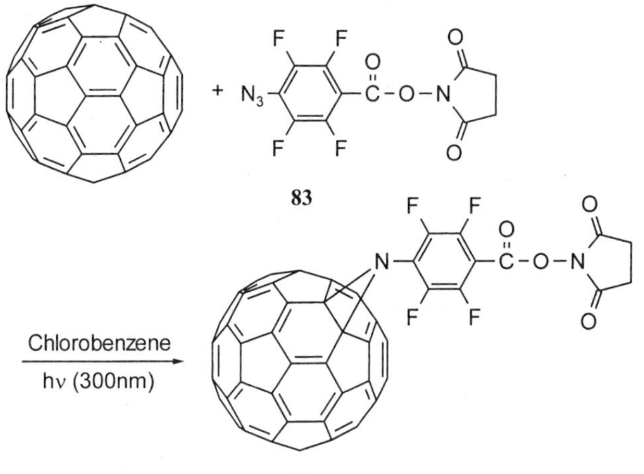

Scheme 33 Formation of aziridinofullerene 84 upon irradiation of C_{60} and perfluorophenyl azide 83.

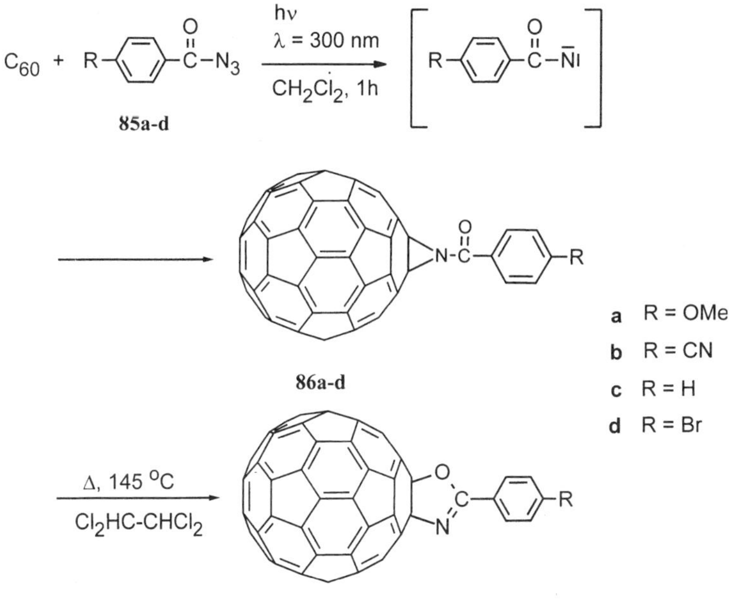

Scheme 34 Photolysis of C_{60} and aroylazides 85a–d in oxygen free 1,1,2,2-tetrachlorethan at $\lambda = 300$ nm.

Photochemical Chamber reactor fitted with RPR-3000 Å lamps. The structures of the [6,6]-ring fused 1,2-dihydrofullerenes have been identified by standard spectroscopic methods.

Irradiation at wavelengths > 380 nm did not result in product formation. Thus, the formation of the acylnitrenes does not occur via excited C_{60}. The reaction of C_{60} and 4-cyanobenzoylazide **85b** in benzene at 300 nm did not lead to the expected fulleroaziridine **86b.** Instead a 6,6-brigded dihydrofullerene was obtained as a main product. Mass spectral analysis and NMR spectroscopic investigations indicates an asymmetrical fullerene structure formed by reaction of the nitrene to C_{60} with incorporation of benzene, similar to the incorporation of benzene in the reaction of 1,8-dehydronaphthalene with C_{60} [261]. The exact structure still remains unknown. As a minor product the synthesis of the fullero-oxazole **87b** was observed. The formation of the corresponding fullerooxazoles **87a–d** was also observed by the thermal rearrangement of aziridinofullerenes.

The same group investigated the [3 + 2] cycloaddition reaction of 2*H*-azirines with C_{60} [262,263]. Upon irradiation 2,3-substituted-2*H*-azirines **88a–c** were added to C_{60} along with formation of mono- and oligoadducts (Scheme 35) [262]. Irradiation of a equimolar solution of C_{60} and **88a–c** led to the production of the mono adducts **89a–c** that have been isolated and identified by standard spectroscopic methods. Use of a 10-fold excess of the azirine led to the formation of polyadducts. UV spectroscopy and NMR investigations showed that **89a–c** have a closed [6,6]-bridged 1,2-dihydrofullerene structure with C_1 symmetry. An open [1,6]-substituted [10]annulene structure can be excluded. In all cases, a fraction of regioisomeric bis-adducts can be obtained, which were not separated.

The formation of the [3 + 2] cycloadduct **89** can be explained by two different reaction pathways (Scheme 36). Irradiation at 300 nm leads to ring opening of the azirine **88** to the nitrile ylide as an intermediate which adds to C_{60} yielding the observed product **89**. No product formation is observed at wavelengths > 400 nm. Instead, under photoinduced electron transfer conditions in the presence of

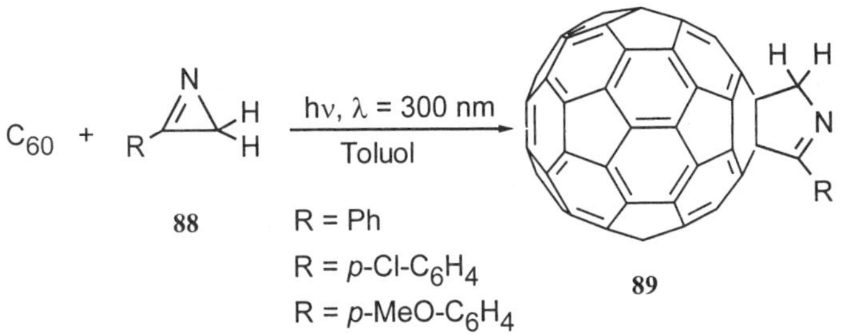

Scheme 35 [3 + 2] photocycloaddition of 2*H*-azirines and C_{60} in toluene at $\lambda = 300$ nm.

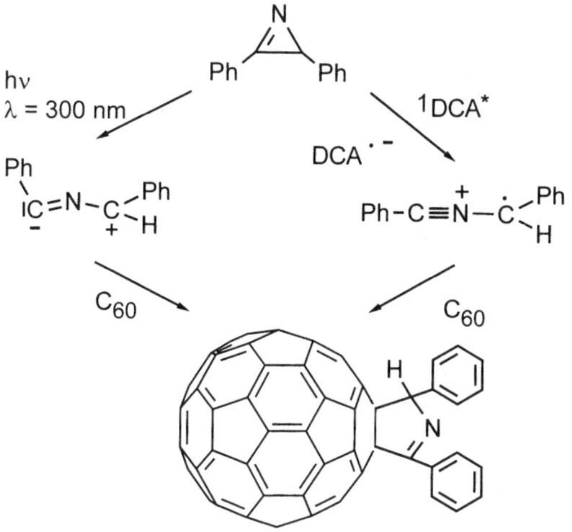

Scheme 36   Mechanism for the [3 + 2] photocycloaddition of 2*H*-azirines and C_{60}.

9,10-dicyanoanthracene (DCA) the [3 + 2] cycloaddition reaction proceeds via the 2-azaallenyl radical cation.

Since amines are good electron donors and C_{60}/C_{70} are excellent electron acceptors the reaction between these two species should readily occur. Indeed, several examples are known for the electron transfer interactions of fullerenes with aromatic and aliphatic amines [32,130,138,224a,224c,224e,227c,263]. The emissions of the fullerene-amine exciplexes being formed in these processes are extremely solvent sensitive. The intense exciplex fluorescence observed in nonpolar solvents such as hexane and cyclohexane are completely quenched in polarizable but still nonpolar solvents such as benzene and toluene. It has been proposed [224a,264] that the quenching is due to the formation of ion pairs in a polarizable solvent environment. Fullerene molecules also undergo efficient photochemical reactions with tertiary amines under ambient conditions to form a complex mixture of addition products. The reaction of C_{60} with *N,N*-dimethylaniline and trimethylamine, leading in the latter case to C_{60}-fused pyrrolidine derivatives upon prolonged irradiation, has already been discussed. Irradiation of C_{60} and triethylamine (TEA) with a Xe lamp led to the formation of a cycloadduct [265]. The structure of the product as a C_{60}-TEA cycloadduct *N*-ethyl-*trans*-2′,5′-dimethylpyrrolidino[3′,4′ :1,2]-[60]fullerene **33** was determined by mass spectral analysis and NMR investigations. According to classical photoinduced electron transfer-proton-transfer mechanism [266], a simple photochemical addition yields the 1-substituted 1,2-dihydrofullerene **90.** Due to the possibility of C_{60} to act again as an electron acceptor, the monoadduct may undergo further electron transfer in-

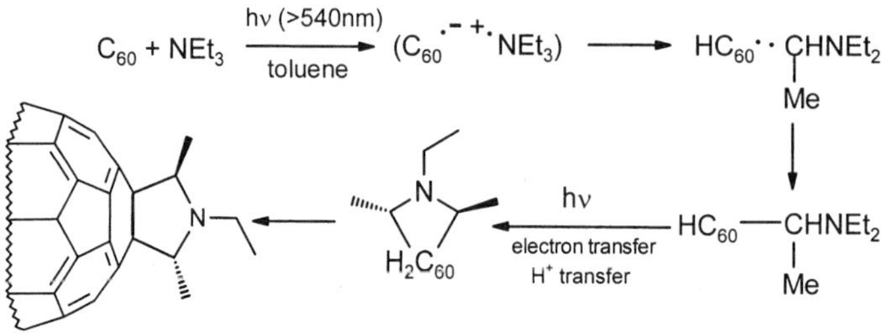

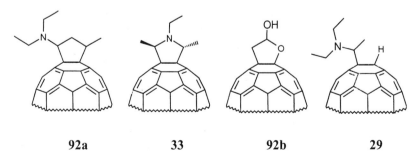

Scheme 37 Proposed mechanism for the formation of the pyrrolidinofullerene EDMP-
C_{60}.

tramolecularly as depicted in Scheme 37, finally leading to compound **33**. The
ring-closed product **91** is formed by a similar process of proton transfer and radical pair combination following the formation of an intramolecular ion pair. The
process leading to the final product **33** may be an oxidation similar to the one observed in the formation of phenanthrene from dihydrophenanthrene in the photocyclization reaction of *cis*-stilbene [267].

Sun et al. further investigated the photoinduced electron transfer reactions
of C_{60} and triethylamine, both in deoxygenated solution and air saturated solution
[79]. Three types of cycloadducts of fullerenes **33** and **92a–b** were obtained,
whereas the formation of the monoalkylated 1,2-dihydro[60]fullerene **29** as described by Liou et al. [230] in the reaction of trimethylamine and *N,N*-dimethy-
laniline with C_{60}, was not observed (see Fig. 32).

Compound **92a** was formed in the deoxygenated solution, compound **33** and
92b were obtained by irradiation of air-saturated solutions. Sun and coworkers
suggest that the simple adduct **29** should be the initial product formed in C_{60}-ter-

Figure 32 Structure of the products obtained upon irradiation of triethylamine and C_{60}.

tiary amine photochemical reactions regardless whether the reaction is performed in deoxygenated or air-saturated solutions. The first step of the reaction with or without dissolved molecular oxygen is assumed to be the formation of the $C_{60}^{\bullet-}/TEA^{\bullet+}$ ion pair by photoinduced electron transfer between the singlet-excited C_{60} and the amine. The formation of the different products in deoxygenated or air saturated toluene solution may be attributed to oxygen effects on the following steps after the initial formation of the radical ion pair. The secondary photochemical reactions leading to the observed products are probably complicated. **29** may undergo both photoinduced intermolecular and intramolecular electron transfers followed by subsequent proton transfer and radical combinations leading to the pyrrolidinofullerene in the presence of oxygen [265]. In the formation of **92a** carbon-nitrogen bond breaking must be involved as described by Gan et al. [268] in the photoaddition reactions of sarcosine esters to C_{60}. The formation of compound **92b** is likely to proceed via the oxidation by dissolved molecular oxygen requiring carbon-nitrogen bond breaking for carbon–oxygen bond formation.

The photochemical reaction of tertiary amines with C_{60} can be used to synthesize alkaloid-C_{60} derivatives [269]. Irradiation of alkaloids bearing a tertiary amino group such as tazettine, gramine, scandine, or 10-hydroxyscandine with C_{60} led to the isolation of alkaloid-C_{60} adducts. Use of tazettine and gramine in the reaction yielded the expected [6,6] adduct. In addition to the pyrrolidinofullerene **94a–b,** a new type of monoadduct **95a–b** with a bis-[6,6] closed structure characterized by UV-vis, FT-IR, ^{1}H-NMR, ^{13}C-NMR, ^{1}H-^{1}H-COSY, ROESY, HMQC, and HMBC spectroscopy was obtained from the reaction with scandine **93a** and its 10-hydroxy derivative **93b** (Scheme 38).

Photoinduced electron transfer from the amine to the fullerene core leading to a radical ion pair is suggested to be the initial step in the reaction mechanism (Scheme 39). Formation of the bis-[6,6] closed adduct proceeds via [3 + 2] cycloaddition of the tertiary amine followed by a [2 + 2] cycloaddition of the vinyl group and the C_{60} double bond adjacent to the previously formed ring connection leading to a structure analogous to 1,2,3,4-$C_{60}H_4$.

Foote et al. used 3-diethylamino-1-propyne (DEAP) **96** as a tertiary amine in the photoreaction with C_{60}. Instead of the expected [2 + 2] cycloaddition of the acetylene unit a acetylene-substituted pyrrolidinofullerene **97** was obtained when C_{60} and DEAP were irradiated with a Xe lamp under air-saturated conditions [270]. Since the reaction involves both oxygen and light, two competing reaction pathways are possible (Scheme 40). Singlet oxygen $^1O_2$ seems to be a key participant in the reaction of DEAP with C_{60} since all tests—solvent deuterium isotope effect, competitive quenching with DABCO and product formation by thermally generated singlet oxygen as from 1,4-dimethylnaphthalene endoperoxide (DMNO$_2$)—indicate the involvement of $^1O_2$ (Type II). Singlet oxygen, formed via the triplet-excited fullerene, may react with the amine to give a radical ion pair through electron transfer. Deprotonation of the methylene carbon α to the amine

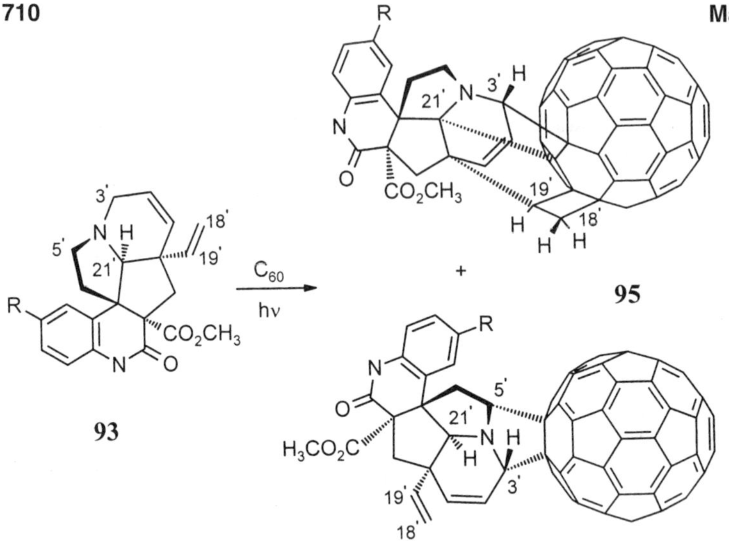

Scheme 38 Photolysis of C_{60} and scandine derivatives 93a–b.

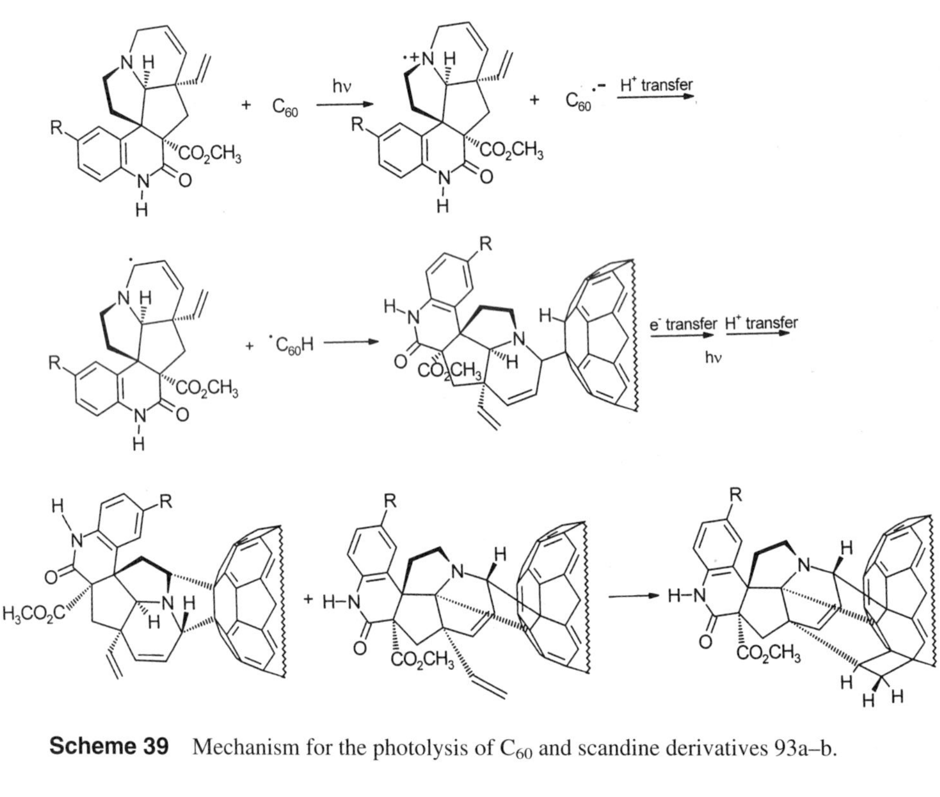

Scheme 39 Mechanism for the photolysis of C_{60} and scandine derivatives 93a–b.

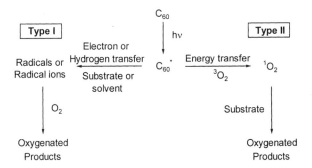

Scheme 40   Competing reaction pathways for the reaction of C$_{60}$ and DEAP.

cation leads to the neutral amine radical, which reacts with the fullerene under C–C bond formation. Repeating this process a second time results in the formation of the observed product (Schemes 41 and 42).

Amino acid derivatives react with C$_{60}$ both thermally and photochemically to yield different products. Gan et al. intensively studied the photochemical reactions of C$_{60}$ and amino acid derivatives such as sarcosine esters, glycine esters, and iminodiacetic esters [268]. The thermal reaction of the sarcosine ethyl ester **98** led to the formation of a 4:1 mixture of two products, the major being compound **99** and the minor being compound **100** containing two ester functionalities (Scheme 43).

If the sarcosine esters are used in the photoreaction, compounds **100** bearing two ester functionalities and **101** containing only one ester functionality, are obtained (Scheme 44). Sarcosine has one more methyl group than glycine that results in the amino group being more basic and also more crowded. Obviously, the electronic effect predominates leading to an increased rate compared to the glycine analogues. The methyl ester gives the best results in terms of reaction rate and yield. It was not possible to distinguish between the methanofullerene and fulleropyrrolidine structure for compound **100** by the spectral investigations (UV-vis, IR, ^{1}H, and ^{13}C

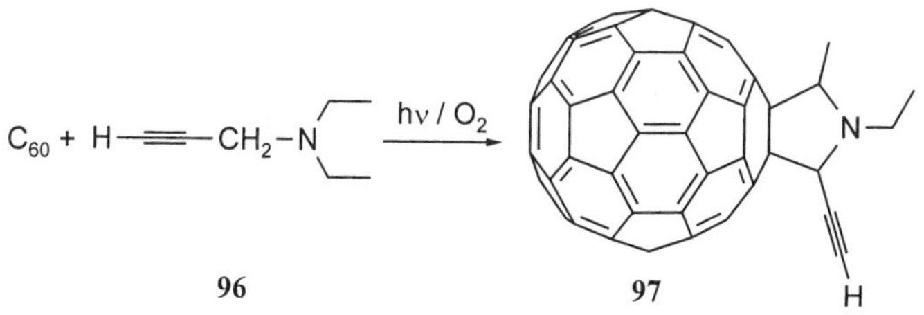

$$C_{60} + H\!\!-\!\!\equiv\!\!-CH_2\!\!-\!\!N \xrightarrow{h\nu\,/\,O_2}$$

96   97   H

Scheme 41   Photolysis of C$_{60}$ and DEAP.

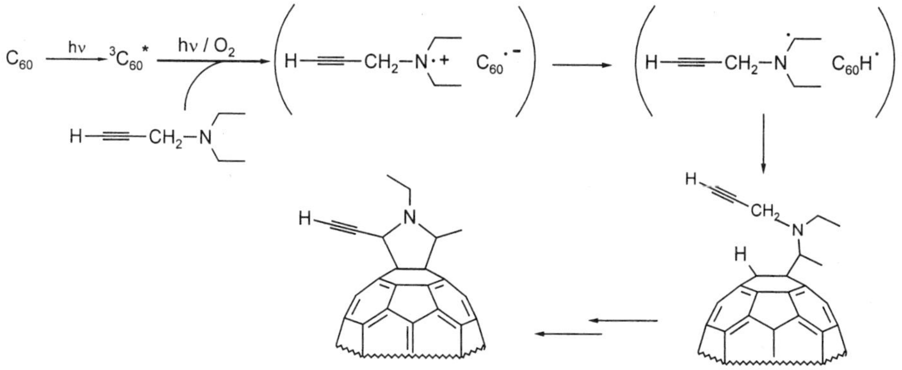

C60
hν
3C60*
hν / O2
H—≡—CH2—N
H—≡—CH2—N·+ C60·−
H—≡—CH2—N C60H·
H
CH2
N
H
N

Scheme 42 Mechanism for the reaction of C_{60} and DEAP.

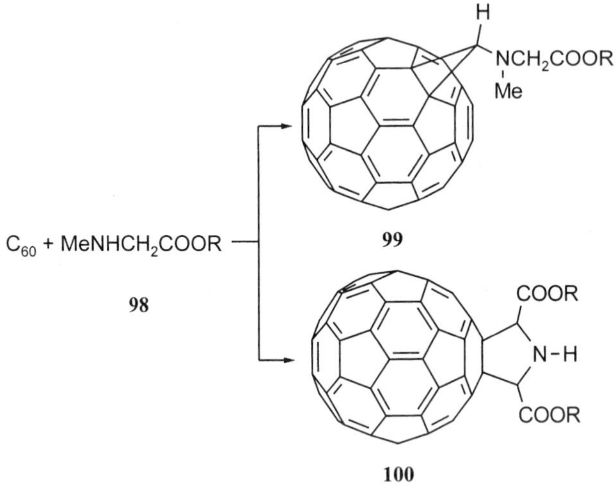

H
NCH2COOR
Me
C60 + MeNHCH2COOR
99
98
COOR
N-H
COOR
100

Scheme 43 Thermal reaction of sarcosine esters 98 with C_{60}.

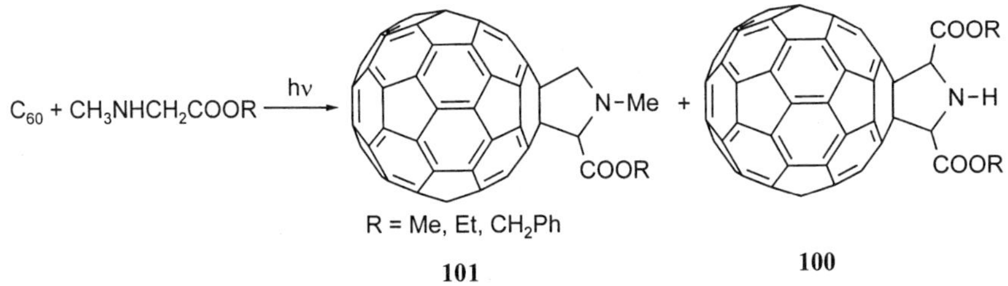

C60 + CH3NHCH2COOR
hν
N-Me
COOR
+
COOR
N-H
COOR
R = Me, Et, CH2Ph
101
100

Scheme 44 Photochemical reaction of sarcosine esters 98 with C_{60}.

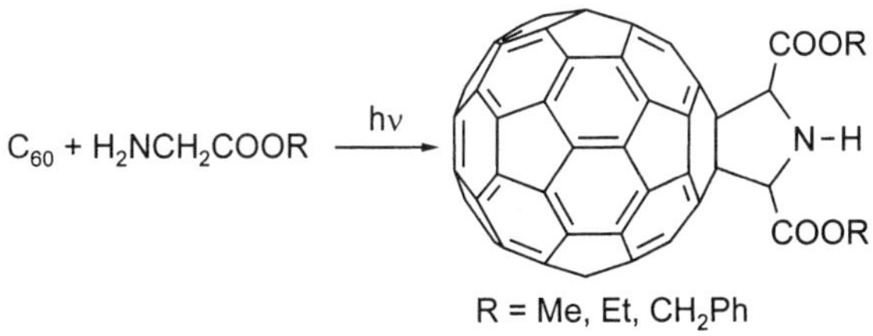

$$C_{60} + H_2NCH_2COOR \xrightarrow{h\nu} 100$$

R = Me, Et, CH$_2$Ph

100

Scheme 45   Photochemical reaction of glycine esters to C$_{60}$.

NMR spectroscopy) executed. Reaction of iminodiacetic methyl ester also results in the formation of compound **100**. These ester derivatives are difficult to hydrolyze in excess mineral acids. The *tert*-butyl ester of compounds **101** can be hydrolyzed with excess trifluoromethanesulfonic acid. In contrast, by the photoaddition of various glycine esters to C$_{60}$ only compounds **100a–c** were isolated (Scheme 45). Therefore, the reaction mechanism for the photochemical and the thermal pathway must be different. Again, the reactivity of the glycine esters increases from the benzyl to the methyl ester probably due to steric hindrance.

Gan et al. first proposed a radical mechanism without singlet oxygen involvement [268a]. Based on further studies they suggested an novel air assisted radical reaction mechanism, in which both C–N bond breaking and bond formation processes are involved. Recently, Foote et al. [270] confirmed the involvement of singlet oxygen by a control experiment. The reaction of the glycine ester with C$_{60}$ was performed thermally under the same reaction conditions using DMNO$_2$ as the singlet oxygen source resulting in the formation of a product with identical ^{1}H-NMR and ^{13}C-NMR spectra to that reported by Gan et al. The reaction mechanism for the sarcosine reaction is outlined in Scheme 46. In the first step three radicals are formed: the aminium radical (a) arising from single electron transfer from the amino group to C$_{60}$ and the two carbon centered radicals (b,c) produced by the interaction with ^{1}O$_2$, which is generated by C$_{60}$ efficiently. Addition of the carbon centered radicals to the fullerene is followed by electron transfer from the amine to the C$_{60}$ fragment to form the zwitterionic radical (step 3). The C–N bond cleavage and the release of glycine at step iv is unusual and may be addicted to the unique properties of C$_{60}$. Moderate heating is probably necessary due to this endothermic C–N bond breaking step. The driving force for the rearrangement in steps vi and xii may be the more stable five-membered ring. Besides the formation of the main products **100** and **101** several other products are possible such as C$_{60}$(CH$_2$), C$_{60}$(HCCOOR), and the amine addition product C$_{60}$(H)(CH$_3$NCH$_2$COOR). Photolysis of aminopolycarboxylates such as tetramethyl ethylenediaminetetraacetate (EDTA) **102** and pentamethyldimethylenetriaminepentaacetate (DTPA) **106** with C$_{60}$ has been shown to be an effective method for the preparation of isomerically pure pyrrolidinofullerene derivatives

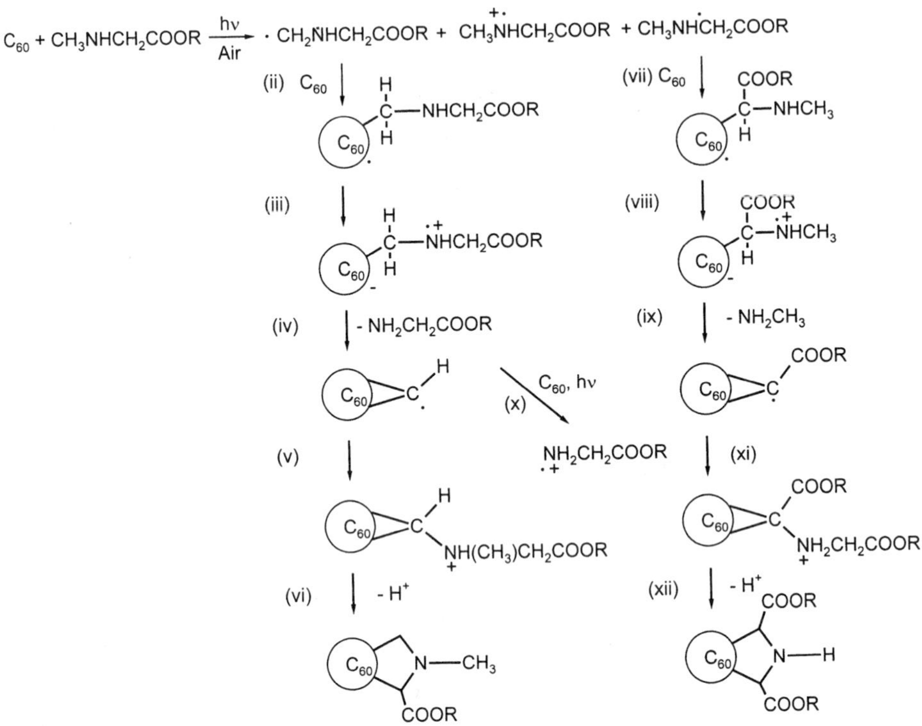

$C_{60} + CH_3NHCH_2COOR \xrightarrow[\text{Air}]{h\nu} \cdot CH_2\overset{+\cdot}{N}HCH_2COOR + CH_3\overset{+\cdot}{N}HCH_2COOR + CH_3N\overset{\cdot}{H}CH_2COOR$

Scheme 46   Mechanism for the photochemical reaction of sarcosine esters 98 with C_{60}.

that can be used as complexones or polydentate ligands [232]. The presence of free acid functionalities leads to decarboxylation and 1-alkylated 1,2-dihydro [60]fullerenes are formed as described by Zhang et al. [231]. In the case of esters, pyrrolidine-fused fullerenes, are the major products. All reactions involve radicals as described for the photoaddition of glycine and sarcosine esters. Unlike primary amino acid esters that undergo complicated bond-breaking and formation processes, amino acid esters with secondary and tertiary amino groups add to C_{60} by simply losing two H atoms. In comparison to the well known synthesis of pyrrolidine ring-fused fullerene adducts in the above reaction the use of aldehydes or ketones can be avoided. Table 12 gives a short overview of fullerene derivatives obtained by the photoreaction of EDTA and DTPA.

Banks et al. reported the formation of the expected [6,6]-closed aziridino-fullerene by the reaction of C_{60} and singlet oxycarbonylnitrenes [271]. In addition, small amounts of unexpected closed [5,6]-closed adducts were postulated [272]. In further investigations, Banks et al. studied the irradiation of C_{60} and ethyl azid-oformate 109 in benzene leading to a 4:1 mixture of two major products (Scheme

Table 12 Products of the Photolysis Aminopolycarboxylates With C_{60}

Starting Material	Products

100

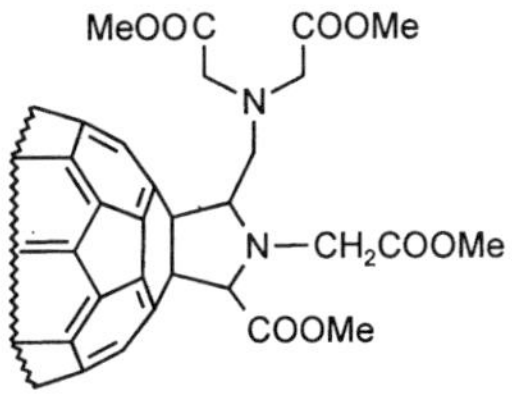

103

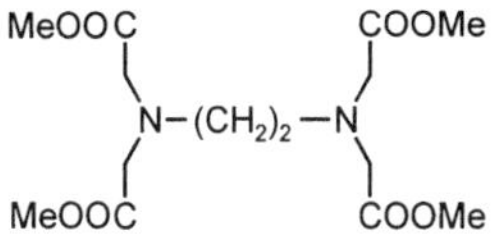

102

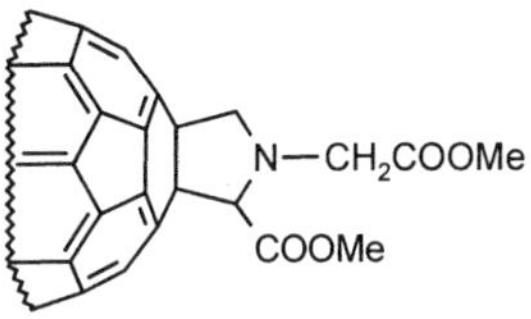

104

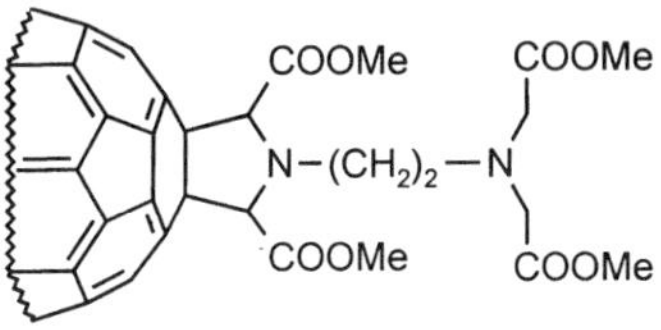

105

100

103

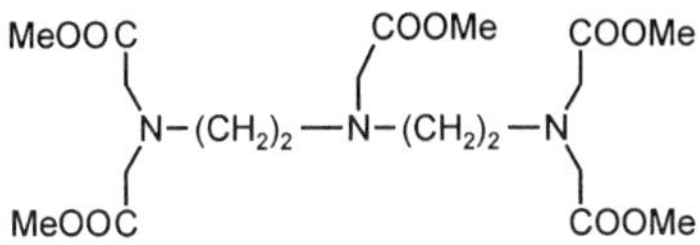

106

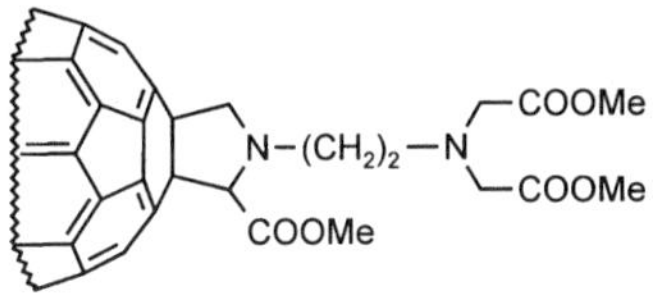

107

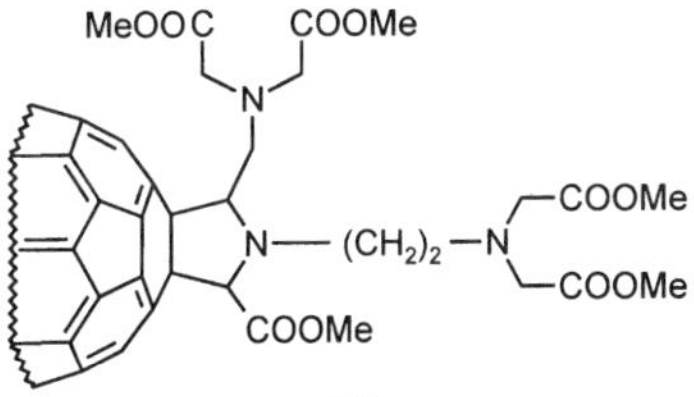

108

Source: Ref. 232.

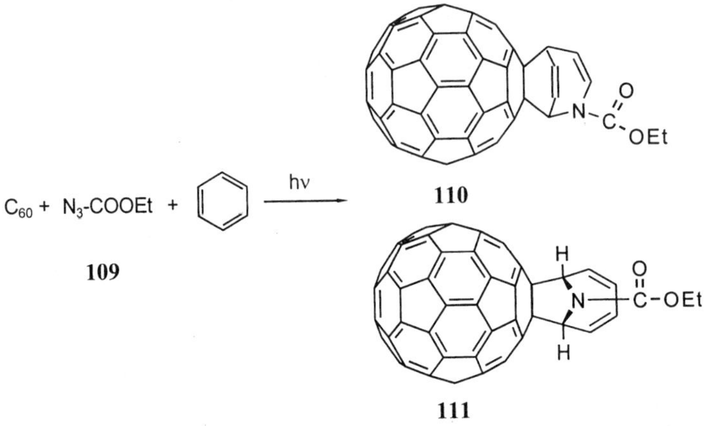

Scheme 47   Photolytic cycloaddition reaction of azepines with C_{60}.

47) [273]. Mass spectral analysis showed that the two products were isomeric and correspond to an adduct of C_{60} combined with benzene and photolytically generated ethoxycarbonylnitrene. Control experiments confirmed the formation of N-ethoxy-carbonylazepine which underwent further photochemical [2 + 4] and [2 + 6] cycloaddition reactions with C_{60} leading to the observed products.

Whereas the major product is the [2 + 4] cycloadduct **110,** the minor isomer **111** is the product of a [2 + 6] cycloaddition reaction. The structures of the two isomers were determined by NMR investigations.

The addition of the azepine to the photoexcited triplet state of C_{60} is suggested to be involved in the reaction. A diradical intermediate as shown in Fig. 33

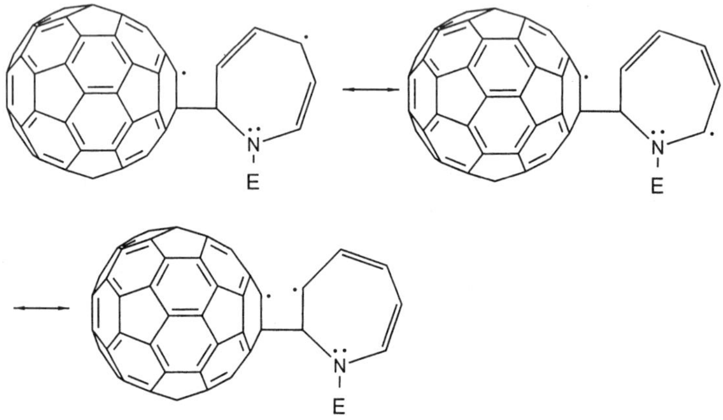

Figure 33   Mechanism for the [2 + 4] and [2 + 6] photocycloaddition of azepine to C_{60}.

may be involved preferentially leading to **110.** No formation of the [2 + 2] cycloadduct was observed probably due to the fact that the formation of a four-membered ring is energetically less favored than the larger rings.

3. Heterocycles Containing Silicon

It has been well established that virtually all silanes that contain a particularly
weak Si–Si bond, such as $Ph_3SiSiMe_3$ and those containing other bulky substituents, are cleaved by 604 nm UV radiation to give silyl radicals which add,
e.g., to fullerenes. In other studies the reaction was performed with siliranes
[274], disilanes [275], trisilanes [276], and cyclosilanes [277]. Trisilane **112** as
a silylene precursor was photolyzed with a low pressure mercury lamp in a
toluene solution of C_{60}. Upon irradiation cleavage of the trisilane **112** leads to
the formation of bis(2,6-diisopropylphenyl)silylene **113** which readily adds to
the fullerene core leading to a [6,6]-closed silicon adduct **114** with C_{2v} symmetry analogous to the methanofullerenes (Scheme 48) [276a]. The formation of
the [6,6]-open structure can be excluded based on NMR spectra which shows 17
signals for the C_{60} skeleton in the ^{13}C NMR expected for an adduct with C_{2v}
symmetry and one signal at $\delta = -72.74$ in the ^{29}Si NMR typical for a silicon
atom on a silirane ring.

   Irradiation of C_{70} under the same conditions leads to a mixture of two [6,6]-
closed isomers **115a** and **115b** in a 2:1 ratio having both C_s symmetry (Fig. 34)
[276b]. Comparison of the NMR data for the two isomers **115a** and **115b** with the
number of possible peaks for the a,b-, c,c-,d,e-, e,e-isomers suggests that **115a** and
115b are the a,b- and c,c-isomers. Theoretical investigations strongly support the
higher reactivity of the a–b double bond located at the pole region of C_{70} and the
preferred formation of the silirane structure rather than the isomeric 1,6-sil-
amethano [10]annulene structure.

   Irradiation of a toluene solution of 1,1,2,2-tetramesityl-1,2-disilirane **116**
resulted in the formation of the [6,6]-closed 1,1,3,3-tetramesityl-1,3-disilolane
117 with C_s symmetry (Scheme 49) [274a]. The reaction is surpressed by addition
of DABCO and 1,2,4,5-tetramethoxybenzene and completely inhibited by addition of rubrene. As rubrene is known to be an effective triplet quencher, there is

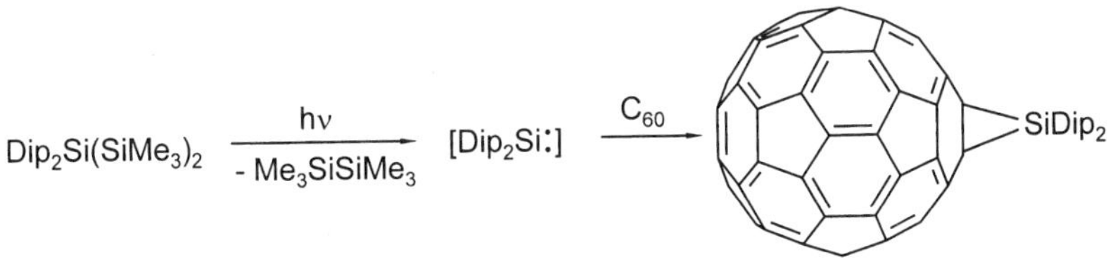

Scheme 48   Irradiation of C_{60} and trisilane 112 (Dip = 2,6-diisopropoylphenyl-).

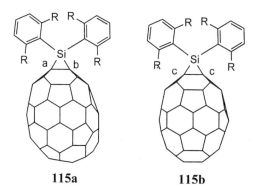

115a   **115b**

Figure 34   Products of the irradiation of C_{70} and trisilane 112.

evidence for the reaction to proceed via an exciplex derived from the silirane **116** and the triplet state of C_{60}.

The corresponding [3 + 2] cycloadduct of C_{70} was also obtained under the same reaction conditions [274b]. The equatorial bonding for this compound having C_2 symmetry, in which either 1,2-adduct or 1,4-adduct structures are conceivable has been elucidated by NMR investigations and confirmed by AM1 molecular orbital calculations showing that these C_2 structures correspond to energy minima. Addition of bis(alkylidene)disilacyclobutanes **118** should yield the corresponding C_{60} annulated disilacyclohexane derivatives **120,** but instead adducts **119a–b** are formed resulting from an unexpected rearrangement of the disilacyclobutane moiety (Scheme 50) [276a]. The structures of **119a–b** were determined by spectroscopic methods, including ^{29}Si-^{1}H HMBC hetero nuclear shift correlation experiments. The reaction mechanism still remains unclear but two alternative reaction pathways are suggested. Since rubrene completely inhibits the reaction, triplet excited C_{60} might be involved in the course of the reaction.

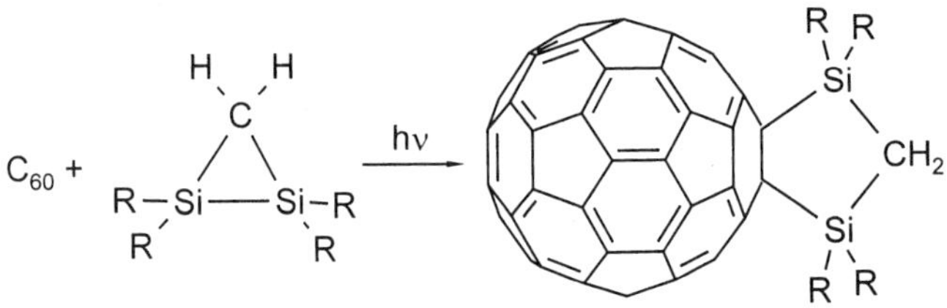

116   **117**

Scheme 49   Photolysis of C_{60} and silirane 116.

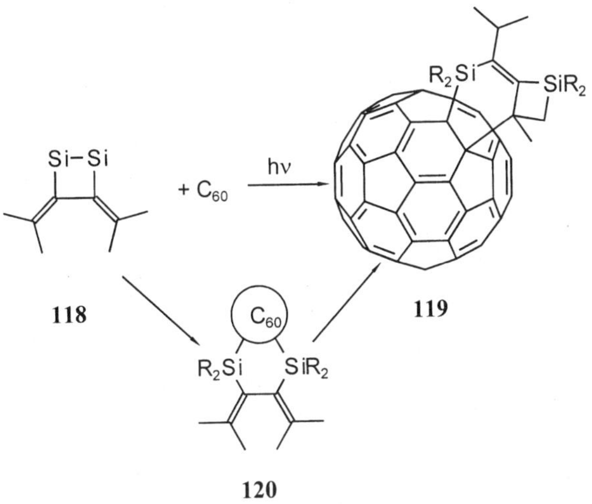

Scheme 50 Irradiation of bis(alkylidene)disilacyclobutanes **118** to C_{60}.

Alternatively, the initially formed C_{60} annulated disilacyclohexane derivative may rearrange under the applied photolytic conditions.

Photolysis of *tert*-butyl-substituted disilanes **121a–d** with C_{60} result in the formation of 1,16-adducts **122** (Scheme 51). The unusual 1,2 products **123** and **125–27,** in which the silyl and the phenyl groups are attached on the 1,2-postions

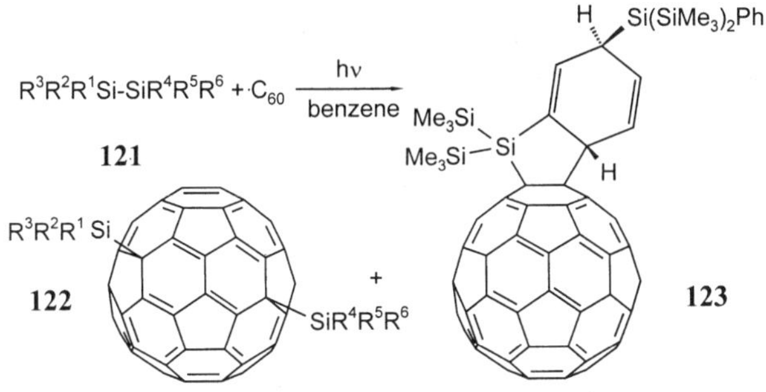

a: $R^1 = R^4 = $ t-Bu, $R^2 = R^3 = R^5 = R^6 = $ 4-Me-C_6H_4
b: $R^1 = R^2 = R^4 = R^5 = $ t-Bu, $R^3 = R^6 = $ Ph
c: $R^1 = R^4 = $ t-Bu, $R^2 = R^3 = R^5 = R^6 = $ Ph
d: $R^1 = R^4 = $ t-Bu, $R^2 = R^3 = $ Ph, $R^5 = R^6 = $ 4-Me-C_6H_4
e: $R^1 = R^2 = R^4 = R^5 = $ Me$_3$Si, $R^3 = R^6 = $ Ph

Scheme 51 Photolysis of disilanes **121a–e** and C_{60}.

of C_{60}, are also obtained in the reactions of **121e** and **124** with C_{60} (Scheme 52) [275b]. A free radical process is suggested for these conversions on the basis of experiments with other disilanes and oligosilanes.

All the reactions may involve attack of a phenylsilyl radical formed upon ir-radiation on C_{60} to give the radical intermediate as shown in Scheme 53. There is probably a rapid equilibrium between this radical and the cyclohexadienyl radical. The product ratio depends on the relative rates of recombination with the other sil-icon radical produced initially. Bulky phenylalkylsilyl radicals react preferentially at the 16-position of the C_{60} polycycle in the initially formed radical. In contrast, trimethylsilyl and silylsilyl radicals prefer the cyclohexadienyl ring.

Further investigations on the photochemical bissilylation of C_{60} with disi-lanes [275c] indicate that $^3C_{60}^*$ and $C_{60}^{\cdot-}$ do not play an important role in these re-actions due to the fact that no absorption band for the C_{60} radical anion is observed in the course of the reaction and the decay of $^3C^*_{60}$ is not accelerated by addition of the disilanes. Another hint for the involvement of the phenylsilyl radical is that no product is formed upon irradiation at >300 nm where the cleavage of the dis-ilane does not take place. The radical reaction pathway is further confirmed by a lower yield obtained by addition of radical scavengers.

The photochemical reaction of the aryl substituted cyclotetrasilane **128a** with C_{60} afforded stable 1:1 adducts **129a** and **130a**, the latter was obtained from a rearrangement of the cyclotetrasilane unit (Scheme 54) [277]. In case of the cy-clotetrasilane **128b**, only the rearranged product was obtained. Reaction of the cy-

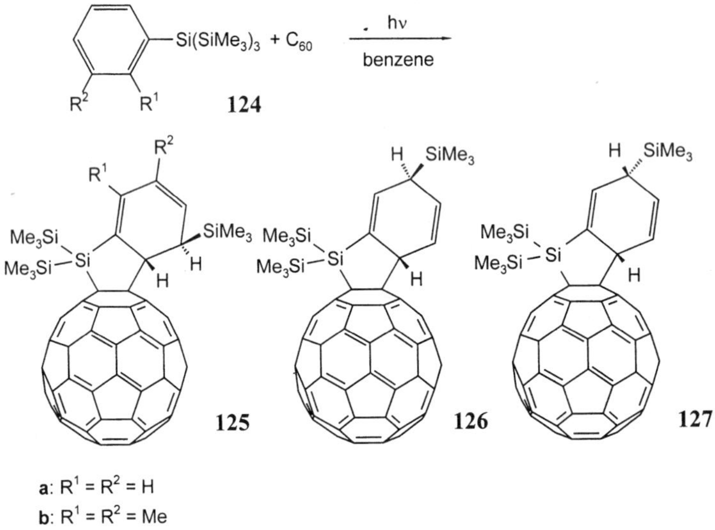

a: $R^1 = R^2 = H$
b: $R^1 = R^2 = Me$

Scheme 52   Photolysis of oligosilanes 124 and C_{60}.

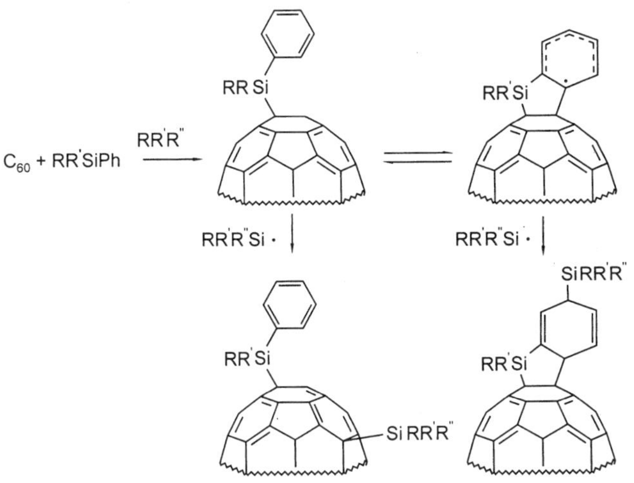

Scheme 53 Possible mechanism for the reaction of C_{60} and disilanes 121.

clotetragermane led to the germanium analogues. Thermal or catalyzed ring open-ing may be expected to provide fullerene-substituted polysilanes and polyger-manes.

As a control experiment the irradiation of the cyclosilanes was performed at $\lambda > 300$ nm in the presence of CCl_4, to give 1,4-dichloro-1,1,2,2,3,3,4,4-oc-taphenyltetrasilane and 1,3-dichloro-1,1,2,2,3,3-hexaphenyltrisilane. Under iden-tical photolytic conditions the cycloadduct 129a did not rearrange to 130a indi-cating that a biradical might be involved in the course of the reaction as depicted in Scheme 55. Compound 129a is quite stable toward ionic and radical reagents,

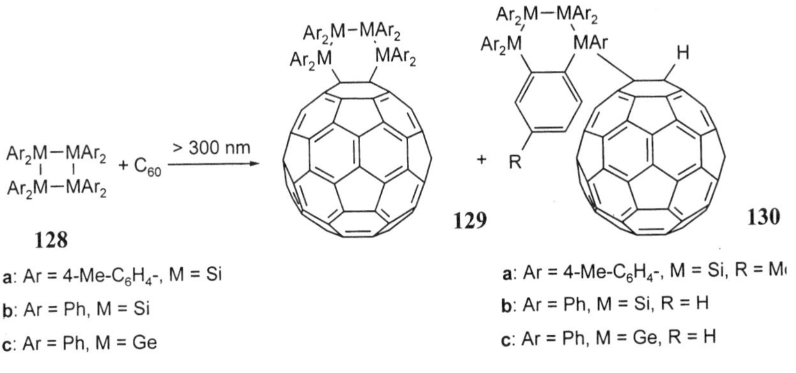

Scheme 54 Photolysis of C_{60} with cyclosilanes 128.

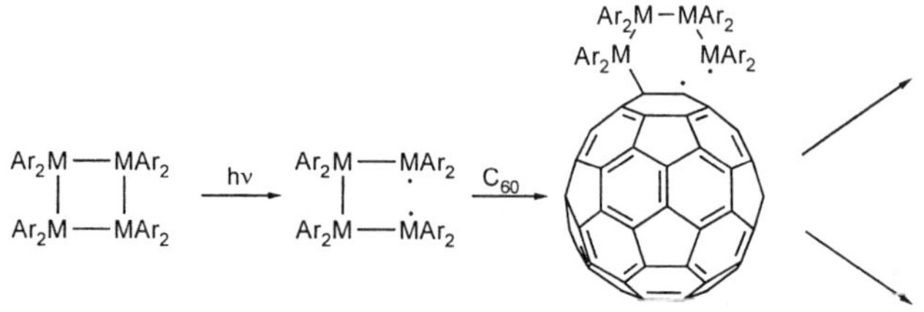

Scheme 55 Mechanism for the photolysis of cyclotetrasilanes to C_{60}.

thus reaction of the silicon unit is not observed by treatment with di-*tert*-butyl peroxide, benzoyl peroxide, chlorine, and MeLi. On the other hand, the silicon-fullerene bond is cleaved upon irradiation in the presence of bromine leading to C_{60} and compound **131** (Scheme 56) [278].

The most likely course of this conversion involves H abstraction by bromine atoms. The resulting radical may undergo homolysis of the fullerene–silicon bond as outlined in Scheme 57. The silyl radical thus formed then undergoes intramolecular cyclization to give **132**. While this type of intramolecular reaction readily occurs with radical species, it is not a common one in silicon ring systems. The Si–Si bond of **132** then must react with bromine followed by hydrolysis to give siloxane **131**.

E.   Photochemistry of C_{60} and C_{70} Derivatives

1.   Addition of Singlet Oxygen

Photophysical investigations of C_{60} and its derivatives have revealed that [60]fullerene and 1,2-dihydro[60]fullerene derivatives produce singlet oxygen in high quantum yields. Consequently, the fullerene compounds are used as photo-

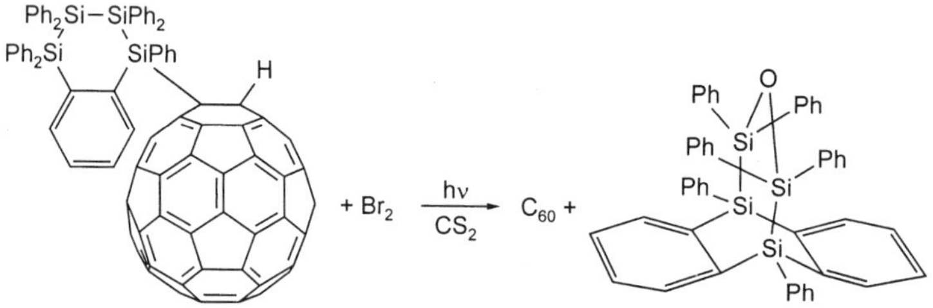

Scheme 56 Reaction of bromine with the C_{60} cyclosilane adduct 129a.

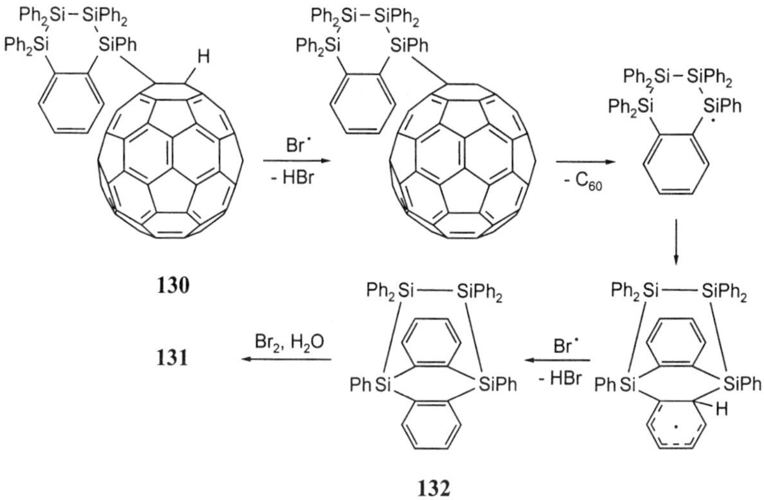

Scheme 57 Mechanism for the reaction of bromine with the C_{60} cyclosilane adduct 129a.

sensitizers in ene and Diels-Alder-type reactions of olefins and dienes to give the photooxygenation products [279]. Rubin et al. reported the photo-oxygenation of C_{60}-fused cyclohexenes **133** leading in almost all cases to the formation of allylic alcohols **135** with endocyclic double bonds after reduction of the corresponding hydroperoxides **134** with PPh_3 (Scheme 58) [280]. In all cases, the fullerene moiety acts as a $^1O_2$-sensitizer.

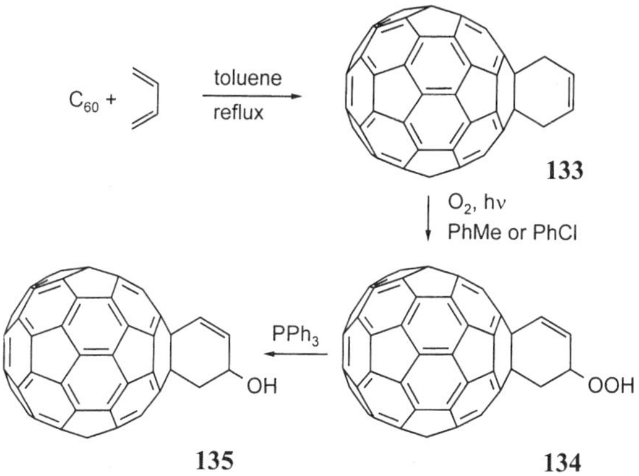

Scheme 58 Photo-oxygenation of C_{60}-fused cyclohexenes.

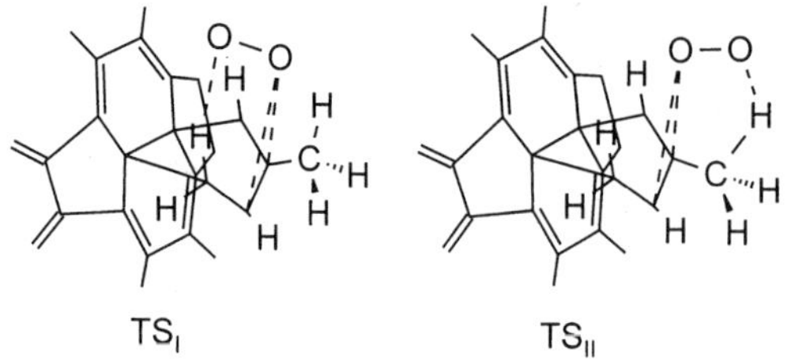

Scheme 59 Transition states in the photo-oxygenation of C_{60}-fused cyclohexenes.

In contrast, reaction of alkyl substituted cyclohexenes yields a majority of products with an exocyclic double bond[281]. The unexpected regioselectivity is explained in conformational and electrostatic terms. The boat conformation of the C_{60}-fused cyclohexenes confers a rigid framework to the six-membered ring in which two of the 3,6 hydrogens are pseudoaxial. Because of favorable interactions between the incoming $^1O_2$ and these axial hydrogens, the *endo*-perepoxide resulting from transition state **TS$_I$** is formed rather than the *exo*-isomer derived from transition state **TS$_{II}$** (Scheme 59). Additional favorable electrostatic or electronic interactions may exist between the negative oxygen of the developing endoperepoxide I and C_{60}, which is strongly electron deficient.

Oxidation of C_{60} cycloadducts with singlet oxygen results in ring-opening or epoxide formation, although the proposed epidioxy intermediates could not be isolated. Foote et al. studied the self-sensitized photo-oxygenation of various fullerenyl cyclobuteneamines synthesized by [2 + 2] photocycloaddition of ynamines to C_{60} or C_{70} (Scheme 60) [239,248]. These fullerene adducts are unique in that they have a photosensitizer (the dihydrofullerene) and a photo-oxidizable group (the enamine) in the same molecule. Photo-oxidative cleavage of enamines is a well-known process and proceeds via an intermediate 1,2-dioxetane [282], which is extremely unstable and cleaves to ketone and amide fragments. The products obtained by the self-sensitized photo-oxygenation are summarized in Table 13.

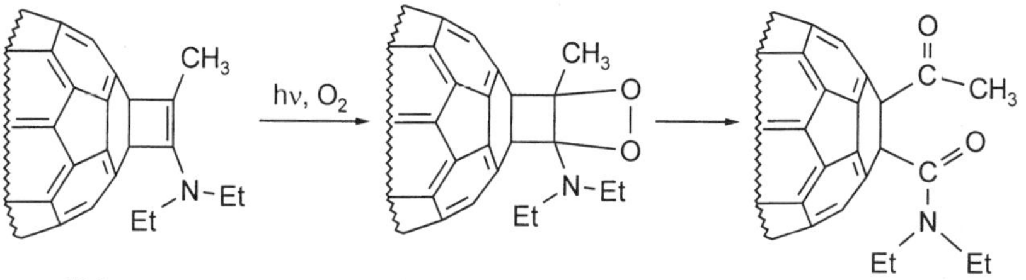

Scheme 60 Self-sensitized photo-oxygenation of fullerenyl cyclobutenamine 136.

Table 13 Products of Self-Sensitized Photo-oxygenation of Cyclobuteneamines

Starting material		Product	

Source: Refs. 239, 248.

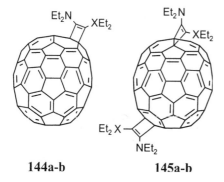

144a-b   **145a-b**

Figure 35 Structures of C$_{70}$ derivatives obtained upon irradiation of C$_{70}$ and ynamines 138 and 139.

The 1,2-dihydro-[60]fullerene-1,2-diamide and thioester amide, obtained by photochemical oxidation of the [2 + 2] photocycloadduct of C$_{60}$ with *N,N,N′,N′*-tetraethylethynediamine **138** and *N,N*-diethyl-2-ethylthioethyneamine **139,** can be further derivatisized to the acid anhydride C$_{62}$O$_3$ upon treatment with *p*-toluenesulfonic acid. Irradiation of a 1:1 mixture of C$_{70}$ with **138** or **139** leads regiospecifically to the addition at the 1,9-bond of C$_{70}$ yielding compound **144a** or **144b** (Fig. 35). Using 2.2 equivalents of the ynamines results in the formation of a mixture of two isomeric bis-adducts **145a–b.** The corresponding C$_{70}$ acetic anhydride can not be obtained via the initially formed [2 + 2] cycloadduct. Instead, photocycloreversion to C$_{70}$ and tetraethyl oxamide is observed. One pot reaction of *N,N*-diethyl-*tert*-butylthioethyneamine with C$_{70}$ and photo-oxygenation of the resultant cycloadduct gives C$_{72}$O$_3$ [239a].

The cycloadducts from the tetraalkoxyethylenes **148–150** (Fig. 36) undergo cycloreversion through the triplet-excited state, giving back C$_{60}$ and C$_{70}$ cleanly

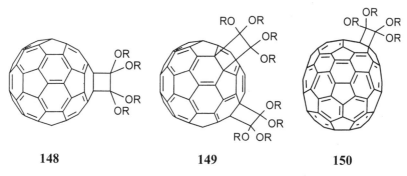

148   **149**   **150**

Figure 36 Structure of dihydrofullerene cyclobutanediketals.

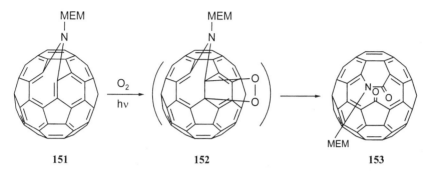

Scheme 61 Self-sensitized photo-oxygenation of N-(MEM)-substituted [5,6]-aza-fulleroid 151.

and efficiently on visible or UV light irradiation. Efficient inhibition of the cycloreversion upon addition of rubrene, a triplet quencher, confirmed that triplet excited states are involved in the reaction.

Wudl et al. [283] obtained the ring-opened N-MEM-ketolactam **153** by self-sensitized photo-oxygenation of N-methoxyethoxymethyl (MEM)-substituted [5,6]azafulleroid **151** (Scheme 61). The reaction is highly regioselective, most likely because the anti-Bredt carbon–carbon double bonds in **151** are more strained.

Eguchi et al. [284] studied the [4 + 2] cycloaddition of C_{60} with heteroaromatic analogues of o-quinodimethane including furan, thiophene, oxazole, thiazole, indole and quinoxaline to give the corresponding heterocycle-linked [60]fullerenes. Among them, the furan and oxazole derivative are found to be oxygen-labile. Intended self-sensitized photo-oxygenation yielded epoxy-γ-lactone **158** from the former and diester **164** from the latter after methanolysis and hydrolysis. According to the reported menthofuran case [285] which proceeds through collapsing of an endoperoxide cycloadduct, it would be reasonable to estimate that the primary product formed is a 4-hydroxybutenolide such as **156** (Scheme 62). In this case, however, the rigid framework of C_{60} seems to force the hydroxyl group on the butenolide ring to be eclipsed with the p-orbital of C_2–C_3 double bond. Therefore, the epoxide **158** may instead be formed. Contact with silica gel might assist the conversion of 4-hydroxybutenolide to 3,4-epoxybutenolide.

Wasserman et al. [286] reported a closely related reaction of an oxazole as shown in Scheme 63. The major product was formed via [4 + 2] cycloaddition of singlet oxygen, followed by rearrangement to 5-cyanopentanoic benzoic anhydride **161** and methanolysis to the cyano ester **162**. In the case of the oxazole-linked [60]fullerene, however, the treatment with methanol for a long time to ensure methanolysis might result in further conversion of the once formed cyano

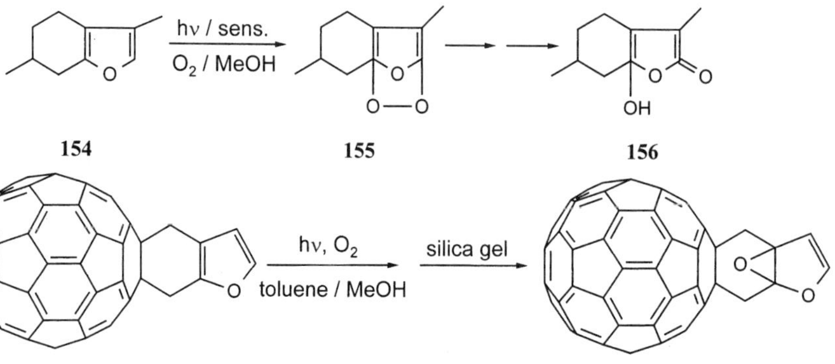

Scheme 62   Self-sensitized photo-oxygenation of the furan derivative 157.

ester into an imidate ester with catalytic action of benzoic acid which should be formed as a byproduct on methanolysis. Although this proposal has not been checked, the final product diester **164** could be produced probably by hydrolysis during silica gel chromatography.

In Diels-Alder adducts of 9,10-dimethylanthracene with C_{60}, the anthracene moiety can be removed by irradiation with light in the presence of oxygen to yield the anthracene-endoperoxide [287]. Irngartinger et al. reported the synthesis of stable endoperoxides **168–170** of fullerene derivatives **165–167** substituted by an anthracenyl group [288]. The anthryl groups of the fullerene derivatives react readily with singlet oxygen to form the 9,10-endoperoxides under irradiation. The results are summarized in Table 14. The fullerene group and the anthryl group

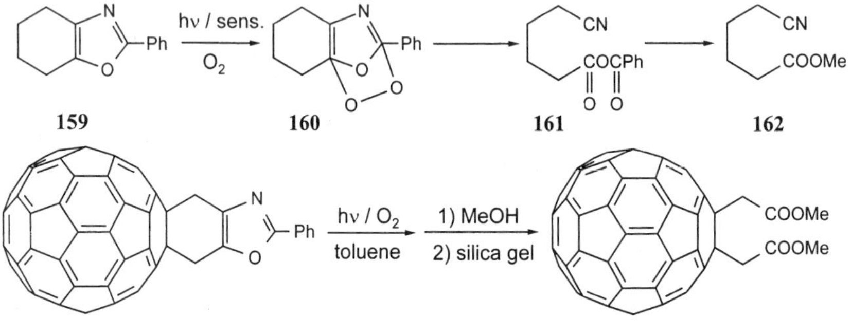

Scheme 63   Self-sensitized photo-oxygenation of the oxazole derivative 163.

Table 14 Products of the Irradiation of Fullerene Derivatives 165-167 Substituted by an Anthryl Group

Starting Material	Product of Photo-oxygenation

moiety seem to be photosensitizers for singlet oxygen generation, since the precursors 4-(9-anthryl)benzaldoxime and 9-anthrylacetic acid are also photo-oxidized in the absence of the fullerenes.

2. Secondary Derivatization Leading to Methanofullerenes/ Fulleroids and Aziridino-Fullerenes/Azafulleroids

A reliable and versatile route to functionalized fullerenes has been found via the addition of diazo compounds [289]. This route offers the possibility of obtaining

171 172 173

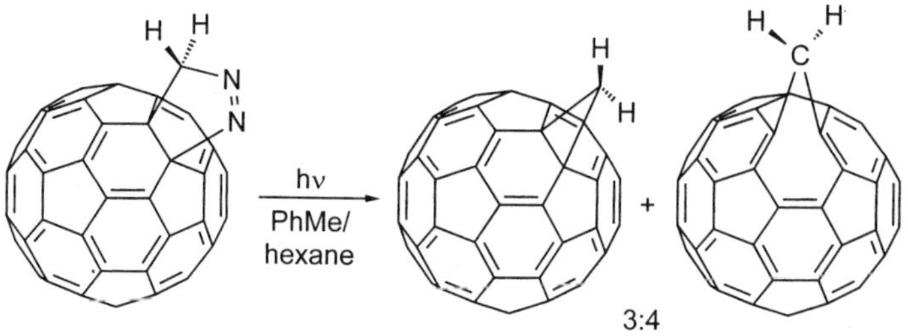

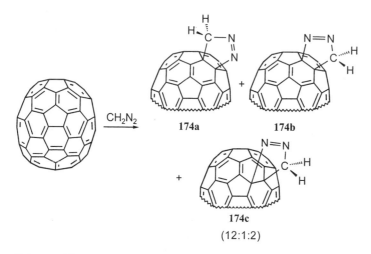

Scheme 64 Irradiation of the diazomethane adduct of C_{60}.

monoaddition products across a 5,6 ring junction (fulleroid) or a 6,6 ring junction (methanofullerene). Parent cyclopropanes, annulenes, and related derivatives of C_{60} and C_{70} have been synthesized by Smith et al. [290]. Irradiation of the C_{60} diazomethane adduct **171** resulted under N_2 extrusion in the formation of the [6,6]-closed cyclopropane $C_{61}H_2$ **172** accompanied by the [5,6]-open fulleroid **173** (Scheme 64). The two isomers can not be converted neither photochemically or thermally [95].

Addition of diazomethane to a toluene solution of C_{70} generated a 12:1:2 mixture of pyrazolines. The structures of the three isomers are shown in Scheme 65. The formation of **174a** and **174b** in unequal amounts appears to be a well-characterized regioselective transformation at a fullerene bond. Photolysis and ther-

174a 174b 174c

(12:1:2)

Scheme 65 Addition of diazomethane to C_{70}.

molysis of the pyrazolines generated four isomeric methylene derivatives of C_{70}. Irradiation of the pyrazoline mixture through Pyrex resulted in the formation of two isomers in a 7:1 ratio, the major arising from pyrazoline **174a** and being the cyclopropane derivative **177** as determined by the obtained NMR data. Although the data for isomer **178** are consistent with the [6,6] cyclopropane structure, the isomeric [5,6] cyclopropane can not be excluded. Upon thermolysis the isomers **175** and **176,** the major again arising from isomer **174a.** The two reaction pathways are summarized in Scheme 66.

Mattay et al. synthesized triazolinofullerenes **180** by the thermal [3 + 2] cycloaddition of aryl and sulfonyl azides **179** with C_{60} [263,291]. Thermal extrusion of N_2 predominantly leads to the formation to the opened [5,6]-bridged azafulleroids **182,** whereas the major products formed by photolysis of triazolinofullerenes are the closed [6,6]-bridged aziridinofullerenes **181** (Scheme 67) [292,293].

In 1995, Wudl et al. reported that thermal or photochemical treatment of fulleroids led to the conversion to methanofullerenes. Irradiation of fulleroid **183** in deoxygenated p-xylene at ambient temperature with light from an Ar laser (480nm, 50 mW, 10 mm path length) gives rise of a quantitative conversion to the methanofullerene **184** (Scheme 68) [238]. Conversion of the methanofullerene to the fulleroid is not observed. The conversion is suggested to proceed via a di-π-methane-rearrangement, described by the following mechanistic scheme (Scheme 69) [294]. In acyclic systems, the di-π-methane-rearrangement is known to proceed preferentially via the singlet state. Rigid (bicyclic) molecules are known to react via the triplet state. Based on this general rule and the fact that intersystem crossing in C_{60} is very efficient, it is considered that **183** rearranges via the triplet

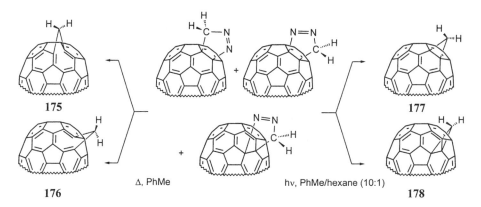

Scheme 66 Products of the thermolysis and photolysis of the mixture of pyrazolino[70]fullerenes.

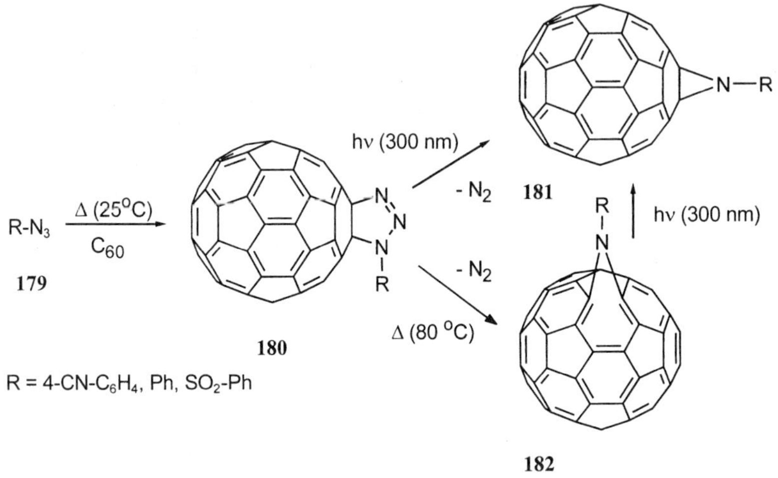

Scheme 67   Synthesis of aziridinofullerenes 181 upon irradiation of triazolinofullerenes 180.

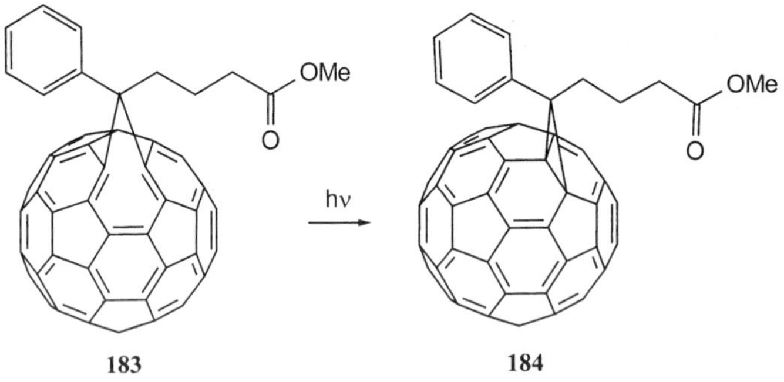

Scheme 68   Conversion of fulleroids to methanofullerenes upon irradiation.

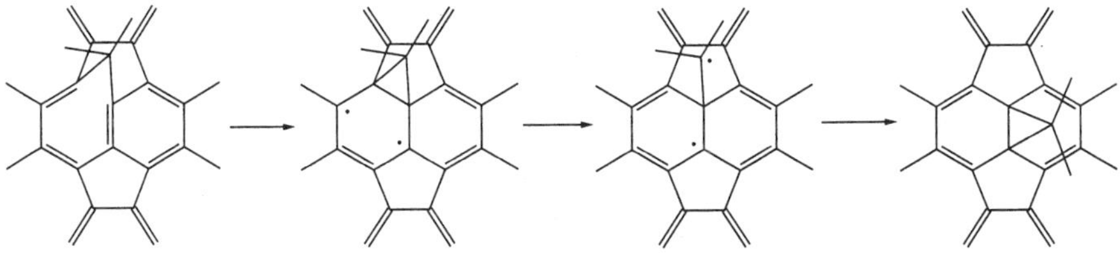

Scheme 69   Mechanism for the conversion of fulleroids to methanofullerenes.

state. In addition, performing the reaction in air-saturated *p*-xylene leads to a decrease of the rate of conversion by a factor more than 20, due to triplet quenching by oxygen. The conversion for [5,6]-$C_{61}H_2$ does not take place, consistent with the general requirement for the di-π-methane-rearrangement that the methane carbon carries substituents. Evidently, the intermediate 1,3-diradical will be stabilized by the phenyl ring.

Wudl et al. also observed the photochemical rearrangement of an azafulleroid, namely [5,6]-*N*-MEM-azafulleroid **151,** leading to a 1,2-isomer, the aziridinofullerene [283]. When the obtained aziridinofullerene is irradiated, no photochemical isomerization with or without oxygen could be detected. This indicates a fundamental equivalence in reactivity between aziridinofullerenes and methanofullerenes. Mattay et al. also reported the rearrangement of aryl-azafulleroids **182** [263]. Irradiation of azafulleroids **182** in a deoxygenated toluene solution with light of 300 nm wavelength showed complete conversion to aziridinofullerenes **181.** Sulfonyl-azafulleroids can also be converted to the corresponding [6,6]-bridged aziridinofullerenes by photochemically induced rearrangement [291]. The photochemically induced azafulleroid-aziridinofullerene rearrangement might be rationalized in terms of a di-π-methane-rearrangement, as reported for the fulleroids, rather than a cleavage-cycloaddition process of a nitrene unit as outlined in Scheme 70.

3. Secondary Functionalization of Fullerene Monoadducts by Cycloaddition Reactions

The aziridinofullerenes and azafulleroids obtained by Mattay et al. were subjected to further reactions to yield fullerene adducts of the second generation. As a model compound 4-methyl-1,2,4-triazoline-3,5-dione (NMTAD) **187,** which is one of the most reactive dienophiles for Diels-Alder reactions, was selected. Sheridan et al. studied the photoinduced [4 + 2] cycloadditions of NMTAD to arenes [295]. With alkenes, however, NMTAD reacts in photoinduced cycloadditions in a [2 + 2] fashion under formation of diazetidines [296]. Both the sulfonyl-azafulleroids (**185**) and aziridinofullerene (**186**) were irradiated at 420 nm to give both the same [2 + 2] cycloadduct **186** (Scheme 71) [291]. This indicates that at a certain stage rearrangements must be involved.

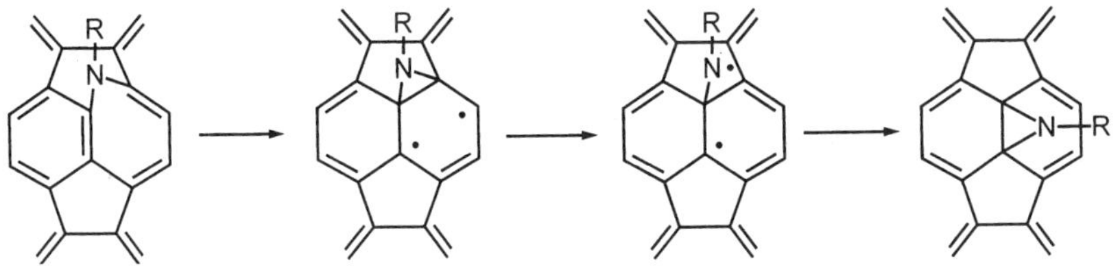

Scheme 70 Rearrangement of azafulleroids to aziridinofullerenes.

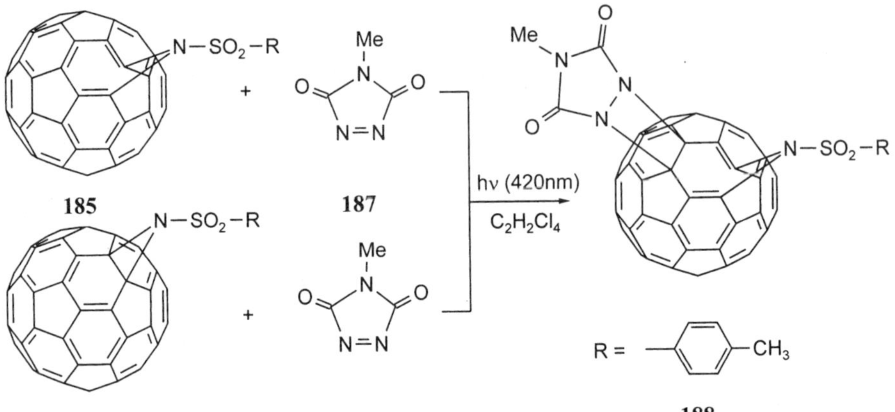

Scheme 71   Addition of NMTAD 187 to sulfonyl-azafulleroids/aziridinofullerenes.

Two routes for the formation of **188** by photochemically induced cycloaddition of NMTAD might be possible (Scheme 72). In the case of route A, first the opened [5,6]-bridged isomer **185** rearranges to the isomer **186**. After the photoinduced [2 + 2] cycloaddition of NMTAD to a *cis*-1 bond of the fullerene cage the azabridge must be rearrange back to the opened form. The other possibility might be a direct addition of NMTAD to the opened isomer **185** at a 3,4-bond at the fullerene cage (route B). As shown by control experiments the addition of NMTAD most probably follows route A.

Hummelen et al. [297] obtained a mixture of **190** and the photodimer **191** (Scheme 73) upon irradiation of the *m*-phenylenbis(arylmethanofullerene) **190** or the corresponding fulleroid **189** indicating that the new compound is formed from **190** not **189**. Formation of **190** from **189** can be explained by a rearrangement as described earlier, which then reacts further to give the final product. The UV–vis data can be assigned to a *cis*-3 regioisomer that is consistent with the formation of photodimer **191** from structure **190**. Structure **191** is in full agreement with the predicted unique dimerization site for two specific (enantiotopic) *cis*-3 C=C bonds in each fullerene moiety of **190**. The three different fullerenes **189, 190, 191** displayed identical behavior when solutions in ODCB were irradiated for 44h at 20°C. In all cases, mixtures of **190/191** of fairly identical composition (58/42 from **189**, 61/39 from **190**, 61/39 from **191**). This confirms the reversibility of the photodimerization process **190→191**. Control experiments showed that the dimerization process is significantly retarded by molecular oxygen, implying a mechanism involving the fullerene triplet excited state. A similar effect was reported for the photopolymerization process in thin solid films of C_{60} [205]. the photopolymerization is suggested to proceed via a [2 + 2] photocycloaddition.

As cyclohexadienes derivatives of C_{60} are not easy accessible, Cheng et al. studied the nickel promoted cycloaddition of bisalkynes **192** to C_{60} in order to obtain such derivatives (Scheme 74) [298]. Reports on the preparation of a cyclohexadiene ring via metal-catalyzed cyclotrimerization of two alkynes with an alkene have appeared in the literature. Indeed, treatment of C_{60} with terminal 1,6-diynes ($HC\equiv CCH_2)_2X$ **192a-h** in the presence of $NiCl_2(PPh_3)_2$, Zn and PPh_3 afforded [2 + 2 + 2] bicyclic hexadiene derivatives **193a-h** in good yields. Among them, the derivatives **193a-c**, **193f** and **193h** readily underwent photoinduced [4 + 4] cycloaddition upon irradiation (350 nm) to give the corresponding bisfulleroids **194a-c**, **194f** and **194h** in excellent yields (Scheme 75). The formation of **194** is likely to proceed via an intramolecular photoinduced [4 + 4] cycloaddition reaction of **193** to give an intermediate bismethanofullerene. This, upon subsequent carbon-carbon bond cleavage and rearrangement, yielded the bisfulleroid product. It should be noted that the methano groups in the intermediate are across [5,6]-ring junctions of the C_{60} moiety. The rearrangement of the intermediate to **194** is expected in view of the fact that a bisfulleroid is generally more stable than the corresponding methanofullerene. Examples of such rearrangement are known [97].

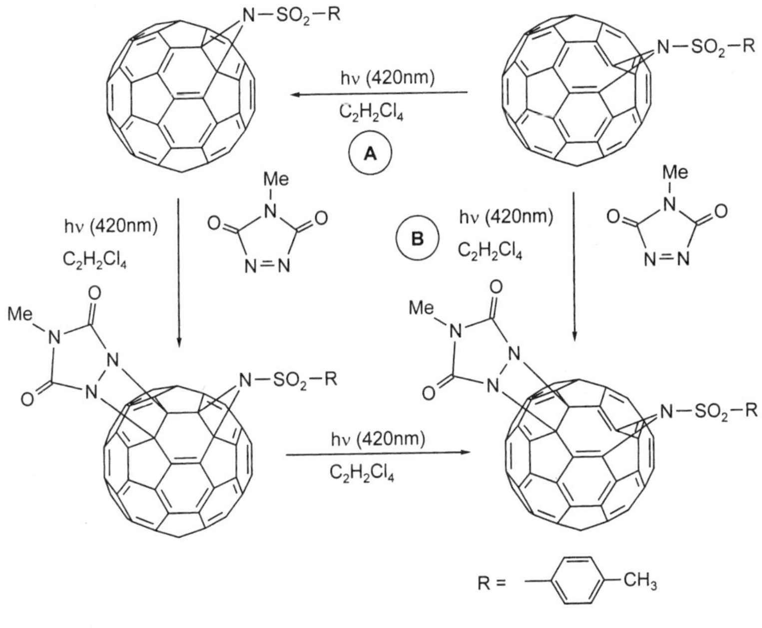

Scheme 72 Reaction pathways of the addition of NMTAD **187** to sulfonyl-azafulleroids/aziridinofullerenes.

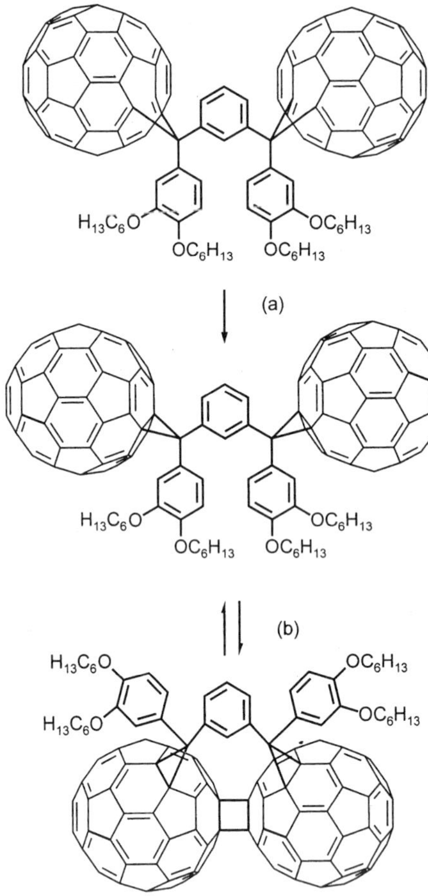

(a) ODCB, Δ
(b) ODCB, hν, 17 °C

Scheme 73 Formation of the photodimer 191 via [2 + 2] photocycloaddition.

Prior to the work of Cheng et al. Rubin et al. reported a similar [4 + 4] photocyclization reaction [299]. Rubin has shown that a cyclohexadiene derivative
underwent a very facile photochemically promoted rearrangement to the stable
bridged bisfulleroid **198** (Scheme 76). This process occurs via the initial [4 + 4]
photoadduct (not observed), which undergoes a thermally allowed [2 + 2 + 2] cycloreversion to afford a bis-methano[12]annulene structure **198.** Alternatively,
compound **198** can be obtained by photolysis of the allylic alcohol **197** under re-
flux and acidic conditions in good yields. The allylic alcohol is obtained by acidic

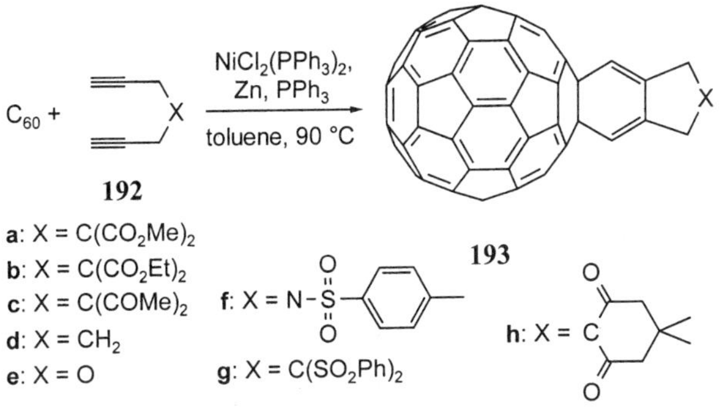
C60 +
X
NiCl2(PPh3)2,
Zn, PPh3
toluene, 90 °C
X
192
a: X = C(CO2Me)2
b: X = C(CO2Et)2
c: X = C(COMe)2
d: X = CH2
e: X = O
f: X = N-S
g: X = C(SO2Ph)2
193
h: X = C
O
O

Scheme 74 Formation of cyclohexadiene-fused C_{60} derivatives via [2 + 2 + 2] cycloadditions.

work-up of the cycloadduct **196** formed by addition of a two-fold excess of 1-((trimethylsilyl)oxy)-1,3-butadiene **195** to C_{60}. The structure of **198** was deduced from the high symmetry displayed by the ^1H and ^{13}C NMR spectra and the characteristic UV–vis spectrum. The X-ray structure of complex **199** (Fig. 37) formed by addition of $CpCo(CO)_2$ confirmed the suggested structure for compound **198.** Interestingly, the C1 and C59 bond of the five-membered ring adjacent to the bis-methano[12]annulene ring of **198** has been broken by oxidative insertion of the cobalt.

Rubin [300] also described the synthesis of C_{62}, a four-membered ring isomer of C_{60} (Scheme 77). The dicarbonyl bridged bisfulleroid **203** seems to be a

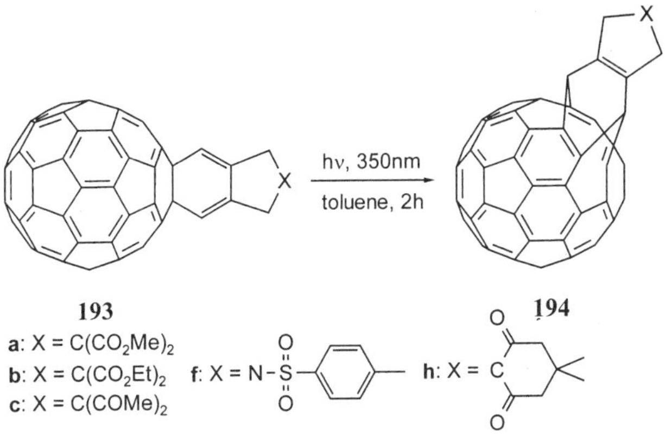
X
hν, 350nm
toluene, 2h
X
193
194
a: X = C(CO2Me)2
b: X = C(CO2Et)2
c: X = C(COMe)2
f: X = N-S
h: X = C
O
O

Scheme 75 Photolysis of cyclohexadiene-fused C_{60} derivatives.

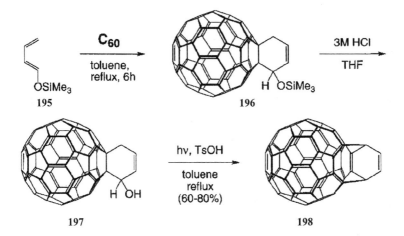

Scheme 76   Double scission of a six-membered ring on the surface of C_{60} via [4 + 4]/[2 + 2 + 2] cycloaddition reactions. (From Ref. 299.)

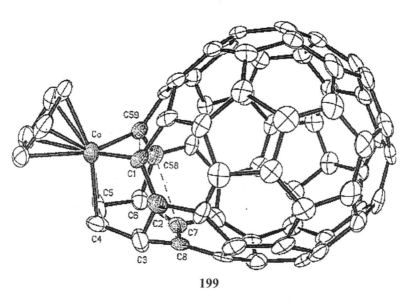

199

Figure 37   X-ray structure of the cobalt complex 199. (From Ref. 299.)

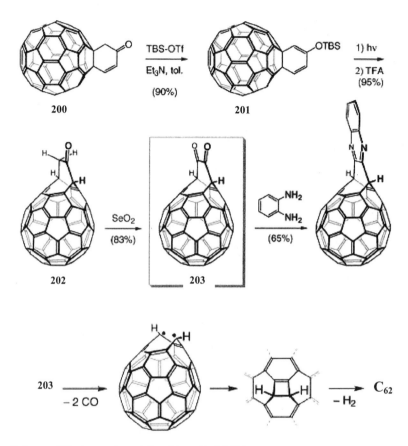

Scheme 77 Formation of C_{62}. (From Ref. 300.)

key precursor to C_{62}. Compound **203** was prepared by a stepwise strategy. Compound **200** [301] was converted to the corresponding silyl enol ether **201.** Visible light irradiation initiated its sequential [4 + 4]/[2 + 2 + 2] rearrangement to the bisfulleroid, which was deprotected with TFA to give the α-methylene ketone **202.** Oxidation of **202** results in the formation of the dicarbonyl compound **203.** Examples of photolytic α-fragmentation of 1,2-dicarbonyl compounds with loss of two CO units have been described [302]. In the case of C_{62}, removal of both carbonyl groups under photochemical or thermal excitation should result in two adjacent radicals combining to the four-membered ring. Subsequently, facile dehydrogenation that is characteristic for hydrofullerenes [303], would give C_{62}.

V. SUMMARY

In this chapter we have described the photophysics and photochemistry of C_{60}/C_{70} and of fullerene derivatives. On the one hand, C_{60} and C_{70} show quite similar photophysical properties. On the other hand, fullerene derivatives show partly different photophysical properties compared to pristine C_{60} and C_{70} caused by pertubation of the fullerene's π-electron system. These properties are influenced by (1) the electronic structure of the functionalizing group, (2) the number of addends, and (3) in case of multiple adducts by the addition pattern. As shown in the last part of this chapter, photochemical reactions of C_{60}/C_{70} are very useful to obtain fullerene derivatives. In general, the photoinduced functionalization methods of C_{60}/C_{70} are based on electron transfer activation leading to radical ions or energy transfer processes either by direct excitation of the fullerenes or the reaction partner. In the latter case, both singlet and triplet species are involved whereas most of the reactions of electronically excited fullerenes proceed via the triplet states due to their efficient intersystem crossing.

ACKNOWLEDGMENTS

Financial support was provided by the Federal Department of Science, Research and Technology (BMBF) and the Deutsche Forschungsgemeinschaft (DFG). We also gratefully acknowledge the valuable contributions of Johannes Averdung, Christina Siedschlag, Ingo Schlachter, Gregorio Torres-Garcia, Waldemar Iwanek, Lothar Dunsch, Osamu Ito, Peter Luger, Werner Abraham, and Dirk M. Guldi. Special thanks go to Heinrich Luftmann and Christian Wolff for their important contributions to Mass Spectrometry and NMR Spectrometry, respectively. Support of Hoechst AG (now Aventis) is also gratefully acknowledged.

REFERENCES

1. Kroto, H. W.; Heath, J. R.; O'Brien, S. C.; Curl, R. F.; Smalley, R. E. *Nature* **1985,** *318,* 162–163.
2. Krätschmer, W.; Lamb, L. D.; Fostiropoulus, K.; Huffmann, D. R. *Nature* **1990,** *347,* 354–358.
3. Hebard, A. F.; Rosseinsky, M. J.; Haddon, R. C.; Murphy, D. W.; Glarum, S. H.; Palstra, T. T. M.; Ramirez, A. P.; Kortan, A. R. *Nature* **1991,** *350,* 600–601.
4. Allemand, P.-M.; Khemani, K. C.; Koch, A.; Wudl, F.; Holczer, K.; Donovan, S.; Grüner, G.; Thompson, J. D. *Science* **1991,** *253,* 301–303.
5. Friedmann, S.; DeCamp, D. L.; Sijbesma, R.; Srdanov, G.; Wudl, F.; Kenyon, G. L. *J. Am. Chem. Soc.* **1993,** *115,* 6506–6509.
6. Boutorine, A. S.; Tokuyama, H.; Takasugi, M.; Isobe, H.; Nakamura, E.; Hélène, C. *Angew. Chem. Int. Ed. Engl.* **1994,** *33,* 2462–2465.
7. Da Ros, T.; Prato, M. *J. Chem. Soc., Chem. Commun.* **1994,** 663–669.

8. Signorini, R.; Zerbetto, M.; Meneghetti, M.; Bozio, R.; Maggini, M.; De Faveri, C.; Prato, M.; Scorrano, G. *J. Chem. Soc., Chem. Commun.* **1996,** 1891–1893.

9. Sun, Y.-P.; Riggs, J. E. *J. Chem. Soc., Faraday Trans.* **1997,** *93,* 1965–1970.

10. Sun, Y.-P.; Riggs, J. E.; Bing, L. *Chem. Mater.* **1997,** *9,* 1268–1272.

11. Taylor, R.; Walton, D. R. M. *Nature* **1993,** *363,* 685–693.

12. Hirsch, A. *The Chemistry of the Fullerenes,* Georg Thieme, Stuttgart, **1994.**

13. Diederich, F.; Isaacs, L.; Phlip, D. *Chem. Soc. Rev.* **1994,** *23,* 243–255.

14. Hirsch, A. *Synthesis* **1995,** 895–912.

15. Hirsch, A. (Ed.) Fullerenes and Related Structures *Top. Curr. Chem.* **1999,** *199.*

16. Birkett, P. R. *Annu. Rep. Prog. Chem., Sect. A* **1999,** *95,* 431–451.

17. Taylor, R. *Synlett* **2000,** 776–793.

18. Reed, C. A.; Bolskar, R. D. *Chem. Rev.* **2000,** *100,* 1075–1120.

19. Foote, C. S. Photophysical and Photochemical Properties of Fullerenes in *Top. Curr. Chem.* **1994,** *169,* 347–363.

20. Arbogast, J. W.; Darmanyan, A. P.; Foote, C. S.; Rubin, Y.; Diederich, F. N.; Alvarez, M. M.; Anz, S. J.; Whetten, R. L. *J. Phys. Chem.* **1991,** *95,* 11–12.

21. Arbogast, J. W.; Foote, C. S. *J. Am. Chem. Soc.* **1991,** *113,* 8886–8889.

22. Ajie, H.; Alvarez, M. M.; Anz, S. J.; Beck, D. R.; Diederich, F. N.; Fostiropoulos, K.; Huffmann, D. R.; Krätschmer, W.; Rubin, Y.; Schriver, K. E.; Sensharma, D.; Whetten, R. L. *J. Phys. Chem.* **1990,** *94,* 8630–8633.

23. Leach, S.; Vervloet, M.; Desprès, A.; Bréheret, E.; Hare, J. P.; Dennis, T. J.; Kroto, H. W.; Taylor, R.; Walton, D. R. M. *Chem. Phys.* **1992,** *160,* 451–466.

24. Sun, Y. P.; Wang, P.; Hamilton, N. B. *J. Am. Chem. Soc.* **1993,** *115,* 6378–6381.

25. Catalán, J.; Elguero, J. *J. Am. Chem. Soc.* **1993,** *115,* 9249–9252.

26. Ma, B.; Sun, Y.-P. *J. Chem. Soc., Perkin Trans. 2* **1996,** 2157–2162.

27. Kim, D.; Lee, M.; Suh, Y. D.; Kim, S. K. *J. Am. Chem. Soc.* **1992,** *114,* 4429–4430.

28. Sibley, S. P.; Argentine, S. M., Francis, A. H. *Chem. Phys. Lett.* **1992,** *188,* 187–193.

29. Wang, Y. *J. Phys. Chem.* **1992,** *96,* 764–767.

30. Sun, Y. P.; Bunker, C. E. *J. Phys. Chem.* **1993,** *97,* 6770–6773.

31. Palewska, K.; Sworakowski, J.; Chojnacki, H.; Meister, E. C.; Wild, U. P. *J. Phys. Chem.* **1993,** *97,* 12167–12172.

32. Williams, R. M.; Verhoeven, J. W. *Chem. Phys. Lett.* **1992,** *194,* 446–451.

33. Watanabe, A.; Ito, O.; Watanabe, M.; Saito, H.; Koishi, M. *J. Phys. Chem.* **1996,** *100,* 10518–10522.

34. Ebbesen, T. W.; Tanigaki, K.; Kuroshima, S. *Chem. Phys. Lett.* **1991,** *181,* 501–504.

35. Tanigaki, K.; Ebbesen, T. W.; Kuroshima, S. *Chem. Phys. Lett.* **1991,** *185,* 189–192.

36. Lee, M.; Song, O.-K.; Seo, J.-C.; Kim, D.; Suh, Y. D.; Jin, S. M.; Kim, S. K. *Chem. Phys. Lett.* **1992,** *196,* 325–329.

37. Wasielewski, M. R.; O'Neil, M. P.; Lykke, K. R.; Pellin, M. J.; Gruen, D. M. *J. Am. Chem. Soc.* **1991,** *113,* 2774–2776.

38. Palit, D. K.; Sapre, A. V.; Mittal, J. P.; Rao, C. N. R. *Chem. Phys. Lett.* **1992,** *195,* 1–6.

39. Fraelich, M. R.; Weismann, R. B. *J. Phys. Chem.* **1993,** *97,* 11145–11147.

40. Hung, R. R.; Grabowski, J. J. *J. Phys. Chem.* **1991,** *95,* 6073–6075.

41. Sension, R. J.; Phillips, C. M.; Szarka, A. Z.; Romanow, W. J.; McGhie, A. R.; Mc-Cauley, Jr., J. P.; Smith III, A. B.; Hochstrasser, R. M. *J. Phys. Chem.* **1991,** *95,* 6075–6078.

42. Kajii, Y.; Nakagawa, T.; Suzuki, S.; Achiba, Y.; Obi, K.; Shibuya, K. *Chem. Phys. Lett.* **1991,** *181,* 100–104.

43. Biczok, L.; Linschitz, H.; Walter, R. I. *Chem. Phys. Lett.* **1992,** *195,* 339–346.

44. Dimitrijevic, N. M.; Kamat, P. V. *J. Phys. Chem.* **1992,** *96,* 4811–4814.

45. Bensasson, R. V.; Hill, T.; Lambert, C.; Land, E. J.; Leach, S.; Truscott, T. G. *Chem. Phys. Lett.* **1993,** *201,* 326–335.

46. Bensasson, R. V.; Hill, T.; Lambert, C.; Land, E. J.; Leach, S.; Truscott, T. G. *Chem. Phys. Lett.* **1993,** *206,* 197–202.

47. McLean, A. J.; McGarvey, D. J.; Truscott, T. G.; Lambert, C. R.; Land, E. J. *J. Chem. Soc., Faraday Trans.* **1990,** *86,* 3075–3080.

48. Zeng, Y.; Biczok, L.; Linschitz, H. *J. Phys. Chem.* **1992,** *96,* 5237–5239.

49. Gevaert, M.; Kamat, P. V. *J. Phys. Chem.* **1992,** *96,* 9883–9888.

50. Closs, G. L.; Gautam, P.; Zhang, D.; Krusic, P. J.; Hill, S. A.; Wassermann, E. *J. Phys. Chem.* **1992,** *96,* 5228–5231.

51. Bennati, M.; Grupp, A.; Mehring, M.; Dinse, K.-P.; Fink, J. *Chem. Phys. Lett.* **1992,** *200,* 440–444.

52. Zhang, D.; Norris, J. R.; Krusic, P. J.; Wassermann, E.; Chen, C. C.; Lieber, C. M. *J. Phys. Chem.* **1993,** *97,* 5886–5889.

53. Steren, C. A.; Levsten, P. R.; van Willigen, H.; Linschitz, H.; Biczok, L. *Chem. Phys. Lett.* **1993,** *204,* 23–28.

54. Steren, C. A.; van Willigen, H.; Dinse, K.-P. *J. Phys. Chem.* **1994,** *98,* 7464–7469.

55. Bennati, M.; Grupp, A.; Mehring, M. *Synthetic Metals* **1997,** *86,* 2321–2324.

56. Terazima, M.; Sakurada, K.; Hirota, N.; Shinohara, H.; Saito, Y. *J. Phys. Chem.* **1993,** *97,* 5447–5450.

57. Goudsmit, G.-H.; Paul, H. *Chem. Phys. Lett.* **1993,** *208,* 73–78.

58. Levanon, H.; Meiklayr, V.; Michaeli, S.; Gamliel, D. *J. Am. Chem. Soc.* **1993,** *115,* 8722–8727.

59. Isaacs, L.; Wehrsig, A.; Diederich, F. *Helv. Chim. Acta* **1993,** *76,* 1231–1250.

60. Isaacs, L.; Diederich, F. *Helv. Chim. Acta* **1993,** *76,* 2454–2464.

61. Lin, S.-K.; Shiu, L.-L.; Chien, K.-M.; Luh, T.-Y.; Lin, T.-I. *J. Phys. Chem.* **1995,** *99,* 105–111.

62. Bensasson, R. V.; Bienvenüe, E.; Fabre, C.; Janot, J.-M.; Land, E. J.; Leach, S.; Leboulaire, V.; Rassat, A.; Roux, S.; Seta, P. *Chem. Eur. J.* **1998,** *4,* 270–278.

63. Guldi, D. M.; Hungerbühler, H.; Asmus, K.-D. *J. Phys. Chem.* **1995,** *99,* 9380–9385.

64. Hamano, T.; Okuda, K.; Mashino, T.; Hirobe, M.; Arakane, K.; Ryu, A.; Mashiko, S.; Nagano, T. *J. Chem. Soc., Chem. Commun.* **1997,** 21–22.

65. Sun, Y.-P.; Lawson, G. E.; Riggs, J. E.; Ma, B.; Wang, N.; Moton, D. K. *J. Phys. Chem. A* **1998,** *102,* 5520–5528.

66. Ma, B.; Riggs, J. E.; Sun, Y.-P. *J. Phys. Chem. B* **1998,** *102,* 5999–6009.

67. Sun, Y.-P.; Guduru, R.; Lawson, G. E.; Mullins, J. E.; Guo, Z.; Quinlan, J.; Bunker, C. E.; Gord, J. R. *J. Phys. Chem. B.* **2000,** *104,* 4625–4632.

68. Armaroli, N.; Marconi, G.; Echegoyen, L.; Borgeois, J.-P.; Diederich, F. *Chem. Eur. J.* **2000,** *6,* 1629–1645.

69. Diederich, F.; Jonas, U.; Gramlich, V.; Hermann, A.; Ringsdorf, H.; Thilgen, C. *Helv. Chim. Acta* **1993**, *76*, 2445–2453.
70. An, Y. Z.; Anderson, J. L.; Rubin, Y. *J. Org. Chem.* **1993**, *58*, 4799–4801.
71. Anderson, J. L.; An, Y.-Z.; Rubin, Y.; Foote, C. S. *J. Am. Chem. Soc.* **1994**, *116*, 9763–9764.
72. Bensasson, R. V.; Bienvenue, E.; Janot, J.-M.; Leach, S.; Seta, P.; Schuster, D. I.; Wilson, S. R.; Zhao, H. *Chem. Phys. Lett.* **1995**, *245*, 566–570.
73. Janot, J.-M.; Bienvenue, E.; Seta, P.; Bensasson, R. V.; Tomé, A. C.; Enes, R. F.; Cavaleiro, J. A. S.; Leach, S.; Camps, X.; Hirsch, A. *J. Chem. Soc., Perkin Trans. 2* **2000**, 301–306.
74. Liddell, P. A.; Sumida, J. P.; MacPherson, A. N.; Noss, L.; Seely, G. R.; Clark, K. N.; Moore, T. A.; Gust, D. *Photochem. Photobiol.* **1994**, *60*, 537–541.
75. Nakamura, Y.; Minowa, T.; Tobita, S.; Shizuka, H.; Nishimura, J. *J. Chem. Soc., Perkin Trans. 2*, **1995**, 2351–2357.
76. Rubin, Y.; Khan, S.; Freedberg, D. I.; Yeretzian, C. *J. Am. Chem. Soc.* **1993**, *115*, 344–345.
77. Beer, E.; Feuerer, M.; Knorr, A.; Mirlach, A.; Daub, J. *Angew. Chem., Int. Ed. Engl.* **1994**, *33*, 1087–1090.
78. Bennati, M.; Grupp, A.; Mehring, M.; Belik, P.; Gügel, A.; Müllen, K. *Chem. Phys. Lett.* **1995**, *240*, 622–626.
79. Lawson, G. E.; Kitaygorodskiy, A.; Sun, Y.-P. *J. Org. Chem.* **1999**, *64*, 5913–5920.
80. Williams, R. M.; Koeberg, M.; Lawson, J. M.; An, Y.-Z.; Rubin, Y.; Paddon-Row, M. N.; Verhoeven, J. W. *J. Org. Chem.* **1996**, *61*, 5055–5062.
81. Torres-Garcia, G.; Luftmann, H.; Wolff, C.; Mattay, J. *J. Org. Chem.* **1997**, *62*, 2752–2756.
82. Folcy, S.; Berberan-Santos, M. N.; Fedorov, A.; McGarvey, D. J.; Santos, C.; Gigante, B. *J. Phys. Chem. A* **1999**, *103*, 8173–8178.
83. Benedetto, A. F.; Bachilo, S. M.; Weismann, R. B.; Nossal, J. R.; Billups, W. E. *J. Phys. Chem. A* **1999**, *103*, 10842–10845.
84. Williams, R. M.; Zwier, J. M.; Verhoeven, J. W. *J. Am. Chem. Soc.* **1995**, *117*, 4093–4099.
85. Ma, B.; Bunker, C. E.; Guduru, R.; Zlang, X.-F.; Sun, Y.-P. *J. Phys. Chem. A* **1997**, *101*, 5626–5632.
86. Catalán, J.; Saiz, J. L.; Laynez, J. L.; Jagerovic, N.; Elguero, J. *Angew. Chem. Int. Ed. Engl.* **1995**, *34*, 105–107.
87. Guldi, D. M.; Maggini, M.; Scorrano, G.; Prato, M. *J. Am. Chem. Soc.* **1997**, *119*, 974–980.
88. Luo, C.; Fujitsuka, M.; Watanabe, A.; Osamu, I.; Gan, L.; Huang, Y.; Huang, C.-H. *J. Chem. Soc., Faraday Trans.* **1998**, *94*, 527–532.
89. Zhou, D.; Tan, H.; Gan, L.; Luo, C.; Huang, C.; Yao, G.; Zhang, P. *Chem. Lett.* **1995**, 649–650.
90. Zhou, D.; Gan, L.; Tan, H.; Luo, C.; Huang, C.; Yao, G.; Zhang, B. *J. Photochem. Photobiol. A: Chemistry* **1996**, *99*, 37–43.
91. Prato, M.; Maggini, M. *Acc. Chem. Res.* **1998**, *31*, 519–526.
92. Irngartinger, H.; Fettel, P. W.; Escher, T.; Tinnefeld, P.; Nord, S.; Sauer, M. *Eur. J. Org. Chem.* **2000**, 455–465.

93. Da Ros, T.; Prato, M.; Lucchini, V. *J. Org. Chem.* **2000,** *65,* 4289–4297.

94. Eiermann, M.; Wudl, F.; Prato, M.; Maggini, M. *J. Am. Chem. Soc.* **1994,** *116,* 8364–8365.

95. Diederich, F.; Isaacs, L.; Phlip, D. *J. Chem. Soc., Perkin Trans.* **1994,** *2,* 391–394.

96. Arias, F.; Echegoyen, L.; Wilson, S. R.; Lu, Q.; Lu, Q. *J. Am. Chem. Soc.* **1995,** *117,* 1422–1427.

97. Diederich, F.; Isaacs, L.; Phlip, D. *Chem. Soc. Rev.* **1994,** 243–255.

98. Creegan, K. M.; Robbins, J. L.; Robbins, W. K.; Millar, J. M.; Sherwood, R. D.; Tindall, P. J.; Cox, D. M.; Smith III, A. B.; McCauley, Jr., J. P.; Jones, D. R.; Gallagher, R. T. *J. Am. Chem. Soc.* **1992,** *114,* 1103–1105.

99. Smith III, A. B.; Strongin, R. M.; Brard, L.; Furst, G. T.; Romanow, W. J. *J. Am. Chem. Soc.* **1993,** *115,* 5829–5830.

100. Suzuki, T.; Li, Q. („Chan'); Khemani, K. C.; Wudl, F. *J. Am. Chem. Soc.* **1992,** *114,* 7301–7302.

101. Averdung, J.; Wolff, C.; Mattay, J. *Tetrahedron Lett.* **1996,** *37,* 4683–4684.

102. Bensasson, R. V.; Bienvenüe, E.; Dellinger, M.; Leach, S.; Seta, P. *J. Phys. Chem.* **1994,** *98,* 3492–3500.

103. Bensasson, R. V.; Hill, T. J.; Land, E. J.; Leach, S.; McGarvey, D. J.; Truscott, T. G.; Ebenhoch, J.; Gerst, M.; Rüchardt, C. *Chem. Phys.* **1997,** *215,* 111–123.

104. Palit, D. K.; Mohan, H.; Mittal, J. P. *J. Phys. Chem. A* **1998,** *102,* 4456–4461.

105. Spielmann, H. P.; Weedon, B. R.; Meier, M. S. *J. Org. Chem.* **2000,** *65,* 2755–2758.

106. Palit, D. K.; Mohan, H.; Birkett, P. R.; Mittal, J. P. *J. Phys. Chem. A* **1997,** *101,* 5418–5422.

107. Coheur, P.-F.; Cornil, J.; dos Santos, D. A.; Birkett, P. R.; Liévin, J.; Brédas, J. L.; Janot, J.-M.; Seta, P.; Leach, S.; Walton, D. R. M.; Taylor, R.; Kroto, H. W.; Colin, R. *Syn. Met.* **1999,** *103,* 2407–2410.

108. Guldi, D. M.; Asmus, K.-D. *J. Phys. Chem. A* **1997,** *101,* 1472–1481.

109. Taki, M.; Sugita, S.; Nakamura, Y.; Kasashima, E.; Yashima, E.; Okamoto, Y.; Nishimura, J. *J. Am. Chem. Soc.* **1997,** *119,* 926–932.

110. Nakamura, Y.; Taki, M.; Tobita, S.; Shizuka, H.; Yokoi, H.; Ishiguro, Y.; Sawaki, Y.; Nishimura, J. *J. Chem. Soc., Perkin Trans. 2* **1999,** 127–130.

111. Schick, G.; Levitus, M.; Kvetko, L.; Johnson, B. A.; Lamparth, I.; Lunkwitz, R.; Ma, B.; Khan, S. I.; Garcia-Garibay, M. A.; Rubin, Y. *J. Am. Chem. Soc.* **1999,** *121,* 3246–3247.

112. Levitus, M.; Schick, G.; Lunkwitz, R.; Rubin, Y.; Garcia-Garibay, M. A. *Photochem. Photobiol. A* **1999,** *127,* 13–19.

113. Arbogast, J. W.; Foote, C. S.; Kao, M. *J. Am. Chem. Soc.* **1992,** *114,* 2277–2281.

114. Mattay, J. *Angew. Chem. Int. Ed. Engl.* **1987,** *26,* 825–845.

115. Mattay, J.; Vondenhof, M. *Top. Curr. Chem.* **1991,** *159,* 219–255.

116. Fukuzumi, S.; Guldi, D. M. Electron Transfer Chemistry of Fullerenes in *Electron Transfer in Chemistry Vol. II, Part. I* (Ed. Balzani, V.), Wiley-VCH, New York, **2001,** pp. 270–337.

117. Samanta, A.; Kamat, P. V. *Chem. Phys. Lett.* **1992,** *199,* 635–639.

118. Staško, A.; Brezová, V.; Biskupic, S.; Dinse, K.-P.; Schweitzer, P.; Baumgarten, M. *J. Phys. Chem.* **1995,** *99,* 8782–8789.

119. Osaki, T.; Tai, Y.; Tazawa, M.; Tanemura, S.; Inukai, K.; Ishiguro, K.; Sawaki, Y.; Saito, Y.; Shinohara, H.; Nagashima. H. *Chem. Lett.* **1993,** 789–792.

120. Sasaki, Y.; Yoshikawa, Y.; Watanabe, A.; Ito, O. *J. Chem. Soc., Faraday Trans.* **1995,** *91,* 2287–2290.

121. Dubois, D.; Moninot, G.; Kutner, W.; Jones, M. T.; Kadish, K. M. *J. Phys. Chem.* **1992,** *96,* 7137–7145.

122. Xie, Q.; Arias, F.; Echegoyen, L. *J. Am. Chem. Soc.* **1993,** *115,* 9818–9818.

123. Echegoyen, L.; Echegoyen. L. E. *Acc. Chem. Res.* **1998,** *31,* 593–601.

124. Brezová, V.; Staško, A.; Rapta, P.; Domschke, G.; Bartl, A.; Dunsch, L. *J. Phys. Chem.* **1995,** *99,* 16234–16241.

125. Ito, O.; Sasaki, Y.; Yoshikawa, Y.; Watanabe, A. *J. Phys. Chem.* **1995,** *99,* 9838–9842.

126. Takeda, K.; Kajii, Y.; Shibuya, K. *J. Photochem. Photobiol. A: Cemistry* **1995,** *92,* 69–72.

127. Steren, C. A.; van Willigen, H.; Biczók, L.; Gupta, N.; Linschitz, H. *J. Phys. Chem.* **1996,** *100,* 8920–8926.

128. Brezová, V.; Gügel, A.; Rapta, P.; Staško, A. *J. Phys. Chem.* **1996,** *100,* 16232–16237.

129. Staško, A.; Biskupic, S.; Dinse, K.-P.; Groβ, R.; Baumgarten, M.; Gügel, A.; Belik, P. *J. Electroanalytical Chem.* **1997,** *423,* 131–139.

130. Caspar, J. V.; Wang, Y. *Chem. Phys. Lett.* **1994,** *218,* 221–228.

131. Sension, R. J.; Szarka, A. Z.; Smith, G. R.; Hochstrasser, R. M. *Chem. Phys. Lett.* **1991,** *185,* 179–183.

132. Palit, D. K.; Mittal, J. P. *Full. Science Tech.* **1995,** *3,* 643–659.

133. Ghosh, H. N.; Palit, D. K.; Sapre, A. V.; Mittal, J. P. *Chem. Phys. Lett.* **1997,** *165,* 365–373.

134. Ghosh, H. N.; Pal, H.; Sapre, A. V.; Mittal, J. P. *J. Am. Chem. Soc.* **1993,** *115,* 11722–11727.

135. Seshadri, R.; Rao, C. N. R.; Pal, H.; Mukherjee, T.; Mittal, J. P. *Chem. Phys. Lett.* **1993,** *205,* 395–398.

136. Ma, B.; Lawson, G. E.; Bunker, C. E.; Kitaygorodskiy, A.; Sun, Y.-P. *Chem. Phys. Lett.* **1995,** *247,* 51–56.

137. Sun, Y.-P.; Ma, B.; Lawson, G. E. *Chem. Phys. Lett.* **1995,** *233,* 57–62.

138. Park, J.; Kim, D.; Suh, Y. D.; Kim, S. K. *J. Phys. Chem.* **1994,** *98,* 12715–12719.

139. Rath, M. C.; Pal, H.; Mukherjee, T. *J. Phys. Chem. A* **1999,** *103,* 4993–5002.

140. Mittal, J. P. *Pure & Appl. Chem.* **1995,** *67,* 103–110.

141. Sasaki, Y.; Fujitsuka, M.; Watanabe, A.; Ito, O. *J. Chem. Soc., Faraday Trans.* **1997,** *93,* 4275–4279.

142. Dawe, E. A.; Land, E. J. *J. Chem. Soc., Faraday Trans. 1* **1975,** *71,* 2162–2169.

143. Lafferty, J.; Land, E. J.; Truscott, T. G. *J. Chem. Soc., Faraday Trans. 1* **1978,** *74,* 2760–2762.

144. Lafferty, J.; Truscott, T. G. *J. Chem. Soc., Chem. Comm.* **1978,** 51–52.

145. Fukuzumi, S.; Suenobu, T.; Kawamura, S.; Ishida, A.; Mikami, K. *Chem. Commun.* **1997,** 291–292.

146. Alam, M. M.; Watanabe, A.; Ito, O. *J. Photochem. Photobiol. A: Chemistry* **1997,** *104,* 59–64.

147. Ito, O.; Sasaki, Y.; Watanabe, A.; Hoffmann, R.; Siedschlag, C.; Mattay, J. *J. Chem. Soc., Perkin Trans. 2* **1997,** 1007–1011.

148. Nojiri, T.; Alam. M. M.; Konami, H.; Watanabe, A.; Ito, O. *J. Phys. Chem. A* **1997,** *101,* 7943–7947.

149. Krasnovsky Jr., A. A.; Foote, C. S. *J. Am. Chem. Soc.* **1993,** *115,* 6013–6016.

150. Yamakoshi, Y.; Sueyoshi, S.; Fukuhara, K.; Miyata, N.; Masumizu, T.; Kohno, M. *J. Am. Chem. Soc.* **1998,** *120,* 12363–12364.

151. Fujisawa, J.; Ohba, Y.; Yamauchi, S. *Chem. Phys. Lett.* **1998,** *294,* 248–254.

152. Alam, M. M.; Sato, M.; Watanabe, A.; Akasaka, T.; Ito, O. *J. Phys. Chem. A* **1998,** *102,* 7447–7451.

153. Nojiri, T.; Watanabe, A.; Ito, O. *J. Phys. Chem. A* **1998,** *102,* 5215–5219.

154. Fukuzumi, S.; Suenobu, Patz, M.; Hirasaka, T.; Itoh, S.; Fujitsuka, M.; Ito, O. *J. Am. Chem. Soc.* **1998,** *120,* 8060–8068.

155. Fukuzumi, S.; Suenobu, T.; Hirasaka, T.; Sakurada, N.; Arakawa, R.; Fujitsuka, M.; Ito, O. *J. Phys. Chem. A* **1999,** *103,* 5935–5941.

156. Konishi, T.; Sasaki, Y.; Fujitsuka, M.; Toba, Y.; Moriyama, H.; Ito, O. *J. Chem. Soc., Perkin Trans. 2* **1999,** 551–556.

157. Staško, A.; Brezová, V.; Rapta, P.; Dinse, K.-P. *Full. Science Tech.* **1997,** *5,* 593–605.

158. Sasaki, Y.; Konishi, T.; Fujitsuka, M.; Ito, O.; Maeda, Y.; Wakahara, T.; Akasaka, T.; Kako, M.; Nakadaira, Y. *J. Organomet. Chem.* **2000,** *599,* 216–220.

159. Konishi, T.; Fujitsuka, M.; Ito, O. *Chem. Lett.* **2000,** 202–203.

160. Scurlock, R. D.; Ogilby, P. R. *J. Photochem. Photobiol. A: Chem.* **1995,** *91,* 21–25.

161. El-Kemary, M.; Fujitsuka, M.; Ito, O. *J. Phys. Chem. A* **1999,** *103,* 1329–1334.

162. El-Kemary, M.; El-Khouly, M.; Fujitsuka, M.; Ito, O. *J. Phys. Chem. A* **2000,** *104,* 1196–1200.

163. Allemand, P.-M.; Srdanov, G.; Koch, A.; Khemani, K.; Wudl, F.; Rubin, Y.; Diederich, F.; Alvarez, M. M.; Anz, S. J.; Whetten, R. L. *J. Am. Chem. Soc.* **1991,** *113,* 2780–2781.

164. Dubois, D.; Kadish, K. M.; Flanagan, S.; Wilson, L. J. *J. Am. Chem. Soc.* **1991,** *113,* 7773–7775.

165. Dubois, D.; Kadish, K. M.; Flanagan, S.; Haufler, R. E.; Chibante, L. P. F.; Wilson, L. J. *J. Am. Chem. Soc.* **1991,** *113,* 4364–4366.

166. Kato, T.; Kodoma, T.; Shida, T.; Nakawaga, Y.; Matsui, S.; Suzuki, H.; Shiromaru, K.; Yanauchi, K.; Achiba, Y. *Chem. Phys. Lett.* **1991,** *180,* 446–450.

167. Nonell, S.; Arbogast, J. W.; Foote, C. S. *J. Am. Chem. Soc.* **1992,** *96,* 4169–4170.

168. Nadtochenko, V. A.; Denisov, N. N.; Rubtsov, I. V.; Lobach, A. S.; Moravskii, A. P. *Chem. Phys. Lett.* **1993,** *208,* 431–435.

169. Michaeli, S.; Meiklyar, V.; Schulz, M.; Möbius, K.; Levanon, H. *J. Phys. Chem.* **1994,** *98,* 7444–7447.

170. Yao, S. D.; Lian, Z. R.; Wang, W. F.; Zhang, J. S.; Lin, N. Y.; Hou, H. Q.; Zhang, Z. M.; Qin, Q. Z. *Chem. Phys. Lett.* **1995,** *239,* 112–116.

171. Lian, Z. R.; Yao, S. D.; Lin, W. Z.; Wang, W. F.; Lin, N. Y. *Radiat. Phys. Chem.* **1997,** *50,* 245–247.

172. Guldi, D. M.; Hungerbühler, H.; Janata, E.; Asmus, K.-D. *J. Phys. Chem.* **1993,** *97,* 11258–11264.

173. Siedschlag, C.; Luftmann, H.; Wolff, C.; Mattay, J. *Tetrahedron* **1997**, *53*, 3587–3592.
174. Siedschlag, C.; Luftmann, H.; Wolff, C.; Mattay, J. *Tetrahedron* **1999**, *55*, 7805–7818.
175. Dunsch, L.; Ziegs, F.; Siedschlag, C.; Mattay, J. *Chem. Eur. J.* **2000**, *6*, 3547–3550.
176. Siedschlag, C. *University of Kiel, PhD thesis*, **1998**.
177. Fedurco, M.; Rapta, P.; Kurán, P.; Dunsch, L. *Ber. Bunsnenges. Phys. Chem.* **1997**, *101*, 1045–1049.
178. Koptyug, I. V.; Goloshevsky, A. G.; Zavarine, I. S.; Turro, N. J.; Krusic, P. J. *J. Phys. Chem.* **2000**, *104*, 5726–5731.
179. Luo, C.; Fujitsuka, M.; Huang, C.-H.; Ito, O. *J. Phys. Chem. A* **1998**, *102*, 8716–8721.
180. Palit, D. K.; Mohan, H.; Birkett, P. R.; Mittal, J. P. *J. Phys. Chem. A* **1999**, *103*, 11227–11236.
181. Staško, A.; Brezová, V.; Rapta, P.; Asmus, K.-D.; Guldi, D. M. *Chem. Phys. Lett.* **1996**, *262*, 233–240.
182. Guldi, D. M. *J. Phys. Chem. A* **1997**, *101*, 3895–3900.
183. Brezová, V.; Staško, A.; Asmus, K.-D.; Guldi, D. M. *J. Photochem. Photobiol. A: Chem.* **1998**, *117*, 61–66.
184. Brezová, V.; Staško, A.; Dvoranová, D.; Asmus, K.-D.; Guldi, D. M. *Chem. Phys. Lett.* **1999**, *300*, 667–675.
185. Guldi, D. M.; Hungerbühler, H.; Asmus, K.-D. *J. Phys. Chem. B* **1999**, *103*, 1444–1453.
186. Guldi, D. M. *J. Phys. Chem. B* **2000**, *104*, 1483–1489.
187. Guldi, D. M.; Prato, M. *Acc. Chem. Res.* **2000**, *33*, 695–703.
188. Nakamura, Y.; Minowa, T.; Hayashida, Y.; Tobita, S.; Shizuka, H.; Nishimura, J. *J. Chem. Soc., Faraday Trans.* **1996**, *92*, 377–382.
189. Sun, Y.-P.; Ma, B.; Bunker, C. E. *J. Phys. Chem. A* **1998**, *102*, 7580–7590.
190. Martín, N.; Sánchez, L.; Illescas, B.; Pérez, I. *Chem. Rev.* **1998**, *98*, 2527–2547.
191. Simonsen, K. B.; Konovalov, V. V.; Konovalov, T. A.; Kawai, T.; Cava, M. P.; Kispert, L. D.; Metzger, R. M.; Becher, J. *J. Chem. Soc., Perkin Trans. 2* **1999**, 657–666.
192. Martín, N.; Sanchez, L.; Herranz, M. A.; Guldi, D. M. *J. Phys. Chem. A* **2000**, *104*, 4648–4657.
193. Maggini, M.; Guldi, D. M.; Mondini, S.; Scorrano, G.; Paolucci, F.; Ceroni, P.; Roffia, S. *Chem. Eur. J.* **1998**, *4*, 1992–2000.
194. Polese, A.; Mondini, S.; Bianco, A.; Toniolo, C.; Scorrano, G.; Guldi, D. M.; Maggini, M. *J. Am. Chem. Soc.* **1999**, *121*, 3446–3452.
195. Imahori, H.; Hagiwara, K.; Akiyama, T.; Taniguchi, S.; Okada, T.; Sakata, Y. *Chem. Lett.* **1995**, 265–266.
196. Imahori, H.; Hagiwara, K.; Akiyama, T.; Aoki, M.; Taniguchi, S.; Okada, T.; Shirakawa, M.; Sakata, Y. *Chem. Phys. Lett.* **1996**, *263*, 545–553.
197. Wasielewski, M. R. *Chem. Rev.* **1992**, *92*, 435–461.
198. Gust, D.; Moore, T. A.; Moore, A. L. *Acc. Chem. Res.* **1993**, *26*, 198–205.
199. Maruyama, K.; Osuka, N.; Mataga, N. *Pure Appl. Chem.* **1994**, *66*, 867–872.

200. Imahori, H.; Yamada, K.; Hasegawa, M.; Taniguchi, S.; Okada, T.; Sakata, Y. *Angew. Chem. Int. Ed. Engl.* **1997**, *36*, 2626–2629.
201. Imahori, H.; Sakata, Y. *Adv. Mater.* **1997**, *9*, 537–632.
202. Armaroli, N.; Diederich, F.; Dietrich-Buchecker, C. O.; Flamigni, L.; Marconi, G.; Nierengarten, J.-F.; Sauvage, J.-P. *Chem. Eur. J.* **1998**, *4*, 406–416.
203. Imahori, H.; Sakata, Y. *Eur. J. Org. Chem.* **1999**, 2445–2457.
204. Guldi, D. M. *Chem. Commun.* **2000**, 321–327.
205. Rao, A. M.; Zhou, P.; Wang, K.-A.; Hager, G. T.; Holden, J. M.; Wang, Y.; Lee, W.-T.; Bi, X.-X.; Eklund, P. C.; Cornett, D. S.; Duncan, M. A.; Amster, J. *Science* **1993**, *259*, 955–957.
206. Wilson, S. R.; Kaprinidis, N.; Wu, Y.; Schuster, D. I. *J. Am. Chem. Soc.* **1993**, *115*, 8495–8496.
207. Karfunkel, H. R.; Hirsch, A. *Angew. Chem. Int. Ed. Engl.* **1992**, *31*, 1468–1470.
208. Hirsch, A.; Lamparth, I.; Karfunkel, H. R. *Angew. Chem.* **1994**, *106*, 453–455.
209. Taylor, R. *Pure Appl. Chem.* **1997**, *69*, 1411–1434.
210. Taylor, R. *J. Chem. Soc., Perkin Trans. 2* **1993**, 813–824.
211. Schuster, D. I.; Lem, G.; Jensen, A. W.; Courtney, S. H.; Wilson, S. R. *Fullerenes, Recent Advances in the Chemistry and Physics of Fullerenes and Related Materials,* Ed. R. Ruoff, K. M. Kadish, The Electrochemical Society, **1995**, *2*, 441–448.
212. Schuster, D. I.; Lem, G.; Jensen, A. W.; Hwang, E.; Safanov, I.; Courtney, S. H.; Wilson, S. R. *Fullerenes, Recent Advances in Chemistry and Physics of Fullerenes and Related Materials,* Ed. R. Ruoff, K. M. Kadish, The Eletrochemical Society, **1996**, *3*, 287–296.
213. Mattay, J.; Torres- Garçia, G.; Averdung, J.; Wolff, C.; Schlachter, I.; Luftmann, H.; Siedschlag, C.; Luger, P.; Ramm, M. *J. Phys. Chem. Solids* **1997**, *58*, 1929–1937.
214. a) Mattay, J.; Siedschlag, Ch.; Torres-Garçia, G.; Ulmer, L.; Wolff, C.; Fujitsuka, M.; Watanabe, A.; Ito, O.; Luftmann, H. Photoreactions with fullerenes, *Recent Advances in the Chemistry and Physics of Fullerenes and Related Materials,* Ed. R. Ruoff, K. M. Kadish, The Eletrochemical Society, **1997**, *14*, 326–337. b) Siedschlag, Ch.; Torres-Garçia, G.; Wolff, C.; Mattay, J.; Fujitsuka, M.; Watanabe, A.; Ito, O.; Dunsch, L.; Ziegs, F.; Luftmann, H. Radical ions in fullerene chemistry, *Recent Advances in Chemistry and Physics of Fullerenes and Related Materials,* Ed. R. Ruoff, K. M. Kadish, The Eletrochemical Society, **1997**, *42*, 296–305. c) Siedschlag, Ch.; Torres-Garçia, G.; Wolff, C.; Mattay, J.; Fujitsuka, M.; Watanabe, A.; Ito, O.; Dunsch, L.; Ziegs, F.; Luftmann, H. *J. Inf. Rec.* **1998**, *24*, 265–270.
215. Lem, G.; Schuster, D. I.; Courtney, S. H.; Lu, Q.; Wilson, S. R. *J. Am. Chem. Soc.* **1995**, *117*, 554–555.
216. Tokuyama, H.; Isobe, H.; Nakamura, E. *J. Chem. Soc., Chem. Commun.* **1994**, 2753–2754.
217. Yamago, S.; Takeichi, A.; Nakamura, E. *J. Am. Chem. Soc.* **1994**, *116*, 1123–1124.
218. a) Mikami, K.; Matsumoto, S. *Synlett* **1995**, 229–230, b) Mikami, K.; Matsumoto, S.; Ishida, I.; Takamuku, S.; Suenobu, T.; Fukuzumi, S. *J. Am. Chem. Soc.* **1995**, *117*, 11134–11141.
219. Fukuzumi, S.; Fujita, M.; Otera, J.; Fujita, Y. *J. Am. Chem. Soc.* **1992**, *114*, 10271–10278.

220. see ref 4, 5 in Mikami, K.; Matsumoto, S.; Tonoi, T.; Suenobu, T.; Ishida, A.; Fukuzumi, S. *Synlett* **1997,** 85–87.

221. Mikami, K.; Matsumoto, S.; Tonoi, T.; Suenobu, T.; Ishida, A.; Fukuzumi, S. *Synlett* **1997,** 85–87.

222. Fukuzumi, S.; Fujita, M.; Otera, J. *J. Chem. Soc., Chem. Commun.* **1993,** 1536–1537.

223. Fukuzumi, S.; Suenobu, T.; Fujitsuka, M.; Ito, O.; Tonoi, T.; Matsumoto, S.; Mikami, K. *J. Organomet. Chem.* **1999,** *574,* 32–39.

224. a) Hirsch, A.; Soi, A.; Karfunkel, H. R. *Angew. Chem., Int. Ed. Engl.* **1992,** *31,* 766–768. b) Kitagawa, T.; Tanaka, T.; Takata, Y.; Takeuchi, K. *J. Org. Chem.* **1995,** *60,* 1490–1491. c) Tanaka, T.; Kitagawa, T.; Komatsu, K.; Takeuchi, K. *J. Am. Chem. Soc.* **1997,** *119,* 9313–9314.

225. Banim, F.; Cardin, D. J.; Heath, P. *Chem. Commun.* **1997,** 60–66.

226. a) Wudl, F.; Hirsch, A.; Khemani, C.; Suzuki, T.; Allemand, P.-M.; Koch, A.; Eckert, H.; Srdanov, G.; Webb, H. M. *Fullerenes: Synthesis, Properties, and Chemistry of Large Carbon Clusters,* Ed. G. S. Hammond and V. J. Kuck, American Chemical Society Symposium Series 481, **1992,** 161. b) Seshadri, R.; Govindaraj, A.; Nagarajan, R.; Rao, C. N. R. *Tetrahedron Lett.* **1992,** *33,* 2069–2072. c) Hirsch, A.; Li, Q.; Wudl, F. *Angew. Chem., Int. Ed. Engl.* **1991,** *30,* 1309–1311.

227. Skiebe, A.; Hirsch, A.; Klos, H.; Gotschy, B. *Chem. Phys. Lett.* **1994,** *220,* 138–140.

228. a) Sun, Y. P.; Ma. B. *Chem Phys. Lett.* **1995,** *236,* 285–291. b) Wang, Y.; Cheng, L.-T. *J. Phys. Chem.* **1992,** *96,* 1530–1532.

229. Kajii, Y.; Takeda, K.; Shibuya, K. *Chem. Phys. Lett.* **1993,** *204,* 283–286.

230. Liou, K.-F.; Cheng, C.-H. *Chem. Commun.* **1996,** 1423–1424.

231. Zhang, W.; Su, Y.; Gan, L.; Jiang, J.; Huang, C. *Chem. Lett.* **1997,** 1007–1008.

232. Gan, L.; Jiang, J.; Zhang, W.; Su, Y.; Shi, Y.; Huang, C.; Pan, J.; Lü, M.; Wu, Y. *J. Org. Chem.* **1998,** *63,* 4240–4247.

233. Okada, K.; Okubo, K.; Oda, M. *J. Photochem. Photobiol. A* **1991,** *57,* 265–277.

234. Suzuki, T.; Li, Q.; Khemani, K. C.; Wudl, F.; Almarsson, Ö. *Science* **1991,** *254,* 1186–1188.

235. a.) Vasella, A.; Uhlmann, P.; Waldraff, C. A. C.; Diederich, F.; Thilgen, C. *Angew. Chem., Int. Ed. Engl.* **1992,** *31,* 1388–1390. b) Komatsu, K.; Kagayama, A.; Murata, Y.; Sugita, N.; Kobayashi, K.; Nagase, S.; Wan, T. S. M. *Chem. Lett.* **1993,** 2163–2166. c) Tsuda, M.; Ishida, T.; Nogami, T.; Kurono, S.; Ohashi, M. *Tetrahedron Lett.* **1993,** *34,* 6911–6912. d) Tokuyama, H.; Nakamura, M.; Nakamura, E. *Tetrahedron Lett.* **1993,** *34,* 7429–7432. e) An, Y.-Z.; Rubin, Y.; Schaller, C.; McElvany, S. W. *J. Org. Chem.* **1994,** *59,* 2927–2929. f) Anderson, H. L.; Faust, R.; Rubin, Y.; Diederich *Angew. Chem., Int. Ed. Engl.* **1994,** *33,* 1366–1368.

236. *The Chemistry of Fullerenes,* Taylor, R., Ed., World Scientific, Singapore, **1995.**

237. Akasaka. T.; Liu, M. T. H.; Niino, Y.; Maeda, Y.; Wakahara, T.; Okamura, M.; Kobayashi, K.; Nagase, S. *J. Am. Chem. Soc.* **2000,** *122,* 7134–7135.

238. a)Jansen, R. A. J.; Hummelen, J. C.; Wudl, F. *J. Am. Chem. Soc.* **1995,** *117,* 544–545. b) Gonzalez, R.; Hummelen, J. C.; Wudl, F. *J. Org. Chem.* **1995,** *60,* 2618–2620.

239. a) Zhang, X.; Foote, C. S. *J. Am. Chem. Soc.* **1995,** *117,* 4271–4275. b) Zhang, X.; Fan, A.; Foote, C. S. *J. Org. Chem.* **1996,** *61,* 5456–5460.

240. a) Vassilikogiannakis, G.; Orfanopoulos, M. *J. Am. Chem. Soc.* **1997,** *119,* 7394–7395. b) Vassilkogiannakis, G.; Orfanopoulos, M. *Tetrahedron Lett.* **1997,** *38,* 4323–4326.

241. Vassilikogiannakis, G.; Chronakis, N.; Orfanopoulos, M. *J. Am. Chem. Soc.* **1998,** *120,* 9911–9920.

242. Schuster, D. I.; Cao, J.; Kaprinidis, N.; Wu, Y.; Jensen, A.; Lu, Q.; Wang, H.; Wilson, S. R. *J. Am. Chem. Soc.* **1996,** *118,* 5639–5647.

243. Vassilikogiannakis, G.; Orfanopoulos, M. *J. Org. Chem.* **1999,** *64,* 3392–3393.

244. Jensen, A. W.; Khong, A.; Saunders, M.; Wilson, S. R.; Schuster, D. I. *J. Am. Chem. Soc.* **1997,** *119,* 7303–7307.

245. Wilson, S. R.; Wu, Y.; Kaprinidis, N. A.; Schuster, D. I. *J. Org. Chem.,* **1993,** *58,* 6548–6549.

246. for a recent review, see: Schuster, D. I.; Lem, G.; Kaprinidis, N. *Chem. Rev.* **1993,** *93,* 3–22.

247. Hatzimarinaki, M.; Vassilikogiannakis, G.; Orfanopoulos, M. *Tetrahedron Lett.* **2000,** *41,* 4667–4669.

248. Zhang, X.; Romero, A.; Foote, C. S. *J. Am. Chem. Soc.* **1993,** *115,* 11024–11025.

249. Tomioka, H.; Ichihashi, M.; Yamamoto, K. *Tetrahedron Lett.* **1995,** *36,* 5371–5374.

250. a) Win, W.W.; Kao, M.; Eiermann, M.; McNamara, J. J.; Wudl, F. *J. Org. Chem.* **1995,** *59,* 5871–5878.

251. Mikami, K.; Matsumoto, S.; Tonoi, T.; Okubo, Y. *Tetrahedron Letters,* **1998,** *39,* 3733–3736.

252. Mikami, K.; Matsumoto, S.; Okubo, Y.; Fujitsuka, M.; Ito, O.; Suenobu, T.; Fukuzumi, S. *J. Am. Chem. Soc.* **2000,** *122,* 2236–2243.

253. Schuster, D. I.; Baran, P. S.; Hatch, R. K.; Khan, A. U.; Wilson, S. R. *Chem. Comm.* **1998,** 2493–2494.

254. Juha, L.; Hamplová, V.; Kodymová, J.; Spalek, A. *J. Chem. Soc. Chem. Commun.* **1994,** 2437–2438.

255. Heymann, D.; Chibante, L. P. F. *Chem. Phys. Lett.* **1993,** *207,* 339–342.

256. Smith, A. B., III; Strongin, R. M.; Brard, L.; Furst, G. T.; Atkins, J. H.; Romanow, W. J.; Saunders, M.; Jiménez-Vázquez, H. A.; Owens, K. G.; Goldschmidt, R. J. *J. Org. Chem.* **1996,** *61,* 1904–1905.

257. a) Prato, M.; Li, Q. C.; Wudl, F.; Lucchini, V. *J. Am. Chem. Soc.* **1993,** *115,* 1148–1450. b) Banks, M. R.; Cadogan, J. I. G.; Gosney. I.; Hodgson, P. K. G.; Langridge-Smith, P. R. R.; Rankin, D. W. H. *J. Chem. Soc., Chem. Commun.* **1994,** 1365–1366. c) Kuwashima, S.; Kubota, K.; Kushida, K.; Ishida, T.; Ohashi, M.; Nogami, T. *Tetrahedron Lett.,* **1994,** *34,* 4371–4374. d) Ishida, T.; Tanaka, K.; Nogami, T. *Chem. Lett.* **1994,** 561–562.

258. Yan, M.; Cai, S. X.; Keana, J. F. W. *J. Org. Chem.* **1994,** *59,* 5951–5954.

259. Poe, R.; Schnapp, K.; Young, M. J. T.; Grayzar, J.; Platz, M. S. *J. Am. Chem. Soc.* **1992,** *114,* 5054–5067.

260. Averdung, J.; Mattay, J.; Jacobi, D.; Abraham, W. *Tetrahedron* **1995,** *51,* 6043–6052.

261. Averdung, J.; Mattay, J. *Tetrahedron Lett.* **1994,** *35,* 6661–6664.

262. Averdung, J.; Albrecht, E.; Lauterwein, J.; Luftmann, H.; Mattay, J.; Mohn, H.; Müller, W. H.; ter Meer, H.-U. *Chem. Ber.* **1994,** *127,* 787–789.

263. Averdung, J.; Mattay, J. *Tetrahedron* **1996,** *52,* 5407–5420.

264. Sun, Y.-P.; Bunker, C. E.; Ma, B. *J. Am. Chem. Soc.* **1994**, *116*, 9692–9699.

265. Lawson, G. E.; Kitaygorodskiy, A.; Ma, B.; Bunker, C. E.; Sun, Y.-P. *J. Chem. Soc., Chem. Commun.* **1995**, 2260–2266.

266. a) Lewis, F. D. *Acc. Chem. Res.* **1986**, *19*, 401–405. b) Barltrop, J. A. *Pure Appl. Chem.* **1973**, *33*, 179–195. c) Bryce-Smith, D.; Gilbert, A. *Tetrahedron* **1977**, *33*, 2459–2490. d) Lewis, F. D.; Ho, T.-I. *J. Am. Chem. Soc.* **1977**, *99*, 7991–7996. e) Lewis, F. D.; Ho, T.-I.; Simpson, J.-T. *J. Org. Chem.* **1981**, *46*, 1077–1082. f) Lewis, F. D.; Ho, T.-I.; Simpson, J.-T. *J. Am. Chem. Soc.* **1982**, *104*, 1924–1929. g) Hubb, W.; Schneider, S.; Dörr, F.; Oxman, J. D.; Lewis, F. D. *J. Am. Chem. Soc.* **1984**, *106*, 708–715.

267. a) Saltiel, J.; Charlton, J. L. *Rearrangements in Ground and Excited States,* ed. P. de Mayo, Academic Press, New York, **1980**, p. 60. b) Mazzucazo, U. *Pure Appl. Chem.* **1982**, *54*, 1705–1721. c) Mallory, F. B.; Mallory, C. W. *Organic Reactions* **1984**, *30*, 1. d) Laarhoven, W. H. *Organic Photochemistry,* ed. A. Padwa, Marcel Dekker, New York, **1989**, *10*, p. 163.

268. a) Zhou, D.; Tan, H.; Luo, C.; Gan, L.; Huang, C.; Pan, J.; Lu, M.; Wu, Y. *Tetrahedron Lett.* **1995**, *36*, 9169–9172. b) Gan, L.; Zhou, D.; Luo, C.; Tan, H.; Huang, C.; Lu. M.; Pan, J.; Wu, Y. *J. Org. Chem.* **1996**, *61*, 1954–1961.

269. a) Guo, L.-W.; Gao, X.; Zhang, D.-W.; Wu, S.-H.; Wu, H.-M.; Li, Y.-J. *Chem. Lett.* **1999**, 411–412. b) Guo, L.-W.; Gao, X.; Zhang, D.-W.; Wu, S.-H.; Wu, H.-M.; Li, Y.-J. *J. Org. Chem.* **2000**, *65*, 3804–3810.

270. Bernstein, R.; Foote, C. S. *J. Phys. Chem. A* **1999**, *103*, 7244–7247.

271. a) Banks, M. R.; Cadogan, J. I. G.; Gosney, I.; Hodgson, P. K. G.; Langridge-Smith, P. R. R.; Rankin, D. W. H. *J. Chem. Soc., Chem. Commun.* **1994**, 1365–1366. b) Banks, M. R.; Cadogan, J. I. G.; Gosney, I.; Hodgson, P. K. G.; Langridge-Smith, P. R. R.; Millar, J. R. A.; Taylor, A. T. *Tetrahedron Lett.* **1994**, *35*, 9067–9070.

272. Banks, M. R.; Cadogan, J. I. G.; Gosney, I.; Hodgson, P. K. G.; Millar, J. R. A.; Langridge-Smith, P. R. R.; Rankin, D. W. H.; Taylor, A. T. *J. Chem. Soc., Chem. Commun.* **1995**, 877–878.

273. Banks, M. R.; Cadogan, J. I. G.; Gosney, I.; Hodgson, P. K. G.; Langridge-Smith, P. R. R.; Millar, J. R. A.; Parkinson, J. A.; Sadler, I. H.; Taylor, A. T. *J. Chem. Soc., Chem. Commun.* **1995**, 1171–1172.

274. a) Akasaka, T.; Ando, W.; Kobayashi, K.; Nagase, S. *J. Am. Chem. Soc.* **1993**, *115*, 10366–10367. b) Akasaka, T.; Mitsuhida, E.; Ando, W.; Kobayashi, K.; Nagase, S. *J. Am. Chem. Soc.* **1994**, *116*, 2627–2628.

275. a) Kusukawa, T.; Kabe, Y.; Erata, T.; Nestler, B. Ando, W. *Organometallics* **1994**, *13*, 4186–4188. b) Kusukawa, T.; Ando, W. *Organometallics* **1997**, *16*, 4027–4029. c) Akasaka, T.; Suzuki, T.; Maeda, Y.; Ara, M.; Wakahara, T.; Kobayashi, K.; Nagase, S.; Kako, M.; Nakadaira, Y.; Fujitsuka, M.; Ito, O. *J. Org. Chem.* **1999**, *64*, 566–569.

276. a) Akasaka, T.; Ando, W.; Kobayashi, K.; Nagase, S. *J. Am. Chem. Soc.* **1993**, *115*, 1605–1606. b) Akasaka, T.; Mitsuhida, E.; Ando, W.; Kobayashi, K.; Nagase, S. *J. Chem. Soc., Chem. Commun.* **1995**, 1529–1530.

277. a) Kusukawa, T.; Kabe, Y.; Ando, W. *Organometallics* **1995**, *14*, 2142–2144. b) Kusukawa, T.; Shike, A.; Ando, W. *Tetrahedron* **1996**, *52*, 4995–5005.

278. Kusukawa, T.; Ohkubo, K.; Ando, W. *Organometallics* **1997**, *16*, 2746–2747.

279. a) Tokuyama, H.; Nakamura, E. *J. Org. Chem.* **1994,** *59,* 1135–1138. b) Or-
      fanopoulos, M.; Kambourakis, S. *Tetrahedron Lett.* **1994,** *35,* 1945–1948.
280. An, Y.-Z.; Viado, A. L.; Arce, M.-J.; Rubin, Y. *J. Org. Chem.* **1995,** *60,* 8330–8331.
281. a) Foote, C. S. *Acc. Chem. Res.* **1968,** *1,* 104–110. b) Schulte-Elte, K. H.; Rauten-
      strauch, V. *J. Am. Chem. Soc.* **1980,** *102,* 1738–1740.
282. a) Zhang, X.; Foote, C. S. *J. Org. Chem.* **1993,** *58,* 5524–5527. b) Foote, C. S.; Dza-
      kpasu, A. A.; Lin, J. W.-P. *Tetrahedron Lett.* **1975,** *15,* 1247–1250. c) Wasserman,
      H. H.; Terao, S. *Tetrahedron Lett.* **1975,** *15,* 1735–1739.
283. Hummelen, J. C.; Prato, M.; Wudl, F. *J. Am. Chem. Soc.* **1995,** *115,* 7003–7004.
284. Ohno, M.; Koide, N.; Sato, H.; Eguchi, S. *Tetrahedron* **1997,** *53,* 9075–9086.
285. Foote, C. S.; Wuesthoff, M. T.; Wexler, S.; Burstain, I. G.; Denny, R.; Schenck, G.
      O.; Schulte-Elte, K.-H. *Tetrahedron,* **1967,** *23,* 2583–2599.
286. Wasserman, H. H.; Druckrey, E. *J. Am. Chem. Soc.* **1968,** *90,* 2440–2441.
287. Lamparth, I.; Maichle-Mössmer, C.; Hirsch, A. *Angew. Chem., Int. Ed. Engl.* **1995,**
      34, 1607–1609.
288. Irngartinger, H.; Weber, A.; Escher, T. *Eur. J. Org. Chem.* **2000,** 1647–1651.
289. Wudl, F. *Acc. Chem. Res.* **1992,** *60,* 157–161.
290. Smith, A. B., III; Strongin, R. M.; Brard, L.; Furst, G. T.; Romanow, W. J.; Owens,
      K. G.; Goldschmidt, R. J.; King, R. C. *J. Am. Chem. Soc.* **1995,** *117,* 5492–5502.
291. Ulmer, L.; Torres-Garcia, G.; Wolff, Ch.; Mattay, J.; Luftmann, H. in *Recent Ad-
      vances in the Chemistry and Physics of Fullerene and Related Materials,* Ed.
      Kadish, K. M.; Ruoff, R. S., The Electrochemical Society, **1997,** *5,* 306–312.
292. Averdung, J.; Torres-Garcia, G.; Luftmann, H.; Schlachter, I.; Mattay, J. *Full. Sci.
      Techn.* **1996,** *4,* 633–654.
293. Schick, G.; Hirsch, A.; Mauser, H.; Clark, T. *Chem. Eur. J.* **1996,** *2,* 935–943.
294. a) Zimmerman, H. E. in *Rearrangements in ground and excited states,* Vol. 3, de
      Mayo, P., Ed.; Academic Press, New York, **1980.** b) Scott, L. T.; Erden, I. *J. Am.
      Chem. Soc.* **1982,** *104,* 1147–1149.
295. a) Kjell, D. P.; Sheridan, R. S. *J. Am. Chem. Soc.* **1984,** *106,* 5368–5370. b) Ham-
      rock, S. J.; Sheridan, R. S. *Tetrahedron Lett.* **1988,** *29,* 5509–5512. c) Kjell, D. P.;
      Sheridan, R. S. *J. Photochem.* **1985,** *28,* 205–213.
296. Koerner von Gustorf, E.; White, D. V.; Kim, B.; Hess, D.; Leitich, J. *J. Org. Chem.*
      1970, *35,* 1155–1165.
297. Knol, J.; Hummelen, J. C. *J. Am. Chem. Soc.* **2000,** *122,* 3226–3227.
298. Hsiao, T.-S.; Santhosh, K. C.; Liou, K.-F.; Cheng, C.-H. *J. Am. Chem. Soc.* **1998,**
      120, 12232–12236.
299. Arce, M.-J.; Viado, A. L.; An, Y.-Z.; Khan, S. I.; Rubin, Y. *J. Am. Chem. Soc.* **1996,**
      118, 3775–3776.
300. Qian, W.; Bartberger, M. D.; Pastor, S. J.; Houk, K. N.; Wilkins, C. L.; Rubin, Y. *J.
      Am. Chem. Soc.* **2000,** *122,* 8333–8334.
301. An, Y.-Z.; Ellis, G. A.; Viado, A. L.; Rubin, Y. *J. Org. Chem.* **1995,** *60,* 6353–6361.
302. a) Chapman, O. L.; Mattes, K.; McIntosh, C. L.; Pacansky, J.; Calder, G. V.; Orr, G.
      J. Am. Chem. Soc. **1973,** *95,* 6134–6135. b) Rubin, Y.; Kahr, M.; Knobler, C. B.;
      Diederich, F.; Wilkins, J. *J. Am. Chem. Soc.* **1991,** *113,* 495–500.
303. Henderson, C. C.; Rohlfing, C. M.; Gillen, K. T.; Cahill, P. A. *Science* **1994,** *264,*
      397–399.

Index

Acidity
  excited-state, 578
Acyloxoamides
  enantioselective photorearrangement,
    393
Addition of amines to aromatics, 209
Addition (1,4) of amines to aromatics,
  211
Addition of arenes to alkenes
  electron transfer pathway, 216–218
Addition of arenes to amines
  electron transfer pathway, 219–223
Alumina
  crystal violet, 567
Anthracene 9-carboxylic acid
  dimerization of
    micelles, 368
Anthracene dimerization
  Nafion membranes, 369

Benzonorbornadienes
  rearrangement of 366

Carboxamide
  enantioselective photoarrangement
    inclusion complex, 408

Chemosensors, 464
Chiral auxiliary strategy
  enantioselective photorearrangement,
    402
Chiral crystals, 385
Cinnamic acid
  dimerization of vesicles, 367
Crystal engineering, 428
  role of hydrogen bond, 436
Cycloaddition
  anthracenes to dienes, 152
  aromatics to nitriles
    sensitized, 20
  chrysenes to alkenes, 157
  cyanonaphthalenes to alkenes
    hydrogen bonding effect, 158
  cyanonaphthalenes to cyclooctene, 138
  cyanonaphthalenes to dienes, 144
  cyanonaphthalenes to
    tetramethylethylene, 137
  cyanonaphthalenes to vinyl ethers,
    136
  cyanophenanthrenes to alkenes, 154
  intramolecular
    cyclohexenones, 405
    naphthalenes, 177–182

[Cycloaddition]
    phenanthrenes, 185, 186
    pyrenes, 187, 188
    naphthalenes to acrylonitrile, 139
    pyrenes to dienes, 157
Cycloaddition of arenes to alkenes
    classification, 130
    electron transfer, 129
Cycloaddition of arenes to dienes, 133
Cycloaddtion-meta, 2
Cycloaddition-ortho, 2, 4
    aromatics to acetylene(s)
        dicarboxylate, 11–13, 22–25
    aromatics to alkenes, 49–72
        intramolecular, 73–79
        mechanism, 80–94
    aromatics to maleimides, 17, 27–29
    aromatics to nitriles, 26
    benzene, 2
    benzene to acetylene, 2
    benzene to acetylene dicarboxylate,
        10
    benzene to maleic anhydride, 2, 7
    benzonitrile, 2
    correlation diagram, 102, 105, 106
    ground-state complex, 4
    intramolecular, 16, 163–175
    maleimides to biphenyl, 19
    phenanthrene to maleic anhydride, 5
    sensitized
        aromatic ketones, 30–49
    toulene to maleic anhydride, 9
    vs. meta, 94
        electron transfer, 98–100
Cycloaddition-para, 3
Cyclodextrin, 461
    energy transfer, 487
    modified with
        azobenzene, 493
        benzophenone, 488
        calixarene, 478
        dansyl, 468
        naphthalene, 464
        N,N-dimethyl aminobenzonitrile,
        479
        peptides, 467, 483

[Cyclodextrin]
    pyrene, 480
    stilbene, 497
    modified with dansyl
        sensors, 469
    rotaxanes, 491

Dimerization
    crystals
        temperature effect, 443
        wavelength effect, 446
    dienes
        solid state, 431
    geometric criteria, 434
    halocoumarins
        crystals, 429
    intramolecular, anthracene, 194–202
        cation effect, 202
        benzene, 189
        naphthalene, 190–192
    styryl coumarins
        crystals, 432
Dimerization of
    alkenes
        zeolite, 367
    alkoxynaphthalene, 147
    anthracene
        cyclodextrin, 484
    cyanoanthracene, 149
    cyanophenanthrene, 147

Enantioselective dimerization
    coumarins
        inclusion complex, 419
    cycloalkenes
        inclusion complex, 418
    thiocoumarins
        inclusion complex, 420
Enantioselective reactions, 385
Ene reaction
    allylic
        alcohols and amines, 253
        silanes, 267
        stannanes, 268
    cis effect, 254
    dienes, 273

[Ene reaction]
  fullerenes, 274
  geminal selectivity, 265, 269
  homoallylic alcohols, 272
  phenylsubstituted alkenes, 256
  regio selectivity, 259
  singlet oxygen
    site selectivity, 250
  solvent effect, 279
  steric effect, 261
  styrenes, 257
  syn/anti selectivity, 253
  zeolite effect, 276
Excited-state acidity
  aromatic hydroxy compound, 579
  carbazole
    pH probe, 595
  crown ethers, 615
  double proton transfer, 581
  hypericin
    pH probe, 591, 604, 610
  intramolecular proton transfer, 580,
    597, 603, 605, 611
  LB films, 615
  monolayers, 61
  naphthols
    pH probe, 583, 599, 608
  polymers and gels, 616
  pyranine
    pH probe, 589, 602
  table, 622–628
  3-hydroxyflavone
    pH probe, 597
Exciton, 527

Fluorine as a steering group, 429
Forster cycle, 578
Fullerene derivatives, 641
  electron transfer, 658, 671
  energy transfer, 665
  fluorescence, 646, 651
  quenching by β-carotene, 665
  radical anion spectra, 660
  reaction with amines, 675, 692
  triplet properties, 654
Fullerene-amino derivatives, 677

Fullerene–C_{60}
  addition of
    alkenes, 687, 698
    aminocarboxylates, 715
    diazomethane, 706
    disilirane, 717
    enamines, 725
    singlet oxygen, 722
  carbon numbering, 681
  cycloaddition of
    diazirine, 695
    dienes, 701
    enones, 696
  cyclodextrin complex, 675
  oxidation, 666
  photophysical data, 639
  radical addition, 670
  reduction via electron transfer, 681
  Schlegal diagram, 681
  singlet oxygen generation, 678
Fullerene-C_{60} and C_{70}
  phosphorescence, 640
Fullerene-C_{70}
  photophysical data, 639

Green fluorescent protein, 604

Heavy atom effect
  zeolite, 366

Intramolecular cycloaddition of arenes,
  132
Intramolecular dimerization of arenes
  polyethylene film effect, 334–337
  zeolite effect, 325–333

Layered double hydroxides, 518
  porphyrins, 560
  pthalocyanines, 562

Molybdates, 517

Nafion membranes
  structure, 321
Naphthols
  bilayers
    excited-state acidity, 585

Norrish-Yang reaction
  chiral chemistry
    zeolites, 373

Olefin-oxygen complex
  zeolite effect, 357
Orthocycloadducts of arenes
  secondary reactions, 107
Oxidation
  olefin-oxygen complex, 357
Oxoamides
  enantioselective photorearrangement
    crystals, 386
    inclusion complex, 392

Phenylglyoxamides
  enantioselective photorearrangement,
    389
Photo-Birch reaction, 213
Photocyclization
  acetanilides
    inclusion complex, 415
  valerolactams
    inclusion complex, 410
Photo-Fries reaction, 359
  polyethylene film effect, 364
  zeolite effect, 362
Photon antenna, 538
Photo-oxidation of alkenes
  effect of vesicles, 538
  electron transfer pathway, 338
  polyethylene film effect, 344
  selectivity
    zeolite effect, 353–357
  zeolite effect, 337, 340–344
Photo-oxidation of dienes
  zeolite effect, 350–352
Polyethylene films
  structure, 322
Polymorphism, 437
Polymorphs
  succinimide crystals
    reactivity difference, 397, 399

Reaction cavity, 448
Room temperature phosphorescence, 462

Schenk reaction, 245
Singlet-oxygen
  addition to dienes, 244
  diffusion, 290
  ene reaction, 245
  ene reaction
    mechanism, 245
    perepoxide intermediate, 246, 249
  generation, 244
  lifetime, 291
  liposomes, 309
  quenching, 293
    by methyl indole, solvent effect,
      296
    by tryptophan, 301
    effect of vesicles, 312
    micellar effect, 300
    reverse micellar effect, 303
    solvent effect, 295
    vesicles, 304
Substitution reactions
  arenes, 207

Thiourea derivatives
  enantioselective photorearrangement,
    393
Titanates, 517
Titania
  ruthenium trisbipyridyl complex, 565
Topochemical photopolymerization
  cinnamic acid derivatives, 400
Topochemical rules
  modification, 441
Transition metal oxides
  structure, 517
Tropolone(s)
  chiral chemistry
    inclusion complex, 404
    zeolites, 371

Vesicles
  structure, 323

Zeolites
  structure, 318

Zirconium phosphate
  anthracene excimer, 522
  azobenzene, 537
  binding of
    arenes, 520, 526
    crystal violet, 533
    ethidium bromide, 529
    methylene blue, 532
    thionin, 532
  electron transfer, 538, 549
  energy transfer, 538
  fluorescin-rhodamine, 540
  non-linear optical properties, 536, 570

[Zirconium phosphate]
  porphyrins, 554
  preparation, 509
  proteins, 555
    redox activity, 557
  pyrene excimer, 524
  ruthenium trisbipyridyl complex,
    544
  structure, 510–513
  uranyl-europium, 543
  viologens, 552
Zirconium phosphonate
  structure, 513